GLOBAL CHANGE IN THE HOLOCENE

Edited by

Anson Mackay, University College London, UK
Rick Battarbee, University College London, UK
John Birks, University of Bergen, Norway and University College London, UK
Frank Oldfield, University of Liverpool, UK

First published in Great Britain in 2003 by
Hodder Education, part of Hachette Livre UK,
338 Euston Road, London NW1 3BH

http://www.hoddereducation.co.uk

British Library Cataloguing in Publication Data
A catalogue record for this book is available from the British Library

Library of Congress Cataloging-in-Publication Data
A catalog record for this book is available from the Library of Congress

ISBN: 978 0 340 81214 3

Production Controller: Anna Keene
Cover Design: Terry Griffiths

Picture credit: photograph of Lake Prestesteinsvatnet and the Fannaråkbreen glacier in the Jotunheimen mountains of central Norway (back cover) by John Birks.

Typeset in 10 on 12 pt Garamond by Phoenix Photosetting, Chatham, Kent

If you have any comments to make about this, or any of our other titles, please send them to educationenquiries@hodder.co.uk

CONTENTS

LIST OF CONTRIBUTORS

Michael G.L. Baillie, School of Archaeology and Palaeoecology, Queen's University of Belfast, Northern Ireland

Keith E. Barber, Palaeoecology Laboratory, Department of Geography, University of Southampton, UK

Richard W. Battarbee, Environmental Change Research Centre, Department of Geography, University College London, UK

H. John B. Birks, Botanical Institute, University of Bergen, Norway and Environmental Change Research Centre, University College London, UK

Hilary H. Birks, Botanical Institute, University of Bergen, Norway and Environmental Change Research Centre, University College London, UK

Raymond S. Bradley, Climate System Research Center, University of Massachusetts, USA

Peter Brimblecombe, School of Environmental Sciences, University of East Anglia, UK

Stephen J. Brooks, Department of Entomology, The Natural History Museum, London, UK

David M. Brown, School of Archaeology and Palaeoecology, Queen's University of Belfast, Northern Ireland

Mark Bush, Department of Biological Sciences, Florida Institute of Technology, USA

Dan J. Charman, Department of Geographical Sciences, University of Plymouth, UK

Martin Claussen, Potsdam Institute for Climate Impact Research, Climate System Department, Germany

Julia E. Cole, Department of Geosciences, University of Arizona, USA

Svein Olaf Dahl, Department of Geography, University of Bergen, Norway

Daniel R. Engstrom, St Croix Watershed Research Station, Science Museum of Minnesota, USA

Virginia Ettwein, Environmental Change Research Centre, Department of Geography, University College London, UK

David A. Fisher, National Glaciology Programme, Terrain Sciences Division, Geological Survey of Canada, Natural Resources Canada, Canada

Sherilyn C. Fritz, Department of Geosciences, University of Nebraska, USA

Ian D. Goodwin, Environmental Geoscience Group, School of Environmental and Life Sciences, University of Newcastle, Australia

Jonathan A. Holmes, Environmental Change Research Centre, Department of Geography, University College London, UK

Phil D. Jones, Climatic Research Unit, University of East Anglia, UK

Vivienne J. Jones, Environmental Change Research Centre, Department of Geography, University College London, UK

Roy M. Koerner, National Glaciology Programme, Terrain Sciences Division, Geological Survey of Canada, Natural Resources Canada, Canada

Stein-Erik Lauritzen, Department of Earth Science, University of Bergen, Norway

Melanie J. Leng, NERC Isotope Geosciences Laboratory, British Geological Survey, Nottingham, UK

Anson W. Mackay, Environmental Change Research Centre, Department of Geography, University College London, UK

André F. Lotter, Laboratory of Palaeobotany and Palynology, University of Utrecht, The Netherlands

Mark Maslin, Environmental Change Research Centre, Department of Geography, University College London, UK

Atle Nesje, Department of Earth Science, University of Bergen, Norway

Frank Oldfield, Department of Geography, University of Liverpool, UK

Jennifer Pike, Department of Earth Sciences, Cardiff University, UK

Jon R. Pilcher, School of Archaeology and Palaeoecology, Queen's Unversity of Belfast, Northern Ireland

Antoni Rosell-Melé, ICREA and Institute of Environmental Science and Technology, Autonomous University of Barcelona, Spain

Louis Scott, Department of Plant Sciences, University of the Free State, South Africa

Stephen Shennan, Institute of Archaeology, University College London, UK

Catherine Stickley, Environmental Change Research Centre, Department of Geography, University College London, UK

Joanna L. Thorpe, Limnological Research Center, University of Minnesota, USA

Roy Thompson, Department of Geology and Geophysics, University of Edinburgh, UK

Paul J. Valdes, Department of Meteorology, Reading University, UK

H.E. Wright Jr, Limnological Research Center, University of Minnesota, USA

Bernd Zolitschka, Geomorphologie und Polarforschung, Institut für Geographie, Universität Bremen, Germany

ACKNOWLEDGEMENTS

The editor and publishers would like to thank the following for permission to use copyright material in this book: *AA Balkema:* Koinig, K.A, Sommaruga-Wögrath, S., Schmidt, R., Tessadri, R. and Psenner, R. 1998. Acidification processes in high alpine lakes. In M.J. Haigh, J. Krĕcĕk, G.S. Raijwar and M.P. Kilmartin (eds) *Headwaters: water resources and soil conservation*. A.A. Balkema Publishers, Rotterdam, pp. 45–54. *American Association for the Advancement of Science:* Douglas, M.S.V., Smol, J.P. and Blake Jr W. 1994. Marked post-18th century environmental change in high-arctic ecosystems. *Science*, 266, 416–419; Shindell, D.T., Schmidt, G.A., Mann, M.E., Rind, D. and Waple, A. 2001. Solar forcing of regional climate change during the Maunder Minimum. *Science*, 294, 2149–2152. *Arizona Board of Regents for the University of Arizona:* Bar-Yosef, O. 2000. The impact of radiocarbon dating on Old World archaeology: past achievements and future expectations. *Radiocarbon* 42, 23–39; Nydal, R. and Gislefoss, J.S. 1996. Further application of bomb ^{14}C as a tracer in the atmosphere and ocean. *Radiocarbon*, 38, 389–407. *Arnold:* Bigler, C., Larocque, I., Peglar, S.M., Birks, H.J.B. and Hall, R.I. 2002. Quantitative multi-proxy assessment of long-term patterns of Holocene environmental change from a small lake near Abisko, northern Sweden. *The Holocene*, 12, 481–496. *The British Academy:* Bayliss, A., Ramsey, B. and McCormac, F.C. 1997. Dating Stonehenge, Science and Stonehenge In B. Cunliff and C. Renfrew (eds), *Proceedings of the British Academy*, 92, pp. 39–59. *Elsevier Science:* Bar-Matthews, M., Ayalon, A., Kaufman, A. and Wasserburg, G.J. 1999. The eastern Mediterranean paleoclimate as a reflection of regional events: Soreq Cave, Israel. *Earth and Planetary Science Letters*, 166, 85–95; Dean, J.M., Kemp, A.E.S. and Pearce, R.B. 2001. Paleo-flux records from electron microscope studies of Holocene laminated sediments, Saanich Inlet, British Columbia. *Marine Geology*, 174, 139–158; deMenocal, P., Oriz, J., Guilderson, T., Adkins, J., Sarnthein, M., Baker, L. and Yarusinsky, M. 2000. Abrupt onset and termination of the African Humid Period: rapid climate responses to gradual insolation forcing. *Quaternary Science Reviews*, 19, 347–361; Gasse, F. 2000. Hydrological changes in the African tropics since the last Glacial Maximum. *Quaternary Science Reviews*, 19, 189–211; Stuiver, M., Braziunas, T.F., Becker, B. and Kromer, B. 1999. Climatic, solar, oceanic and geomagnetic influences on Late Glacial and Holocene atmospheric $^{14}C/^{12}C$ change. *Quaternary Research*, 35, 1–24; Tiljander, M., Ojala, A., Saarinen, T., Snowball, I. 2002. Documentation of the physical properties of annually laminated (varved) sediments at a sub-annual to decadal resolution for environmental interpretation. *Quaternary International*, 88, 5–12; Wasson, R.J. and Claussen, M. 2002. Earth system models: a test using the mid-Holocene in the southern hemisphere. *Quaternary Science Reviews*, 21, 819–824. *Geological Survey of Finland:* Ojala, A. 2001. Varved lake sediments in southern and central Finland: long varve chronologies as a basis for Holocene palaeoenvironmental reconstructions, PhD Thesis, University of Turku. Geological Survey of Finland, Espoo. *International Peat Society:* Lappalainen, E. (ed.) 1996.

Global peat resources. International Peat Society, Finland. *International Union of Geological Sciences*: cover page of *Episodes,* 9(1). *John Wiley and Sons:* Thomas, D.S.G. 1997. *Arid Zone Geomorphology: process, form and change in drylands.* John Wiley and Sons, Chichester, pp. 732. *Kluwer Academic Publishers:* Bradbury, J.P., Cumming, B. and Laird, K. 2002. A 1500-year record of climatic and environmental change in Elk Lake, Minnesota. III: measures of past primary production. *Journal of Paleolimnology*, 27, 321–340; Frolich, C. 2000. Observations of irradiance variations. *Space Science Reviews*, 94, 15–24 (Fig 6); Halsey, L.A., Vitt, D.H. and Bauer, I.E. 1998. Peatland initiation during the Holocene in continental western Canada. *Climatic Change,* 40, 315–342; Laird, K.R., Fritz, S.C. and Cumming, B.F. 1998. A diatom-based reconstruction of drought intensity, duration, and frequency from Moon Lake, North Dakota: a sub-decadal record of the last 2300 years. *Journal of Paleolimnology*, 19, 161–179. *Macmillan*: Neff, U., Burns, S.J., Mangini, A., Mudelsee, M., Fleitmann, D. and Matter, A. 2001. Strong coherence between solar variability and the monsoon in Oman between 9 and 6 kyr ago. *Nature,* 411, 290–293. *Nature Conservation and Tourism Branch of the South West Africa Administration:* Brain, C.K. and Brain, V. 1977. Microfaunal remains from Mirabib: some evidence of palaeo-ecological changes in the Namib. *Madoqua,* 10, 285–293. *Springer-Verlag:* Claussen, M. 2002. Does landsurface matter in weather and climate. In P. Kabat, M. Claussen, P.A. Dirmeyer, *et al.* (eds) *Vegetation, water, humans and the climate: a new perspective on an interactive system.* Springer-Verlag, Berlin (Figs A.22, A.25, A.28). *University of Washington Press:* Post, A. and LaChapelle, E.R. 1998. *Glacier ice.* University of Washington Press, Seattle, 2000.

The authors thank the EU Framework IV programme for funding an Advanced Study Course at UCL, which provided the original idea for this book (ENV4-CT97-4013). AWM also thanks the Cartography Unit, Department of Geography for undertaking to redraw many of the figures. Finally, we also thank all the external referees who kindly agreed to provide feedback on each chapter.

PREFACE

Both the number and diversity of studies of Holocene climate variability have expanded dramatically over the last decade, as new regions around the world are investigated, new archives exploited, and the range of proxy types expanded. Moreover, the questions that are now posed with respect to understanding patterns in variability have become increasingly sophisticated. The impetus for this book initially stemmed from an EU Advanced Study Course, held at UCL, on the subject of Holocene Climate Variability. While the course brought together a large selection of scientific experts, it quickly became apparent that there were few advanced textbooks on the subject of high resolution Holocene climate variability suitable for upper undergraduate students, postgraduates and postdoctoral scientists.

The book can be loosely divided into six sections, which cover what we believe to be the main foci of current research. The first section provides an introduction to the Holocene, the principal forcing factors driving climate and a description of the range of modelling techniques used during this time, ending with case studies that highlight recent progress in linking human cultural development to past climate variability. In all of aspects of reconstructing past environments, the development of robust chronologies together with an assessment of accuracy, calibration and methodological limits are of the utmost importance, and in the second section of the book special attention is paid not only to radiocarbon dating, but also to annually resolved chronologies derived using dendrochronology and freshwater and marine laminations. Increasingly, it has been possible to derive quantitative estimates of past climate variability using multivariate techniques and stable isotope analysis, and chapters on each of these topics form the basis of the third section of the book.

The fourth section of the book introduces the reader to the use of instrumental and documentary records, data that have been collected and collated for several centuries. However, these records rarely extend further back than 1000 years, and it is here that natural archives come into their own. The range of archives is large, but we have chosen to focus on archives that are able, under suitable conditions, to provide time-series from the sub-centennial to sub-decadal resolution or better. In selecting our archives, we have made an attempt to choose examples from a wide range of environments, including tropical corals, marine sediments, peatlands, lake sediments, speleothems, glaciers and ice-cores. Within these archives, the range of proxy climate indicators is even more extensive and the examples chosen here (section five: diatoms, ostracods, chironomids, pollen, plant macrofossils and alkenones) have been perhaps some of the most extensively researched, but this list is far from exhaustive. However, we have tried to ensure that the range of case studies in both of these sections reflects the diversity of work being carried out around the world.

The final section brings together applications and case studies that we felt deserved special merit. Increasingly, the nature of our field requires multidisciplinary approaches to be taken, but rarely has the potential of such studies been critically examined. This is the focus of the chapter on multi-proxy reconstructions. Furthermore, the majority of published studies are based on archives and proxies from northern hemisphere landmasses, especially Europe and North America. We felt it appropriate therefore to address some of the imbalance, and we have included chapters specifically on Holocene climate variability from lowland tropical forests, arid regions in middle latitudes and sea-level changes in coastal regions around the world. Our final case study exemplifies the use of climate system models to simulate Holocene climate change, especially in northern latitudes and northern Africa.

Holocene research is an area of enormous activity and of concern, because of the need to establish the magnitude of natural variability prior to the onset of the Anthropocene in the last 200 years. This has made Holocene climate research a rapidly developing and challenging field. This book attempts to capture the direction of current research and to provide an overview of the important questions and research approaches.

Anson Mackay, Rick Battarbee,
John Birks & Frank Oldfield,
November 2004

CHAPTER

1

INTRODUCTION: THE HOLOCENE, A SPECIAL TIME

Frank Oldfield

1.1 The Holocene in Temporal Perspective

For anyone raised in the tradition of field-based Quaternary studies in northwestern Europe, the transition from the end of glacial times to the beginning of the Holocene is one of the most notable and readily detectable of stratigraphic boundaries. Almost everywhere, it is marked by evidence for a dramatic shift in surface processes, denoting a major climate change that in turn triggered a whole sequence of responses in both abiotic and biotic ecosystem components. Evidence for the precise age, suddenness and synchroneity of the transition has gradually accumulated over the last 70 years until now, we have remarkably precise chronological control on its timing, the pace of change and the extraordinary spatial and temporal coherence of response over large areas of the globe. Indeed, were it not that colleagues dealing with **contemporary transformations** of the Earth System had placed their concerns so firmly under the heading of 'Global Change', that term could serve perfectly for the opening of the Holocene.

The isotopically-inferred temperature record in the GRISP/GISP ice cores from Central Greenland points up the sharp contrast between late Pleistocene and Holocene in terms both of mean values and of the amplitude of variability (Dansgaard *et al.*, 1993). To a large extent, the shift to the Holocene appears to be a rapid switch in mode from low mean temperatures and extreme variability on all time-scales from decadal to millennial, to one of much higher mean values and lower variability. Thus, if the record from Central Greenland were the only template for our interpretation of Holocene environmental change, we would be considering a period of rather remarkable invariance relative to that which preceded it. But the empirical evidence from other parts of the world, as well as our knowledge of the changing patterns of, and interactions between, external forcings and feedbacks reveal this as a serious oversimplification.

The orbitally driven changes that appear to have triggered the Pleistocene–Holocene transition were relatively gradual. Moreover, orbitally driven changes in solar irradiance have continued throughout the Holocene and they have had different expressions at different latitudes. Only in the second half of the Holocene, roughly the last 6000 years, have they been broadly comparable to those prevailing today. Thus, at the opening of the Holocene, the effects of smoothly changing external forcing were mediated by internal system dynamics to generate a range of abrupt and synchronous changes in many parts of the world, but not all the responses were immediate and speedily accomplished. Ice takes time to melt and the great northern hemisphere polar ice did not disappear overnight. Nor did it simply wane smoothly and continuously everywhere. In consequence, eustatic sea-level too took several millennia to reach its mid-Holocene levels. Not only did many physical responses to Holocene warming and deglaciation take place over several thousands of years, biotic responses too were not completed instantaneously. Migration, soil development, competition and succession all played a part in modulating ecological responses to the major changes in the Earth System that marked the opening of the Holocene. We may therefore think of this latter shift as the beginning of a longer, complex period of transition as well as a sharp boundary between Earth System regimes. Depending on our research focus and on where and how we look, it was both.

Did these transitional changes during the first half of the Holocene play out against the backdrop of a global climate as relatively invariant as the Central Greenland temperature record suggests? Undoubtedly not. High latitude temperature variability may have been reduced, but there were still major changes, especially during the early Holocene. Elsewhere, at lower latitudes and notably in tropical regions, hydrological variability was extreme over the same period, with dramatic changes in lake level well documented in Africa and South America. To some extent, the climatic variability that is recorded during the first half of the Holocene may be attributed to the sequence of changes taking place in the wake of deglaciation and to the way in which the changes interacted with the prevalent patterns of orbital forcing, but these factors alone fail to account for all the changes observed. Changes in ocean currents and land biota also appear to have influenced climate, at least on a continental scale.

Even during the second half of the Holocene, when orbitally driven external forcing was broadly similar to today, ice had melted to a minimum and eustatic sea-level had recovered, there is strong evidence for significant climate variability in all areas and on all time-scales. Such variability, as well as having had important effects on past hydrological regimes and ecosystems, is of outstanding interest at the present day. It is against this background variability that we must seek to detect and characterize the imprint of human-induced climate change resulting from ever increasing atmospheric greenhouse gas concentrations. Moreover, future climate change will be, in part, an expression of similar variability as it plays out in the future and interacts with the effects of any human-induced climate change.

The Holocene period thus emerges not as a bland, pastoral coda to the contrasted movements of a stirring Pleistocene symphony; rather we now see it as a period of continuous change, the documenting and understanding of which becomes increasingly urgent as our concerns for future climate change grow. All the foregoing serves to reinforce the importance of the Holocene as a major research challenge; but there is an additional element that may be of even greater importance, for it is during the Holocene, and especially the later part, that human activities have begun to reshape the nature of the Earth System not only through systemic impacts on the composition and concentrations of atmospheric trace gases, but through the cumulative effects of land clearance, deforestation, soil erosion, salinization, urbanization, loss of biodiversity and a myriad other impacts that have transformed our environment at an ever accelerating rate. These processes began many thousands of years ago at local and regional levels in long settled areas of the globe, but over the last two centuries and at an accelerating rate in the last few decades, the impacts have become global and the implications for rapidly increasing human populations a cause for growing anxiety. It follows from all the above that the themes of this book are of major relevance to our present-day environmental concerns (cf. Oldfield and Alverson, 2003).

1.2 THE DEMISE OF THE 35-YEAR MEAN

One of the cornerstones of climatology 50 years ago was the notion of the 35-year mean. This purported to encapsulate an adequate first approximation to the climate of a station or region. At the same time, it was recognized that climate had changed in the past, as witness the sequence of glaciations and the climate oscillations they implied. It is doubtful whether any conflict was perceived between these two perspectives as they were the concerns of quite different scholarly communities. Reconciling the notion of the 35-year mean with the realization that climate had changed was, in any case, quite easy if one took the view that past change entailed a switch between distinctive episodes, each of relative constancy. The 'post-glacial' period in northwest Europe for example, was one that could be divided into a suite of phases – Pre-Boreal, Boreal, Atlantic, Sub-Boreal and Sub-Atlantic. From around 500 BC, we had been in the cool, wet Sub-Atlantic phase and, by implication, the 35-year mean could serve to describe the climate regime typical of that period for any given location. It took the work of scholars like Gordon Manley (1974) and Hubert Lamb (1963) to bridge the gap between instrumental records and the longer time-scales of climate change. In so doing, they helped to show that climate variability was continuous on all time-scales, that short-term changes were nested within longer-term trends and that there was no such thing as a mean value that could serve for any time interval other than that for which it was calculated.

Put another way, change is the norm. This has important implications for almost every aspect of environmental science, for it shifts our perspective away from any static descriptor of a relatively constant state to an acknowledgement that for any place and over any time-scale there has been an envelope of variability which changes with the time-span which it represents. Characterizing and understanding the processes contributing to and modulating past variability on a wide range of time-scales constitutes a major scientific challenge, but one that is of vital interest at the present day and for the future.

1.3 LESSONS FROM THE PAST

When the threat of future greenhouse warming was first identified and clearly stated (see summary in Oeschger, 2000), it was tempting to turn to the past for analogues. There had been warmer worlds in the past; what were they like? Could they provide a partial template for a possibly warmer world of the future? Quite quickly, this rather simple way of using hindsight was seen to be seriously flawed. We cannot hope to find analogues with any useful degree of realism by turning to periods when the external boundary conditions and the very configuration of the planet were different. Instead, palaeo-scientists began to interrogate the past record of environmental change with questions about processes, rates of change, long-term Earth System dynamics, non-linear responses to external forcing, feedback mechanisms involving the hydrosphere and biosphere and a myriad other similar issues (see e.g. Alverson *et al.*, 2000, 2003).

In adopting this much more realistic research agenda, the main focus in palaeo-science has been on the late Quaternary period. Indeed, the record of the last four glacial cycles spanning the last 430,000 years from Vostok in Antarctica (Petit *et al.*, 1999) has come to serve as an almost universal template for this type of research. The Holocene represents no more than the last 2.7 per cent of this time interval. What are the special qualities of the period that make it of compelling interest? What are the key questions we can address by using the record from the Holocene and what are the key issues that improved knowledge of the Holocene may help us to resolve?

1.4 THE SPECIAL INTEREST OF THE HOLOCENE

The realization that the isotopically inferred temperature record from Central Greenland was not a template for all aspects of Holocene climate everywhere has been noted above. Nevertheless, the contrast between late Pleistocene and Holocene variability in the ice core record has strongly influenced thinking in the research community. It has, for example, added special point to questions about climate variability in warm, interglacial intervals. Evidence for climate variability in the Eemian interglacial (Marine Isotope Stage 5e) has evoked a good deal of interest, but continuous, well dated, fine resolution records from the Eemian are rare. It is to the Holocene itself that we must turn for the bulk of our evidence for 'warm climate' variability. The paragraphs that follow seek to highlight some of those qualities of the Holocene that make the record of environmental change during the period of such special interest and value.

1.4.1 Boundary Conditions, External Forcing and Internal Feedbacks

As already hinted at above, the Holocene as a whole is the period for which we have the most information about climate variability during warm, interglacial times. Significant changes in temperature that were certainly synchronous between Greenland and Europe have been well documented for the early Holocene (Alley *et al.*, 1997; von Grafenstein *et al.*, 1998). Even more dramatic in human terms were the widescale changes in lake levels, plant cover and soil moisture that took place at lower latitudes and continued at least until around 4000 years ago (Gasse and Van Campo, 1994). Less dramatic, but nevertheless highly significant, changes in hydrology have also been recorded throughout the second half of the Holocene (see e.g. Verschuren *et al.*, 2000).

The pattern of orbitally driven solar forcing changes relatively slowly and continuously, but over the last 6000 years, which is to say during the second half of the Holocene, it has not differed greatly from the pattern that prevailed during the centuries immediately before the human-induced increase in atmospheric greenhouse gas concentrations began. By the middle of the Holocene, other aspects of the Earth System that influence climate significantly – polar ice cap and sea-ice extent, sea-level and major terrestrial biomes, for example – had all achieved states within an envelope of variability not significantly different from that typical of the last millennium. Thus the main patterns of forcing and feedbacks that characterized the period immediately before human activities began to modify the atmosphere significantly were, broadly speaking, in place by the middle of the Holocene. Anything that we can learn about variability and environmental change since then thus has special relevance for understanding the processes operating now and in the most recent past.

As Bradley (pp. 10–19 in this volume) points out, solar irradiance reaching the outer edge of the earth's atmosphere varies on many time-scales and is modulated by processes some of which are quite independent of orbital changes. The role of these shorter-term variations in solar activity as drivers of global climate has recently received increasing attention. In part, this is due to the fact that for the Holocene period it appears possible to reconstruct a detailed and well dated proxy record of variations in received solar irradiance by measuring deviations in the relationship between the decline in radiocarbon concentrations with age in tree-rings and true dendrochronological age (Stuiver *et al.*, 1991). Where records of past climate variability have been sufficiently well and independently dated, this opens up the possibility of exploring the extent to which the climate changes recorded are coherent with inferred variations in solar activity. Our growing knowledge of the Holocene thus provides key information for testing hypotheses about climate forcing.

1.4.2 Modes of Variability

One of the ways in which climatologists make sense of climate variability on a global scale is by identifying and characterizing relatively distinct modes of variability. The El Niño Southern Oscillation (ENSO) and the North Atlantic Oscillation (NAO) are well-known examples. Other modes currently recognized include a decadal oscillation in the North Pacific and an Arctic Oscillation that interacts with the NAO. One of the key findings of recent research on late Holocene records is that these modes of variability are remarkably protean. Over a period of decades and centuries, their amplitudes, frequencies and spatial domains change (see e.g. Cole and Cook, 1998; Markgraf and Diaz, 2001). This knowledge presents both a contemporary caution and a future challenge. In our present state of knowledge it reduces the confidence with which predictions of the long-term incidence and effects of these modes in the future can be made. At the same time, it challenges us to discover the factors responsible for the decadal- and century-scale variability. Only by understanding these and incorporating them in model simulations will there be any realistic chance of improving future predictability. Once more, the Holocene period is the crucial time interval for exploring these issues, though longer-term insights into the nature of ENSO variability, for example, also shed important light on the range of possible behaviours ENSO may assume (Tudhope *et al.*, 2001).

1.4.3 Continuity and Overlap with the Present Day

Many of the archives and proxies that form the toolkit of the palaeo-scientists can bring the Holocene record of variability right up to the present day. Tree-rings are still being formed, lake sediments continue to accumulate, corals and speleothems still grow. This allows Holocene research to reap multiple benefits. The insights gained contribute to our understanding of present-day ecosystems and environmental processes that have been in part conditioned by their antecedents. By bringing records of climate change from the centuries well before significant human impact right through to the short period of instrumental records (Jones and Thompson, pp. 140–158 in this volume), palaeoclimatology makes a crucial contribution to resolving the questions of detection and attribution raised by global warming in recent decades. The same kind of temporal overlap allows direct comparison between recent instrumental records of the amplitude, duration and recurrence intervals of extreme events and their longer-term history (Page *et al.*, 1994; Knox, 2000).

The above examples stress only one facet of the importance of continuity and overlap, for the points made would count for little were it not possible to translate proxy records of environmental change into inferences sufficiently quantitative to permit comparison with direct measurements. This requires calibration (Birks, pp. 107–123 in this volume) and, in this regard, the period of overlap between past proxy records and present-day measurements is crucial. Calibration, whether achieved by comparing directly measured sequences with proxy records covering the same time interval, or by linking proxies to a spatial array of contemporary measurements spanning a range of variability, is at its most robust for situations where the processes, biological communities or geochemical signatures in which the proxy signals reside lie within or as close as possible to the variability encompassed by the calibration process. For time intervals in which past biological communities lack present-day analogues, or abiotic proxies have values that can only be matched by significant extrapolation of a calibration function, the inferences become less secure and the statistical uncertainties much greater. Once more, the Holocene, and especially the late Holocene, have important advantages. As we move further back in time, confidence in quantitative reconstructions of climate often decline quite steeply (see e.g. Bigler *et al.*, 2002).

1.4.4 Chronology

Not only are Holocene palaeoarchives much more abundant than those for earlier time intervals, especially in continental situations, they can generally be dated much more accurately and precisely. Radiocarbon dating, whether used conventionally or by 'wiggle-matching' (van Geel and Mook, 1989), tephra analysis, varve counting and use of annual speleothem growth increments are all most effective for the Holocene and immediately pre-Holocene period. Tree-rings, with relatively few exceptions, are pretty well limited in their use both for direct chronologies and as climate proxies, to the late Holocene (Baillie and Brown, pp. 75–91 in this volume). Refinements in chronology are vital for addressing many of the most urgent questions in palaeoenvironmental research and recent advances should not blind us to the need for ever better chronologies.

1.4.5 Testing Models

All the above advantages ascribed to the Holocene reinforce its value for testing climate models. Rigorous tests require that the empirical basis for the 'ground truth' against which model simulations are compared be as secure and well constrained as possible. Because models provide the only way of developing scenarios of future conditions other than expert opinion or simple extrapolation, testing their output against known conditions in the past is of prime importance. If models are unable to replicate a known set of conditions or sequence of events in the hind-cast mode, they can have little credibility as predictors of future changes. This realization has led to a whole range of model–data interactions using palaeo-research both to improve parameterization and to provide the basis for testing. The synergy between data acquisition and model refinement is well illustrated in Claussen (pp. 422–434 in this volume), mainly using output from EMIC (Earth Models of Intermediate Complexity). Model–data comparison also plays an important role in ascribing recent climate variability to different forcing mechanisms (e.g. Crowley and Kim, 1999), in testing time-slice simulations performed by more complex global circulation and coupled ocean–atmosphere models (Kohfeld and Harrison, 2000; Valdes, pp. 20–35 in this volume), in exploring the implications for past climates of proxies such as stable isotope signatures (Hoffmann, 2002) and reconstructing biomes representing both present and past climate conditions (Prentice *et al.*, 1992, 1993). In all these roles, well-dated and calibrated proxy records from the Holocene are of major importance.

1.5 PREREQUISITES FOR RESEARCH INTO HOLOCENE VARIABILITY

In the above section we seek to identify some of the key reasons why research on Holocene environmental history is of such outstanding importance. Below, we consider briefly what is required for carrying out effective research on the history of environmental change during the Holocene. These are the 'tools' that allow us to extend the record of past change beyond the instrumental period. Clearly, in many parts of the world, documentary records span a much longer period than do instrumental records and these have been used with increasing skill for reconstructing past climate variability and the impacts on human populations and ecosystems of extreme events such as major floods and persistent droughts (Brimblecome, pp. 159–167 in this volume). Below, we concentrate on the evidence available from environmental archives rather than documentary sources.

1.5.1 Environmental Archives

Many of these have already been referred to in the foregoing paragraphs. In essence they are environmental contexts that preserve one or more types of decipherable record of past environmental conditions. They are as diverse as trees, both living and dead, lake and marine sediments, peat bogs, corals, glaciers and ice fields, and speleothems. In many cases a single archive will contain several possible proxies that can be translated, through calibration, into well validated information on the nature of past environmental conditions. Over the last decade, the main thrust of this type of research has been in reconstructing some aspect of past climate.

1.5.2 Proxies

Proxies are components within an archive that can be extracted, identified and quantified in such a way that their implications for past environmental conditions can be reliably and consistently inferred. The current emphasis on climate reconstruction has led to an incredible diversity of proxy climate signatures. In most cases, these refer to temperature, either annual or, more often, seasonal. Overall, there is a tendency for the majority of proxies to reflect spring and/or summer temperatures: for example, biological proxies are usually calibrated to growing-season conditions and glacier melt layers reflect summer warmth. Reconstructions of palaeo-precipitation are less common, but they can be found in sources as contrasted as the stratigraphic record from ombrotrophic peat bogs (Barber and Charman, pp. 210–226 in this volume) and both the geomorphological and stratigraphic evidence for lake level variations. Biological remains often record a range of influences. Aquatic organisms such as diatoms or chironomids, respond to lake chemistry as well as water temperature. Pollen-producing plants that provide the source of the palynological record in peats and sediments reflect human activities such as deforestation and agriculture as well as changes in climate. In these cases, calibration to some aspect of climate is, in effect, imposing a filter on the full range of information intrinsic to the biological record.

Just as many sub-fossil remains contain a range of potentially calibratable signatures, some types of proxy can be identified within a wide and diverse range of contexts. The demonstration that the ratio between the stable isotopes of oxygen ($\delta^{18}O$) and hydrogen (δD) in rain water reflects air temperature has led to the use of stable isotope records as (often rather indirect) palaeoclimateic proxies in a wide range of archives – ice cores, tree-rings, carbonates, both marine and fresh water, speleothems and, more recently, sedimentary plant cellulose and diatom silica. The use of stable isotopes thus constitutes a versatile methodology applicable to a wide range of archives and components, both biotic and abiotic, within them (Leng, pp. 124–139 in this volume).

All the proxies currently available pose challenges in interpretation and no single one alone can be relied on universally to provide a complete and secure record of past climate change. This has led to the frequent use of what is often termed a multi-proxy approach to climate reconstruction (e.g. Lotter, pp. 373–383 in this volume). Where used, this allows the mutually independent records to act as constraints on each other, either reinforcing inferences where they are in agreement, or posing new questions where they differ significantly.

1.5.3 Chronology

The need for chronological control on all records of past environmental change has already been stressed. Ideally, chronological control in Holocene records should achieve decadal or sub-decadal resolution, with annually or seasonally resolved records highly desirable at least for the last millennium. This degree of control permits close comparison between sites and archives, greatly reduces the scope for miscorrelation, allows characterization of short-term, transient responses to events such as volcanic eruptions and provides evidence that can be smoothly linked to instrumental time series. Even where absolute dates are not known, a finely resolved 'floating' chronology makes possible calculations of rates of change as well as the precise sequence of ecosystem responses to perturbations. Tree-rings (Baillie and Brown, pp. 75–91 in this volume), varved lake (Zolitschka, pp. 92–106 in this volume), or marine sediments and speleothems (Lauritzen, pp. 242–263 in this volume) all provide the opportunity for annual dating, but in the many studies, this level of accuracy and precision is unattainable. Nevertheless, increasingly effective use of AMS radiocarbon dating, often coupled with tephra recognition, is providing an ever firmer chronological framework for Holocene research (Pilcher, pp. 63–74 in this volume).

1.6 CONCLUDING OBSERVATIONS

Some of the special qualities as well as the crucial significance of research into Holocene environmental variability have been identified above. This volume seeks to provide an authoritative guide to these, through chapters on methodologies, selected archives and proxies, as well as illustrative examples of applications and case studies. The special attention devoted to the Holocene record from lower latitudes reflects the relative paucity of evidence from these parts of the world (Bush, pp. 384–395 in this volume; Scott, pp. 396–405 in this volume), in comparison with the wealth of information available from North America and Europe.

The interaction between Holocene and immediately pre-Holocene environmental change and a crucial step in the development of human societies is the subject of the chapter by Wright and Thorpe (pp. 49–62 in this volume) and broader issues surrounding the role of climate change in societal development are addressed by Shennan (pp. 36–48 in this volume). These chapters give some sense of the interactions between past climate variability and social organization. Human societies have never been so insulated from environmental processes as to escape vulnerability to major shifts in climatic regime, extreme events or persistent droughts. Recognizing the role of environmental change in human affairs does not imply a return to simple-minded environmental determinism, for the impacts of environmental change on human societies are mediated through all manner of cultural and socio-economic processes. Nevertheless, they are far from insignificant.

This still applies today, especially for the poorest societies, which, in many cases, are the ones deemed most likely to be negatively impacted by future climate change (Houghton *et al.*, 2001). Even for the most technologically advanced societies, evidence from the past can highlight future threats of extreme gravity, as for example, in the case of rapidly dwindling groundwater resources in parts of the western United States. There, as in many other parts of the world, these are largely fossil waters formed during moister periods either in the first half of the Holocene or earlier. They are now being mined at a pace greatly in excess of any foreseeable rate of recharge. The palaeo-record thus provides both a glimpse into history and a dire warning for the future. Possible future impacts are also implicit in the last two chapters. As Claussen (pp. 422–434 in this volume) shows, where models and data are used together to shed light on climate system dynamics, feedbacks between the different components of the Earth System often give rise to non-linear responses disproportionate to the original forcing – an important point to bear in mind when we consider the future implications of the huge global experiment initiated by greenhouse gas enrichment of our atmosphere. Goodwin (pp. 406–421 in this volume) focuses on sea-level variability during the late Holocene and cites empirical evidence pointing up the links between climate variability and sea-level over the last two millennia, a period when the amplitude of global climate variability was much less than even the most modest projections for the next century. Much more work is needed to quantify possible links between recent secular variations in climate and sea-level, but the evidence so far gives no grounds for complacency in a world of densely settled shorelines and numerous coastal conurbations.

Almost all the major changes in human societies that have formed mileposts on the way from hunter-gatherer to the complex civilizations existing today have taken place during the Holocene period, the time since the end of the last Ice Age. Throughout the Holocene, there have been major environmental changes in every part of the world, but the rate of change has accelerated during the last 50 to 100 years, largely as a result of human activities. It is essential that we place our concerns for the future of the environment in the context of the changes that have occurred during the Holocene as a whole. We need to know how quickly ecosystems have responded to changes in the past and how resilient they may be to changes forced by current and future threats, for, if our Holocene past is anything to go by, we may expect 'surprises' – environmental responses well beyond any change in the external processes that provoked them and extremes well outside the range documented during the short period of instrumental records. The history of environmental change during the Holocene provides us with essential knowledge about the way the climate system and the biosphere work, rapidly growing insight into processes and rates of environmental change, a temporal context within which to place many of our present-day observations and a test bed for models designed to look into the future – for if the models cannot create scenarios to match the reality of the past, we must remain sceptical of their power to shed realistic light on the future.

CHAPTER

2

CLIMATE FORCING DURING THE HOLOCENE

Raymond S. Bradley

Abstract: The role of several important factors that have played a role in Holocene climate change is examined. These forcing factors operate on different time-scales: lower frequency (millennial-scale) climate changes associated with orbital forcing, century-scale variability associated with solar forcing, and annual- to decadal-scale variability associated with volcanic forcing. Feedbacks within the climate system may involve non-linear responses to forcing, especially if critical thresholds are exceeded. In addition, there may be distinct regional climate anomaly patterns that result from certain types of forcing. Other anomalies that appear in Holocene paleoclimatic records may be unrelated to external forcing factors, but reflect conditions entirely within the climate system. General circulation model simulations play an important role in helping to understand how these various factors interact to produce the observed changes in Holocene climate.

Keywords: Orbital forcing, Solar forcing, Volcanic forcing

Why did climate change during the Holocene? The paleoclimatic records that are discussed at length in other chapters reflect, to a large extent, the composite effects of external factors operating on the climate system, plus feedbacks within the climate system that were triggered by these factors. Climate forcing (external factors that may cause climate to change) can be considered on several time-scales, ranging from very long-term (multi-millennial) to interannual. The resulting climate in any one region is the consequence of variability across all time-scales, but breaking the spectrum of climate variability down, from lower to higher frequencies, provides a useful way of assessing different forcing mechanisms. Here some of the main forcing factors during the Holocene are considered, beginning with factors operating at the lower frequency end of the spectrum.

2.1 Orbital Forcing

On the very longest, multi-millennial time-scales, the main factors affecting Holocene climate change are related to orbital forcing (changes in obliquity, precession and eccentricity). These changes involved virtually no change in overall global insolation receipts (over the course of each year) but significant re-distribution of energy, both seasonally and latitudinally. Representing the time- and space-varying nature of orbitally driven insolation anomalies is difficult in a single diagram, but Plate 1 shows these changes schematically for each month, with each panel representing the time-varying anomaly pattern over the last 10,000 years, with respect to latitude. In the Early Holocene, precessional changes led to perihelion at the time of the northern hemisphere summer solstice (today it is closer to the winter solstice). This resulted in higher summer insolation in the Early Holocene at all latitudes of the northern hemisphere (ranging from ~40°W/m^2 higher than today at 60°N to 25°W/m^2 higher at the Equator). Thus, July insolation (radiation at the top, or outside, the atmosphere) has slowly decreased over the last 12,000 years (See Plate 1). Anomalies during southern hemisphere summers were smaller, centred at lower latitudes, and they were opposite in sign (that is, insolation increased over the course of the Holocene). For example, January insolation anomalies were ~30°W/m^2 below current values at 20°S in the Early Holocene.

What impact did such changes have on climate in different regions? Unfortunately, it is not a simple matter to translate insolation anomalies of solar radiation entering the atmosphere into radiation receipts at the surface, and it is even more difficult to then infer the effect of such changes on climate. Radiation passing through the atmosphere is reflected and absorbed differently from one region to another (depending to a large extent on the type and amount of cloud cover). Furthermore, surface albedo conditions also determine how much of the radiation reaching the surface will be absorbed. There may also be complexities induced in the local radiation balance. For example, Kutzbach and Guetter (1986) found that a 7 per cent increase in solar radiation at low latitudes, outside the atmosphere, at 9 ka BP was associated with 11 per cent higher net radiation at the surface due to a decrease in outgoing long-wave radiation (because of increased evaporation and higher water vapor levels in the atmosphere, which absorb long-wave radiation). However, this amplification of solar radiation effects was not as important at higher latitudes (where precipitation amounts are much less a function of solar radiation anomalies). Differences in surface properties may lead to changes in regional-scale circulation; for example, differential heating of land versus ocean (with the same insolation anomaly) could lead to land–sea circulation changes. Indeed, this effect served to drive an enhanced monsoon circulation over large parts of the northern continents in the Early Holocene, leading to wetter conditions and consequent changes in vegetation (see below). Finally, on an even larger scale, differential radiation anomalies from the Poles to the Equator may have led to changes in temperature gradients and consequent changes in the overall strength of atmospheric circulation, with associated shifts in the Hadley and extra-tropical circulation (Rind, 1998, 2000).

To assess such complexities under a constantly varying insolation regime requires general circulation model simulations. Numerous studies have examined the effects of orbital forcing for selected time intervals during the Holocene, initially using atmosphere global circulation models (GCMs; with fixed sea-surface temperatures: e.g. Kutzbach and Guetter, 1986; Hall and Valdes, 1997), then models with interactive ocean–atmosphere systems (e.g. Kutzbach and Liu, 1997; Hewitt and Mitchell, 1998) and, more recently, fully coupled ocean–atmosphere–biosphere models, where the interactions between the atmosphere and land surface hydrology and vegetation is treated explicitly (Brovkin *et al.*, 1998; Kutzbach, 1996; Coe and Bonin, 1997; Broström *et al.*, 1998). Most of these studies focus in particular on northern Africa where Early Holocene conditions were much wetter than the Late Holocene, and the transition between these states was quite abrupt, at around 5500 calendar years BP (deMenocal *et al.*, 2000a). All model simulations demonstrate that increased summer insolation in northern hemisphere summers, in the Early Holocene, caused a stronger monsoon circulation and increased precipitation in sub-Saharan Africa. However, unless vegetation and hydrological feedbacks are incorporated into the models, the precipitation amounts simulated are well below those that would have been necessary to support the lakes and vegetation changes that are known (from the paleoclimatic record) to have occurred (Foley, 1994).

These conclusions have been obtained from model simulations, generally centred at specific time intervals (snapshots at 3000-year intervals) through the Holocene (e.g. Kutzbach, 1996; Joussaume *et al.*, 1999). With complex GCMs, given computational constraints, it is not feasible to run long, multi-millennia simulations to examine transient changes. However, the paleoclimatic record suggests that the gradual changes in insolation were not matched by equally gradual changes in surface climate over North Africa. Rather, the transition from arid to humid conditions was abrupt, both at the onset of wetter conditions (~14,800 calendar years BP and at its termination ~5500 calendar years BP). Both transitions correspond to summer (June–July–August (JJA)) insolation levels of ~4 per cent greater than today (outside the atmosphere) at 20°N. To investigate this, transient model simulations have been made for the last 9000 years, using a much lower resolution zonally-averaged model, (but with a coupled ocean–atmosphere system and vegetation feedbacks) (Claussen *et al.*, 1998, 1999; Ganopolski *et al.*, 1998a). These point to the importance of vegetation feedbacks as critically important; vegetation changes abruptly amplified the linear orbital influence on precipitation over North Africa, to produce an abrupt, non-linear change at ~5440 BP, corresponding to the paleoclimatic field evidence (Fig. 2.1). Although the North African case may be an extreme example of the role of vegetation feedbacks, other model simulations also suggest that vegetation changes at pronounced ecotones (such as the tundra–boreal forest interface) may also play a strong role in modifying initial forcing factors (Foley *et al.*, 1994; TEMPO, 1996; Texier *et al.*, 1997).

2.2 SOLAR FORCING

Orbital forcing involves the redistribution of incoming solar energy, both latitudinally and seasonally. Thus there are differential effects on the climate system that can lead to circulation changes, and there may be different responses to the forcing in the northern and southern hemispheres. Changes in solar irradiance (energy emitted by the sun) might be expected to affect all parts of the earth equally. However, this is not so because the response to solar irradiance forcing is amplified regionally, as a result of feedbacks and interactions within the atmosphere (Rind, 2002).

Until quite recently it was assumed (based on measurements in dry, high-altitude locations) that total irradiance did not vary, at least not on inter-annual to decadal scales – hence the term 'solar constant' was coined to describe the energy that is intercepted by the atmosphere when the sun is overhead (1368°W/m^2) (National Research Council, 1994; Hoyt and Schatten, 1997). Satellite measurements over the last ~25 years tell a different story – total solar irradiance (TSI; that is, integrated over all wavelengths) varies by ~0.08 per cent over a Schwabe solar cycle (average length of ~11 years), with maximum values at times of maximum solar activity (when there are many sunspots and bright solar faculae; Lean, 1996; Fröhlich and Lean, 1998) (Fig. 2.2). Furthermore, irradiance changes at very short (ultraviolet) wavelengths vary even more over a solar cycle (Lean, 2000). Such changes have significance because an increase in UV radiation causes more ozone (O_3) to be produced in the upper stratosphere; ozone absorbs radiation (at UV wavelengths of 200–340 nm) so heating rates in the upper atmosphere are increased during times of enhanced solar activity. This then affects stratospheric winds (strengthening

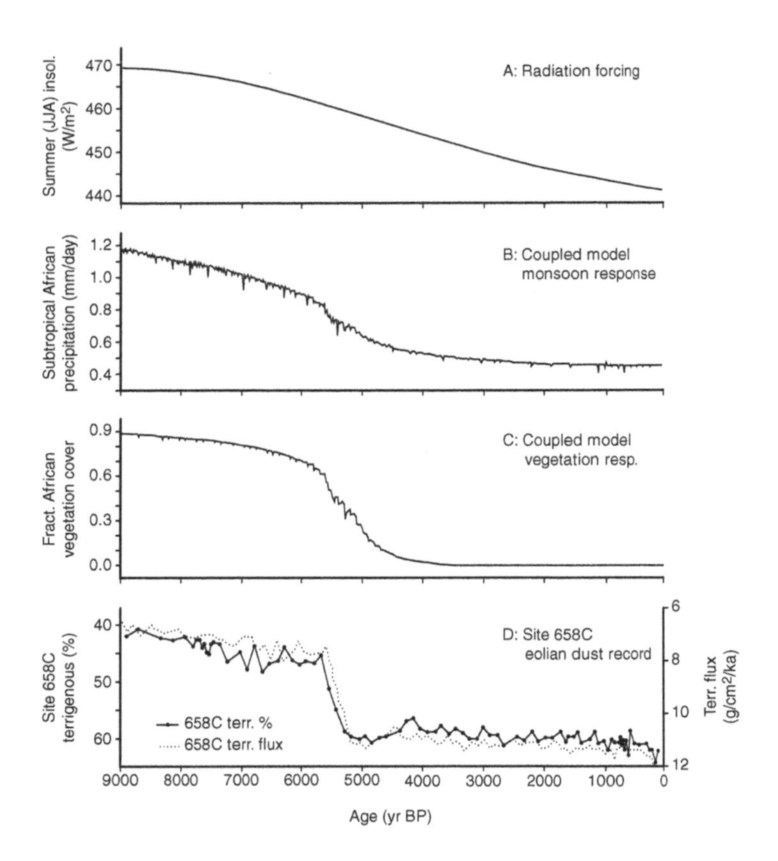

Figure 2.1 Model simulations (Claussen *et al.*, 1999) of the response of North African precipitation (b) and fractional vegetation cover (c) compared to radiation forcing – summer (JJA) radiation at 20°N (a) and the record of eolian (Saharan) dust in a sediment core from off the west coast of North Africa (deMenocal *et al.*, 2000a). The model incorporates vegetation feedbacks that seem to be important in generating a non-linear response to orbital forcing, at ~5500 BP. Reprinted from deMenocal *et al.* (2000a), with kind permission from Elsevier Science.

stratospheric easterlies), which can in turn influence surface climate via dynamical linkages between the stratosphere and the troposphere (Shindell *et al.*, 1999; Baldwin and Dunkerton, 2001; O'Hanlon, 2002). Model simulations of these effects indicate that there is a poleward shift of the tropospheric westerly jet and a poleward extension of the Hadley circulation, by ~70 km from solar minimum to solar maximum, in the summer hemisphere (Haigh, 1996; Larkin *et al.*, 2000). Although such changes are small, if irradiance changes in the past were larger and more persistent than solar cycle variability, the effects may have been quite significant.

How much has irradiance changed over longer time-scales? Satellite measurements are too short to shed light on longer-term irradiance changes, so these must be inferred from other lines of evidence. Lean *et al.* (1992) examined variations in brightness of stars similar to our own sun, concluding that present-day solar activity is at relatively high levels. By analogy with the range

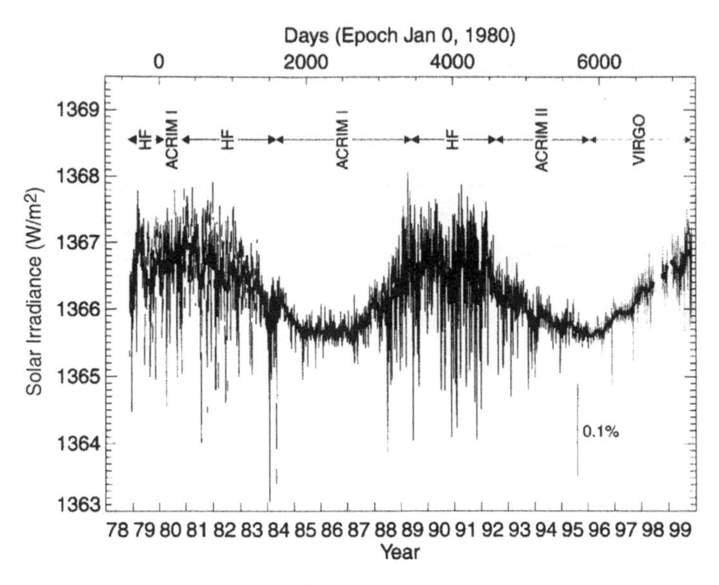

Figure 2.2 Total solar irradiance as recorded by satellites since 1979 (Fröhlich, 2000). This energy is distributed only over the illuminated half of the earth which intercepts it over a circular area ('circle of illumination'). Considering the area of a sphere (4 πr^2) versus that of a circle (πr^2), the average energy impinging at the top of the atmosphere is 1368/4, or ~342°W/m^2 (Hoyt and Schatten, 1997). Hence a variation of 0.08% over an ~11-year (Schwabe) solar cycle is equivalent to a mean forcing of 0.27°W/m^2; this is further reduced (by ~30%) due to planetary albedo effects (scattering, reflection) to ~0.2°W/m^2. It should be born in mind, however, that the earth can not reach radiative equilibrium in relation to forcing over an 11-year solar cycle, as the ocean has great thermal inertia, which smoothes out the effects of rapid changes in external forcing. Reprinted from Fröhlich (2000) with kind permission from Kluwer Academic Publishers.

of brightness in stars like the sun, with or without activity cycles, they inferred that the historical range of TSI varied by ~0.24 per cent, from the time of minimal solar variability at the end of the 17th century (the 'Maunder Minimum', ~AD 1645–1715) to the present (i.e. the mean of the most recent solar cycle). Shorter-term (~11 years) variability of ~0.08 per cent is superimposed on the lower frequency changes (Fig. 2.3). Simple comparisons with long-term temperature estimates suggest that changes in TSI of ~0.24 per cent were associated with mean annual surface temperature changes over the northern hemisphere of 0.2–0.4 °C (Lean *et al.*, 1995) and such changes have also been simulated in GCM and energy balance solar forcing experiments (Rind *et al.*, 1999; Crowley, 2000; Shindell *et al.*, 2001). Indeed, Crowley (2000) concluded that much of the low frequency variability in northern hemisphere temperatures over the last millennium (prior to the onset of global anthropogenic effects) could be explained in terms of solar and volcanic forcing. It is also interesting that distinct patterns of regional temperature change may be associated with solar forcing, as seen in both empirical and modelling studies, due to complex interactions between the circulation in the stratosphere and the troposphere (Shindell *et al.*, 2001; Waple *et al.*, 2001). Prolonged periods of reduced solar activity, like the Maunder Minimum, are associated with overall cooler conditions, but cooling is especially pronounced over mid- to high-latitude continental interiors, and warmer

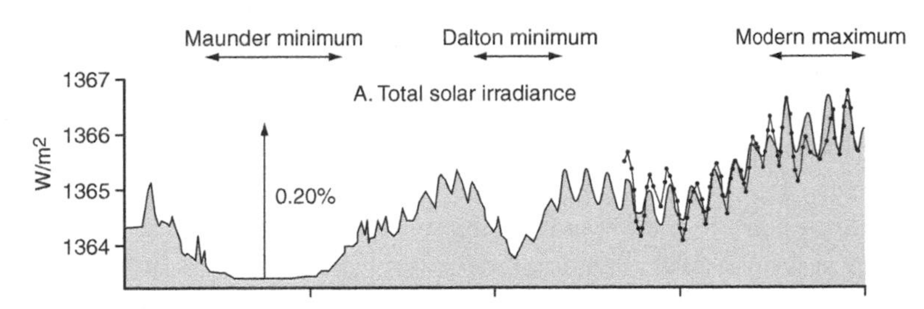

Figure 2.3 Estimated variability in total solar irradiance (TSI) over the last 400 years (Lean, 2000); cf. Fig. 2.2 for the most recent Schwabe solar cycles.

temperatures occur over mid- to high-latitudes of the Atlantic. Such a pattern is characteristic of a shift in the North Atlantic Oscillation (NAO) towards lower index conditions, whereby the pressure gradient between Iceland and the Azores is reduced, leading to less advection of warm, moist air from the Atlantic into western Europe, and cooler temperatures over Eurasia. Shindell *et al.* (2001) simulated winter temperatures over eastern North America and western Europe that were cooler by 1–2 °C during the late Maunder Minimum, compared to a century later when solar irradiance was higher (Plate 2). Such changes are consistent with the paleoclimatic evidence (Pfister *et al.*, 1999; Luterbacher *et al.*, 2001).

If solar irradiance has changed by ~0.24 per cent over the last 350 years, how much change occurred during the Holocene? Long-term changes in solar activity can be estimated from changes in cosmogenic isotopes preserved in natural archives. Cosmic rays intercepted by the upper atmosphere produce cosmogenic isotopes – such as ^{10}Be and ^{14}C – which eventually enter the terrestrial environment at the earth's surface. During times of high solar activity the flux of cosmic rays to the atmosphere is reduced, leading to a reduction in the production rate of these isotopes. Thus, variations in cosmogenic isotopes are inversely related to solar activity. If we make the assumption that solar activity is correlated with TSI changes (as observed in the recent instrumental period – see Beer *et al.*, 1996), long-term changes in ^{14}C (seen as departures from expected age, in tree-rings) or ^{10}Be (in ice-cores) can be used as an index of solar irradiance changes over time. Unfortunately, other factors have also affected the production rate of cosmogenic isotopes over the Holocene, and these must be accounted for in order to isolate the effects of solar variability. In particular, magnetic field variations have had a large impact on production rates (a weaker field being associated with higher production levels). Also, changes in the rate at which radiocarbon was sequestered in the deep ocean (due, for example, to thermohaline circulation changes) may also have affected atmospheric concentrations in radiocarbon over the Holocene. To examine this question for the last millennia, Bard *et al.* (2000) used a box model of ocean carbon variations driven by ^{10}Be variations measured in an ice-core to assess whether ^{14}C variations over the last 1000 years had been affected by ocean circulation changes. On this time-scale, such effects appear to have been minimal, suggesting that ^{14}C can be used to assess solar variability over the last thousand years and perhaps longer. A similar result was found by Beer *et al.* (1996) for the last 4000 years. In the last millennium, both ^{10}Be and ^{14}C indicate that solar activity was high from ~AD 1100 to 1250, decreased to minima in the 15th century and at the end of the 17th century, then increased in the 20th century to levels that were similar to those of the 12th century (Fig. 2.4). Detailed ^{10}Be records are not yet available for the Holocene, but deviations of ^{14}C from background levels (adjusted for magnetic field changes) reveal a large number of solar activity anomalies comparable to the Maunder Minimum (as well as episodes of enhanced solar activity) throughout the Holocene (Stuiver and Braziunas, 1993b) (Fig. 2.5). Using the Maunder Minimum as a guide, the variability of $\Delta^{14}C$ suggests that TSI may have varied by ± 0.4 per cent from modern levels during the Holocene.

If climatic conditions during the Maunder Minimum were driven by solar forcing alone, there ought to have been many similar climatic episodes during earlier periods. In fact, there are many Holocene paleoclimatic studies that claim solar forcing has driven observed changes, based largely

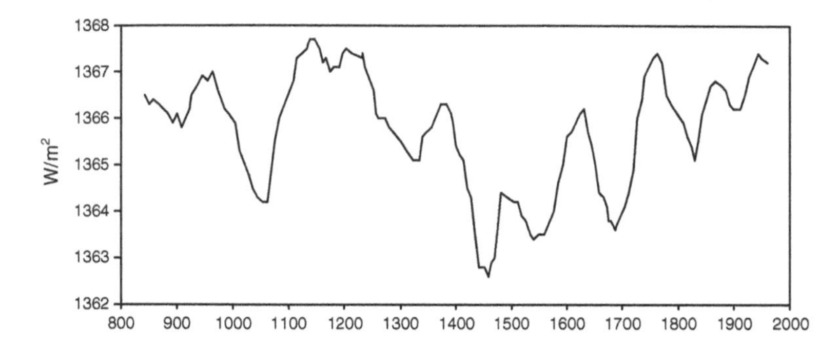

Figure 2.4 Total solar irradiance from AD 843 to 1961, estimated from ^{10}Be variations, recorded in an Antarctic ice-core, scaled to the estimates of Lean *et al.* (1995) (cf. Fig. 2.3). Other estimates of the magnitude of change in TSI from the Maunder Minimum to the present are higher – up to 0.65%, which, if correct, would simply amplify the scale of change shown here (data from Bard *et al.*, 2000).

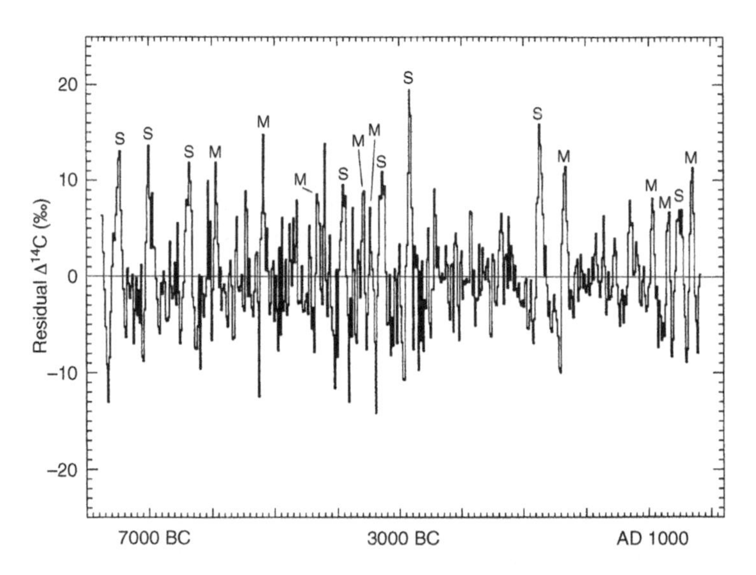

Figure 2.5 Solar activity variations (lower values indicating higher solar activity/enhanced irradiance), as recorded by radiocarbon variations in tree-rings of known calendar age (after taking geomagnetic field effects into account) M = Maunder Minimum-like events; S = Sporer Minimum-like events. Reprinted from Stuiver *et al.* (1991), with kind permission from Elsevier Science.

on comparisons of proxy records with the ^{14}C anomaly series. In particular, paleoclimatic records of precipitation variability across the tropics, from northern South America and Yucatan (Black *et al.*, 1999; Haug *et al.*, 2001; Hodell *et al.*, 2001) to East Africa and the Arabian Peninsula (Verschuren *et al.*, 2000; Neff *et al.*, 2001), show strong correlations with solar activity variations recorded by ^{14}C anomalies (e.g. Fig. 2.6). Furthermore, solar variability may also have played a role in mid-continental drought frequency on both short and long time-scales (Cook *et al.*, 1999; Yu and Ito, 1999; Dean *et al.*, 2002). Other studies have also identified potential links between solar activity variations and climate changes in the Holocene (Magny, 1993b; van Geel *et al.*, 2000). Bond *et al.* (2001) argue that temporal variations in the abundance of ice-rafted debris in North Atlantic sediments vary with the same frequency (~1450–1500 years) as ^{14}C anomalies, and Stuiver *et al.* (1991) also noted the similarity between a ~1470-year periodicity in ^{14}C data and a similar periodicity in oxygen isotopic data from GISP2. It remains to be seen how robust these relationships are, and what plausible mechanism might link solar activity/irradiance variations with climate in such diverse parts of the globe, from the Arabian Sea to the mid-continental USA, to Greenland. One possibility (seen in some model simulations) involves solar variations influencing the Hadley circulation (intensity and/or extent), which then leads to tropical and subtropical precipitation anomalies, and further teleconnections to extra-tropical regions.

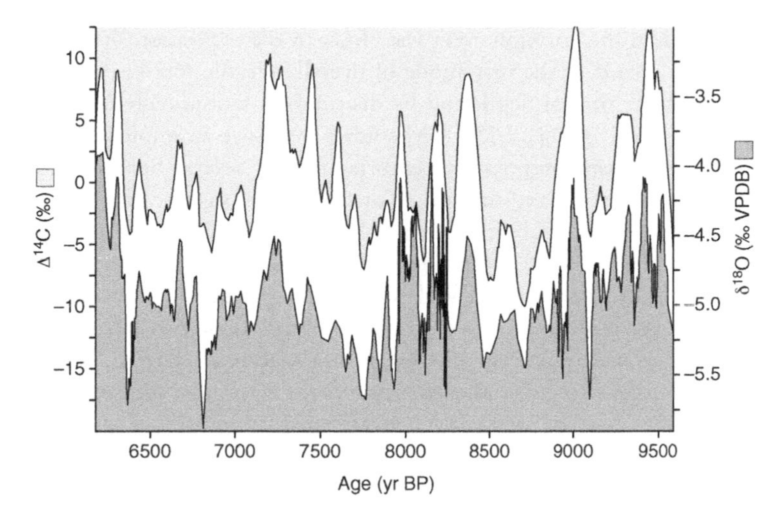

Figure 2.6 The relationship between $\delta^{18}O$ in an Early Holocene speleothem from Oman (representing rainfall, with lower values indicating wetter conditions) and $\Delta^{14}C$, representing solar activity variations (lower values indicating higher solar activity/enhanced irradiance). Reprinted from Neff *et al.* (2001), with kind permission from Macmillan Publishers Ltd.

2.3 VOLCANIC FORCING

It is well known, from studies of instrumental records, that explosive volcanic eruptions can have short-term cooling effects on overall hemispheric or global mean temperatures (Bradley, 1988; Robock, 2000). These result from direct radiative effects, with the volcanic aerosol reducing energy receipts at the surface, plus associated circulation changes that may result from such effects. Such circulation changes (often involving an amplification of the upper Rossby wave pattern) lead to large negative temperature anomalies in some regions, but other areas may become warmer. For example, it has been noted that warming in high-latitude continental interiors, in winter months, was commonly associated with major eruptions during the 20th century (Groisman, 1992; Robock and Mao, 1992, 1995).

Most temperature effects are not detectable after a few years, so individual explosive eruptions only contribute short-term variability to the spectrum of Holocene climate. However, if eruptions were more frequent in the past, or if they happened to occur in clusters of events, it is possible that the cumulative effect of eruptions could have persisted for longer, resulting in decadal- to multi-decadal-scale impacts. Such effects would be enhanced if the initial cooling led to feedbacks within the climate system, such as more persistent snow and sea-ice cover, which would raise the surface albedo and possibly alter the atmospheric circulation. Sulphate levels in ice-cores from Greenland provide an index of explosive volcanism in the past, albeit possibly biased towards high-latitude eruption events (Fig. 2.7). The GISP2 record suggests that there were indeed periods of more frequent events in the past, such as in the period 9500–11,500 calendar years BP (Zielinski *et al.*, 1994) Furthermore, in the early Holocene, there were many more large volcanic signals greater than that recorded after the eruption of Tambora (1815) which was the largest eruption in recent centuries (registering 110 ppb of volcanic sulphate at the GISP2 site in central Greenland) (Fig. 2.7). Of course, a larger sulphate signal might simply mean the eruption event was closer to the deposition site so we do not have a definitive long-term record of the magnitude of overall volcanic forcing, or more specifically, the record of atmospheric optical depth and its distribution latitudinally (cf. Roberston *et al.*, 2001). Nevertheless, energy balance and GCM studies that have attempted to parameterize the effects of explosive volcanism over recent centuries (where several lines of evidence can be combined to resolve the location and magnitude of each event) suggest that explosive volcanism has contributed to the natural variability of hemispheric and global mean temperatures over this interval of time (Crowley and Kim, 1999; Free and Robock, 1999; Crowley, 2000; Ammann *et al.*, 2003). Explosive volcanism, together with solar forcing, explains most of the variability of temperatures over the last millennium, so it seems likely that these two factors have also played a significant role in overall Holocene forcing, with volcanism having been of particular importance at certain times. Recent studies, for example, suggest that unusually cold conditions

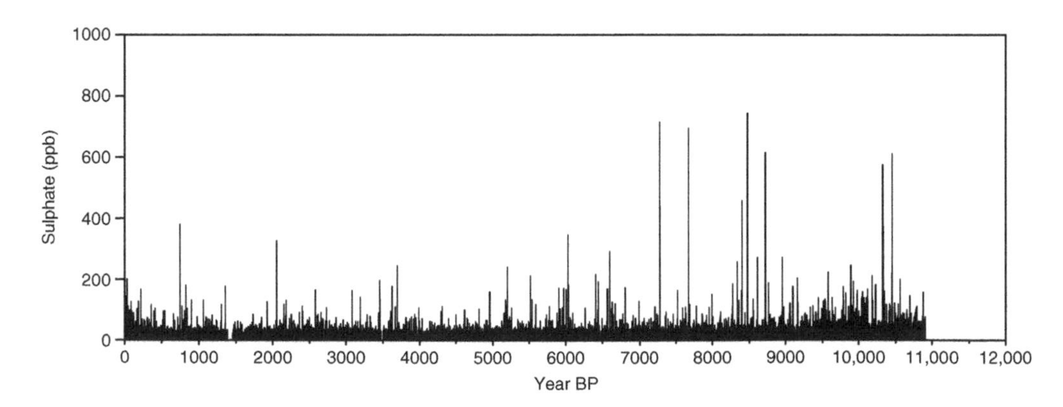

Figure 2.7 The Holocene record of volcanic sulphate (anomalies from background variations) recorded in the GISP2 ice-core from Summit, Greenland (Zielinski *et al.*, 1994).

occurred in western Europe and southern Alaska in the latter half of the solar Maunder Minimum, because this period also coincided with an episode of exceptionally active explosive volcanism (Bauer *et al.*, 2002; Cubasch, 2002; D'Arrigo *et al.*, 2002, Shindell *et al.*, 2002).

2.4 DISCUSSION

Most paleoclimatic studies that cite a relationship to particular forcing mechanisms do so either via simple correlations in the time domain (curve-matching) or in the frequency domain (finding a spectral peak that corresponds to something similar in a particular forcing factor; e.g. Black *et al.*, 1999; Bond *et al.*, 2001). There has generally been little interaction between those working at the frontiers of understanding how forcings affect the climate system, in a mechanistic or dynamic sense, and those observing the paleoclimatic record (though see Friis-Christensen *et al.*, 2000). Modelling provides a link between these two approaches, particularly when simulations involve coupled ocean–atmosphere GCMs (with a realistic stratosphere), incorporating vegetation and land–surface (hydrological) feedbacks. With such tools, it may be possible to comprehend the complex interactions that are driven by what often appears to be a simple forcing function (most commonly exemplified by a simple plot of insolation anomalies for a particular latitude and month!). With such models, the spatial climatic response to particular forcing factors can be determined, and perhaps thresholds and feedbacks within the system can also be identified (as suggested by Fig. 2.1) to help explain the observed Holocene paleoclimatic record.

Finally, it must be recognized that not all paleoclimatic variability seen in the Holocene can (or should) be ascribed to specific external forcings. Perhaps the best example of this is the ~8200 calendar year BP 'event', seen in many paleoclimatic archives, that resulted from catastrophic pro-glacial lake drainage at the margins of the Laurentide Ice Sheet (Alley *et al.*, 1997; Barber *et al.*, 1999a; Baldini *et al*, 2002). This rapid flooding of the North Atlantic with freshwater clearly had a significant regional impact, unrelated to any external forcing. There are also internal modes of climate system variability (e.g. El Niño Southern Oscillation, North Atlantic Oscillation, Pacific Decadal Oscillation) that likely vary on both long and short time-scales (though we know relatively little about their long-term behaviour). Furthermore, stochastic resonance in the climate system – by which a weak quasi-periodic forcing signal may be amplified into a non-linear, bi-stable climate signal – may have brought about relatively abrupt changes in the past by pushing the system across critical thresholds (Lawrence and Ruzmaikin, 1998; Ruzmaikin, 1999; Rahmstorf and Alley, 2002).

CHAPTER

3

AN INTRODUCTION TO CLIMATE MODELLING OF THE HOLOCENE

Paul J. Valdes

Abstract: An increasingly important motivation for palaeoclimate studies of the Holocene is the evaluation of computer climate models. These models play a central role in future climate change prediction and it is important that they are fully tested in climate regimes other than the present. In addition, the models can provide valuable insights into the interpretation of the proxy climate indicators. This chapter will describe the hierarchy of climate models, ranging from very simple box models to the most sophisticated general circulation models. The fundamental principles of the models will be discussed, as well as their strength and weaknesses. The Mid-Holocene, 6000 years before present, will provide one example of the use of such models

Keywords: Coupled models, General circulation models, Holocene, Monsoons, Palaeoclimate

Understanding climate change and its impact on human society is of fundamental societal interest and importance. A lot of the recent growth in research on this subject has been motivated by concerns about future climate change. There is a growing consensus that climate is changing due to mankind's activities, but the magnitude of future change, especially its regional and seasonal patterns, remains very uncertain (Houghton *et al.*, 2001). Predictions of future climate change are almost entirely based on computer climate models, and it is essential that these models are accurate and well tested. In most cases, these models have been developed within the context of present-day climate only, and thus the past represents a truly independent test of the models.

The Holocene represents a particularly valuable test of such models because it is a period with a huge wealth of high quality data. Moreover, it can be argued that warm, interglacial periods are of special interest because many climate feedback processes are potentially sensitive to the sign of the climate change. For instance, a small warming of tropical sea-surface temperatures will produce a much larger effect than a small cooling (due to the moisture-holding capacity of the atmosphere being non-linear). Thus although glacial climates can be used as a test of the models, the warmer Holocene may test components of the model which are more relevant for warm climate conditions.

However, it is important to emphasis that the Holocene should NOT be used as a direct analogue of future climate change (Mitchell, 1990). This is because the main cause of the long-term Holocene climate changes is associated with changes in the Earth's orbit. This produces a seasonal change in incoming solar radiation, which generally results in a seasonal change in climate. Thus Holocene warmth should not necessarily be thought of as universal warmth. In many regions it was probably summer warmth and winters were often colder. By contrast, increases in radiatively active gases (such as CO_2 and CH_4) cause an annual mean change in the radiation balance of the Earth and hence future climate change is predicted to exhibit a large annual mean change in climate (Houghton *et al.*, 2001). Thus the forcing mechanisms are very different and the Holocene should not be thought of as a direct analogue of future change. Instead, studying Holocene climate change helps to develop our understanding of climate change processes and allows us to evaluate and improve the climate models used for future climate change predictions.

A slightly different role of Holocene climate studies is the use of very high temporal resolution proxy climate data (ideally, annual resolution). This can be used to help in the attribution of the causes of recent climate change (Stott *et al.*, 2001). A key question is whether the last 150 years of climate variability is entirely natural, or is part of it due to human activity. Methods have been developed to 'fingerprint' the anthropogenic component using climate model estimates of natural variability. The results have led to strong statements about climate change, but such statements are very dependent on whether models are correctly simulating the temporal and spatial patterns of natural climate variability. High quality palaeo-data can help evaluate whether the models are able to simulate the natural component successfully (Collins *et al.*, 2002).

If we are to rigorously test climate models, it is essential that great care must be taken when interpreting the proxy record of climate. We do not have direct observational data beyond a few centuries, so we have to rely on indirect proxies for how the climate system was varying. The climatic interpretation of the geological data must be carefully and systematically performed. In all cases, the relationship between the proxy and the climate is imprecise and appropriate errors bars much be calculated. It should always be remembered that the climatic interpretation of the proxy dataset is a form of model. This type of 'data model' (e.g. a transfer function) is an empirical model which may well have weaknesses, uncertainties and errors associated with it (see Birks, pp. 107–123 in this volume). In this sense, it is not just computer climate models that require testing. It is essential that we always consider the testing of the 'data model'. There are several examples where climate model testing has shown potential errors in the data interpretation.

In addition, it is vital that there is a good understanding of how the proxy is related to climate. Some proxies may be more sensitive to summer warmth than the annual mean, and we must not assume that these are the same (even if we have good correlations for the modern climate). In general, the climate models are able to simulate all of the important aspects of climate, including a full seasonal cycle of surface temperatures and precipitation so it is possible to perform model–data comparisons using the most appropriate climate variable.

Furthermore, some proxies may depend on several climate variables and it is difficult to separate the different effects. For instance, vegetation can depend on temperature and precipitation so that a pollen distribution can be interpreted as indicating a warmer or wetter climate or both. Similarly, oxygen isotope data are sensitive to the mean temperature, the seasonal cycle and the circulation (and hence source) of the moisture. In such cases, it is difficult for the proxy to give a unique reconstruction of climate and hence it is difficult to test the models rigorously. In these circumstances, a very effective tool is to use 'forward' modelling techniques in which the climate model is used to directly predict the proxy data. These methods have been widely used for predicting lake levels (e.g. Coe, 1997), vegetation distributions (e.g. Harrison *et al.*, 1998) and oxygen isotopes (e.g. Jouzel *et al.*, 2000; Werner *et al.*, 2001).

3.1 MODELLING CLIMATE

There are many different types of model and before showing some results for the Holocene, it is important to describe briefly how they work. '**Data models**' such as transfer functions are almost entirely empirical. They rely on present-day training-set data to devise a quantitative link between climate and the proxy data. An understanding of the mechanisms which relate climate to the proxy are not directly needed, although an understanding will often be essential for good use of a proxy.

Computer climate models start from a different perspective. These models are physically based, process-orientated, numerical models. Their starting point is NOT knowledge of the present climate. Instead, they start from the basic physical laws that govern the climate system. This includes Newton's laws of motion, thermodynamic laws, ideal gas law and conservation of mass and moisture. These physical laws apply for all time periods and if we were truly able to solve such equations perfectly, then we would be able to have high confidence in predictions of future climate change. In practice, we solve these equations using computers and advanced numerical methods but have to make many approximations and assumptions, some of which may be influenced by present-day climate regimes (Trenberth, 1992). All models divide the world into a set of grid boxes which are typically a few hundred kilometres in size. There are many important climate processes (e.g. clouds and convection) that work on scales smaller than this. We cannot ignore such processes so instead we have to estimate (parameterize) the effects of sub-grid scale processes. The parameterization schemes are also developed based on physics but they always include some empirical estimates based on present-day conditions.

In addition, we are still learning about many potentially important climate processes (and palaeoclimate research has been a major source of our knowledge of important climate change processes). Some of these processes are only just being incorporated into the models and some are difficult to simulate from first principles (this is especially true for biological processes related to vegetation cover and the carbon cycle). The result is a computer model of the climate system which is based on first principles but which contains uncertainty. One method for quantifying such uncertainty is to test them in climate regimes different to those used in their development.

Within this category of physically based climate models, there are a number of different types but they can be broadly categorized into four sub-classes.

3.1.1 Box Models

As the name suggests, box models split the Earth System into a small number of boxes and often include a limited set of equations. They are a very effective tool for establishing and quantifying the importance of a particular processes. Examples include those by Maasch and Saltzman (1990) who used a three-box model (representing global ocean temperature, global ice volume and global carbon dioxide concentration) to examine the long-term variability of the climate system. The model was able to reproduce many aspects of Milankovitch time-scale variability including the change in amplitude and frequency that occurred at around 700,000 years BP. The strength of this type of model is the ability quickly to quantify processes and their interactions, especially on very long time-scales. It is also much easier to understand all of the details of the interactions. Their weakness is that the models will generally not be able to give detailed simulations of the Earth System, and it is possible to 'tune' (i.e. choose approximations and empirical constants) the models to over-emphasize a particular process.

3.1.2 Energy Balance Models (EBMs)

In terms of the degree of complexity and detail, the next group of models are based on **energy balance**. The fundamental physics is that energy is conserved. The Earth System is driven by incoming solar radiation and is cooled by the emission of longwave (infrared) radiation (Crowley and North, 1992). If the climate is in equilibrium, the incoming and the outgoing radiation must exactly balance. Any imbalance will result in warming or cooling of the system, at a rate which is proportional to the thermal capacity of the climate system.

The amount of absorbed solar radiation depends on the incoming flux of solar energy (which depends on the solar output and the distance from the Sun), and the reflectivity (albedo) of the system. The albedo is influenced by the surface properties (especially vegetation type and ice and snow cover) and cloud cover. These latter aspects are part of the climate system and represent potentially important internal feedbacks.

The amount of outgoing long-wave radiation depends on the temperature, atmospheric composition (CO_2, water vapour, CH_4, etc.), surface conditions and cloud cover. The former is controlled by the basic physics of the system (Stefan-Boltzman equation), whereas the latter depends on the climate system and again represents potentially important internal feedbacks.

The most advanced versions of energy balance models (e.g. Gallée *et al.*, 1991; Sakai and Peltier, 1996; Weaver *et al.*, 2001) include two or more dimensions (either latitude–height or longitude–latitude) and couple the energy balance to ice sheet and ocean models. They have been very effective tools for understanding the details of climate change over the last 100,000 years, and have shown that these depend on feedbacks between the climate and the ice sheets and carbon cycle. These models also have predictive powers for future climate change and have also played an important role in understanding the causes of climate change during the last century or two (Crowley and Kim, 1999).

The weakness of this type of model is that, in general, the only predicted variable is temperature. However, some models (e.g. Weaver *et al.*, 2001) have included a representation of the hydrological cycle and have coupled this model to a dynamic ocean model. Most energy balance models do not have much internally generated variability so that temporal variability is from the forcing only. In addition, all of these models have to make large approximations about processes not explicitly included in the energy balance equations, such as the circulation of the atmosphere.

3.1.3 Earth System Models of Intermediate Complexity (EMICs)

Another group of models bridge the gap between the energy balance models and the full complexity, general circulation models (described in 3.1.4). These intermediate complexity models (Claussen *et al.*, 2002) include the physics of energy balance but also include more detailed representations of other key climate physics. The division between EMICs and EBMs is not precise. In general, one of the key aspects is that there is an attempt to explicitly represent the **circulation of the atmosphere and/or ocean**. An example of an EMIC is the Potsdam model (Petoukhov *et al.*, 2000). This is described in more detail by Claussen (pp. 422–434 in this volume). An additional aspect of many EMICs is that they include an attempt at representing more than just the atmosphere and ocean. The models include (or are in the process of developing) sub-components to represent the terrestrial vegetation, carbon cycle and land ice sheets, as well as the atmosphere and ocean.

The great benefit of this type of model is their computational speed, and the incorporation of the processes important for the study of the Earth System as a whole. It is possible to run these models for many millennia and to perform many simulations so that the detailed responses and a detailed understanding can be developed. They have been important tools for examining the stability of the thermohaline circulation (e.g. Ganopolski *et al.*, 1998b; Ganopolski and Rahmstorf, 2001) and the stability, sensitivity and feedbacks of North African climates and vegetation (e.g. Claussen *et al.*, 1999; Claussen, pp. 422–434 in this volume). They are also an important tool for studying long time-scale cycles, such as those observed by Bond *et al.* (1997).

The weaknesses of this type of model mainly lie in the heavily approximated nature of the governing equations. This is particularly true for the atmosphere and ocean. In addition, the models are often very coarse resolution (e.g. the grid boxes representing the Earth System can be 50° longitude × 10° latitude) which causes problems when evaluating such models against the data. Moreover, EMICs are not heavily used in future climate change predictions for the next century. Nonetheless, they represent a useful and important part of the hierarchy of climate models.

3.1.4 General Circulation Models (GCMs)

The most complex models are those which attempt to solve the **full dynamical equations** that govern the climate system. These solve the basic laws of physics by simulating the time variations of the atmosphere and ocean on relatively small spatial and temporal scales (the models often have time steps of 1 hour or less). The typical grid that such models use is approximately 3° (equivalent to an average resolution of about 250 km). The models have the potential for variability on time-scales from diurnal up to decadal and beyond. They simulate a full spectrum of variability, both forced and internally generated (such as the El Niño Southern Oscillation (ENSO) or the North Atlantic Oscillation (NAO)). A typical grid of a state-of-the art model is shown in Fig. 3.1. The models also include approximations (normally called **parameterizations**) of many climatically important processes which act on spatial scales smaller than the explicit grid. The approximations result in imperfect simulations of the present-day mean climate and climate variability (see Houghton *et al.*, 2001).

The strength of these models is that they represent our best attempts at modelling climate. Originally such models included only the atmosphere and had crude representations of the ocean and virtually no representation of other sub-components of the Earth System. More recently, this has changed so that the latest versions of these models can be referred to as Earth System models and include the same set of components as EMICs. However, full complexity GCMs have a major weakness in that they cannot be used to simulate long-term changes. The

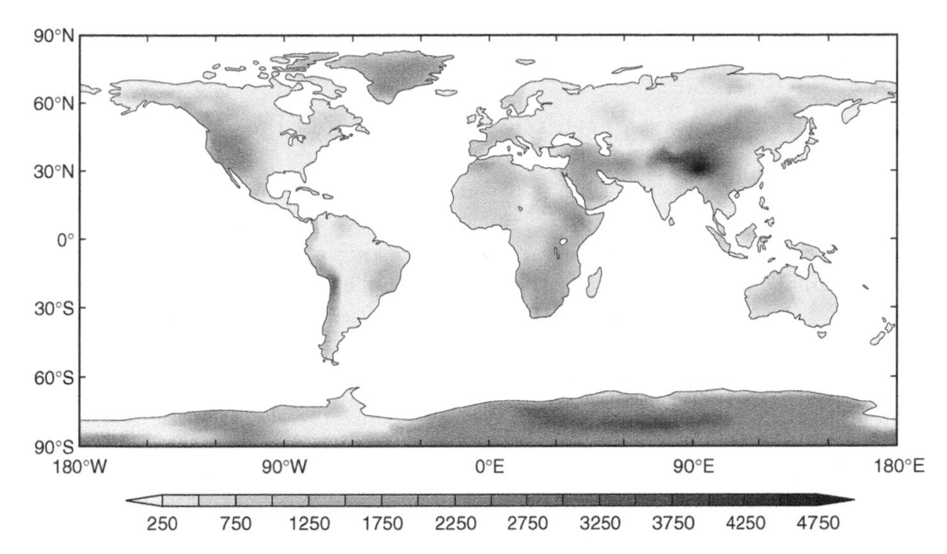

Figure 3.1 The land sea mask and orography of a typical climate model (in this case the Hadley Centre model). The grid size is 3.75° in longitude and 2.5° in latitude. The shading shows the mean orographic height (in m). The contour interval is 250 m. Note that at this resolution, small mountain ranges (such as the Alps) are not resolved. Sub-grid scale parameterization attempt to improve upon this.

longest that most of these models can be run is 1000 years, and more usually they are only used to simulate a few decades. Hence these simulations are often referred to as 'snapshot' time-slice simulations. They simulate the typical climate for a specific time period in the past but do not give detailed information about the temporal variation of climate. Thus to test such models requires detailed data at specific time slices. The most well studied period corresponds to the Mid-Holocene, 6000 years BP.

The models were developed without use of any knowledge of past changes and hence the palaeo-record represents a true independent test of the models. However, it is important to distinguish two different aspects of testing. A model can be wrong because of **errors in the boundary conditions** for the model, or due to **errors in the internal representation of the underlying climate physics.** The latter is the most important in terms of future climate change because any errors in the model physics would potentially apply for all climate regimes, whereas errors in the boundary conditions may only be important aspects of model–data disagreements for the past.

Boundary conditions correspond to the external forcing of the model. For a complete Earth System model, this would only include such things as the solar output, the Earth's orbital parameters, volcanic output and on longer time-scales, continental drift. This is the ultimate dream of an Earth System modeller but is not yet possible. Instead we have to give the model additional forcing conditions, which are in part internal to the climate system. This includes land ice sheets (important in the Glacial and Early Holocene), and carbon dioxide and methane concentrations. In addition, if an atmosphere-only model is used we must specify the sea-surface temperatures. Global reconstructions exist for the Last Glacial Maximum (CLIMAP,

1981) but there are no such reconstructions for the Holocene. Vegetation is another potentially important boundary condition for models.

More recently, atmosphere–ocean models have started to be used for Holocene climates (e.g. Kutzback and Liu, 1997; Hewitt and Mitchell 1998; Otto-Bliesner, 1999; Braconnot *et al.*, 2000; Liu *et al.*, 2000). These models predict the circulation of both the atmosphere and ocean, and thus sea-surface temperature data can be used to evaluate the model, rather than as input to the model. This is an important new development because there is no longer any need to create global datasets as input to the model's. Instead they can be used as a test of the model and uniform global coverage is not essential (although probably still desirable). An additional feature of the coupled atmosphere–ocean models is that they better simulate the variability of climate on interannual- to century-scale, and such variability can be tested with high temporal resolution data (Otto-Bliesner, 1999).

Even more recently, models have also started to include interactive vegetation so that changes in the physical climate system (the atmosphere and ocean) can interact with the biological system (e.g. Kutzbach, 1996; Texier *et al.*, 1997; Braconnot *et al.*, 1999; Kutzbach *et al.*, 1999; Doherty *et al.*, 2000). For the Holocene, this is an important additional interaction, which helps understand some of the changes in the monsoon system during these periods. It now means that the only boundary conditions needed for Holocene studies consist of the true external forcings plus well-known and accurate internal boundary conditions such as CO_2 and CH_4. For the Early Holocene, the ice sheet configuration is also needed but does have uncertainty.

The following sections show examples of palaeoclimate modelling of the Mid-Holocene, 6000 years BP. Many other periods within the Holocene have been studied but a lot of recent work has focussed on the Mid-Holocene. The main reason for choosing this period is that the orbital forcing is large but the ice sheets are the same as in the present day. This ensures that the model simulations are simple to perform, and there is less uncertainty about the appropriate boundary conditions. This hopefully increases the value of the model evaluations.

3.1.4.1 Mid-Holocene Climates and the Palaeoclimate Model Intercomparison Project

From a modelling perspective, the Mid-Holocene (6000 years BP) represents one of the most heavily studied periods. It has been the focus of a major international project called the Palaeoclimate Model Intercomparison Project (PMIP) (Joussaume and Taylor, 1995). The aim of the first phase of this project was to investigate and quantify the uncertainties in atmospheric GCMs. About 18 different climate modelling groups participated and ran simulations for the present-day and Mid-Holocene using exactly the same changes in boundary conditions. Thus differences between the model-simulated changes in climate gives us an estimate of the modelling uncertainty due to the design of the models (the models differed in their spatial resolution, and the details of their parameterization schemes). Comparison between the models and observations gives an estimate of the errors with the models, or the errors in the boundary conditions.

The boundary conditions were chosen to be simple and do not represent an attempt to precisely simulate the Mid-Holocene. The only boundary condition change was to set the orbital parameters to those appropriate for 6000 years BP (Fig. 3.2). This corresponds to higher

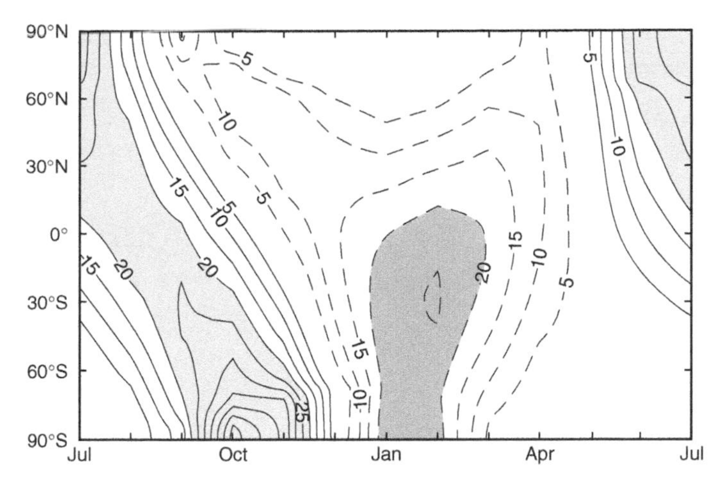

Figure 3.2 The mean change in incoming solar radiation at the top-of-the-atmosphere for the Mid-Holocene (6000 yr BP), as a function of latitude and time of year. The contour interval is 5°W/m^2 and negative values are dashed. Values greater than 20°W/m^2 are lightly shaded, and values less than –20°W/m^2 are darkly shaded. Positive values (i.e. during northern hemisphere summer) indicate that the Holocene receives more solar radiation than present day.

incoming solar radiation during northern hemisphere summers and reduced incoming solar radiation during northern hemisphere winters. The global, annual mean changes are negligible. All other boundary conditions were held constant at their present-day values. Thus there was no change in sea-surface temperature (SST) and vegetation cover, and the ice sheets were assumed to the same as present. The reason for holding SST constant was that there are no global reconstructions of SST for the Mid-Holocene, and the regional datasets suggested that the changes were generally small. Similarly, global vegetation reconstructions did not exist. Methane and carbon dioxide concentrations were also held constant.

All modelling groups ran integrations for the present-day and the Mid-Holocene. The simulations varied in length but had to be at least 10 years duration, preferably longer. The climate was then created by averaging the 10 or more year simulations. Because the models have different present-day climates, the focus was on the modelled changes in climate (i.e. Mid-Holocene minus the present day) for each model. It is also important to be aware that GCMs generate internal variability and some of the differences between the Mid-Holocene and the present day arise because of this variability, and is not due to the forcing of the climate. Statistical tests are used to identify those aspects of model changes which have been caused by the changes in boundary conditions.

Figure 3.3 shows the resulting modelled change in surface air temperature during December–January–February (DJF) and June–July–August (JJA). It is an average over all the models participating in PMIP. The dominant change in JJA is a warming of all of the continents by as much as 2 °C in the centre of the Eurasian continent. The only exception is small local cooling in the regions associated with an enhanced monsoon. In the DJF season, cooling is almost universal but does not reach the same magnitude as the JJA warming. Note

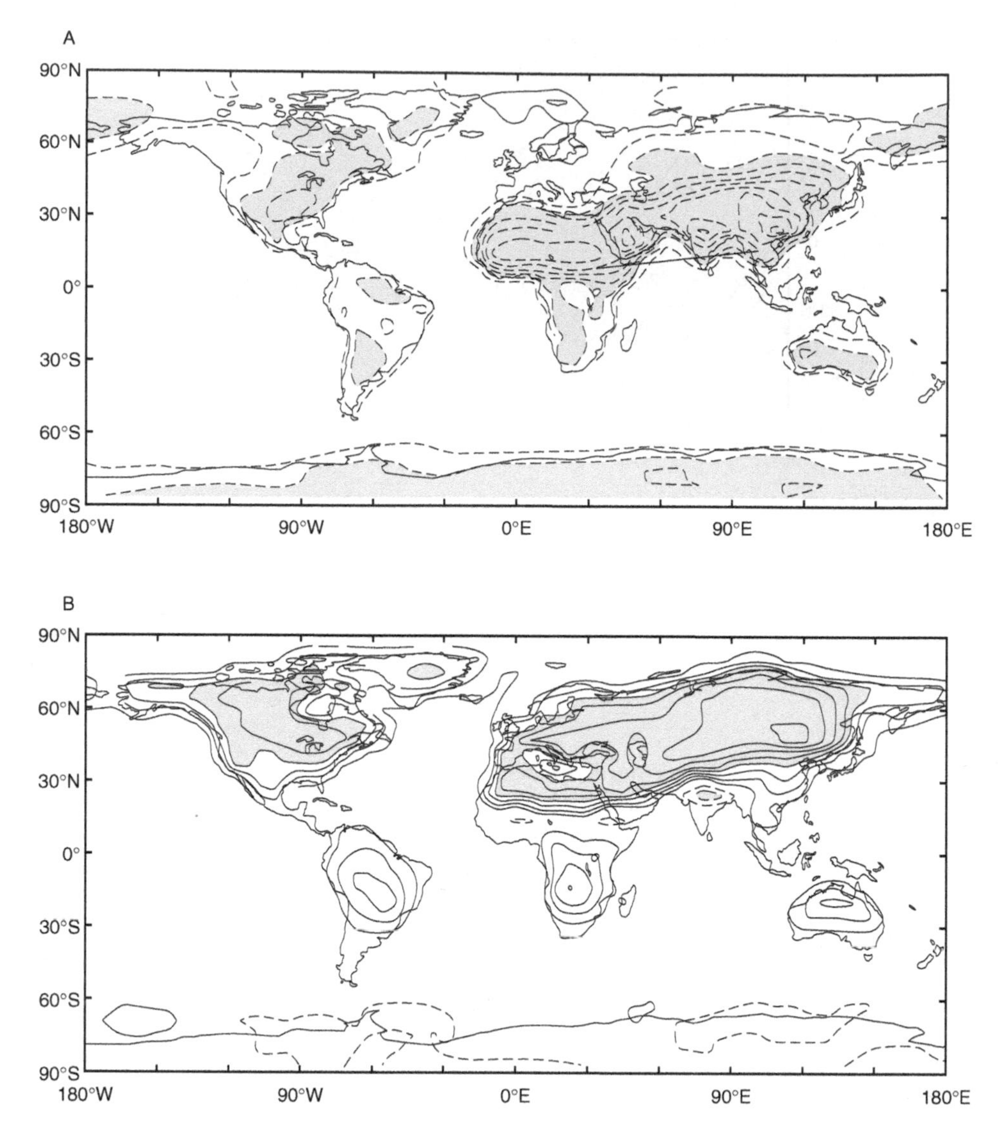

Figure 3.3 Model simulated change (Mid-Holocene – present day) in surface air temperature for (a) December–January–February (DJF) and (b) June–July–August (JJA) for the average of all PMIP model simulations. The contour interval is 0.25 °C, and negative contours are dashed. Light shading corresponds to temperatures less than –0.5 °C and darker shading corresponds to temperatures greater than +1 °C. Positive values indicate that the Mid-Holocene was warmer than the present day. The individual model simulations have been regridded onto a common grid (corresponding to about 5° × 5°).

that over the oceans and away from sea ice, the models predict no change but of course this is because of the choice of boundary conditions.

It is interesting to ask the question of whether all models showed this pattern or whether some models showed different results. Figure 3.4 shows the intermodel standard deviation for DJF

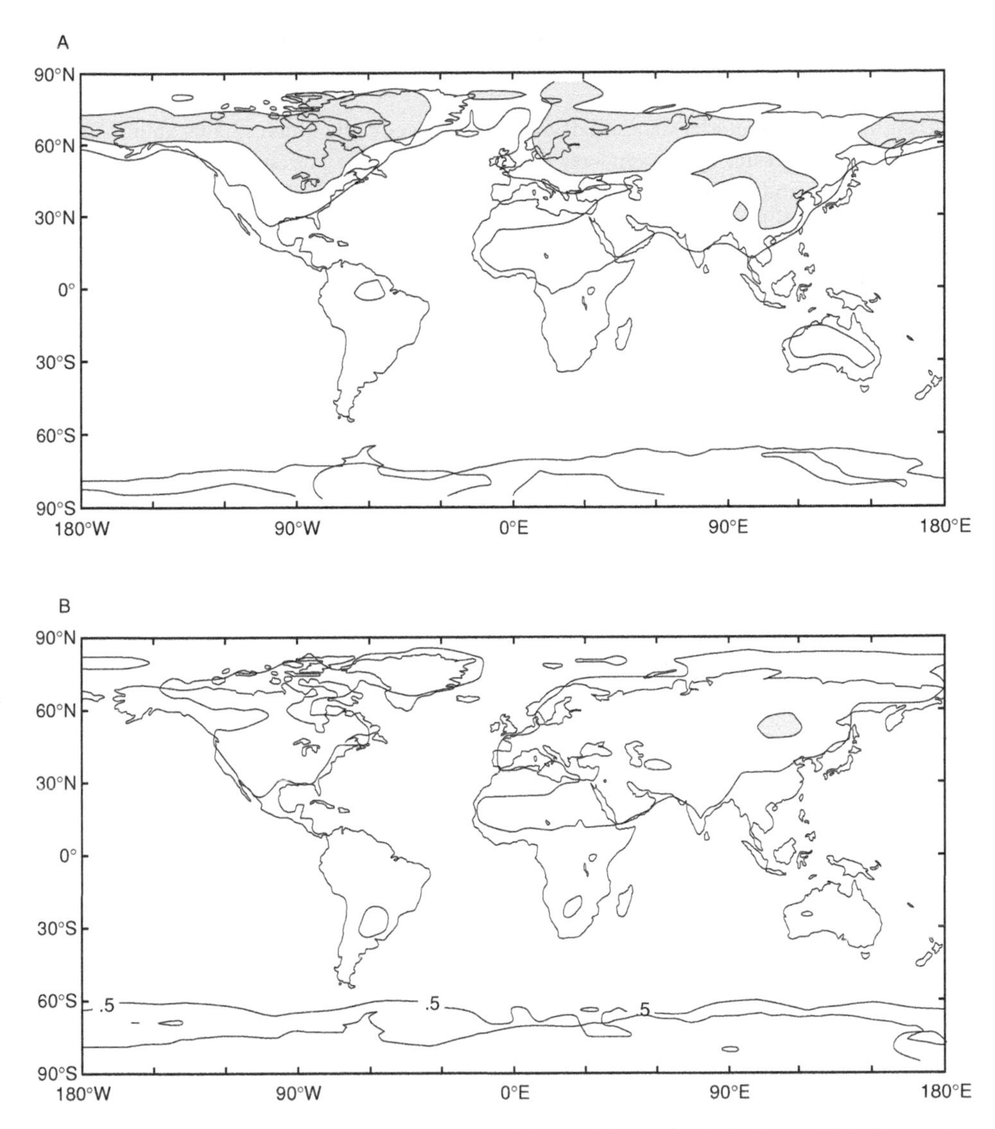

Figure 3.4 As for Fig. 3.3, but showing the model–model standard deviation. This figure shows a measure of the consistency of the models. Differences between the models are due to differences in the parameterisations within the models. The contour interval is 0.5 °C and values greater than 1 °C are shaded.

and JJA averages. This can be thought of as a measure of the uncertainty associated with the models. In DJF, the standard deviation can be very large and even exceeds the mean change. For instance, in North East Europe, the mean temperature change is very small (~0.1 °C) but the intermodel variability is very high (the range of predicted change varies between +7.3 °C and –4.0 °C). This would indicate that we should have relatively little confidence in model-predicted changes in this region. However, it should be cautioned that this region also experiences high amounts of interannual variability and so that some of the model–model differences may be associated with this. The PMIP database does not allow evaluation of the statistical significance of individual model simulations.

In contrast, the JJA intermodel standard deviation shows generally much smaller values, indicating that there is much greater agreement between the models. The largest model–model variability occurs over Asia and is substantially smaller than the mean change. There is also some model–model variability in the regions of cooling associated with enhanced monsoonal precipitation.

An alternative way of examining the robustness of the climate model predictions is to calculate the degree of consistency between the models. A very crude method is to calculate the percentage of the models which agree with the sign of the mean change. For JJA, most continental regions have a high degree of consistency, with all models predicting the warming. Thus any uncertainty is only associated with the magnitude of warming. In DJF, the models give a good consistency of the sign of change for the tropics and southern hemisphere but much smaller agreement throughout the northern hemisphere mid-latitudes. Over North America, Eastern Europe and North Russia, 75 per cent of the models predict the same sign of change. However, as already noted, over Western Europe there is little consistency, with as many models predicting a cooling as a warming.

Figure 3.5 shows the PMIP all model mean changes in precipitation for DJF and JJA, and Fig. 3.6 shows the standard deviation. The most striking feature is the enhanced northern hemisphere summer monsoon for North Africa and Southeast Asia. This is the result of the warmer landmasses, enhancing the land–sea temperature gradient. Almost all models show this increase in precipitation. However, for other parts of the globe there is a lack of any real consistency. In the mid-latitudes, there are as many models showing drying as showing a moistening.

A similar story also applies to DJF. The largest changes and greatest consistency occur in the tropics. The cooler land conditions result in a reduction of tropical rainfall over the land but an increase over the ocean. However, in other regions the consistency among models is poor.

This lack of consistency in precipitation is disappointing but not surprising. Precipitation has much greater variation of spatial and temporal scales and is known to be much harder to model (and indeed to measure). For model–data comparisons, the results can be interpreted in two ways. In the low consistency regions, it can be argued that the data can identify which of the models simulate the correct sign of the change. However, it is important to remember that the boundary conditions were not complete representations for the Mid-Holocene so there is a danger that a model could be rejected because of an error in boundary conditions. Alternatively a model may get the right answer but for the wrong reason (although model–data comparisons over Europe have suggested than none of the models got it right) (Masson *et al.*, 1999). In this sense, it is probably best to focus model–data studies on regions where the models showed good consistency.

In the tropics, the models were very consistent in predicting an increased monsoon. This is indeed seen in a number of proxies including lake-level estimates and pollen reconstructions. However, the proxy data argue for a much bigger change in the North African monsoon region (Joussaume *et al.*, 1999). The proxy data would suggest a mean change in precipitation of about

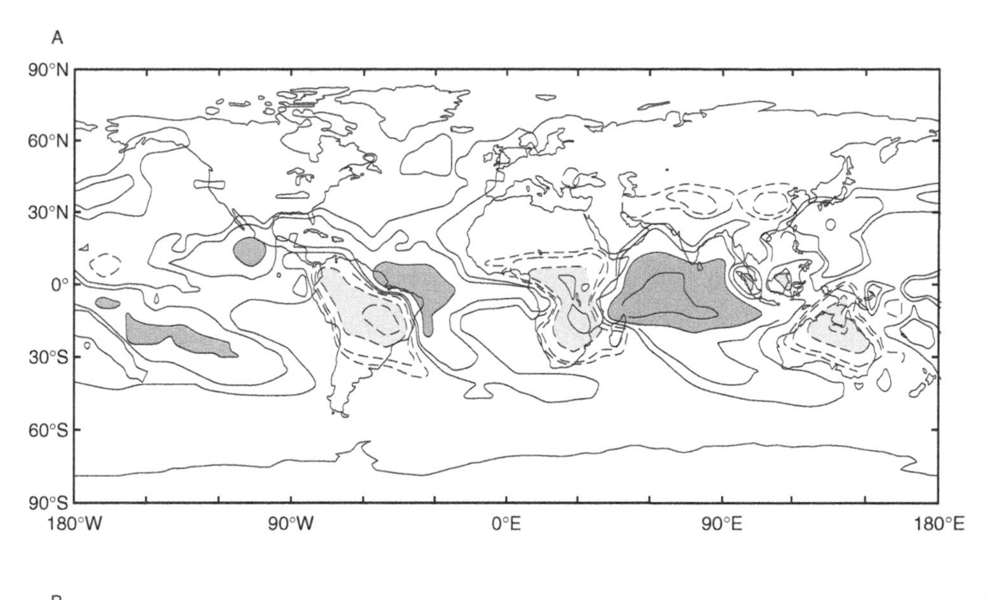

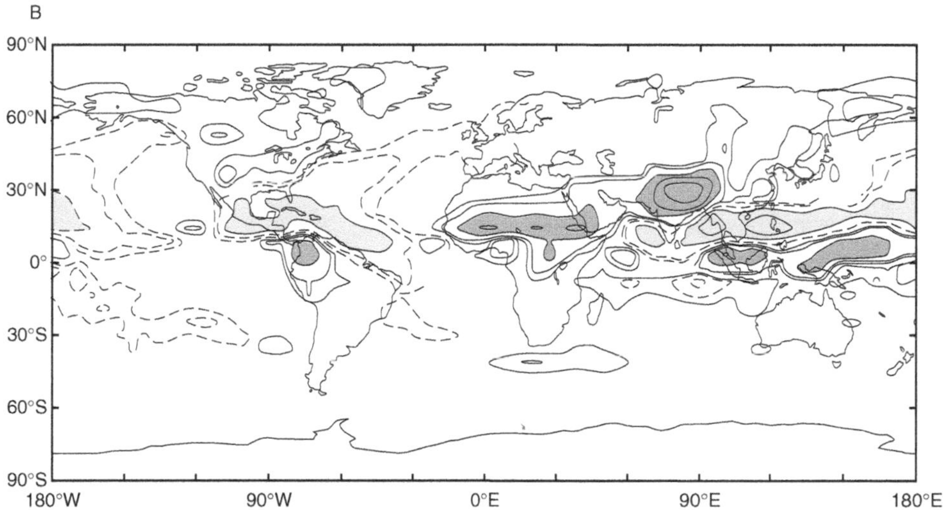

Figure 3.5 As for Fig. 3.3, but showing changes in precipitation. The contours correspond to –2, –1, –0.5, –0.2, 0.2, 0.5, 1, 2 mm/day and the negative contours are dashed. Light shading indicates precipitation less than –0.5 mm/day and darker shading indicates precipitation greater than +0.5 mm/day. Positive values indicate that the Mid-Holocene was wetter than present day.

200–300 mm per year and extending to 25°N or beyond. Most models give much smaller changes and much less northward extension.

So what are the causes of this model–data disagreement? Is this showing that climate model parameterizations are inadequate for predicting Holocene climate change? If so, it would have major implications for future climate change. However, there is an alternative. The poor model–data comparisons may be showing us that the PMIP boundary conditions were too great

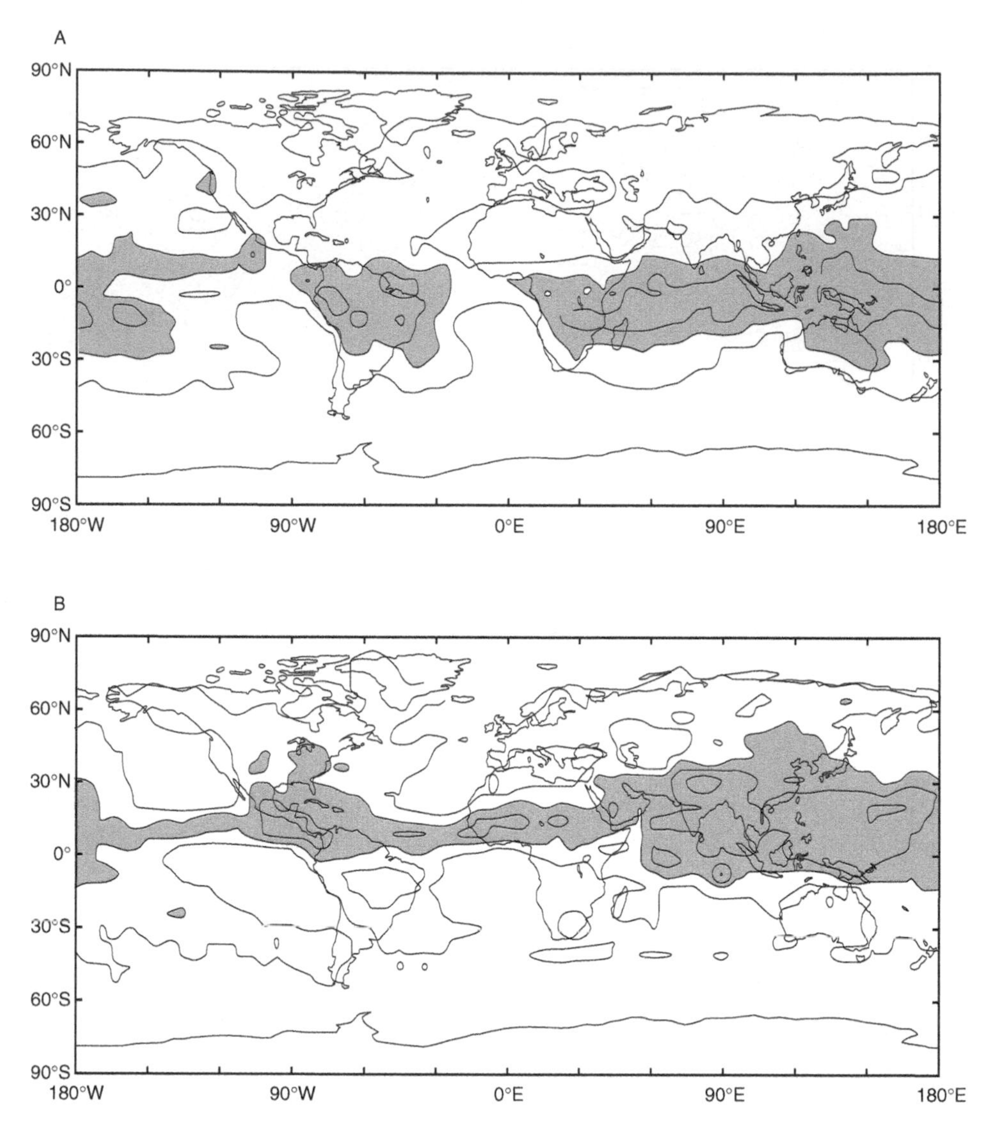

Figure 3.6 As Fig. 3.4, but showing the model–model standard deviation of precipitation. The contours are 0.2, 0.5, 1 and 2 mm/day and the shading indicates value greater than 0.5 mm/day.

a simplification. To investigate this, it is essential to use a more complete modelling structure, which includes changes in SST and surface vegetation.

The simplest ocean model to include is a so-called 'slab ocean' model in which the ocean is represented as a 50-m slab of water with a prescribed horizontal transport of heat in the ocean. Only the upper ocean thermodynamics are represented, and there is no representation of the dynamical changes or changes in the deep ocean. Although very simple, the model may be able to represent the first-order changes in ocean SST during the Holocene.

Plate 3 compares the simulation of June–July–August–September (JJAS) mean change in precipitation from the Hadley centre atmospheric model (Pope *et al.*, 2000) coupled (Plate 3a) to prescribed present-day SST (as in PMIP), to that using a simple slab ocean model (Plate 3b), and a fully coupled dynamic ocean model (Plate 3c). There are a number of important aspects of the changes. First, the ocean temperature changes (not shown) are generally very small. They are close to the limit of accuracy of most palaeo-oceanographic proxies. However, the changes in temperature are sufficient to influence the terrestrial climate, especially over the North African region. For the interactive ocean simulations, the precipitation has significantly increased, and there is a lengthening of the wet season. This is because the winter temperatures are cooler than present and this continues into the spring and early summer seasons. This results in an enhanced temperature gradient and hence increased monsoonal flow and precipitation, especially at the start of the wet season (Kutzbach and Liu, 1997; Hewitt and Mitchell, 1998).

A more complete ocean model is a **dynamic ocean GCM**. This works using similar principles to an atmospheric GCM and is the current tool used for future climate change prediction. Dynamic ocean GCMs are able to simulate changes in ocean circulation, including the deep ocean. When these models are coupled to atmospheric GCMs, the resulting simulations of present-day climate are reasonable, but far from perfect. Some models have resorted to using artificial correction factors (normally referred to as '**flux correction**') to improve the present-day simulations. Alternatively, some models accept a poor simulation of the present day. A few models are considered to give an 'adequate' simulation of the present day, but this is a subjective assessment and depends on the problem being considered. In terms of temperatures, even good models can have errors exceeding 5 °C in specific regions (for more details see chapter 8 in Houghton *et al.*, 2001).

Several fully coupled models have now been run for the Mid-Holocene (Kutzbach and Liu, 1997; Hewitt and Mitchell, 1998; Otto-Bliesner, 1999; Braconnot *et al.*, 2000; Liu *et al.*, 2000). The results for the North African monsoon region are broadly comparable and all models show an enhancement of precipitation. An example is shown in Plate 3c. Analysis shows that changes in the ocean circulation play a role in this. However, no model successfully simulates the full extent of North African moistening during the Holocene. It can be seen that, although the precipitation is enhanced, there is relatively small northward expansion. Some other mechanism must also be important.

3.1.4.2 Vegetation–atmosphere–ocean interactions

Charney (1975) suggested that the land surface conditions could have important feedbacks on the climate system in the North African region. He suggested that an increase in albedo would result in a cooling of the atmosphere and enhanced descent. This would act to reduce rainfall, and hence vegetation cover, which would then act to increase the albedo, thus creating a positive feedback loop. Changes in evapo-transpiration and cloud cover have also been shown to play a role in decreasing the surface moist static energy.

Claussen (1994) has modelled this process using an atmospheric GCM coupled to a biome vegetation model. He found that for present-day conditions, there were two steady states of the coupled system. One corresponds to the present conditions, with a large desert, high albedo and low rainfall. The other state corresponds to a vegetated Sahara, lower albedo and higher rainfall. The positive feedbacks are sufficiently strong to maintain a 'green' Sahara. Claussen and Gayler

(1997) further showed that for a Mid-Holocene orbital configuration, the only equilibrium state corresponded to the 'green' Sahara. This is discussed in more detail in Claussen (pp. 422–434 in this volume).

Finally, the above results were mainly using a vegetation model coupled to an atmospheric GCM. Ganopolski *et al.* (1998a) and Braconnot *et al.* (1999) have shown that there is a strong interaction (synergy) among the atmosphere, ocean and vegetation. A fully coupled model of the ocean–atmosphere–vegetation system produces changes which are more than just a simple addition of the individual components. They argue that a full understanding of the Mid-Holocene climate requires a representation of all of these components. A new phase of PMIP will perform comparisons using these coupled models.

3.2 CLIMATE VARIABILITY AND RAPID CLIMATE CHANGE EVENTS

Until recently, much of the focus of Holocene modelling has been on attempting to understand the long-term changes in the mean climate. Such work is evaluating whether the models have the correct sensitivity to changes in the forcing. As was discussed in the introduction to this chapter, another important role of palaeoclimate studies is to evaluate if the models are able correctly to represent variability on shorter time-scales. This work can be subdivided into three different branches.

The first examines the natural variability of the present-day climate by studying the variability during the past one or two millenniums, including events such as the Little Ice Age or Medieval Warm Period. The focus of the work is on comparing the model-simulated spectrum of variability with the proxy data (from a variety of archives including tree-rings). So far, GCM studies have mainly focussed on long simulations with no changes in external forcing (e.g. Collins *et al.*, 2002). The only variability is internally generated. When this variability is compared to the proxy data, the models predict similar variability to observed but slightly too little variability. However, this is not interpreted as an intrinsic failure of the models. Rather it may suggest that externally generated variability (e.g. from solar variability or volcanoes) is an important part of the causes of variability during the last few millennia. EBMs have also been run for this period using the proxy estimates of climate forcing (Crowley and Kim, 1999). The results suggest that both volcanic and solar variability is important and should be included in Holocene variability studies (see Bradley, pp. 10–19 in this volume).

A second role of high temporal resolution data is to examine whether the models are capable of simulating some of the important patterns of variability, such as ENSO or the NAO. The emphasis of many of these studies is to understand the mechanisms that control ENSO or NAO variability. For instance, Clement *et al.* (1999, 2000) showed that changes in the seasonal cycle (due to long-term orbital changes) could modify the amplitude of ENSO variability.

A third role of high resolution data is to examine rapid climate change events, such as the 8.2 kyr event. This is currently understood as a consequence of a freshwater pulse into the North Atlantic. This causes changes in the North Atlantic ocean circulation (especially the thermohaline circulation (THC)), which then changes climate both regionally and, potentially, globally. Model simulations of this type of event fall into two categories. There is a long history of idealized simulations in which a freshwater pulse is added to the present-day ocean–atmosphere model (e.g.

Manabe and Stouffer, 1997, 1999). The model results show that the speed and amount of freshwater can both be important in determining how the THC responds. However, Ganopolski and Rahmstorf (2001) have shown that the stability of the THC can depend on the initial state of the ocean so that a glacial ocean will have a different response compared to the modern. Because the 8.2-kyr event is during the Early Holocene, the deep oceanic state may still be close to glacial conditions. Thus to fully understand this Early Holocene event, we must perform more realistic types of simulation.

Recently, Renssen *et al.* (2001) have performed such simulations using a model of intermediate complexity. Among the important results from this simulation is that the response of the coupled system can be very sensitive to small changes in the initial conditions, in essence a form of climate 'chaos'. They found that for the same freshwater input scenario, the behaviour of the THC could vary substantially. In some cases the modelled circulation recovered quickly, whereas in other cases the modelled THC remained weak for a millennium. If this result is correct, there are two important implications. Firstly, even if we have a perfect model, it will be difficult to reproduce the details of the observed variability. Instead, modellers will have to follow the same approach used in weather forecasting in which an ensemble of simulations are performed and the probability of a particular event is assessed. Second, if the climate system is truly chaotic, then it further confirms the idea that we really cannot use the past as an analogue for future changes as each event will be truly unique and non-repeatable.

3.3 SUMMARY

The emphasis in this chapter has been on the role of palaeoclimate data to evaluate critically our understanding of climate change processes, and particularly our ability to simulate them using state-of-the-art ocean–atmosphere–vegetation models. Palaeoclimate studies have significantly advanced our knowledge of the basic mechanisms of climate, and the work on the Holocene (especially in North Africa) has been a major force for improving the models by including a more complete representation of the Earth System.

Climate models are only representations of the real world. They are not reality, but just an expression of our level of physical and biological understanding of the processes that control the Earth System. However, it is also very important to remember that the climate interpreted from proxy data may also not be reality. Joint model–data studies can act both to evaluate the models and improve our interpretation of past climate change processes.

In the coming years, the major challenges are to understand the sub-orbital time-scale variability within the Holocene. This includes interannual variability, rapid change events and the suggested 1500-year variability in climate. Such features represent significant challenges for our understanding and for the models. They will only be answered by the combined use of intermediate and full complexity models, and with modellers closely working with the data gatherers. It is certain that the next decade will be an exciting time for Holocene climate research.

CHAPTER

HOLOCENE CLIMATE AND HUMAN POPULATIONS: AN ARCHAEOLOGICAL APPROACH

Stephen Shennan

Abstract: In recent decades archaeologists have been extremely sceptical of invoking environmental factors to account for cultural and social change. The increasing availability of archaeological and climatic data with high degrees of temporal resolution has meant that the study of such questions now has a much more substantial database. However, even with improved chronological resolution, identifying causal connections remains difficult. This chapter examines three case studies where climatic factors have been used to explain demographic decline, the circum-Alpine Neolithic in the later 4th millennium, the East Mediterranean in the late 3rd millennium and the arid southwest of North America AD 400–1400, and emphasizes the importance of understanding patterns of costs and benefits to people in particular local situations. Although the issues involved are complex, the impact of climate change on past human societies and economies should not be ignored or considered insignificant.

Keywords: Alpine Neolithic, Climate change, Demography, East Mediterranean Early Bronze Age, Southwest North American Anasazi

Until very recently, most archaeologists have been extremely reluctant to invoke climate change as an explanation for any of the major changes they have observed in the Holocene archaeological record. It was assumed that, in most parts of the world at least, climatic conditions have been relatively constant since the beginning of the Holocene and that the changes which have occurred in human societies over the last 10,000 years must always have been the result of factors internal to those societies themselves. When climatic changes have been invoked as relevant explanatory factors, they have not been taken very seriously, for a variety of reasons. One is the belief that to propose that climate change might have had an effect on human societies represents '**environmental determinism**' that can therefore be dismissed out of hand (cf. Jones *et al.*, 1999c). More important is the fact that evidence for climate change was often only vaguely dated, so that it was hard to demonstrate a convincing chronological link between evidence for climate change, on the one hand, and the archaeological patterns to which it was supposed to be relevant, often equally vaguely dated, on the other (cf. Buckland *et al.*, 1997). Finally, and perhaps most importantly, such attempts rarely specified convincing mechanisms by which the hypothesized climate changes would have led to the claimed human impact (cf. McIntosh *et al.*, 2000).

While the scepticism with regard to supposed environmental determinism has remained, in the last few years archaeologists have become much more ready to admit that climatic factors might have influenced human societies during the Holocene, and case studies have begun to emerge. Several factors have led to this change in attitude. First is the rise of concern about the impact of climate change on present and immediately future human societies. Just as the concern about the impact of human populations on the world's resources which emerged in the late 1960s led to archaeological interest in the role of population in prehistory, so current concerns about climate have prompted interest in its effects on human societies in the past. More important though are the increasing quantities of data which have demonstrated the dynamic nature of Holocene climate and have enormously improved our knowledge of the chronological resolution of those dynamics. It has become possible to identify fine-grained changes and to show that they can operate at short time-scales. In addition, the ability to establish the synchronicity of environmental changes in different regions and using a variety of data sources has made it easier to infer when they are reflecting a climatic signal (Magny, 1993a). The result of all this is a greatly improved potential to link climatic variations to the archaeological record. At the same time, the dating of the archaeological record has also been improving as a result of the increasingly extensive and careful use of radiocarbon dating and of dendrochronology (see Pilcher, pp. 63–74 in this volume). The result is the emergence of realistic new agendas concerning the impact of climate change on human communities, the factors affecting social and economic change and the mechanisms involved in climatic impacts, both in terms of climate and human responses.

The particular aspect on which this chapter will focus is the potential link between climate change and human demographic patterns. Identifying patterns of population growth and decline offers the archaeologist a powerful source of information about past human adaptations and their success. Population growth is an indicator not of population pressure, as the archaeological conventional wisdom has it, but of the availability of new resources, which may stem from a variety of factors, including technological innovation and, of central concern here, environmental change. It is population stability that indicates pressure on resources, while population decline is an indication of failing adaptations.

Boone (2002) recently concluded on theoretical grounds that human population history is one of periods of rapid growth interrupted by frequent crashes caused by both density-dependent and external factors. Anthropological genetics has recently begun to reveal evidence of such expansions and crashes (Richards *et al.*, 1996). Moreover, there is increasing archaeological evidence for population fluctuations now that more detailed archaeological surveys are being carried out and the chronological resolution of archaeological data is being improved. Population patterns have important consequences for many areas of human social and economic life. Population growth often leads to increased regionalization and territoriality in human societies (David and Lourandos, 1998), as well as changes in human social institutions (Johnson and Earle, 1987). Furthermore, if a population expands, its cultural inventory and social institutions will expand with it. Some of the major expansions of specific languages and cultures in different parts of the world – for example, the expansion of Bantu languages in Africa – are likely to be explicable in this way. The impact of demographic decline is equally important (Shennan, 2000, 2001).

Accordingly, to the extent that climate change has had a significant impact on human resources, it should be visible in the archaeological record of population fluctuations. To establish whether or not it has had such an impact in a particular case we need independent evidence of climatic patterns and human population patterns. Moreover, we cannot simply be satisfied with documenting a correlation between the two. We also need to be able to specify the probable processes through which the connection operated. The case studies which make up the rest of this chapter have attempted to do precisely this.

4.1 CASE STUDIES

4.1.1 The Circum-Alpine Late Neolithic

The high degree of chronological resolution provided by the dendrochronological dates from the Late Neolithic lake villages of the European circum-Alpine region provides an excellent opportunity to look at prehistoric population patterns in detail. Figure 4.1 shows the dates of tree-felling activity identified in the wood used for piles and house construction at Neolithic settlements dating between 4000 and 2400 BC around the shores of Lake Constance, on the Swiss–German border (Billamboz, 1995). The most striking feature is the fluctuations in occupation they suggest. A similar pattern is seen in the results of pollen analysis from the same region, which show a series of phases of cereal cultivation, separated by gaps (Rösch, 1993). The dendrochronologically dated wood from the lake villages also provides evidence for patterns in the development of the woodland itself, in which repeated patterns of clearance are succeeded by regeneration and the development of secondary forest.

Precisely the same sort of situation is to be found during the later Neolithic on the much smaller lakes of Chalain and Clairvaux in the western foothills of the Alps. The left-hand column of Fig. 4.2 shows the number of settlements around the two lake shores, from 4000 to 1600 BC. Pétrequin and colleagues (Arbogast *et al.*, 1996; Pétrequin, 1997; Pétrequin *et al.*, 1998) see this pattern as a series of demographic cycles of population growth and decline which is also reflected in the dendrological evidence of woodland clearance.

Not everyone accepts that such patterns can be interpreted in demographic terms. The fluctuating numbers of settlements on the circum-Alpine lakes are correlated, although by no means perfectly, with lake-level fluctuations. The number of settlements known from low lake-level phases is much greater than for higher levels. One possible explanation therefore is that because the settlements which would have been occupied in times of high lake-levels were never submerged, they were poorly preserved, so the wood to provide dendrochronological dates has not survived. Accordingly, the fluctuations in settlement numbers could also be a result of

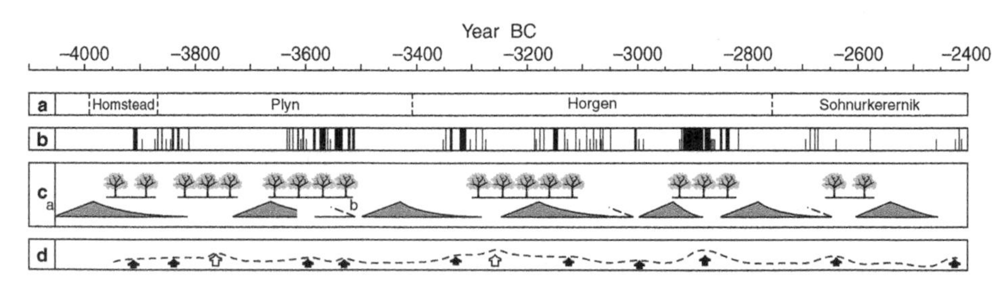

Figure 4.1 Later Neolithic occupation around Lake Constance, showing (a) the cultural sequence, (b) felling dates of dendrochronologically dated timbers from settlements, (c) woodland history and (d) inferred demographic history (after Billamboz, 1995). a. forest regeneration; b. clearance.

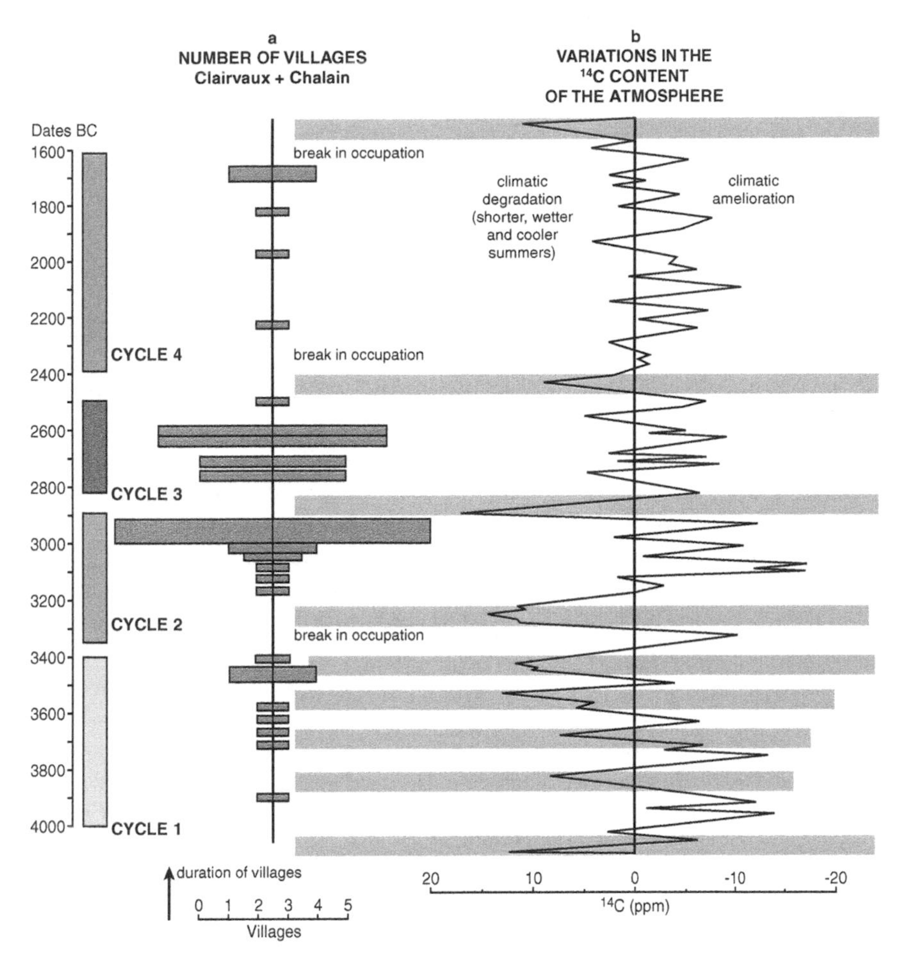

Figure 4.2 Later Neolithic occupation in the Chalain/Clairvaux area of the French Jura, showing (a) the demographic sequence and (b) variations in the ^{14}C content of the atmosphere as a climate proxy (after Pétrequin, 1997).

fluctuations in preservation (Gross-Klee and Maise, 1997). However, the fact that variations in human impact are apparent in both the pollen diagrams and the evidence for woodland history suggests that population fluctuations were at least a significant factor.

It is now generally agreed that the lake-level variations are a result of climatic variations, with cooler, wetter phases corresponding to high lake levels and warmer, drier phases to low levels; lake-level changes are synchronous across different circum-Alpine regions and lake-level rises are correlated with lowering of the timberline and glacier growth (Magny, 1993a). The pattern of rises and falls also correlates closely with the high resolution record of variations in $\delta^{14}C$ concentrations in the atmosphere, as measured from tree-rings (Fig. 4.2b), related to levels of solar activity. The variations in solar energy affect patterns of atmospheric circulation, leading to changed weather conditions. It appears that at this latitude in the North Atlantic area, low levels of solar energy (high $\delta^{14}C$) result in a reinforced westerly circulation in the summer, with an increase in the frequency and intensity of depressions, producing wet and cool conditions, while flows of arctic air to the south occur in winter, conditions corresponding to the Little Ice Age of the 16th to 18th centuries AD.

How would this have affected population levels? Maise (1998) used historical records from Western and Central Europe to show that the cooler, wetter conditions of the Little Ice Age led to frequent crop failures, because the growing season was shortened. Mortality rates increased as a result. The same problems are likely to have arisen in the region in prehistory. In fact, by the 18th century the broader food supply network that had been established meant that the impact of this sort of crisis was not as severe as it had been in earlier times (Fischer, 1996). Calculations based on evidence from Neolithic settlements on Lake Zürich dating to around 3700 BC suggested that the subsistence economy was vulnerable, in terms of the margin between what would have been needed by the population and what the economy could provide. A climatic downturn leading to cool, wet conditions and a shorter growing season would have produced severe problems (Gross *et al.*, 1990) for local communities, with significant demographic consequences, in terms of increased mortality rates and decreased fertility. As Fig. 4.2 shows, it appears that in the Chalain–Clairvaux area at least the periods of low solar activity and cooler, wetter summers (high $\delta^{14}C$) were indeed associated with lower numbers of settlements, while the converse was true of higher solar activity conditions. It seems likely that periods of favourable climate led to local population expansion, quite possibly involving not just the growth of local communities but also, on the basis of other aspects of the archaeological evidence, the immigration of groups from other areas where population was growing. Equally, when conditions deteriorated and people could foresee the likely consequences of failed harvests, rather than waiting to experience them they may have decided to abandon the area to find better conditions. In fact, the Chalain–Clairvaux area was probably always relatively marginal for agricultural populations, occupied during favourable conditions and largely, if not entirely, abandoned in unfavourable ones.

However, it is important to be aware that climatic oscillations were not the only factor affecting local populations. Examination of the settlement numbers and the $\delta^{14}C$ values in Fig. 4.2 shows that after *c.* 2500 BC there were periods of favourable climate which do not correspond with increased lake edge occupation. The reasons for this remain unclear.

4.1.2 The East Mediterranean and the Near East in the 3rd Millennium BC

The second case study raises some different issues. Apart from the fact that the climatic issues in question involve aridity, rather than the problems posed by cool, wet conditions, the societies in the Near East and the East Mediterranean were organized very differently from those in West–Central Europe at around the same time. In the latter case all evidence suggests that communities were small villages, independent from one another and responsible for their own subsistence. While they had trade and kinship relations with one another, there is no indication that individual communities were parts of larger-scale political entities. This is very different from the Near East, where settlements numbering thousands of people already existed and there are indications of large-scale hierarchical political entities which exercised political control from a major centre over lesser settlements in their hinterland and exacted tribute from them, a process which often involved the storage and redistribution of large amounts of food. If we can indeed make the case here, as we have done for later Neolithic circum-Alpine Europe, that there was a connection between climate change and socio-economic change, especially population patterns, what did it involve and why did it take the form it did?

Peltenburg (2000) has recently reviewed the situation in the Near East and East Mediterranean at this time, taking as his starting point the region of Upper Mesopotamia, where the dense, hierarchical settlement distribution, including major centres, of the mid 3rd millennium BC was replaced by a pattern with a much lower degree of archaeological visibility. The changes involved the contraction of many large sites, evidence of destruction followed by urban decline at some important settlements, abandonment of many sites and a process of settlement dispersal, with an eventual increase in the number of small sites.

The strongest argument for linking these changes in settlement pattern and apparent demographic decline to environmental factors is that made by Weiss *et al.* (1993; but see also Rosen, 1998), who identified a process of abandonment and increasingly arid conditions around the site of Tell Leilan, and postulated a long-term process of climatic deterioration in the region. There is certainly good evidence for an abrupt late 3rd millennium BC arid event in the eastern Mediterranean, both from speleothems (Bar-Matthews *et al.*, 1998) and from deep-sea cores (Cullen *et al.*, 2000). Evidence is also available from a number of other regions, including Egypt, where there is documentary evidence for progressive decreases in the Nile flood levels in the late 3rd millennium (Peltenburg, 2000), at a time of decentralization and disturbances. Cyprus too shows major changes at this time, with extensive abandonment of late Chalcolithic settlements and far fewer in the succeeding Early Bronze Age phase, while those that did exist in the later period were not of the same size as earlier ones. Peltenburg (2000) concluded that there was a sequence starting with the development of drier conditions, leading on to the depletion of woodland and erosion, which was followed by the settlement abandonments that occurred at the end of the Chalcolithic and were associated with a major break in the cultural sequence.

In Greece, similar patterns of change may be identified, with a major decrease in the number of sites in the EH III period compared with the preceding EH II. In the southern Argolid, which has been the subject of a detailed archaeological survey, the number of sites decreased from 28 to two. This is also a time of major erosion in the region. As Peltenburg (2000) points out, the team that carried out the survey saw the changes as arising from the increasing intensity of human landuse, especially the introduction of the plough, rather than as the result of climatic factors, so the question arises whether it is is possible to distinguish such causal factors from one another. Clearly, one way of doing so is to collect more data from the region concerned, but looking for evidence of similar processes affecting several different regions is also a possibility. Do we have good enough evidence to suggest that settlement pattern collapse in the different regions mentioned really was contemporary, or are we attempting to bring into relation processes which occurred at different times in different places? As we saw at the beginning of this chapter, the problem of establishing convincing linkages between the timing of socio-economic and demographic change, on the one hand, and climate change on the other, is one of the reasons archaeologists have been so sceptical of climatic explanations.

Examination of the currently available radiocarbon dates (see Fig. 4.3) suggests that in Upper Mesopotamia the contraction and dispersal phase began *c.* 2200 BC; dates for the corresponding developments in the East Mediterranean and the Levant also point to *c.* 2200 BC. As Peltenburg (2000) points out, the societies in these different regions were markedly different in character, in particular in their degree of centralization, population concentration and hierarchical differentiation, but all experienced an apparently similar process of collapse and dispersal at the same time, suggesting that the events were in some way linked. The fact

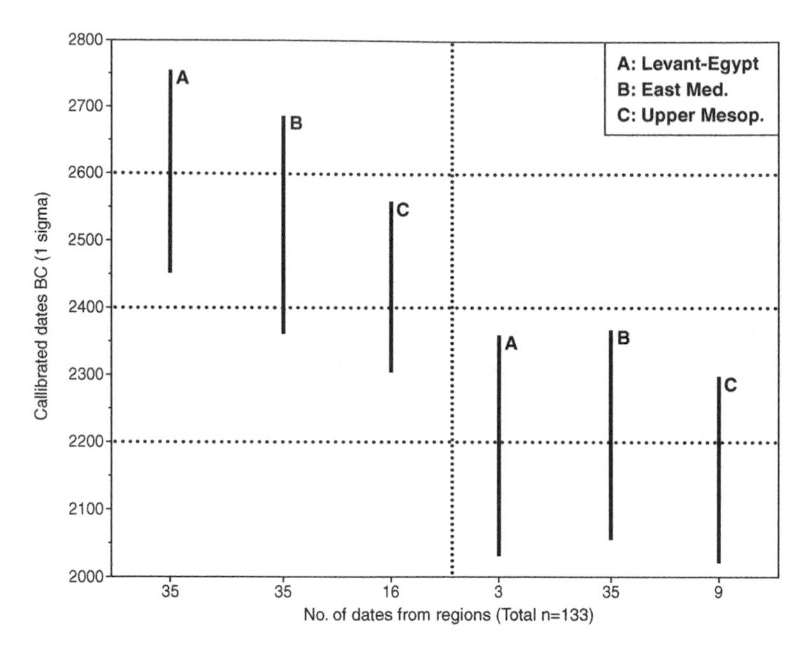

Figure 4.3 Averaged radiocarbon dates from the nucleated (left) and dispersal (right) phases of the late 3rd millennium BC in the Near East and Aegean (after Peltenburg, 2000).

that indications of environmental deterioration and increased aridity are common to all of them, as well as being attested in speleothems and deep-sea cores, points to the significance of this factor. While the more centralized forms of organization may have been able to buffer local populations from short-term agricultural downturns through the organization of storage and redistribution, and thus maintain population densities higher than would have been possible without them, the environmental deterioration eventually went beyond what local forms of centralized organzation could cope with. Indeed, these may themselves already have become weaker, perhaps as a result of the gradual breakdown of the exchange networks which provided local elites with prestige goods (Peltenburg, 2000). In other words, local subsistence problems may have compromised the capacity of particular regions to produce surpluses sufficient for their elites to be involved in exchange, thus having inter-regional effects additional to those produced directly by local environmental deterioration and affecting their credibility. However, we should be careful not to assume that all societies would have reacted in the same way to such stresses; for example, some of the Early Bronze Age centres seem to have survived the drastic changes around 2200 BC.

The process has been examined in detail for the southern Levant by Rosen (1995). Here the late 4th and earlier part of the 3rd millennium BC had seen the growth of large population centres, until the EB III phase, when a process of abandonment began, which by the end of the 3rd millennium had led to the collapse of local centralized societies and the abandonment of most urban centres, together with a drastic decline in population. Evidence suggests that environmental change, and in particular a shift towards more arid conditions, with decreasing rainfall, played an important role in this process. The key mechanism in Rosen's account centres on the role of floodwater farming. Hydrological studies suggest that in the late 4th and earlier 3rd millennium BC, increased precipitation led to aggrading streams regularly flooding valley bottoms and providing the basis for floodwater farming. This would have been essential to support the large concentrations of population that existed through the first part of the 3rd millennium as it would have given greater yields in general than dry farming and, in particular, would have provided a buffer against dry years. The evidence for the existence of such water-assisted farming is in the form of phytoliths. In dry conditions, cereals produce phytoliths of only a small number of cells whereas in wet conditions they can be up to several hundred cells in size and such phytoliths have now been found at 4th and 3rd millennia BC sites in Israel (Rosen, 1995).

When conditions became drier, aggradation and flooding in the valley bottoms were replaced by down-cutting, as the levels of streams dropped, and floodwater farming became impossible. However, as Rosen (1995) points out, there were large urban centres in the area from the Middle Bronze Age to the Hellenistic Period coping perfectly well with the environmental and agricultural conditions associated with the abandonment of the Early Bronze Age urban centres. What was the difference? Rosen (1995) postulates that Early Bronze Age agriculture was divided into two sectors, one concerned with the subsistence production of cereals and the other with the cash-cropping of vines and olives, controlled by local elites. This involved large-scale specialization, organized by those elites, including the redistribution of food staples, indicated by the existence of major storage facilities. Such stored staples would have helped populations to survive short-term food shortages caused by droughts, but at the expense of of reducing the control of local farmers over their own resources and decisions. The end of floodwater farming led to the collapse of this redistributive system and its effects were magnified by the fact that the traditional mechanisms for coping with drought employed by individual subsistence farmers had been subverted by the role of the state redistribution system.

Of course, this still does not explain why they did not start to use canal irrigation in response to the problems, a technique already known in other regions and which was used used successfully by later populations. One possibility that Rosen (1995) suggests is that cosmological ideas may have affected the response, in that people may have perceived the growing environmental problem as stemming from their relation to the gods rather than being anything to which they could respond in what we would regard as a more practical manner. The building of large temples at a number of sites in the EB III period, immediately before the collapse, may be relevant here. Nor should we forget the possibility that, in its initial stages at least, the increasing drought might have been useful to local elites as it would have intensified the dependence of local populations on the storage and redistribution mechanisms that they controlled.

4.1.3 Climate and Demography in the North American Southwest, AD 400–1400

The relationship between climate and demography in the North American Southwest has long been of interest to archaeologists working in this arid region but the following account will be mainly based on Larson *et al.* (1996). The period between AD 400 and 1400 saw subsistence intensification and increasing social complexity at various times over this period before major parts of the region were abandoned altogether. While some archaeologists have seen climatic factors as playing a major role in the changes in human settlement and society observed over that time, others have reacted against the environmental determinism they see in such approaches and have emphasized internal social factors. For example, they have pointed to the role of growing social inequality and centralized decision-making and have argued that when the archaeological and climatic evidence are brought together the idea that socio-economic change was a response to changing climate is not convincing (Plog, 1990). Larson *et al.* (1996) challenge such arguments on the basis of new high-resolution climate data and propose that climatic change had a major impact on socio-economic and demographic patterns, especially between AD 900 and 1300.

The basic argument is that as previous hunting and gathering subsistence strategies were increasingly replaced by reliance on domesticated crops between AD 900 and 1100, populations increased due to the availability of these new resources, but as a result local populations, known by archaeologists as the Anasazi, became increasingly vulnerable to climatic instability and its effects on their primary resources. This in turn was compensated by the development of various buffering tactics, both agricultural, for example in the form of fields ingeniously designed to collect and retain moisture (Dominguez, 2002), and social, including dependence on storage and also the exchange of food between settlements. It appears to be precisely the periods with evidence for the greatest temporal variability in water availability that show the greatest evidence for increased storage and exchange among the Vermillion Cliffs and northern Black Mesa Anasazi studied by Larson *et al.*(1996). However, there were times when the climatic situation became so bad that these mechanisms were no longer sufficient and regions had to be abandoned. There is extensive evidence for migrations in the southwest in the 12th century AD (Hegmon *et al.*, 2000), which themselves caused problems in areas where new groups attempted to establish themselves (see e.g. Billman *et al.*, 2000 for evidence of cannibalism in connection with drought-driven population movements and hostilities).

Figure 4.4 shows reconstructed population patterns based on archaeological survey data for several local areas of the western Anasazi. While specific patterns may be open to question in

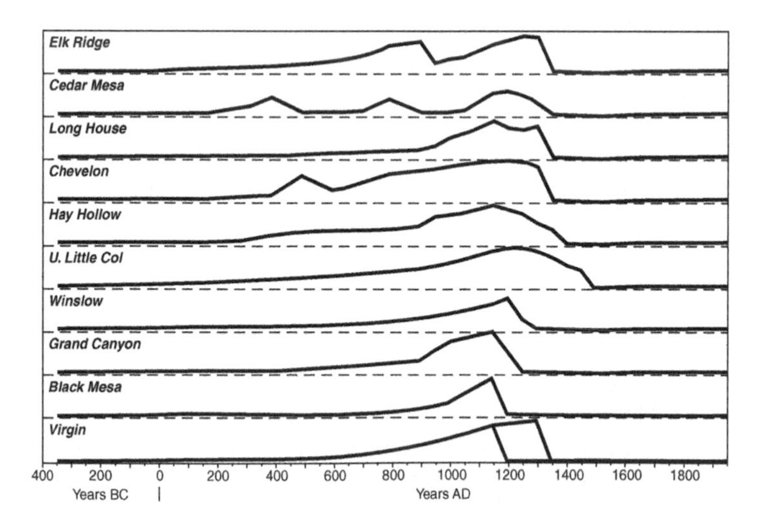

Figure 4.4 The relative change in population levels among the western Anasazi, 400 BC to AD 1800 (after Larson et *al.*, 1996).

detail, and it is highly likely that if data with greater chronological resolution could be obtained there would be far more fluctuations than shown here, the trends shown can be considered reliable and indicate the demographic success associated with the employment of agricultural subsistence strategies.

Larson *et al.* (1996) proposed that Anasazi farmers would have experienced especially severe problems during periods of prolonged drought but also at times when there was very high moisture variance from year to year, frequently leaving communities short of the means to meet their dietary needs. In order to establish whether Anasazi farmers were indeed affected by such conditions it was necessary to find a means of reconstructing past climate variability. This was done using the Palmer Drought Severity Index (PDSI), an empirical water balance index which gives a measure of soil moisture and which can be reliably reconstructed in the past using tree-ring data. Reconstructed values of PDSI were tested against data which had been held back from the initial equation relating tree-ring widths to PDSI and also against historical records; they were found to show a good fit.

Figure 4.5 shows the reconstructed index for the period AD 900–1300, where negative values correspond to dry periods and positive values to wet ones. The marked fluctuations are very apparent. The data can also be used to reconstruct patterns of climatic variability, with the result shown as variations in the standard deviation of the index from year to year for overlapping 50-year periods (see Fig. 4.6d). Very high variability at the end of the 10th century was followed by extremely low variability in the early-mid 11th, succeeded in turn by very high variability in the late 11th century, with fairly high variability in the 12th century, increasing markedly in the 13th. The indications of exchange intensity and storage capacity for the Anasazi of the Black Mesa area shown in Fig. 4.6a–c and show a strong association between these measures and increased climatic variability, in keeping with Larson *et al.*'s (1996) hypothesis

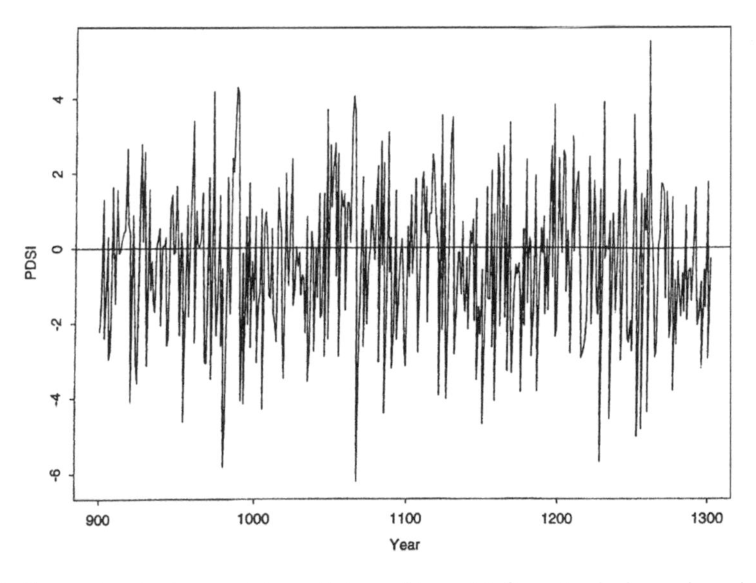

Figure 4.5 Reconstructed annual June Palmer Drought Severity Index values between AD 900 and 1300 for the western Anasazi area (after Larson *et al.*, 1996).

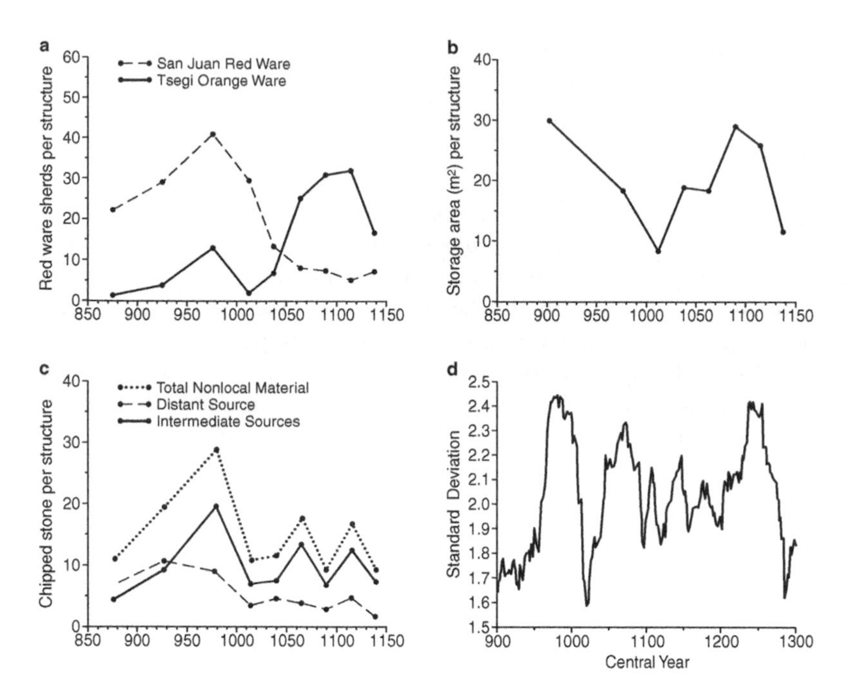

Figure 4.6 (a) Total red and orange ware per structure, (b) storage area per structure, (c) non-local chipped stone raw materials per structure for the Peabody Eastern Lease Area on northern Black Mesa AD 900–1150 and (d) reconstructed standard deviations of June PDSI values for overlapping 50-year periods (after Larson et *al.*, 1996).

that exchange and storage were used to buffer subsistence shortfalls arising from climatic variability.

It appears that the northern part of Black Mesa was abandoned between AD 980 and 1010, then reoccupied, before being permanently abandoned around AD 1130. Both periods of abandonment correspond with severe climatic conditions. Moreover, even during the period of reoccupation there are indications from Black Mesa burials of nutritional stress, especially among children and young women, that render the subsequent abandonment unsurprising. In other words, this particular adaptation to local conditions failed and the population emigrated. As Larson *et al.* (1996) point out, and indeed as we saw in the case of the Early Bronze Age Levant, environmental pressures do not automatically bring into existence new and more successful adaptations on the part of local populations. In this case the solution was abandonment, which meant that at least some of the people survived even if the local adaptation did not. However, even in cases like this it is necessary to qualify the environmental determination of the process. A recent simulation study of the Long House Valley, for which a population reconstruction is shown in Fig. 4.4, suggests that it would not in fact have been necessary to abandon the area completely from the subsistence point of view; a small population could have successfully survived there on the resources likely to have been available (Dean *et al.*, 1999a). It is likely to have been social factors which meant that everyone decided to leave.

4.2 CONCLUSION

Clearly, in a sense the case studies are biased. They are examples where there is convincing evidence of regional demographic decline, a process implying negative effects on the well-being of the communities concerned. Once the existence of these detrimental historical patterns had been established the next step was to examine why they might have occurred and to suggest a climatic mechanism. The beneficial parts of these historical sequences – the periods where there is evidence for population increase – were emphasized far less. More importantly, there has been no attempt to find cases where there has been no change to the human socio-economic and demographic systems in the face of major climate changes. This would certainly be at least as interesting to investigate, if not more so, because it would be necessary to establish the mechanisms by which stability was maintained in the face of potential disruption.

Even in cases like those which were selected, coming to causal conclusions is a complex process, requiring the independent characterization of changes in human societies, in this case especially demographic decline, and of climatic changes which might have a relevant impact. Establishing a clear chronology for both the human and climatic patterns is essential. Nor can we necessarily assume that there will be a straightforward chronological correlation even when there is a causal connection between the two. There may be lags in the impact of climatic factors on those aspects of the Earth System which are relevant to human activities, while human perceptions of those impacts and, even more, decisions based on those perceptions, may vary in their speed. For example, people may be quick to take up new opportunities and slow to accept that things are deteriorating and that they must react to them, or vice versa.

What is clear from the case studies is that it is not appropriate to think of the impact of climatic change on human societies in terms of environmental determinism. It has to be conceived in terms of perceived costs and benefits in the context of relevant constraints. For example, the costs of abandoning an area in the face of problems may be quite low if the adjacent areas are not very densely populated, but extremely high if they are already heavily occupied by other groups. In the latter circumstances people may simply have to suffer the effects of the new circumstances until a new demographic equilibrium is established. Furthermore, the social context cannot be neglected. When societies become differentiated the costs, benefits and constraints are not the same for everybody. This emerged very clearly in Rosen's (1995) Levant case study, where the interests of local elites and the general population did not coincide and the latter had to accept the situation created by the former. Up to a point it was in their interests to do so as the storage and redistributive system created and run by the elite had proved highly successful in the past, nevertheless, if local farmers had defected from it earlier the ensuing collapse might not have been so dramatic. Equally clear is the fact that problems do not automatically call into being appropriate solutions in the form of innovations. The factors affecting the innovation process are still extremely poorly understood. Fitzhugh (2001) has suggested that people are more likely to innovate, and therefore to be less risk-averse, when it is clear that the current way of doing something cannot possibly produce a satisfactory solution in the circumstances, but this too needs qualifying. It certainly did not lead to the later Early Bronze Age farmers of the Levant adopting canal irrigation. Here again the social context may be relevant, in that the local elites who controlled things may not have seen it as in their interest to do so, because they were still getting what they wanted from the existing system. However, these complexities do not indicate that the impact of climate change on past human societies and economies should be ignored or considered insignificant. Humans are part of socio-natural systems and their past activities should be analysed in that light.

Acknowledgements

I am grateful to Arlene Rosen and an anonymous referee for advice and references and to Anson Mackay for the invitation to contribute to this project.

CHAPTER

5

CLIMATIC CHANGE AND THE ORIGIN OF AGRICULTURE IN THE NEAR EAST

H.E. Wright Jr and Joanna L. Thorpe

Abstract: Calibration of the dozens of radiocarbon dates for Epi-Paleolithic and Neolithic times of early settlements and plant domestication in the eastern Mediterranean region by Bar-Yosef allows correlation of archeological trends with the Late Glacial climatic chronology based on annual layers in the Greenland ice-cores, especially with the Younger Dryas cool episode. This chapter evaluates the evidence for the Younger Dryas seen in pollen diagrams from the Taurus-Zagros Mountains north of Mesopotamia (Lake Van and Lake Zeribar) and from the Levant (Ghab Marsh and Lake Huleh), based on the calibration of radiocarbon dates. The paleoclimatic chronology of the speleothems from Soreq Cave in Israel is also utilized. The view is supported that the improved climatic conditions before the Younger Dryas episode permitted population expansion, and that the dry conditions of the Younger Dryas led to plant domestication and incipient agriculture.

Keywords: Calibrated radiocarbon dates, Levant, Neolithic, Younger Dryas

Ideas about the relation of climatic change to the origin of agriculture go back to Ellsworth Huntington, who accompanied Raphael Pumpelly on explorations in Central Asia, but it was V. Gordon Childe (1952) who popularized the subject. In his time the prevailing opinion was that during the glacial period North Africa experienced a pluvial climate, which sustained the temperate vegetation that had been excluded from Europe north of the Alps by the cold climate that produced the Scandinavian ice sheet. Childe assumed that with the retreat of the ice sheet the forests became diminished and converted to treeless landscapes with isolated 'oases', where animals, plants and human hunters and gatherers congregated. This juxtaposition led to domestication. Another factor in the early concepts of plant domestication was the concept of Vavilov that cereal grains were domesticated first in areas of their natural ranges, which was known to be in the hills of southwest Asia.

When geological studies in Africa showed no concrete evidence that the well known pluvial periods were contemporaneous with glaciation in Europe (Flint, 1959), the foundation of Childe's oasis theory was weakened. Further evidence came from paleoecological studies in the western Mediterranean region, including the pollen stratigraphy of the sediments of a crater lake in Italy, which implied that steppe vegetation rather than forest prevailed during the glacial period (Frank, 1969). On the archeological side, skepticism about the validity of the theory led R.J. Braidwood to initiate archeological surveys and excavations in the foothills of the Zagros Mountains of Kurdistan to identify the beginnings of sedentary cultures and domestication of plants and animals (Braidwood and Howe, 1960). At this time the chronology of the transition from Paleolithic nomadic hunting cultures to sedentary communities could only be estimated by extrapolation backward on the basis of king lists and pottery styles, or, assuming some relationship to climatic change, extrapolation forward from the time of retreat of the Scandinavian ice sheet, as based on the Swedish varve chronology.

During the early stages of Braidwood's excavations at Jarmo, the development of radiocarbon dating made possible the direct estimation of the time of domestication. He took the initiative to encourage natural scientists to join in the field programme to obtain direct evidence for environmental change during the time of cultural transition. One of the first results showed that glaciation had indeed affected the Zagros Mountains, indicating that the climate in the past was colder than the present (Wright, 1961). Although the time of glaciation could not be dated there directly, it was postulated that ice retreat in the Zagros was contemporaneous with that in the Alps of southern Europe, where radiocarbon dating had shown that glaciers had retreated almost to their modern limits well before the cultural transition then dated at Jarmo (Fig. 5.1). Thus it was concluded that the climatic change in the Near East occurred well before the time of plant domestication, and therefore that the two events were not related (Wright, 1960).

This conclusion reinforced the opinion of Braidwood (1952, 1960) that environmental change was not a factor in the origin of agriculture. Because it was based on the long-distance correlation of paleoclimatic events rather than on well dated local environmental events contemporaneous with the cultural transitions, a causal connection was indeed speculative. Other approaches prevailed, e.g. the idea that gradual cultural evolution led to sedentary communities based on more diversified use of existing resources, and that sedentary populations then expanded to peripheral areas less well endowed, where they increased the resources by domestication of cereal grains (Flannery, 1965; Binford, 1968).

The opportunity for more specific and well dated paleoenvironmental reconstruction in the Near East came with the application of pollen analysis to lake sediments – the technique that had proven so effective in climatic reconstructions in Europe. Pollen analysis and radiocarbon dating

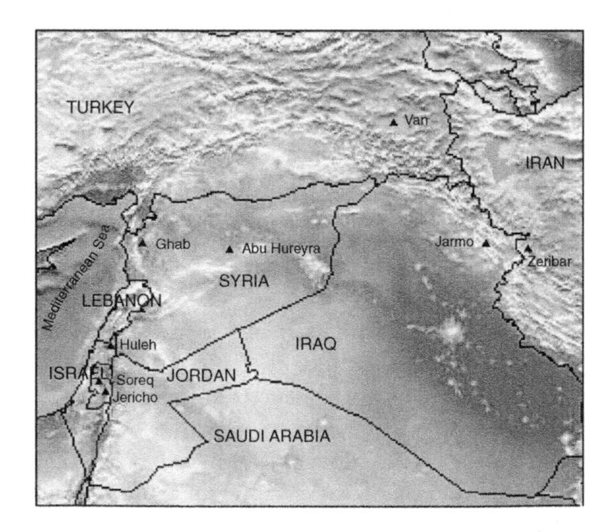

Figure 5.1 Map of the Near East showing location of sites mentioned in the text.

of a long sediment core from Lake Zeribar in the Zagros Mountains of southwestern Iran (Fig. 5.2) showed that cold, dry steppe changed to oak-*Pistacia* savanna at the same time as the end of the glacial period in Europe (van Zeist and Wright, 1961; van Zeist, 1967; van Zeist and Bottema, 1977). This work prompted the generalization that the domestication of plants, which by this time was well dated at a number of sites in the Zagros foothills as well as in the Levant, was contemporaneous with the change in climate from dry to temperate, and it was postulated that the climatic change set the stage for domestication (Wright, 1968, 1976, 1993). It was first speculated that the annual grasses that became domesticated were not common in the cold steppe of the glacial period, but that the change to more temperate climatic conditions allowed them to immigrate from unidentified refuges. A postulated refuge in the Atlas Mountains of Morocco has not been confirmed, and subsequent studies have shown that wild cereal grains did in fact occur in the Levant as early as 19,000 radiocarbon years ago (Kislev *et al.*, 1992). Expansion rather than immigration was apparently the process. A variant of the environmental hypothesis involves the actual speciation of annual grasses, including cereal grains, in response to increased seasonality of climate (McCorriston and Hole, 1991).

The identification and dating of many Epipaleolithic and early Neolithic sites in the Levant and in the Euphrates valley of Syria have elaborated the conclusion that village life had already been established while food gathering was still the economic base. It may have been the expansion of

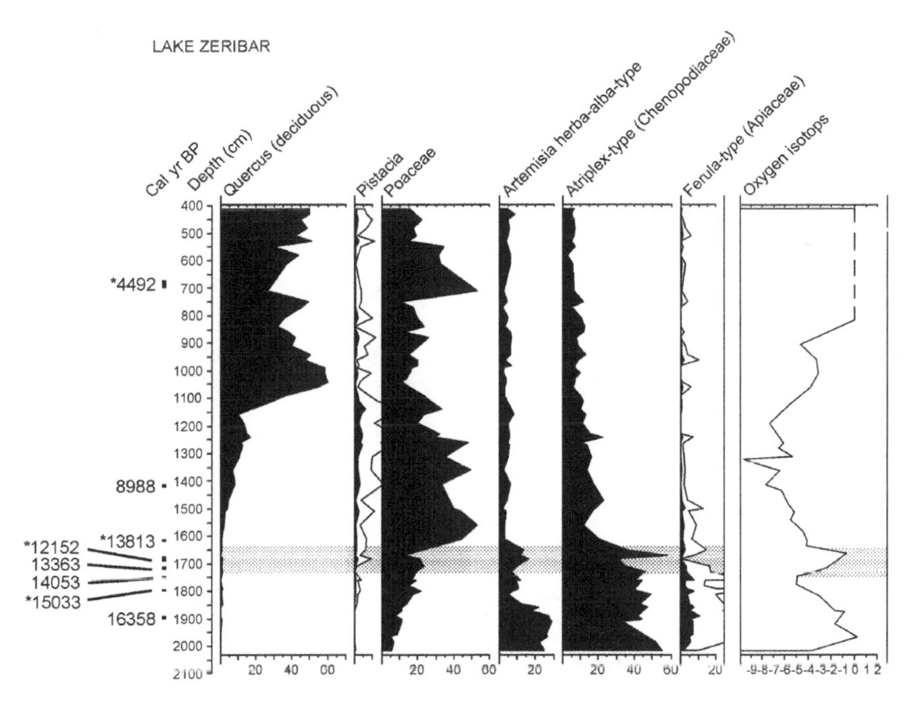

Figure 5.2 Lake Zeribar pollen diagram, adapted from van Zeist and Bottema (1977). Extracted from the European Pollen Data Base. The radiocarbon dates marked with a star are based on terrestrial macrofossils, the others on bulk calcareous sediment. Shaded zone represents the Younger Dryas.

large-seeded annual grasses at the time of the initial climatic amelioration that made them more conspicuous on the landscape and more extensively utilized by foragers.

While these contrasting views about the situation in the Near East were being explored and debated, the details of the climatic transition at the end of the glacial period in the North Atlantic region were being defined, especially with the interpretations of annually layered cores from the Greenland ice sheet. These cores contained evidence for short-term fluctuations, notably the so-called Younger Dryas cold interval, which interrupted the gradual warming that had prevailed during the early stages of deglaciation. The Younger Dryas cold interval had long been known from pollen studies in Europe, but the ice-core findings stimulated the search for its occurrence beyond the North Atlantic region. Near Eastern archeologists picked up on the idea that the Younger Dryas should have been registered in the eastern Mediterranean as a dry episode and that the stress of dry conditions led to the actual cultivation of cereal grains by the early villagers.

The occurrence of two critical climatic episodes – one leading to increased plant resources and a second marking the onset of dry climatic conditions – had been recognized in a comprehensive archeological synthesis by Moore (1985), but among the first to make explicit correlation of the second interval with the Younger Dryas were the reviews of Bar-Yosef and Belfer-Cohen (1992) and Moore and Hillman (1992). The chronological comparison between the archeological sequence and the ice-core data (Bar-Yosef, 2000) has been aided by conversion of all the relevant archeological dates to calendar years, made possible by calibration of the radiocarbon dates with the tree-ring chronology, because the ice-core dates, based on layer counting, are believed to represent calendar years (Fig. 5.3). Similarly, all dates used in the present review have been calibrated to calendar years (Tables 5.1 and 5.2).

What was lacking in this scenario that the stress of the dry climate of the Younger Dryas led to plant domestication was the actual local paleoecologial and chronological evidence that the Younger Dryas interval actually affected the Near East. Efforts to identify precisely this interval

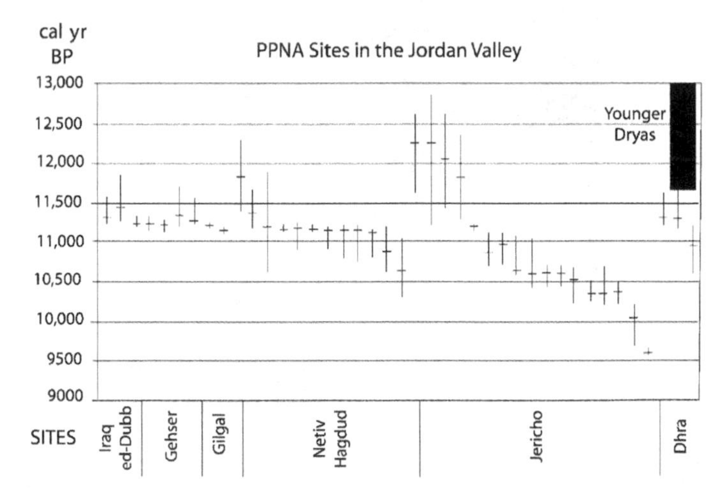

Figure 5.3 Calibrated radiocarbon dates for early Neolithic sites in the Jordan Valley in relation to the Younger Dryas (after Bar-Yosef, 2000). Reprinted with kind permission from the Arizona Board of Regents for the University of Arizona.

	Depth (cm)	Material	Raw date	Calibrated date
Lake Zeribar				
CURL 5788	670–700	Sedge seed	4010 ± 75	4492
Y 1432	1410–1420	Bulk	8100 ± 160	8988
CURL 5789	1610–1620	Macro	11,850 ± 120	13,813
CURL 5790	1675–1690	Macro	10,300 ± 50	12,152
Y-1687	1710–1720	Bulk	11,450 ± 160	13,363
CURL 5791	1745–1750	Macro	12,050 ± 55	14,053
CURL 5792	1790–1800	Macro	12,750 ± 110	15,033
Y-1686	1890–1900	Bulk	13,650 ± 110	16,358
Y-1451	2535–2545	Bulk	22,600 ± 500	
GrN-7627	3415–3430	Bulk	37,350 ± 1250	
GrN-7950	4015–4030	Bulk	42,600 ± 3600–2500	
	Depth (cm)		**Raw date**	**Calibrated date**
Ghab Marsh (Niklewski and van Zeist, 1970)				
	129–137		10,080	11,353
	169–171		23,030	28,300
	645–655		45,650	c. 50,000
Ghab Marsh (Yasuda et al., 2000)				
	130		3450 ± 90	3691
	210		4910 ± 90	5644
	252		5010 ± 110	5736
	315		6620 ± 100	7502
	352		7750 ± 100	8464
	405		8680 ± 100	9613
	425		9970 ± 100	11,187
	510		12,890 ± 160	15,259
	590		14,820 ± 180	17,723
	Depth (cm)	**Raw date**	**Corrected date**	**Calibrated date**
Lake Huleh				
	1120 1140	10,440 ± 120	9400	10,372
	1235–1242	11,540 ± 100	10,960	12,879
	1475–1500	15,560 ± 220	14,540	17,415
	1600–1625	17,140 ± 220	16,040	18,915

Table 5.1 Key radiocarbon dates for the stratigrahies of Lake Zeribar, Ghab Marsh, and Lake Huleh, calibrated on the basis of Stuiver and Reimer (1993) and Beck *et al.* (2001). The original dates for Huleh were corrected by Cappers (2001)

	Huleh	Ghab	Monticchio	GISP2	Soreq	Van	Zeribar
			Younger Dryas				
Raw ^{14}C	11–10.3	12–10	11–10				
Corrected	10.5–9.5						
Calendar	12.4–10.8	13.1–11.4	13–11.4	12.8–11.6	13.2–11.4	11.6–10.5	13.5–12
			Bølling-Allerød				
Raw ^{14}C	?–11	?–12	12.5–11				
Corrected	?–10.5						
Calendar	?–12.4	?–13.1	14.7–13	14.8–12.8	?–13.2		15.7–13.5
			Warming before Bølling				
Raw ^{14}C	15.6–?	17.8–?					
Corrected	14.6–?						
Calendar	17.5–?	21.2–?			18–?		18–?

Table 5.2 Radiocarbon and calibrated (calendar) dates for Lake Zeribar in the Zagros Mountains and for Ghab Marsh and Lake Huleh in the Levant. Dates for Lake Van in the Taurus Mountains are based on varves, and for Soreq Cave on U-Th. Monticchio in southern Italy and the GISP2 ice-core from Greenland are included for comparison

in available pollen diagrams from the eastern Mediterranean area were made by Bottema (1995) and Rossignol-Strick (1995), but the results were not very convincing, either because the sequence was not well dated or the stratigraphic resolution of the pollen diagrams was too coarse. This chapter evaluates the paleoclimatic interpretation and chronology of the few sites in the Near East where the transition from glacial to post-glacial conditions are well displayed – Lake Van and Lake Zeribar in the Taurus-Zagros Mountains and Ghab Marsh, Lake Huleh and Soreq Cave in the Levant.

5.1 TAURUS-ZAGROS MOUNTAINS

The particular problem of independent paleoclimatic evidence for the Younger Dryas interval in the Near East has recently been reduced by the analysis of a core of annually laminated (varved) sediments from Lake Van (Fig. 5.4) in eastern Turkey (Wick *et al.*, 2003). Close-interval geochemical analysis showed a short phase of increased salinity in the lake, interpreted as a product of dry climatic conditions, and a detailed pollen study of the same samples indicated a pronounced maximum of pollen types that pointed to cold, dry climatic conditions. Radiocarbon dating was not possible in the sediments, but varve counts down from the surface identified the Younger Dryas as the sharp fluctuation 11.6 to 10.5 ka varve years BP (Table 5.2), *c.* 1000 years younger than a similar fluctuation in the Greenland chronology (12.8–11.6 ka cal BP), indicated also by stratigraphic studies of lake sediments in Europe (Ammann *et al.*, 2000). The 1000-year discrepancy can be attributed to the admitted difficulty in counting thousands of varves down from the surface (Wick *et al.*, 2003).

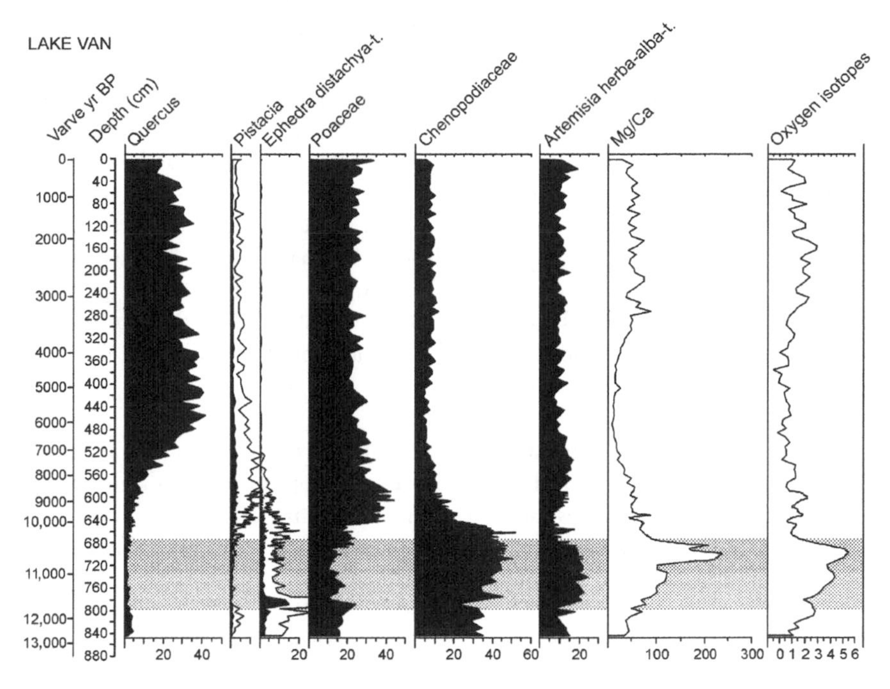

Figure 5.4 Lake Van pollen and geochemical stratigraphy. Shaded zone represents the Younger Dryas (after Wick *et al.*, 2003).

Recent re-examination of the 1963 cores from Lake Zeribar in Iran (Fig. 5.2) have identified a similar sharp interval in records of stable-isotopes (Stevens *et al.*, 2001), diatoms (Snyder *et al.*, 2001) and plant macrofossils (Wasylikowa, unpublished data) at the same stratigraphic level as an *Artemisia*-Chenopod maximum in the pollen diagram of Van Zeist and Bottema (1977), all consistent with an interpretation of dry climatic conditions attributed to the Younger Dryas interval. Dating of the Lake Zeribar sequence was uncertain because of errors inherent in dates on calcareous lacustrine sediment, but the correlation of this interval with the Younger Dryas is strengthened by new radiocarbon dates based on accelerator mass spectrometry of terrestrial macrofossils rather than on decay counting of calcareous sediment. Estimated dates for the span of the Younger Dryas are about 13.5–12 ka cal BP, compared to12.8–11.6 ka for the Greenland ice-core (Table 5.2). The underlying *Artemisia* minimum at 15.7–13.5 ka cal BP at Zeribar may represent the Bølling-Allerød of the European chronology.

The conclusion is that the Younger Dryas dry interval was indeed recorded in the Taurus and Zagros Mountains at these two sites, but the evidence for it in the Levant, which is affected by a somewhat different climatic regime, is not so clear.

5.2 THE LEVANT

5.2.1 Ghab Marsh

Ghab Marsh is located in the long depression that is the structural continuation of the Jordan Valley rift. The pollen diagram published by Niklewski and van Zeist (1970) covers the time since at least 45,000 yr BP (depth 650 cm), but the only relevant date in that study is 10,080 ^{14}C yr BP on a mollusk shell at a level (130 cm) now believed to be well after the beginning of the Holocene, which is represented by the strong increase in *Quercus* and *Pistacia* pollen along with a major decline of both Chenopodiaceae and *Artemisia* (Fig. 5.5). Such a horizon is interpreted as a reliable signature of the Younger Dryas–Holocene transition by Rossignol-Strick (1995) after correcting the radiocarbon dates on calcareous lake sediments by comparing them with pollen profiles from the better-dated marine cores in the eastern Mediterranean, although the marine records are of low stratigraphic resolution. Hillman (1996) also suggested that the chenopod maximum at Ghab represents the Younger Dryas, based on the proposal that a decline in woodland plant remains between 11,000 and 10,000 ^{14}C yr BP at Abu Hureyra, an archeological site 180 km east of the Ghab basin, indicates forest retreat in the northern Levant during the Younger Dryas.

A new AMS date (CURL-5966) on a terrestrial seed is 23,030 ^{14}C yr BP at 484 cm. The chronology used here is based on this date and the rejection of the 10,080 date as being too old because of the hard-water effect. Instead, it is assumed that the abrupt rise in oak pollen and increase in *Pistacia* marks the beginning of the Holocene at 10 ka ^{14}C BP. Dates between 10 ka and the AMS date of 23,030 ^{14}C yr BP are interpolated.

With this chronology an earlier sharp decrease in *Artemisia* and Chenopods and increase in oak (345 cm) is dated about 17.5 ka ^{14}C BP (21.2 ka cal BP), a date that in Europe is just after the last

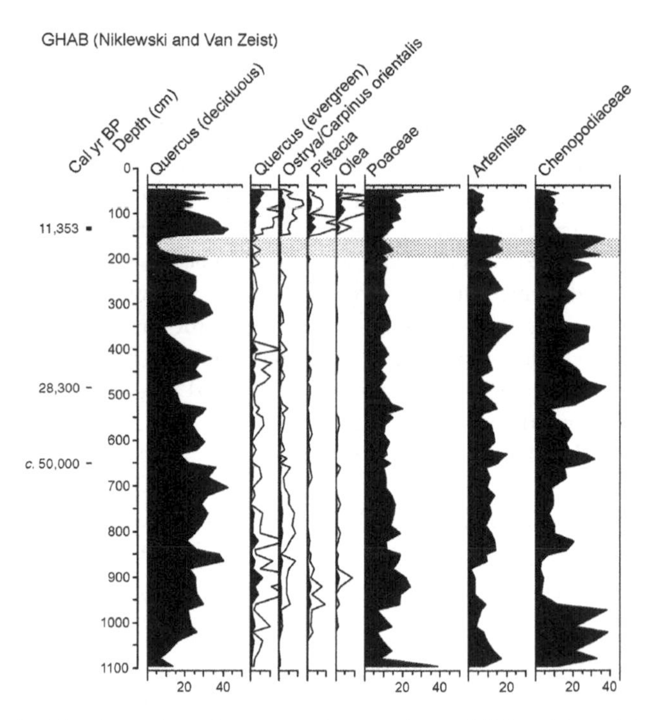

Figure 5.5 Ghab Marsh pollen diagram, adapted from Niklewski and Van Zeist (1970). Extracted from the European Pollen Data Base. The calibrated radiocarbon date of 11,353 yr BP is considered to be too old because of the hard-water effect. Shaded zone represents the Younger Dryas.

glacial maximum. The oak values remain high until the Younger Dryas, which is dated by interpolation as about 11.7–10 ka ^{14}C BP. The latter part of this oak zone would include the Bølling-Allerød. The similarity to the sequence after the glacial maximum in the southern Levant (Huleh and Soreq) is discussed later.

A new pollen study for Ghab by Yasuda *et al.* (2000) includes nine ^{14}C dates on freshwater mollusks back to 14,820 ^{14}C yr BP (Fig. 5.6). The authors assume that no hard-water errors are involved, because the dates they indicate for the Younger Dryas by interpolation between 470 and 430 cm are 11.5 to 10.1 ka ^{14}C BP, similar to the generally accepted radiocarbon dates elsewhere for this interval. However, their identification of the Younger Dryas in the pollen stratigraphy is based not on the Chenopod maximum percentage of total pollen, as Rossignol-Strick had done, but rather on the *Artemisia*/Chenopod ratio, which had been suggested as an index of aridity by Van Campo and Gasse (1993). However, a Chenopod maximum with respect to *Artemisia* may rather reflect local expansion of Chenopod taxa that favour saline soils exposed at a time of low lake level. *Artemisia* may be a better indicator of regional aridity, for *Artemisia* does not have a preference for saline soils. At both Van and Zeribar, where the Younger Dryas is identified by geochemical profiles, the Chenopod and *Artemisia* curves rise and fall together.

Actually the zone designated by Yasuda *et al.* (2000) as Younger Dryas contains instead a maximum percentage of *Quercus* pollen, which in the Ghab core of Niklewski and van Zeist (1970) is accompanpied by a major expansion of *Pistacia*, emphasized by Rossignol-Strick

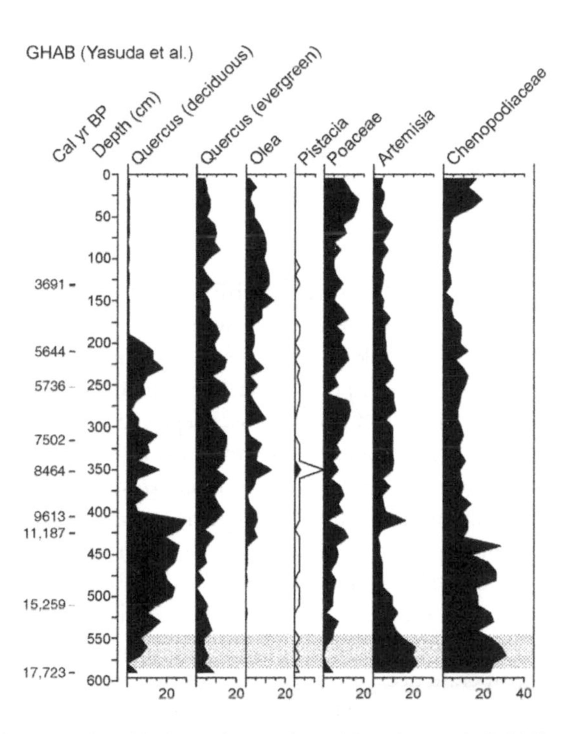

Figure 5.6 Pollen diagram for Ghab, redrawn from Yasuda *et al.* (2000) with calibrated dates. The shaded zone with maximum of Artemisia is considered here to represent the Younger Dryas. The overlying oak maximum at 400–580 cm then marks the Early Holocene. The listed calibrated radiocarbon dates from Yasuda *et al.* (2000) must therefore be too old (see text).

(1995) as a sure sign of the Early Holocene here and elsewhere. The underlying pollen zone (590–470 cm) has higher maxima of both *Artemisia* and Chenopods and is a better indication for the Younger Dryas, suggesting that these ^{14}C dates are much too old rather than being without error.

Yasuda *et al.* (2000) attribute the sharp reduction in oak pollen at their dated level of about 9000 ^{14}C BP to Neolithic forest clearance, and they suggest that the occurrence of pollen of 'cultivated grasses' (not described) during their Younger Dryas interval indicates that plant cultivation started at a time of unfavourable climatic conditions, a conclusion that may be consistent with the archaeological evidence elsewhere but not justified from their Ghab pollen stratigraphy, neither on the basis of signs of cultivation nor on chronologic and paleoclimatic interpretations.

Although the interpretation of the Ghab pollen diagram by Yasuda *et al.* (2000) may be in error with respect to chronology or the record of human impact, the Younger Dryas may indeed be represented, and the site is otherwise important because the earlier study of Niklewski and van Zeist (1970) indicates that the Levant has maintained a population of oak since at least 45,000 years ago, even during the Younger Dryas and certainly during the Bølling-Allerød interval that preceded it. Thus oak was available to expand when the climate became favourable in the Early Holocene.

5.2.2 Lake Huleh

Lake Huleh is in the Jordan Valley in northern Israel. The latest pollen diagram (Baruch and Bottema, 1999) has ten ^{14}C dates over 16 m of sediment, ranging from about 17 to 3 ka ^{14}C BP (Fig. 5.7). The dates were considered too old by an average of 1000 years because of hard-water error and were corrected individually on the basis of the ^{13}C measurement (Cappers, 2001). *Quercus* pollen has values of at least 20 per cent throughout the core, indicating the continuous presence of oak in the region.

The sequence starts at about 16 ka ^{14}C (corrected) BP with an *Artemisia*/Chenopod assemblage in which grasses were abundant as well, implying more temperate climatic conditions for this time than at the interior sites, where oak was a very minor component of the pollen assemblage and may have resulted from distant transport. When the climate changed at Van and Zeribar at the end of the Younger Dryas (10 ka ^{14}C BP), grasses replaced *Artemisia* and Chenopods, and oak and *Pistacia* produced an Early Holocene savanna such as can be seen today in the Zagros foothills on the dry side of the oak forest. At Huleh, in contrast, oak was much more abundant in the area earlier in combination with *Artemisia,* Chenopods and grasses, and when the dry-land plants decreased at about 14.6 ka ^{14}C (corrected) BP, then oak and *Pistacia* expanded into the grass steppe, producing an oak-*Pistacia* savanna at a time when the interior sites were still treeless. As the humidity further increased at Huleh, the oak filled in the savanna, reaching 70 per cent at 11 ka ^{14}C (corrected) BP. Such a strong increase in oak can correlate with the high values of oak and other temperate deciduous trees at Monticchio in southern Italy, the only well dated high-resolution pollen site in the Mediterranean lowlands, where the Allerød is dated at 12.5–11 ka ^{14}C BP (Watts *et al.*, 1996). At Huleh at 11 ka ^{14}C (corrected) BP, oak then decreased from 70 to 30 per cent and grasses rose to 30 per cent in an assemblage attributed to the Younger Dryas. The prominence of *Artemisia* and Chenopods in the Younger Dryas at the interior sites is not apparent at Huleh, because of the less arid conditions of the southern

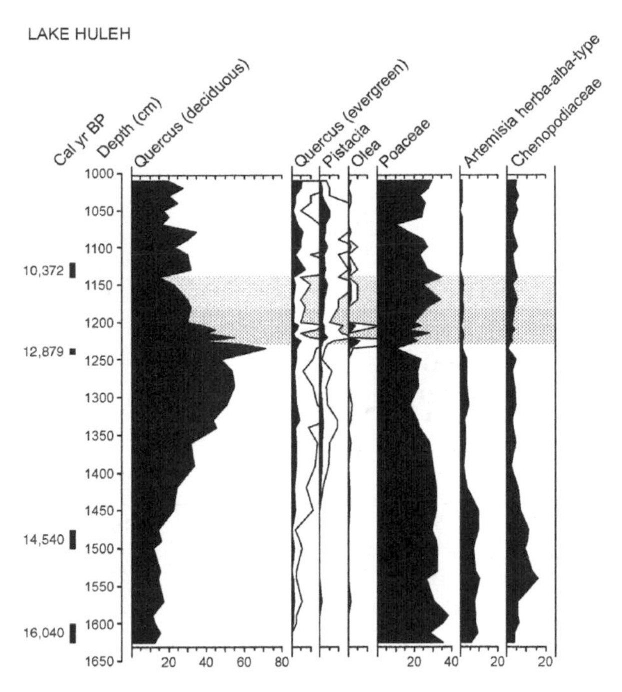

Figure 5.7 Pollen diagram for Huleh, redrawn from Baruch and Bottema (1999), with radiocarbon dates corrected by Cappers *et al.* (2001) and calibrated. The shaded zone represents the Younger Dryas.

Levant. The episode was really represented only by the opening of the oak forest to a savanna structure, with the dry-land shrubs not represented.

After the Younger Dryas, the oak values at Huleh remained at *c.* 30 per cent rather than rising to Allerød levels, perhaps because by that time anthropogenic forest disturbance was becoming significant, as represented by the rise in evergreen oak and the continued presence of olive.

5.2.3 Soreq Cave

Support for the paleoecological interpretation for Lake Huleh comes from the stable-isotope analysis of speleothems in Soreq Cave in central Israel (Bar-Matthews *et al.*, 1999), well dated in calendar years by the U-Th method (TIMS). The sequence indicates high $\delta^{18}O$ values during the glacial period, interpreted as a reflection of a cold dry climate. This was followed at 18–13.2 ka BP during early deglaciation by a drop of 3‰ to values comparable to those of the Early Holocene (Fig. 5.8). The implied warming was interrupted by several reversals. The one at 16.5 ka is correlated with Heinrich event 1, and the subsequent resumption of warming reached a climax that can be correlated with the Bølling-Allerød interstadial of Europe, dated in the Greenland GISP-2 ice-core as 14.8–12.8 calendar years. If the corrected dates of 14.7–10.5 ka ^{14}C BP inferred for the long oak pollen rise at Huleh are calibrated to 17.5–12.4 ka cal BP, correlation may be made with this long warming trend at Soreq (18–13.2 ka cal BP). Similarly for the Younger Dryas, if the corrected dates for Huleh (10.5 to *c.* 9.5 ka ^{14}C BP) are calibrated to 12.4–10.8 ka cal BP, they can be compared to the Younger Dryas at Soreq as based on a $\delta^{18}O$ increase of 1‰ from 13.2 to 11.4 ka cal BP. They can also be compared to 12.8–11.6 ka cal BP

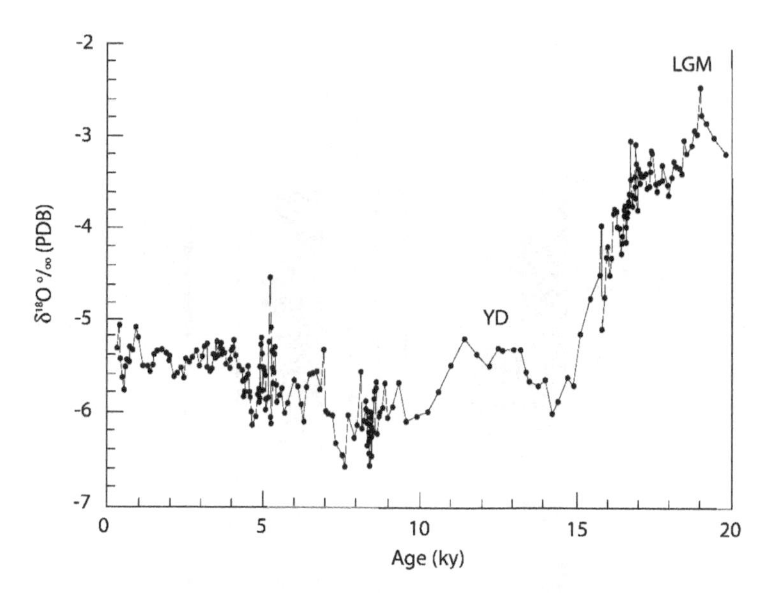

Figure 5.8 $\delta^{18}O$ stratigraphy for speleothems from Soreq Cave. Dates are based on U-Th analyses and are therefore presumed to be calendar dates. LGM = Last Glacial Maximum; YD = Younger Dryas. Reprinted from Bar-Matthews *et al.* (1999) with kind permission from Elsevier Science.

for the Younger Dryas in Greenland. These comparisons imply that the hard-water corrections at Huleh are approximately correct.

According to these correlations, the warming trend after the Last Glacial Maximum at Huleh and Soreq started about 18 ka cal BP and led into the Bølling-Allerød interstadial after H-1. A similar trend is seen at Ghab, with the warming after the glacial maximum starting abruptly at 17.8 ka ^{14}C BP (21.2 ka cal BP). In contrast, the transition to the Late Glacial in GISP2 and Europe in general started abruptly with the Bølling-Allerød at 14.8 ka cal BP. It is not clear why the apparent warming in the Levant should precede the warming in the North Atlantic by several thousand years. The latter was still under the climatic influence of the slowly waning ice sheets and the effects of their wastage on the thermohaline circulation, whereas the Levant was more distant from the ice sheets and (except for the strong perturbation of the Younger Dryas episode) was more directly affected by the steady increase in summer insolation as well as in atmospheric CO_2 (Kutzbach *et al.*, 1993). Closer to moisture sources, the higher summer insolation, which reached a maximum at about the Late Glacial/Holocene boundary, resulted in the enhanced monsoonal precipitation that at its maximum brought the lakes to northern Africa, the increased flow of the Nile River, and the sapropels to the Mediterranean Sea floor (Street-Perrott and Perrott, 1993; Rossignol-Strick, 1985).

5.2.4 Conclusion

Of the three paleoclimatic sites in the Levant (Ghab, Huleh and Soreq), dating control for the published Ghab diagrams of Yasuda *et al.* (2000) is problematical because of hard-water errors. The last oak minimum at 170 cm on the earlier Ghab core of Niklewski and van Zeist (1970) represents the Younger Dryas. The preceding oak minimum at 350 cm has a date of 17.8 ka ^{14}C

BP (21.2 ka cal BP) according to the chronology adopted here and can be correlated with the Last Glacial Maximum as represented at Soreq Cave by the $\delta^{18}O$ peak at 18 ka BP. In this case the broad oak maximum at Ghab between the two minima (170 and 350 cm) would similarly represent the long warming increase that culminated in the Bølling-Allerød.

The next lower oak minimum at 480 cm at Ghab has a date of 22 ka ^{14}C BP (27 ka cal BP), which might be correlated with the oxygen-isotope peak identified as Heinrich event 2 (25 ka) at Soreq. With this framework the corrected dates at Huleh and Ghab when calibrated may thus be used to correlate the pollen sequence with European Late Glacial climatic events. Thus at Huleh the steady major increase in oak following the decrease in *Artemisia* and Chenopods would represent the deglacial warming that culminated in the Bølling-Allerød, rather than the Early Holocene warming as proposed by Rossignol-Strick (1995), and at Ghab the long oak interval between the minima at 350 and 170 cm would represent this same deglacial period.

With these revised chronologies the apparent problem of the opposing vegetational sequences of the Ghab and Huleh records, broadly discussed by e.g. Hillman (1996), is no longer so much of a problem. The histories of the northern and southern segments of the Levant are comparable, and the difference from the interior sites of Zeribar and Van make sense, because they are even farther from the Atlantic. The climatic trend that might be critical for subsistence of human populations is the warmer, moister conditions associated with the end of the glacial maximum and the warming trend leading to the Bølling-Allerød expansion of oak forest, rather than with the similar conditions following the Younger Dryas, as had been earlier supposed for the Zagros region (Wright, 1968). On the other hand, the still earlier speculation may have been closer to the truth – that the critical climatic change in the Zagros Mountains as represented by retreat of the mountain glaciers was contemporaneous with that in the Swiss Alps, where the glaciers had retreated nearly to their modern limits by the time of the Bølling-Allerød (Wright, 1960). The end of the glacial period is more properly considered as the beginning of the Late Glacial (i.e. Bølling-Allerød) rather than the end of the Younger Dryas, which was a *c.* 1000-year perturbation in a long warming trend that started with the retreat of the glaciers and had a temporary culmination in the Bølling-Allerød. This was the time of initial afforestation in Europe by the expansion of birch and pine (and of temperate beetles). The emphasis epitomized by the so-called Meltwater pulses 1 and 2, derived from estimations of ice-sheet volumes based on marine records (Fairbanks, 1989), illustrates the equal climatic importance of the beginning of the Bølling-Allerød and the beginning of the Holocene. The relevance of this conclusion to the origin of agriculture may now be considered.

5.3 THE ROLE OF CLIMATE IN PLANT DOMESTICATION

In the early days of archaeological field investigations on the development of village life and the origin of agriculture, a friendly rivalry developed between Braidwood, who had chosen the foothills of the Zagros Mountains for exploration and excavation because it was the principal area where wild cereal grains were still common, and Kenyon, who focussed on the southern Levant, where many excavations of Paleolithic sites provided a background for extending the record to younger periods. Once the prime sites of Jarmo and Jericho had been identified and were in the midst of revealing the nature of their respective cultures, the development of radiocarbon dating placed the focus on the timing of settlement and of plant domestication.

In the half century that has subsequently elapsed, a very large number of Natufian (Epi-Paleolithic) and early Neolithic sites have been investigated in the southern Levant. In addition, the key locality of Abu Hureyra in northern Mesopotamia has yielded abundant and well dated evidence of the transition between these two cultural phases, largely because of the meticulous recovery and interpretation of the botanical remains. It now appears that the Epi-Paleolithic phase that marked the transition from the nomadic life of hunters and foragers to year-round settlements, with the accompanying increase in population, developed in an environment in which plant resources had expanded as a result of the more favourable climatic conditions that terminated the cold and dry regime of the glacial period. Specifically, this was the time equivalent of the Bølling-Allerød Late Glacial phase well known in Europe. The areas where this occurred were principally in the forest/steppe in both the northern and southern Levant and the Zagros foothills, within the natural range of cereal grains as well as nut trees. Then when the climatic interruption of the Younger Dryas reversed the favourable conditions, the Natufians, already committed to a sedentary life, began to cultivate the cereal grains they had been collecting, leading to the genetic changes in the seed morphology that permits their identification as domesticated. The large number of radiocarbon dates for the archaeological sequence in the Levant, when calibrated to the calendar time scale set by the counts of annual layers in the Greenland ice-cores (Fig. 5.7) (Bar-Yosef, 2000), indicates that the sedentary cultures of the Natufian or Epi-Paleolithic had developed during the favourable climatic regime of the Bølling-Allerød, and that they persisted for most populations during the Younger Dryas, when the first signs of plant domestication appeared, e.g. at Jericho in the Jordan Valley. With the renewal of favourable climatic conditions at the end of the Younger Dryas, marked by an increase in atmospheric CO_2 and the expansion of annual grasses (including cereals), the number of Neolithic food-producing settlements expanded and subsequently spread from the nuclear Levantine area to other parts of the Near East and ultimately to Europe.

Thus the original premise of Childe (1952) that the agricultural revolution in the Near East was impelled by climatic change can be confirmed on the basis of local paleoecological evidence, but it proves to be more complex than he envisioned. An early improvement of climate and plant resources attendant on the early phase of ice-sheet wastage led to the Natufian (Epi-Paleolithic) settlements and expansion of populations. A set-back in food availability during the dry time of the Younger Dryas phase resulted in the cultivation of the cereal grains as necessary to retain the sedentary lifestyle, and the Neolithic was born. The climatic improvement at the end of the Younger Dryas, marking the inception of the Holocene, saw the expansion of the farming culture from its Near Eastern nucleus.

CHAPTER

RADIOCARBON DATING AND ENVIRONMENTAL RADIOCARBON STUDIES

Jon R. Pilcher

Abstract: Since the discovery of natural radiocarbon in the 1950s, its study has diversified into medical, environmental and dating applications. In this chapter the current state of dating methods is reviewed with the advantages of the different methods. Matters of accuracy, calibration and the limits of the method both in age range and in accuracy are considered from the user's viewpoint. Examples of environmental applications of radiocarbon measurement are described before returning to consider the environmental information that is appearing as a bonus from the efforts to calibrate radiocarbon dates.

Keywords: AMS dating, Calibration, Carbon-14, Radiocarbon, Scintillation counting

6.1 Radiocarbon Measurement as a Dating Method

Simple accounts of the method for radiocarbon date users can be found in Pilcher (1991a, 1991b, 1993) and Bowman (1990). It is worth pointing out at the start that the natural levels of carbon-14 (^{14}C) in the biosphere are about one part in 10^{12} (one part in a million, million). A radiocarbon date estimate and most environmental radiocarbon measurements are estimating the levels of radioactivity between this low level and, in effect, the zero concentration reached after some 100,000 years (see Section 6.1.3). Measurement of such low levels of radioactivity will never be easy and never cheap.

There are two fundamentally different ways of measuring the radiocarbon content of a sample; either by measuring the amount of radioactivity it produces, or by measuring the proportion of ^{14}C atoms in the sample. The former, more traditional, method follows long established methods of radioactive counting. Because the amount of ^{14}C is so small and also because the beta radiation produced by ^{14}C is low energy, the sample must be converted into a gas or liquid medium that can be contained within the measurement system. In the first experiments on radiocarbon dating by Libby (1965), the sample carbon was deposited as a solid on the inner wall of a proportional counter. This resulted in the loss of radioactivity by adsorption within the carbon. Attention then turned to using the sample as a gas and, from the 1960s to 1980s, gas systems predominated, but with the development of highly stable and accurate liquid scintillation systems, designed largely for medical applications, most laboratories changed over to liquid scintillation counting. The chemical preparation process for liquid scintillation counting starts with the conversion of sample carbon to carbon dioxide gas by combustion in clean oxygen and the trapping of the carbon dioxide as lithium carbide by passing the gas over red-hot lithium metal. The release of acetylene gas from the lithium carbide by adding water is followed by the polymerization of acetylene to benzene on a catalyst. Benzene is an ideal liquid for radiocarbon dating as about 78 per cent of the molecule is carbon (all derived from the sample) and it behaves well in a scintillation counter. A scintillant is added to the sample so that light is emitted as a result of the energy of radioactive decay. The scintillation counter measures the pulses of light and counts those that derive from ^{14}C decay. Precautions have to be taken at all stages of the chemistry to prevent external contamination and to prevent cross-contamination from one sample to another.

6.1.1 Refinement of the Method

From the mid-1980s a number of radiocarbon laboratories sought to reduce the errors of radiocarbon dating. Those errors inherent in the statistics of random radioactive decay can never by removed but can be reduced by increasing the amount of radioactivity counted, either by increasing sample size or by increasing counting time or a combination of the two. Both exact a price. Increased sample size carries the price of destruction of precious samples, larger excavations and often increased stratigraphical uncertainty. The price of longer counting times is an increase in unit cost as fewer can be measured in a year with the added pressure on machine stability imposed by the long counting times. To reduce the statistical error from ± 50 years to ± 25 years requires four times the counting time – equivalent in cost to measuring four samples. All this, however, only deals with the statistical error from random radioactive decay. A series of international inter-laboratory tests over the 1980s and 1990s (International Study Group, 1982; Scott *et al.*, 1990a, 1990b; Rozanski *et al.*, 1992; Gulliksen and Scott, 1995) proved that there were systematic errors in many laboratories that occurred in many different stages of this complex process. The second task of those laboratories seeking ‘high precision’ was to track down and eliminate or mitigate these errors by meticulous repetition and testing, mostly using known-age samples. It was this work in laboratories such as Belfast, Seattle and Groningen that eventually produced the high-precision measurements that were (and are still) used for calibration measurement.

High-precision scintillation counting is still the method of choice where the highest accuracy is essential and sample size is not limiting. By using equipment of exceptional stability and very long counting times, some laboratories have achieved high precision on sample sizes as small as 0.5 g carbon (McCormac *et al.*, 2001). More usual sample sizes for high-precision measurement are in the range 5–10 g of carbon. This translates to field sample sizes of 25 g charcoal, 150 g wood and up to1 kg wet peat or lake sediment. Thus sample size is often the greatest limitation to high-precision dating, particularly in dealing with samples of low carbon content. As an example, a typical freshwater lake sediment might have a water content of 75 per cent and a carbon content of 20 per cent dry weight. Pre-treatment to remove mobile humic acid will remove 50 per cent of the carbon. If the sediment is sampled with a Livingstone corer with a 5 cm diameter chamber, a 20 cm length of core would be required. This depth of sediment, however, might represent many hundreds of years of sediment accumulation, nullifying the value of high-precision dating.

The second method called **accelerator mass spectrometry** (AMS), involves the separation and counting of ^{14}C atoms. The equipment for this process is from the world of high energy physics. Mass spectrometers are widely used to separate isotopes such as oxygen-16 (^{16}O) and oxygen-18

(^{18}O) used in the study of past ocean temperatures. Unfortunately, ^{14}C is much more difficult to measure than the oxygen isotopes and this cannot be done in an ordinary mass spectrometer, because of the presence of similar mass elements such as nitrogen-14 (^{14}N). To separate ^{14}C from other isotopes, the atoms of carbon have to be accelerated to a very high speed using a van de Graff generator. The sample preparation for AMS dating involves burning the sample to carbon dioxide and the reduction of this to graphite. The graphite, a solid form of carbon (as in pencil lead), is compressed into a pellet that is used as the AMS target. Because the AMS samples are so small, contamination is proportionally a much bigger hazard. The carbon in a fingerprint or in a speck of dust floating in through the window could add a significant proportion of modern carbon to the AMS sample.

Inside the AMS equipment, charged ions are knocked off the solid sample target made of graphite and accelerated by the van de Graff chamber. The speeding ions pass through a powerful magnetic field that bends their path. The heavier ions, with more momentum, travel in a straighter path than the lighter ^{12}C and ^{13}C atoms. Hence each isotope ends up in a different detector that counts the amount of each isotope. Technology from the world of atomic physics removes ^{14}N and the ions such as $^{12}CH_2$ which have the same atomic weight as ^{14}C.

Because AMS counts the ^{14}C atoms rather than waiting for the ^{14}C atoms to decay, the measurement can be rapid with a dedicated laboratory capable of some 2000 samples a year. The ideal sample size is 1–5 mg, but many laboratories handle samples in the microgram range, particularly where the samples are being separated into chemical constituents by gas chromatography (e.g. see McNichol *et al.* 2000). The advent of AMS has had a major impact on many fields such as on lake sediment studies where the diverse origins of the components of lake sediment have always added uncertainty to conventional dating. By dating individual plant components much of the uncertainty of dating the matrix can be removed. Lowe and Walker (2000) warn, however, of the hidden pitfalls in dating macrofossils and Turney *et al.* (2000) give an example where leaf fragments in lake matrix appear to be contemporary but *Carex* (sedge) seeds appear to have entered the lake basin from older deposits. This refined interpretation would have been impossible before the advent of AMS dating.

6.1.1.1 What Can Be Dated?

Anything with carbon in it within the age range of the method CAN be dated. Whether a particular sample is WORTH dating is a more complex question. Carbon dating will only tell the mean age of the carbon in the sample. It will tell nothing about samples of mixed age material or about contaminated material. The radiocarbon laboratory measures what is there. It is up to the submitter to decide whether the carbon in that sample will provide a reasonable estimate of the age of the object or event that is being studied. Familiar examples from the literature include dating archaeological sites by dating hearth charcoal. The carbon in the charcoal dates from the years in which those rings of the tree were growing – this may predate the event (in this case a cooking fire) by several hundred years. Wood that has been stored in a museum may have been treated by preservative resins (polyethylene glycol, for example) of petrochemical origin. The average age of wood plus resin will be considerably older than the use of the wood that the archaeologist wishes to establish. Samples of particularly low organic content are very susceptible to contamination. Small fragments of coal, lignite or even mineral graphite can contaminate lake sediment samples with a low organic content. Thus the most important aspects of radiocarbon dating for the Quaternary scientist are understanding the site and its catchment, understanding the site stratigraphy and having detailed information about the exact nature of the material supplied to the dating laboratory.

6.1.1.2 How Accurately Can It Be Measured?

There is a random element in all radioactive decay and in the particle counting in an AMS system. This poses a limit on the precision of measurement that is typically about 3‰ in routine measurements using scintillation counting and AMS dating. High-precision liquid scintillation counting using large samples can achieve a precision of ± 2‰ (± 12 years in 2000) and, under ideal conditions, AMS can achieve ± 2.5‰ (± 20 years in 2000). The date provided by the laboratory will always carry a ± figure after the age measurement. The two are an integral part of the measurement and should never be separated! It used to be a convention in the radiocarbon business that the ± figure quoted was only the statistical error from the random decay process. Once the inter-laboratory tests (TIRI) showed the range of other errors inherent in the dating process, most laboratories tried to estimate a realistic error based of reproducibility – this is often about 1.2–1.5 times the statistical error.

6.1.2 Calibration

Uncalibrated dates are not calendrical dates. This is because the basis of the radiocarbon method makes the assumption that the natural production of radiocarbon has always remained constant. This assumption is untrue and calibration makes a correction for past variations in natural radiocarbon. Most scientific measurements involve calibration. This is normally treated as an integral part of the laboratory procedure and is hidden from the end-user. Because there was no calibration in the early days of radiocarbon dating, users got used to uncalibrated dates and were confused when calibrated dates started to appear.

Calibration is now (or at least should be) an integral part of radiocarbon date measurement. The reasons for using uncalibrated dates are purely historical. It is no more valid now to quote dates in radiocarbon years than in Martian years – the rest of the world works in calendar years to which calibrated radiocarbon years are a good approximation. It is, however, still important to quote the primary radiocarbon measurement with its identifying laboratory number and to state which calibration has been used. A good review of the history of calibration is given by Damon and Peristykh (2000).

6.1.2.1 How Can I Calibrate My Dates?

Dates supplied to users by radiocarbon laboratories normally now come with uncalibrated and calibrated results. To carry out your own calibration, to use a different calibration curve or to calibrate old, uncalibrated dates, the web-based calibration programme at http://www.calib.org/ is easy to use. There is still a difficult decision that the user has to make and that is which calibration curve to use. In an ideal world the most recent calibration should automatically be the best – it contains the biggest and most recent data set. However, as we find out more about regional variations in radiocarbon, it appears that in some cases regional calibrations may be more appropriate. For example, for samples in the UK and the more oceanic parts of western Europe, it is probably more appropriate to use the 1986 calibration based on Irish oak than the INTCAL98 calibration (Stuiver *et al.*, 1998) based on a combination of Irish, German and North American data (see for example van der Plicht *et al.*, 1995).

6.1.3 What Are the Limits of Radiocarbon Dating?

Variously, 35,000, 50,000 and even 60,000 years are quoted as the limits of radiocarbon dating. A date of 40,000 years is eight half-lives (Fig. 6.1). There is very little original ^{14}C present by then – and plenty of scope for non-original ^{14}C in various forms of contamination to be there instead. By concentrating the minute amounts of ^{14}C in old samples before making the measurement of radioactivity, the limits can be extended back to 75,000 years. Stuiver *et al.* (1978) describe the dating of last glaciation interstadials by concentrating the ^{14}C some sixfold using a thermal diffusion column before measuring the samples in a gas counter. The thermal diffusion process took about 5 weeks for each sample and this method has been little used since then. The theoretical limits of AMS measurement are possibly as much as 100,000 years which is a radiocarbon concentration of 6.7×10^{-18}, but at the moment a practical limit is still at best 50,000 years. Measurements on geologically old materials such as coal, oil and anthracite that should contain no ^{14}C invariably have a radiocarbon concentration of about 10^{-15} – equivalent to ages in the 50 ka region (Gove, 2000). Research is underway in several centres to produce an AMS target production facility using the sort of ultraclean technology developed for studies such as DNA fingerprinting. These may allow AMS dating to be pushed back past the 50 ka barrier without using isotope enrichment (Bird *et al.*, 1999). There is also a different sort of limit at the younger end. Calibration wiggles in the last 400 years prevent the separation of dates in this range except using a wiggle-match technique (see below).

6.1.3.1 How Can I Get a Closer Estimate of Real Age?

There are two ways of getting closer to an estimate of real age than is provided by the best single radiocarbon date. The first uses the **wiggles** (short-term variations) in the calibration curve. A series of stratigraphically linked samples (ideally annual samples such as tree-rings) are dated at high precision and the resulting section of calibration curve matched to the appropriate regional

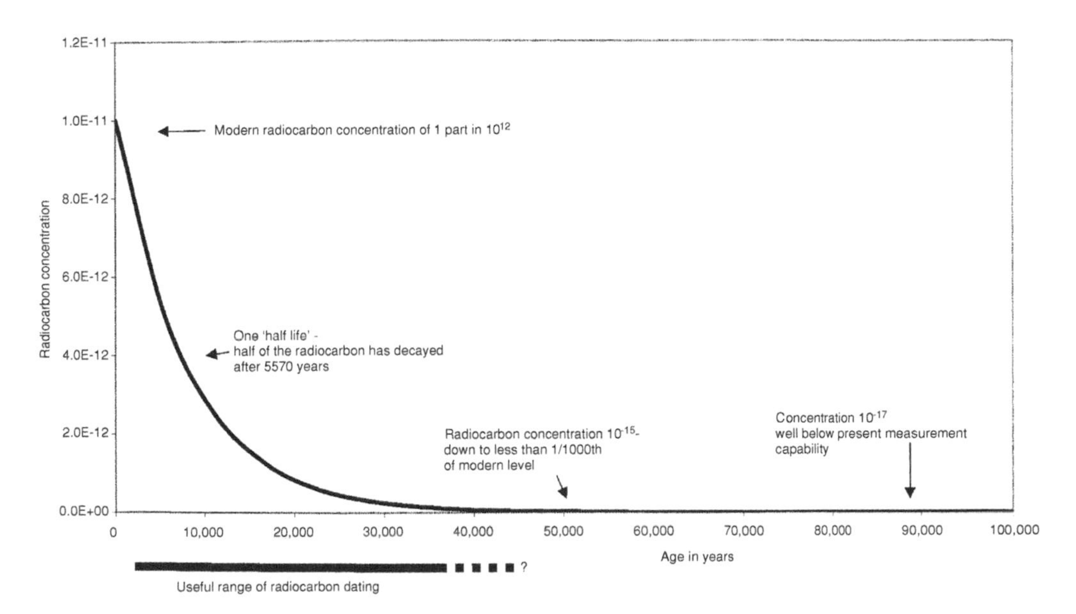

Figure 6.1 Radiocarbon decay curve illustrating why there is a limit to the age of samples that can be radiocarbon dated. The curve is based on a half-life for radiocarbon of 5570 years.

calibration curve (an example using wood is given by Mallory *et al.*, 2002; for examples using peat, see Kilian *et al.*, 1995, 2000). There are various techniques for matching the series from simple graphical approaches to the use of the Bayesian statistics mentioned below. As an example of wiggle-matching, samples were taken to date a tephra horizon in an Irish bog derived from an eruption of Hekla in Iceland (called Hekla 4). The single radiocarbon date covering the tephra layer was measured with a precision of ± 24 but because it fell on a flat part of the radiocarbon calibration (see Fig. 6.2), the calibrated date range was 2457–2205 BC. This analysis was combined with four others in a stratigraphic sequence and the results matched to the calibration curve as shown in Fig. 6.2. Note how the reversal of radiocarbon dates at *c.* 2875 BC exactly mirrors the wiggle in the calibration. Using this additional information allowed a date range of 2395–2279 BC to be specified for Hekla 4 (Pilcher *et al.*, 1996). It might be of interest in the context of regional radiocarbon variations mentioned later that this wiggle-match was carried out with a calibration curve based on local wood (Pearson and Stuiver, 1986). If the same exercise is carried out using a combined international data set, INTCAL98 (Stuiver *et al.*, 1998), the fit of the wiggle-match is not nearly as good.

The second technique utilizes **Bayesian statistics**. The idea behind Bayesian statistics is simple and attractive to the radiocarbon user. We know some information about a sample or a set of samples before we carry out most radiocarbon measurements. For example, we might know the stratigraphic order of a series of samples, we might know the end date of a sequence or perhaps the spacing between samples. Bayesian statistics provides a way of converting this anecdotal information into probabilities that can then be entered into the date calculation. The problem, however, is that the mathematics of Bayesian calculation is complex, forcing most of us to use it as

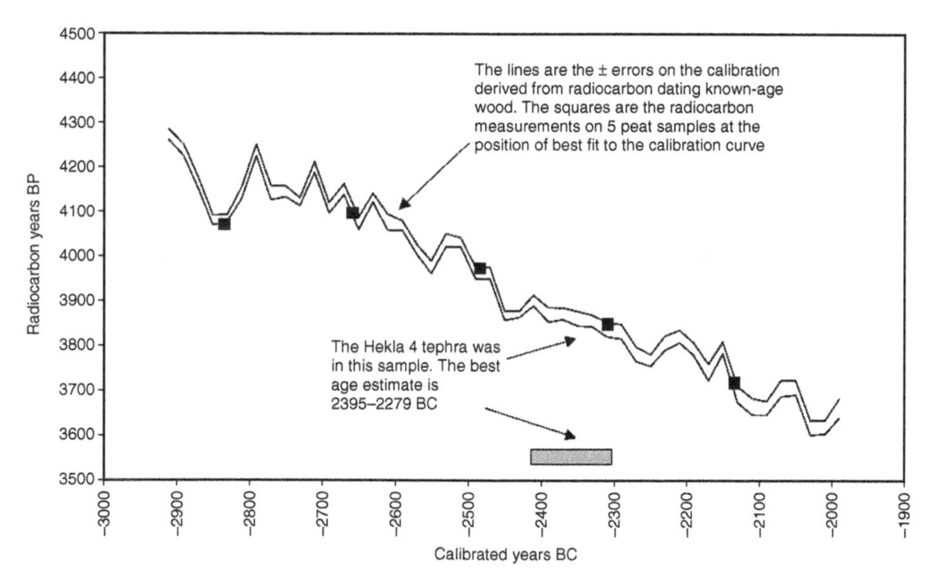

Figure 6.2 Wiggle-matching using five peat samples from Ireland to refine the dating of the Hekla 4 eruption in Iceland. One sample contains the volcanic ash. The position of the five samples is fixed on the *y*-axis by their radiocarbon measurements, but they are free to move together on the *x*-axis until a position of best fit to the calibration curve is found. Note how the oldest sample picks up the big radiocarbon wiggle at about 2880 BC and helps to fix the position of the whole series.

a black box. Steier and Rom (2000) point out, using a simulation, that the assumptions built into the Bayesian model can lead to increased errors while giving apparently improved precision. As Scott (2000) points out, Bayesian statistics is not a cure-all!

Bayliss *et al.* (1997) show the use of a Bayesian approach to the dating of phases in the construction of Stonehenge using the prior information based on stratigraphic context. They combined high-precision dates on antlers used to dig the ditch, AMS dates on bone collagen, with stratigraphic information on the sequence of events as interpreted by the archaeologists. Fig. 6.3 shows how they summarize the results and group them according to the prior information used in the Bayesian calculation.

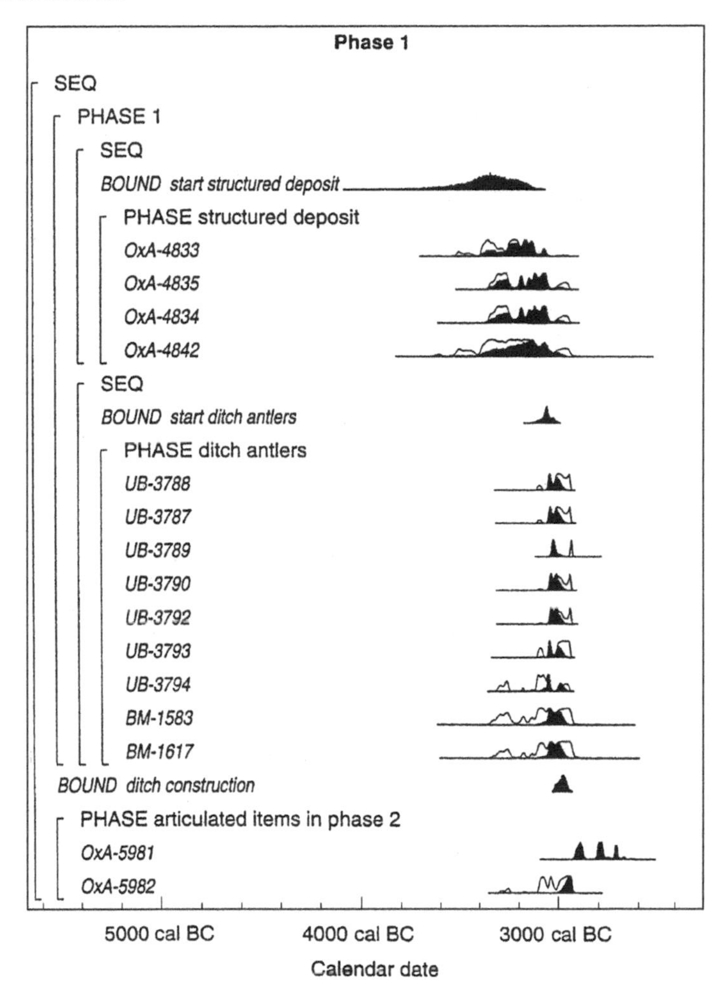

Figure 6.3 Bayesian statistics used to constrain the dating of Phase I of Stonehenge. The stratigraphic information (Phases) are combined with the AMS dates (labelled OxA-) and liquid scintillation dates (labelled UB-). For each date measurement the probability distribution based on simple calibration is given as an open curve and the probability distribution taking into account the additional stratigraphic information is shown in black. Note the narrow range established for the ditch construction. Redrawn from Bayliss *et al.* (1997) with kind permission from The British Academy.

6.1.3.2 Limits of Accuracy

There is good reason to believe that we are already close to the limits of accuracy of radiocarbon dating. There is scope to push back beyond the 50,000-year barrier as described above and there is scope to improve the chemistry of sample preparation, but the limits imposed by the random statistics will remain. This means that there are certain questions that can not be answered. A good example is brought to light by the Hekla 4 dating mentioned above. There is an event in the tree-ring record that indicates a severe climatic downturn starting in the year 2354 BC and culminating in 2345 BC. Other such events have been linked by fairly convincing circumstantial evidence to volcanic eruptions (Baillie and Munro, 1988). Our best estimate of the Helka 4 date in the range 2395–2279 BC falls **close to** the tree-ring date. This does not prove there is any connection between the two or that the climate downturn was caused by the dust veil from the volcanic eruption. Only when the exact calendrical date of Hekla 4 is known will the picture become clearer. Work is ongoing to try to find the volcanic ash from Hekla 4 and other tephra layers in the NGRIP ice core from Greenland (Dahl-Jensen *et al.*, 1997). A major research programme is now underway to achieve a secure calendrical chronology for the NGRIP core that will eventually allow any volcanic events recorded in the ice to be dated precisely. To spend more effort trying to tackle this by radiocarbon dating, however good, would never answer the question.

6.1.3.3 Is My Sample Really Worth Dating?

It is worth considering the limitations of radiocarbon dating before embarking on a dating exercise. Not only are there some questions that radiocarbon dating will not solve, but there are also many situations for which the prior knowledge of the date is as good as the predicted ^{14}C result. Many archaeological periods in northwest Europe are so well dated that additional radiocarbon dating contributes little to total knowledge. This brings us to the accept/reject attitude to radiocarbon dating. The sample submittor has in mind a date based on prior knowledge. If the radiocarbon date lies close to this, it is accepted and if it lies outside the prior knowledge range an excuse is found to reject the result – the sample was contaminated or there was a laboratory error. Such samples should not be dated. It is only where samples of exceptional integrity are found where the sample carbon really does belong to the age under study and where there is a real possibility that the date will contribute to knowledge, that the sample should be dated. It could also be argued that in most cases single dates are of limited value. Sequences of dates with good stratigraphic control have the potential to yield much more useful information.

A good example of both the value and problems of carefully controlled dating is work on the iceman from an Alpine pass on the Austrian–Italian border (Rom *et al.*, 1999). Here careful use of materials of known composition, dating of tissues from the body itself and the replication of results by three laboratories yielded a date with a narrow calibrated confidence range (3360–3100 BC). Even in this carefully controlled example, two samples confidently assigned as property of the iceman turned out to belong to other periods when the Alpine pass was in use.

6.2 RADIOCARBON AS AN ENVIRONMENTAL TRACER

6.2.1 Dead Carbon as Tracer – Where Does Fossil Fuel CO_2 Go To?

As modern biological material contains ^{14}C from the atmosphere and fossil fuels do not, the ^{14}C can be used as a tracer for following the pathways of the additional CO_2 that we are adding to the atmosphere by modern industrial activity. As the most significant greenhouse gas, the fate of CO_2 in the atmosphere, biosphere and oceans is critical to an understanding of, and prediction of, future climate change. Other aspects of the global carbon cycle such as the fate of dissolved inorganic carbon in groundwater are amenable to study by using natural ^{14}C as a tracer (see special volume of *Radiocarbon*, vol 38, Number 2, 1996).

6.2.2 Bomb Carbon as a Tracer – Ocean Experiments

The radiation from nuclear weapons explosions is of high enough energy to form ^{14}C from atmospheric ^{14}N in the same way as the natural production occurs by the action of high energy cosmic rays in the atmosphere. This ^{14}C is injected into the upper atmosphere in atmospheric weapons tests. The spate of weapons testing, climaxing in the autumns of 1961 and 1962 during the Cold War, followed by the Test Ban Treaty of 5 August 1963, provided a sharp peak of ^{14}C activity (Fig. 6.4) that has since been used in a range of experiments on carbon movement in the Earth System.

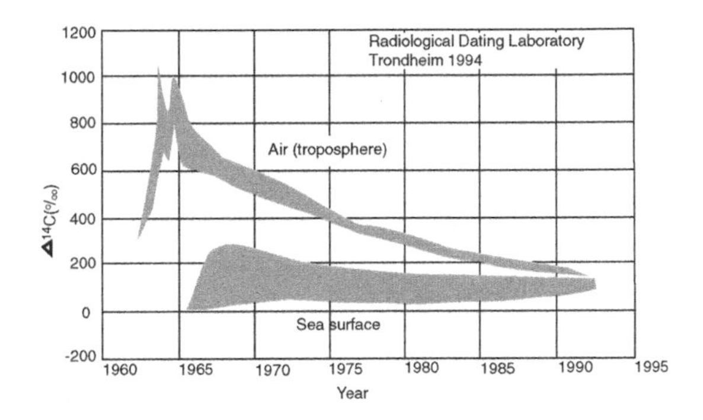

Figure 6.4 Radiocarbon activity in air and sea surface measured from 1960 to 1993 by the Trondheim radiocarbon laboratory. The atmosphere reacted rapidly to the peak of weapons testing in 1961 and 1962 and to the subsequent Test Ban Treaty in 1963. The seawater responds and recovers more slowly. Figure redrawn from Nydal and Gislefoss (1996), with kind permission from the Arizona Board of Regents for the University of Arizona.

A recent volume of *Radiocarbon* was devoted to ocean carbon studies using ^{14}C (^{14}C Cycling and the Oceans. *Radiocarbon* 38, 1996). Here, Nydal and Gislefoss (1996) summarize applications of bomb ^{14}C as a tracer in oceans and atmosphere and show the take-up of the ^{14}C into ocean water at various depths with time. In the same volume, Key (1996) describes the World Ocean Circulation Experiment (WOCE) that used ^{14}C as well as many other measurements to further understanding of ocean circulation.

6.2.3 Is There an Environmental Story in the Natural Radiocarbon Variations?

As the early efforts at radiocarbon calibration showed that atmospheric radiocarbon concentrations have not remained constant over the last few millennia (Suess, 1965), there has been interest in the environmental significance in these fluctuations. First, we must look at what controls the level of ^{14}C available to plants. The driving force behind ^{14}C production is the flux of high-energy cosmic rays from distant space. It is assumed that this flux is reasonably constant (with the exception of supernovae, see below). The amount of this cosmic energy that arrives in the Earth's atmosphere, however, depends on the Earth's magnetic field and on the Sun's activity – both deflect cosmic rays away from the Earth and both are variable on the time-scales we are considering.

Once ^{14}C has been produced, the amount available for absorption by plants (i.e. the samples used for constructing the calibration curve) will depend on the global carbon cycle. For example, the biosphere absorbs carbon on a short time-scale, and is temperature-dependent. The oceans absorb carbon on a longer time-scale and this will be affected by variations in ocean currents. Any additional 'dead' CO_2 from natural or man-made sources will cause temporary dilutions of the atmospheric ^{14}C concentration.

There are several components to variation in atmospheric radiocarbon concentrations as seen in the calibration curves. It was clear from the first long sequence of measurements by Hans Suess (1965) on *Sequoia* wood that the level of atmospheric radiocarbon has varied and that the relationship of radiocarbon dates to calendar years was not linear. The long-term trend found by Suess is retained in all the more recent data sets and has been extended beyond the limits of tree-ring-dated wood by measuring the radiocarbon age of varves and corals (INTCAL98 – Stuiver *et al.*, 1998). This long-term trend is assumed to be caused by a gradual change in the Earth's dipole moment. The Earth's magnetic field has changed in the past and, in geological time, has many times reversed North–South – most recently at *c.* 735,000 years ago at the Brunhes–Matuyama transition, a useful chronological marker in ocean sediments.

If this long trend is removed from the calibration data there are still many variations, not least of which are the flat portions of the calibration that are such a hindrance to dating the last glacial–interglacial transition (13,000–11,000 BP) or, for example, the European Iron Age (800–400 BC).

If the calibration data set had been homogeneous and had been based on annual samples, further analysis of the variation using Fourier analysis would have been relatively simple. However, the calibration is made up of a combination of annual, decadal, 20-year wood samples and coral and sediment samples of variable age span. With such a data set the limits of signal analysis are constrained by the lowest resolution sampling. Even with these limitations there are some clear cycles apparent in the calibration data. The clearest is the 88-year Gleissberg cycle whose origin is certainly solar (Damon and Peristykh, 2000). There is also a 207-year cycle that is probably solar and a 512-year cycle that may be related to the oscillations in the ocean circulation.

Where single-year wood samples have been analysed these portions of the calibration can be analysed at higher resolution and show clear evidence of the 10.4-year Schwabe cycle and the 21.3-year Hale cycle, both solar and well known from other environmental time series. These cycles are related to sun-spot frequency and relate to some fundamental oscillation in the sun.

So far we have modes of variation related to the Earth's magnetic field, to solar activity and to ocean currents. Damon and Peristykh (2000) present evidence for the enhancement of atmospheric ^{14}C levels by the high-energy particles from a supernova explosion dated to AD 1006. The ^{14}C increase started in 1008 and faded way by 1015. Damon and Peristylkh consider this to be consistent with the arrival of energetic particles from the supernova. If this is true then other such events during the last 100,000 years are certain to have affected the atmospheric ^{14}C levels. Beck *et al.* (2001) present radiocarbon dating of a Uranium/Thorium dated speleothem that appears to show very large-scale fluctuations in natural radiocarbon between 45,000 and 33,000 years ago which are greater than could be produced by solar variations or the Earth's magnetic field. They attribute this to substantial fluctuations in the carbon cycle.

Finally, we have to consider the possibility of ^{14}C dilution by the addition of 'dead' carbon to the atmosphere. It is well known that this has been happening since 1850 due to the burning of fossil fuel. In fact it was the analysis of ^{14}C that showed that the additional CO_2 in the atmosphere was dead carbon and pointed the finger at fossil fuel burning. What is much less clear is the possible role of natural injections of dead carbon into the atmosphere. One of the possible sources of this carbon is the huge reserve of methane trapped in shallow ocean sediments in the form of clathrates. Seismic or other geological disturbance could potentially release significant amounts of dead carbon suddenly into the atmosphere. Baillie (1999a, 1999b) describes historical accounts suggestive of big out-gassing events that happen to coincide with depletions in atmospheric ^{14}C. While such events are certainly possible, release of deep ocean (^{14}C-depleted) carbon is also possible as a result of upwelling of deep ocean water. Changes in ocean regimes such as El Niño are likely to lead to changes in the exchange of ocean and atmospheric carbon.

6.2.4 Marine and Southern Hemisphere Offsets

It has been known for some time that marine samples have radiocarbon dates that are older than contemporary terrestrial samples, with the 'offset' often quoted as 400 years. The extent of the marine offset actually varies from place to place according to the speed and degree of ocean mixing and to the amount of upwelling of deep ocean water. Typical values are 400–500 years for the North Atlantic and as much as 1000 years for the North Norwegian Sea. It is unlikely that this offset has remained constant and one would predict that major perturbations would have occurred around the time of the Younger Dryas when large volumes of 'old' water were added to the North Atlantic from melting glacial ice. Lowe and Walker (2000) discuss this in some detail and quote an offset of 700–800 years for the North Atlantic in Late Glacial times. A further complication is that the atmospheric concentration variations will also impinge on the marine radiocarbon on time-scales that depend on rates of mixing between surface and deep water. Stuiver and Braziunas (1993a) tackle this complexity from a theoretical standpoint and use their calculations to produce calibration curves for marine samples for the last 10,000 years. Once more data are available the offset will become a tool for the study of variations in ocean currents in the past. Marine corrections are available at the calib website (http://radiocarbon.pa.qub.ac.uk/calib or http://depts.washington.edu/qil/calib).

At present the atmospheric ^{14}C concentration in the southern hemisphere is slightly lower than that from the northern hemisphere. This hemispheric offset of about 25–40 years has been assumed to be a constant related to the relative areas of land and sea in the two hemispheres. A meticulous replicated, recent study using wood from Ireland and New Zealand has shown that the hemispheric offset is variable through time (McCormac *et al.*, 2002). Thus, ultimately, there will have to be separate radiocarbon calibrations for the northern and southern hemispheres. At present this is limited by the progress in extending southern hemisphere tree-ring chronologies. The replicated northern and southern measurements extend back to AD 950 and the offset can be seen to vary from about 20 to as much as 80 years. The full environmental implications of this are still to be assessed, but at present it is assumed that ocean currents must play a significant role in determining the offset.

6.3 THE FUTURE

What does the future hold for radiocarbon dating? As stated above, it is unlikely that date measurement precision will improve much beyond the 2‰ level attainable at present. The advent of the smaller (and considerably cheaper) 0.5 MeV AMS machines will greatly extend the availability and perhaps reduce the cost of AMS measurement over the next 10 years. Extension of European tree-ring chronologies into the Late Glacial will extend the northern hemisphere calibration curve and further calibration extension will be achieved by measuring the radiocarbon activity of coral and speleothem samples dated by Uranium series. The big advances will not be in dating applications but in using radiocarbon for environmental studies and the extension of the southern hemisphere calibration will be a part of this. This will further understanding of ocean ventilation and ocean circulation and feed the global change models for the future high CO_2 world.

CHAPTER

7

DENDROCHRONOLOGY AND THE RECONSTRUCTION OF FINE-RESOLUTION ENVIRONMENTAL CHANGE IN THE HOLOCENE

Michael G.L. Baillie and David M. Brown

Abstract: Dendrochronology by establishing the year-by-year chronology for the Holocene has set the ultimate chronological standard. All other sources of information on fine resolution climate change must ultimately fit themselves to deductions from the tree-ring calendar. This chapter reviews the principal long chronologies and explores the types of fine-resolution information becoming available from them, be it archaeological, tectonic, volcanic or climatic. A strong indication is given that the most profitable direction for future research lies in a multi-proxy approach that combines information from various well-dated proxies in order to paint the broadest possible picture of short-term events in the past.

Keywords: Abrupt changes, Archaeology, Climate reconstruction, Index chronologies, Vulcanism

Dendrochronologists are faced with what is almost an over-abundance of riches, namely information of some description about every year in a tree-ring chronology. To provide perspective, the longest chronology currently available is the Hohenheim oak chronology which contains an annual record back to 10,480 BP (Friedrich *et al.*, 1999). To show how rapidly new chronologies are developing, the Hohenheim workers are now indicating the existence of a pine chronology which overlaps with the oak chronology and extends the total annual record back to almost 12,000 BP. The annual descriptive information available from the tree-rings is normally some measure of mean ring width, perhaps an annual growth index value, though it could be maximum late-wood density or some measure of radioactive or stable-isotope concentration; it could even be a measure of trace element concentration. Moreover, the existence of area chronologies implies much more than single records. Although there may be one master chronology for an area, presenting a general picture, there are going to be ring-width records from many individual trees at any point in the record. The sources of those trees may be localized at a sub-regional level, or there may be differing sub-chronologies from differing ecological settings such as forests or bog surfaces, mineral soils, even exposed lake margins.

Not only are there ring-width chronologies but, when a period becomes of interest, dendrochronologists can go back to the timber specimens themselves and examine the individual growth rings for changes in character, or for the existence of unusual growth or even physical damage. This chapter will look at the range of tree-ring information which is becoming available, at how it influences dating and chronology in the Holocene, and how it can contribute to an understanding of fine-resolution climate change.

7.1 THE CHRONOLOGIES

Dendrochronology has expanded so dramatically in the last few decades that it is essentially impossible to keep up with the literature; there are now many thousands of pieces of writing. This reflects the fact that dendrochronologists have sampled trees and constructed chronologies in every part of the temperate world, from Mongolia to Ireland and from northern Canada to Tasmania. Only the tropics remain relatively unstudied, mostly due to the difficulty in confirming ring counts in tropical species, and the poor survival of fallen logs on tropical forest floors. It may be worth taking a brief look at the reasons behind chronology construction as these have dictated the nature of the chronologies and the chronological coverage which currently exists.

The simplest dendrochronological unit is the living-tree site chronology. There are now thousands of site chronologies. It is implicit that a chronology should have some level of replication and thus a site chronology tends to contain information from ten or more trees growing in close proximity. These can be short, e.g. the six Irish oak chronologies covering the last two centuries (Pilcher and Baillie, 1980), intermediate, e.g. the numerous 500- to 700-year juniper and ponderosa pine chronologies from Western North America (Holmes *et al.*, 1986), or extremely long, e.g. the 3620-year record for *Fitzroya cupressoides* from Southern South America (Lara and Villalba, 1993). These chronologies contain almost exclusively continuous ring patterns of still living or recently dead trees. The advantage of such chronologies, from the point of view of environmental reconstruction, is that their constituents have not changed, throughout they represent the same trees in the same location. Such chronologies have been built largely for the attempted reconstruction of climatic or environmental variables. We will look at one major example below.

Separate from the climatically inspired site chronologies are the composite chronologies. The first of these had been developed by Andrew Douglass using yellow pines from the American Southwest. By overlapping living-tree sequences with patterns from timbers preserved in progressively older Pueblo ruins, Douglass had constructed a chronology which by 1929 ran back continously to AD 701 (Robinson, 1976). Elsewhere, composite oak chronologies were constructed for a variety of purposes, most notably in Germany for the elucidation of Holocene valley development (Becker and Schirmer, 1977), and in Ireland for radiocarbon calibration purposes (Pilcher *et al.*, 1984). However, irrespective of initial motivation, in each case the chronologies were eventually used for a suite of purposes including direct dating of archaeological timbers, radiocarbon calibration and environmental/ecological reconstruction. Composite chronologies are constructed from the ring patterns of living trees, timbers from historic buildings and archaeological sites and naturally preserved trees from bogs and river gravels. In short, composite chronologies are constructed using timbers from a variety of sources which may exhibit differing responses to climate; a potential problem if long-term reconstructions are to be attempted. Some long chronologies, e.g. bristlecone pine, use relict timbers from exactly the same location as the living trees and should preserve a more coherent growth response through time than might be anticipated with oaks (Ferguson and Graybill, 1983).

In some areas, progress in chronology construction has been limited by combinations of factors. In New Zealand, cedar and silver pine chronologies have been constructed from living and recently dead trees back for 850 and 1150 years, respectively (Xiong and Palmer, 2000). Further extension is limited by three factors: the lack of really long modern examples, the lack of archaeological sources and failure (so far) to locate suitable sources of earlier timbers which might overlap with the existing chronologies. In such situations further extension requires a combination of luck and exhaustive sampling; the biggest single factor being luck. The reason for making this point is that researchers, hoping to see extended chronologies from potentially key climatic areas such as New Zealand, should not assume that extended chronologies will be easily forthcoming. If at any time, any event, for example major storm or extensive fire, substantially depleted the number of trees which might survive down to the present, then we can expect to run into difficulty when attempting to extend the chronology across that event, and the result could easily be an unbridgeable gap in that chronology. That said, New Zealand offers the only possibility for the construction of long floating chronologies in the period 30,000–50,000 BP due to the unique survival in the North Island of long-lived logs of Kauri pine in swamp (bog) contexts.

In summary, while there are very large numbers of living-tree chronologies, covering parts of the last millennium, there are many fewer chronologies covering the whole of the last two millennia, perhaps only some 35 in total. Table 7.1 lists the longest chronologies (lengths rounded) but is not intended to be exhaustive.

From Table 7.1 it is evident that there are only something like eight really long chronologies running back seven millennia or more. It is important to note that the list is an approximation

Name	Species	Area	Length (yrs)	Type
Hohenheim	Oak	Germany	10,480	C2
Göttingen	Oak	Germany	9700	C2
Methuselah	Bristlecone	USA	8700	C1
Sweden	Pine	Sweden	7400	C2
Finland	Pine	Finland	7500	C2
Ireland	Oak	Ireland	7400	C2
England	Oak	England	7000	C2
Yamal	Larch	Siberia	7000	C1
Campito	Bristlecone	USA	5400	C1
Indian Garden	Bristlecone	USA	5200	C1
Tasmania	Huon pine	Australia	4000	C1
White Chief	Foxtail pine	USA	4000	C1
Fitzroya	Fitzroya	Chile	3600	LT
Giant Forest	Sequoia	USA	3200	LT
Converse Basin	Sequoia	USA	3200	LT
Cirque Peak	Foxtail	USA	3000	C1

Table 7.1 Most of the world's longest continuous tree-ring chronologies, rounded to nearest century. Types of chronology: C1 = composite single site; C2 = composite multiple site; LT = living tree chronology

and does not include important long chronologies under construction. For example, Kuniholm, who has been constructing chronologies in the eastern Mediterranean, has a continuous chronology back to AD 300 and a long prehistoric section spanning circa 700–2700 BC, tightly constrained by multiple radiocarbon determinations (Kuniholm, 1996). Table 7.1 does, however, give an approximation of the state of the art. From Table 7.1 it is apparent that in terms of really long chronologies we could simplify the global coverage down to Siberian Larch, Fennoscandian pine, European oak and American bristlecone pine. From a global environmental reconstruction point of view, it is unfortunate that these are all situated in the northern hemisphere, however, it is some consolation that they are well spaced around the hemisphere.

7.1.1 Local Chronologies and Data Presentation

The chronologies mentioned above all have a regional aspect. Dendrochronologists face the problem of how to present their data which range from the ring widths of individual trees, through site and regional chronologies, to potentially supra-regional groupings. The simplest and easiest form is to take trees from a site or region and produce mean ring-width values for each year – the 'mean' chronology. Unfortunately, there are problems with mean chronologies and this has encouraged the widespread use of index chronologies. First, why are means a problem? Imagine that a European dendrochronologist obtains many oak ring patterns for a 500-year period such as AD 1001–1500. What can happen, because of changes in the history of timber use through time, is that most timbers running up to say 1350 are long-lived, and therefore narrow-ringed. After the Black Death with its reduction in human population, large numbers of previously coppiced oaks are abandoned to grow back into timber trees. These trees, growing from established root stock are fast-grown and wide-ringed. The resultant 1001–1500 mean master chronology would show a dramatic increase in annual mean width after 1350. Obviously such a step would be an artefact of a change in timber source and is not overtly a reflection of any change in climate.

Chronologies with steps are highly suspect when attempting to reconstruct past conditions and, as a result, dendrochronologists have tended to de-trend individual ring patterns by fitting a curve to each ring pattern and producing growth indices which vary around 100 per cent (put simply, the yearly values are converted to percentages of the fitted curve; for a recent elaboration of the process; see Cook and Peters, 1997). The resultant index chronologies have no long-term trend and are essentially year-to-year and short-term variations around a baseline. Thus the same data can be presented in mean and index forms and the presentation used may allow differing interpretations. Index chronologies are essentially misleading in that they present a 'tidy' version of reality; literally they make diagrams more presentable. Figure 7.1 shows an index chronology for all Irish oaks across the period AD 420–660 plotted with a mean chronology for the same constituent trees across the same period. It is observed that the year-to-year variations are the same, so that each chronology is equally good from a dating point of view, however, the message conveyed to the observer is very different in the two cases. What is interesting is that in the mean chronology the lowest point is in the AD 540s. The same impression is lost in the index chronology even though the growth indices do point to the 540s being unusually poor growth. For anyone interested in this period, whether environmentalist, archaeologist or historian, it might be important to compare the two chronologies. For example, in Fig. 7.1 the low index values at and after 635 give an impression that this growth reduction is of an equivalent magnitude to that in the 540s. However, when the mean chronology is

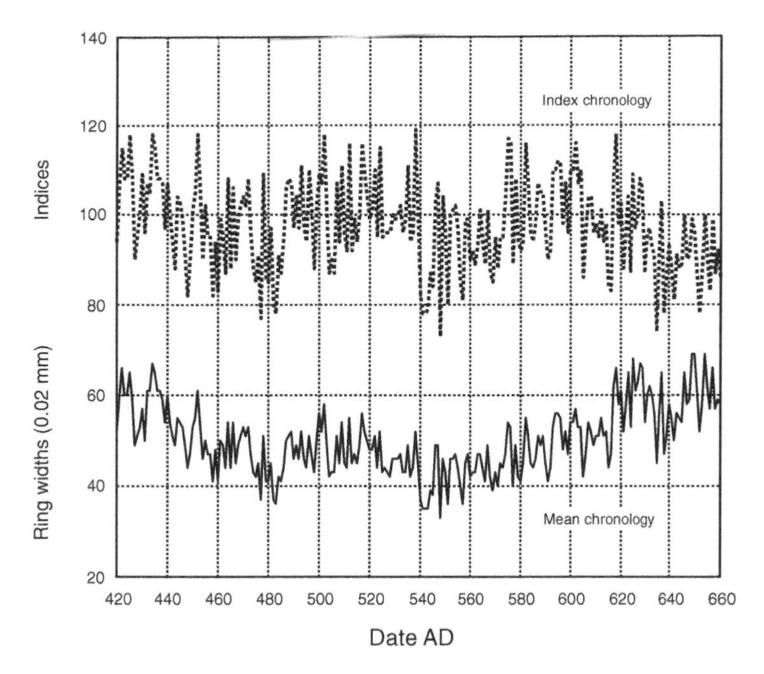

Figure 7.1 The Irish oak tree-ring series for the years AD 420–660 presented in an index format with long-term trend removed from component ring patterns (above) and as a raw mean-ring-width chronology, demonstrating the significantly different appearance of these presentations of the same data after AD 600.

examined, it is clear that the 540s represent an absolute growth minimum, whereas the reduction at 635 is from a higher base, on an overall rising trend in growth, and is presumably of less significance. In Fig. 7.2, each of the data sets in Fig. 7.1 has been subjected to a standard five-point smoothing routine in order to clarify the trends in the data. Here a much clearer picture emerges with the 540 event standing out clearly in both presentations. The reader can ask which is 'best' or 'most correct', the simple answer is that there may not be a best or most correct presentation (see Fig. 7.3); all the dendrochronologist can do is state clearly what is being presented and attempt to avoid distorting the data to make a point. Irrespective of issues concerning data presentation, we now have continuous records at annual resolution which can be used to make observations about changing growth conditions during the Holocene.

7.1.2 Backbone of Chronology

With the completion of several long chronologies, especially those in Europe, dendrochronologists have started to provide the archaeological world with real calendar dates for at least some of their prehistoric sites. This has had the effect of forcing archaeologists and palaeoecologists to move from consideration of raw radiocarbon time-scales to calibrated radiocarbon age ranges which are compatible, within the limits imposed by radiocarbon, to tree-ring ages. Because the main oak chronologies were constructed specifically to refine the radiocarbon calibation curves, it follows that dendrochronology has been the principal driving force behind a move to real dates for most of the Holocene. This same trend can be seen in the increasingly widespread use of tephrochronology. Here, microscopic layers of volcanic ash

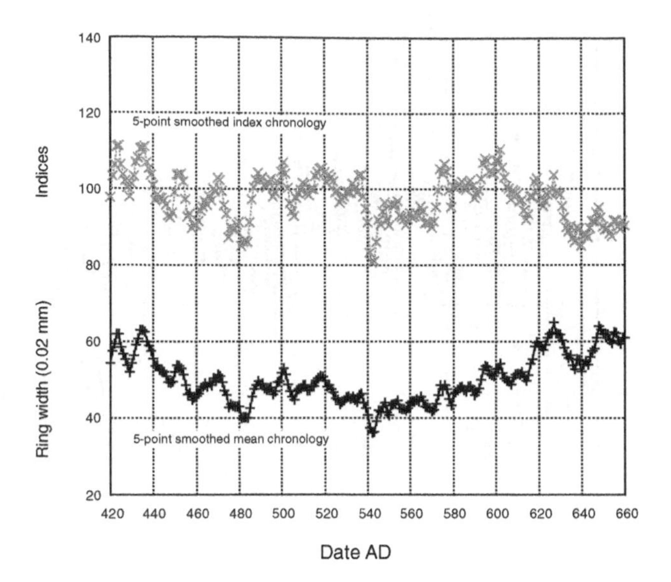

Figure 7.2 The index and mean chronologies for Irish oak, as in Fig. 7.1, presented as five-point smoothed curves. This draws attention to the notable growth reduction at AD 540 which stands out clearly in both representations compared with Fig. 7.1.

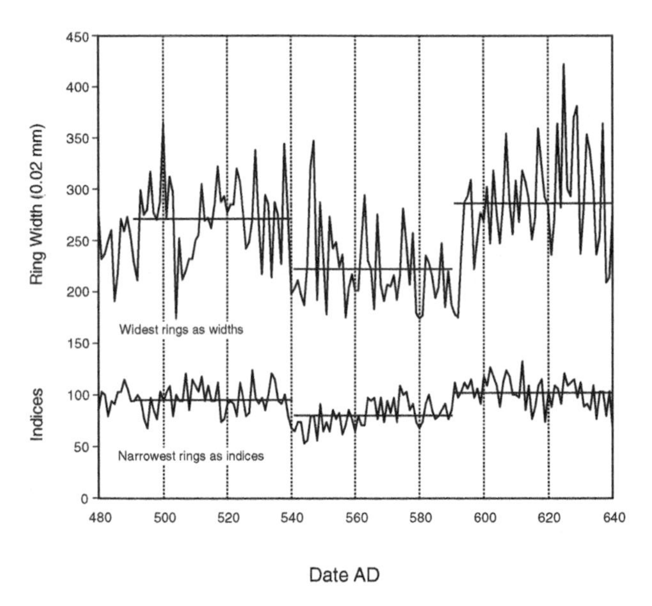

Figure 7.3 An alternative representation of tree-ring data (not normally used). In this case the 'widest rings in any tree in any year' is isolated, similarly the very narrowest ring. The resultant plot represents the entire envelope of oak growth. The narrowest rings are represented as indices for clarity. In this plot we see a curious 50-year reduction in the envelope of Irish oak growth across a period AD 540–590 previously picked out by Gibbon as one which involved 'corruption of the atmosphere' from AD 542–594. Horizontal bars represent successive 50-year mean values.

occurring widely in peat profiles can be tightly constrained in real time by the use of multiple stratified, high-precision (i.e. ± 20 years) radiocarbon determinations (Pilcher *et al.*, 1995). This more refined approach to the dating of peat profiles, coupled with a widespread acceptance of the importance of the well-dated ice-core records from Greenland (e.g. Clausen *et al.*, 1997), means that whole suites of information can, in theory, begin to be drawn together with tight time control. It has to be remembered that, of these techniques, the only one which is truly calendrical in nature is the tree-ring record.

7.2 DATING ENVIRONMENTAL PHENOMENA

From an environmental viewpoint, dendrochronology allows direct dating of many phenomena from earthquakes to volcanic eruptions, subsidence, lake level changes and even tsunami. Numerous tree-ring studies have been applied to physical effects on trees. For example, advancing glaciers can push over, damage, or kill trees in their paths. Where remains of such trees have survived, due to burial by ice or moraines, dates can be obtained for the physical events (Kaiser, 1993; Luckman, 1995). Trees damaged but not killed will tend to heal scars within their ring patterns, allowing precise dating of all manner of physical phenomena. Trees pushed over, say in earthquakes or by storms, tend to exhibit asymmetric wide growth rings (reaction wood) where the tree has attempted to right itself; the initiation of the reaction wood giving a close date for the phenomenon.

Stahle *et al.* (1992) show how baldcypress trees at Reelfoot Lake, Tennessee exhibit an anomalous growth surge in the years following the great New Madrid earthquakes of 1811–1812, which apparently formed the lake. On average these trees show an increase in ring width in excess of 400 per cent; the increase stated to 'reflect the radically improved water budget of these drought sensitive trees' (Stahle *et al.*, 1992). Probably one of the most elegant recent examples of the dating of a physical process is the ability of dendrochronologists to suggest a specific day and indeed time for a massive Richter 9 earthquake affecting the Oregon and Washington coast in the year 1700. Around 1700, the Cascadian Subduction Zone experienced a major earth movement. The evidence for this was the widespread occurrence of 'ghost forests' of dead cedar trees standing in saltwater; obviously the land had subsided. Yamaguchi *et al.* (1997) drew together two disparate pieces of evidence to make a plausible suggestion of the exact time of the earthquake. They refined the exact dating of the dead trees by taking samples from roots with intact bark to confirm a 1699 date for the last year of growth (on the exposed trunks the bark and outer rings had weathered). With this confirmation that the trees died sometime between the autumn of 1699 and the spring of 1700, this made sense of Japanese records of a significant, but previously unexplained, tsunami in January 1700. As the evidence for the giant earthquake is obvious in the inundated trees, as the tsunami record falls into the relevant time window and as the speed of a tsunami is relatively well known, Yamaguchi *et al.* (1997) could reasonably suggest that the most likely time and date of the earthquake was around 21:00 h on 26 January 1700.

Now the reader may be questioning why such geophysical dating exercises are being laboured in a chapter on reconstructing fine-resolution climate change in the Holocene. We would argue that it is important to see the wider context of environmental events. Earthquakes may well be signposts of enhanced volcanic activity in the Holocene. If we take the great New Madrid earthquakes of 1811 and 1812, various ice-core records show volcanic acidity signals in 1809/10 and 1815/16 (Zielinski *et al.*, 1994; Clausen *et al.*, 1997; Cole-Dai *et al.*, 2000), which we know are implicated in climate change. Briffa (1999) states, 'Over the hemisphere as a whole, 1810–1820 was one of the coolest decades of the millennium, in part due to the effects of the large explosive volcanic eruptions in 1809...and 1815'. Such packages hint at broad relationships between tectonic episodes and environmental change. The example is not unique; the ice acidity and cold temperatures in 1698–1699 (Briffa *et al.*, 1998), coupled with the Cascadian earthquake of 1700 may suggest another. It is clear that as the application of dendrochronology spreads in time and space we can expect to find more and more interactions with other disciplines which can provide comparable chronological resolution.

7.3 RECONSTRUCTING CLIMATE/ENVIRONMENT

It is implicit that if tree-ring patterns cross-match from tree to tree and from area to area then there must be some controlling climatic influence. Douglass could see, even in his earliest work in the American Southwest, that the extremely narrow, sometimes absent, rings in his yellow pine record represented drought years in an area where growth was controlled by available moisture. Thus two types of information come directly from dendrochronology – the dates for the archaeological timbers and environmental conditions deduced from the ring patterns of the same timbers. Numerous exercises have followed this course. For example, Rose *et al.* (1981), studying a complex set of ruins in the American Southwest (a pueblo) were able to make the following related observations. In a period of favourable climatic conditions 'the pueblo grew to nearly a hundred times its original size in the first decades of the 1300s...The settlement reached its greatest size around 1330, comprising 24 room blocks constructed around ten...enclosed plazas...'. In this case the tree-ring dates for the actual constructional timbers directly record the enhanced building activity brought on by the improving food supply indicated by the 'wet' proxy of better tree-ring growth. Inevitably, when the trees record a pattern of severe droughts after about 1335, the town's population began to decline leading to virtual abandonment around a decade later. There are now numerous studies on regional droughts in the American Southwest which broadly link droughts and site abandonment. For example, Larson and Michaelson (1990) note the Anasazi being affected by major droughts in AD 1000–1015 and AD 1120–1150, the latter leading to the abandonment of the southwestern Great Basin. Ahlstrom *et al.* (1995) tackle the complexities of a migration between the Mesa Verde region and the Rio Grande in the late 13th century AD, traditionally believed to have been due to regional droughts. What becomes clear is that there are now numerous interlocking studies for the American Southwest for the period AD 1000–1500. It is then of interest to observe that comparisons can be made with other regions. Information on drought frequency for central California exists for the whole of the last two millennia, deduced from sequoia growth (Hughes and Brown, 1992). Similarly, a record of fire scars in giant sequoia from Sierra Nevada, California, allows related deductions on drought (Swetnam, 1993). With all this available information, the nearest thing to an overall synthesis is probably the attempt by Hughes and Diaz (1994) to define if there really was a 'Medieval Warm Period', and if there was, where and when it occurred. The essential point is that with the tight chronology provided by dendrochronology, data from disparate sources can now be combined into broader pictures.

Before leaving droughts there is one surprising example which should be mentioned. It seems counter intuitive that trees which grow in a 'frequently-flooded riparian habitat' could actually be drought-sensitive. Yet workers studying baldcypress in the southeastern USA have found that growth of such trees is directly correlated with precipitation. The chronologies have been used to provide a reconstruction of the Palmer Hydrological Drought Index for areas in southeastern Virginia. In an intriguing application, Stahle *et al.* (1998) believe they can link drought to the fate of two early English settlements known as the Roanoke and Jamestown Colonies. As Stahle *et al.* (1998) state, 'The Roanoke colonists were last seen by their English associates on August 22, 1587, the same summer when the tree-ring data indicate the most extreme growing-season drought in 800 years'.

They also note that 'the settlers of Jamestown had the monumental bad luck to arrive in April, 1607, during the driest seven-year period in 770 years' (Stahle *et al.*, 1998). In both cases, the English settlers had been expecting to trade for food with native American populations but this idea was frustrated by the droughts which restricted food supplies generally. Thus dendrochronology is capable, under the right circumstances, of being able to add a new dimension to our understanding of past historical events. Obviously the results from this latter study can be added to those from the Southwest and California to begin to establish broader patterns of American drought and climate change. Elsewhere links have been made between what was happening in the American Southwest, in the decades running up to AD 1350, with more global goings on in the run-up to the outbreak of the Black Death in the Old World (Baillie, 1999a, 1999b). In particular, it transpires that in the 1340s there was a global downturn in tree growth which must, almost certainly, be due to a reduction in global temperature.

7.4 ACCUMULATED DATES

With the establishment of long chronologies, particularly those for oak in Europe and pine in the American Southwest, dating of archaeological sites has become routine; already a whole corpus of dates exists. In Europe, Hollstein (1980) drew together all the German sites and buildings he had dated in a working lifetime, while in the American Southwest, Robinson and Cameron (1991) were able to produce a list of 1300 tree-ring-dated sites spanning the last two millennia. Overall, literally thousands of dendrochronological dates exist. One interesting fact emerging from this routine application of dendrochronology to archaeology is the new perspective that is provided as the individual dates gradually accumulate into patterns of past human activity; we now have evidence for building episodes in the past. Tree-ring dates do not suffer from the innate 'smearing' effects normally associated with radiocarbon dating, where a short event in real time, investigated with multiple radiocarbon determinations, almost inevitably leads to a blurred picture. Dendrochronology allows us to see a sharper picture of human building activity which can then be related to increasingly well-dated environmental stories.

Thus, dendrochronologists are becoming used to seeing patterns of human building activity in real time. For example, the dating of around 50 prehistoric wetland sites in Ireland reveals that these are not uniformly, or even randomly, distributed in time, rather there are tight clusters of building activity involving oak timbers (Baillie, 1995). Figure 7.4 shows the distribution of Irish sites clustering around 1500 BC, 950 BC, 400 BC and 100 BC. Although the reason for these clusters is not yet fully understood, the fact that some of the clusters contain habitation sites on lake margins and in wetlands does tend to suggest relatively dry episodes when lake-levels were low and bog surfaces were accessible. The fact that tree-ring dates are directly comparable from

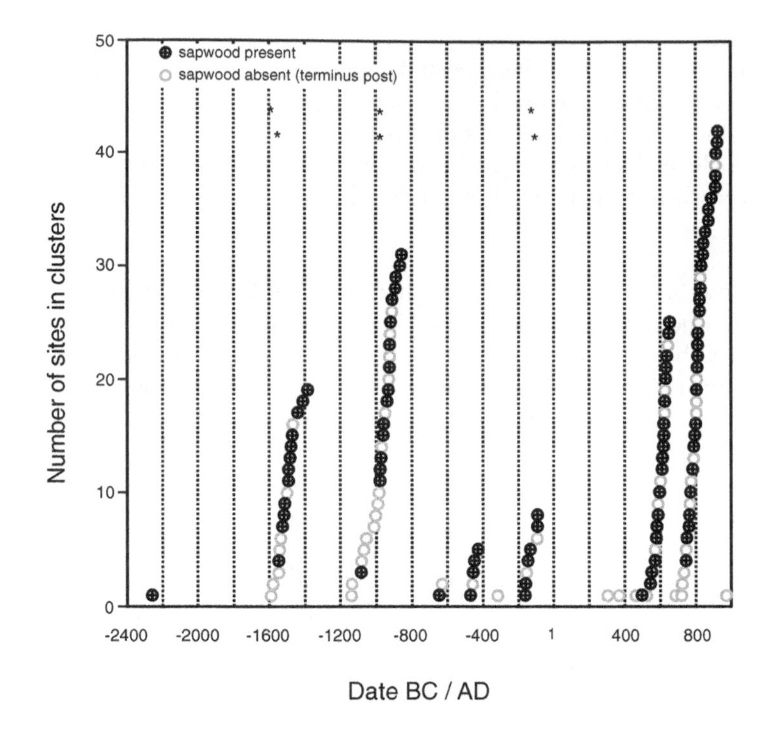

Figure 7.4 The non-random pattern of dates for Irish, oak-bearing, archaeological sites developed from the application of routine dendrochronological dating. Sapwood presence confers absolute dates; absence provides *terminus post quem* dates only. The six stars represent the first six prehistoric sites to be dated.

region to region means that additional clues can be drawn from other areas. The finding of a substantial Bronze Age bog settlement at Siedlung Forschner in the Federsee marsh, Germany, with one major building phase dating to 1511–1480 BC, again suggests a wet area being accessible during the 1500 BC Irish phase (Billamboz, 1992). Billamboz also notes a floating chronology representing three phases, spanning 50 years, from a large, well defended, Late Bronze Age site at Wasserburg Buchau, Federsee. Although not yet precisely dated it 'shows a continuous extension in circumference in the same way as Siedlung Forschner 500 yr before' (Billamboz, 1996). It would be surprising if this second construction phase in the Federsee was not also coincident with the 10th century building phase in Ireland.

The value of these long-distance linkages is that they again add colour to the environmental story which can be derived from the tree-ring chronologies themselves.

7.5 CLIMATE RECONSTRUCTION FROM TREE-RINGS

The examples noted above form a backdrop to environmental reconstruction rooted in the historical development of dendrochronology. A different approach was developed from the 1970s and is well paraphrased in H.H. Lamb's foreword to Fritts' seminal work *Reconstructing Large-scale Climatic Patterns from Tree-Ring Data* (1991):

> Progress had to wait on the development of the necessary computing capacity and the devising of multivariate analysis techniques for handling the data from very large numbers of carefully selected trees and sites....Later, another vital step was to use the relationship discovered between tree-growth and weather to take the tree-ring data and 'back-cast' (i.e. reconstruct) the climate of previous periods within the present century and then test the result by comparison with the known reality (Fritts, 1991).

This approach moved away from the more qualitative 'low growth approximates to drought', to a more quantitative reconstruction of actual values for parameters such as temperature, precipitation and pressure over large spatial grids. This move was driven by the needs of climate modellers to have numerical quality data with which to feed their models. Fritts noted that there were four main causes of climatic variation over time-scales of years to centuries, namely solar output, volcanic forcing, ocean circulation and atmospheric transparency. At its simplest, the work of dendroclimatologists, over the last few decades, has been to reconstruct aspects of these variables from very large spatial grids of tree-ring chronologies following calibration of the grids with modern climate data.

7.5.1 A Volcano Story

In order to demonstrate the application of this grid approach, we can use an example where aspects of summer temperature have been reconstructed for the northern hemisphere for the last 600 years. This work used a grid of 383 temperature-sensitive site chronologies from right around the northern hemisphere, broken down into eight regional averages (Briffa *et al.*, 1998). In this case, maximum late-wood density was used in preference to ring width as tests indicated that density of the late-wood was strongly correlated with summer temperature. Figure 7.5 shows the northern hemisphere temperature reconstruction (termed NHD1). It is

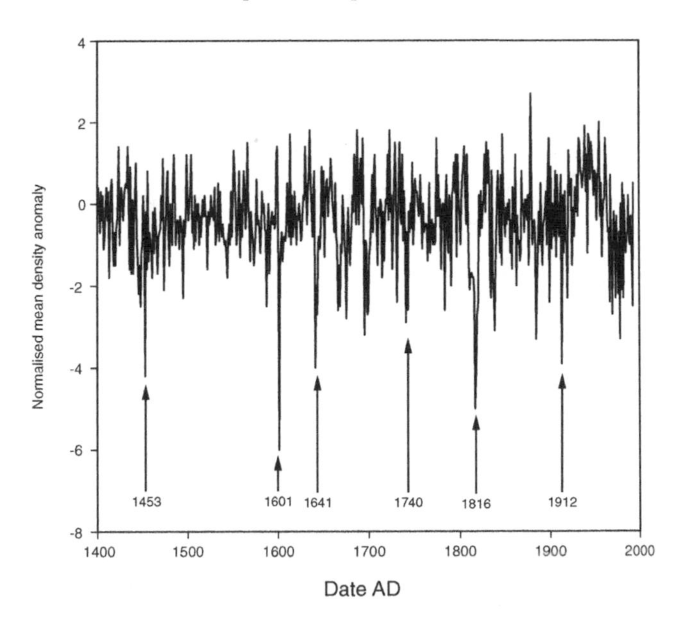

Figure 7.5 Northern hemisphere summer temperature as deduced from density record NHD1 (after Briffa *et al.*, 1998) with the dates for a few well-known volcanic eruptions (arrows).

clear from Fig. 7.5 that most of the large temperature depressions coincide with the dates of known volcanic eruptions deduced historically or from the Greenland ice-core records. Some of the dates are well known, such as 1601 following the eruption of Huaynaputina, Peru (1600); 1816–1818 following Tambora, Lesser Sunda (1815) and 1884 following Krakatau, West of Java (1883). Obviously these temperature-sensitive trees indicate a definite response to the lowering of temperatures following significant volcanic dust veil events.

However, in order to demonstrate how dendrochronology can add to development of a wider picture, we can look at the response of European oak to these same events, introducing a more temperate deciduous species into the equation. Figure 7.6 shows NHD1 with the average growth index derived from a grid of eight regional oak chronologies which span a transect across northern Europe from Ireland to Poland. Comparison of the two charts indicates that some of the cooling events picked out by the temperature-sensitive NHD1 also show as reduced ring width in the oak series, others do not. For example, in the 17th century the oaks present low growth value in 1601, 1635, 1652, 1667, 1675 and 1685, with 1685 the lowest growth in the century. Of these major growth reductions only 1601, 1667 and 1675 are coincident with significant pine density reductions. A similarly patchy agreement occurs throughout the 600-year record. Thus adding in the European oak story changes the picture implied by NHD1. Not all volcanoes show immediate effects on temperate deciduous trees. It seems logical that the

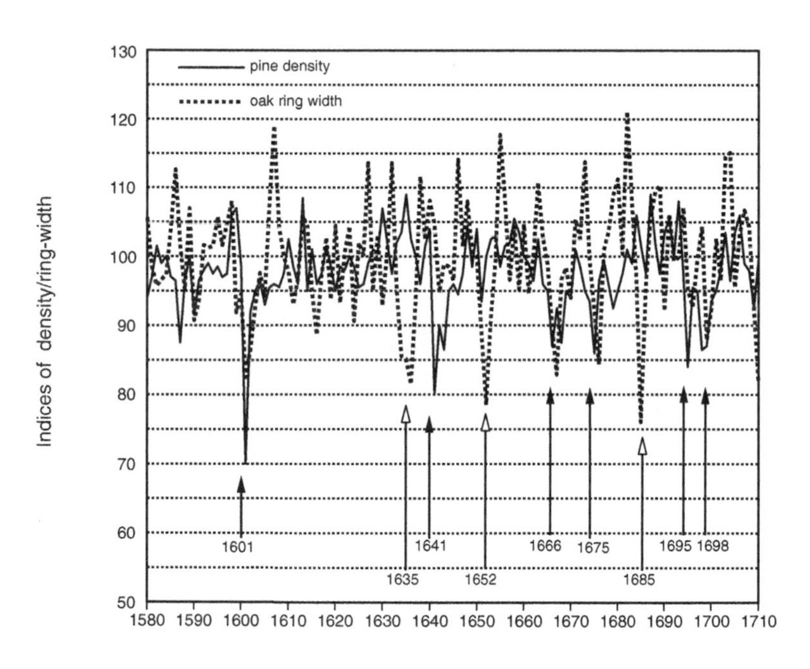

Figure 7.6 A section of the summer temperature record, from Fig. 7.5, plotted with an eight-site European oak ring-width chronology (transect Ireland to Poland) for comparison. Solid black arrows indicate volcanic eruptions, open arrows represent oak growth downturns not associated with the NHD1 temperature response.

important events are those where the effects show up in both records and involve more temperate areas.

The power of dendrochronology lies in its ability to add further data. Taking only the 1600–1601 event, we observe that in South America there is a dramatic downturn in the growth ring for 1599 (Boninsegna and Holmes, 1985). Because, by convention, rings in the southern hemisphere are dated to the year in which they start growth, this implies that these trees in Argentina were affected sometime between October 1599 and May 1600. In contrast, cedars in New Zealand show a dramatic lagged effect in 1602 (Xiong and Palmer, 2000). These observations are interesting because European oak values are all below 100 per cent for all the years 1599 to 1605, implying that the so-called '1600' event may be more complex than a single volcano in 1600; something hinted at by Briffa *et al.* (1998) who wonder if Huaynaputina may have been part of a package of volcanic activity.

Hopefully this demonstrates that the temperature signal provided by the hemispheric density grid, is only telling part of the story. While temperatures at high-latitude, high-altitude sites may well be depressed by explosive volcanism, the parallel effects in temperate oak would seem to suggest that, in the last 600 years, only the 1600, 1740 and 1815 events were of genuinely widespread environmental significance.

This volcano-related example serves to show the power of dendrochronology to describe, and refine, pictures of past climate which are simply not available from human documentation. Further exercises can be anticipated for any period back to circa 5000 BC given the number of chronologies becoming available (see Table 7.1).

7.5.2 Cycles

It is apparent that there is a great deal of underlying cyclicity in the tree-ring records (e.g. Fig. 7.2). From the earliest days of dendrochronology this was a source of interest, especially to Douglass who was first and foremost an astronomer (Douglass, 1919). However, studies of cycles in tree-ring series have had a mixed history. For example, in the early 1970s, LaMarche and Fritts (1972) attempted to discover if there was a solar signal in tree-rings and applied power spectrum analysis to the question. The results were inconclusive in that they found 'no convincing evidence of consistent relationships between ring width and solar variation'; however, they made the important observation that 'the power spectrum analyses do not rule out the possibility of solar-related oscillations in tree-ring series which undergo frequent phase shifts or reversals'. More has been done since and the problem that arises in almost every study is that cyclic behaviour is time transgressive. It becomes apparent that cycles are present in the system, but they change and interact with one another through time. As a brief excursion we can look at the same 600-year period that has already been discussed.

The authors, courtesy of colleagues worldwide (see Acknowledgements), have access to a major grid of eight world chronologies constructed using pine (Polar Urals, Sweden), oak (Europe); oak, pine, juniper (Aegean); bristlecone pine (USA); Fitzroya (South America); cedar (New Zealand); huon pine (Tasmania). For simplicity, one oak chronology represents all European oaks and the bristlecone pine chronology is the mean of nine such chronologies. So, the resultant overall 'world 8 ' chronology is a proxy for the response of a huge number of trees worldwide (outside the tropics). Again, each of the chronologies has been indexed and smoothed with a standard five-point smoothing routine. Figure 7.7 shows the chronology from 1400 to near present. In it we see notable changes in the character of the curve through time. Most dramatic is the change from cycles of approximately 18 years to cycles of around 11 years beginning at AD 1600. It is possible, though not quite so clear, that the short cycles revert to longer periods at 1740. Given that the event at 1600 has already been highlighted as occurring in both the oak and pine records, and as being related to a volcanic dust veil, it is permissible to wonder if such events can be responsible for dramatic alterations in the world climate system as reflected in tree-growth. In other words, can an abrupt environmental event, triggered, say, by a dust veil, flip the world system from one state to another instantly? Would the logic be that the cooling effect of the dust veil, if severe enough, can alter ocean circulation or move the polar front? Would the logic also be that we live in a world where any given climate state exists for only a century or two at best? If this were the case then attempting to reconstruct past climate by calibration of tree-rings against modern weather records might be fraught with difficulty. If, as seems apparent in Fig. 7.7, the world tree-growth system – which implies the world's climate system – can flip from one state to another, is it fair to ask if there ever was a 'normal' period. Aspects of the same changing picture can be observed back for at least 7000 years through the dendrochronological record. Figure 7.8 shows an emerging 37-year cyclicity in European oaks in the later 3rd millennium BC, a period already known to have involved significant and apparently global environmental change (Dalfes *et al.*, 1997). Worse, it isn't that there are two states and the system flips between them, the tree-ring record hints at a long series of episodes, each different in character. If Figs 7.7 and 7.8 are anything to go by, and we would suggest that they are entirely typical, then no period of stability lasts longer than a couple of centuries.

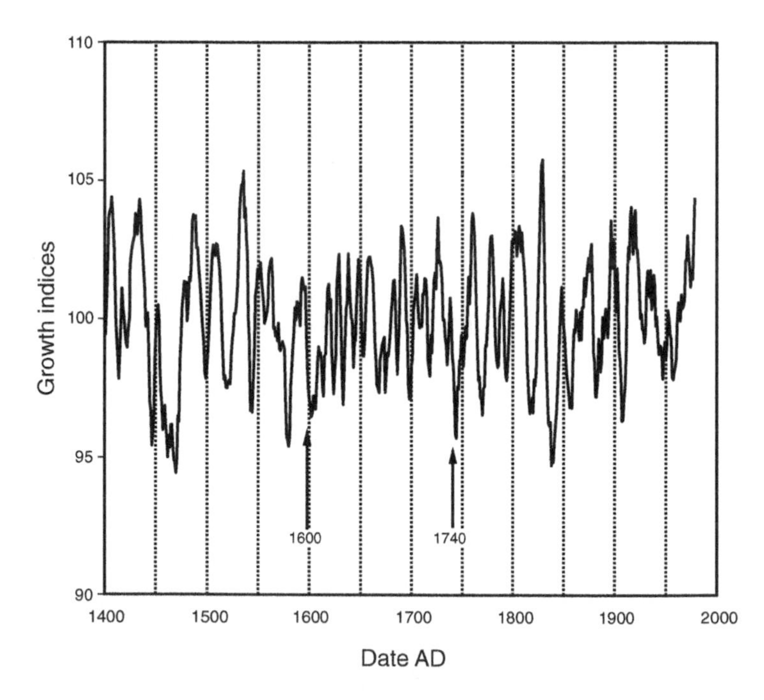

Figure 7.7 The mean of eight world chronologies (all five-point smoothed) for the period AD 1400–2000 showing notable changes in underlying cyclic behaviour around 1600 and 1740.

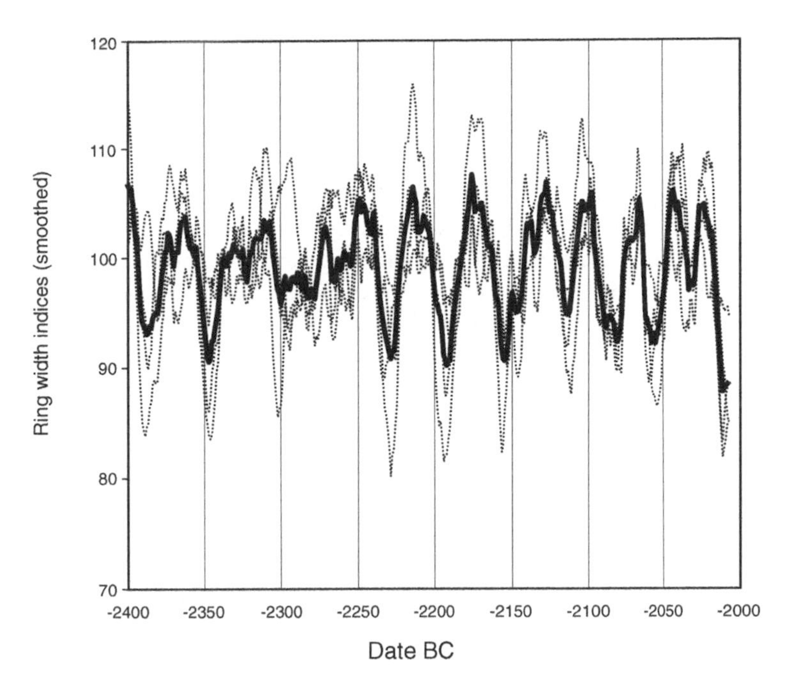

Figure 7.8 The mean of four European oak chronologies showing a clearly developing 37-year cycle in the later 3rd millennium BC. The component chronologies (Ireland, West and East England and North Germany) are indicated as broken lines.

7.5.3 Abrupt Events

It is impossible to leave a discussion of events such as that at AD 1600 without looking briefly at the general question of abrupt environmental events. Dendrochronologists tend to be the people to observe these simply because they are the only workers with multiple annual resolution records. Much has been made of the abrupt event at AD 540, evident in Fig. 7.1 (Baillie, 1995, 1999b; Keys, 1999). This environmental event is synchronous with widespread famines, affecting populations at least from Europe to China, and with the outbreak of a major plague. Opinions differ on the likely cause of the environmental downturn which can be clearly seen in chronologies from around the world. There is little doubt that the event is complex, and at least two-staged, and may have involved an interaction with a comet or cometary debris (Baillie, 1999b). Keys (1999), on the other hand, opts for a supervolcano, with an explosive power of 5 million megatons, on the site of Krakatau, taking place in February 535. Of interest in the context of this chapter is that new tree-ring information from Mongolia indicates, yet again, that the most severe part of the event was not around 535–536 but in the early 540s (D'Arrigo *et al.*, 2001), something apparent in chronologies from Europe (Baillie, 1994, 1995). It is interesting to see what they say about the event:

> In AD 536 the Index value is 0.645, relative to the long-term mean of 1.0. Standard deviation (SD) is 0.204 over the full length of the chronology. This low growth value signals the onset of an unusually cold decade (AD 536–545) in which the mean ring-width index is 0.670 (SD 0.240), with a minimum of 0.37 in AD 543. AD 538 shows a brief recovery with an index value of 1.223. As noted, this two stage pattern is also evident in the

> European oak and other tree-ring series and may signify a delayed climatic response to one event (as is typical for many volcanic eruptions – e.g. Stothers (2000) or possibly two separate events (Baillie, 1994, 1999b). (D'Arrigo *et al.* 2001)

Again, as more tree-ring information accumulates a better and better regional, and global, picture begins to emerge to supplement that already available from Europe and the Americas. Suffice to say that it is dendrochronology which has supplied the focus for study of this and other abrupt events. These abrupt environmental events should act as foci for scholars in different disciplines and it is anticipated that their full effects and causes will eventually be elucidated. One essential aspect of this refined-chronology research which has been slow to be addressed is the linking of the annual tree-ring records with the near-annual ice-core records. Something which can be done by exploiting the simultaneous occurrence of tree-growth downturns and volcano-related acid in the two records (Baillie, 1996).

7.6 CONCLUSION

European tree-ring workers have already provided an absolute, annual, time frame back to 11,919 BP (Friedrich *et al.*, 1999). Thus, dendrochronology is providing a chronological and environmental backbone for the Holocene. The information from other proxies such as ice-cores, varves and pollen form the environmental ribs which must ultimately be fitted to that backbone. Dendrochronology has the particular advantage of allowing fresh environmental information to be obtained which is already precisely dated, as shown in the AD 1600 case. However, most workers have hardly come to terms with what this implies. Let us look briefly at the 6th century example. As noted, the AD 536–545 event includes aspects from 535–536 (Scuderi, 1990; Baillie, 1994) through to 543–545 (D'Arrigo *et al.*, 2001; Baillie, 1994). This is a **minimum** span of 7 years' effect if we accept the latter authors, or 8 years if we accept Keys' 535 dating (Keys, 1999). Stothers (1999) has written about this event but has the dilemma that he accepts the concept of volcanic perturbation of 'the climate system in an important way for up to 5 years'. Dendrochronologists are seeing definite effects in trees for at least 2 years longer. Either volcanologists are wrong on how long the effects of volcanoes can last, or what happened around AD 540 was a two-stage event. Thus, dendrochronologists are setting the agenda on environmental understanding associated with abrupt change in the later Holocene where historical information exists. When we move to prehistory and consider what was happening at dates between 2354 and 2345 BC, or at 3200/3199 BC (Baillie, 1995, 1999b), only dendrochronology currently has anything to say.

The inescapable conclusion is that, until the advent of dendrochronology, most rapid change in the Holocene, which may have had dramatic effects on human populations, was largely invisible due to poor chronological control and the blurring of point events by chronological smearing. We can now see these events through their effects on trees. Inevitably, if we really want to study these events, real effort is required to improve chronology in all related areas of environmental reconstruction.

Acknowledgments

The authors are grateful to numerous colleagues for supplying unpublished European oak data of the last two millennia, namely Ian Tyers, Cathy Groves, Jennifer Hillam, Esther Jansma, Niels

Bonde, Thomas Bartholin, Dieter Eckstein, Hubert Leuschner and Tomasz Wazny. For the unpublished world data sets we are particularly indebted to Ed Cook (Tasmanian Huon pine); Keith Briffa (Fennoscandian and Polar Urals pine); Peter Kuniholm (Aegean oak and pine); and Xiong Limin (New Zealand cedar). The data collection relevant to this article was funded under EU grant ENV4–CT95–1027, Climatology and Natural Hazards 10K.

Further Reading

Cook, E.R. and Kairiukstis, L.A. (eds) 1990. *Methods of tree-ring analysis: applications in the environmental sciences.* Kluwer Academic Publishers, Dordrecht.

Dean, J.S., Meko, D.M. and Swetnam, T.W. (eds) 1996. Tree-rings, environment and humanity. Proceedings of the International Conference, Tucson, Arizona, 17–21 May 1994. *Radiocarbon.*

CHAPTER 8

DATING BASED ON FRESHWATER- AND MARINE-LAMINATED SEDIMENTS

Bernd Zolitschka

Abstract: Annually laminated or varved sediments are deposited and preserved under certain environmental conditions in lakes and in the ocean. Clastic varves are formed in proglacial and periglacial lakes in arctic and alpine regions, whereas organic varves dominate in mesotrophic to eutrophic lakes in midlatitudes and in the tropics. Evaporitic varves are deposited in saline lakes and along shallow coasts in arid regions. However, many transitional forms exist between these three varve types. In the marine environment, organic and clastic varves can be formed in oxygen minimum zones and in areas with high sedimentation rates. After detailed investigation of fabric and composition of varved sediments using thin sections and scanning electron microscopy as well as identification of the processes that cause cyclic deposition, varve counts can be transformed into 'absolute' or 'floating' chronologies. Counting errors can be quantified by multiple counts of two or more investigators on several cores from the same site, while systematic chronological errors can only be determined by comparison with chronologies from other records or with other independent dating methods. Future developments of varve studies are directed towards image analyses of scanned thin sections, radiographs and polished sediment surfaces, all of which provide fast, repeatable, non-destructive, objective and high-resolution measurements making varve chronologies even more reliable.

Keywords: Chronology, Dating, Lake sediments, Laminated sediments, Marine sediments, Varves

Different types of archives have to be compared and correlated in order to expand high-resolution reconstructions of Holocene environmental change from a local view, achieved from one individual site, to a regional or global view. The success of such attempts is strongly dependent on a precise time control. Generally, two different dating methods are used: **radiometric dating** and incremental dating. The former depends on the decay of radiogenic isotopes, e.g. ^{14}C dating (see Pilcher, pp. 63–74 in this volume), which is related to time by their respective half-lives and through models of production and deposition of the respective element. Therefore, these dating methods provide a time control in, for example, radiocarbon years but not in calendar years. A calendar-year chronology, which can be correlated without calibration to historical or instrumental data, can only be obtained by incremental dating methods.

Incremental dating is based on cyclical accumulation of biological or lithological material with time. As the annual climatic cycle is the strongest part of the entire spectrum of climate variability, seasonal variations in temperature and precipitation with amplitudes twice as high as long-term climatic fluctuations are the major control mechanisms of this rhythmicity and produce a kind of 'internal clock'. Such annual increments are either of organic origin and recorded by biological processes in trees (dendrochronology) (cf. Baillie and Brown, pp. 75–91 in this volume), in corals (see Cole, pp. 168–184 in this volume) or in organic lacustrine and marine sediments (this chapter). Annual layers can also be formed by physical transport processes and deposition controlled by seasonal change of climatic conditions. This is the case for ice core records and for minerogenic sediments in lakes and oceans (this chapter). As varved sediments are much more widespread than any other palaeo-archives and occur in different environmental settings, they are important records for past environmental conditions. They also provide one of the highest possible time resolution in calendar-years recorded in sediments.

8.1 HISTORY AND DEFINITION OF VARVES

Already in the 19th century, pale-dark laminations of lacustrine sediments were regarded as a source of chronological information (Heer, 1865). Half a century later a major step was taken by using sediments from proglacial deposits in Sweden to determine the time elapsed since the end of the last glacial (DeGeer, 1912). It was also DeGeer who coined the term 'varve', the Swedish word for 'layer', which is used today as a synonym for annually laminated sediments. Soon after DeGeer published his data from glacial varves, laminated records of non-glacial lakes were studied and their annual character was described (Whittaker, 1922; Perfiliev, 1929). The potential of varves for dating and understanding of past environments was recognized immediately and studies were carried out applying these possibilities to instrumental meteorological records (Schostakowitsch, 1936), palaeomagnetic studies (McNish and Johnson, 1938) and pollen records (Welten, 1944).

After 1950, the introduction of the radiocarbon dating method distracted from varves as a dating tool, which were at that time regarded as an imprecise and subjective time control. Studies of varves in the 1960s and 1970s were restricted to Mesozoic and Palaeozoic sediments, partly because of much easier access and subsampling possibilities (Anderson and Kirkland, 1960; Richter-Bernburg, 1964). These studies allowed to derive a relative chronology and to establish sedimentation rates of hitherto unknown precision.

Varves from Quaternary marine sediments were studied intensively starting in the 1960s, when improved sampling techniques made available continuous and high-quality sediment cores from the ocean floor (Seibold, 1958; Gross *et al.*, 1963; Olausson and Olsson, 1969).

Improvements in the recovery of undisturbed and continuous soft sediments from modern lakes by applying piston coring techniques and *in situ* freezing techniques intensified the study of lacustrine varves since the 1970s. This was amplified by new methods of sample preparation and investigation (Merkt, 1971; Renberg, 1981b; Pike and Kemp, 1996; Lotter and Lemke, 1999). As the need for high-resolution environmental records grew during the 'global change' discussion, the value of data from varved records was increasingly recognized. During the 1990s, laminated marine sediments also became the focus of intensified research (Hughen *et al.*, 1996; Schulz *et al.*, 1996).

The original **definition of varves** – still presented in many textbooks and encyclopaedias – is that of proglacial varves (DeGeer, 1912). However, studies of finely laminated sediments from various non-glacial lakes in the late 1970s and in the 1980s (Sturm, 1979; Renberg, 1981a; O'Sullivan, 1983; Anderson *et al.*, 1985a; Saarnisto, 1986) extended the meaning of varves to all types of annually laminated sediments. Furthermore, improved coring techniques and intensified search for finely laminated sediments on the ocean floor during the last decade added another depositional environment – marine varves. Thus, the term 'varve' includes annually deposited:

1. minerogenic lake sediments in a glacial hydrological regime (proglacial or glacial lakes)
2. minerogenic lake sediments in snowmelt- or rainfall-dominated hydrological systems (together with glacial varves they are also termed as clastic varves)
3. organic lake sediments dominated by algal remains
4. evaporitic sediments dominated by precipitation of minerals from the water column
5. deep-sea sediments composed of minerogenic and/or organic materials.

All types of annually laminated sediments or varves have a similar structure: they consist of at least two sediment layers (laminae) with different composition, texture, structure and thickness representing certain periods or events during the year of deposition. These laminae can either be distinguished visually or microscopically.

8.2 FORMATION OF VARVES

Varves are the result of seasonal climatic variations controlling the biological productivity of a lake (organic varves) as well as the flux of minerogenic particles from the catchment into a lake (clastic varves). Thus, laminae with a different composition are produced during the course of one year leading to the formation of characteristic varve couplets (two laminae), triplets (three laminae) or even a larger number with up to 19 laminae, the maximum number of layers investigated to date (Dean *et al.*, 2001). Larger numbers of laminae are more common in clastic environments.

While this seasonal variability is necessary for the formation of varves, their preservation is dependent on two other factors: absence of bottom currents in the lake basin and oxygen deficiency of the deeper water body (hypolimnion). Bottom currents frequently disrupt a continuous sediment profile by causing erosional hiatuses, while anoxic conditions preclude the sediment surface from being mixed by bottom-dwelling organisms through bioturbation. However, there is no need of permanent anoxia for varves to be preserved. For example, varves have been reported from meromictic lakes without oxygen in their hypolimnion (Anderson *et al.*, 1985a) as well as from eutrophic and dimictic (holomictic) lakes with only seasonally anoxic conditions. Because the mode of circulation in the water body of a lake is dependent on size and shape of the lake basin and of the catchment area, varved sediments are most common in deep and wind protected, eutrophic lakes (Ojala *et al.*, 2000).

The prevalence of anoxic conditions as well as high sedimentation rates that exceed the rate of bioturbation (Domack *et al.*, 2001) cause the preservation of marine varves. Such conditions occur in drowned glacial valleys (fjords) like Saanich Inlet (Nederbragt and Thurow, 2001) or other basins, for example the Gulf of California (Gorsline *et al.*, 1996) and the Cariaco Basin off Venezuela (Hughen *et al.*, 1996). Oxygen-minimum zones may also occur in upwelling areas like the Arabian Sea off the coast of Pakistan (Schulz *et al.*, 1996).

8.2.1 Lacustrine Varves

As noted above, three different varve types can be distinguished according to their composition: clastic, organic and evaporitic varves. Each type is characteristic for certain environmental conditions in the lake and the catchment area during the time of deposition. However, in reality, varves are composed of a variety of materials and therefore in most instances many transitions occur between these three types.

8.2.1.1 Clastic Lacustrine Varves

Clastic varves predominate under cold climatic conditions, e.g. in high latitudes or alpine regions. Intensive physical weathering and the lack of a dense vegetation cover provide a high amount of minerogenic detritus, which can be eroded and transported into the lake. In general, sediment transfer from the catchment into the lake is correlated with the amount of runoff. In continental climatic regimes, runoff is governed by the melting of snow and ice through solar insolation during the summer (Hardy *et al.*, 1996; Moore *et al.*, 2001). Under oceanic climatic conditions, runoff is controlled either by the melting of snow and ice through advective heat transport associated with increased rainfall (Hicks *et al.*, 1990) or solely by precipitation (Anderson *et al.*, 1985b).

Varves are formed when a discontinuous, i.e. highly seasonal, stream loaded with suspended sediment enters a lake with a stratified water body (Sturm, 1979). The inflowing stream water is positioned as overflow, interflow or underflow according to its density in relation to the density of the lake water. Both are dependent on water temperature and the amount of suspended and dissolved substances. Overflows and interflows usually cause a wide distribution of suspended matter across the lake, whereas underflow deposits may be locally restricted to a sediment fan. Unlike overflows and interflows, underflows may also cause erosion.

After having entered the lake, flow velocity of the stream is transformed into turbulence causing a reduction of transport capacity for suspended sediment particles. As a consequence, coarse particles are deposited first out of the sediment plume, whereas fine particles remain in suspension until turbulent conditions cease. This is usually the case some time after runoff and inflow has stopped. Thus clastic varves are composed of a pale coarse-grained basal lamina and a darker fine-grained (clay) top lamina. Additional coarse-grained laminae are frequent and may be related to several successive runoff events in the course of 1 year. The formation of clastic varves is therefore strongly dependent on discontinuous discharge.

The first and most well known investigation of clastic varves was carried out with the study of glacial and proglacial sediments in Sweden (DeGeer, 1912). Therefore, these deposits are often referred to as 'classic varves' producing the 'Swedish time-scale' (Wohlfarth *et al.*, 1995). In addition to the classic sites in Sweden, proglacial varves have been studied in many other locations, like glacial Lake Hitchcock, New England, USA (Ashley, 1975) and modern sites from Spitsbergen, Sweden (Cromack, 1991), the Swiss Alps (Ohlendorf *et al.*, 1997) and Alaska, USA (Smith, 1978). The formation of snowmelt-controlled varves is also frequent and occurs in non-glacial environments in the Canadian High Arctic (Lamoureux, 1999) and in the Pleistocene of Germany (Brauer, 1994). This type of clastic varves is referred to as 'periglacial varves' (Plate 4).

8.2.1.2 Organic Lacustrine Varves

Physical or clastic varves are predominantly formed in oligotrophic lakes. In mesotrophic and eutrophic lakes with higher levels of available nutrients, increased organic productivity leads to the formation of biological (biogenic) or organic varves. This varve type has been reported from mid-latitudes throughout the world. Good examples exist for Elk Lake, Minnesota, USA (Anderson, 1993), Lovojärvi, Finland (Saarnisto *et al.*, 1977), Lake Gosciaz, Poland (Goslar, 1992), Meerfelder Maar, Germany (Brauer *et al.*, 2001), Holzmaar, Germany (Zolitschka *et al.*, 2000) and Hämelsee, Germany (Merkt and Müller, 1999), Soppensee, Switzerland (Lotter, 1989) and Lake Suigetsu, Japan (Kitagawa and van der Plicht, 1998). But organic varves have also been recognized in the Canadian Arctic (Hughen *et al.*, 2000) and in the inner tropics, like in Lake Magadi, Kenya (Damnati and Taieb, 1995), Lake Malawi, Tanzania (Johnson *et al.*, 2001) and Lake Ranu Klindungan, Indonesia (Scharf *et al.*, 2001).

The catchment of lakes with organic varves is vegetated which reduces the availability and the transport capability of clastic material into the lake. Furthermore, the prevalence of chemical weathering releases nutrients from the bedrock that are either incorporated into plant organic matter, buffered in soils or washed out and transported as dissolved substances into the lake. Mesotrophic, eutrophic or even polytrophic conditions in a lake cause seasonal algal blooms starting with diatoms in spring, followed by green and blue-green algae during summer and terminated by another diatom bloom in autumn. While green and blue-green algae are easily decomposed during deposition, siliceous diatom frustules are much more resistant to dissolution and are therefore well preserved in the sedimentary record.

Collectively, they form spring laminae and less frequently autumn laminae. The second part of a typical organic varve couplet is made of detritus, either composed of amorphous organic matter from the lake's plankton, or of littoral or terrestrial organic matter. The latter two components are washed into the central part of the lake basin during increased runoff periods related to autumn or winter rains (Plate 5).

This basic pattern of organic varve formation is only observed in absence of carbonaceous rocks in the catchment area of the lake. If dissolved carbonates are brought in via runoff, another lamina can be formed composed of calcite crystals produced by autochthonous biochemical precipitation (Kelts and Hsü, 1978). The ions of Ca^{2+} and HCO_3^- in the lake water stay in solution until saturation is obtained. This is partly achieved by the seasonal temperature increase in the upper water layer – the epilimnion – during the insolation maximum in summer. As $CaCO_3$ is less soluble with increasing temperature, calcite may precipitate due to this effect alone. Even more important is the (temperature-related) increase of photosynthetic activities in the epilimnion through phytoplankton blooms.

Related to increased biological activity is enhanced photosynthesis causing a withdrawal of CO_2 out of the lake water and thus a pH rise up to pH 9 resulting in a reduced solubility of $CaCO_3$. This combination of temperature increase and phytoplankton activities initiates the precipitation of calcite crystals during the warmest season. A calcite lamina forms as a distinct pale layer and makes the varve couplet easy to distinguish (Plate 6).

8.2.1.3 Evaporitic Lacustrine Varves

Evaporitic varves are formed as the salinity of a lake increases through evaporation, resulting in the saturation of specific mineral compounds (salts), which then precipitate out of the lake's water column. This type of varve occurs under arid or semiarid climatic conditions. In addition to calcite, which can precipitate either biogeochemically in mid-latitudes or physico-chemically under semi-arid conditions, laminae can also be composed of calcium sulphate (gypsum) and sodium chloride (halite), which precipitate only under arid climates. Either algal mats or terrigenous clastic deposits are interrupting these precipitates, causing a characteristic pale and dark cycle (Fig. 8.1).

Although there are some examples for recent deep-water evaporitic varve formation from the Dead Sea, Israel (Heim *et al.*, 1997) and Lake Van, Turkey (Lemcke and Sturm, 1997), most evaporitic varves are formed – unlike other varve types – in shallow lakes, as in lagoons from the Christmas Islands, Australia (Trichet *et al.*, 2001). Because of highly saline conditions, bioturbation does not occur in this environment. Only currents can disrupt the sedimentary record. Overall, evaporitic varves are less frequently found than organic or clastic varves today.

8.2.1.4 The Lacustrine Reality

Lacustrine varves are the result of several depositional processes and contain features of more than one of the varve types mentioned. Only clastic varves may occur without any organic contribution because lakes in arctic or high alpine regions are usually oligotrophic with minimal vegetation in the catchment (Plate 4). As soon as organic productivity increases, organic matter interrupts the pure clastic depositional cycle and a transitional varve type is formed or the varve type changes completely from clastic to organic (Plate 5), which can, for example, be observed

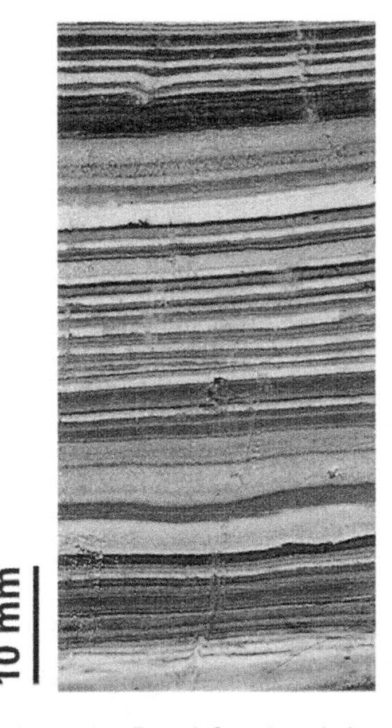

Figure 8.1 Evaporitic laminae from the Dead Sea, Israel, (core DS 7-1 SC, 2.7–2.8 m). Pale laminae consist of either aragonite (white) or gypsum (grey) crystals, whereas the dark laminae consist of detritic silts and clay (Heim et *al.*, 1997). Macroscopic photograph of the split core obtained from http://www.gfz-potsdam.de/pb3/pb33/wiav.html.

at the Weichselian/Late Glacial transition at Lake Holzmaar, Germany (Zolitschka *et al.*, 2000). However, most organic varves have an additional runoff-related clastic component and, with carbonaceous bedrock in the catchment area, they may also contain precipitated calcite laminae (Plate 6). Similarly, evaporitic varves may include organic as well as clastic components (Fig. 8.1).

Varves can be attributed to certain types, which allows drawing a general picture about the climatic conditions at the time of their formation. The succession of varve types at one site demonstrates that conditions controlling the formation of annual laminations vary through time generating a characteristic set of sublaminae to every year. However, a closer look demonstrates that every lake behaves individually and that even varves from the same region may look different despite similar regional environmental settings.

8.2.2 Marine Varves

Much like lacustrine varves, marine varves are composed of a variety of different components, with certain combinations of laminae forming one annual cycle. Clastic marine varves are the result of a varying terrigenous sediment flux to coastal marine basins. A major runoff peak produces a distinct coarse-grained lamina followed by a fine-grained unit out of suspension similar to clastic varve formation in lakes. Such a site was investigated in the Palmer Deep, Antarctic Ocean (Domack *et al.*, 2001). In more productive marine environments, organic components, mostly diatoms, occur (Fig. 8.2), as in Saanich Inlet, British Columbia (Dean *et al.*, 2001), Cariaco Basin off Venezuela (Hughen *et al.*, 1996) and the Gulf of California (Kemp *et al.*, 2000). In addition to diatoms, non-siliceous organic matter may be added forming a succession of organic and clastic laminae, as is the case for the Arabian Sea (Schulz *et al.*, 1996).

8.3 INVESTIGATION OF VARVES

With the study of finely laminated sediments an attempt is made to establish a precise time control for high-resolution palaeoenvironmental reconstructions. To achieve this goal it is necessary to recover continuous and undisturbed sediment records. Most coring equipment available today is capable of producing such records. However, to obtain a complete stratigraphy including the sediment–water interface, usually more than one technique for sediment recovery has to be applied. Piston-coring and percussion-coring systems are adequate for sections below the sediment–water interface, while freeze-corers provide best results for surface samples. Top-quality surface cores obtained with freeze-corers or gravity corers are necessary to compare and calibrate the sedimentary record with modern processes, which are crucial for any process-related study.

Because individual core sections are usually not longer than 1 or 2 m, it is essential to obtain more than one core from every lake. With at least two parallel and overlapping cores, a complete record can be analysed by correlating the cores to skip the gaps between successive sections.

8.3.1 Sample Preparation for Different Analytical Methods

Sample preparation and analytical techniques are the next step to construct a varve chronology. This depends on a detailed analysis of the sediment (micro)structures. There are several methods

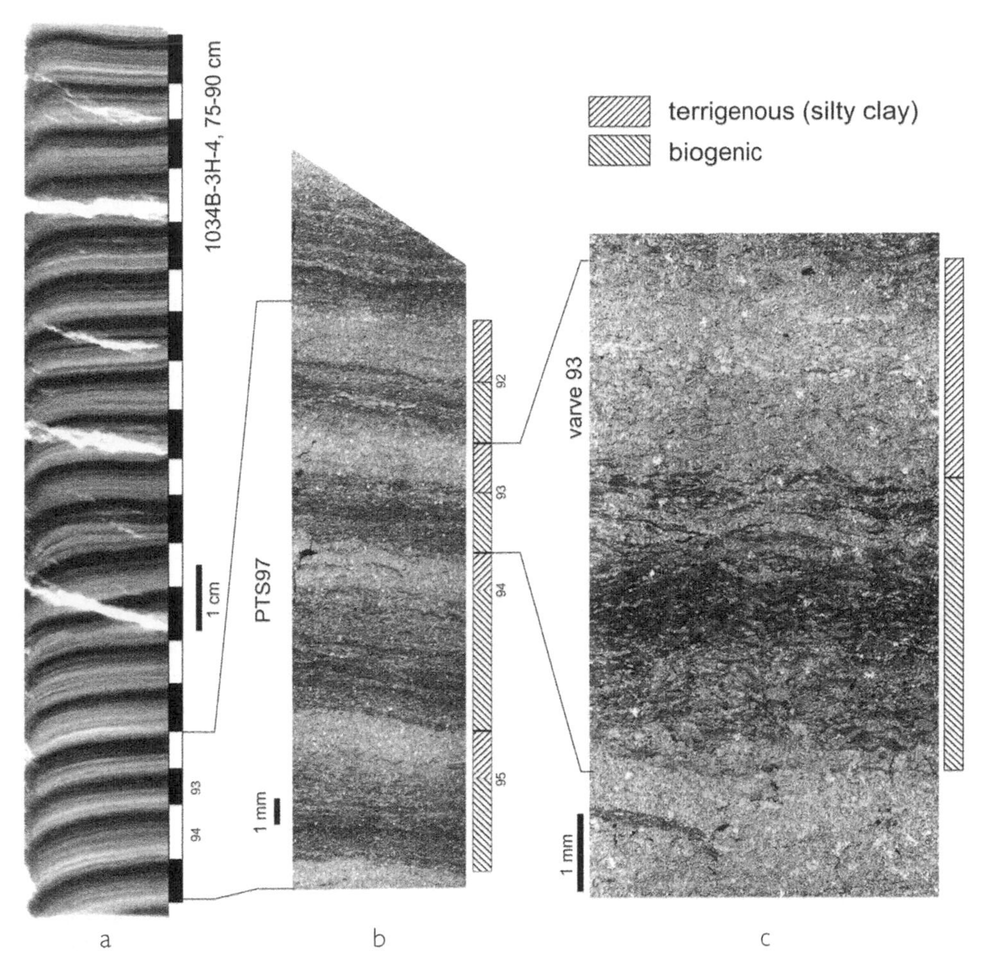

Figure 8.2 Well-laminated Late Holocene siliceous mud (interval 169S-1034B-3H-4, 75–90 cm) from Saanich Inlet (ODP Leg 169S), British Columbia, Canada. The black and white bars indicate a total of 19 individual years. (a) X-ray image of the 13-cm long section; (b) backscattered electron image (BSEI) of varve numbers 92–95 from a polished thin section (PTS #97); (c) BSEI of an individual varve (varve number 93). Reprinted from Dean *et al.* (2001) with kind permission from Elsevier Science.

in use including: (1) direct analysis of the split core surface, (2) analysis of enlarged photographs, (3) application of X-ray techniques, or (4) drying and impregnation of soft sediments to harden the sample for preparation of sediment slices, polished surfaces and thin sections.

8.3.1.1 Direct Analysis of the Split Core Surface

If varves are relatively thick (>1 mm) and easy to distinguish with pronounced pale and dark laminae (Fig. 8.1), the surface of a split sediment core can be macroscopically investigated aided with a magnifying glass or a stereomicroscope. Depending on the type of sediment, varve counts may have to be carried out immediately after splitting the core before the surface is getting oxidised and the sedimentary structures are obscured. In contrast, with iron-rich sediments it may be helpful to wait until the surface is completely oxidized thus making laminations easier to be determined. For multiple counting, individual layers or sets of layers are marked with coloured pins. A precondition for directly analysing the split core – as well as for all more advanced methods explained below – is a smooth and clean sediment surface. This is obtained by cutting the core either with a knife or a wire and cleaning the face with a cover-slip or a razor blade or by using an osmotic knife for clay-rich sediments (Pike and Kemp, 1996).

8.3.1.2 Analysis of Enlarged Photographs

Enlarged photographs of the split core surface can also be used to magnify sedimentary structures and to document this information together with varve counts directly made from the photograph. As with direct counts on the split core surface, this is only practicable if the varve structures are clearly visible. Conventional photographical techniques are increasingly replaced by video or digital cameras (Petterson *et al.*, 1999) and semi-automatic image analysis. Digital images can also be obtained from automated systems like digital image scanners (Nederbragt and Thurow, 2001; Zolitschka *et al.*, 2001).

8.3.1.3 Application of X-ray Techniques

Visually indistinct or macroscopically invisible sediment laminations and other structures like bioturbation or slumps often become apparent using X-ray techniques. Such radiographs (X-ray images) should not be obtained from the whole or split core because the decreasing sediment thickness at the edges causes uneven exposures. Better results are achieved if X-rays are applied to sediment slabs of constant thickness (1 cm or less), which can later be used for impregnation (Figs 8.2–8.5). Radiographs can also be scanned with a flatbed scanner and transformed into

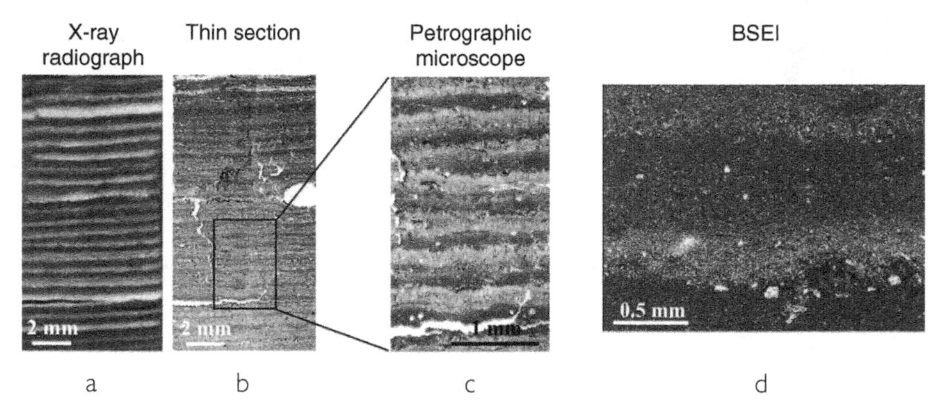

Figure 8.3 Laminated sediments from c. AD 200 of Lake Nautajärvi, Finland. Organo-clastic varves are presented as (a) X-ray radiographs made from a 2-mm thick impregnated sediment slab, flatbed scan of a (b) thin section prepared from the same impregnated sediment slab and as (c) an enlarged section of the thin section viewed through a petrographic microscope. For thin section images dark laminae represent deposition of detritic clastic material during the spring snow melt runoff, whereas the pale laminae represent deposition of various types of organic matter from summer to winter. For the radiograph this relation is the opposite of the Back Scattered Electron Image (BSEI) (d). The latter image is close to one individual varve with pale dots representing minerogenic matter. Reprinted from Ojala (2001) with kind permission from the Information Department of the Geological Survey of Finland.

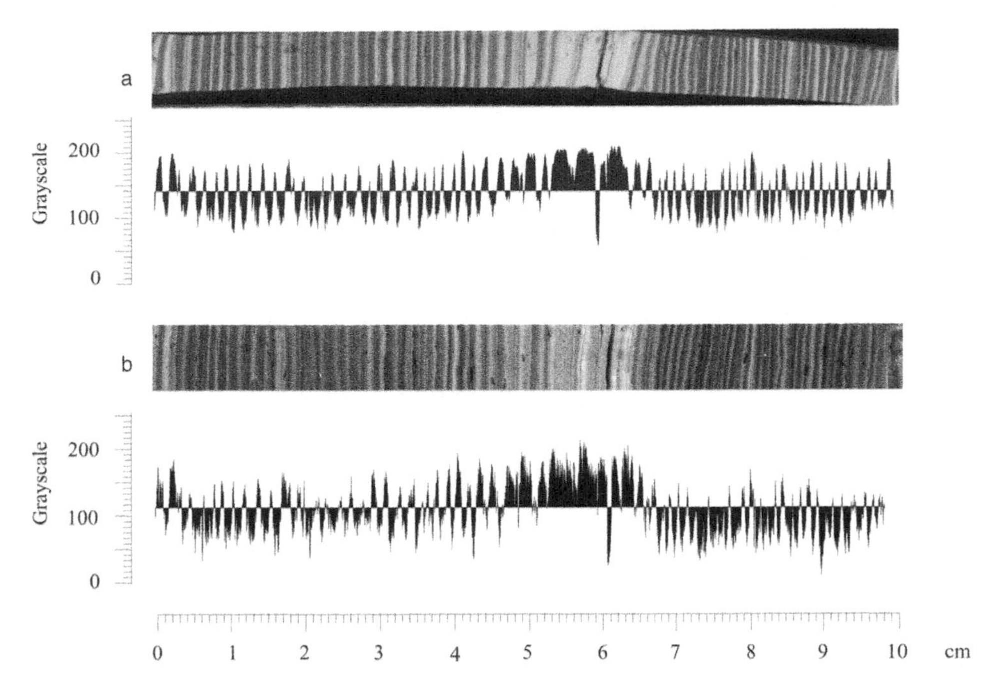

Figure 8.4 (a) Radiograph and (b) flatbed scan of a polished impregnated sediment slab from the same section (15–25 cm below sediment surface) of Lake Korttajärvi, Finland. For both images grey-scale curves were generated demonstrating the annual signature of this record. Reprinted from Tiljander *et al.* (2002) with kind permission from Elsevier Science.

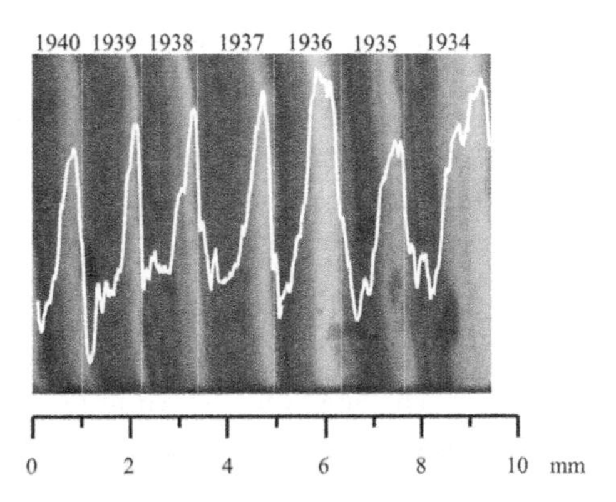

Figure 8.5 Radiograph combined with an X-ray density curve from Lake Korttajärvi, Finland for the years AD 1934–1940. Minerogenic material (pale laminae) is deposited during the spring snowmelt floods and organic matter (dark laminae) is deposited from summer through winter. Reprinted from Tiljander *et al.* (2002) with kind permission from Elsevier Science.

grey-scale or X-ray density records (Figs 8.4 and 8.5) using standard image-analysing software (Tiljander *et al.*, 2002).

8.3.1.4 Impregnation of Soft Sediments and Applications

As soft sediments are difficult to handle, the preparation and impregnation of sediment slabs is a standard procedure for analysing finely laminated sediments. Impregnated sediments have been used to prepare (a) thin sections. However, in recent years more and more applications have used sediment slabs for (b) X-ray techniques, (c) flatbed scanning of polished sediment surfaces as a means to obtain a digital image of sediment textures and (d) for scanning electron microscopy (SEM). Prior to impregnation, wet sediments are subsampled and dried. An advanced method to circumvent disturbance of the sediment during subsampling is the use of the electro-osmotic effect (Francus and Asikainen, 2001). Water removal, necessary because impregnation media (usually epoxy resins) do not harden in the presence of water, is usually achieved by freeze-drying (Merkt, 1971) or by fluid displacement (Lamoureux, 1994; Pike and Kemp, 1996). Only the latter method leaves the sediment in a wet condition throughout the entire process and thus minimises textural damages like cracks or shrinkage. Once impregnated, soft sediment slabs can be cut and polished like rock samples.

a. Originally, laminated sediments were impregnated to prepare **thin sections** (Plates 4-6) which allow the investigation of the internal structure and composition of varves (Merkt, 1971). Petrographic microscopy using normal (Plate 5 and Fig.8.3) and polarized light (Plate 4) allows analysing the internal fabric and composition of laminations and thus provides a way to distinguish between annual and non-annual laminae. Furthermore, compositional data and the seasonal succession of algal remains also reveal the annual character and are typically not visible from the split core surface. This is especially the case for sediments with low sedimentation rates or without a clear differentiation between laminae.
b. In addition to **X-rays** of fresh sediment samples, it becomes more and more common to use 1–3 mm thick even slices of impregnated sediments (Figs 8.2–8.5). Advantages of this method are the absence of time pressure as impregnated sediments do not dry out or oxidize and the precise thickness and smooth surface of the impregnated samples (Ojala and Francus, 2002).
c. **Flatbed scans** of polished and impregnated sediment slabs (Plate 6 and Figs 8.3 and 8.4) can easily be obtained by commercially available scanners (Tiljander *et al.*, 2002). However, results are only useful if the analysed sediment record shows clearly visible laminations. They cannot be used to investigate the composition of varves in detail unless additional methods are consulted.
d. Like microscopic thin section investigations, **SEM analyses** (Figs 8.2 and 8.3) also provide compositional information but can be carried out with much higher resolution (Dean *et al.*, 2001; Ojala and Francus, 2002). The SEM can be used in two different modes: secondary electron images (SEI) and backscattered electron images (BSEI). SEI produces topographic images from a fractured piece of sediment, broken either parallel or perpendicular to the bedding plane, whereas BSEI provides compositional and porosity data from polished thin sections or sediment slabs. SEM with an energy dispersive spectrum (EDS) analyser additionally provides qualitative and quantitative information of elemental composition (Dean *et al.*, 1999b).

8.3.2 Varve Determinations and Counting

Prior or parallel to varve counting and measuring it is necessary to develop a conceptual model how the (idealized) varve is formed and composed at the specific site of investigation (Lotter and Lemke, 1999; Lamoureux, 2001). As described above, this depends on the limnological conditions of the lake as well as on the environmental conditions of the catchment area. The model should characterize specific features, which are then used to determine boundaries between consecutive varves and to identify laminae that do not represent years, for example turbidites or homogenites. This becomes especially important if varves are not well defined and cannot be distinguished easily. The development of such a process model is often supported by sediment trap studies (Leeman and Niessen, 1994a; Retelle and Child, 1996) or by detailed micro-palaeontological methods, e.g. pollen or diatom analysis (Peglar *et al.*, 1984; Card, 1997; Hausmann *et al.*, 2002). In contrast to standard pollen and diatom studies, here the succession of pollen types or diatom species are examined during 1 year.

The preferred approach to establish a varve chronology should include replicate varve counts along three or more transects by at least two different researchers (Lamoureux, 2001). Image analysis (Saarinen and Petterson, 2001) can be used if sediments show a distinct pattern of lamination with negligible disturbances. However, automated counting techniques are restricted, if applicable at all, to the very few perfectly developed and very well defined varves (Lotter and Lemke, 1999). But computer-based interactive utilities can assist the investigator with varve measurements (Francus, 1998).

Usually, the formation of varved records responds to complex forcing factors. Only for simple clastic systems with a well-understood depositional model it is advisable to generate a macroscopic varve count by measuring pale and dark cycles from the open core face, from photographs or radiographs. Generally, SEM analyses would be preferable for in-depth studies of varves. However, this technique cannot provide a continuous record over many hundreds or even thousands of years with reasonable effort. Long records can only be achieved by patching together images from flatbed scans, digital cameras or radiographs. For better inspection, these images are often transformed into grey-value records (Figs 8.4 and 8.5) using image analysis software tools (Petterson *et al.*, 1999; Dean *et al.*, 2002; Ojala and Francus, 2002; Tiljander *et al.*, 2002). If varves are composed of a clear pale and dark annual cycle and thick enough to be captured by the resolution of the image-capturing process, the grey-value record can be used to establish a varve count as well as a record of varve thickness. This can be carried out in a semi-automated way (Francus *et al.*, 2002b).

However, digital images cannot provide insight into the composition of varves, which is critical for the verification of their annual character. This is only achieved for longer records by thin section analysis with petrographic microscopes. With the help of magnifying factors between 20× and 1000× varves can be distinguished from each other, they can be counted and their thickness can be measured (Plates 4, 5 and Fig. 8.3). The kind of method applied depends on the type of sediment studied, the technical capabilities of the laboratory and the personal preference of the analyst.

A thorough knowledge of sediment fabric and composition as well as of sedimentary processes are important prerequisites not only to develop palaeoclimatic and palaeoenvironmental models of varve formation but also to verify that laminations are produced due to seasonal cycles. This information is essential to establish a lamination-based chronology. Such a varve chronology, however, has to be cross-checked against other, independent dating methods (**multiple dating approach**).

Varve chronologies may be regarded as 'absolute', if they are continuous and extend to the present. However, in many cases, inconsistencies of the chronology occur, hiatuses have been detected or the connection to modern sedimentation is not possible. These varve chronologies are termed '**floating**'.

8.3.3 Error Estimations

Errors occurring within varve chronologies can be attributed to three different sources (Sprowl, 1993):

1. *Errors related to technical problems*: This kind of error is caused by incomplete recovery of sediment cores – sediment gaps as a result of coring artefacts – or by missing or incorrect core correlation. Technical errors can be avoided if appropriate coring techniques are applied and two or more overlapping core series are studied.
2. *Errors related to depositional events*: Such errors may occur over larger areas of the lake basin as the result of erosion (turbidites, volcanic ash layers) or locally restricted as the result of deposition of plant or animal macrofossils, drop stones, faecal pellets or of degassing channels. These influences can be avoided if several parallel cores from the same lake are investigated, correlated and cross-dated. However, in some rare cases, like basin-wide deposition of tephra layers that disrupt the underlying sediment, no compensation for this type of error is possible.
3. *Errors related to sedimentation processes*: Very high or very low sediment accumulation rates make individual varves difficult to distinguish. In addition, changes in lake-level or autochthonous productivity have significant effects on anoxic conditions at the lake floor and thus for varve preservation. For example, in periods with low lake-levels, predominant oxic conditions may prevent the conservation of varves.

Such systematic errors are difficult to circumvent, possibly by a careful selection of the coring site or by certain analytical techniques (SEM techniques for very thin laminations) and must be regarded as an inherent problem to most varve chronologies.

If varves are well developed and not disrupted, replicate counts show only little variation between different counts and observers (Ojala and Saarnisto, 1999), the counting error will be low. But if sediments are disturbed or not very distinct, counting errors increase. An easy way to estimate counting errors is the root of the mean squared error of multiple counts. As this value mostly gives a false precision, it is suggested to additionally provide the maximum and minimum deviation from the mean varve count (Lotter and Lemke, 1999).

Generally, the determination of errors as variance between different investigators and/or replicate counts gives only a measure for the studied section. It does not provide any indication about the chronological error of the record as the result of undetected systematic errors. Verification of the chronology, i.e. the exclusion of systematic errors, can only be achieved by comparison with other records (Sprowl, 1993) or with other independent dating methods applied to the same record. Such a multiple dating approach has been carried out using ^{137}Cs

and ^{210}Pb isotopes (Lotter *et al.*, 1997b; Hughen *et al.*, 2000; Johnson *et al.*, 2001), AMS ^{14}C dating (Goslar, 1992; Hajdas *et al.*, 1995) or tephrochronology (Merkt *et al.*, 1993; Zolitschka *et al.*, 1995). Additionally, correlation with palaeomagnetic records from different lakes using inclination and declination data may help to elucidate problems in some chronologies (Stockhausen, 1998; Saarinen, 1999).

8.4 FUTURE DEVELOPMENTS OF VARVE STUDIES

Future perspectives provided by high-resolution palaeoenvironmental reconstructions based on varved sediment records will strengthen their regional and temporal coverage. Furthermore, improvements of image analysis techniques driven by the rapid development of computer software and hardware will increase the precision of varve chronologies. Developments in image analysis techniques have been enormous for the last decade, which highlights their possibilities for the future. Image analysis is fast, non-destructive, objective and performs high-resolution measurements (Saarinen and Petterson, 2001). Thus it is not only capable for core documentation but also for varve counting, varve thickness measurements and, to some extent, for studying varve composition (Francus, 1998; Francus *et al.*, 2002a).

As laminated sediments provide a sequence of pulses of flux, generated by episodes of primary production and terrigenous input, they may also be used to establish time series for palaeoenvironmental studies. Of special interest is the possibility to determine quantitative influx rates (Zolitschka, 1998; Petterson *et al.*, 1999).

Many varved sediment records have been analysed from arctic and midlatitudes as well as from marine sites. Little is known about high-resolution records from the tropics. Tree-ring chronologies basically fail to produce good environmental archives at low latitudes due to a lack of seasonality. Tropical ice-cores are very rare and these ice caps are vanishing rapidly. Until now lakes with annually laminated sediments have not been studied to a large extent in the tropics. However, new records have been developed (Walker *et al.*, 2000; Johnson *et al.*, 2001; Scharf *et al.*, 2001) and promise new exciting results from an area where more information about terrestrial environmental change is urgently needed.

The majority of varved records is restricted either to the last few decades or centuries with varve formation being the result of anthropogenic eutrophication (Lotter *et al.*, 1997b; Lüder and Zolitschka, 2001) or to the Holocene, sometimes including the Late Glacial, with climatic forcing related to the interglacial warming initiating natural eutrophication processes (Bradbury and Dean, 1993; Ralska-Jasiewiczowa *et al.*, 1998). Records reaching from the present beyond the Late Glacial are extremely rare and still have chronological inconsistencies (Kitagawa and van der Plicht, 1998; Allen *et al.*, 1999; Zolitschka *et al.*, 2000). Therefore, future attention will be given to find and investigate palaeoclimate archives with a complete glacial–interglacial cycle on land. Such records could potentially help to further explore Late Pleistocene climatic variations and to provide a high-resolution terrestrial counterpart to marine records.

A fine-resolution reconstruction of climate change during the Holocene and beyond is vital to predict the climate of the near future. Only the inter-archive and inter-regional comparison of palaeoenvironmental data will lead to meaningful interpretations of past conditions and thus will finally help to understand the mechanisms of future rapid climatic changes. This can only be achieved with a precise 'absolute' chronology in calendar-years obtained for as many records as possible – annually laminated or varved lacustrine and marine records are prime sources for such data.

Acknowledgements

The author gratefully acknowledges Antti Ojala (Geological Survey of Finland), Jean Dean and Alan Kemp (University of Southampton, UK) for providing images of sediments from Finnish lakes and marine Saanich Inlet, respectively. In particular, I am grateful to Dirk Enters for major improvements on an earlier version of the manuscript. Thanks also go to the editors of this textbook and to Scott Lamoureux and an anonymous reviewer for very helpful comments on the manuscript.

Further Reading

Suggested for further reading are several monographs about laminated sediments covering various methodological aspects and interdisciplinary approaches to palaeoclimatic and palaeoenvironmental interpretation. Special issues are available for individual lake sites like Elk Lake, Minnesota (Bradbury and Dean, 1993), Lake C2, Canadian High Arctic (Bradley, 1996), Baldeggersee, Switzerland (Wehrli, 1997), Lake Gosciaz, Poland (Ralska-Jasiewiczowa *et al.*, 1998) and for marine deposits from Saanich Inlet, British Columbia, Canada (Bornhold and Kemp, 2001). Additionally, there are four volumes dedicated to methods and applications of varved sediment records: an early compilation introducing the formation and composition of clastic varves (Schlüchter, 1979), methods and case studies of laminated lacustrine sediments from Europe (Saarnisto and Kahra, 1992), structure, formation, dating and significance of varved sediments for the study of human impacts on the environment (Hicks *et al.*, 1994) and application of mainly marine laminated sediments as palaeo-indicators (Kemp, 1996). A series of up-to-date methodological handbooks (Last and Smol, 2001a, 2001b) adds to many of the aspects discussed in this chapter.

CHAPTER

9

QUANTITATIVE PALAEOENVIRONMENTAL RECONSTRUCTIONS FROM HOLOCENE BIOLOGICAL DATA

H. John B. Birks

Abstract: There are three main approaches to reconstructing quantitatively Holocene environments from fossil biological assemblages: (1) indicator species, (2) assemblage or modern analogues (including response surfaces) and (3) multivariate transfer functions. Their assumptions, strengths and limitations are discussed. The basic concepts, requirements and numerical procedures involved in transfer functions are reviewed, and the properties of weighted averaging and weighted averaging partial least squares regression are outlined. A reconstruction of Holocene mean July temperatures in northern Norway from pollen-stratigraphical data is presented. Possible future developments are discussed.

Keywords: Climate reconstruction, Error estimation, Transfer functions, Weighted averaging, Weighted averaging partial least squares

Many Holocene palaeoecological studies aim to reconstruct features of the past environment from fossil assemblages preserved in sediments. Although fossil assemblages are usually studied quantitatively with individual pollen, chironomids, etc. being identified and counted, environmental reconstructions may be qualitative and presented as 'cool', 'warm', etc. The need for quantification in Holocene research is increasing, largely in response to demands for quantitative reconstructions of past environments as input to, or validation of, simulations by Earth System models of past, present and future climate patterns. There are three main approaches to reconstructing quantitatively past environments from biostratigraphical data (Birks, 1981, 1995, 1998; Birks and Birks, 1980): (1) indicator species approach, (2) assemblage approach and (3) multivariate indicator species approach involving mathematical transfer functions. All require information about modern environmental requirements of the taxa found as fossils. The basic assumption is methodological uniformitarianism (Rymer, 1978; Birks and Birks, 1980), namely that modern-day observations and relationships can be used as a model for past conditions and, more specifically, that organism–environment relationships have not changed with time, at least in the Late Quaternary.

This chapter discusses the multivariate indicator species approach involving transfer functions as a means of quantitatively reconstructing Holocene climates from fossil assemblages. It presents the basic concepts of transfer functions, the assumptions, requirements and data properties of the approach, the relevant numerical procedures, an application, and a discussion of the limitations of the approach. In order to put this approach into its palaeoecological context, I briefly discuss the indicator species and assemblage approaches. I conclude by suggesting possible areas for future development. I draw extensively on reviews by Birks (1981, 1995, 1998) and ter Braak (1995).

9.1 INDICATOR SPECIES APPROACH

Fossil occurrences of a species with known modern environmental tolerances provide a basis for environmental reconstructions. Assuming methodological uniformitarianism, the past environment is inferred to have been within the environmental range occupied by the species today. This approach requires information about what environmental factors influence the distribution and abundance today of the species concerned. The commonest means of obtaining such information is to compare present-day distributions of species with selected climatic variables of potential ecological and physiological significance, such as mean temperature of the coldest month or maximum summer temperature (Dahl, 1998). If the geographical trend of an eco-climatic variable covaries with the species distribution in question, a cause-and-effect relationship is often assumed. For example, Conolly and Dahl (1970) related the modern distribution of *Betula nana* in the British Isles to the 22 °C maximum summer temperature isotherm for the highest points in areas where *B. nana* grows today. Fossil records show that it occurred widely in lowland Britain during the Late Glacial, where maximum summer temperatures are 30 °C today. Conolly and Dahl proposed therefore that there was a depression of 8 °C in maximum summer temperatures in the Late Glacial.

In some instances it is more realistic to consider species distributions in relation to two or more variables (Hintikka, 1963). This 'bivariate' approach was pioneered by Iversen (1944) in his classic work on *Viscum album*, *Hedera helix* and *Ilex aquifolium*. On the basis of detailed observations, Iversen delimited the 'thermal limits' within which they flowered and produced seed. He showed that *Ilex* is intolerant of cold winters but tolerant of cool summers, *Hedera* is intolerant of winters with mean temperatures colder than –1.5 °C but requires warmer summers than *Ilex*, and *Viscum* is tolerant of cold winters but requires warmer summers than either *Ilex* or *Hedera*. These shrubs are ideal indicator species because their pollen is readily identifiable to species level, it is not blown great distances so interpretative problems arising from far-distance transport rarely arise, and their berries are rapidly dispersed by birds. Their distributions are likely to be in equilibrium with climate. From fossil pollen occurrences, Iversen used this approach to suggest that Mid-Holocene summers were 2–3 °C warmer and winters 1–2 °C warmer than today in Denmark.

This approach has been extended to several species simultaneously to identify areas of climatic overlap for pollen (e.g. Grichuk, 1969; Markgraf *et al.*, 1986; McKenzie and Busby, 1992; Pross *et al.*, 2000), plant macrofossils (Sinka and Atkinson, 1999), chironomids (Dimitriadis and Cranston, 2001), molluscs (Moine *et al.*, 2002) and beetles (Elias, 1997; Atkinson *et al.*, 1987), the so-called mutual climatic range approach. This approach assumes that spatial correspondence between species distributions and selected climatic variables implies a causal relationship. As discussed by Birks (1981) and Huntley (2001), it can be unwise to assume such relationships exist. When climatic factors have been studied in detail (e.g. Forman, 1964; Pigott, 1970, 1981, 1992; Pigott and Huntley, 1991) many factors may be operative at different spatial scales. Moreover, only a few indicator taxa are usually considered and little or no attention is given to the numerical frequencies of the different taxa in the fossil assemblages. An alternative approach, considering the composition and abundance of the whole assemblage, is the assemblage approach.

9.2 ASSEMBLAGE APPROACH

This considers the fossil assemblage as a whole and the proportions of its different fossil taxa. It has been widely used in an intuitive non-quantitative way for many decades. For example, pollen assemblages are interpreted as reflecting tundra, pine forest, or deciduous forest. Past environmental inferences are based on the present-day environment in which these vegetation types occur. More recently it has been put on a more quantitative basis, the so-called **modern analogue technique** (MAT) and related response surface approach.

The basic idea of MAT is to compare numerically, using a dissimilarity measure (e.g. squared chord distance) (Overpeck *et al.*, 1985), the fossil assemblage with modern assemblages. Having found the modern sample(s) that is most similar to the fossil sample, the past environment for that sample is inferred to be the modern environmental variable(s) for the analogous modern sample(s). The procedure is repeated for all fossil samples and a simultaneous reconstruction of several environmental variables can be made using modern analogues. The environmental reconstruction(s) can be based on the modern sample that most closely resembles the fossil assemblage or, more reliably, it can be based on a mean or weighted mean of, say, the 10 or 25 most similar modern samples, with weights being the inverse of the dissimilarities so that modern samples with the lowest dissimilarity have the greatest weight in the reconstruction. Examples of MAT for reconstructing Holocene climates include Bartlein and Whitlock (1993) and Cheddadi *et al.* (1998). MAT has been extended by Guiot (1990, Guiot *et al.*, 1992, 1993a) to reconstruct several climatic variables from pollen assemblages for the last interglacial–glacial cycle.

Response surfaces are three-dimensional graphical representations of the occurrence and/or abundance of individual taxa in modern environmental space (Huntley, 1993). The x and y axes represent environmental variables and the z axis represents the occurrence or relative abundance of the taxon of interest. Modern pollen–climate response surfaces have been constructed to illustrate relative abundances of modern pollen varying along major climatic gradients (e.g. Bartlein *et al.*, 1986). The surfaces are fitted by multiple regression (Bartlein *et al.*, 1986) or locally weighted regression (Bartlein and Whitlock, 1993). Palaeoenvironmental reconstructions from Holocene assemblages (e.g. Allen *et al.*, 2002) are made by 'stacking' modern surfaces to produce synthetic assemblages for a series of grid nodes, usually 20 × 20 nodes, in modern climate space. These synthetic assemblages are then compared to fossil assemblages by a dissimilarity measure, usually the squared chord distance as in MAT. Climate values for the 10 grid nodes with synthetic pollen spectra most similar to the fossil assemblages are used to infer the past climate. The final inferred value is a mean of the climate values weighted by the inverse of the squared chord distances (Prentice *et al.*, 1991). Environmental reconstructions are thus done by MAT but the modern data consist of fitted pollen values in relation to modern climate and not the original pollen values. The fitted values naturally smooth the data to varying degrees depending on the smoothing procedures used (ter Braak, 1995). Much inherent local site variability that is assumed to be unrelated to broad-scale climate is removed (Bartlein and Whitlock, 1993).

Limitations in using MAT and response surfaces are the need for high-quality, taxonomically consistent modern data sets from comparable sedimentary environments as the fossil data and covering a wide environmental range, the absence of any reliable error estimates for reconstructed values, the problems of defining 'good', 'poor' and 'no analogues', selecting an appropriate dissimilarity measure and the occurrence of no analogues and multiple analogues (Birks, 1995). No analogues arise when no modern assemblages are similar to the fossil assemblage (Huntley, 1996). Multiple analogues arise when the fossil assemblage is similar to several modern samples that differ widely in climate (Huntley, 1996, 2001), for example assemblages dominated by pine pollen that can be derived from northern, central, or Mediterranean Europe, all of which have very different climates today. Constraints can be built into the analogue-matching procedure to help minimize the multiple analogue problem by restricting possible analogues to be from the same modern biome as the biome reconstructed from the fossil assemblage (Huntley, 1993; Allen *et al.*, 2000).

The multivariate indicator species approach involving transfer functions attempts to overcome some of these problems.

9.3 MULTIVARIATE TRANSFER FUNCTION APPROACH

The idea of quantitative environmental reconstructions involving transfer functions is summarized in Fig. 9.1. There is one or more environmental variable $\mathbf{X}_0$ to be reconstructed from fossil assemblages $\mathbf{Y}_0$ consisting of m taxa in t samples. To estimate $\mathbf{X}_0$, we model the responses of the same m taxa today in relation to the environmental variable(s) (**X**). This involves a modern 'training set' or 'calibration set' of m taxa at n sites (**Y**) studied as assemblages preserved in surface sediments (e.g. surficial lake muds, ocean sediments), with associated modern environmental variables (**X**) for the same n sites. The modern relationships between **Y** and **X** are modelled numerically and the resulting transfer function is used to transform the fossil data $\mathbf{Y}_0$ into quantitative estimates of the past environmental variable(s) ($\mathbf{X}_0$). The various stages are schematically shown in Fig. 9.2.

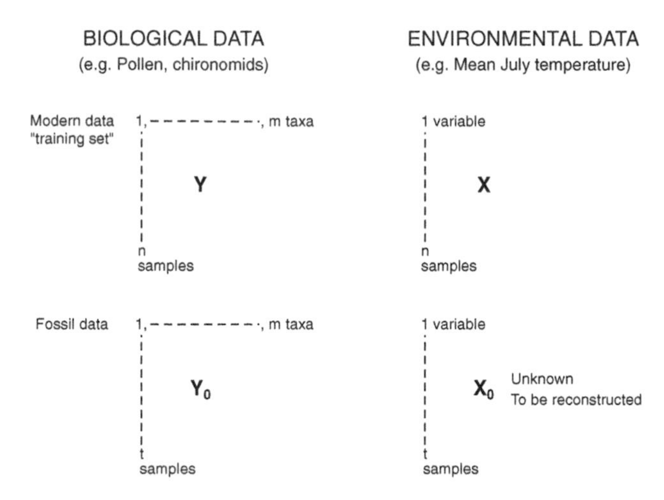

Figure 9.1 The principles of quantitative palaeoenvironmental reconstruction showing X_0, the unknown environment variable (e.g. July temperature) to be reconstructed from fossil assemblages Y_0, and the essential role of a modern training set consisting of modern biological (Y) and environmental (X) data.

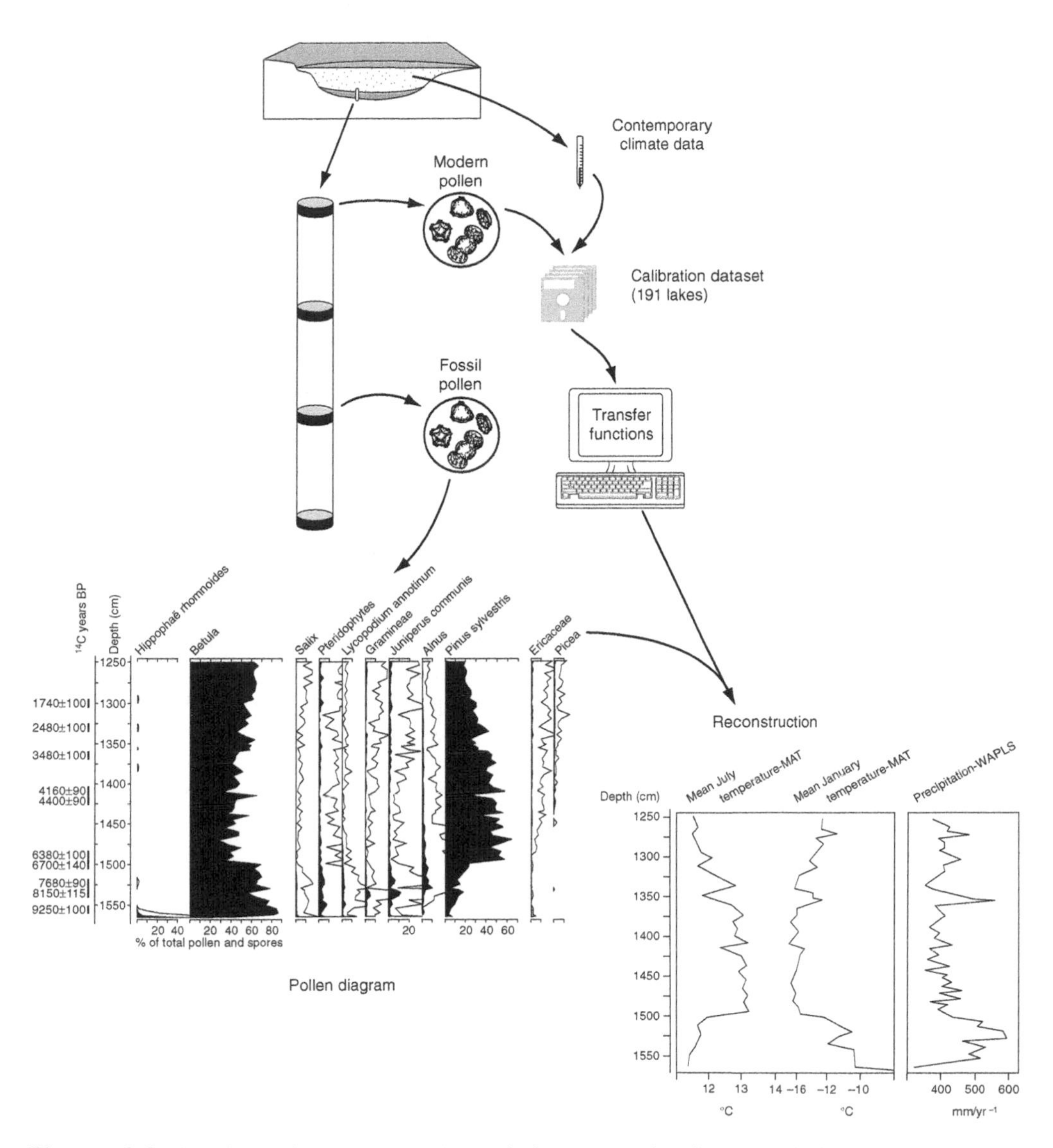

Figure 9.2 A schematic representation of the stages involved in deriving a quantitative reconstruction of past climate from pollen-stratigraphical data using a modern calibration or training set. Modified from an unpublished diagram by Steve Juggins.

Since Imbrie and Kipp (1971) revolutionized Quaternary palaeoecology by presenting, for the first time, a numerical procedure for quantitatively reconstructing past environments from fossil assemblages, several numerical techniques have been developed for deriving transfer functions (Birks, 1995). Some have a stronger theoretical basis, either statistically, ecologically, or both, than others. Some (e.g. weighted averaging partial least squares (WA-PLS) and its simpler relative, two-way weighted averaging (WA) regression and calibration) fulfil the basic requirements for quantitative reconstructions, perform consistently well with a range of data, do not involve an excessive number of parameters to be estimated and fitted and are thus relatively robust statistically and computationally economical.

As ter Braak (1995, 1996) and Birks (1995, 1998) discuss, there is a major distinction between models assuming a linear or monotonic response and a unimodal response between organisms and their environment, and between classical and inverse approaches for deriving transfer functions. It is a general law of nature that organism–environment relationships are usually non-linear and taxon abundance is often a unimodal function of the environmental variables. Each taxon grows best at a particular optimal value of an environmental variable and cannot survive where the value of that variable is too low or too high (ter Braak, 1996). Thus all taxa tend to occur over a characteristic but limited environmental range and within this range to be most abundant near their environmental optimum. The distinction between classical and inverse models is less clear (ter Braak, 1995).

In the classical approach taxon responses (Y) are modelled as a function of the environment (X) with some error:

$$Y = f(X) + \text{error}$$

The function f ( ) is estimated by linear, non-linear and/or multivariate regression from the modern training set. Estimated f ( ) is then 'inverted' to infer the unknown past environment from Y_0. 'Inversion' involves finding the past environmental value that maximizes the likelihood of observing the fossil assemblage in that environment. If function f ( ) is non-linear, which it almost always is, non-linear optimization procedures are required (e.g. Birks *et al.*, 1990; Line *et al.*, 1994). Such procedures are not without programming problems (e.g Birks, 2001b) and can be computationally demanding (e.g. Vasko *et al.*, 2000; Toivonen *et al.*, 2001).

In the inverse approach this difficult inversion is avoided by estimating directly the function (g) from the training set by inverse regression of **X** on **Y**:

$$X = g(Y) + \text{error}$$

The inferred past environment (X_0), given fossil assemblage (Y_0) is simply the estimate

$$X_0 = g(Y_0)$$

As ter Braak (1995) discusses, statisticians have debated the relative merits of both approaches. Inverse models perform best if the fossil assemblages are similar in composition to samples in the central part of the modern data, whereas classical models may be better at the extremes and under some extrapolation, as in 'no-analogue' situations. In the few comparisons of these two major approaches, inverse models (e.g. WA or WA-PLS) nearly always perform as well as classical models of Gaussian or multinomial logit regression in a non-Bayesian (ter Braak *et al.*, 1993; ter Braak, 1995; Birks 1998) or a Bayesian framework (Vasko *et al.*, 2000; Toivonen *et al.*, 2001) but with a fraction of the computing resources of classical approaches.

I will only consider WA and WA-PLS, as they currently represent simple, robust approaches for quantitative reconstructions. Ter Braak *et al.* (1993) concluded 'until such time that such sophisticated methods mature and demonstrate their power for species–environment calibration, WA-PLS is recommended as a simple and robust alternative.'

9.3.1 Assumptions

There are five major assumptions in quantitative palaeoenvironmental reconstructions (Imbrie and Webb, 1981; Birks *et al.*, 1990).

1. The *m* taxa in the modern data (Y) are systematically related to the environment (X) in which they live.
2. The environmental variable(s) to be reconstructed is, or is linearly related to, an ecologically important determinant in the system of interest.
3. The *m* taxa in the training set (Y) are the same biological entities as in the fossil data (Y_0) and their ecological responses to the environmental variable(s) of interest have not changed over the time represented by the fossil assemblage. Contemporary spatial patterns of taxon abundance in relation to X can be used to reconstruct changes in X through time.
4. The mathematical methods adequately model the biological responses to the environmental variable(s) of interest and yield transfer functions with sufficient predictive power to allow accurate and unbiased reconstructions of X.
5. Other environmental variables than the ones of interest have negligible influence, or their joint distribution with the environmental variable in the past is the same as today.

9.3.2 Data Properties

Modern training sets (e.g. pollen, diatoms, etc.) contain many taxa (e.g. 50–300), whereas there may be 50–200 samples. Data are usually quantitative and commonly expressed as percentages of the total sample count. They are thus closed, multivariate compositional data with a constant-sum constraint. They often contain many zero values (up to 75 per cent of all entries) for sites where taxa are absent. The data are complex, showing noise, redundancy and internal correlations, and often contain outliers. Taxon abundance is usually a unimodal function of the environmental variables.

Modern environmental data usually contain fewer variables (*c.* 1–10) than the corresponding biological data. Environmental data rarely contain zero values. Quantitative environmental variables often follow a log-normal distribution and commonly show linear relationships and high correlations between variables (e.g. mean July temperature, number of growing-day degrees). There is thus often data redundancy.

9.3.3 Requirements

There are nine major requirements for quantitative reconstructions (Birks, 1995).

1. A biological system is required that produces abundant identifiable and preservable fossils and is responsive to the environmental variable(s) of interest today at the spatial and temporal scales of study.
2. A large high-quality training set is available. This should be representative of the likely range of past environmental variables, have consistent taxonomy and nomenclature and be of comparable quality (counting techniques, size, sampling methodology, preparation procedures, etc.) and from the same sedimentary environment (e.g. lakes).
3. Fossil data sets used for reconstruction should be of comparable taxonomy, quality and sedimentary environment as the training set.
4. Good independent chronology is required for the fossil data sets to permit correlations, comparisons and, if required, assessments of rates of biotic and environmental change.
5. Robust statistical models are required that can model the non-linear relationships between modern taxa and their environment and take account of the numerical properties of the biological data.

6. Reliable estimation of prediction errors is required. As the reliability of the reconstructed environmental values can vary from sample to sample, depending on composition, preservation, etc., sample-specific prediction errors are needed.
7. Critical evaluations and validations of all reconstructions are essential as any statistical procedure will produce a result. What matters is whether the result is ecologically sensible and statistically reliable.
8. The numerical methods are theoretically sound statistically and ecologically, easy to understand, robust, perform well with large and small data sets and taxon-poor and taxon-rich assemblages and are not too demanding in terms of computer resources.
9. The relevant computer programs are available to the research community.

9.3.3.1 Two-way Weighted Averaging (WA)

The basic idea behind WA (ter Braak, 1996) is that at a site with a particular environmental variable x, taxa with optima for x close to the site's value of x will tend to be the most abundant taxa present, if the taxa show a unimodal relationship with x. A simple and ecologically reasonable estimate of a taxon's optimum for x is the average of all the x values for sites at which the taxon occurs, weighted by the taxon's relative abundance. The estimated optimum is the weighted average of x. Taxon absences have no weight. The taxon's tolerance can be estimated as the weighted standard deviation of x. An estimate of a site's value of x is the weighted average of the optima for x for all the taxa present. Taxa with a narrow tolerance for x can, if required, be given greater weight than taxa with a wide tolerance. The underlying theory of WA and the conditions under which it approximates Gaussian logit regression and calibration are discussed by ter Braak (1996), ter Braak and Looman (1986) and ter Braak and Barendregt (1986).

Because the computations involved in WA are simple and fast, computer-intensive bootstrapping (Efron and Tibshirani, 1993) can be used to estimate the root mean square error of prediction (RMSEP) for inferred values of x for all modern samples, the whole training set and individual fossil samples (Birks *et al.*, 1990; Line *et al.*, 1994). The idea of bootstrap error estimation is to do many bootstrap cycles, say 1000. In each, a subset of modern samples is selected randomly but with replacement from the training set to form a bootstrap set of the same size as the original training set. This mimics sampling variation in the training set. As sampling is with replacement, some samples may be selected more than once in a cycle. Any modern samples not selected form a bootstrap *test* set for that cycle. WA is then used with the bootstrap training set to infer the variable of interest for the modern samples (all with known observed modern values) in the bootstrap *test* set. In each cycle, WA is also used to infer the environmental variable, x_0, for each fossil sample. The standard deviation of the inferred values for both modern and fossil samples is calculated. This comprises one component of the overall prediction error, namely estimation error for the taxon parameters. The second component, due to variations in taxon abundance at a given environmental value, is estimated from the training set by the root mean square of the difference between observed values of x and the mean bootstrap of x when the modern sample is in the bootstrap *test* set. The first component varies from fossil sample to fossil sample, depending on the composition of the fossil assemblage, whereas the second component is constant for all fossil samples. The estimated RMSEP for a fossil sample is the square root of the sum of squares for these two components (Birks *et al.*, 1990).

WA has gained considerable popularity in palaeoecology in the last decade for various reasons.

1. It combines ecological realism (unimodal species responses and species-packing model) with mathematical and computational simplicity, rigorous underlying theory and good empirical power.
2. It does not assume linear species–environment responses, it is relatively insensitive to outliers and it is not hindered by multicollinearity between variables or by the large number of taxa in training sets.
3. Because of WA's computational simplicity, it is possible to use bootstrapping to derive RMSEP for all samples.
4. WA performs well in 'no-analogue' situations (Hutson, 1977; ter Braak *et al.*, 1993). In such situations, environmental inferences are based on the WA of the optima of taxa in common between the modern and fossil assemblages. As long as there are reliable optima estimates for the fossil taxa of high numerical importance, WA inferences are often relatively realistic. WA is thus a multivariate indicator species approach rather than an analogue-matching procedure.
5. WA appears to perform best with noisy, species-rich compositional data with many taxa absent from many samples and extending over a relatively long environmental gradient. WA does, however, have two important weaknesses (ter Braak and Juggins, 1993).
   a. WA is sensitive to site distribution within the training set along the environmental gradient of interest (ter Braak and Looman, 1986).
   b. WA disregards residual correlations in the biological data, namely correlations that remain in the biological data after fitting the environmental variable of interest that result from environmental variables not considered directly in WA. The incorporation of partial least squares (PLS) regression (Martens and Næs, 1989) into WA (ter Braak and Juggins, 1993) helps overcome the second weakness by utilizing residual correlations to improve optima estimates.

9.3.3.2 Weighted Averaging Partial Least Squares Regression (WA-PLS)

The relevant feature of PLS is that components are selected not to maximize the variance of each component within **Y** as in principal components analysis but to maximize covariance between components that are linear combinations of the variables within **Y** and **X**. In the unimodal equivalent, WA-PLS, components are selected to maximize covariance between the vector of weighted averages of **Y** and **X**. Subsequent components are chosen to maximize the same criterion but with the restriction that they are orthogonal and hence uncorrelated to earlier components (ter Braak *et al.*, 1993). Ter Braak and Juggins (1993) show that, with a small modification, WA is equivalent to the first PLS component of suitably transformed data. In WA-PLS, further orthogonal components are obtained as WA of the residuals for the environmental variable, in other words the regression residuals of x on the components extracted to date are used as new sample scores in the basic WA algorithm. A joint estimate of x is, in PLS, a linear combination of the WA-PLS components, each of which is a WA of the taxon scores, hence the name WA-PLS. The final transfer function is a WA of updated optima, but in contrast to WA, optima are updated by considering residual correlations in the biological data. In practice, taxa abundant in samples with large residuals are most likely to have updated optima (ter Braak and Juggins, 1993).

The main advantage of WA-PLS is that it usually produces models with lower RMSEP and lower bias than WA (ter Braak and Juggins, 1993). There are two reasons for this:

1. All WA-based models suffer from 'edge effect' problems that result in inevitable overestimation of optima at the low end of the gradient of interest and underestimation at the high end of the gradient. As a result there is a bias in the inferred values and in the residuals. WA-PLS implicitly involves a weighted inverse deshrinking regression that pulls the inferred values towards the training set mean. WA-PLS exploits patterns in the residuals to update the transfer function, thereby reducing errors and patterns in the bias.
2. In real life there are often additional environmental predictors that influence the biological assemblages. WA ignores structure resulting from these variables and assumes that environmental variables other than the one of interest have negligible influence. WA-PLS uses this additional structure to improve estimates of the taxa 'optima' in the final transfer function. For optimal performance, the joint distribution of these environmental variables in the past should be the same as in the modern data (ter Braak and Juggins, 1993).

The main disadvantage of WA-PLS compared to WA is that great care is needed in model selection. As more components are added, the WA-PLS model seems to fit the data better as the root mean square error (RMSE) decreases and becomes 0 when the number of components equals the number of samples. RMSE is an 'apparent' statistic of no predictive value as it is based on the training set alone. An independent test set is needed to evaluate different models as the optimal model is the model giving the lowest RMSEP for the test set. In real life there are usually no independent test sets and model evaluation is based on cross-validation to derive approximate estimates of RMSEP. In leave-one-out cross-validation, the WA-PLS modelling procedure for 1, ..., p components, where p is less than n (usually 6–10), is applied n times using a training set of size (n-1). In each of the n models, one sample is left out and the transfer function based on the (n-1) modern samples is applied to the one test sample omitted from the training set, giving a predicted x for that sample. This is subtracted from the known value to give a prediction error for that sample. Thus, in each model, individual samples act in turn as a test set, each of size 1. The prediction errors are accumulated to form a 'leave-one-out' RMSEP. The final WA-PLS model to use in reconstruction is selected on the basis of low RMSEP, small number of 'useful' components (a 'useful' component gives a RMSEP reduction of ≥ 5 per cent of the one-component model) (Birks, 1998), and low mean and maximum bias (ter Braak and Juggins, 1993; Birks, 1995). Sample-specific errors are estimated by cross-validation and Monte Carlo simulation (Birks, 1998).

Like WA, WA-PLS performs surprisingly well and considerably better than direct analogue-matching procedures when none of the fossil assemblages are similar to the modern data (ter Braak *et al.*, 1993; ter Braak, 1995). For very strong extrapolation beyond the modern training set, WA may perform better than WA-PLS. Like WA, WA-PLS is an indicator species approach but where all taxa are used in the transfer function and estimates of the relevant taxon parameters (beta regression coefficients or 'optima') are derived from the modern training set rather than from modern autecological observations.

9.4 EVALUATION AND VALIDATION OF PALAEOENVIRONMENTAL RECONSTRUCTIONS

This has received surprisingly little attention. It is important as all reconstruction procedures will produce results. How reliable are the results?

The most powerful validation is to compare reconstructions, at least for the recent past, against recorded historical records (e.g. Fritz *et al.*, 1994; Lotter, 1998). An alternative approach compares reconstructions with independent palaeoenvironmental data, for example by comparing pollen-based climate reconstructions with plant macrofossil data (Birks and Birks, pp. 342–357 in this volume), pollen-based climate reconstructions with stable-isotope stratigraphy (Hammarlund *et al.*, 2002), chironomid-based climate reconstructions with plant macrofossil data (Brooks and Birks, 2000), etc. Such comparisons are part of the importance of multi-proxy approaches (Lotter, pp. 373–383 in this volume). Without historical validation or independent palaeoenvironmental data, evaluation must be done indirectly using numerical criteria (Birks, 1998).

There are four useful numerical criteria (Birks, 1998):

1. sample-specific RMSEP for individual samples
2. 'goodness of fit' statistics assessed by fitting fossil samples 'passively' onto the ordination axis constrained by the environmental variable being reconstructed for the modern training set and evaluating how well individual fossil samples fit onto this axis in terms of their squared residual distance (Birks *et al.*, 1990)
3. 'analogue' measures for each individual fossil sample in comparison with the training set. A reconstructed environmental variable is likely to be more reliable if the fossil sample has modern analogues within the training set (ter Braak, 1995)
4. the percentages of the total fossil assemblage that consist of taxa (a) that are not represented at all in the training set and (b) that are poorly represented (e.g. < 10 per cent occurrences) in the training set and hence whose transfer function parameters (WA optima, WA-PLS beta-coefficients, etc.) are poorly estimated and have high associated standard errors in cross-validation (Birks, 1998).

Hammarlund *et al.* (2002) and Bigler *et al.* (2002) illustrate these evaluation criteria in multi-proxy studies on Holocene climatic change in northern Sweden.

9.4.1 Computer Software

As transfer functions and reconstruction diagnostics are computer-dependent, the relevant DOS/Windows software is:

Two-way WA	– WACALIB (Line *et al.*, 1994), CALIBRATE (Juggins and ter Braak, 1997a).
Sample-specific errors in WA	– WACALIB
WA-PLS and leave-one-out cross-validation	– CALIBRATE, WAPLS (Juggins and ter Braak, 1997b).
Sample-specific errors in WA-PLS	– WAPLS
Analogue statistics	– ANALOG (unpublished program by J.M. Line and H.J.B. Birks), MAT (Juggins, 1997).
Reconstruction goodness-of-fit statistics	– CANOCO 4.5 (ter Braak and Šmilauer, 2002).
Other reconstruction diagnostics	– CEDIT (program by Onno van Tongeren supplied with CANOCO 4.0).

For details of availability, see http://www.campus.ncl.ac.uk/staff/stephen.juggins (CALIBRATE, WAPLS, MAT) and http://www.microcomputerpower.com (CANOCO, CEDIT), or e-mail John Birks (WACALIB, ANALOG) (John.Birks@bot.uib.no).

9.4.2 Limitations

The major limitation is the quality and internal consistency of the modern and fossil data sets. Such sets require a detailed and consistent biological taxonomy and, for the modern data sets, reliable and representative environmental data. The creation of modern training sets with detailed and consistent taxonomy ideally requires that all the biological analyses be done by the same analyst who must be skilled in the relevant taxonomy. Many training sets covering broad geographical areas and ecological gradients have, by necessity, to be constructed from samples analysed by different analysts (e.g. Seppä and Birks, 2001). In such cases, taxonomic workshops, standardization of methodology, quantitative analytical quality control and agreed taxonomic and nomenclatural conventions are essential. Such harmonization between data sets and analysts is time-consuming and unattractive and as a result is underfunded or even bypassed as national, continental and global computer databases rapidly develop. Even less attention is often given to the quality and representativeness of the modern environmental data used. Particular problems arise in deriving reliable climate data for mountainous areas (e.g. Korhola *et al.*, 2001). Standardized procedures of interpolation, lapse-rate corrections, etc. must be used throughout.

Transfer functions generally work well in the Late Glacial (e.g. Birks *et al.*, 2000) where environmental changes are large and exceed the inherent sample-specific errors of prediction for individual fossil samples (*c.* 0.8–1.5 °C) (e.g. Brooks and Birks, 2001). In the Holocene these errors are near to the likely range of temperature change and interpretation is correspondingly more difficult (e.g. Korhola *et al.*, 2000). An important area for future research (see below) concerns reducing these errors by adopting a Bayesian approach for environmental reconstruction (Korhola *et al.*, 2002) or by using local training sets (Birks, 1998).

An inherent limitation of all unimodal-based reconstruction methods using WA estimation (WA, WA-PLS) is the 'edge effect' that results in distortions at the ends of the environmental gradient (ter Braak and Juggins, 1993). Although the implicit inverse regression in WA-PLS helps reduce edge effects, it has its own problems by 'pulling' the predicted values towards the mean of the training set, resulting in an inevitable bias with overestimation at low values and underestimation at high values. At present there seems no way to reduce the truncation of taxon responses and hence under- or overestimation of optima, except by using shorter environmental gradients, linear-based methods and local training sets (Birks, 1998).

A further problem arises in interpreting quantitative palaeoenvironmental reconstructions. They are invariably rather 'noisy' resulting, in part at least, from the inherent sample-to-sample variability in the biostratigraphical data used. Non-parametric regressions such as locally weighted regression smoothing (LOWESS) provide useful graphical tools for highlighting 'signal' or major patterns in time-series of reconstructed environmental variables. LOWESS (Cleveland, 1993) models the relationship between a dependent variable (e.g. pollen-inferred July temperature) and an independent variable (e.g. age) when no single functional form such as a linear or quadratic model is appropriate. LOWESS provides a graphical summary that helps assess the relationship and detect major trends within 'noisy' data or reconstructions. LOWESS can also be used to provide a 'consensus' reconstruction based on several reconstructions (e.g. different transfer functions or different proxies) (Birks, 1998). Examples include Bartlein and Whitlock (1993), Lotter *et al.* (1999, 2000) and Fig. 9.3. More sophisticated smoothers such as SiZer (Chaudhuri and Marron, 1999) have considerable potential in palaeoecology because they assess which features seen in a range of smooths are statistically significant (see Korhola *et al.*, 2000).

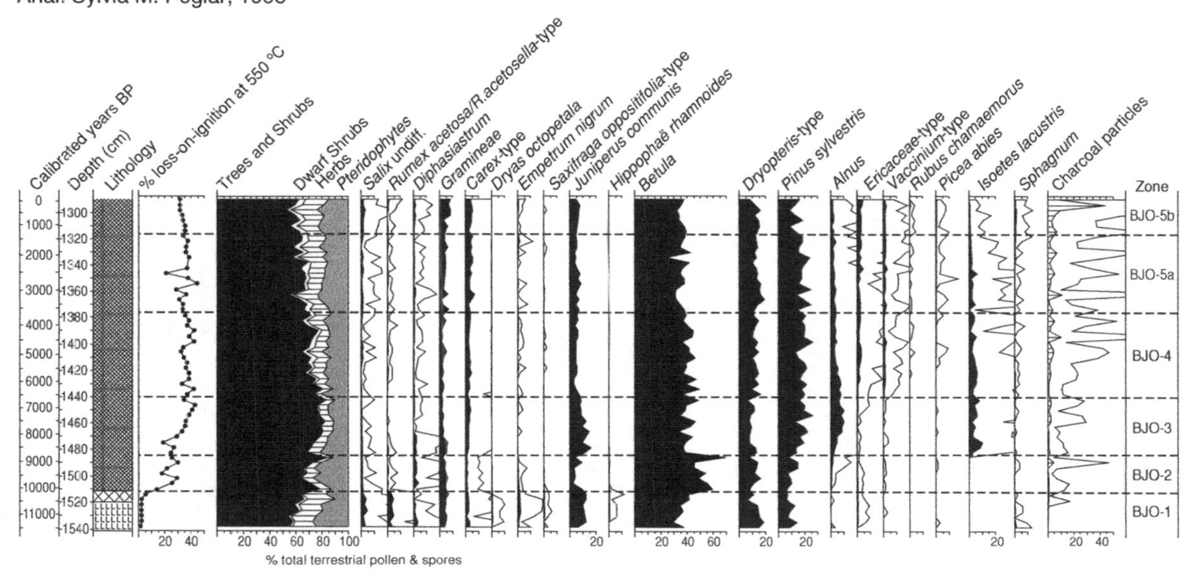

Figure 9.3 Summary pollen diagram from Björnfjelltjörn, north Norway showing major taxa only. The pollen and spores values are expressed as percentages of total land pollen and spores. The unshaded curves are ×10 exaggeration. An age-scale based on eight AMS ^{14}C dates and an associated age–depth model for the core is also shown.

9.5 AN APPLICATION

To illustrate using different environmental reconstruction techniques in Holocene palaeoclimatology, WA, WA-PLS and MAT are applied to the Holocene pollen stratigraphy from Björnfjelltjörn, northern Norway (Fig. 9.4). This is a small 3 ha lake at 510 m above sea level near the Norwegian–Swedish border east of Narvik at 68°26'N latitude and 18°04'E longitude. A 2.9 m long core of sediment was obtained from the deepest (12.9 m) part and the core analysed for pollen, spores and plant macrofossils by Sylvia M. Peglar. Eight AMS ^{14}C dates were obtained and an age-depth model developed using a weighted non-parametric regression procedure. The site occurs just above present-day *Betula pubescens* forest limit and lies within the low-alpine region. Vegetation cover is patchy with much bare rock, dwarf-shrub heath dominated by *Betula nana* and heaths (e.g. *Vaccinium myrtillus, Arctostaphylos alpinus*) and snow-beds dominated by *Salix herbacea*. Present-day mean July temperature is 10.5 °C.

Holocene mean July temperatures were reconstructed using MAT, WA and WA-PLS and a 191-sample training set of surface-mud samples from throughout Norway and northern Sweden (Seppä and Birks, 2001). The training set covers 7.7–16.4 °C mean July temperature, has 152 pollen and spore taxa, and a RMSEP of 1.07 °C (WA), 1.33 °C (MAT) and 1.03 °C (WA-PLS 3 components).

The reconstructions (Fig. 9.5) are based on MAT, WA, WA-PLS and all 98 terrestrial fossil pollen and spore taxa. For comparability the *y*-axis is plotted on the same scale throughout and the reconstructions are plotted on a calibrated (cal) age scale (years BP) based on the age-depth model for the core. In addition the WA-PLS reconstruction and associated sample-specific errors are shown on Fig. 9.5. LOWESS smoothers are fitted to each reconstruction to highlight major trends. The MAT suggests little change throughout the Holocene, whereas WA suggests a rise of about 1 °C between 5000 and 9000 cal years BP. WA-PLS suggests larger climatic shifts, with mean July temperatures about 1.5 °C warmer than today between 6000 and 9000 cal years BP and marked changes in the Late Holocene. Sample-specific errors are about 1.0–1.2 °C, but are largely relative to the magnitude of change in the reconstructions. Numerical evaluation of the individual reconstructions suggest that they are reliable on statistical criteria, with low residual distances, good analogues, almost all fossil taxa well represented in the training set and consistent sample-specific errors.

A consensus reconstruction (Fig. 9.5) is derived by fitting a LOWESS smoother through all reconstructed values (MAT, WA, WA-PLS and also PLS). This consensus highlights warmer July temperatures than today from about 4500 to 9000 cal years BP. It should be validated using an independent proxy, in this case plant macrofossils that are not used in deriving the pollen-based temperature reconstructions. This is presented by Birks and Birks (pp. 342–357 in this volume).

9.6 CONCLUSIONS AND POSSIBLE FUTURE DEVELOPMENTS

The major conclusion is that quantitative Holocene palaeoenvironmental reconstructions are possible from biostratigraphical data but we are probably near the resolution of current data and methods, with sample-specific errors of 0.8–1.5 °C. The transfer function approach is dependent on modern and fossil data of high taxonomic and analytical quality. The acquisition

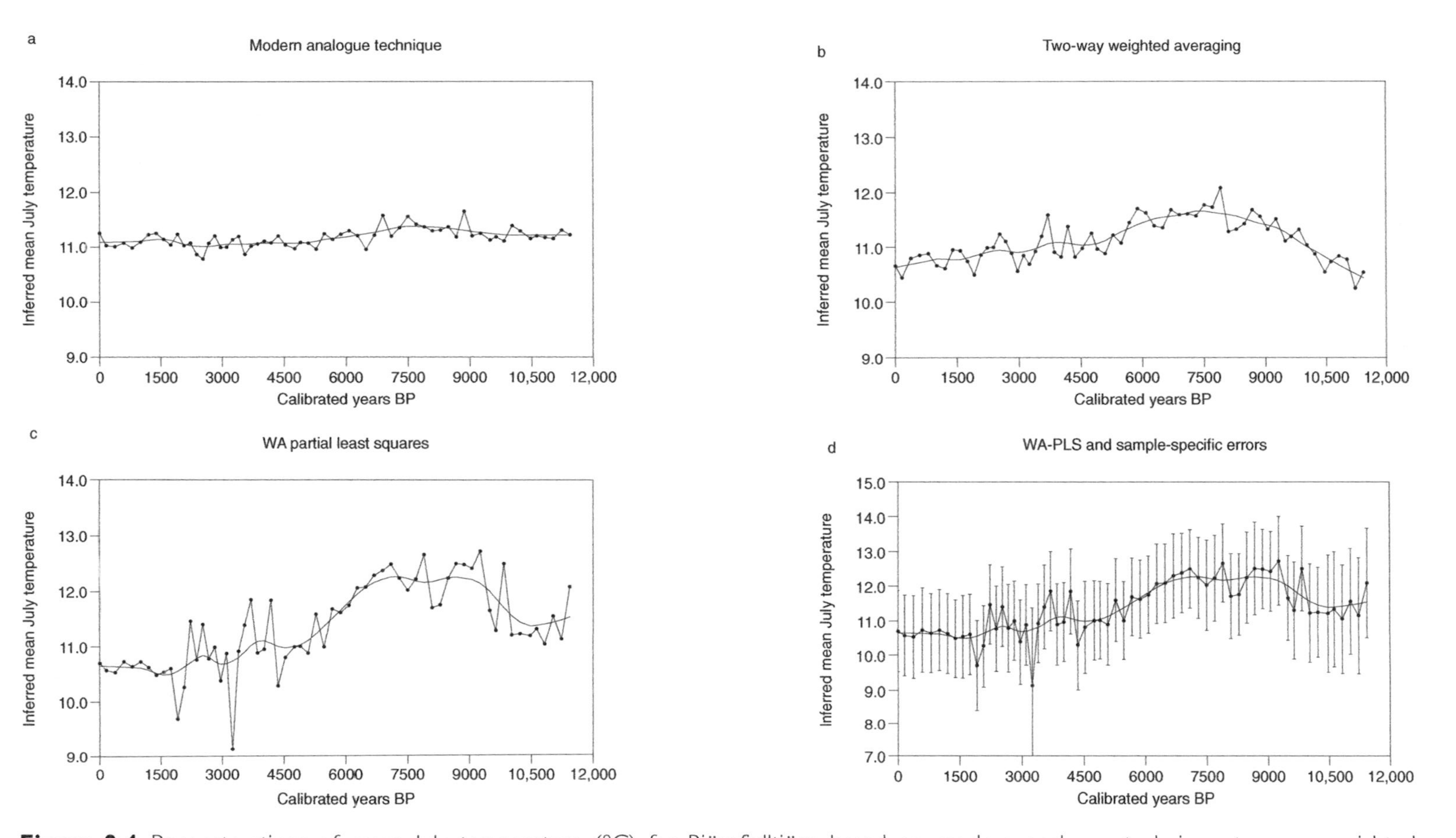

Figure 9.4 Reconstructions of mean July temperature (°C) for Björnfjelltjörn based on modern analogue technique, two-way weighted averaging, and weighted-averaging partial least squares (WA-PLS). The reconstructed values are joined up in chronological order and a LOWESS smoother (span = 0.2, order = 2) has been fitted to highlight the major trends. In the lower right-hand plot, the sample-specific errors of reconstruction for the WA-PLS reconstruction are also shown. All reconstructions are plotted against the age scale (calibrated years BP) of Figure 9.3.

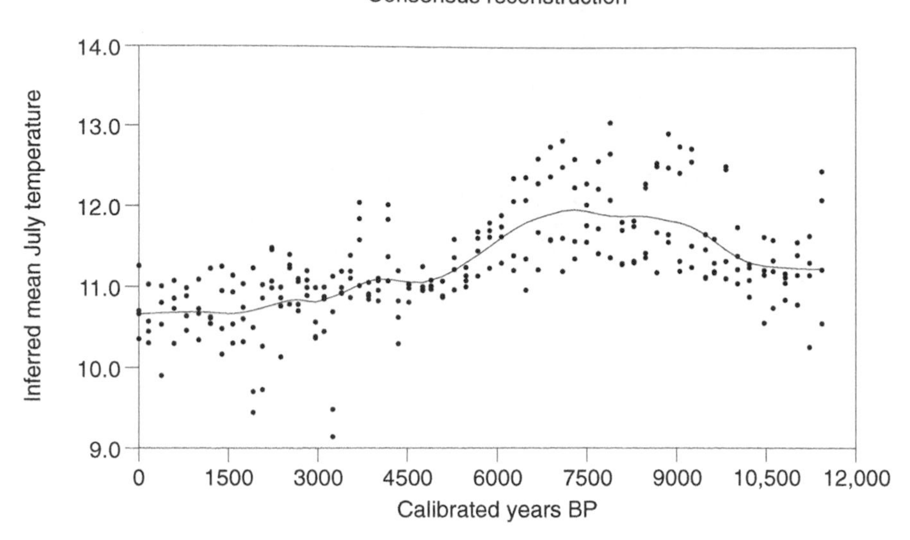

Figure 9.5 Consensus reconstruction for mean July temperature (°C) at Björnfjelltjörn based on modern analogue technique, two-way weighted averaging, weighted-averaging partial least squares regression (see Fig. 9.4), and partial least squares regression. The consensus reconstruction is shown as the fitted line and is a LOWESS smoother (span = 0.2, order = 2) fitted through all the reconstructed values.

of such data is time-consuming. A further challenge in Holocene palaeoenvironmental research is to refine transfer function methodology, reduce sample-specific errors, and distinguish 'signal' from 'noise' in reconstructions. Such improvements may come about in three ways: (1) improvements in the quality and reliability of the modern environmental data; (2) improvements in numerical methods for deriving transfer functions reconstructions; and (3) improvements in interpreting palaeoenvironmental reconstruction time-series.

It is difficult to see how to improve the modern climatic data used in training sets given the availability and quality of modern climate data and the complex patterns of climate variation over small areas, especially with complex topography. Geographical information systems and spatially-explicit statistical modelling and interpolation procedures (e.g. Fotheringham *et al.*, 2000) have the potential for improving climatic data for modern transfer functions.

Improvements in numerical methods may come from artificial neural networks and adopting a Bayesian framework. Applications of artificial neural networks in palaeoceanography (Malmgren and Nordlund, 1997; Malmgren *et al.* 2001) and palaeolimnology (Racca *et al.* 2001, 2003) show their potential in Holocene research. The Bayesian approach does not rely on an explicit model of relationships between variables but on the modification of some prior belief about the specific value of a variable on the basis of some additional information (Robertson *et al.*, 1999). In Bayesian terms, this is known as the prior probability. This can be refined with additional information provided by the modern training set, for example, to give the conditional probability. Once the conditional probability density function has been obtained, it can be combined with the prior probability density function to provide a posterior probability density function using Bayes' theorem (Robertson *et al.*, 1999). Recent studies (Vasko *et al.*, 2000; Toivonen *et al.*, 2001; Korhola *et al.*, 2002) indicate potential advantages of developing transfer functions within a Bayesian framework, although the current computing demands are beyond the computing facilities generally available to palaeoecologists.

Despite considerable advances in transfer function methodology and in developing organism–environment training sets, our abilities to interpret and compare time-series of palaeoenvironmental reconstructions have hardly developed beyond visual comparisons of time-series (Bennett, 2002). There is great scope for applying robust approaches for comparing time-series (e.g. Burnaby, 1953; Malmgren, 1978; Schuenemeyer, 1978; Malmgren *et al.*, 1998). Newly developed techniques for spectral and cross-spectral analysis and of unevenly spaced time-series that are so frequent in palaeoecology (Schulz and Stattegger, 1997; Schulz and Mudelsee, 2002) have the potential for permitting a critical and statistically rigorous interpretation of Holocene palaeoenvironmental time-series. These are major challenges for the future.

Acknowledgements

I am grateful to Sylvia M. Peglar for providing pollen data from Björnfjelltjörn and the modern pollen data set, to Arvid Odland for providing modern climate data, to Hilary Birks, Steve Juggins, Andy Lotter, Heikki Seppä and Cajo ter Braak for helpful discussions about palaeoenvironmental reconstructions, and to Anne Birgit Ruud Hage, Sylvia Peglar and Christopher Birks for help in preparing the manuscript.

CHAPTER

10

STABLE-ISOTOPES IN LAKES AND LAKE SEDIMENT ARCHIVES

Melanie J. Leng

Abstract: Stable-isotope ratios in the marine, terrestrial and lacustrine environment are important in palaeoclimate reconstruction. In lakes, isotopes can be used over a range of time-scales from interannual to millennial. In the recent past, it is often one of the few methods that can be successfully used where anthropogenic influences may be the predominant effect on faunal and floral change. However, to understand and interpret isotopic data from various components within lake sediments requires a knowledge of the processes that control and modify the signal, their effects need to be quantified and a robust calibration is necessary to establish the relationship between the measured signal and the isotopic composition of the host waters. This chapter is aimed at giving an insight into this process, describing the potential as well as the problems of using isotope data from various types of lake sediment components.

Keywords: Palaeoclimate, Lake sediments, Lakes, Stable-isotopes

10.1 General Principles: Isotopes in the Modern Environment

The use of stable-isotope ratios in palaeoclimate studies of lake sediments is based on the relationship between the $^{18}O/^{16}O$ (and $^{2}H/^{1}H$) ratios in waters (precipitation, groundwater, rivers, streams and lakes) and climate, especially temperature. $^{18}O/^{16}O$ and $^{2}H/^{1}H$ ratios are referred to as delta values (ie. $\delta^{18}O$ and δD see next Section). On a global scale mean annual values for $\delta^{18}O$ (and δD) form broadly parallel zones corresponding to temperature (Fig. 10.1). In general, the isotopically heavier values occur at the equator and the lighter values at the poles and $\delta^{18}O$ values of precipitation change by +0.2 to +0.7‰/°C. At the local scale isotopes in precipitation are controlled by:

1. origin of the air mass or the isotopic composition at the source
2. amount of prior rainfall precipitated by the air mass (reflecting continentality, altitude and sometimes referred to as the 'amount effect')
3. the temperature at which precipitation occurs (which may vary seasonally).

On reaching the Earth's surface the isotope ratios of water may subsequently undergo change, and in lakes this is dependent on water residence time and climate. Lake-waters with long residence times, especially in arid environments, undergo the most change through evaporation:

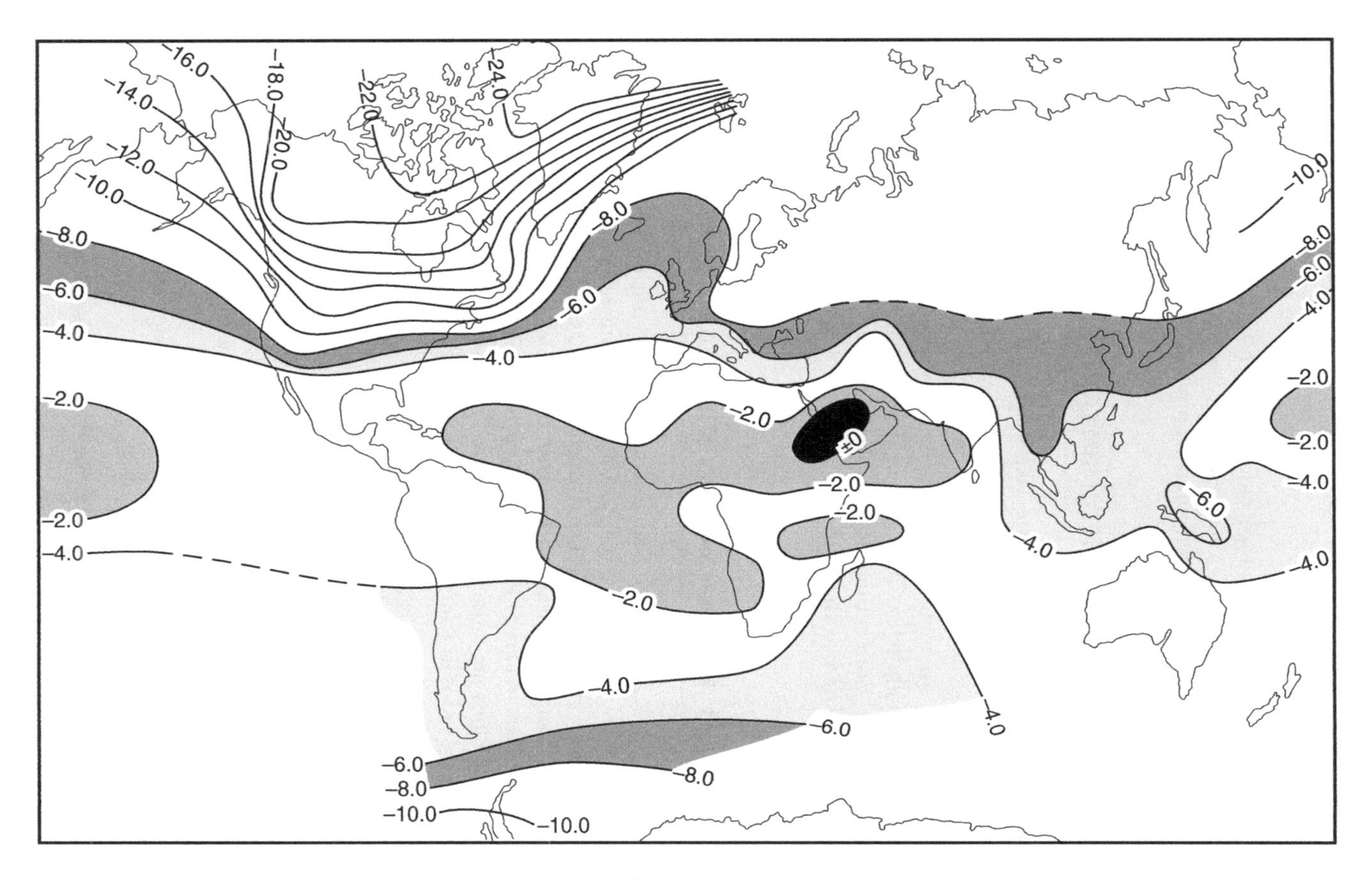

Figure 10.1 Modern-day global distribution of mean $\delta^{18}O$ in precipitation (based on Yurtsever and Gat, 1981).

preferential loss of lighter isotopes in the vapour leads to 'isotopic enrichment' of the remaining lake-water. In oxygen isotope studies of lake sediments, it is vital to assess the hydrological setting, in order to ascertain whether the oxygen isotope composition of lake-waters change in response predominantly to air temperature variation or variations in precipitation/evaporation (see Clark and Fritz, 1997).

In addition to $^{18}O/^{16}O$ and $^{2}H/^{1}H$ isotope ratios, the $^{13}C/^{12}C$ ratio of total dissolved inorganic carbon (TDIC) in lakes can be used as a tracer since the isotope composition (mostly bicarbonate (HCO_3^-) at pH 7–9) is controlled by environmental processes.

10.2 THE SEDIMENTARY RECORD

The $^{18}O/^{16}O$ and $^{13}C/^{12}C$ ratios from the lake-waters get incorporated into many primary precipitates and preserved in lake sedimentary records. These precipitates are either **authigenic** (also referred to as endogenetic) or **biogenic**. In general, authigenic refers to calcite (marl) precipitated in response to algal and macrophyte photosynthesis, while biogenic includes organic matter but generally refers to skeletal structures such as ostracod and mollusc shells as well as diatom frustrules. Both authigenic and biogenic precipitates have isotope ratios ($^{18}O/^{16}O$ and $^{13}C/^{12}C$) that are related to the isotopic composition of the lake-waters from which they precipitated.

10.3 NOTATION AND STANDARDIZATION

$^{18}O/^{16}O$, $^{2}H/^{1}H$ and $^{13}C/^{12}C$ ratios from lake-waters, carbonates, organic materials and diatom silica are normally measured by mass spectrometry. A mass spectrometer ionizes a gaseous sample, and then forms a spectrum of charged molecules which are separated, on the basis of their different masses, by passage through electrical and magnetic fields. Detailed descriptions can be found in Bowen (1988), Coleman and Fry (1991), Clarke and Fritz (1997), Hoefs (1997) and Criss (1999). Because the isotope ratios are more easily measured as relative differences, rather than absolute values, we refer to the ratios in terms of delta values (δ) and this is measured in units of per mille (‰). The δ value is defined as:

$$\delta = (\text{Rsample}/\text{Rstandard}) - 1 \,.\, 1000$$

Where R = the measured ratio of the sample and standard respectively. Because a sample's ratio may be either higher or lower than that of the standard, δ values can be positive or negative. The δ value is dimensionless, so where comparisons are made between samples (e.g. where $\delta_A < \delta_B$) the δ value of A is said to be 'lower' than that of B (and B 'higher' than A). Where reference is made to absolute ratios, A may be said to be 'depleted' in the heavier isotope compared to B (and B 'enriched' compared to A).

In the laboratory it is necessary to use 'working standards' with values calibrated against recognized standard materials, thus all values are quoted relative to the latter according to Craig (1957) and Deines (1970). For waters (oxygen and hydrogen) and silicates (oxygen) we use V-SMOW (Vienna–Standard Mean Oceanic Water) an average ocean water, while for carbonate and organic material we use V-PDB (Vienna–PeeDee Belemnite). See discussions in Bowen

(1988), Coleman and Fry (1991), Clarke and Fritz (1997), Hoefs (1997) and Criss (1999). The data are presented as per mille (‰) deviations from the relevant international standard (i.e. ‰ V-PDB).

Isotopes are useful as environmental tracers because of **isotope fractionation**. Isotope fractionation occurs in any thermodynamic reaction due to differences in the rates of reaction for different molecular species. The result is a disproportional concentration of one isotope over the other on either side of the reaction. Isotope fractionation is dependent on temperature of the reaction (see details in Clark and Fritz, 1997).

10.4 WATER ISOTOPES IN THE CONTEMPORARY ENVIRONMENT

10.4.1 Rainfall and Lake-Water

When water evaporates from the surface of the ocean, the water vapour is enriched in the lighter water isotopes (^{16}O, ^{1}H), because of their lower vapour pressures, which causes isotopic fractionation. As the vapour rises, rain forms when the dew point is reached. During removal of rain from a water mass, the residual vapour is continually depleted in the heavy isotopes, because rain leaving the system is enriched in ^{18}O and ^{2}H. As the air mass moves pole-ward and becomes cooler, additional rain formed will contain less and less ^{18}O and ^{2}H. The flux of moisture from the oceans and its return via rainout and runoff is, on an annual basis and global scale, close to dynamic equilibrium. Only major climatic shifts will change the isotopic composition of the oceans, such as the amount of storage of freshwater in glacial reservoirs between glacial and interglacial times. On a local scale, starting with evaporation from the ocean surface, rainout, re-evaporation, snow and ice accumulation and melting, runoff, different climatic regimes – are all different steps that will affect the isotopic composition of the water at a given point. The isotopic composition of precipitation (rain, snow) in most parts of the world falls on to the 'Global Meteoric Water Line' (GMWL) (Craig 1961), which is represented by the equation:

$$\delta D = 8\delta^{18}O + 10 \text{ (}\delta D \text{ and } \delta^{18}O \text{ values versus VSMOW)}$$

This line is an average of many local or regional meteoric water lines whose slopes and intercepts may differ from the GMWL due to varying climatic and geographic parameters. In general, however, warmer, tropical rains have higher δ values, while colder, polar precipitation has lower δ values. Lake-waters plotting on or close to the GMWL are isotopically the same as precipitation, whereas lake-waters that plot off the GMWL on a local evaporation line (LEL) have undergone kinetic fractionation. Molecular diffusion from the water to the vapour is a fractionating process due to the fact that the diffusivity of $^{1}H_2{}^{16}O$ in air is greater than $^{2}H^{1}H^{16}O$ or $H_2{}^{18}O$. With evaporation the isotopic composition of the residual water in the lake and the resulting water vapour become progressively more enriched (Fig. 10.2), in both cases the kinetic fractionation of ^{18}O exceeds that of ^{2}H. Experimental work (summarized in Gonfiantini, 1986) shows that water becomes progressively enriched in ^{18}O and ^{2}H during evaporation, the rate of which is dependent on humidity. If humidity is 0 per cent then this enrichment follows a Rayleigh distillation (Clarke and Fritz, 1997). Where humidity is about 100 per cent there is no net diffusivity as diffusion occurs in both directions. On the $\delta^{18}O$ versus δD diagram (Fig. 10.2), water that deviates from the MWL along a LEL with a lower slope (s) depends largely on

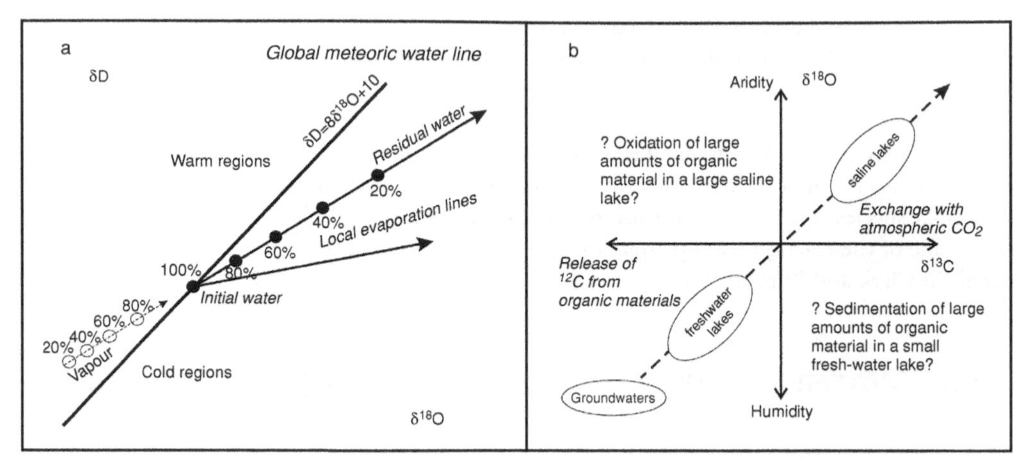

Figure 10.2 Major controls on the (a) $\delta^{18}O$ v δD of precipitation and lakewaters and (b) $\delta^{13}C$ v $\delta^{18}O$ of lakewaters. In (a) evaporation defined by LELs is shown to increase the isotopic composition of both the residual lakewater and resultant water vapour. The slope of the LEL is dependent on local humidity.

relative humidity (h) but also on temperature and windiness. The slope of the line can be an estimate of the relative humidity. Gat (1971) calculated that for s = 0.25, h = 4; whereas if h = 0.75, s = 5. Not until h = 0.9 does the slope approach 8.

The aim in investigating the isotopic composition of modern lake-waters is to establish their relationship to modern climate. This is an important question as the isotopic composition of lake-water can be recorded in precipitates such as authigenic calcites (McCrea, 1950; O'Neil *et al.*, 1969). Thus, if authigenic and/or biogenic precipitates are preserved in the lake sediments then measurement of their isotopic composition can be useful for palaeolimnology and, if the relationship between lake-water composition and climate is understood, also for palaeoclimatology. However, to calibrate lake-water response to climate it is necessary to acquire a knowledge of the baseline conditions for isotopes (O, H and C) in the meteoric and surface-water parts of the hydrological cycle. In most parts of the world, $^{18}O/^{16}O$ and $^{2}H/^{1}H$ ratios in precipitation are routinely measured by the International Atomic Energy Agency (http://isohis.iaea.org), ground-waters can be measured or calculated (as they tend to reflect the weighted mean isotopic composition of annual rainfall) and lake-waters can be measured. In addition to oxygen and hydrogen, carbon in the form of bicarbonate is usually present in lake-waters. The carbon is derived from soils, i.e.

$$CO_{2(aq)} + H_2O = H_2CO_3$$

$$\text{and } H_2CO_3 = H^+ + CO_3^{2-}$$

In lake-waters the carbon gets incorporated into authigenic and biogenic carbonates, i.e.

$$Ca^{2+} + CO_3^{2-} = CaCO_3$$

Carbon isotopes are not normally measured in hydrological studies but understanding the primary controls on changes in $\delta^{13}C$ (i.e. $^{13}C/^{12}C$ ratios from bicarbonate in the lake-waters) are paramount to understanding palaeoclimate records as both carbon and oxygen isotope records from lacustrine calcites are routinely measured and interpreted in terms of climatic variation (e.g. Talbot, 1990; Drummond *et al.*, 1995; Li and Ku, 1997).

Modern lakes tend to be classified as either 'open' or 'closed'. Open refers to lakes that have an inflow and an outflow, while closed refers to lakes that have only an inflow, and water losses are primarily through evaporation. In general, groundwater-fed open lake-waters should have a $\delta^{18}O$ and δD composition similar to mean weighted values for precipitation, and fall on a MWL. Evaporating lakes will have $\delta^{18}O$ and δD values which lie on a LEL with a slope determined by local climate (Table 10.1, Fig. 10.2).

	General characteristics of lakes where temperature change estimates are required	**General characteristics of lakes where precipitation/evaporation (relative humidity) estimates are required**
Lake-water volume	Small	Large
Residence time	Short ('open' lake)	Long ('closed' lake)
Geographical setting	High altitude/latitude, temperate	Arid and semi-arid, any altitude
Typical $\delta^{18}O$ ranges in core sediments	Often negative values, small range of 1–2‰	Negative to positive values, large swings (>10‰)

Table 10.1 Features of lakes likely to produce temperature or precipitation/evaporation reconstructions from isotopic composition of primary precipitates within a lake sediment.

10.5 OXYGEN AND CARBON ISOTOPE RATIOS IN PRIMARY PRECIPITATES

10.5.1 Oxygen Isotopes

Carbonates, diatom silica and cellulose etc. precipitated in lake-waters, preserve a record of the $^{18}O/^{16}O$ ratio of the water at a given temperature. Urey recognized the potential of the ^{18}O content of carbonates to record temperature information (Urey *et al.*, 1951), which led to the $\delta^{18}O_{CaCO3}$–palaeotemperature scale (Epstein *et al.*, 1953):

$$t = 16.5 - 4.3(c - w) + 0.14(c - w)^2$$

Where t = temperature, c = $\delta^{18}O_{calcite}$, w = $\delta^{18}O_{water}$. Several other determinations of equilibrium fractionation have been made (on inorganic and organic carbonates) such as Craig (1965), O'Neil *et al.* (1969) and Erez and Luz (1983), a popular one to use is that of Hays and Grossman's (1991) fit of O'Neil *et al.*'s (1969) data:

$$t = 15.7 - 4.36(c - w) + 0.12(c - w)^2$$

The temperature-dependent mineral-water fractionation means that the oxygen isotope composition of carbonates, diatom silica etc. have potential as palaeo-thermometers in the sedimentary record and are commonly used in marine studies. However, calculation of temperature is only possible if the composition of the water is known, i.e. in ocean waters (between glacial and interglacial cycles) or in open (low residence), freshwater lakes where the isotopic composition of the water is the same as the mean weighted composition of meteoric water. These characteristics are most common in high-altitude/latitude lakes (Table 10.1) (e.g. Shemesh *et al.*, 2001; Teranes and McKenzie, 2001). However, using oxygen isotope ratios to reconstruct palaeo-temperatures is complicated by the fact that there are actually two temperature effects recorded by $\delta^{18}O$ of primary precipitates from lake-waters:

1. the mean annual air temperature recorded by the $\delta^{18}O$ of precipitation (Figs 10.1 and 10.2a) which is controlled by the local temperature ~$\delta^{18}O$ relationship (Dansgaard, 1964)
2. the water temperature at which the precipitation took place, which controls the equilibrium fractionation between water and mineral (equations above).

To derive palaeo-temperature data from primary precipitates one must first ascertain that the lake-water temperature signal is not indecipherable due to other effects. Lake-waters that are evaporating do not readily provide temperature information, because the $\delta^{18}O$ of the lake-water is changed through preferential evaporative loss of the lighter oxygen isotope (^{16}O) and this is usually the predominant process. In the majority of lakes it is easy to rule out temperature as the principal driving force as variations in $\delta^{18}O$ records due to changes in water balances are often much larger (several ‰) than potential temperature effects (Fig. 10.3). However, where

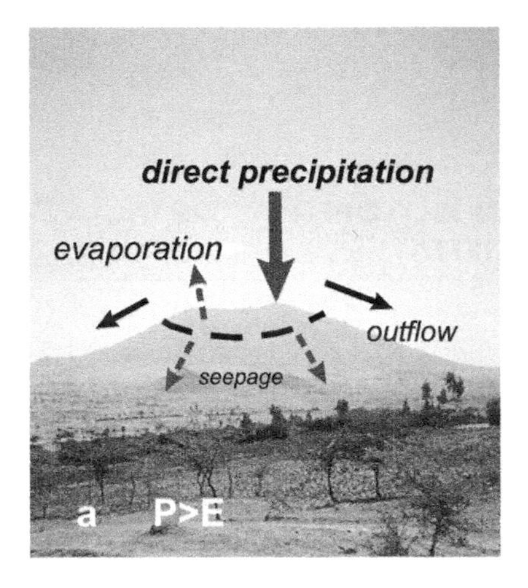

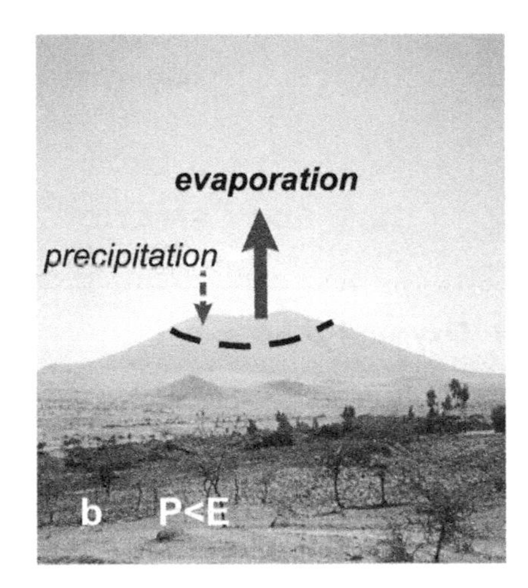

Figure 10.3 Lakes have the potential of recording either temperature or precipitation/ evaporation through the isotopic composition of lake-waters which gets incorporated into primary precipitates. In caldera lakes where direct precipitation>evaporation (a) and the lake-water has a short residence time (due to an outflow), the lakewater isotopic composition changes with the isotopic composition of precipitation. This type of lake has the potential of recording past air temperature information. In lake (b) precipitation<evaporation, and lake-water is lost primarily through evaporation. The isotopic composition of the lake-water is a function of aridity as evaporation of the lighter isotopes causes the lake-waters to become isotopically 'enriched'. Dependent on the climatic regime, caldera lakes have the potential to switch between these scenarios.

evaporation is insignificant, as in the case of some high-altitude/latitude lakes (Rosqvist *et al.*, 1999; Shemesh *et al.*, 2001), it is possible to derive temperature directly although the $\delta^{18}O$ value of the lake-water (i.e. through the relationship between the isotopic composition of precipitation and temperature, $\delta^{18}O_{precip}$~t) must be known. However, there are inferences that can be made. The isotopic fractionation during the precipitation of calcite is about –0.25‰/°C (Craig, 1965) and between –0.2 and –0.5‰/°C for diatom silica (e.g. Brandriss *et al.*, 1998; Shemesh *et al.*, 1992). Variations in $\delta^{18}O_{precipitation}$ on a global scale range between +0.2 and +0.7‰/°C, being greatest for marine and polar regions (Dansgaard, 1964). This range takes into account numerous processes (e.g. altitude, latitude and continentality or rainout, and seasonal effects, summarized in Clarke and Fritz, 1997). The combined effect of the fractionation due to temperature of precipitation and calcite-water fractionation therefore suggests a change in $\delta^{18}O_{calcite}$ of –0.05 to +0.45‰/°C in mean annual temperature. In temperate regions with $\delta^{18}O_{precipitation}$ variation of ~ 0.60‰/°C this would amount to +0.35‰/°C (Siegenthaler and Eicher, 1986).

Information on temperature from a lake record can only be gained if there were no changes in the temperature or isotopic composition of the seawater undergoing evaporation or changes in the long-distance trajectory of air-masses as both affect the isotopic composition of precipitation (δp). A number of recent studies have demonstrated significant short-term deviation from the classic Dansgaard relationship in the sediment records and have highlighted important shifts in the Holocene climate system. Isotopic studies of lake sediments that attempt to interpret oxygen isotopic records solely in terms of temperature change at the site potentially ignore the possible effects of changes in δp that are related to more distant climatic processes (Edwards *et al.*, 1996; von Grafenstein *et al.*, 1999; Teranes and McKenzie, 2001). δp can really only be unravelled from temperature in either deep lakes where bottom-water temperatures are constant (von Grafenstein *et al.*, 1999) or in lakes where there is an independent temperature estimate.

The above section shows that the relationship between temperature, δp and precipitation still has considerable uncertainties, and there is some debate over additional variables that may effect isotopic fractionation, such as water chemistry (Zeebe, 1999). It is usually very difficult to derive quantitative palaeo-temperature/δp data with any certainty (see further discussion in Section 10.7).

10.5.2 Carbon Isotope Ratios

Inorganic carbon isotope ratios mainly from HCO_3^- in the lake-waters are useful as tracers as they get fractionated during several important carbon-cycle processes and ultimately get incorporated into authigenic and biogenic carbonates. The bicarbonate is derived from interaction of groundwaters with rocks and soils in the catchment. In general there are three processes that control the inorganic carbon isotope composition of the TDIC. These processes are:

1. the isotope composition of inflowing waters
2. CO_2 exchange between atmosphere and water
3. photosynthesis/respiration.

These processes are all in some way connected to climate, water residence time and nutrient availability.

Groundwaters and river-waters in general have $\delta^{13}C$ values that are typically low; values between −10 and −15‰ have been reported from Northern Europe (Andrews *et al.*, 1993, 1997). In lake-waters the carbon pool is often changed by biological productivity, where plants preferentially take up ^{12}C during photosynthesis. During periods of enhanced productivity, the carbon pool may become enriched in ^{13}C and hence have higher $\delta^{13}C$. Long lake-water residence times enable exchange with isotopically heavier atmospheric CO_2, which leads to values about +2‰ at final equilibrium (Usdowski and Hoefs, 1990). This type of information derived from carbon isotopes is often complementary to $\delta^{18}O$. $\delta^{13}C$ and $\delta^{18}O$ often show a close correlation, and are often used to estimate the degree of hydrological 'openness' of a lake through time (Talbot, 1990; Li and Ku, 1997). A summary of the predominant process effecting closed basin lake-waters are given in Fig. 10.2.

10.6 SAMPLE PREPARATION

For isotope analysis of lacustrine sediments, cores are normally sampled in small slices (e.g. 0.1, 0.5 and 1 cm) at stratigraphic intervals of 0.5, 1, 2, 4, 8, 16 cm etc. dependent on the resolution required. Sub-samples are then taken, dependent on the amount of the required material present. In marl-rich sediments where both the fine-grained authigenic calcite and biogenic components are required, only a small amount of material will be needed for the authigenic sample. However, a large sample will probably be required to extract sufficient of the biogenic components. There are numerous methods for the initial raw sediment preparation and concentration dependent on the different proportions of the various components. Figure 10.4 shows recommended raw sediment preparation techniques. Subsequent extraction for preparation of gas for mass spectrometry are detailed elsewhere (Bowen, 1988; Coleman and Fry, 1991; Clarke and Fritz, 1997; Hoefs, 1997; Criss, 1999). It is worth emphasizing that isotopic differences between samples to be measured are often extremely small. Great care is taken to avoid isotopic fractionation during chemical and physical treatment of the sample. The cleaned samples are then converted to a gas by an extraction method, which is necessary for mass spectrometry. The clean-up stage and extraction must provide a 100 per cent yield because loss of any fraction will lead to isotopic fractionation since the different isotopic species have different reaction rates. The gas produced must be pure, as contamination with other gases having the same molecular mass and similar physical properties can be a serious problem in mass spectrometry.

10.7 INTERPRETATION OF OXYGEN ISOTOPES IN LAKE SEDIMENTS

10.7.1 Authigenic Carbonate

Authigenic carbonates are precipitated by photosynthetic utilization of CO_2 and resultant calcium carbonate supersaturation. In temperate and high-latitude regions, authigenic carbonates are precipitated mainly in the summer months during periods of maximum phytoplankton productivity (Leng *et al.*, 1999b; Teranes and McKenzie, 2001). In mid-latitude and tropical lakes, phytoplankton growth may occur throughout the year although other mechanisms such as supersaturation may also cause continuous carbonate precipitation, e.g. the Dead Sea aragonite (Niemi *et al.*, 1997). More commonly, in the tropics, carbonate precipitation is related to phytoplankton blooms associated with annual lake-water mixing and

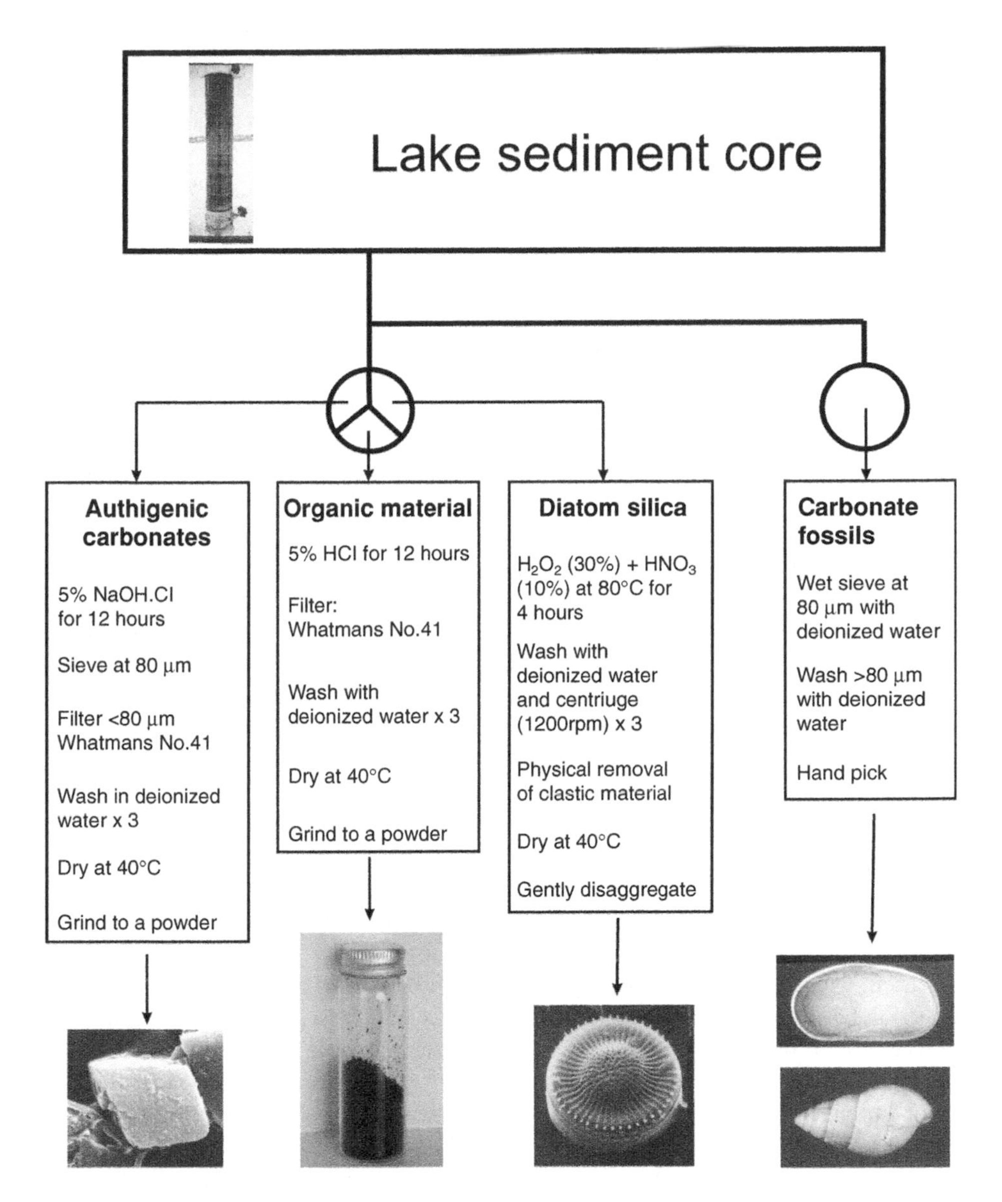

Figure 10.4 A summary of some of the more common components of lake sediments and clean-up procedures to produce sample materials ready for stable-isotope extraction techniques and mass spectrometry.

nutrient availability (cf. Lamb *et al.*, 2002). The advantage of using authigenic carbonate is that it provides an integrated climate signal for the whole sample which may be 1, 10 or even 100 years, depending on sedimentation rate. However, there are some potential problems. It is difficult to distinguish between authigenic and allogenic (derived from the terrestrial environment) carbonates, especially in karstic regions, so there is always the possibility of contamination of the isotope signal from a washed-in component. Allogenic input is generally only associated with lakes with riverine or stream input. In addition, there are several various carbonate minerals that could precipitate out in a lake and each mineral has its own mineral-water fractionation (Friedman and O'Neil, 1977). In freshwater systems, calcite ($CaCO_3$) usually forms but with increasing evaporation other forms occur such as aragonite ($CaCO_3$) and dolomite ($CaMg(CO_3)_2$) as at Lake Bosumtwi (Talbot and Kelts, 1986). Aragonite is about 0.6‰ more positive (contains more ^{18}O) than calcite formed under the same conditions, while dolomite is 3–4‰ more positive. Separating different carbonate minerals is not easy, although by limiting reaction times dolomite dissolution can generally be excluded (Al-Asam *et al.*, 1990).

Authigenic calcite is probably the most common mineral used for isotope analysis in lake sediments. In general, studies fall into two types: either closed lakes which tend to be more common at the low altitudes; or open high-altitude lakes (Fig. 10.5). There are, however, a whole range in lake types and different water residence times, from hundreds of years in large closed lakes where the only loss is through evaporation, to open lakes which essentially allow a continuous throughflow of water. Large closed-lake systems tend to lose water predominantly through evaporation and have enriched $\delta^{18}O_{lake\text{-}water}$ values. Fluctuations in the isotope composition of authigenic carbonate (which can have amplitudes of >10‰) are mainly a function of long-term changes in the precipitation/evaporation ratio. These types of lakes are often classified as marl or salt lakes and have abundant authigenic carbonate, and there are many examples. They are particularly common at lower latitudes (Li and Ku, 1997) especially in the tropics (Ricketts and Johnson, 1996; Holmes *et al.*, 1997; Ricketts and Anderson, 1998; Lamb *et al.*, 2000) and the Mediterranean (Frogley *et al.*, 1999; Leng *et al.*, 1999b; Reed *et al.*, 1999; Roberts *et al.*, 2001) although are also known from the Antarctic (Noon *et al.*, 2003).

In contrast, high-altitude lakes which have a degree of throughflow typically have oxygen isotope compositions that vary by no more than a few ‰. These variations are generally ascribed to variations in temperature and, as described above, this is an effect on both the isotopic composition of precipitation and the fractionation during carbonate/diatom precipitation (Fig. 10.5). The lake-water is generally similar to mean weighted annual precipitation. These types of lakes are common in Northern Europe and at high altitudes where the $\delta^{18}O$ values of authigenic carbonate preserve a record of changing isotopic composition of local precipitation. In lakes where there is a limited catchment and short water residence time, it has been assumed that carbonate precipitation occurs at approximately the same time during the annual cycle when the range of the surface-water temperatures is about the same each year. Variations in the $\delta^{18}O$ of the carbonate are then related directly to past changes in the isotopic composition of mean annual precipitation and therefore air temperature (McKenzie and Hollander, 1993).

However, authigenic calcites can show down-core variations in isotope ratios which may not always be climatically influenced. For example, variations in amount of hydrothermal water input, unrelated to climate variation, have been demonstrated as a cause of large changes in authigenic calcites from a caldera lake in Ethiopia (Lamb *et al.*, 2000).

10.7.2 Biogenic Materials

The use of oxygen and carbon isotopes as environmental tracers can also be used in biogenic materials. However, knowledge of growth period, habitat etc. are essential in interpreting this type of data. Carbonate studies tend to be restricted to ostracod and snail shell carbonates

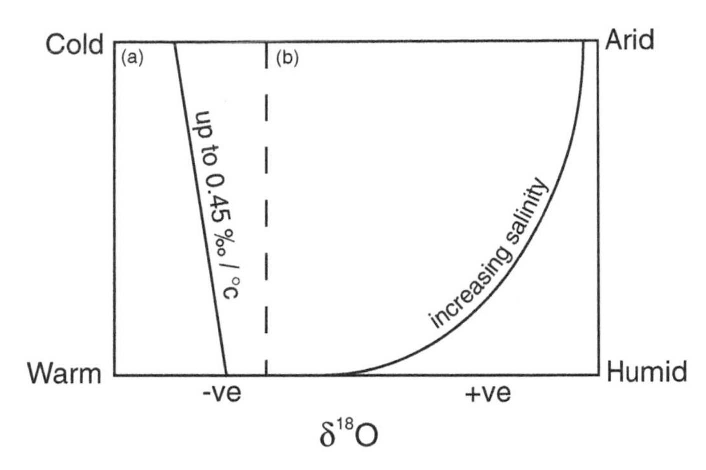

Figure 10.5 Simplified hypothetical response of open-closed lakes to climate change as seen in a lake sediment core isotope curve. (a) Open spring-fed freshwater lake with a short water residence time, the isotopic composition of the lake-water will respond predominantly to changing air temperature. (b) Closed lake with a long lake-water residence time, the isotopic composition of the authigenic components within the lake-water will be a function of the precipitation/evaporation balance (assuming constant lake-water volumes).

although other forms, such as Chara (Coletta *et al.*, 2001) and the larval shell of the molluscan glochidia (Griffiths *et al.*, 2002) also have potential. Outlined below are the 'pros and cons' associated with the various types of biogenic materials, with examples of where they have been successfully used.

10.7.2.1 Ostracod Shells

The oxygen and carbon isotope composition of ostracod calcite is commonly used for reconstructing past climates (Holmes, 1996). There are distinct advantages (but also some disadvantages) compared to other forms of carbonate (Ito, 2003). The advantages include the multiple moults that ostracods undergo to reach the adult stage (analysis of individual shells from the sample period can give an indication of seasonal variation within the lake), the short duration of calcification after each moult, the shell surface morphology (which can be used as an indication of preservation) and the relative ease with which they can be isolated from the bulk sediment. The disadvantages include the potential for ostracods to be washed in to lakes from rivers/streams and may have calcified their shells in an environment other than where they are found. Ostracods grow their shells seasonally so a good knowledge of the ecology, especially of habitat preferences and moult stages is required because individuals reach the adult stage over a period of time and thus a random selection of individual shells will span an unknown period (Heaton *et al.*, 1995). In lakes that have large seasonal variation in temperature or lake-water $\delta^{18}O$, single or small numbers of shells do not represent a known calcification period so this type of data is difficult to interpret (Lamb *et al.*, 1999; Bridgewater *et al.*, 1999). Ostracod shells are relatively small, normally weighing only a few micrograms. Techniques that are commonly used for ostracod isotope analysis normally require a minimum of a few micrograms of $CaCO_3$, so some very small species such as *Cypria ophtalmica* which are ubiquitous but weigh much less than this (H.I. Griffiths, personal communication) cannot be analysed as individuals.

Taking only a small number of shells from a seasonally changing lake obviously has the potential of producing ambiguous data unless a statistically significant population is used. However, ostracod calcite has been used successfully (e.g. Holmes *et al.*, 1997; Frogley *et al.*, 1999; Griffiths *et al.*, 2002), although a good knowledge of the habitat preferences and moult stages is essential.

To use $\delta^{18}O_{ostracod}$ in a quantitative, rather than qualitative, way depends on, amongst other things, knowledge of the isotope fractionation between ostracod calcite and water. Studies of $^{18}O/^{16}O_{ostracod\text{-}water}$ fractionation have shown that ostracods precipitate their shell calcite out of equilibrium with lake-waters. In laboratory cultures, Xia *et al.* (1997a, 1997c) found that the calcite valves of *Candona rawsoni* had $\delta^{18}O$ values higher than expected for equilibrium, and that the amount of offset from equilibrium differed under different culture conditions. In monitoring studies of lakes in southern Germany, von Grafenstein *et al.* (1996) recorded $\delta^{18}O_{ostracod}$ values higher than expected for equilibrium, with different offsets for different taxa. Studies on spring-fed ponds in southern England enabled Keating *et al.* (2002) to determine ostracod calcite-water fractionation for three species. They found that the ostracods had $\delta^{18}O$ values 2.5-3.0‰ higher, probably dependent on pH, than those calculated for equilibrium using the traditional water-calcite oxygen isotope fractionation equations while the carbon appears to be in isotopic equilibrium only under certain pH conditions.

Despite problems, ostracod calcite isotope data can be used in a very sophisticated manner if modern calibrations can assess the relationship between $\delta^{18}O_{ostracod}$, $\delta^{18}O_{lake\text{-}water}$ and temperature. For example, through analysis and modelling of the isotopic hydrology of a lake in southern Germany, von Grafenstein *et al.* (1996, 1999) were able to show that the oxygen isotope composition of the lake-water is mainly controlled by the isotopic composition of local precipitation, which is empirically linked to air temperature. Therefore the isotopic composition of ostracods from the lake provides a direct link to air temperature variation.

10.7.2.2 Mollusc Shells

The isotopic composition of fossil fresh-water mollusc shells is generally under-utilized relative to other forms of carbonate. They are widespread in Quaternary lacustrine deposits although tend not to occur continually, and are often composed of thermodynamically unstable aragonite, which can convert to calcite and effectively 're-set' the isotope signal. However, it is generally thought that the oxygen and carbon stable-isotope values of snail-shell carbonate ($\delta^{18}O_{snail}$, $\delta^{13}C_{snail}$) reflect the isotopic composition of lake-water (Fritz and Poplawski, 1974; Leng *et al.*, 1999a). Some studies have raised concerns with inter-species fractionation differences (e.g. Abell and Williams, 1989), although records have been published using multiple species where variation has been considered insignificant (Gasse *et al.*, 1987; Abell and Hoelzmann, 2000). Some of these records use only one or two individual shells from a stratigraphic level (e.g. Bonadonna and Leone, 1995; Zanchetta *et al.*, 1999), although others have shown considerable ranges, albeit from unspecified numbers of shells, from individual sample levels (Abell and Hoelzmann, 2000). In a small lake in southern Turkey, Jones *et al.* (2002) demonstrated that analysis of individual shells provides a range in data that is related to both seasonal and annual changes in the $\delta^{18}O$ and $\delta^{13}C$ of the lake-water, as samples from lake-sediment cores represent a number of years, dependent on sedimentation rates. This study also shows that co-existing species yield different $\delta^{18}O$ and $\delta^{13}C$ values, probably due to habitat differences within the same lake. It cannot be assumed that, although different species may be precipitating carbonate in equilibrium with the water in which they are living, the isotopic composition of the lake-water is constant between microhabitats. If multi-species records are necessary, due to the nature of the fossil mollusc record, overlapping species should be chosen to allow comparison of the species through time.

Some of the larger species of freshwater snails form carbonate in a regular manner over a fixed and relatively short period. These are important in providing the possibility for achieving continuous, interseasonal information about the changing isotopic composition of the lake-water. The gastropod *Melanoides tuberculata* has great potential. This snail is widespread in modern (Abell, 1985) and Quaternary deposits throughout Africa and Asia and is ubiquitous in both fresh and highly evaporated lakes. In one study of whole-shells and incremental growth, Leng *et al.* (1999a) analysed both modern and fossil *Melanoides* from two lakes in the Ethiopian Rift Valley. $\delta^{18}O$ values in the modern shells show that the snail carbonate precipitates in equilibrium with modern waters. $\delta^{18}O$ values in fossil shells show changes over the lifetime of the organism associated with enhanced monsoonal conditions in the early Holocene.

10.7.2.3 Biogenic Silica

Biogenic silica is deposited by a variety of freshwater organisms including diatoms and sponges. It is especially useful in acidic lakes with no authigenic or biogenic carbonates. The oxygen isotope composition of biogenic silica is potentially useful as a palaeo-thermometer, because it is often found in greatest abundance in high altitude, open (low residence time), freshwater systems where the isotopic composition of the lake-water is the same as the composition of meteoric water. Biogenic silica has been used as a proxy for temperature change in many studies (Rietti-Shati *et al.*, 1998; Rosqvist *et al.*, 1999; Leng *et al.*, 2001), although the temperature-dependence of oxygen isotope fractionation between diatom silica and water is still controversial. The fractionation has been estimated previously from analyses of diatoms from marine and freshwater sediments, coupled with estimates of the temperatures and isotopic compositions of coexisting waters during silica formation (Labeyrie, 1974; Juillet-Leclerc and Labeyrie, 1987; Matheney and Knauth, 1989). Realistic fractionations vary somewhat because data from calibration studies are limited and based on bulk samples (Labeyrie and Juillet, 1982; Wang and Yeh, 1985; Juillet-Leclerc and Labeyrie, 1987; Shemesh *et al.*, 1995). Published estimates of the average temperature dependence for typical ocean temperatures range from –0.2 to –0.5‰/°C (Juillet-Leclerc and Labeyrie, 1987; Shemesh *et al.*, 1992) and these estimates are often used in lake-based studies although little may be known about the effect of the changing $\delta^{18}O_{water}$ which is important in lakes. This was partially addressed by analysis of diatoms cultured in the laboratory (Brandriss *et al.*, 1998), which showed a diatom-temperature coefficient of –0.2‰/°C (i.e. has a negative slope). Further questions have been raised by Schmidt *et al.* (1997), who analysed the oxygen isotope composition of diatom frustrules collected live from the oceans. They found no regular correlation between temperature and the oxygen isotope fractionation between diatom silica and water. These results led to the hypothesis that the temperature-dependent oxygen isotope fractionation preserved in biogenic opaline sediments may have been established during diagenesis rather than acquired during growth.

Certain diatom records have been shown to be sensitive to other aspects of climate, such as amount or source of precipitation. For example, abrupt shifts of up to 18‰ in $\delta^{18}O_{diatom\ silica}$ have been found in a 14 ka-long record from two alpine lakes on Mount Kenya (Barker *et al.*, 2001), which cannot be entirely temperature-related given the current knowledge of the diatom-temperature fractionation. Instead the variation have been interpreted as a moisture balance effect related to changes in Indian Ocean sea surface temperatures through the Holocene. Episodes of heavy convective precipitation are linked to enhanced alkenone-based sea-surface temperature estimates. In another study in Northern Scandinavia, oxygen isotope ratios in diatoms have been linked with changes in source of precipitation (Shemesh *et al.*, 2001). Lake-waters in the region have undergone limited evaporation so the stratigraphic $\delta^{18}O_{diatom}$ record is primarily controlled by changes in the summer isotopic composition of the lake water. The overall 3.5‰ depletion in $\delta^{18}O_{diatom}$ since the Early Holocene is interpreted as an increase in the influence of the Arctic polar continental air mass that carries depleted precipitation.

10.7.2.4 Organic Matter

Studies of organic material in lake sediments have demonstrated that the carbon isotope ratio can vary considerably (Coleman and Fry, 1991). There are many reported processes that control this variation. The most predominant are the effects associated with different sources of organic material and their different photosynthetic pathways and productivity (Ariztegui *et al.*, 1996). Different sources of organic material may have different $\delta^{13}C$ values (e.g. Meyers and Lallier-Verges, 1999), but there is considerable overlap in the $\delta^{13}C$ of various plants (Fig. 10.6), so carbon/nitrogen (C/N) ratio analysis of the same material can help to define the source of the organic matter (Silliman *et al.*, 1996). The C/N ratio of organic matter are normally used to help distinguish between algal and higher plant carbon sources. Typical values <10–12 are common for lacustrine algae, values between 10 and 20 for submergent and floating aquatic macrophytes or a mixed source, and values >20 for emergent macrophytes and terrestrial plants (Tyson, 1995). Where there is a limited source of terrestrial carbon, the $\delta^{13}C$ of organic carbon can be a reliable proxy for palaeoproductivity, which may be due to the response of aquatic plants to increased nutrients via enhanced inwash during wetter periods (Hodell and Schelske, 1998; Battarbee *et al.*, 2001a).

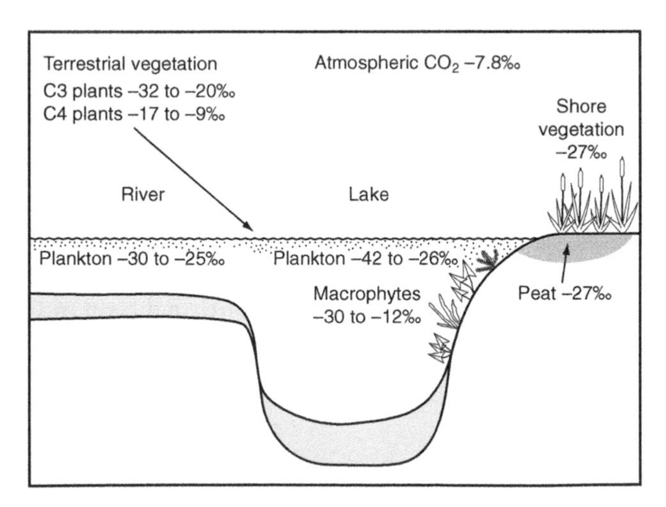

Figure 10.6 A summary of $\delta^{13}C$ composition of different types of organic materials found in lake sediments (from Coleman and Fry, 1991).

In recent years, measurement of carbon isotope ratios from individual compounds (biomarkers) within a lake sediment have allowed more detailed analysis of the composition of organic material. Both terrestrial and aquatic organic matter sources can be independently assessed and the changing proportion of input from terrestrial, aquatic and algae can be identified (Ficken *et al.*, 1998). In biomarkers from terrestrial plant remains washed into lakes, $\delta^{13}C$ values have been shown to exhibit large shifts between glacial and interglacial times associated with shifts from C4 to C3 plants (Huang *et al.*, 1999).

The oxygen isotope analyses on the fine-grained cellulose fraction of lacustrine organic matter is a relatively new tool for reconstructing past hydrological conditions (e.g. Wolfe *et al.*, 2000, 2003; Anderson *et al.*, 2001; Saucer *et al.*, 2001). Lake-water $\delta^{18}O$ histories have been directly inferred from the $\delta^{18}O$ of cellulose in sediment cores on the assumption that the isotopic fractionation between cellulose and lake-water is unaffected by changes in temperature and plant species. Assuming a lacustrine origin and constant isotopic fractionation between cellulose and lake-water, interpretation of $\delta^{18}O$ in lake-sediment cellulose is reduced to distinguishing between changing source of precipitation or precipitation/evaporation.

10.8 SUMMARY

This chapter is aimed at giving the reader a glimpse in to the use of isotopes as a palaeoenvironmental tool in the lacustrine environment. It is by no means exhaustive, but gives an insight into the most common materials used and isotope ratios analysed. The underlying message is that to understand and interpret isotope data from the various components within a lake sediment requires a knowledge of the processes that control and modify the signal, their effects need to be quantified, and a robust modern calibration is necessary to establish the relationship between the measured signal and the isotopic composition of the host waters. This is particularly important as many systems even within a geographical area are site specific. A robust calibration may not be easy; the materials may not occur in the contemporary lake for example or the lake may be in an isolated geographical region making a rigorous contemporary study impossible. Where such a calibration is not possible assumptions have to be made, but should be based on evidence from both a multi-proxy approach using isotope signals from different materials as well as using other palaeolimnological techniques.

Acknowledgements

My thanks go to the following people for their comments on earlier drafts: Carol Arrowsmith, Tim Heaton, Jonathan Holmes, Angela Lamb and an anonymous reviewer.

CHAPTER

11

INSTRUMENTAL RECORDS

Phil D. Jones and Roy Thompson

Abstract: This chapter reviews the instrumental climate record considering the homogeneity of the basic data, the hemispheric and global temperature series and some of the longest temperature, precipitation and pressure records from each of the continents. Instrumental records are not only essential to all aspects of climatology, but are the observational data against which all palaeoclimatic evidence must be calibrated. Calibration, and verification with independent data withheld, proves the usefulness of the indicator as a proxy for past periods. The strong spatial coherence of most climatic series shows that observational data does not have to be adjacent to the proxy source, but can be some distance due to the typical 400–1300 km correlation decay lengths. This coherence also applies to climate observations from different elevations, allowing remote, often high-elevation proxy information to be calibrated against nearby low-elevation sites with long records. The chapter concludes by discussing our present knowledge of the last thousand years, placing the 150-year instrumental record in a much longer context.

Keywords: Hemispheric temperature series, Global temperature series, Instrumental climate record, Proxy

Instrumental observations of the weather can be aggregated to build up detailed records of climate change as far back as the 17th century. These carefully and painstakingly produced climate series have been developed over many years. During the past two decades much progress has been made in improving the fidelity of the records by removing systematic biases and also by gathering together and analysing daily data rather than relying on monthly averages. Instrumental records now find a multitude of uses in studies of global change, in addition to their central role of documenting the temporal and seasonal changes of the last 350 years. For example, the size and extent of geographical regions, over which records of climate change can be expected to be coherent, can be assessed through the spatial coherence of instrumental time-series. Another particularly valuable application is the calibration of proxy records. Such calibrations can be either spatial (cf. Section 11.3.3 and Birks, pp. 107–123 in this volume), when the method derives its power from the present-day geographical variation of climate, or temporal (cf. Section 11.3.3), when the calibration methodology utilizes the strong interannual signal. Instrumental records also provide estimates of climatic variability and of extreme values: two crucial factors needed for characterizing the natural climate system and its environmental impact. Furthermore, local instrumental time-series, when spatially averaged, allow the large-scale climate to be described and reliable global averages to be constructed.

Out of the myriad of available climate parameters, global surface air temperature plays a key role. Over the last 150 years, instrumental records demonstrate that it has risen by about 0.6 °C. This rate of global temperature increase has been highly unusual and has led to recent decades experiencing warmer average temperatures than at any other time during the last millennium. Recent warming since 1950 has been greater at night, leading to a decrease in the diurnal temperature range. Climatic parameters are also naturally interrelated so a number of climatic elements have co-varied with the global temperature increase. For example, snow-cover extent as monitored over the last 30 years has fallen, while evaporation rates have risen. Most alpine glaciers in the world have dramatically receded since the mid-19th century. Instrumental records have also revealed additional widespread changes to other climatic parameters. For example, northern hemisphere land areas between 55 and 80°N have been shown to have experienced 20th century increases of around 80 mm in annual precipitation totals.

Europe is particularly well endowed with instrumental series. The climatic parameter that provides the longest record is mean air temperature. The earliest continuous monthly record starts in 1659 with the renowned Manley (1974) series for central England. Precipitation series are available from 1697 (Kew), while the longest continuous surface-pressure sequence commences in 1722 at Uppsala in Sweden. The earliest North American climate series begin by the mid-18th century. By the 1780s, recordings of the three main climatic parameters (temperature, precipitation and pressure) were available from many locations in Europe as well as from North America and India. Many more instrumental series had commenced by the 1830s and 1840s with excellent records starting up in Australia, South America, Japan, China and Southeast Asia. By the 1850s, instrumental records had become more and more commonplace, and their geographical coverage increased so much over both land and sea, that global mean air and/or sea-surface temperatures can be reliably estimated from this time onwards. Today the vast observational network of surface stations, ships and buoys and upper-air stations launching radiosondes, largely established for weather forecasting, coupled with satellite soundings and imagery has led to millions of weather observations being made each day. Data are typically recorded by thousands of synoptic and climatological stations on land; hundreds of aircraft, ships and moored buoys; hundreds of radiosondes and from many satellites since the mid-1970s. Re-analysis studies, in which daily observations since 1948 have been assimilated into a comprehensive global representation of the daily state of the atmosphere, provide a synthesis of this great wealth of instrumental weather data (see also Section 11.1.3).

11.1 INSTRUMENTAL CLIMATE DATA

11.1.1 Homogeneity

The vast majority of instrumental data collected to describe climate are primarily information about the weather. In this respect climate data are often referred to as second-hand weather data. Although day-to-day changes in the weather are large, the exact details of the way observations are taken (housing of instruments, observation times and the immediate local environment) are much more important for their climatic usage. Determining the average climate of a location and possible long-term changes requires that the data be collected in a consistent manner. For climatic purposes, considerable care must be taken to ensure the homogeneity of climatic time-series (Peterson *et al.*, 1998a).

A climatic time-series is said to be homogeneous if the variations are due solely to the weather (Conrad and Pollak, 1962). A myriad of potential problems must be considered when assessing and correcting series for potential inhomogeneities as climatic series collected over centuries necessarily involve changes in instrumentation, sites and observers. Bradley and Jones (1985) discuss all the issues. These range from relocation of the sites, improvements to the exposure of the instruments (screen design), changes in observation times and the methods used to calculate daily and monthly means and changes to the environment around the station. For example, urban growth (particularly in the local neighbourhood of 4–10 km^2) may lead to an artificial warming trend while the growth of vegetation, such as trees in the vicinity of the gauge may impair the efficiency of the precipitation catch.

Numerous methods have been developed to assess homogeneity (see the review by Peterson *et al.*, 1998a) ranging between reasonably objective to totally subjective techniques. One often-used approach is to plot the differences between overlapping time-series for neighbouring sites, to identify significant steps in the difference sequences. No one technique has been shown to be superior to all others, superior in the sense that it locates and corrects all inhomogeneities without making adjustments to series where they are not needed.

11.1.2 Climate Parameters/Indices

Standard meteorological parameters include temperature, precipitation, pressure (Section 11.1.3), humidity, sunshine and cloudiness. The importance of each depends, to a large extent, on the region of interest and the type of study, but in most cases it is impossible to consider one particular parameter in isolation from the others. Precipitation and temperature averages, for example, will show variability from year to year and decade to decade, variability that can often be well explained by changes in the regional circulation.

Perhaps the most important climatic index is the global temperature average (GLT) (Jones *et al.*, 1999b, 2001a). GLT and the two hemispheric components (NHT and SHT) are the yardstick against which we judge the present and assess past climate. Past glacial periods were colder than today and recent interglacials experienced similar temperature levels to today's values. GLT, however, while important, is not of much practical value at the local and regional scale. Temperature series from a region do not follow the course of GLT over the 20th century, nor should they be expected to. Similarly, GLT cannot be reconstructed from a single observational site or proxy. Series may be intercompared and differences discussed, particularly where neighbouring series are involved, but the potential spatial coherence of climate parameters needs to be remembered (see also Section 11.1.5). Land and sea-surface temperature data sets (e.g. Jones *et al.*, 1997a) show that there are about 60–80 spatial degrees of freedom over the globe on the seasonal time-scale. This is a considerably smaller number than the over 1000 sites used to monitor temperatures over land and the thousands of sea-surface temperature measurements made each month.

Observational data only cover about 80 per cent of the Earth's surface, with the major areas lacking data being the mid-to-high latitudes of the southern hemisphere and large parts of Antarctica and the central Arctic. Despite these missing areas, the network is considered reasonable (Jones, 1995). Spatial availability from proxy climate reconstructions is, however, considerably poorer (Mann *et al.*, 1998; Jones *et al.*, 2001b).

Precipitation is the most important variable from a societal and an economic perspective. Precipitation is particularly vital for agriculture and water resources. Direct measurements are only taken over land areas, but unlike temperature, the network is less than adequate because the spatial variability of precipitation is considerably greater than that of temperature. To achieve the same level of accuracy in a regional average, there should in general be between five and 10 times as many precipitation measurements made than for temperature. Precipitation variability is much more important in tropical and sub-tropical regions, with seasons defined by the presence or absence of rain. With precipitation measurements being inadequate in much of the world, large-scale assessments of change can, in some cases, be more reliably estimated from riverflow (discharge) measurements. With a relatively simple catchment model, it is possible to reconstruct long-term changes in precipitation or discharge from the other variable, at the monthly time-scale (see, e.g. Jones and Lister, 1998).

The final principal group of climate parameters are measures of humidity, sunshine, cloudiness and windiness. They are important, as together with temperature, their measurement is needed to produce estimates of monthly potential evapotranspiration, necessary to complete the hydrological cycle. Measurement standards for these variables differ between countries more than for temperature and precipitation. Homogeneity assessments are more difficult because of a sparser and shorter observational record. Despite the many problems, New *et al.* (1999, 2000) have developed gridded data sets (0.5 ° × 0.5 °) on a monthly basis for the following variables: mean temperature, diurnal temperature range, precipitation, vapour pressure, cloudiness, rain-day and groundfrost-day counts. They developed fields for each variable in an absolute form for the 1961–1990 period, before developing the time-series as anomalies from this widely used base period. These data sets are being extensively used in many aspects of climatology, including vegetation modelling and climate model validation.

11.1.3 Circulation

11.1.3.1 Surface Air Pressure

The climate parameters of the previous section are all intimately associated with the atmospheric circulation. A key measurement used to deduce circulation is that of atmospheric pressure, first studied by Evangelista Torricelli in 1643 following his invention of the barometer. Continuous series of monthly averages of station air pressure in Europe are extant from the mid-1750s, while daily sequences of similar lengths have recently been digitized and homogenized for a few sites by Camuffo *et al.* (2000b). These long station records have been used to reconstruct monthly mean gridded pressure values (Jones *et al.*, 1987, 1999a). Grids are available for Europe (back to 1780), North America (to 1858) and the whole globe since 1871 (Basnett and Parker, 1997).

11.1.3.2 Circulation Indices

The circulation of the atmosphere displays a rich, ever-changing pattern of spatial and temporal variations. A convenient method of summarizing the patterns of change is through circulation indices. A wide variety of indices has been produced ranging from simple time-series of pressure differences, through daily series of subjectively classified local weather types (e.g. Lamb, 1972b) to statistical measures using principal components analysis (Yarnal, 1993). Sir Gilbert Walker, while studying tropical climate fluctuations after the disastrous failure in 1899 of the monsoon rains in India, noticed that when pressure was high in the Pacific Ocean it tended to be low in the Indian Ocean, and vice versa. He coined the term Southern Oscillation to describe this east–west seesaw in pressure between the Pacific and Indian Oceans (Walker and Bliss, 1932). He also named two

Northern Oscillations – one in the Atlantic and another in the Pacific. Quantitative measures of all three of his large-scale ocean–atmosphere phenomena continue to be widely used today. Another early index was Rossby's (1939) 'zonal index' of the strength of westerly flow. It was defined as the mean pressure difference between 35 °N and 55 °N.

11.1.3.3 Re-analysis Assimilations

Retrospective analysis of weather observations using state-of-the-art four-dimensional data assimilation techniques combined with powerful, modern weather-forecasting models is producing very valuable climate databases. Ambitious re-analysis activities are underway at a number of organizations (Trenberth, 1995). Continuing improvements in re-analysis assimilation techniques, in combination with downscaling studies, now provide the potential for detailed (i.e. daily) 40-year records of climate change to be constructed for localities from all around the world (Kalnay *et al.*, 1996). Observational changes such as the introduction of satellite data inevitably disrupt re-analysis products. Other problems are large spatial data gaps in radiosonde records from the mid-to-high latitudes of the southern hemisphere. Retrospective re-analysis, however, has great potential for providing very valuable data sets to researchers in the climate community. A particularly appealing aspect of re-analysis is that the four-dimensional assimilation system can transport information from data-rich to data-poor regions (Kistler *et al.*, 2001), providing internally consistent fields of the important parameters.

11.1.4 Typical Extent and Nature of Long Surface Climate Records

The three panels of Fig. 11.1 show time-series of some of the longest annual series of temperature (Fig. 11.1a), precipitation (Fig. 11.1b) and pressure (Fig. 11.1c) variations, each with typical series from each continent. Apart from the annual precipitation series for Bombay, all the series are shown at approximately the same relative scale to enable comparisons of both the differences of interannual variability and in the timing and magnitude of trends.

For temperature (Fig. 11.1a), the long Central England temperature record is shown. Series from other continents begin later. Orcadas (located on the South Orkney Islands) has the longest record near the Antarctic continent. Continuous instrumental recording on the continent only began in the Peninsula region in the late-1940s and elsewhere at a few stations from the mid-1950s. Except for Bahia Blanca, long-term warming is evident. Interannual variability is lower in the tropics.

The precipitation series (Fig 11.1b) mostly show larger interannual variability than temperature as measured by the coefficient of variation. Variability is greater in the tropics and subtropics than in more poleward latitudes. All series also show marked decadal to bi-decadal-scale variability, which will have important consequences locally and some series show longer time-scale trends. Dakar in Senegal is typical of much of the Sahelian zone of Africa with lower totals after the late-1960s (Hulme, 1996). Perth, Australia shows the decline in precipitation totals evident over parts of Southwestern Australia (Allan and Haylock, 1993). The Bombay record illustrates that Indian monsoon failures have been less prevalent since the 1950s than in the period up to 1920.

For pressure (Fig. 11.1c), the differences in interannual variability between middle and high latitudes and the tropics are more marked than for temperature. As the mass of the atmosphere is fixed, any changes should be counterbalanced by opposite changes elsewhere. The series for

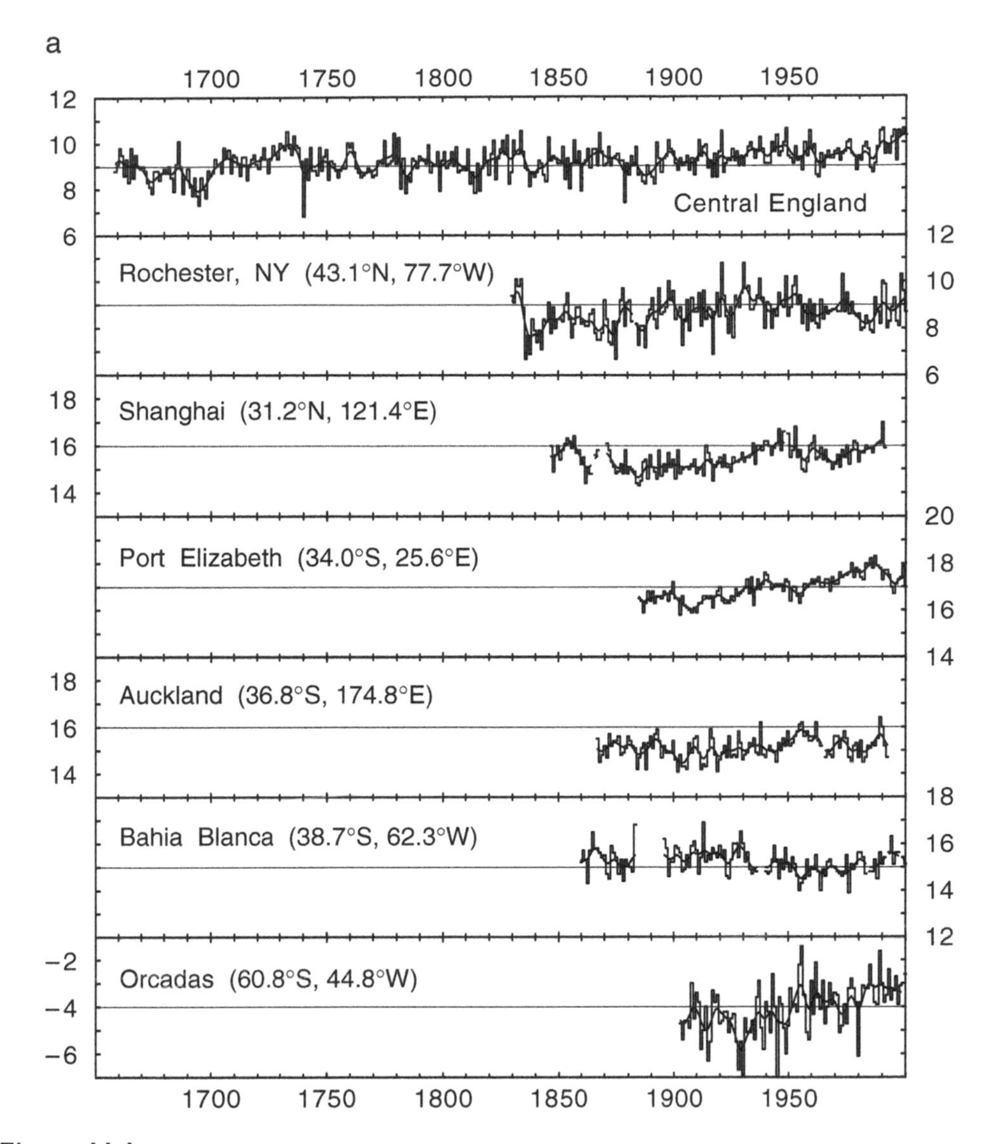

Figure 11.1a

Tahiti and Darwin are near the centres-of-action of the Southern Oscillation (these two series are used as the most common index of the phenomenon) so tend to co-vary in an opposite sense (Trenberth, 1995). Madras tends to co-vary with Darwin, despite the large distance between the two sites. Uppsala and Gibraltar also tend to vary in an inverse sense, and this is more marked still for the winter season. This inverse relationship is linked to the North Atlantic Oscillation (generally defined as the difference in pressure between stations on the Azores and Iceland).

11.1.5 Spatial Coherence of Climate Parameters

Wide variations are found in the spatial coherence and geographical patterns of climatic anomalies according to the parameter of interest and the time-scale. A simple but effective means of visualizing

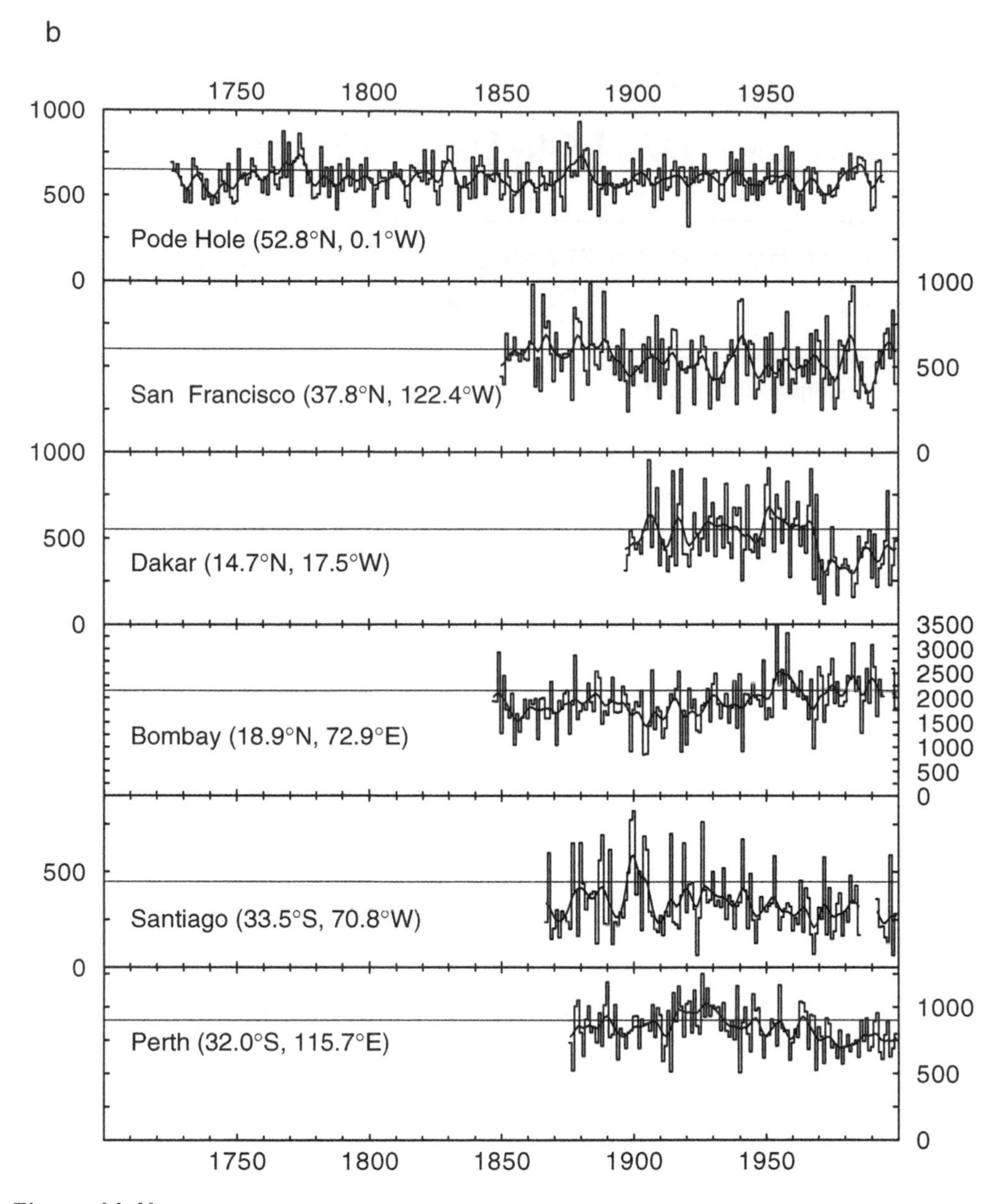

Figure 11.1b

the spatial coherence is through the use of scatterplots (e.g. New *et al.*, 1999, 2000), where correlation coefficients between pairs of randomly selected stations are plotted against distance (Fig. 11.2). A convenient scale length, which can be used to summarize the overall spatial coherence, is the distance at which the correlation coefficient drops to about 0.7 (r^2~0.5). For daily air pressure in mid-latitudes the scale length is some 1300 km, reflecting the general size of pressure systems. Temperature anomalies, although also spatially coherent, are generally not quite as extensive as those of pressure and have scale lengths of around 900 km (in Fig. 11.2, r~0.5 (r^2~0.25) at this

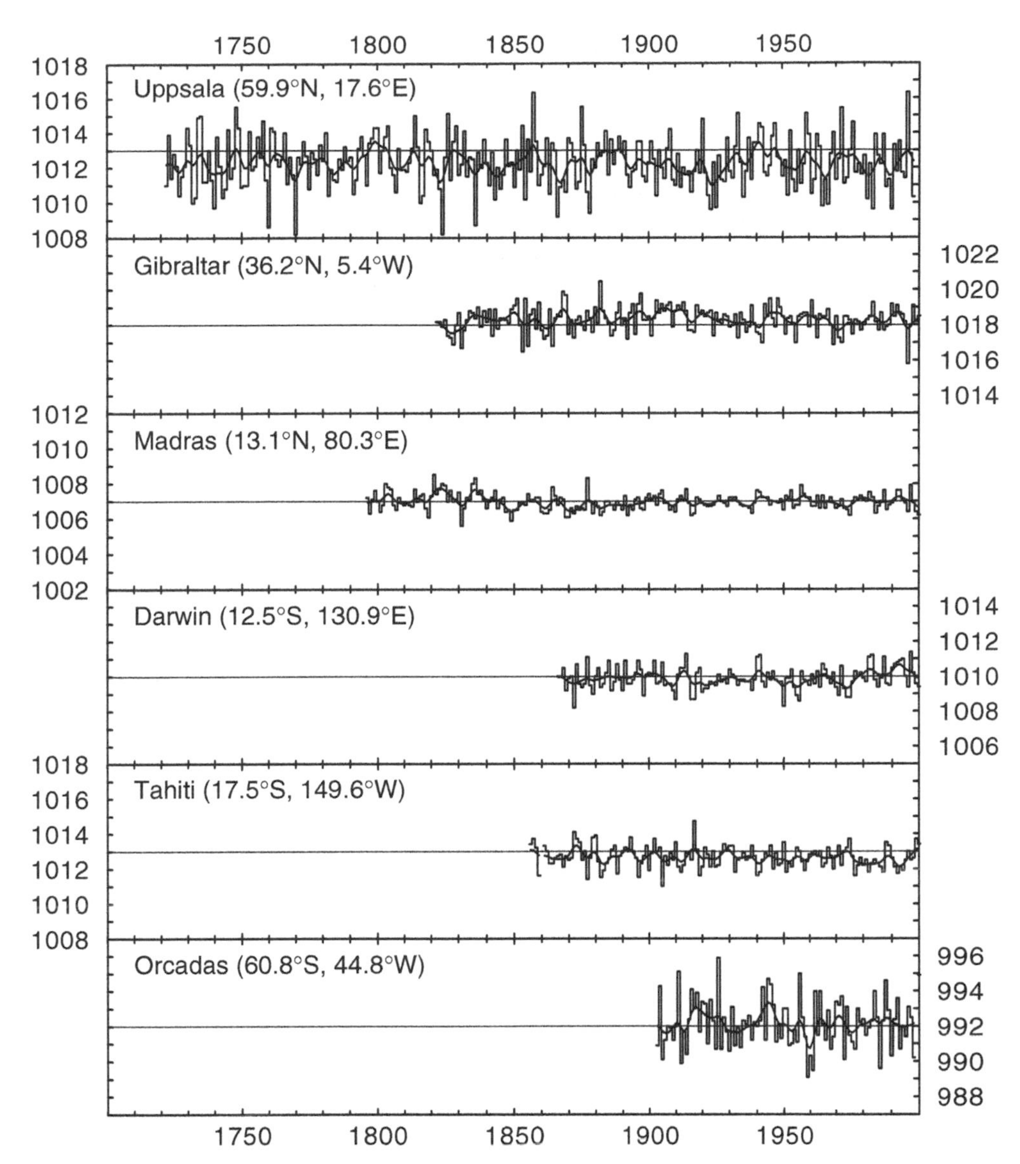

Figure 11.1 Annual time-series of selected stations for (a) average temperature (°C), (b) precipitation total (mm) and (c) average pressure (hPa). The raw calendar year values are shown together with a Gaussian-smoothed series to highlight variations on decadal time-scales.

distance). The spatial coherence of air temperature between upland stations, or between upland and lowland stations, is found to be very similar to that between lowland stations (Fig. 11.2; Agustí-Panareda *et al.*, 2000). This finding means that proxy records of climate change preserved by mountain lake sediments and upper timberline trees can be taken to be representative of the climate change of neighbouring lowlands. Rather shorter-scale lengths of around 400 km are typical for

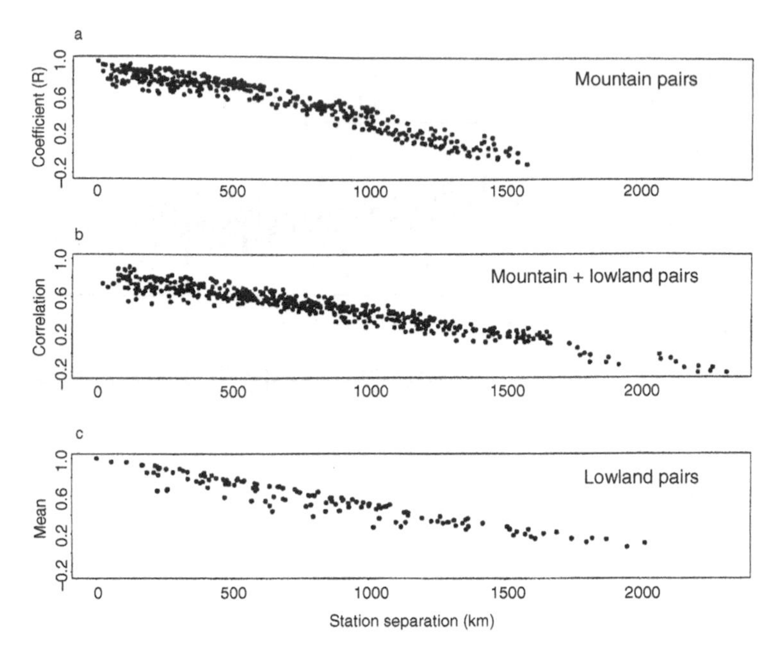

Figure 11.2 Scatterplots of correlation coefficient (r) for daily mean air temperature between pairs of European stations vs. distance. (a) Mountain stations, (b) mountain and lowland stations and (c) lowland stations. The decay of correlation coefficient with distance is very similar for all three datasets.

precipitation, radiation and wind-speed, and even shorter, of around 300 km, for relative humidity (Agustí-Panarada and Thompson, unpublished results). A latitudinal dependence of the spatial coherence is well known (e.g. Hansen and Lebedeff, 1987). So, for example, scale lengths for both daily pressure and temperature anomalies drop from high to low latitudes, as the pressure systems become smaller and less well defined. The reverse is partly true on seasonal-to-interannual time-scales due to the El Niño Southern Oscillation phenomenon. Knowledge of correlation decay lengths is important in determining the effective number of spatial degrees of freedom, a parameter essential for assessing errors due to uneven spatial sampling in large-scale averages (Jones *et al.*, 1997a).

11.2 ANALYTICAL METHODS

11.2.1 Spatial Averaging

Numerous studies have shown that climate data analyses are much more straightforward if the data are regularly spaced. Surface pressure and all operational analyses, and re-analyses, are routinely interpolated to latitude/longitude grids. Before about 1960, gridded pressure fields were derived by reading values off manually-drawn contour maps and then from the mid-1960s to the 1980s by using Cressman (1959) interpolation techniques. Operational analyses involve observations being entered into data assimilation schemes with a weather-forecasting model. Incremental improvements to the models mean that jumps in the analysed fields occur, hence the need for re-analyses to produce a consistent set of fields for the last 40 years with a constant version of the model.

11.2.2 Time-series

Time-series data frequently occur in environmental studies and in the words of the poet John Masefield (1926), can be succinctly defined as 'One damn thing after another'. We can regard a time-series as being made up of three types of variation, namely:

1. a trend, T, which describes the long-term behaviour of the series
2. a seasonal (cyclic) component, S, which describes the variation in a cycle of fixed duration (e.g. 1 year)
3. an error term, ε, describing essentially the random fluctuations of the series about the trend and the seasonal components.

Thus if $\mathbf{X_t}$ denotes the value of the series at the time point **t**, we may write

$$X_t = T_t + S_t + \varepsilon_t \qquad 1$$

The simplest method of extracting a linear trend in a data series is to regress the time-series data against time. Irregular fluctuations can be removed from the time-series by smoothing. Many of the time-series illustrated in this chapter have the raw values plotted together with a data-adaptive Gaussian filter, highlighting variations on decadal- and longer time-scales. Cyclic changes need very careful identification. They can be studied in either the time or the frequency domain. A correlation coefficient approach can be used very effectively in the time domain to produce correlations between successive observations, referred to as autocorrelations. True cyclic fluctuations stand out in a correlogram as distinctive oscillations. In contrast, random time-series (white noise) yield autocorrelations close to zero. In the frequency domain the basic tool of spectral analysis is the periodogram. An important theoretical result, the spectral representation theorem states that any stationary time-series can be represented as a sum of sinusoidal cycles. A particularly convenient sequence of sinusoidal frequencies to use is one made up of harmonics, as originally described by Fourier. The relative contribution that each cycle makes to the series is referred to as the spectrum of the time-series (Fig. 11.3). The only true periodicities in climatic series are those of the Earth's orbital parameters (the Croll-Milankovitch periods of 10, 4.1 and 2.2 × 10^4 years and the annual and daily cycles plus their harmonics). Thus on Holocene time-scales there is a marked absence of periodicities and spectral power in the millennial, centennial and decadal bands as previously noted by Mitchell (1976) (see Bradley, pp. 10–19 in this volume).

Complex demodulation is a frequency-domain approach (Bloomfield, 1976; Thompson, 1995) which allows the trend and slow changes in amplitude and phase of the cyclic component(s) to be extracted simultaneously from a time-series. It also allows a climatic time-series to be divided into the three components of equation 1 of trend, seasonal cycle and irregular fluctuations. An example of complex demodulation is shown in Fig. 11.3c where the seasonal components (S_t), the annual and 6-month cycles, with their slowly changing amplitude and phase, are superimposed on the trend.

11.2.3 Extremes

Most of this chapter considers climate as being composed of monthly averages or monthly totals. For climate reconstruction during the Holocene and even for the last thousand years, monthly data are adequate. The shortest resolution used with some climate proxies (e.g. trees, corals and

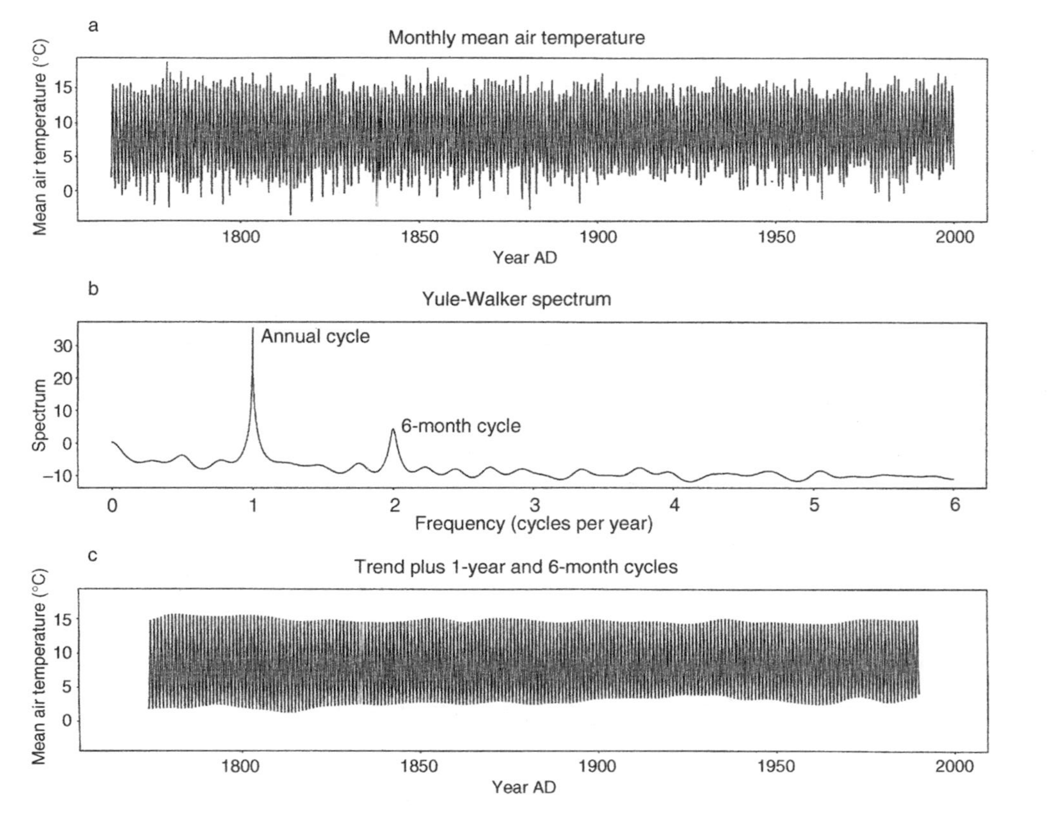

Figure 11.3 (a) Monthly air temperature for Edinburgh since January 1764. (b) Spectral analysis of the air temperature series. In this smoothed version of the frequency spectrum, as calculated using an autoregressive method, the power (in decibels) is plotted against frequency (in cycles per year) between 0 and the Nyquist frequency. The spectrum is dominated by the annual cycle. A 6-month cycle, caused by European summers being longer than European winters, also stands out. The annual and 6-month cycles are seen to be superimposed upon low background power, which declines from long to short periods. (c) Decomposition of the Edinburgh air temperature series. The irregular variations have been removed to leave the long-term trend plus the six-month and annual cycles. The mean air temperature can be seen to have gradually risen over the last 236 years and the annual cycle to have been weaker during the 1920s and 1930s. These changes are mainly a reflection of Edinburgh winters having become less cold.

historical documentary data) is seasonal. Whilst this time-scale can still involve extremes (e.g. warmest/coldest summers) the general public perception of extremes is warm/cold/wet days and dry spells. To fully assess changes in the frequency of extremes requires daily time-series: information that will only be available from instrumental records. For Europe, several daily time-series that span over 200 years have recently been developed (e.g. Jones *et al.*, 2002). Widespread daily data sets back to the 1950s have been digitized in many countries of the world, but many national meteorological agencies charge for data at this resolution. An international initiative (Global Climate Observing System (GCOS)) aims to put together a network of about 1000 stations around the world, with daily series of maximum and minimum temperature and precipitation measurements freely available.

On hemispheric scales analyses indicate that since the early 1950s most regions show greater warming at night than during the daytime (Easterling *et al.*, 1997), leading to a reduction in the diurnal temperature ranges. Analyses of trends in temperature and precipitation extremes show few consistent trends (e.g. Frich *et al.*, 2002), the most consistent being a trend towards more intense daily precipitation totals and fewer cold nights. Many recent studies (e.g. for Canada by Bonsal *et al.*, 2001) conclude that recent warming trends have resulted more from there being fewer cold days than from there being more warm days. As most analyses are restricted to the last 50 years, the longer European records are particularly important as they allow recent trends to be considered in the context of the last two centuries.

11.3 APPLICATIONS OF INSTRUMENTAL RECORDS IN HOLOCENE RESEARCH

11.3.1 Climate Normals

The New *et al.* (1999, 2000) data sets, discussed above, are based on a reference period of 1961–1990. Averages for 30-year periods are often referred to in climatology as climatic normals. The 1961–1990 period is both the most recent complete 30-year period and the one for which we have the most climate data, both from a spatial perspective and from a simple count of the number of sites with data. Jones *et al.* (1999b) show maps of average surface temperatures for the 1961–1990 period, for the four mid-season months of January, April, July and October. They show that the average temperature of the world during this period was 14.0 °C, with the northern hemisphere being warmer than the southern hemisphere (14.6 °C cf. 13.4 °C).

11.3.2 Hemispheric and Global Temperature Trends Since 1850

Figures 11.4 and 11.5 show seasonal and annual temperature series for the northern and southern hemispheres, based on land and marine values (see Jones *et al.*, 1999b, 2001a for more details of their construction). Interannual variability is greatest over the northern hemisphere, particularly in the winter season, when a stronger dynamical circulation is maintained by strong temperature gradients, and weakest in summer. In contrast, the southern hemisphere shows similar levels of variability in all four seasons. Both hemispheres show long-term warming of about 0.6 °C since 1851. As little change occurred in the second half of the 19th century, warming between 1901 and 2000 is also about 0.6 °C. The southern hemisphere shows similar trends between the seasons, but in the northern hemisphere long-term summer warming is less than in the other seasons. This aspect may be a real feature, but it equally well may be due to reduced coverage in the mid-19th

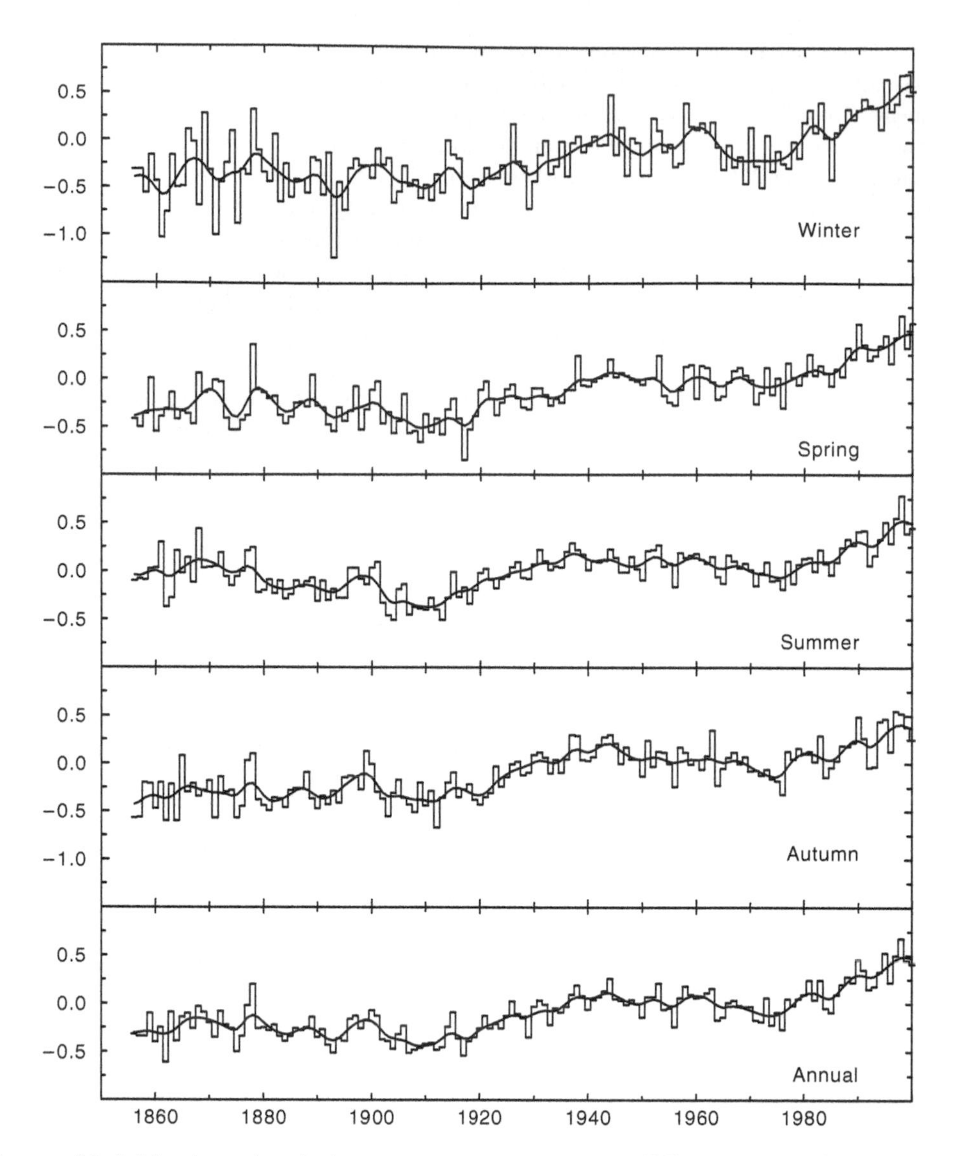

Figure 11.4 Northern hemisphere average temperatures (°C) by season for the period 1856–2001, relative to 1961–1990. The smooth line highlights variations on the decadal time-scale. Standard northern hemisphere seasons are used: Winter (December–January–February), Spring (March–April–May), Summer (June–July–August), Autumn (September–October–November).

century. At that time the available data came from more mid-to-high latitude regions of the northern hemisphere. Instrumental exposure problems (before the introduction of Stevenson screens during the 1870s) might have caused direct insolation to influence thermometers on north wall locations (Parker, 1994).

In neither hemisphere does the 0.6° warming occur in a linear fashion. In both, the warming occurs in two distinct periods, 1915–1940 and since 1975. For the northern hemisphere, the warming in the

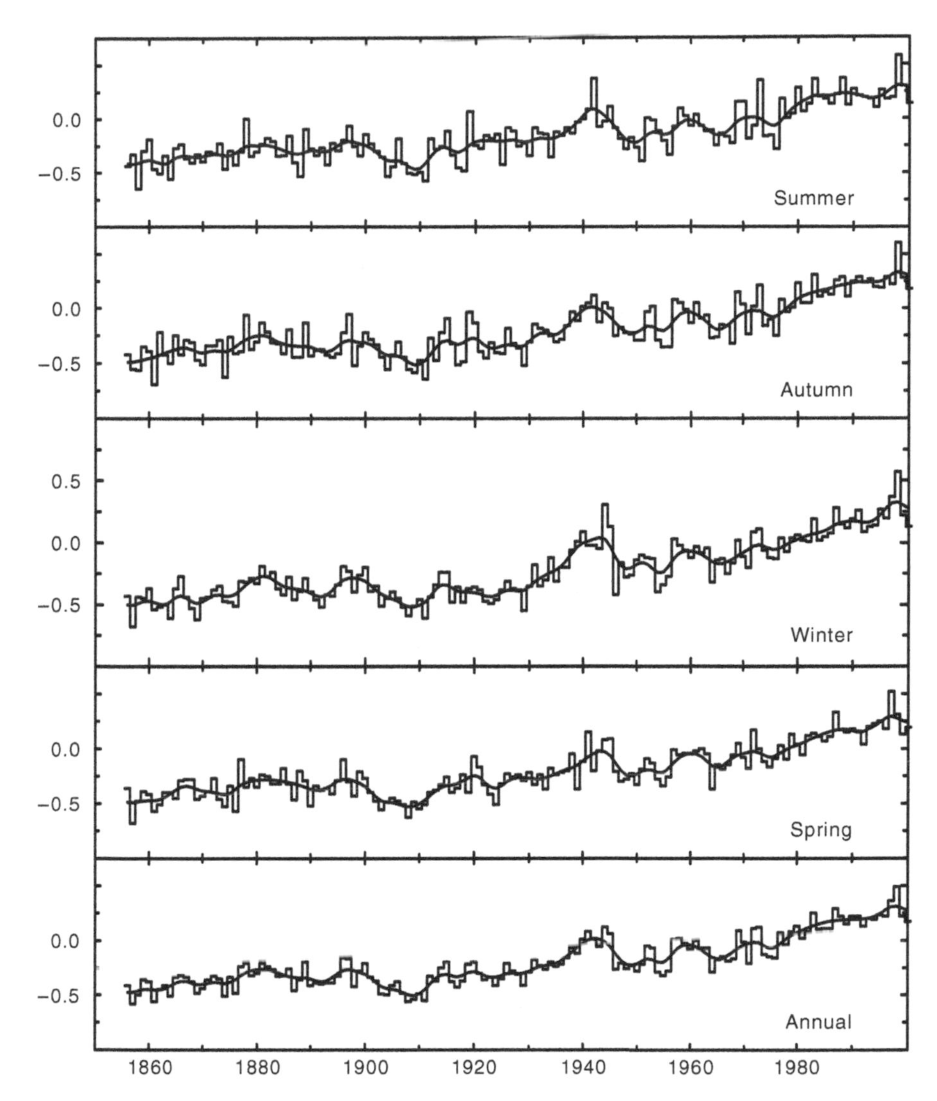

Figure 11.5 Southern hemisphere average temperatures (°C) by season for the period 1856–2001, relative to 1961–1990. The smooth line highlights variations on the decadal time-scale. Standard southern hemisphere seasons are used: Summer (December–January–February), Autumn (March–April–May), Winter (June–July–August), Spring (September–October–November).

two epochs is similar with a slight cooling between them. For the southern hemisphere, the recent warming is slightly greater and there is no indication of a cooling between 1940 and 1975.

Wigley *et al.* (1997) discuss some of the possible causes of the low- and high-frequency variability in the hemispheric and global temperature series. Solar output changes may explain some of the temperature rise in the early 20th century. Solar output, however, has not changed over the last 30 years so cannot explain the recent increase in temperature. These results are confirmed by more recent analyses reviewed by Mitchell and Karoly (2001).

Both hemispheres exhibit considerable high-frequency variability, much of which is common to the two hemispheres. The strong interhemispheric correlation on interannual time-scales suggests that both respond to common forcing. On these time-scales, the common forcing relates principally to the response to the Southern Oscillation Index (SOI) which is a measure of whether the atmosphere/ocean system is in an El Niño, La Niña or neutral state. Distinctive patterns of temperature and precipitation response occur during El Niño and La Niña phases in the tropical Pacific Ocean. El Niño events lead to more of the tropics and some mid-latitude regions being warmer than average so El Niño events tend to be warm with the approximately opposite situation occurring during La Niña events. About 25 per cent of the high-frequency variance of hemispheric temperature can be explained by the SOI, which is simply the normalized pressure differences between Darwin and Tahiti (annual series shown in Fig. 11.1c).

The other factor influencing global temperatures on high-frequency time-scales is explosive volcanic eruptions where ejecta can form a veil over the Earth at an elevation of between 25 and 30 km. For the 2–3 years after a major tropical/sub-tropical eruption, the dust veil in the upper atmosphere reflects and absorbs sunlight and so the troposphere and surface are cooled while the stratosphere is warmed. The cooling below the dust veil is particularly evident in the northern hemisphere summer, because the veil has its greatest influence when insolation is highest and the greater northern hemisphere land area has a lower heat capacity compared to the ocean dominated southern hemisphere. The most recent eruption of Pinatubo in June 1991 caused significant surface cooling during 1992 and 1993, particularly in the northern hemisphere.

11.3.3 Basis for Transfer Functions/Calibration of Proxy Data

The recent transformation of Holocene palaeoclimatology from a rather qualitative, descriptive subject into a quantitative science has been made possible by the development of new multivariate statistical techniques. These statistical tools have allowed precise calibration of proxy data by providing the means for matching them to modern instrumental climate records. Calibration is needed both to extract the palaeoclimatic signal, especially when the signal is weak or embedded in noise, and to ascertain to which component of the climate system the palaeo-signal corresponds.

A good example of the use of time-series of instrumental data is in deciphering the climatic signal preserved by clastic varves (Wohlfarth *et al.*, 1998). Multiple regression analysis of the 90-year sequence, from 1860 to 1950, from the Ångermanälven Estuary in north-central Sweden revealed strong correlations between spring/summer precipitation and varve thickness. Monthly temperatures and varve thickness, however, showed no, or only weak, correlations. The clearest climatic relationships were found between the precipitation in the mountains (the source region for the river discharge) and the estuarine varves. Instrumental time-series have similarly proved to be invaluable for unravelling the climatic information preserved by tree-rings. Temporal regression methods provide the preferred techniques for reconstructing climate from tree-rings. Samples from the limits of a tree's distribution give the clearest climatic signals, as growth tends to be limited by one factor only. Best-subset regression methods point to summer temperatures as limiting the growth of latewood density series at high latitudes or at the elevational timberline, but to precipitation as the main control at the lower or semiarid/temperate timberline. As with any regression-based calibration method it is important to assess the relationship using some data withheld when developing the regression.

Another approach to using modern instrumental records as a basis for calibrating proxy data is through the use of spatial transfer functions. Here spatial statistics are used to relate the present-day distribution of biota to the current climate. Imbrie and Kipp (1971) first developed this powerful technique as a palaeoceanographic tool, and Birks (pp. 107–123 in this volume) reviews recent developments and numerical improvements to the method.

Many environmental factors affect species abundance, and all regression/multiple factor-based methods of environmental reconstruction have to face the general problem of co-linearity. Jones and Briffa (1995) have carried out a detailed study of growing-season temperatures over the former Soviet Union (fSU). Their analyses of daily mean temperatures at 223 stations, 23 reaching back to 1881, can be used to illustrate typical spatial and temporal co-linearities in climate data and hence to cast light on important differences between the temporal validation and the spatial calibration approaches to palaeoclimatic reconstruction. Their stations are typically around 300 km apart and time-series from neighbouring stations correlate well with each other. Very consistent spatial patterns of the start, end, duration and number of degree-days above 5 °C are found right across the fSU (i.e. their regional isopleths are nearly parallel). Duration and degree-days, for example, have a spatial correlation of +0.94. By contrast, some correlations in the temporal domain can be very weak. For example, while correlations between degree-days and May through September temperatures (MJJAS) are typically high (> +0.9) those between duration and degree-days are typically only +0.3 or less. There are thus surprisingly marked differences between the behaviour of climate variables in the spatial and temporal domains. Palaeo-reconstructions should thus preferentially use the time-series approach, as it has a greater potential to discriminate between climatic parameters. Whichever method is used, it must be remembered that the approach is statistical and it is vital to assess the strengths of relationships using a verification period with data withheld from the calibration of the transfer functions.

11.3.4 Retrodiction of Derived Climatic Elements (Local-Scale)

Environmentally relevant parameters such as evaporation (e.g. Barber *et al.*, 1999b), growing season duration (e.g. Agustí-Panareda *et al.*, 2000), water temperature or ice-cover (e.g. Agustí-Panareda *et al.*, 2000), which may only have been monitored for a small number of years, can be hindcast over much longer periods using long instrumental climate records. The recent period is used to establish a relationship between the environmental variable and the climate at nearby stations. Figure 11.6 shows an example of retrodiction of water temperature in a Scottish highland stream. The retrodiction uses the multiple regression relationship of equation 2, which is based on water temperature monitoring since 1988.

$$\mathbf{W}_i = 0.83\,\mathbf{A}_i + 0.06\,\mathbf{A}_{i-1} - 0.04\,\mathbf{P}_i^{0.5} \qquad 2$$

Where **W** is water temperature anomaly (°C) about the annual cycle (as based on a two-term Fourier series):
i is the month
A is air temperature (°C)
P is precipitation (mm/month).

Here the stream water temperature (**W**) is modelled as a linear combination of monthly mean air temperature (**A**) and precipitation (**P**) as measured at a lowland observatory some 100 km away.

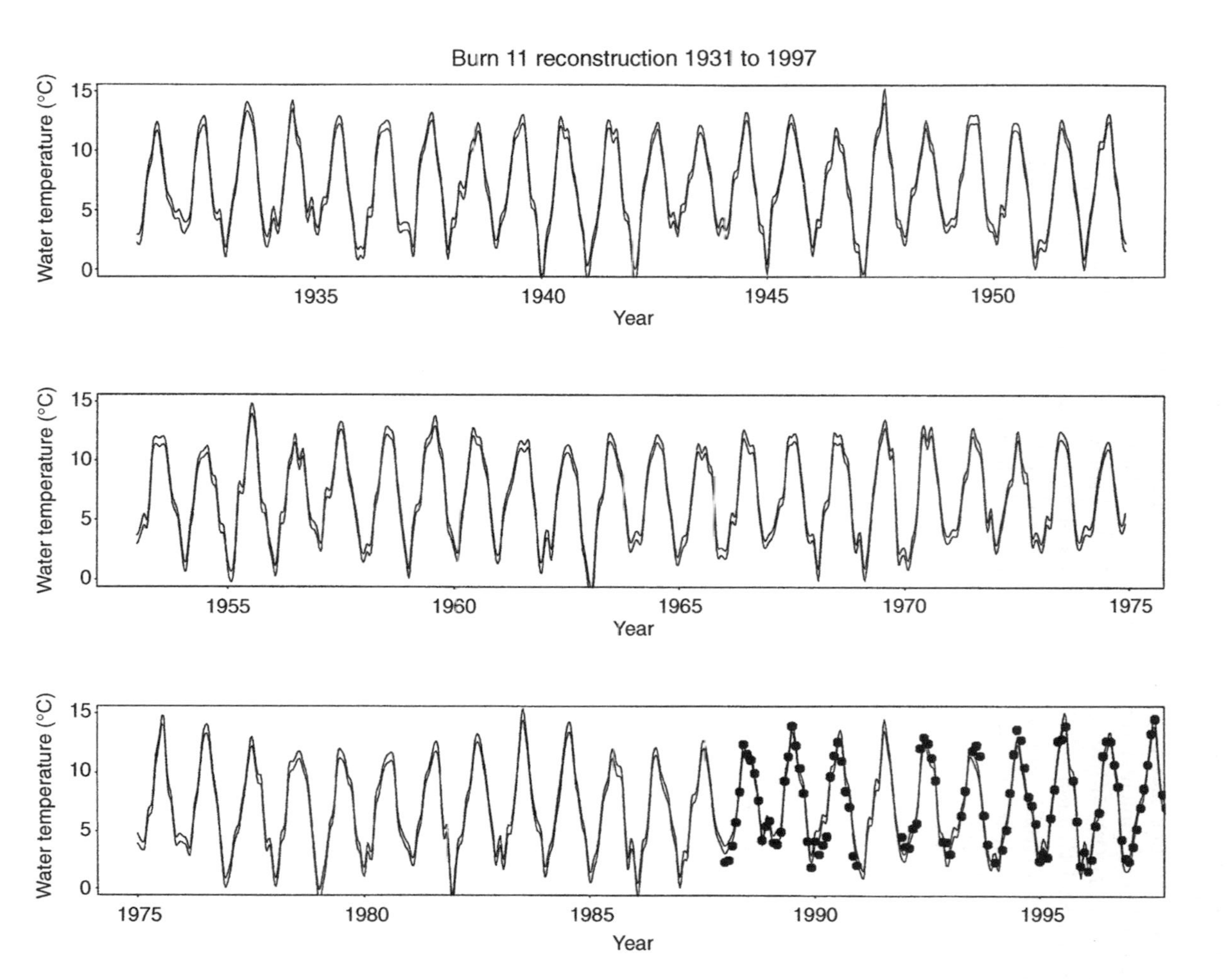

Figure 11.6 Monthly water temperatures for a Scottish stream (Burn 11). Observations since 1988 plotted as black circles (data provided by R. Harriman). The two lines plot confidence limits about the water temperature retrodiction obtained using equation 2 for the period 1931–1997.

While it is important not to over-interpret the coefficients of regression models, equation 2 can be seen to make good physical sense. Consider the various terms:

- air temperatures are positively correlated with stream temperatures as is to be expected
- an antecedent term is found to add skill to the reconstruction, presumably a memory of the air temperature of the previous month is transferred to the stream water through the ground temperature
- precipitation has a negative coefficient, as high discharge leads to lower water temperatures
- the square root transformation is helpful as it makes the monthly precipitation data more normally distributed (Thompson, 1999).

The regression model of equation 2 has been used with lowland weather records to generate the retrodiction of Fig. 11.6.

11.3.5 One Millennium of Air Temperature Variations

In climate change studies attempts are made to explain the course of temperature changes since 1850 in terms of natural forcing factors (solar and volcanic) and anthropogenic influences (greenhouse gases and sulphate aerosols). Although the majority of scientists believe that the 20th century temperature rise can only be explained by involving anthropogenic factors (IPCC, 2001), acceptance is not universal. Periods of past warming are often cited as evidence that the climate has changed naturally by amounts as large as during the 20th century.

Changes to boundary conditions (Croll-Milankovitch forcing and the presence of ice sheets) mean that only periods in the Late Holocene are relevant to the instrumental period and to the 21st century. The period since about AD 1000 is the period for which we have the most knowledge about the pre-instrumental past. Evidence comes from many parts of the world from natural archives (tree-rings, corals, ice-cores, laminated sediments) and documentary material. The extent and reliability of this evidence is discussed in detail in many of the chapters in this book. In our chapter, we have sought to stress the importance of instrumental records for the calibration and verification of these recent proxy data.

Plate 7 shows the results of several compilations of average temperatures for the northern hemisphere (Jones *et al.*, 1998; Mann *et al.*, 1998, 1999; Crowley and Lowery, 2000; Briffa *et al.*, 2001). All compilations stress the uncertainties in the series, which are considerably larger than for the instrumental period. Two of the compilations limit their seasonal windows to the extended summer (May to September) while the other two reconstruct the calendar year. Differences between the series are much greater than for the different compilations of instrumental records (Peterson *et al.*, 1998b), but the general course of temperature change is reasonably consistent. Temperatures during the first 400 years were relatively stable and slightly warmer than the average for the whole millennium. During the 15th century, temperatures cooled, with the coldest century occurring in the 17th. After a milder 18th century, the 19th century marked a return to cooler temperatures. The 20th century experienced the largest rise within a century and temperatures during the last 20 years are warmer than in any period of the millennium. Although the above papers have received much attention, the basic conclusions about northern hemisphere temperature trends since 1400 have not substantially changed since the earlier study of Bradley and Jones (1993).

11.4 SUMMARY

The most comprehensive reviews of what instrumental and palaeoclimatic records tell us about the past have been undertaken over the last 12 years by the Intergovernmental Panel on Climate Change (IPCC, 2001). Our review considers the instrumental record in the context of the Holocene. Instrumental data provide the backbone for all of palaeoclimatology as they enable us to quantify objectively how the climate has varied and the spatial differences in climate between regions.

Instrumental records cover a whole range of climatic parameters. The first question that must always be asked of any climatic series is whether it is homogeneous. The second and third questions concern how the homogeneity was assessed and what adjustments were made. Temperature may be the most widely studied parameter, because it is the yardstick against which we measure a climatic state, but precipitation is generally more important in practice. Pressure is the third most recorded climatic variable and is a favourite of climatologists. Until recently, however, it has only been considered by a small percentage of palaeoclimatologists. Palaeoclimatologists need to consider more the implications of their reconstructions in a climate context (i.e. what a temperature and a precipitation reconstruction means for the circulation).

The rise in instrumental temperatures since 1850 is the basis for much global change research. The 0.6 °C rise has recently been shown to be unprecedented over the last 1000 years, but the uncertainties in the multiproxy compilations are large. The instrumental record tells us that improvements will come by incorporating more records (both from more regions but also from more proxy variables) if we are to significantly reduce uncertainties in the air temperature changes during the last millennium.

Acknowledgements

PDJ has been supported for several years by the US Department of Energy, under Grant No. DE-FG02-98ER62601. Climate change work by RT is supported by the EU Fifth Framework contracts ENV4-CT97-0642 and EVK1-CT-1999-00032. Ron Harriman of the Freshwater Fisheries Laboratory kindly provided the unpublished water temperature data. Anna Agustí-Panareda generated Fig. 11.2.

CHAPTER

12

DOCUMENTARY RECORDS

Peter Brimblecombe

Abstract: Documents provide a unique account of the human experience of climate. This can be a series of detailed observations of the weather that span many years through to almost casual comments that carry indirect information about the climate. These records of the weather often relate to a specific time and place so are not averaged over days or weeks or related to the climate of a region. Records of the positions of glacial margins or the condition of crops, for example, may give more general information. The record is one that spans as much as 5000 years, but is fraught with difficulties in interpretation. The human observer can be biased and in some cases deliberately misleading. Issues of verification and analysis make interpretation of the documentary record particularly challenging.

Keywords: Content analysis, Indices, Reliability, Verification

In some ways it is possible to see how documentary records of climate represent the most direct that we have because at best they can be an accurate log of personal observations. In the extreme the documents may be an accurate set of instrumental observations as described by Jones and Thompson (pp. 140–158 in this volume). This ideal situation is far from universal. Most of the older historic documents do not give direct measurements of an immediately quantifiable nature and are plagued with bias and errors. Much of the methodology of using documents is concerned with using qualitative information and assessing the reliability of the data itself.

Human observers display bias for all sorts of reasons, even if only because the records are more frequent in regions with a high population of recorders. Documents containing geophysical observations were not usually written to serve our purposes of obtaining an accurate record of past climate. Often climate was included as a justification or explanation of particular historical events. If a battle was lost or a harvest yielded less profit than expected we are hardly surprised to find the vicissitudes of weather were much exaggerated. The breadth of historical in such materials was much expanded by writers such as Fernand Braudel and Emmanuel Le Roy Ladurie in the 20th century.

Despite difficulties, the quality of the documentary record is unique. Often it has specificity in time or place, or is the type of observation that is never possible in other proxy records of climate. It is, after all, a record of the human experience of climate and in that way is valuable, almost because of the difficulties thrown into its interpretation. This chapter will review some of the types of record and the forms this information takes. It will try to give examples that expose the range of materials available for study.

12.1 SPAN OF THE RECORD

The documentary record of climate is a surprisingly lengthy one. This is especially true if we think beyond paper archives. Writing goes back some 5000 years and there are extensive Babylonian records of great value. There are records of many activities related to climate, such as precisely dated records dealing with overseeing the supply of water to the city of Larsa for the period 1898–1877 BC (Walters, 1970) or Egyptian descriptions of travel in the ancient Mediterranean, such as the *Report of Wen-Amon* (Goedicke, 1975). Such early non-classical materials are scattered, difficult to read and not widely known.

Written evidence has a different time span for various regions of the world as shown in Table 12.1, suggested by Ingram *et al.* (1978). However, in some ways these do not take a very broad view of what documentation might actually represent. In the case of the Americas and Australia, this span is limited to that of European (and in Australia, English) explorers. It is reasonable to consider the possibility of other types of documents. These might be petroglyphs that would not record the details of climate itself, but rather the presence of human habitation, the local animal populations or types of foodstuffs, which could give clues about the early climate.

Egypt	3000 BC
Middle East	2500 BC
China	2500 BC
Southern Europe	500 BC
Northern Europe	0
Japan	AD 500
Iceland	AD 1000
North America	AD 1500
South America	AD 1550
Australia	AD 1800

Table 12.1 Earliest dates for written evidence of climatic phenomena from different parts of the globe (after Ingram *et al.*, 1978)

Recent records create a parallel problem, because of the rapid change in the way things have become recorded through the 20th century. The regular use of journals to record observations was consistent over hundreds of years, and these materials were often accumulated in a semi-regular and systematic way by local government or archives. In the 20th century, as observations were obtained through instrumental observation, the documentary record often faded. It is likely that even diarists were less regular and systematic in their observations, although there is also a problem with access to the most recent materials of this type, as they may still have to find their way into archives.

12.2 RECORD TYPES

The most obvious forms of documentary records are written observations of climate that are recorded in: inscriptions, annals, chronicles, public records, military records (e.g. ships logs), mercantile accounts and private papers. These sometimes mutate into formalized records of a strictly meteorological character that essentially represent scientific observation. These would be characterized by a regularity of observation and regularity in content. Perhaps the earliest such records are the formalized weather diaries of Bacon and Merle in Medieval England. Here daily observations of the weather were maintained for a substantial period and these early diaries are essentially non-instrumental. In the 17th and 18th centuries, some of the diarists gradually introduced instrumental observations into their accounts and there is a period of overlap with the instrumental record before the introduction of formalized meteorological offices. Weather diaries became less frequent and became rare, as hinted above, by the late 19th century. Thus the analysis of this non-instrumental record for London, using early weather diaries, undertaken by the Victorian meteorologist Mossman (1897) for the period 1713–1896, already seemed quite dated.

It is important not to restrict our concept of documents to these written observations of weather and much use has been made of other observations such as flowering dates, freezing dates, crop prices, harvest failure etc. Approaches to using such data require considerable care and often great ingenuity.

However, more literary materials can also provide insight into climate. The *Tawatodsamad Klongdun* is an ancient Thai poem that dates to about 1464 and describes a year of exile to a rural area (for detailed weather analysis see Aranuvachapun and Brimblecombe, 1979). Although there are problems with the authorship of this poem, it was written at a period with a rising concern about realistic description of the external world. The *Tawatodsamad Klongdun* was probably written in the middle part of Thailand which has a rather moderate climate away from such violent weather events as typhoons. Unlike so many early climate observations, this poem treats rather ordinary weather events, which the poet uses as metaphors for his isolation, such that storms become part of the turmoil of life, rain become synonymous for tears and heat for passion. Separated from his spouse the poet writes (Verse 58 – June):

> 'In month seven noise fills the sky
> Heavy rain falls like drops of gold
> Since last month I cry continually because
> I'm not with my only love
> Time passes slowly, I pine away.'

The frequency of thunderstorms described in the verses, the high water-levels and the sharp transition to the wet season at the beginning of May are all very clear in the writing and coincide closely with the modern dates for such events. It suggests that, despite a view that European climate has changed much since the Little Ice Age, in Southeast Asia the seasonality of the 15th century may be relatively similar to the present day.

Purely fictional materials can also offer interesting climate observations, provided one remains conscious of their underlying goal. The Sherlock Holmes stories of Conan Doyle are an interesting example of this (Brimblecombe, 1987) or the novels of Raymond Chandler in Los Angeles give particular views of cities with striking air pollution problems that concern the actions of the fictional detectives (Brimblecombe, 1990). Like so many writers of the time, Conan Doyle described the vivid, typically yellow colour of London fog. There are relatively few detailed observations of the fog droplets in scientific writings, but they suggest that coal tars were to be found in the droplets (Lewes, 1910). This idea is also found in Conan Doyle's *The Adventure of the Bruce-Partington Plans*: 'we saw the greasy heavy brown swirl still driving past us and condensing into oily drops onto the window panes'. There is even a hint of quantitative information as these detective stories are generally dated and the frequency of weather types can be assigned to months. Fogs, for example, show the expected seasonal variation. This is not so much a measure of weather as a sign of the expected acuteness of an observer such as Arthur Conan Doyle writing about an extraordinary observant fictional detective.

Some documentary materials are not written. Maps, for example, provide a rich source of climate information, although they must be treated with the same caution as other historical records. They have often been used to assess changes in hydrological features, coastlines or glaciers (Engeset *et al.*, 2000). Maps of coastlines can tell an interesting story of the impacts these changes have on navigation and even the development of meteorology. In 15th century Sandwich, an increasingly difficult harbour meant the port was increasingly difficult to use and it seemed to provoke the local government to try to provide a primitive meteorological service for mariners (Brimblecombe, 1995).

Drawings and paintings can also provide evidence of past climate. Some may just give hints of the type of crops grown or the animals that were reared or hunted. Other material can be more quantitative. Landscape paintings can provide evidence of the extent of flooding or the position of glacial margins. One of the best known examples is the retreat of the Franz Josef Glaciers of the Southern Alps of New Zealand (Grove, 1988), whose retreat can even be noticed from its appearance on postage stamps. Paintings often seem one-off pieces of evidence, but some painters undertake series of works, such as the paintings of Krakaota sunsets that spanned many years (Austin, 1983). Through the 20th century, aerial, and later satellite, photographs have provided a wealth of material (Chinn, 1999).

In urban settings there are also valuable images. The Venetian painter Antonio Canal (known as Caneletto, 1697–1768) painted exceedingly accurate pictures of Venice using a camera obscura. These contain evidence of the high water mark embedded in the *commune marino* the thin green line of algae that denotes the high water mark along the canals. This can be used as evidence of changes in sea level in the Venetian lagoon (e.g. Camuffo, 1993).

12.3 DATA RECORDED

Here I take documentary records to be non-instrumental, although quite clearly many documents contain instrumental observations. The nature of instrumental records is discussed by Jones and Thompson (pp. 140–158 in this volume).

12.3.1 Perceptible Meteorology

Some meteorological parameters are rather easy to observe without instruments. Visibility or fog frequency can be extracted from recorded observations fairly readily. The change in the occurrence of fogs in London has been established from a range of diary observations (Brimblecombe, 1995). The difficulty with these data is that fogs are not classified in the same way by different observers. One might, for example, use the term fog only when visibility is very low and call it a mist when it is much thinner. It is possible to gain some idea of this by examining overlapping parts of a record. Despite the problems, the record of fog frequency in London gives a satisfying agreement with the expected air pollution concentrations, which would be an important cause of increased fogginess in the city.

Thunderstorms, or at least the occurrence of audible thunder, is easy to record and good records of this exist along with such parameters as 'state of the ground'. Mossman (1897) showed this for London and there is a recent study of Chicago revealing that a city site averaged 4.5 more thunderstorm days per year, an increase of 12 per cent, than the adjacent rural site (Changnon, 1999).

However, it is often important to establish some of the key meteorological parameters such as temperature, rainfall amount and pressure. The latter is a particularly difficult one to establish prior to the invention of the barometer and has to be done on the basis of pressure patterns expected under particular weather types. This is illustrated by the work of Lamb and his co-workers in establishing the weather at the time of the Spanish Armada (CRU, 1987).

12.3.2 Temperature

It is possible to gain some idea of temperature from qualitative descriptions of seasons identified as abnormally cold or hot. Hubert Lamb used an approach such as this to establish the concept of the **Little Ice Age**. The difficulty is that such approaches are fraught with error and our perception of an abnormally hot season might be distorted through the experience of a single extremely hot day. Writers such as Lamb and Alexandre converted qualitative observations into indices such as the **winter severity index**. The index is C-M, where C is the number of unmistakably cold winter months in a decade and M the number of unmistakably mild winter months. Unfortunately, the correlation between the indices established by Lamb for England and Alexandre for Belgium are very low for AD 1100–1400 and has been attributed to the unreliable data set used by Lamb (Ingram *et al.*, 1978).

In the last decades, great care has been paid to the careful assembly and verification of large groups of proxy records of climate that now amount to many hundreds of thousands of items (Pfister *et al.*, 1999). These offer a hope of greater reliability. Temperatures over much of Europe have been assessed using semi-quantitative indices on the basis of proxy information. Winter temperatures of the Benelux countries, eastern France, western Germany, Switzerland and northern Italy have been assessed by classifying the winters on the basis of proxy information on frost, freezing of water bodies, duration of snow-cover and untimely activity of vegetation. It is likely that severe winters were somewhat less frequent and less extreme during AD 900–1300, than in the 9th century and from 1300 to 1900; this time has often been called the **Medieval Warm Period**. Warm and stable winter climates allowed sub-tropical plants to grow well into central Europe (Pfister *et al.*, 1998), but by the 14th century severe winters were being experienced (Pfister *et al.*, 1996).

Human observers tend to act as a high pass filter, and typically not be good at sensing seasonal averages. However, water bodies with their large thermal inertia or the soil that undergoes slower changes at depth, seem able to reflect average temperatures more effectively. In the Netherlands, lock-keepers' logbooks record the state of the canals, so it is possible to establish the freezing dates (van den Dool *et al.*, 1978). The number of days which the canal is frozen can be correlated with the average winter temperature and provided the layout of the canal does not change this can be used to establish the average winter temperature for years when no instrumental records were available. Similar studies have been done for the freezing of estuaries in Hudson Bay (Moodie and Catchpole, 1976) and data concerning ice-on, ice-off and ice duration on the Angara river in Siberia (Magnuson *et al.*, 2000).

Vegetation growth and flowering dates can be a good proxy for climate and ecosystems may be sensitive to climate change (Walther *et al.*, 2002). Long Japanese records of blossoming dates seem to be related to spring temperature. In Koszeg, Hungary, the length of grapevine sprouts on 24 April have been measured since 1740. These measurements can be correlated with temperatures and allow spring temperatures in the 18th century to be reconstructed (Strestik and Vero, 2000). There are probably more sophisticated approaches than length, and it seems that the square root of length gave somewhat better results. In Norwich, England, a single branch of a chestnut tree and a daisy clump was photographed every year from 1914 to 1942 which can be related to temperature and precipitation records (Willis, 1944), in line with a much longer set of phenological records from Norfolk (Sparks and Carey, 1995). Glaciers also have the potential to record long-term thermal or precipitation (e.g. Grove, 1988).

12.3.3 Wetness

Lamb utilized a **summer wetness index** that was derived in a similar way to his **winter severity index**, i.e. W-D, where W is the number of summer months each decade with evidence of frequent rains and D the number of summer months with evidence of drought. Lamb's evaluation of this index over the period AD 1200–1350 agrees well with Totow's analysis of the manorial account rolls of Winchester Cathedral (Ingram *et al.*, 1978). Other factors such as harvest yields, ripening dates, auction accounts and commodity prices can also be used to reconstruct past climate.

Classical writers and archaeological records have always suggested that the North Africa of the Roman Empire had a moister climate than at present. A detailed analysis of vegetation distribution based on fossil pollen maps and historical records seems to support this long-held belief (Reale and Dirmeyer, 2000). This analysis has been combined with a global circulation model (GCM) which has suggested albedo changes from the Imperial vegetation cover could have altered the atmospheric circulation over northern Africa and the Mediterranean. The model output suggests a northward shift of the ITCZ (inter-tropical convergence zone) substantially increasing precipitation over the Sahel, the Nile valley and northwestern Africa (Reale and Shukla, 2000). This work emphasizes the ability to couple documentary observations with sophisticated climate modelling.

12.3.4 Storms and Extreme Events

Storms and sea surges (e.g. Gotschalk, 1971, 1975, 1977) are often well recorded because they have the potential to cause such extensive economic damage. This has often prompted interest in climate history on the part of insurance companies. Severe storms on the coast or at sea tend to be well recorded, especially where there are shipwrecks (Forsythe *et al.*, 2000) or loss of life. Because these are violent events there is a tendency to focus on the extreme nature of storms and surges, so there is a bias that can make the data hard to analyse.

Hailstorms, for example, are easily observed and provoke considerable historical comment within documents (Camuffo *et al.*, 2000a). There is an excellent record of 2500 British hailstorms beginning with an event during AD 1141. They can be readily expressed in terms of intensity on the TORRO international scale, which is based on damage caused and the size of the hailstones. Intense hailstorms in Britain occur between May and September, with a typical swath length of 25 km. In only a single 17th century case was the hailstorm severe enough to cause loss of life (Webb *et al.*, 2001).

Hurricanes and large storms can cause significant storm surges (>3 m in some cases) removing sediments from the beach and near-shore environment, and depositing over-wash fans across back-barrier marshes, lakes and lagoons. In some cases the deposits preserve a record of over-wash deposition that can be related to historic storms and suggest the occurrence of storms for which there was no record (Donnelly *et al.*, 2001). Storms and sea surges may not always cause inundation, in some cases they can lead to extremely low water. In historical records it is possible to mistake these events for periods of drought (Aranuvachapun and Brimblecombe, 1978).

12.4 RELIABILITY AND DATA VERIFICATION

As emphasized in earlier sections it is very important that the reliability of documentary materials is assessed before being used in climate reconstruction. Many early attempts to reconstruct the history of climate have foundered because of unreliable sources. There is a particular problem with compilations of past climate observations collected prior to the 1950s before source verification became a more widely used technique (Ingram *et al.*, 1978).

In source verification it is important to ask questions about the author and their status (Bell and Ogilvie, 1978), especially questions as to whether we expect them to be reliable and unbiased. Greater value is often given to first-hand observations actually made by the author and recorded soon after the event. If the writer did not experience the event then reports are likely to be more reliable if they came from the author questioning witnesses or at least if the report is contemporaneous with the event. Perhaps the author had access to first-hand reports that are no longer available, and if reliably recorded, then these are of value. Independent confirmation by other authors is always useful, though it is important that such additional reports about the same event are not 'double counted'.

The problems that typically arise in early materials are that they may be slavishly derivative and provide no independent information, inaccurately copied or interpreted, abridged or fabricated. Such errors are not random and therefore not self-correcting. There are particular problems with the dating of events, which can lead to particular difficulties including the multiplication of a given event. If a storm, for example, which occurs in a given year is erroneously assigned to another year by a later chronicler, a subsequent compiler might eventually come to believe there were two storms. Sometimes these problems are exacerbated by different calendar styles or the use of reginal years. Another common problem occurs when the significance of events of purely local character become magnified such that it is ultimately seen as a widespread calamity.

A rather different problem arises with gaps in the data sources. One can imagine two kinds of gaps, one in which a period or a region is missing from a series and another where there are no records at all. In some cases there is internal evidence that a period is missing from the record, but in other cases there may not be any clues that elements are missing. Difference indices of the type adopted by Lamb and others are particularly sensitive to gaps in the data, so Ingram *et al.* (1978) suggested a ratio-index might suffer from less distortion, e.g. the winter severity index might be written (C-M)/(C+M) rather than C-M. It is possible to check on gaps in records as undertaken by Camuffo and Sturaro (2004) in an analysis of the frequency with which solar eclipses have been recorded. Although the capture was low for the earliest periods, after 1200 it became about 70 per cent, which is high considering the likelihood of cloud cover.

12.5 ANALYSIS OF DOCUMENTARY DATA

The aim of many climate studies is to convert the data in documents into quantities that resemble modern meteorological measurements. Thus it is desirable to convert freezing or flowering dates into average temperatures, river flows to rainfall, etc. This often means starting with data that are essentially qualitative and converting them into a more parametric measure.

The first stage of interpretation follows closely the questions of verification mentioned in the section above, but takes place on a more detailed level, although a comprehensive approach to interpretation is not possible to define. Indeed, depending on the imagination of the analyst, some general questions have been proposed by Ingram *et al.* (1978): Why were the data recorded? How was the writer's purpose likely to bias the collection of data and its selection? What is the significance of adjectival terms of degree (e.g. superlative words such as 'greatest', 'most') to be interpreted?

This can be quite subtle – in the case of the record of London fogs, the number of fogs recorded in two contemporaneous records is much affected by the width of the columns in the observer's notebook (Brimblecombe, 1982). Wide columns allowed plenty of detail to be recorded, while a narrow column means the only the most obvious features of the days weather could be recorded, hence fogginess was probably entered only when the fog was very intense.

Content analysis is very useful as a systematic procedure for converting lexical and other qualitative information into quantitative measures (Carney, 1972; Krippendorff, 1980) and ultimately valid climatic measures. It derives from techniques that have led to basic innovations in the social sciences (Deutsch *et al.*, 1971) and media interpretation. An excellent illustration of this technique is the analysis of daily journals kept by employees of the Hudson Bay Company at their forts from 1715 to 1871 (Catchpole and Moodie, 1975; Moodie and Catchpole, 1976). Although dates of freezing and thawing are not recorded frequencies counts of specified symbols or themes within the text can be analysed in terms of the intensity or direction of assertions. If there is more than one variable, then contingency analysis can be applied to the extracted data. In general, analysis proceeds by then taking the numerical representations of the observations within the documents and correlating them with periods when there is an overlap with instrumental measurements. This can be complex if there are many sets of data to be compared, but techniques such as principal components analysis allow these problems to be resolved (e.g. Craddock, 1976; Smith *et al.*, 1996).

12.6 CONCLUSION

The documentary record of climate represents insight into the human experience of climate. It can provide observations of sufficient quality to convert into more conventional meteorological observations, but the analysis has to be undertaken with great care. The human observer has a natural bias towards that which is immediate and local.

CHAPTER

HOLOCENE CORAL RECORDS: WINDOWS ON TROPICAL CLIMATE VARIABILITY

Julia E. Cole

Abstract: Corals provide palaeoclimatic information on the behaviour of the tropical oceans that is available from no other proxy. They allow accurate dating by several means, and they incorporate geochemical tracers of climate that are relatively simple to measure to produce high-quality climate records. Records may span centuries at monthly to annual resolution, or they may provide shorter high-resolution windows into the deeper past. They have been most useful in the tropical Pacific and Indian Oceans, where a collection of c. 20 records is now available. This paper summarizes the methods used in coral palaeoclimate work and describes a selection of Holocene palaeoclimate applications where corals have made significant contributions.

Keywords: Coral, Climate, Holocene, Palaeoclimate, Palaeoceanography, Reef

13.1 THE MOTIVATION FOR CORAL PALAEOCLIMATOLOGY

Records of climate from corals provide critical pieces of the Holocene climate puzzle. As one of the only direct sources of high-resolution information about tropical ocean–atmosphere variability, they offer a unique perspective into past climate change, with important insights into the operation of influential tropical systems such as El Niño Southern Oscillation (ENSO) and monsoons. Such tropical ocean–atmosphere systems contribute substantial variance to modern climate, and coral records are providing new insights into their natural range of variability. The climatic response to greenhouse forcing will almost certainly involve perturbations to these systems, and understanding their sensitivity to past changes will help in anticipating their future responses. Moreover, existing reconstructions of large-scale temperature change over the past millennium are deficient in tropical records (e.g. Mann *et al.*, 1998), and more attention to developing annual records of tropical variability will help to fill this important gap. This chapter describes how high-fidelity climate reconstructions can be derived from corals and how they have been applied to understand climate variability during the Holocene. There is special urgency to expand the network of coral climate records quickly, as corals throughout the tropics are threatened by ongoing environmental and ecological degradation (Wilkinson, 1999).

13.2 NATURE AND DISTRIBUTION OF ARCHIVE

13.2.1 Sampling

Coral reefs are widely distributed along tropical coastlines wherever relatively clear, warm waters are found. The massive, long-lived colonies used in palaeoclimate reconstruction are typically found on reefs but may also occur individually in areas where reef growth is marginal or non-existent. They inhabit depths ranging from tens of metres up to the low tide level. (Non-photosynthesizing corals live throughout the deep oceans, but their palaeoclimatic applications will not be discussed here; see Adkins *et al.*, 1998 for an example.) Selection of massive coral colonies usually focuses on those that experience conditions representing the open ocean. Collection typically involves drilling one or more cores vertically through the most rapidly growing portion of the skeleton, using hydraulic or pneumatic diver-operated equipment. The boreholes seal off as the coral continues to grow (J.E. Cole, personal observation). Cores are slabbed and density or fluorescent bands are used to determine optimal (rapidly growing, time-transgressive) sub-sampling transects for geochemical analysis (Plate 8).

Corals build their skeletons from aragonite ($CaCO_3$) and incorporate multiple geochemical tracers of environmental variability, either as different isotopes of common elements (C and O) or as trace 'contaminants' that are incorporated in the skeletal lattice. Coral geochemistry is intimately involved with coral biology, and geochemical tracers may behave differently in different species. Although published climate reconstructions have utilized several different taxa, most studies in the Pacific and Indian Oceans rely on massive corals of the genus *Porites*, which are straightforward to sub-sample and yields reliably high-quality climate records. Other genera (e.g. *Pavona*, *Platygyra*, *Hydnophora* and *Diploastrea*) have been sampled where they appear to be the oldest or only available, or for short demonstration records. In the Caribbean and Atlantic, long-lived *Porites* are unavailable and other massive (e.g. *Montastrea*, *Siderastrea* and *Diploria*) taxa have been used. Few of these have been subject to rigorous calibration experiments.

Coral skeletons exhibit structural complexity that can complicate the development and interpretation of climate records. The scale, architecture and ease of sampling varies greatly among taxa (Plate 8). The most commonly used taxa, massive *Porites* species, can be sampled in bulk along a linear transect of rapid growth rate with minimal complications, because these taxa have relatively fine-scale skeletal architecture (Plate 8a). Those with coarser architecture (Plate 8b) may need to be sampled along specific skeletal elements (Leder *et al.*, 1996). Optimal sampling strategies have not been explicitly defined for many coral taxa used in palaeoclimate work.

13.2.2 Chronology

Coral records that are collected live, or with a known date of death, can be assigned annually precise ages using several methods. Many corals produce annual density bands in their skeletons as a consequence of changing rates of extension and calcification (Knutson *et al.*, 1972). The timing of dense band formation varies from place to place, and it may happen in response to environmental cues such as temperature extremes or to endogenous rhythms such as spawning (Wellington and Glynn, 1983; Barnes and Lough, 1993). Annual fluorescent bands, visible under ultraviolet light, may also form from the uptake of seasonally available organic materials delivered by river runoff. Annual cycles also occur in commonly measured geochemical parameters in response to seasonal temperature variability (oxygen isotopes and trace metals). Seasonality in carbon isotopes is commonly observed and may occur as a photosynthetically mediated response to light availability (Fairbanks and Dodge, 1979) or as a result of changing internal carbon allocation associated with reproductive rhythms (Gagan *et al.*, 1996). Fossil coral material is readily dated by radiometric methods, including radiocarbon and U-series methods (Edwards *et al.*, 1987; Bard *et al.*, 1990), and annual cycles can be identified and used within these radiometrically dated windows (e.g. Gagan *et al.*, 1998; Correge *et al.*, 2000; Tudhope *et al.*, 2001).

13.3 ENVIRONMENTAL RECONSTRUCTION

13.3.1 From Banding and Growth Parameters

Aspects of coral skeletal structure may preserve environmental information, particularly in environments where corals experience conditions stressful for growth at least seasonally. Changes in growth rate, in density band structures, or in colony morphology can reflect local stresses.

Density bands in particular have been explored for their potential as the 'tree-rings of the ocean'; analyses have ranged from the straightforward interpretation of growth rate derived from annual band thickness to more sophisticated analyses that include distinguishing rates of calcification and extension. Not all corals have annual density banding, and some that do nevertheless show little variability in growth rate. However, if bands are clear and calibrate well with instrumental climate variations, then climate reconstruction is possible from band thickness or related parameters. As an example, extensive work along the Great Barrier Reef by Janice Lough and colleagues has demonstrated a correlation between coral calcification rate and temperature, broadly replicated along the reef (Lough and Barnes, 1997, 2000). Anomalous (non-annual) density bands may also indicate coral stress: in Florida, such 'stress bands' correlate with cold air outbreaks (Hudson *et al.*, 1976).

Fluorescent bands may also preserve environmental information under certain settings (Isdale, 1984). These bands most likely originate from river-borne organic materials that fluoresce under ultraviolet light (Boto and Isdale, 1985). Corals that experience a moderate degree of exposure to river runoff may preserve a record of discharge in the varying skeletal fluorescence patterns. On the Great Barrier Reef, inshore corals near the Burdekin river indicate that the intensity of the fluorescence can be correlated with river discharge (Isdale *et al.*, 1998). Other studies have used annual fluorescent bands to develop or refine annual chronologies (Tudhope *et al.*, 1995; Cole *et al.*, 2000).

Coral colony structure may also record information about subtle sea-level changes. When a massive coral colony reaches the ocean surface, it may then grow outward for many more years, forming a discoid 'microatoll'. The coral can accrete skeleton at the existing level of the low tide, so the height of the coral growing surface records changes in sea-level during the microatoll's lifetime. Such changes have been used to address subtle sea-level variability associated with ENSO in the central Pacific and to explore the history of 20th century sea-level in a region remote from tide gauges (Woodroffe and McLean, 1990; Smithers and Woodroffe, 2001).

13.3.2 From Coral Geochemistry

Corals incorporate many geochemical tracers of environmental variability in the aragonite structure of their skeletons, including isotopes of oxygen and carbon and various trace elements that are thought to be taken up as substitutes for calcium in the aragonite lattice or adsorbed onto the aragonite skeleton (e.g. Sr, U, Mg, Ba, Cd, Mn). Oxygen isotopes and Sr are the most widely used tracers for climate reconstruction; dozens of records have been published, many of which predate instrumental climate data in their region. A selection of these (chosen primarily because they have been made available through the World Data Center for Paleoclimatology) is summarized in Table 13.1 and Figs 13.1 and 13.2. With wider usage comes greater recognition of the strengths and limitations of all tracers. As coral records become increasingly valued for their

Region	Site	Location	Reference	No. of years	Samples/ year	Range
Tropical Pacific	1. Gulf of Chiriqui, Panama	7°N, 82°W	Linsley *et al.* 1994	276	10	1708–1984
	2. Urvina Bay, Galapagos. Ecuador	0°, 91°W	Dunbar et *al.* 1994	347	1	1607–1953
	3. Clipperton Atoll, French Polynesia	10°N, 109°W	Linsley et *al.* 2000a	100	12	1894–1994
	4. Kiritimati Atoll, Kiribati	2°N, 157°W	Evans et *al.* 1999	56	12	1938–1994
	5. Palmyra Atoll, USA	5°N, 162°W	Cobb et *al.* 2001	112	12	1886–1998
	6. Maiana Atoll, Kiribati	1°N, 173°E	Urban et *al.* 2000	155	12	1840–1995
	7. Tarawa Atoll, Kiribati	1°N, 172°E	Cole et *al.* 1993	96	12	1894–1990
	8. Nauru	0°, 166°E	Guilderson and Schrag 1999	98	4	1897–1994
	9. Laing Island, Papua New Guinea	4°S, 145°E	Tudhope et *al.* 2001	109	4	1884–1993
	10. Madang Lagoon, Papua New Guinea	5°S, 145°E	Tudhope et *al.* 2001	112	4	1880–1993
	11. Cebu, Phillippines	10°N, 124°E	Pätzold 1986	117	1	1864–1980
South Pacific	12. Rarotonga, Cook Islands	21°S, 159°W	Linsley et *al.* 2000b	270	12	1727–1997
	13. Espiritu Santo, Vanuatu	15°S, 167°E	Quinn et *al.* 1996	172	4	1806–1980
	14. New Caledonia	22°S, 166°E	Quinn et *al.* 1998	335	4	1658–1993
	15. Great Barrier Reef, Australia	15–22°S, 145–153°E	Lough and Barnes 1997	237	1	1746–1982
	16. Abraham Reef, GBR, Australia	22°S, 153°E	Druffel and Griffin 1999	349	1	1638–1986
Eastern Indian	17. Ningaloo Reef, Australia	22°S, 114°E	Kühnert et *al.* 2000	116	6	1879–1994
	18. Houtman Abrolhos, Australia	28°S, 113°E	Kühnert et *al.* 1999	199	6	1795–1994
Western Indian	19. Mahe, Seychelles	5°S, 56°E	Charles et *al.* 1997	148	12	1847–1995
	20. Malindi Reef, Kenya	3°S, 40°E	Cole et *al.* 2000	194	1	1801–1994
Red Sea	21. Ras Umm Sid, Egypt	28°N, 34°E	Felis et *al.* 2000	244	6	1751–1996

Table 13.1 Location and description of selected long coral records (mostly available at the World Data Center for Paleoclimatology, http://www.ngdc.noaa.gov/paleo/corals.html. Numbers refer to map locations in Figure 13.1. These records are plotted in Figure 13.2.

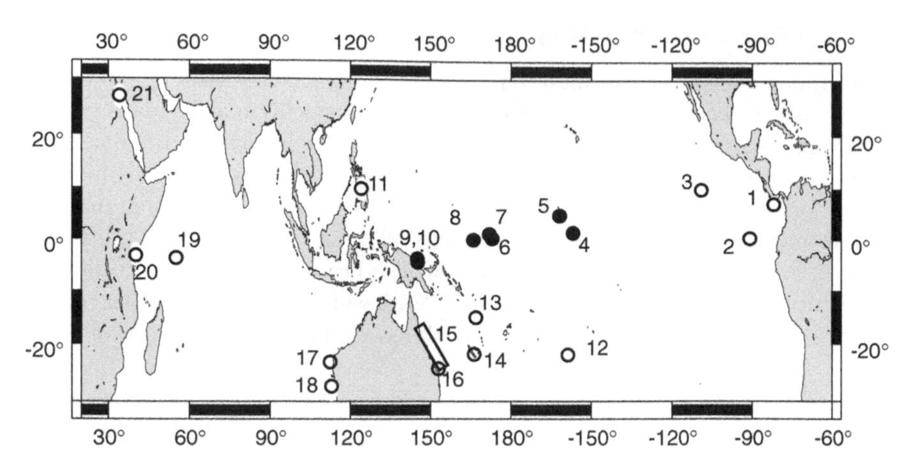

Figure 13.1 This map indicates the locations of century-scale coral records plotted in Fig. 13.2 and available online at the World Data Center for Paleoclimatology (http://www.ngdc.noaa.gov/paleo/corals.html). Solid circles indicate records that are plotted in both Figs 13.2 and 13.4. The diagonal bar along the east coast of Australia represents the range of sites used by Lough and Barnes (1997) for their calcification temperature reconstruction.

contributions to understanding recent climate change, interest and attention has also focussed more intensely on understanding the limitations of these tracers. Their strengths and weaknesses are discussed in the sections below.

13.3.2.1 Oxygen Isotopic Content

The most widely used tracer of climate in corals is the oxygen isotopic composition of skeletal aragonite. Following nearly 50 years of work on the behaviour of oxygen isotopes in carbonates, this is one of the best understood and best calibrated palaeoceanographic tools available. In simplest terms, the oxygen isotopic content (abbreviated $\delta^{18}O$)[1] of a coral skeleton reflects three parameters: seawater temperature, seawater $\delta^{18}O$ and a biological offset from seawater. The early work of Epstein *et al.* (1953) demonstrated that in molluscan calcite, the shell $\delta^{18}O$ decreases by ~0.22‰ per 1 °C increase. This slope also applies to coralline aragonite; coral calibration studies have obtained values ranging from 0.18 to 0.26‰/1°C. In the tropics, the main influence on seawater $\delta^{18}O$ is rainfall. Biological fractionation depresses coral $\delta^{18}O$ values by about 2–5‰ from equilibrium with seawater (Weber and Woodhead, 1972). This offset is large compared to a typical isotope climate signal in corals, and it may vary among individuals (Linsley *et al.*, 1999), but it appears to stay constant in time. Thus long histories of coral $\delta^{18}O$ are interpreted in terms of sea-surface temperature (SST) and rainfall change. Numerous such records that correlate well with local or regional climate have been developed in the tropical Indo-Pacific (Table 13.1, Figs 13.1 and 13.2).

[1] The $\delta^{18}O$ of a sample is defined as follows:

$$\delta^{18}O = (R_{samp} - R_{std})/R_{std}) * 1000$$

where R_{samp} is the $^{18}O/^{16}O$ ratio in a sample and R_{std} is that ratio is a standard. For carbonates, the standard of reference is the Pee Dee Belemnite (PDB); for water, Standard Mean Ocean Water (SMOW) is used.

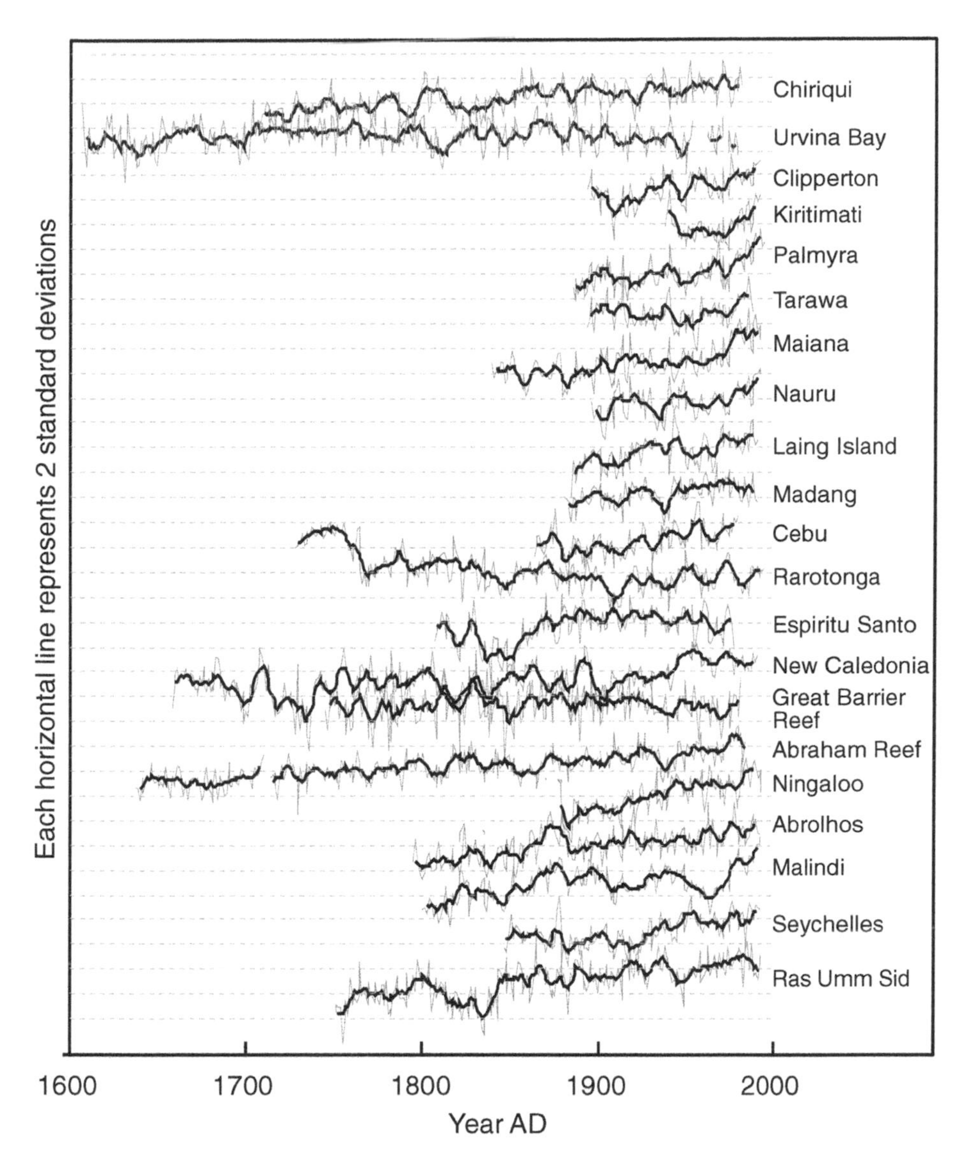

Figure 13.2 Time-series of century-scale coral records that are mostly available at the World Data Center for Paleoclimatology (http://www.ngdc.noaa.gov/paleo/corals.html). All records are derived from $\delta^{18}O$ data except for Rarotonga (Sr/Ca) and the Great Barrier Reef (calcification). All records were converted to annual means (calendar year) and normalized to the mean and standard deviation over a mostly common interval, 1924-80. Heavy lines show a 7-year running mean to emphasize decadal variability; grey lines are the annual normalized values. Horizontal dashed lines represent 2 standard deviation units, and the vertical axis represents warmer/wetter conditions upwards and cooler/drier conditions downwards.

Changes in seawater $\delta^{18}O$ result from changing precipitation–evaporation balance and parallel salinity changes. In the tropical Pacific, a typical relationship between seawater $\delta^{18}O$ and salinity is ~0.2–0.3‰ $\delta^{18}O$ per 1‰ salinity (Fairbanks *et al.*, 1997; Morimoto *et al.*, 2002; J.E. Cole, unpublished data), so salinity changes of less than ~0.25‰ are not detectable with typical measurement precision for coral $\delta^{18}O$. However, in many parts of the tropics, rainfall variability is both sufficient to cause such changes and is diagnostic of large-scale climate patterns (Cole and Fairbanks, 1990; Linsley *et al.*, 1994). The relationship between seawater $\delta^{18}O$ and rainfall amount has the potential for substantial non-linearities. The $\delta^{18}O$ of the rainwater can vary from nearly equal to seawater (~0‰, in which case it will not alter seawater $\delta^{18}O$) to –10‰ or below, depending inversely on the amount of rainfall (Yurtsever and Gat, 1981). The depth of the surface mixed layer will effectively concentrate or dilute the rainwater $\delta^{18}O$ signal, and this depth depends partly on rainfall amount (Lukas and Lindstrom, 1988). Despite these complications, a study using average mixed layer depth and rainfall $\delta^{18}O$ accurately predicted the isotopic signal observed in central Pacific corals (Cole and Fairbanks, 1990). Advection of water masses with contrasting salinities (Picaut *et al.*, 1996) can also influence coral $\delta^{18}O$ records from sites near strong salinity contrasts. Several recent studies have presented $\delta^{18}O$ records of salinity that take advantage of the gradient along the edges of the western Pacific warm/fresh pool, whose position and extent varies with ENSO (Le Bec *et al.*, 2000; Urban *et al.*, 2000; Morimoto *et al.*, 2002).

13.3.2.2 Strontium

Strontium/calcium ratios in corals provide a record of temperature change that is not sensitive to salinity. The temperature dependence of Sr/Ca in corals has been long recognized (Smith *et al.*, 1979), but the low sensitivity of Sr to SST change means that very precise techniques are needed to distinguish the relatively subtle temperature changes in tropical oceans. Beck *et al.* (1992) first demonstrated that such records could be developed using thermal ionization mass spectrometry (TIMS) with analytical accuracy as good as ± 0.03 per cent (equivalent to 0.05 °C; more typical values are ± 0.2 per cent, equivalent to 0.3 °C). However, TIMS measurements involve time-consuming sample preparation and analytical procedures. Schrag (1999) describes an accurate, high-throughput method using inductively coupled plasma atomic emission spectrophotometry (ICP-AES), which allows over 180 analyses per day with an external precision of better than 0.2 per cent (0.3 °C). Other ICP-based methods may provide comparable precision and throughput (e.g. Le Cornec and Correge, 1996).

With analytical issues resolved, the main questions remaining about coral Sr/Ca records involve understanding how well coral Sr/Ca reflects temperature changes. Several studies have focussed on uncertainties in the interpretation of coral Sr/Ca data, including variability of seawater Sr/Ca. Seawater Sr/Ca variations have typically been disregarded on Quaternary time-scales as the residence time of Sr in seawater is approximately 12×10^6 years (Schlesinger, 1997). However, Stoll and Schrag (1998) demonstrate that pulse changes of 1–3 per cent are possible during times of rapid sea-level change, sufficient to amplify coral palaeotemperature reconstructions by 1–3 °C during times of most rapid sea-level rise during the Quaternary. For modern samples, de Villiers *et al.* (1994) argue that the variability in Sr/Ca on tropical reefs would cause palaeotemperature errors of no more than 0.2 °C. However, measurements near the northern limit of hermatypic coral growth, at Nanwan Bay, Taiwan, indicate that during a year, seawater Sr/Ca varies enough to cause a temperature artifact of 0.7 °C (Shen *et al.*, 1996). Documenting this variability at other sites may improve temperature reconstructions.

As more data on coral Sr/Ca become available, identifying a single calibration for temperature reconstruction seems to become more difficult. Several papers discuss the variations in calibration curves (e.g. Shen *et al.*, 1996; Alibert and McCulloch, 1997; Cohen *et al.*, 2002). A notable observation is that the Sr/Ca–SST relationship of inorganic aragonite is about 30 per cent as steep as that observed in most Sr/Ca-based coral paleotemperature studies (Cohen *et al.*, 2001). This discrepancy implicates biological activity as a significant influence on coral Sr uptake. In a thought-provoking pair of papers, Anne Cohen and colleagues (Cohen *et al.*, 2001, 2002) propose a mechanism that may explain calibration irregularities in Sr/Ca. They argue that photosynthesis, by enhancing calcification rate, distorts the palaeotemperature relationship by introducing a kinetic component to coral Sr uptake. This effect is seen in comparisons of daytime and nighttime skeletal elements (sampled at the micron scale) and of photosynthetic and non-photosynthetic *Astrangia poculata*. Nighttime skeletal elements, and non-photosynthetic *A. poculata*, show a Sr/Ca–SST relationship similar to that of inorganic aragonite, while daytime elements and mature photosynthetic *A. poculata* show enhanced sensitivity of Sr/Ca to SST change.

Cohen *et al.* (2002) argue that based on their results, Sr/Ca is an unreliable temperature proxy in photosynthetic corals. However, this argument is probably premature as it ignores the significant ways in which their study differs from a standard palaeoclimatic application (Schrag and Linsley, 2002). The numerous examples of strong correlation between coral Sr/Ca and SST over a broad range of time-scales are unlikely to be driven solely by coincident photosynthetic activity, which responds to many factors besides temperature. However, a better understanding of the biological mechanisms that influence Sr uptake in coral should lead to improved paleoclimate reconstructions.

13.3.2.3 Radiocarbon

The radiocarbon content of coral skeletons provides unique records of past ocean circulation. Although carbon isotopes in coral skeletons are strongly influenced by biological processes, these influences can be removed by correction for ^{13}C content; the residual, denoted $\Delta^{14}C$, reflects the radiocarbon content of overlying seawater (Druffel and Linick, 1978). Changes in coral $\Delta^{14}C$ are typically due to the variable influence of water masses with contrasting radiocarbon signatures. Deep waters contain less radiocarbon than surface waters, due partly to the decay with age of ^{14}C into ^{14}N. This contrast was dramatically enhanced by the addition of bomb-produced radiocarbon to the atmosphere during nuclear testing that commenced in the early 1960s (Broecker *et al.*, 1985). In the tropics, the strongest $\Delta^{14}C$ gradients are seen in regions where newly upwelled waters mingle with waters fed by subtropical gyres, where bomb ^{14}C has equilibrated between atmosphere and seawater.

In the tropical Pacific, coral radiocarbon records of the mid-late 20th century show clear patterns related to the gradients of ^{14}C in surface waters (Fig. 13.3). Year-to-year variations track the weakening of upwelling during El Niño years and its strengthening during La Niña's (Druffel, 1981, 1985; Guilderson *et al.*,1998). Such records show how currents change in strength and position as ENSO varies. As circulation histories of unprecedented length and resolution, they add tremendous detail to our knowledge of surface ocean radiocarbon variability, compared to spot seawater measurements (Moore *et al.*, 1997) and they have also been used to ground-truth models of ocean circulation (Toggweiler *et al.*, 1991; Guilderson *et al.*, 2000). Because they provide direct records of circulation, they can be used to infer

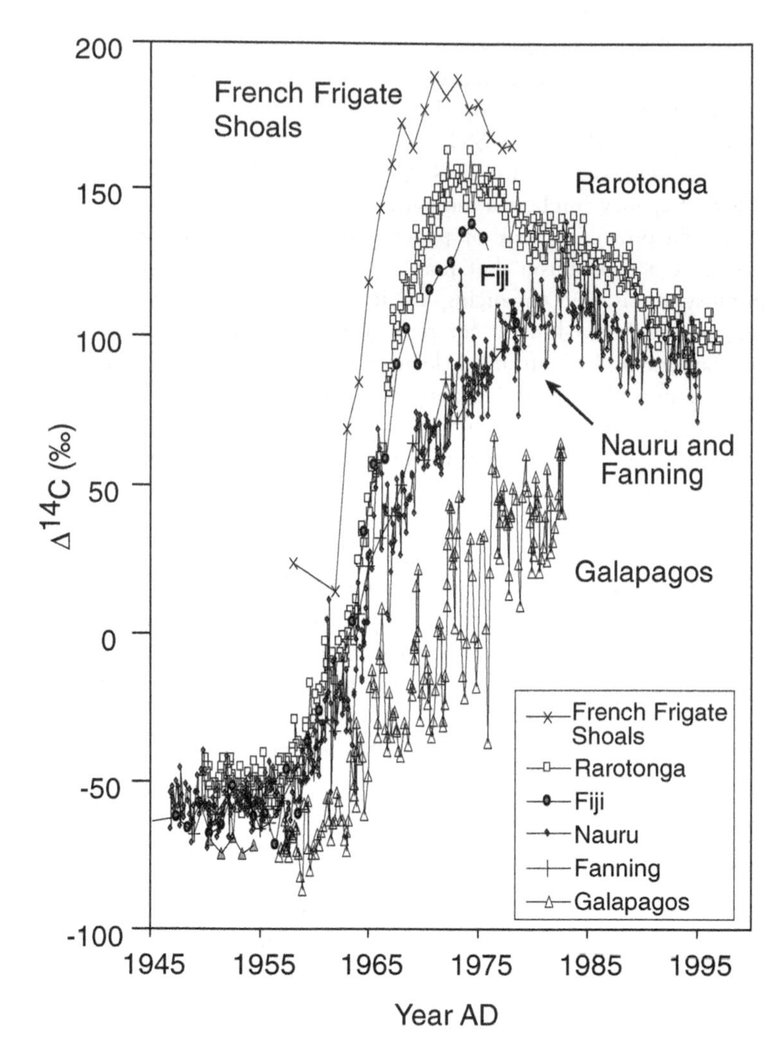

Figure 13.3 Radiocarbon time-series from tropical Pacific corals illustrating the bomb-produced increase in surface water ^{14}C since the early 1960s. The rate of ^{14}C increase associated with atmospheric nuclear testing varies from site to site depending on hydrography. Upwelling in the eastern Pacific (Galapagos) brings low-^{14}C water to the surface where it mixes with surface waters that have equilibrated with atmospheric ^{14}C levels. Central equatorial records (Fanning and Nauru) reflect a mixture of low-^{14}C upwelled water and higher ^{14}C sub-tropical waters. Radiocarbon values in these records peaked in the early 1980s. The most rapid and complete rises are seen in the sub-tropical gyres (Rarotonga, Fiji, French Frigate Shoals), where surface waters equilibrated rapidly with the atmospheric rise and where peak values were reached by the early 1970s. Interannual variability aside from the bomb-produced rise is generally associated with the variable contribution of upwelled water from the eastern Pacific, which weakens during El Niño events. Data are from several references (Druffel, 1987; Guilderson and Schrag, 1998; Guilderson et *al.*, 1998; Toggweiler et *al.*, 1991).

mechanisms of observed surface variability on decadal time-scales inaccessible to instrumental records (Guilderson and Schrag, 1998). Coral radiocarbon records can provide circulation histories that pre-date bomb ^{14}C inputs (Druffel and Griffin, 1993, 1999; Druffel, 1997). However, greater analytical precision is required to distinguish circulation signals, as water mass ^{14}C contrasts are not artificially enhanced.

13.3.2.4 Other Trace Elements

A number of trace elements have been proposed as palaeoenvironmental tracers and have seen limited application to environmental reconstructions. Cd showed early promise as an upwelling tracer (Shen *et al.*, 1987) but additional work showing unexplained large variances led to speculation that biological or early diagenetic processes created unacceptable biases; this issue has not been resolved. In certain contexts, Mn has been proposed as a tracer of wind direction (Shen *et al.*, 1992b). A long Mn record from a Galapagos coral was interpreted in terms of volcanic activity and seasonal upwelling (Shen *et al.*, 1991). Ba appears to be taken up in corals in proportion to its abundance in seawater; processes that change seawater Ba/Ca include upwelling of nutrient-rich water and runoff of Ba-rich river water. Thus Ba/Ca can potentially be used to reconstruct these processes (Lea *et al.*, 1989; Shen *et al.*, 1992a; Fallon *et al.*, 1999; R.B. Dunbar *et al.*, unpublished data).

Other trace elements have been proposed as palaeothermometers unaffected by salinity. Mitsuguchi *et al.* (1996) suggested that Mg/Ca ratios can provide accurate and easily measured temperature reconstructions, although other studies have revealed unusual variability unrelated to temperature (e.g. Fallon *et al.*, 1999). A significant adsorptive component of coral Mg may bias this tracer (Mitsuguchi *et al.*, 2001). Other studies have documented temperature correlations with coral U/Ca ratios (Min *et al.*, 1995; Shen and Dunbar, 1995; Sinclair *et al.*, 1998; Fallon *et al.*, 1999). Seasonal variations in other trace elements have been measured in coral skeletons (e.g. B, Na) (Sinclair *et al.*, 1998; Fallon *et al.*, 1999; Mitsuguchi *et al.*, 2001), but the palaeoclimatic utility of these elements has not been demonstrated.

13.3.2.5 Stable Carbon Isotopic Content

Coral $\delta^{13}C$ showed early promise as a palaeoclimate recorder (Fairbanks and Dodge, 1979; McConnaughey, 1986) but its variability is controlled by a mixture of climatic and biological parameters that is challenging to unravel in a palaeoclimate context. Several papers summarize these controls (Swart, 1983; Swart *et al.*, 1996; Grottoli and Wellington, 1999). Optimism regarding the palaeoclimatic utility of coral $\delta^{13}C$ (defined as for $\delta^{18}O$; see footnote 1) was fueled by studies showing correlation between coral $\delta^{13}C$ and light intensity in a variety of contexts, from seasonal and interannual cycles to depth-dependent gradients (e.g. Fairbanks and Dodge, 1979; McConnaughey, 1989; Grottoli, 1999). Carbon isotopes respond to light intensity via changes in symbiont photosynthesis, which preferentially utilizes ^{12}C and leaves ^{13}C for incorporation into the skeleton; thus higher light intensities tend to correlate with higher $\delta^{13}C$. However, longer records often reveal striking inconsistencies in this relationship, including non-reproducibility of records both within a single coral core and between nearby cores whose $\delta^{18}O$ records match well (e.g. Guilderson and Schrag, 1999). Carbon isotopes also reflect the balance between photosynthesis and respiration, which can be influenced by food availability (Grottoli and Wellington, 1999; Grottoli, 2002).

13.3.2.6 Non-Climatic Influences on Coral Geochemistry: Biology, Growth Rate and Skeletal Alteration

Factors other than climate have the potential to influence records from many tracers described above, particularly the stable-isotopes and trace elements. Possible causes for concern include biological influences, variable growth rates and skeletal alteration. Although they have the potential to complicate coral palaeoclimate reconstruction, these influences can be minimized with judicious sampling strategies and appear to be minor in most records.

No coral climate tracers are immune from biological influences; none are incorporated in a purely thermodynamic, inorganic way. Biological effects on the most commonly used tracers, $\delta^{18}O$ and Sr/Ca, are clear (see above sections), but they do not appear to compromise palaeoclimatic utility on the shorter (annual–interannual) time-scales that are easily calibrated to instrumental records. Most studies assume longer-term variations are climatically driven as well. If true, such variability should be replicable across corals within a climatic regime, whereas biological influences would be more likely to vary among individuals. Eliminating biological influences provides a strong argument for replication of coral records.

Coral geochemistry may also be influenced by calcification rate through kinetic processes. This effect should arise at the temporal and spatial scale of the formation of individual crystals; lacking information on this scale, most studies exploring this relationship use annual measures of linear extension and calcification as crude proxies for the instantaneous rate. For $\delta^{18}O$, McConnaughey (1989) has argued that the influence of growth rate is negligible at extension rates faster than 5 mm/year. For Sr/Ca, this issue has been addressed in several studies, with conflicting results (de Villiers *et al.*, 1995; Shen *et al.*, 1996; Alibert and McCulloch, 1997). In both tracers, the strongest growth rate dependence was implied by samples taken in a way that best palaeoclimate practice would avoid (along a slow-growing side transect in a coral colony). However, subtle differences between samples taken from colony 'hills' and 'valleys' starting at the coral top (Alibert and McCulloch, 1997; Cohen and Hart, 1997) suggest that even samples taken in a reasonable way (from the top down along rapidly growing transects) can exhibit growth-related biases.

Finally, the coral geochemical signal can be corrupted by diagenetic alteration of the primary aragonite (Enmar *et al.*, 2000; Müller *et al.*, 2001). The recrystallization of primary aragonite and the deposition of secondary aragonite cements in the interstitial spaces of the coral skeleton both add material that is higher in $\delta^{18}O$ and Sr than the primary biogenic aragonite, reflecting values in isotopic equilibrium with seawater. A core affected by these processes will produce a record biased towards cooler temperatures in the older part of the core. However, recrystallization of aragonite to calcite is easily assessed by X-ray diffraction, and the presence of secondary aragonite cements can be identified by optical microscopy (or even in hand samples if the problem is severe). Such tests should be routine in selecting cores to sample.

13.4 CALIBRATION: THE CHALLENGE TO QUANTIFY RECORDS AND UNDERSTAND MECHANISMS

Calibration experiments are important tools in the attempt to understand palaeoclimate records from corals. Most palaeoclimate studies include some attempt at statistically relating coral data to available instrumental records. However, these efforts face many challenges:

1. Instrumental data are often either extremely short or are not spatially well matched with the coral data (e.g. gridded SST data may not reflect local reef temperatures).
2. Long records of relevant seawater chemistry (Sr/Ca, $\delta^{18}O$, etc.) are virtually non-existent, so variations must be inferred or neglected in long-term studies.
3. Precise temporal matching of coral and instrumental data is difficult, as extension rates can vary throughout a year.
4. Capturing a full seasonal cycle requires very high resolution; a sample resolution of 12/year can still be smoothed by coral growth or the sampling process so that the full seasonal range is not captured.
5. Sampling strategies that fail to account for the skeletal complexity of the coral may inadvertently yield smoothed geochemical profiles and thereby bias relationships between tracers and environmental variability.

Focussed calibration studies (e.g. Gagan *et al.*, 1994; Leder *et al.*, 1996; Swart *et al.*, 1996; Wellington *et al.*, 1996) control for these factors by controlling or closely monitoring the temperature and seawater $\delta^{18}O$, staining the coral skeleton frequently to establish firm time lines, and sampling at very high resolution (up to 50/year). Ideally, calibration should allow comparison of coral and instrumental data over the time-scale of interest to the reconstruction (e.g. interannual or even decadal); in practice it has proven too labour intensive to maintain closely monitored calibration experiments for many years. More attention needs to be given to calibration studies on different species of coral, as tracer–climate relationships differ among taxa and sampling methods need to be customized for each species.

Few studies have attempted to open the 'black box' of coral biology to address the physiological mechanisms of tracer incorporation during calcification. Our best efforts to understand the relationship between coral records and climate will have to involve statistical calibrations, coupled with a growing knowledge of processes that allows us to avoid sampling those skeletal parts where the climate signal is most compromised by biological processes.

13.5 APPLICATIONS

Coral palaeoclimate records have provided unique and significant insights into past changes in tropical climate (Gagan *et al.*, 2000). Such records have yielded reconstructions that calibrate well with existing instrumental data where they overlap and significantly extend our knowledge of tropical climate phenomena. This section describes several examples of significant contributions of coral records to new insights into Holocene climate variability.

13.5.1 The 20th Century in the Context of the Recent Past: Data From Corals

From coral palaeoclimate records that begin significantly before 1900, one can begin to infer how the tropical oceans may have varied prior to the significant anthropogenic buildup of greenhouse gases in the atmosphere. Virtually all coral palaeoclimate records show a trend towards warmer conditions in the 20th century compared to earlier times (Fig. 13.2). If interpreted strictly in terms of temperature, these data imply a 1–3° warming from the 18th and 19th centuries in all records except the Galapagos, where upwelling maintains cool temperatures (Cane *et al.*, 1997). The trend in $\delta^{18}O$ records probably reflects the superposition of freshening and warming.

A recent paper has used multiple tracers from several coral cores to address the long-term trend in climate on the Great Barrier Reef (Hendy *et al.*, 2002). Using samples representing 5 years each from eight cores, this study examined $\delta^{18}O$, U/Ca and Sr/Ca records to determine changes in SST and salinity over the past four centuries. Their results indicate a much stronger trend in $\delta^{18}O$ than in the temperature-sensitive metals, indicating significant salinity reduction during the 'Little Ice Age' of the 15th to 19th centuries. Their conclusions are consistent, they argue, with increased pole-equator temperature gradients, intensified Hadley circulation and expanded glaciers during the Little Ice Age.

Others have argued for systematic differences in tropical climate variability. Urban *et al.* (2000) note a tendency for more decadal ENSO variability in the mid- to late 19th century relative to the 20th century, when background conditions were cooler and/or drier in the central tropical Pacific. How should ENSO variance respond to a background state that includes stronger trades, higher salinity and cooler temperature? Does ENSO variability help to drive, or simply respond passively to, these background shifts? Gaining a clearer picture of the tropical ocean behaviour prior to the 20th century is a challenge that requires not just more and longer records, but also will benefit from forced coupled model experiments that can identify physically consistent scenarios.

13.5.2 ENSO and Decadal Variability

Geochemical records from modern long-lived coral colonies in the equatorial Pacific yield records of variability that correlate extremely well with instrumental ENSO indices (Fig. 13.4) (Cole *et al.*, 1993; Dunbar *et al.*, 1994; Guilderson and Schrag, 1999; Evans *et al.*, 1999; Urban *et al.*, 2000; Cobb *et al.*, 2001). Several studies have noted a connection between changing variability on seasonal–interannual time-scales and longer (decade-century) time-scales (Cole *et al.*, 1993; Dunbar *et al.*, 1994; Tudhope *et al.*, 1995, 2001; Urban *et al.*, 2000; Cobb *et al.*, 2001). Changes in ENSO intensity consistent across the Pacific are seen clearly in the 20th century, and in isolated records before that time. Longer records are sorely needed to explore century-scale variations and global linkages.

Over longer time-scales, instrumental records suggest a bi-decadal pattern of variability in Pacific SST and related atmospheric variables that resembles ENSO in its spatial pattern, although with a broader latitudinal spread of anomalies about the equator (Zhang *et al.*, 1997; Garreaud and Battisti, 1999; Minobe, 1999). Indices of this phenomenon tend to disagree before 1925, likely a reflection of sparse data (Labeyrie *et al.*, 2003). Palaeoclimate records from the Pacific can provide multiple iterations of Pacific decadal variability and help to clarify its preferred time-scale and spatial extent. However, this potential is only starting to be explored. Among the first papers to show coherent decadal variability among coral and climate records are those focusing on ENSO-sensitive areas. Decadal variance is strongly expressed in central Pacific coral records, particularly in the mid- to late 19th century (Urban *et al.*, 2000). A series of decade-scale ENSO oscillations between 1840 and 1890 include cool and warm extremes lasting up to 10 years, with significant extratropical consequences (Cole *et al.*, 2002). Records in the western Indian Ocean are coherent across interannual–decadal time-scales with Pacific ENSO variability (Cole *et al.*, 2000; Cobb *et al.*, 2001), implying that this variability originates in the tropical Pacific. Other papers have focussed on similarities between coral records and extratropical Pacific variability (Linsley *et al.*, 2000a, 2000b; Evans *et al.*, 2001). However, the definitive palaeoclimatic synthesis of Pacific decadal variability has yet to be written, and coral records hold promise for expanding the record of this phenomenon.

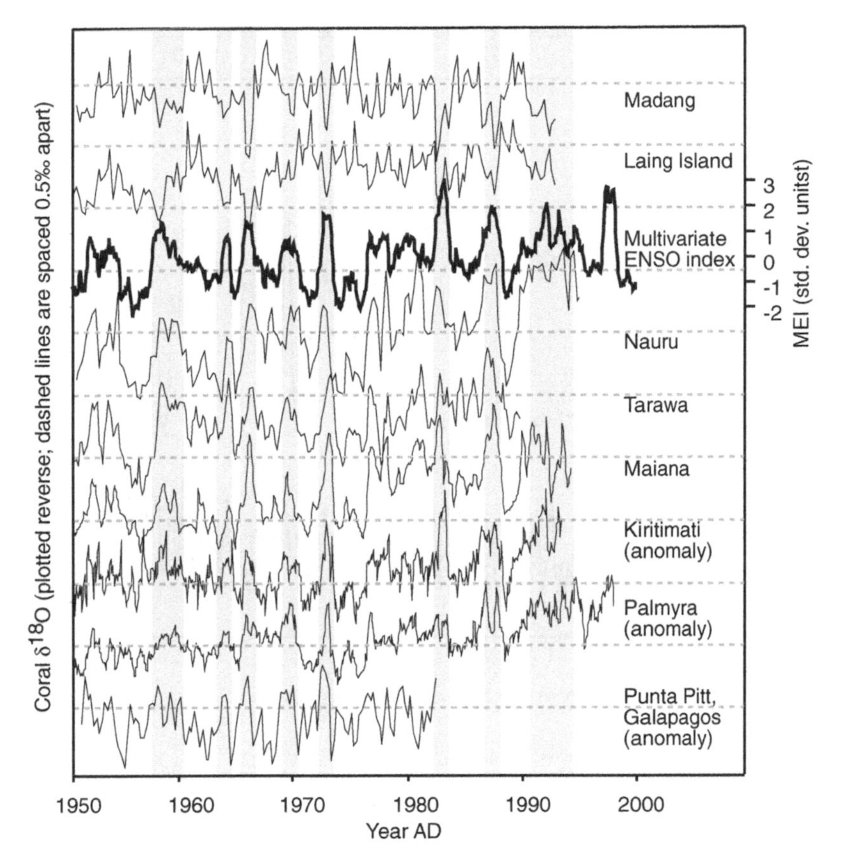

Figure 13.4 Coral records of ENSO, demonstrating close agreement with an instrumental ENSO index, the Multivariate ENSO Index (MEI) (Wolter and Timlin, 1998). Locations of all records are shown in Fig. 13.1 and full-length records are plotted in Fig. 13.2. Note that the Galapagos record shown here is that from Punta Pitt (Shen *et al.*, 1992a), not Urvina Bay. Western Pacific records (plotted above the MEI) show drying and cooling during El Niño records, whereas the eastern and central Pacific records show warmer/wetter conditions. El Niño events (derived from visual inspection of the MEI) are shaded.

13.5.3 Holocene ENSO

The behaviour of the tropical Pacific in the Mid-Holocene has been inferred from several palaeoclimatic and modelling studies, recently summarized (Clement *et al.*, 2000; Cole, 2001). Coral records contribute to this picture mainly in the western Pacific, where a coherent picture is emerging of reduced interannual climate variability (Fig. 13.5). Coral records from the Great Barrier Reef (5400 yr BP) (Gagan *et al.*, 1998) and New Guinea (6500 yr BP) (Tudhope *et al.*, 2001) show no indication of the hydrologic extremes that characterize interannual ENSO fluctuations in modern records, and the former study indicates mean SST warmer than modern. These records are consistent with pollen records from northern Australia that show an absence of drought-adapted taxa before 4000 yr BP (McGlone *et al.*, 1992; Shulmeister and Lees, 1995). A warmer western Pacific would accord with a more La Niña-like background state, but could also conceivably result from a background warming of the entire Pacific. In the eastern

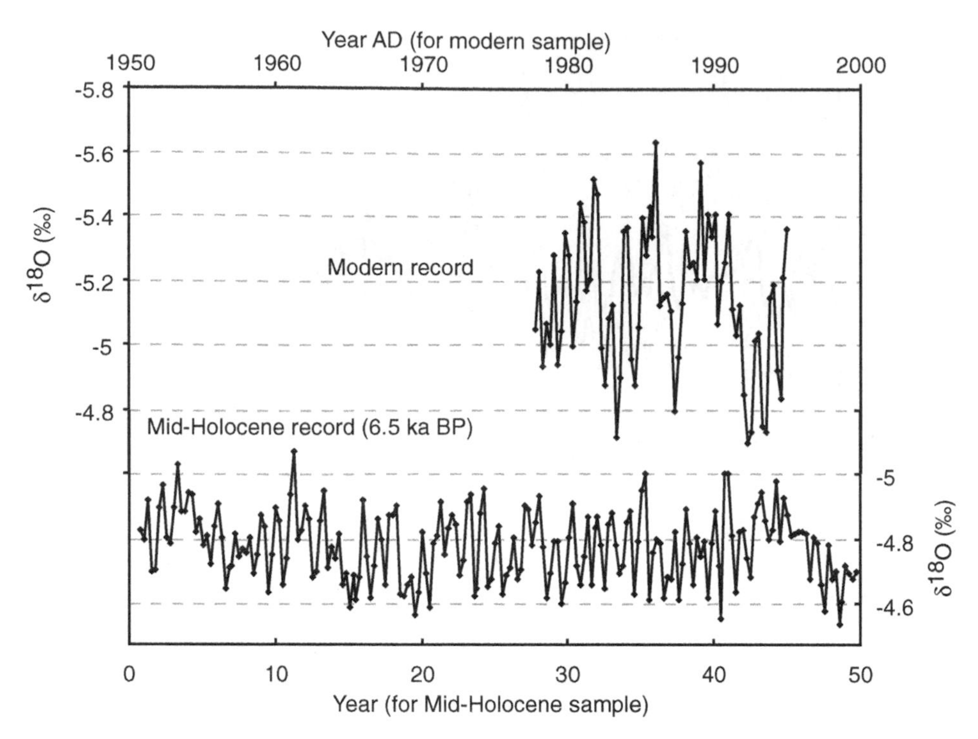

Figure 13.5 Comparison of modern and Mid-Holocene (6500 yr BP) coral records from the Huon peninsula, Papua New Guinea (Tudhope *et al.*, 2001). Both are plotted on the same vertical scale, offset for clarity. The Mid-Holocene record indicates dominantly seasonal variability, with little of the interannual variance that ENSO brings to the modern record today. Longer records from New Guinea (Madang and Laing Island, plotted in Figs 13.2 and 13.4) confirm the dominance of interannual variability in the region today. The Mid-Holocene record is consistent with other palaeoclimate evidence for reduced ENSO variance before about 5000 yr BP (see text discussion).

Pacific, data from Galapagos and Ecuador lakes also suggest reduced variance in the Mid-Holocene (Rodbell *et al.*, 1999; Riedinger *et al.*, 2002), but coastal middens contain molluscan assemblages and fish otoliths that imply warming (Sandweiss *et al.*, 1996; Andrus *et al.*, 2002). Climate model simulations tend to support the idea of a less variable, more La Niña-like state in the Mid-Holocene (Otto-Bliesner, 1999; Clement *et al.*, 2000; Liu *et al.*, 2000). However, no single climate scenario posed for the Mid-Holocene fully agrees with all existing studies, and the behaviour and sensitivity of ENSO remains one of the critical outstanding questions in Holocene palaeoclimatology. More data from corals will contribute to resolving this issue, in coordination with other palaeoclimate data sources.

13.5.4 Climate Field Reconstructions

A stimulating new application for coral climate records is the reconstruction of large-scale climate fields from widely distributed coral data (Evans *et al.*, 2000, 2002). This approach uses an empirical orthogonal function (EOF) analysis to recover common spatiotemporal patterns of variance in annual coral time series. This exercise performed on a dataset of 13 Indo-Pacific coral records identifies three leading spatial patterns of tropical variability, corresponding roughly to ENSO, a tropical warming trend in which the eastern Pacific cools, and a Pacific decadal ENSO-like pattern (Evans *et al.*, 2002). These same three patterns are recovered when SST data from the 13 coral sites are used. One outcome of this type of analysis would be fields of past SST that could be used to drive atmospheric model experiments or to develop indices of phenomena such as ENSO. A more extensive database of coral SST reconstructions is needed to fully recover tropics-wide fields of past variability.

13.6 CONCLUSIONS AND FUTURE WORK

Climate records from corals are already contributing significant new insight into Holocene climate change, both individually and as components of multi-proxy studies. Studies that aim at a better understanding of tracer incorporation and variability should help to improve record development and interpretation, but much can be said now from existing data. Multi-proxy estimates of global and hemispheric temperatures have begun to incorporate multi-century coral SST reconstructions. Coral records have been used to infer oceanographic forcings behind observed decadal climate variations. And coral records contribute a significant part of the evidence for Mid-Holocene weakening of ENSO. Few other sources of proxy information in the tropics can provide the sub-annual resolution, high quality of record and chronological accuracy that corals offer.

Many potential applications for Holocene coral palaeoclimate research are sample-limited. Long (multi-century) chronologies have been more difficult to develop than expected from early work, because long-lived corals are rare in the regions most of interest to study. This problem may be addressed by ongoing studies that explore the palaeoclimate record preserved in slow-growing species, e.g. in the Pacific, *Diploastrea heliopora* (Bagnato *et al.*, 2001; Watanabe *et al.*, 2003; J.E. Cole, unpublished data) and in the Atlantic, *Siderastrea siderea* (Smith, 1995; Guzman and Tudhope, 1998) and *Diploria labyrinthiformis* (Cardinal *et al.*, 2001). These early studies tend to support the preservation of climatic signal in the coral geochemistry, but all note complications and inconsistencies, and they are not unanimous in their conclusions regarding optimal sampling strategies (even for a single species). Analytical techniques that allow for fine-scale sampling of microstructural elements will resolve the best ways to extract climate information from these slowly accreting skeletons.

Coral records can also address palaeoenvironmental questions beyond the strictly climatic that may not require great record lengths. As we attempt to understand the human impact on reefs, records of coral geochemistry that span the period of intense anthropogenic modification of the environment may shed useful light on recent trends. Records of a century or so in length that track nutrient, runoff, heavy metal, or other anthropogenic inputs can reveal how settlement, land-use change, industrialization, agriculture and other human endeavours have changed the reef environment (e.g. Fallon *et al.*, 2002). And records of coral growth and calcification may be able to discern whether rapidly rising atmospheric CO_2 levels are reducing the ability of corals to grow their skeletons, as predicted (Kleypas *et al.*, 1999; Langdon *et al.*, 2000). As human activity impinges on reef health in ever-expanding ways, new applications of coral palaeoenvironmental records will help us to understand how climate and reef environments vary naturally and how humans are impacting these intricate systems.

Acknowledgements

Many thanks to the editors, particularly Anson Mackay, for organizing the UCL Advanced Study Course on Holocene Climate Reconstruction. Thanks also to those who placed their data in the World Data Center for Paleoclimatology, which enabled their inclusion in this paper. I am also grateful to NSF's Earth System History, Palaeoclimatology and CAREER programmes for support of my coral palaeoclimate research. This chapter was improved by discussions with Sandy Tudhope and Warren Beck, and I am grateful to Heidi Barnett for help with the colour plate.

CHAPTER

EVIDENCE OF HOLOCENE CLIMATE VARIABILITY IN MARINE SEDIMENTS

Mark Maslin, Jennifer Pike, Catherine Stickley and Virginia Ettwein

Abstract: Marine sediments have provided critical insight into the climatic variability of the Holocene. They provide continuous palaeoclimatic records for the last 10,000 years with a temporal resolution anywhere from centennial to annual. Marine sediments provide information on changes in the oceans as well as the adjacent continents. For example the following parameters have been reconstructed for the Holocene using marine sediments: (i) ocean conditions: surface and deep circulation patterns, current strengths, sea-surface temperature, sea-surface salinity, iceberg quantity and source, up-welling intensity, productivity, surface-water carbon dioxide content and even water column oxygen content; (ii) continental conditions: global ice volume, failure of ice sheets, wind strength, amount and distribution of dust, aridity, vegetation composition and even erosion rates. Marine sediments have also provided vital palaeo-records to understand the causes of both climatic thresholds and cycles within the Holocene. Examples in this chapter include Holocene Dansgaard–Oeschger cycles, Amazon Basin and African continental hydrology and Asian monsoons.

Keywords: Amazon Basin, Marine sediments, Millennial climate cycles, South China Sea, Tropical North Atlantic

Until a few decades ago it was largely thought that significant large-scale global and regional climate changes occurred at a gradual pace within a time-scale of many centuries or millennia. Climate change was assumed to be scarcely perceptible during a human lifetime. The tendency for climate to change abruptly has been one of the most surprising outcomes of the study of Earth history. In particular, palaeoclimate records from marine sediments have demonstrated that our present interglacial, the Holocene (the last ~10,000 years), has not been as climatically stable as first thought. It seems there are both long-term climate trends (e.g. Haug *et al.*, 2001; Maslin *et al.*, 2001) as well as significant thresholds (e.g. deMenocal *et al.*, 2000b). It has also been suggested that Holocene climate is dominated by millennial-scale variability, with some authorities suggesting a 1500-year cyclicity (O'Brien *et al.*, 1996; Alley *et al.*, 1997, 2001; Bond *et al.*, 1997, 2001; Mayewski *et al.*, 1997; Bianchi and McCave, 1999; deMenocal *et al.*, 2000b; deMenocal, 2001). Evidence from marine sediments has shown that Holocene climate changes can occur extremely rapidly within a few centuries, but frequently within a few decades, and involve regional-scale change in mean annual temperature of several degrees Celsius. In addition many of these Holocene climate changes are stepwise in nature, and may be due to thresholds in the climate system (see Oldfield, pp. 1–9 in this volume).

This chapter is dedicated to the memory of Luejiang Wang, friend and colleague who continues to be sorely missed.

In this chapter we will examine the climate reconstructions that can be made using marine sediments, including bioturbated and laminated sediments. We will summarize some of the key palaeoceanographic records of Holocene climate variability and the current theories for their cause. Figure 14.1 illustrates the Holocene and its climate variability in the context of major global climatic changes that have occurred during the last 2.5 million years.

14.1 THE IMPORTANCE OF THE OCEANS IN REGULATING CLIMATE

Climate is created from the effects of differential latitudinal solar heating. Energy is constantly transferred from the equator (relatively hot) towards the poles (relatively cold). There are two transporters of such energy: the atmosphere and the oceans. The atmosphere responds to an internal or external change in a matter of days, months, or may be a few years (Fig. 14.2). The oceans, however, have a longer response time. The surface ocean can change over months to a few years, but the deep ocean takes decades to centuries. From a physical point of view, in terms of volume, heat capacity and inertia, the deep ocean is the only viable contender for driving and sustaining long-term climate change on centennial to millennial time-scales.

The oceans also have a major impact on the carbon dioxide in the atmosphere. There are four main oceanic processes that help regulate atmospheric carbon dioxide. The first process is ocean temperature, which controls the amount of atmospheric carbon dioxide that can dissolve in the ocean. Second is ocean productivity, as carbon dioxide is extracted by marine organisms to form organic matter via photosynthesis. This biological material can rain out of the surface-waters and some of it is sequestered in marine sediments (Fig. 14.3). At present we are trying to estimate from marine palaeoclimate records how much the ocean productivity varied in the past, particularly in the up-welling zones off the western coasts off South Africa and South America, and its effect on atmospheric carbon dioxide. The third process is the balance between down-welling (deep and intermediate water formation) and up-welling. Surface-water contains both dissolved carbon dioxide and organic matter, when this is down-welled, the carbon is stored in the deep ocean. Most importantly, most deep and intermediate water formation occurs in the high latitudes and the sinking surface-water is extremely cold, and cold water can contain more dissolved gas than warm water. However, oxygen is also dissolved in the water and this over time reacts with the organic matter breaking it down and releasing more carbon dioxide. Hence when the water is finally up-welled it can be extremely rich in carbon dioxide. The fourth process is the amount of calcium carbonate created by corals and planktonic foraminifera and nannofossils, as this regulates the amount of atmospheric carbon dioxide that can dissolve in the ocean.

Since the processes of oceanic heat transfer and carbon dioxide regulation occur on decadal to century time-scales, historic records are too short to provide any record of the ocean system prior to human intervention. Hence we must turn to marine sediments to provide information about oceanic-driven climate change. Such archives can often provide a continuous record on a variety of time-scales. They are the primary means for the study and reconstruction of the stability and natural variability of the ocean system prior to anthropogenic influences.

14.2 THE IMPORTANCE OF PALAEOCEANOGRAPHY

Marine sediments can provide long continuous records of Holocene climate at annual (sometimes intra-annual) to centennial resolution. Marine sediments can also provide information on both

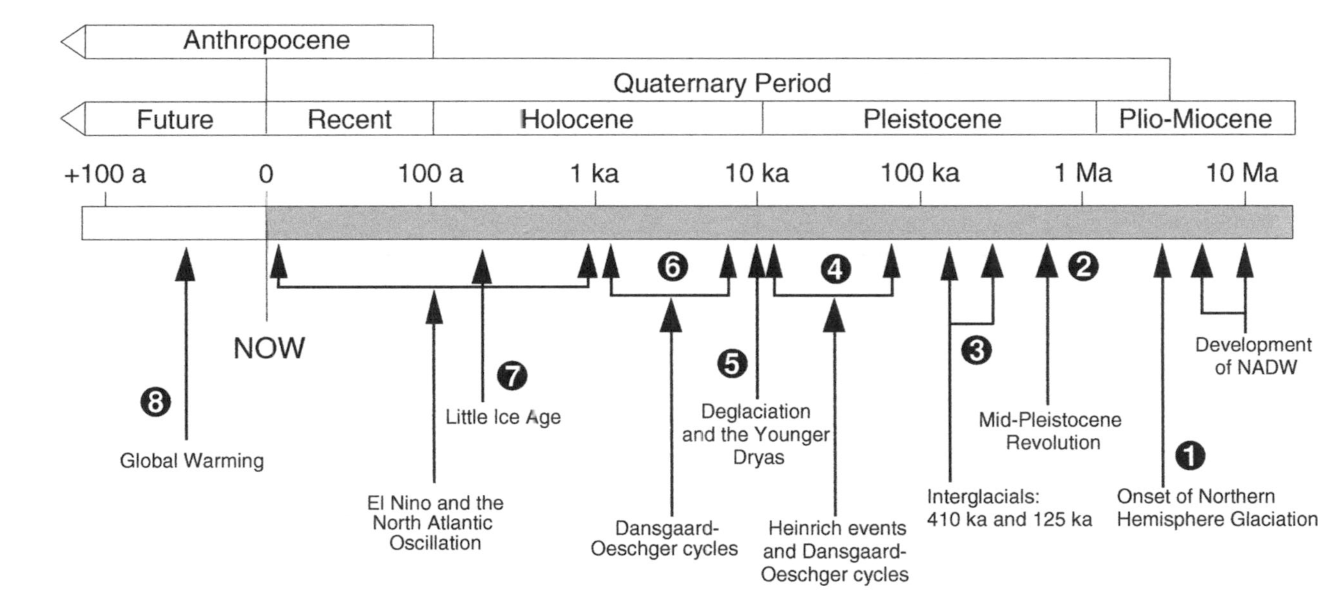

Figure 14.1 Log time-scale cartoon illustrating the most important climate events identified in this study during the Quaternary Period and their relationship to the Holocene (based on Maslin *et al.*, 2001): (1) onset of northern hemisphere glaciation (3.2 to 2.5 Ma) ushering in the strong glacial–interglacial cycles, which are characteristic of the Quaternary Period; (2) Mid-Pleistocene revolution when glacial–interglacial cycles switched from 41 ka, to every 100 ka. The external forcing of the climate did not change, thus, the internal climate feedbacks must have altered; (3) the two closest analogues to the present climate are the interglacial periods at 420–390 ka (oxygen isotope stage 11) and 130–115 ka (oxygen isotope stage 5e, also known as the Eemian); (4) Heinrich events and Dansgaard–Oeschger cycles (see text); (5) deglaciation and the Younger Dryas events; (6) Holocene Dansgaard–Oeschger cycles (see text); (7) Little Ice Age (AD 1700) the most recent climate event, which seems to have occured throughout the northern hemisphere; (8) anthropogenic global warming.

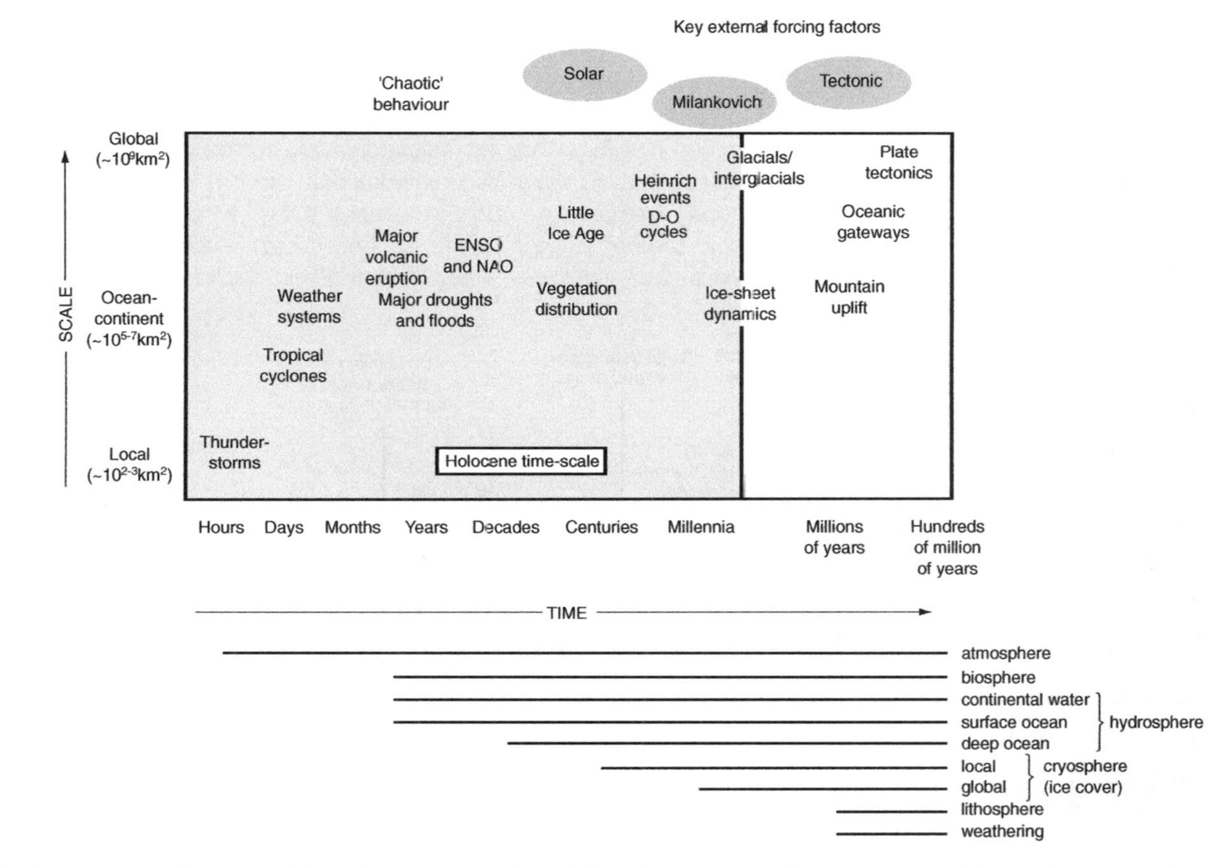

Figure 14.2 Comparison of the spatial and temporal scale of climatic changes with the external forcing factors, and the response of the different parts of the climate system.

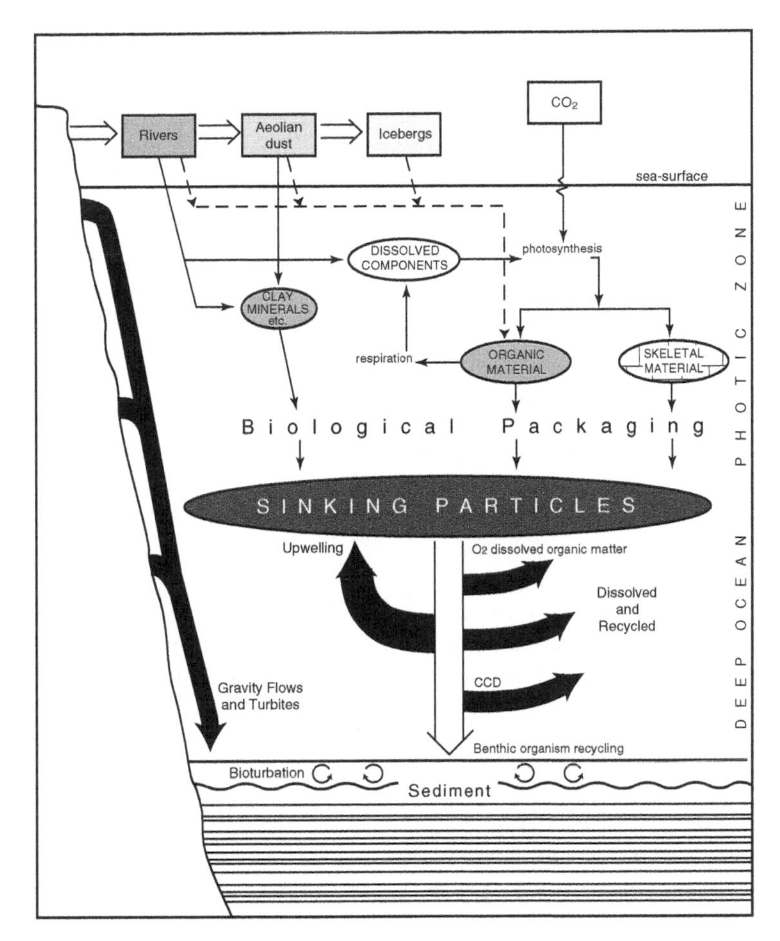

Figure 14.3 Formation of marine sediments, sources and depositional pathways. Adapted from Open University (1989). CCD = Calcium carbonate compensation depth.

changes in the ocean as well as on the surrounding continents. For example, palaeoceanographers have been able to reconstruct the following oceanic parameters using marine sediments: surface and deep circulation patterns, current strengths, sea-surface temperature (SST), sea-surface salinity (SSS), iceberg activity and origin, up-welling intensity, productivity, surface-water dissolved carbon dioxide content and even water column oxygen content. Marine sediments also act as a storage of vital information concerning the continents for example: global ice volume, failure of ice sheets, wind strength, amount and distribution of dust, aridity, vegetation composition and even erosion rates.

14.2.1 Composition of Marine Sediments

Marine sediment is composed of two main components: biogenic and lithogenic (Open University, 1989). The **biogenic component** primarily originates in the surface-water; however, there can be some contribution from the ocean bottom-dwelling community and biogenic material can be transported and deposited in to the oceans by rivers, wind and icebergs (Fig. 14.3). The biogenic component can be sub-divided into three categories:

- organic
- calcium carbonate
- opal-A (biogenic silicate).

Organic matter includes individual molecules, pollen grains, organic-walled microfossils (e.g., dinoflagellate cysts), opaque organic matter (AOM) and wood fragments. The major contribution to the calcium carbonate sediment load comes from coccolithophores (nannofossils) and planktonic foraminifera, with a small contribution from benthic foraminifera and ostracodes. The major contributor to the silicate or opal sediment load are diatoms, with a smaller contribution from silicoflagellates, radiolarians and siliceous sponge spicules. Occasionally, phytoliths (silicate nodules that are formed in land plants, mainly grasses), and freshwater diatoms, transported by either wind or river, are also found in marine sediments.

The **lithogenic component** is primary composed of clays, though larger material up to boulder size can be deposited, for example as iceberg melt. The majority of the lithogenic component originates on the continents where rock and soil are eroded and deposited in the ocean via rivers, wind or icebergs (Fig. 14.3).

One important component of marine sediments that is not usually acknowledged but contains a great deal of chemical and isotopic information, is the pore water within the sediment. This is sea water trapped within the sediment as more sediment accumulates above. Analysis of the oxygen isotopes and chlorine composition of trapped pore waters can provide a valuable insight into past variations in ice volume and the composition of the ocean during the last glacial period (Schrag *et al.*, 1996; Burns and Maslin, 1999).

14.2.2 Distribution of Marine Sediments

Marine sediment can be classified into five main categories based on their primary composition:

1. carbonate
2. siliceous
3. ice-rafted
4. red clay
5. terrigenous.

The distribution of these various sediment types is shown in Fig. 14.4a. The primary controls on the type of sediment are:

- proximity to the continent, which controls whether the ice-rafted debris and terrigenous (mainly riverine) input dominate the sediment
- amount and type of productivity, which controls whether carbonate or siliceous organisms are the dominant component.

In areas far away from the continents where productivity is low, the main input is wind-blown clay, which results in a red clay sediment with extremely low sedimentation rates. Whether carbonate or siliceous material dominates the sediment is complicated but is based on the type of productivity regime and depth of the calcium carbonate compensation depth (CCD). Productivity is controlled by many factors such as light, temperature, salinity and nutrients.

a

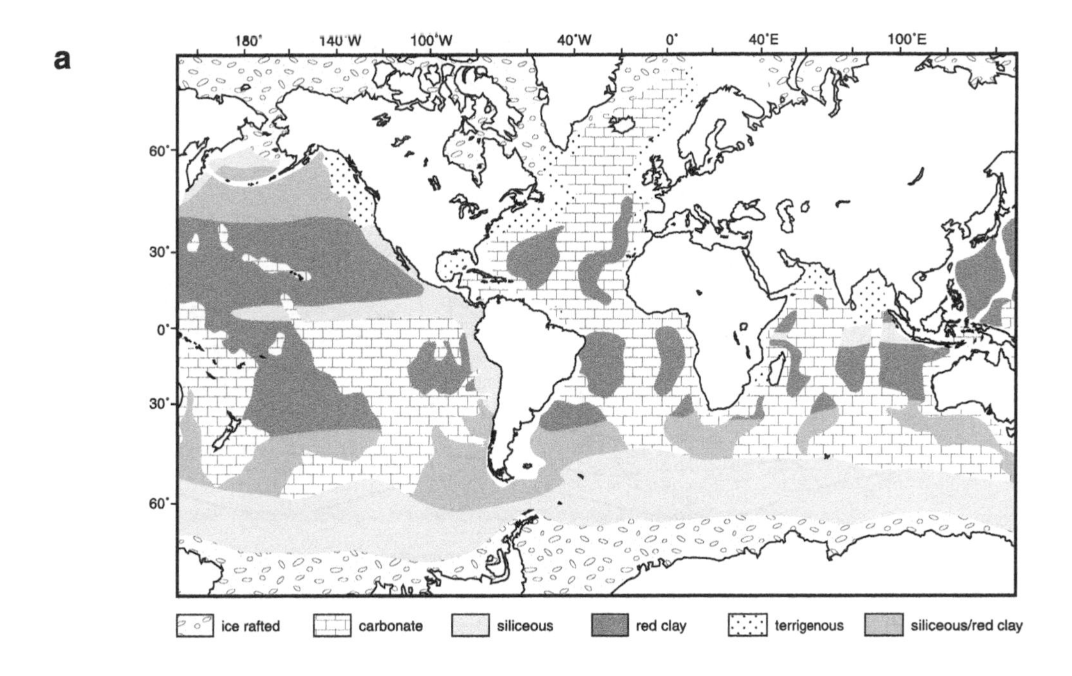

b

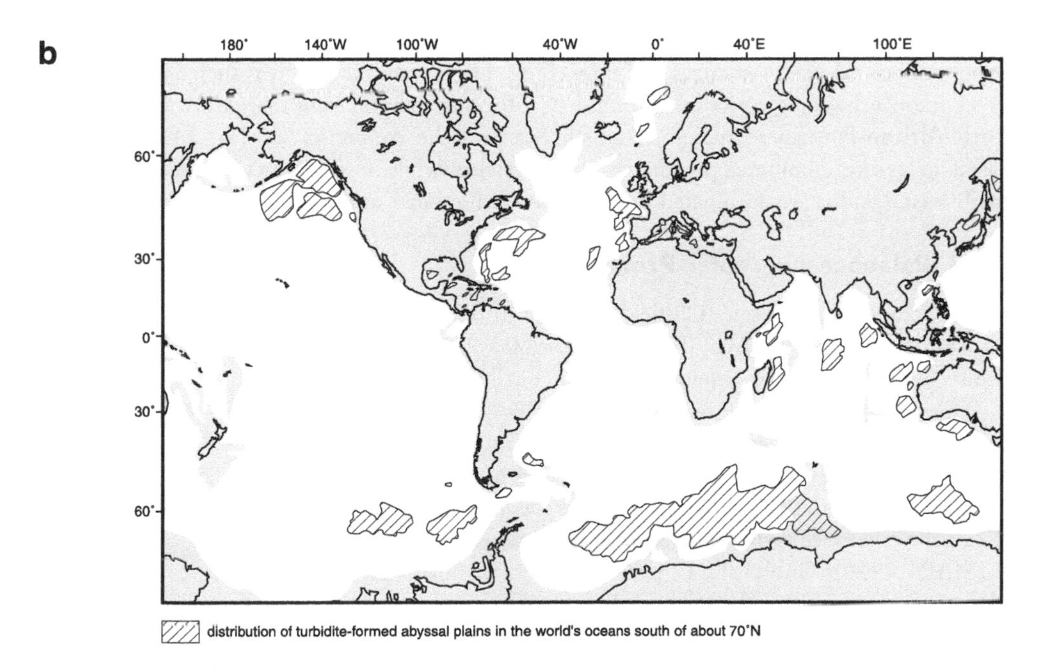

Figure 14.4 Parts (a) and (b). For full caption see page 192.

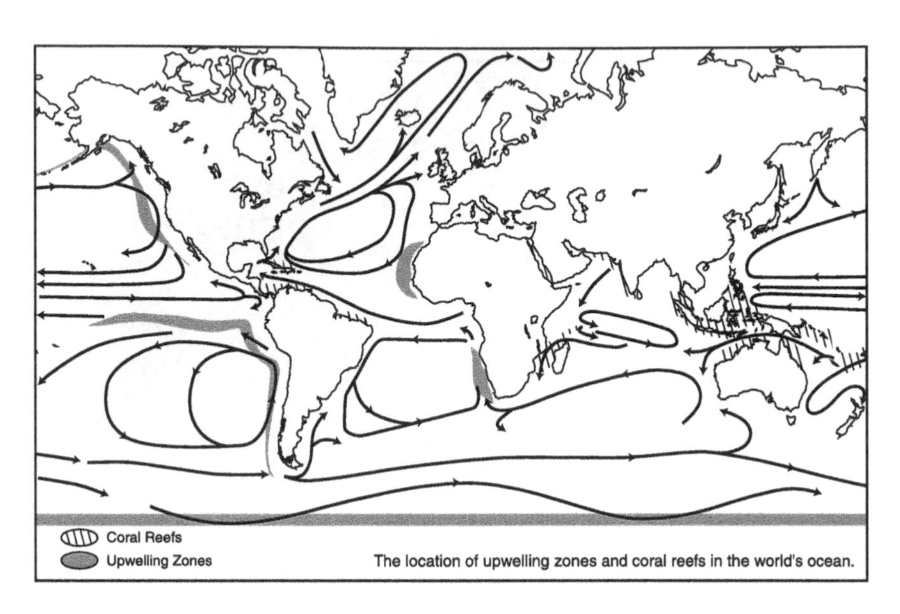

Figure 14.4 (a) Global distribution of marine sediment types. Adapted from Open University (1989). Adapted from Open University (1989); (b) global distribution of continental shelf and abyssal plains, both of which can be highly unstable depositional environments with significant erosion and redistribution of marine sediments; (c) surface ocean circulation and the location of the major high productivity upwelling zones and coral reefs around the world. Adapted from Open University (1989).

Hence both latitude and continental configuration are important because of nutrient availability brought about via season changes and up-welling. Highly productive areas such as the centres of the six main up-welling zones in the world (Californian margin, Peru margin, Angola-Benguela, North African-Portuguese margin, Arabian Sea and the Antarctic, see Fig. 14.4c) and most coastal areas are dominated by diatom productivity, while less productive areas such as the North East Atlantic are dominated by both coccolithophores and foraminifera.

14.2.3 Palaeoceanographic Proxies

In palaeoceanography, proxy (Oldfield, pp. 1–9 in this volume) means a measurable 'descriptor' that stands in for a desired (but unobservable) 'variable' such as past ocean temperature, salinity, nutrient content, productivity, river outflow, wind strength, dust and carbon dioxide etc. These are referred to as '**target parameters**' – things we really want to know about the past ocean and its adjacent continent. The assumption that there is a proxy for these parameters also suggests that these proxies can be calibrated in some way to provide a quantitative estimate of the changes in the target parameter. The classic example of this in palaeoceanography is the relationship between relative species abundance of planktonic foraminifera and SST (e.g. CLIMAP, 1976). Table 14.1 provides a brief overview of the proxies available to palaeoceanographers and the target parameters they represent. A much more comprehensive list of parameters and proxies can be found in Appendix 1 of Wefer *et al.* (1999). Moreover, many of the proxies are discussed in much greater detail in the other chapters in the same volume (Fischer and Wefer, 1999).

Proxies	Target parameters
1. Sedimentology	
Colour:	
Total reflectivity	Per cent calcium carbonate
Red / Blue ratio	Organic matter or clay content
Fabric, texture, structure etc	Depositional environment
Grain size	Sediment type and formation, bottom water current strength
Palaeomagnetics (polarity, inclination, declination and remanent intensity)	Stratigraphy and palaeo-location
Magnetic parameters (Mag Sus, ARM, HIRM, SIRM etc.)	Terrestrial input (iceberg, aeolian, riverine) or relative biological input (Si vs $CaCO_3$)
Crystal structure (XRD)	Composition of sediment, e.g. clay types
Elemental composition (e.g. XRF and ICP-ms)	Composition of sediment e.g., total Ba = productivity, Ti/Al = wind strength proxy, or Ma/Ca in foraminifera tests = SST
2. Stable-isotopes	
$\delta^{18}O$ $CaCO_3$ (foraminifera + nannofossils)	Stratigraphy, global ice volume, temperature and salinity
$\delta^{18}O$ SiO_2 (diatoms)	Stratigraphy, global ice volume and temperature
$\delta^{18}O$ pore water	Bottom water temperature, water masses
$\delta^{13}C$ $CaCO_3$ (foraminifera + nannofossils)	Carbon storage, ocean circulation and productivity
$\delta^{13}C$ total organic	Productivity and terrestrial carbon input
$\delta^{13}C$ biomarkers (individual molecules)	Continental and marine vegetation type, surface water pCO_2
$\delta^{15}N$ total organic	Productivity, nutrient availability, denitrification
$^{87}Sr/^{86}Sr$	Weathering rates and water mass mixing
$^{143}Nd/^{144}Nd$	Riverine input
3. Radiogenic isotopes	
^{14}C (radiocarbon)	Dating, ocean circulation and atmospheric production
Ur/Th series	Dating
$^{40}Ar/^{39}Ar$	Dating
4. Microfossil assemblages	
Planktonic foraminifera	Stratigraphy, sea surface temperature and productivity
Benthic foraminifera	Stratigraphy, surface water productivity, sediment original depth and formation, bottom water oxygenation
Coccolithophores (nannofossils)	Stratigraphy, sea surface productivity
Ostracodes	Water masses and productivity
Diatoms	Stratigraphy, productivity and water masses
Silicoflagellates	Stratigraphy, sea water temperature
Radiolarians	Stratigraphy, palaeolatitude
Dinoflagellate Cysts (Dinocysts)	Stratigraphy and water masses
Pollen	Continental climate and vegetation type
Phytoliths	Continental climate and vegetation type
5. Biomarkers	
Proteins (and DNA)	Species analysis of planktonic foraminifera
Lipids (including waxes and alkenones)	Sea surface temperature, salinity and productivity + continental vegetation type
Pigments (e.g. Chlorins)	Water column oxygen content, productivity

Table 14.1 Brief overview of palaeoceanographic proxies and target parameters.

14.2.4 Resolution

One of the main considerations when trying to reconstruct climate from marine sediments is the time resolution, particularly when studying the rather short period of the Holocene. For example, sedimentation rates in the deep-ocean are on average between 2 and 5 cm/kyr, with highly productive areas producing a maximum of 20 cm/kyr. This limits the temporal resolution to about 200 years per cm (50 years per cm for productive areas). In normal marine sediments an active biological benthic community can mix up the top 20 cm of sediment. This mixing process is called **bioturbation** and reduces the resolution further still to maybe 1000 years per cm. This would provide 10 data points for the Holocene.

In contrast, on the continental shelves, in marginal seas and other specialized natural sediment traps such as anoxic basins and fjords, sedimentation rates can exceed 10 m/kyr (e.g. deep-sea Fans, Maslin *et al.*, 1998; and fjords, Jennings and Weiner, 1996) providing temporal resolution of over 1 year/cm. For example, Pike and Kemp (1997) analysed annual and intra-annual variability within the Gulf of California from laminated sediments containing a record of seasonal accumulation linked to seasonal changes in the structure and nutrient content of the water column. Furthermore, time-series analysis highlighted a decadal-scale variability in *Thalassiothrix*-dominated diatom mat deposition, associated with Pacific-wide changes in surface-water circulation. Such variability is suggested to be influenced by solar-cycles. Often, more localized conditions are recorded in undisturbed laminated marine sediments which form in dysoxic to anoxic environments, where macro-biological activity cannot disturb the sediment. For instance, Pearce *et al.* (1998) revealed seasonal-scale variability during the Late Quaternary at the Mediterranean Ridge (ODP Site 971) from a laminated diatom-ooze sapropel. They inferred changes in the monsoon-related nutrient input to the Mediterranean Basin via the Nile River as the main cause of the variations in the laminated sediments, which suggests a wide influence of changes in seasonality (Kemp *et al.*, 1999, 2000). Other potentially extremely high resolution studies will come from Saanich Inlet, a Canadian fjord (Blais-Stevens *et al.*, 2001) and Prydz Bay in Antarctica (O'Brien *et al.*, 2001), both recently drilled by the Ocean Drilling Program.

A potential drawback to such high-resolution locations is that they may contain highly **localized** environmental and climate information, where some records from laminated sediments can provide **global** climate information. The record of bioturbation from Santa Barbara Basin, California Margin, can be correlated to the Dansgaard–Oeschger cycles seen in the Greenland ice-cores, providing information about the timing of the transmission of climate events around the globe (Behl and Kennett, 1996). Additional problems associated mainly with continental margins are reworking, erosion and redistribution of the sediment by mass density flows such as turbidities and slumps. Figure 14.4b shows the major abyssal plains and the continental shelf which are prone to frequent slope failure and generate the vast majority of the world's turbidites.

14.2.5 Laminated Marine Sediments

Laminated marine sediments contain palaeoceanographic records comparable to other high-resolution proxy climate records such as ice-cores, coral growth rings, tree-rings and even satellite time-series, and they can provide ocean-climate data over time-scales ranging from seasons to millennia. Analysis of laminations has the potential to extend modern instrumental records of seasonality back through the Holocene.

An example of the highest resolution proxy ocean-climate records that can be obtained from laminated sediments is the study of Early Holocene diatom-rich laminae from the Gulf of California, northwest Mexico (Pike and Kemp, 1997). Sediments from the slopes of the Guaymas Basin are laminated throughout the Holocene. A scanning electron microscope-led study revealed that these sediments preserve up to five laminae per year, including four separate diatom assemblages associated with quite different water column conditions, and one terrigenous-rich lamination. Laminae are approximately 1 mm thick, giving a sedimentation rate of approximately 2 mm/yr or 200 cm/kyr.

Holocene laminated sediments from Saanich Inlet, a Canadian fjord, have a sedimentation rate of between 4.96 and 6.79 mm/yr, or 496–679 cm/kyr. Laminations began to form approximately 7000 ^{14}C years ago as the fjord surface-waters became more stratified and primary production increased (Blais-Stevens *et al.*, 2001). Digital sediment colour analysis of approximately 6000 years of the laminated sediments has shown that variations in varve (annual lamina couplet) thickness can be correlated to global climate, but without any obvious one-to-one correlation between wet/dry and warm/cold periods (Nederbragt and Thurow, 2001). The climate around Saanich Inlet was relatively wet during both the Roman Warm Period (2100–1750 yr BP) and the LIA (500 yr BP onwards).

Laminated marine sediments have also been used to analyse high-resolution land–ocean interactions. Monsoonal laminated sediments from the Arabian Sea (Schulz *et al.*, 1998) show the influx of terrigenous sediment from riverine sources alternating with diatom-rich laminations, associated with alternations between the summer and winter monsoon – rain over land and coasts. Analysis of the Late Quaternary laminations from the Cariaco Basin, north of Venezuela, showed that these sediments contain a record of the variation in the strength and position of the Intertropical Convergence Zone during the Late Quaternary, and hence rainfall over northern South America (Hughen *et al.*, 1996).

Not all laminated marine sediments are seasonal. Any cyclical oceanographic or climatic phenomenon has the potential to produce laminations or layered sediments. Flushing of the intermediate and deep waters in the California Margin basin has produced interannual laminations in Santa Monica Basin. The flushing and reoxygenation of the bottom waters of the basin during strong El Niño events leads to laminations controlled by the redox state of the bottom waters (Christensen *et al.*, 1994). Thick laminations in Holocene sediments from the Antarctic Peninsula and Ross Sea, Antarctica, can be correlated with centennial-scale climatic events such as the Little Ice Age and Medieval Warm Period, and longer time-scale events such as the Mid-Holocene Climatic Optimum (Leventer *et al.*, 1993, 1996).

14.3 HOLOCENE CLIMATIC TRENDS AND THRESHOLDS

The best way to demonstrate how marine sediments are helping us understand climate variability in the Holocene is to examine some examples. In this section we have chosen three very different areas, the South China Sea, the Amazon Basin and the tropical North Atlantic. These examples are in no way comprehensive as there are many more excellent marine Holocene studies, but these were chosen to show how marine sediments can pick out small regional changes, large continental climate trends and climatic thresholds.

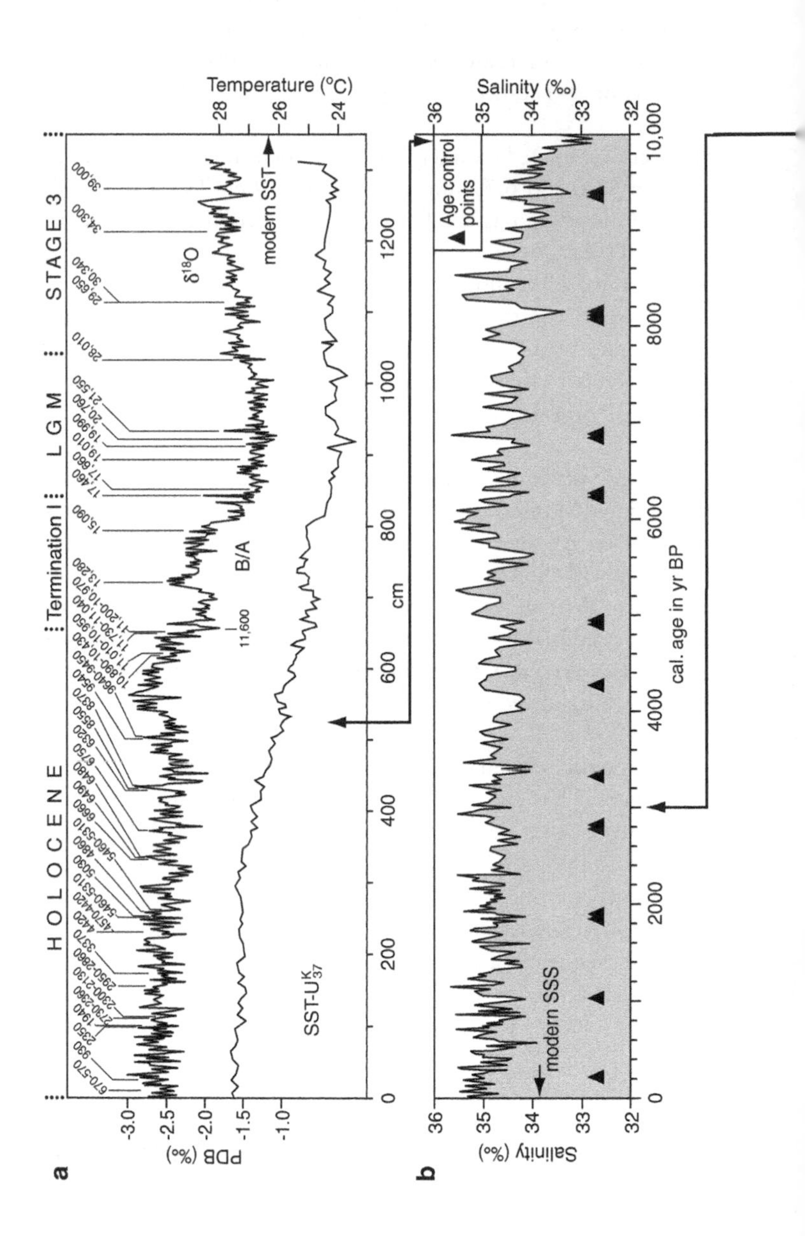
a
PDB (‰)
-3.0
-2.5
-2.0
-1.5
-1.0
HOLOCENE
Termination I
LGM
STAGE 3
δ18O
modern SST
B/A
11,600
SST-U$^{K}_{37}$
Temperature (°C)
28
26
24
cm
0
200
400
600
800
1000
1200
b
Salinity (‰)
36
35
34
33
32
Age control points
modern SSS
cal. age in yr BP
0
2000
4000
6000
8000
10,000

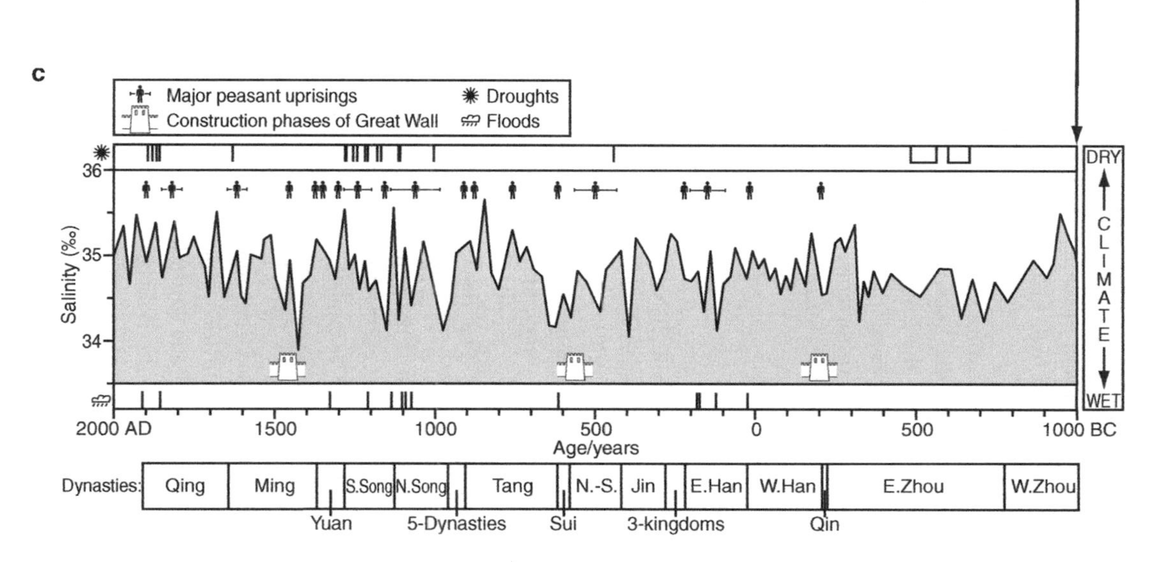

Figure 14.5 South China Sea: (a). *G. ruber* (white) planktonic foraminifera oxygen isotope record and sea-surface temperatures estimated using the $U^{K'}_{37}$ biomarker technique; (b) calculated sea-surface salinity using the above two records for the Holocene. Adapted from Wang *et al.* (1999); (c) comparison of the reconstructed sea-surface salinity for the last 3000 years with documentary evidence from China of periods of droughts, floods, peasant uprisings, construction phases of the Great Wall and changes in the ruling Dynasties. Redrawn from Wang *et al.* (1999).

14.3.1 South China Sea

Wang *et al.* (1999) produced a fascinating study from a marine core recovered from the South China Sea (Fig. 14.5). Using the biomarker $U^{K'}_{37}$ as a proxy for SST they were able to remove the temperature effect from the oxygen isotope record to leave a record of SSS for the last 10,000 years (Fig. 14.5a). This SSS record provides an insight into the Chinese monsoons as the salinity drops in the South China Sea during heavy monsoon periods. As detailed records have been kept by the Chinese over the last 3000 years, Wang *et al.* (1999) were able to compare their SSS reconstruction with the occurrence of both droughts and floods (Fig. 14.5c). Over the last 3000 years there do not seem to be any climatic trends in the climate of the South China Sea and the adjacent Chinese main land but there are a lot of high-frequency fluctuations that seem to occur with a cyclicity of 775, 102 and 84 years. The 775-year periodicity Wang *et al.* (1999) interpret as part of the Dansgaard–Oeschger cycles originating in the North Atlantic which are discussed later. The 84-year cycle corresponds with the known Gleissberg solar periodicity, demonstrating the importance of the sunspot cycles on the Asian monsoons during the Holocene. Wang *et al.* (1999) also correlate the changes in the Asian monsoon over the last 3000 years with changes in the Chinese Dynasties, periods of peasant uprising and construction phases of the Great Wall of China (Fig. 14.5c); suggesting in part that climate variability can be a very strong political force.

14.3.2 Amazon Basin

One of the advantages of marine sediments is that they can provide high-resolution continuous records of continental climate. Moreover, when combined with long lake sequences, it is possible to start piecing together the climate of large regions. Currently a controversial area of Holocene climate research is the moisture history of the Amazon Basin. This is an important area of research as the Amazon Basin may be the source of two very important greenhouse gases, water vapour and methane. Two marine records have been produced using different proxies to examine the moisture history of the Amazon Basin. The first used planktonic foraminiferal oxygen isotopes recovered from Amazon Fan (ODP 942C) as a proxy for the relative mixing of fresh Amazon River water and the Atlantic Ocean (Maslin and Burns, 2000). From this record, Maslin and Burns (2000) were able to make an estimate of the percentage changes in the outflow of the Amazon River and hence the changes in rainfall over the Amazon Basin (Fig. 14.6b and c). The second marine record was produced from the laminated sediments recovered from the Cariaco Basin (ODP 1002). Haug *et al.* (2001) used the variation of the elements iron (Fe) and titanium (Ti) as proxies for increased clay and silt depositions during rainy periods (Fig. 14.6a). Hence the elemental analysis provided a detailed monsoon index, which shows a clear Younger Dryas, Medieval Warm Period and Little Ice Age. If these two records are compared to the long lake records from Peru and the Bolivian Altiplano (Fig. 14.6d), it can be seen that the moisture history of South America is more complicated than any one record can reconstruct. Marine records, however, provide a valuable insight into the regional trends through the Holocene in the Amazon Basin.

14.3.3 Tropical North Atlantic

We have chosen this example of a very sharp climatic threshold in the Holocene as it leads on to the discussion of millennial-scale climate 'variability' in the Holocene. Reconstructions of the climate of North West Africa using deep-sea sediment recovered from ODP Site 658C by deMenocal *et al.* (2000b) are shown in Fig. 14.7b. Using the percentage of terrigenous material

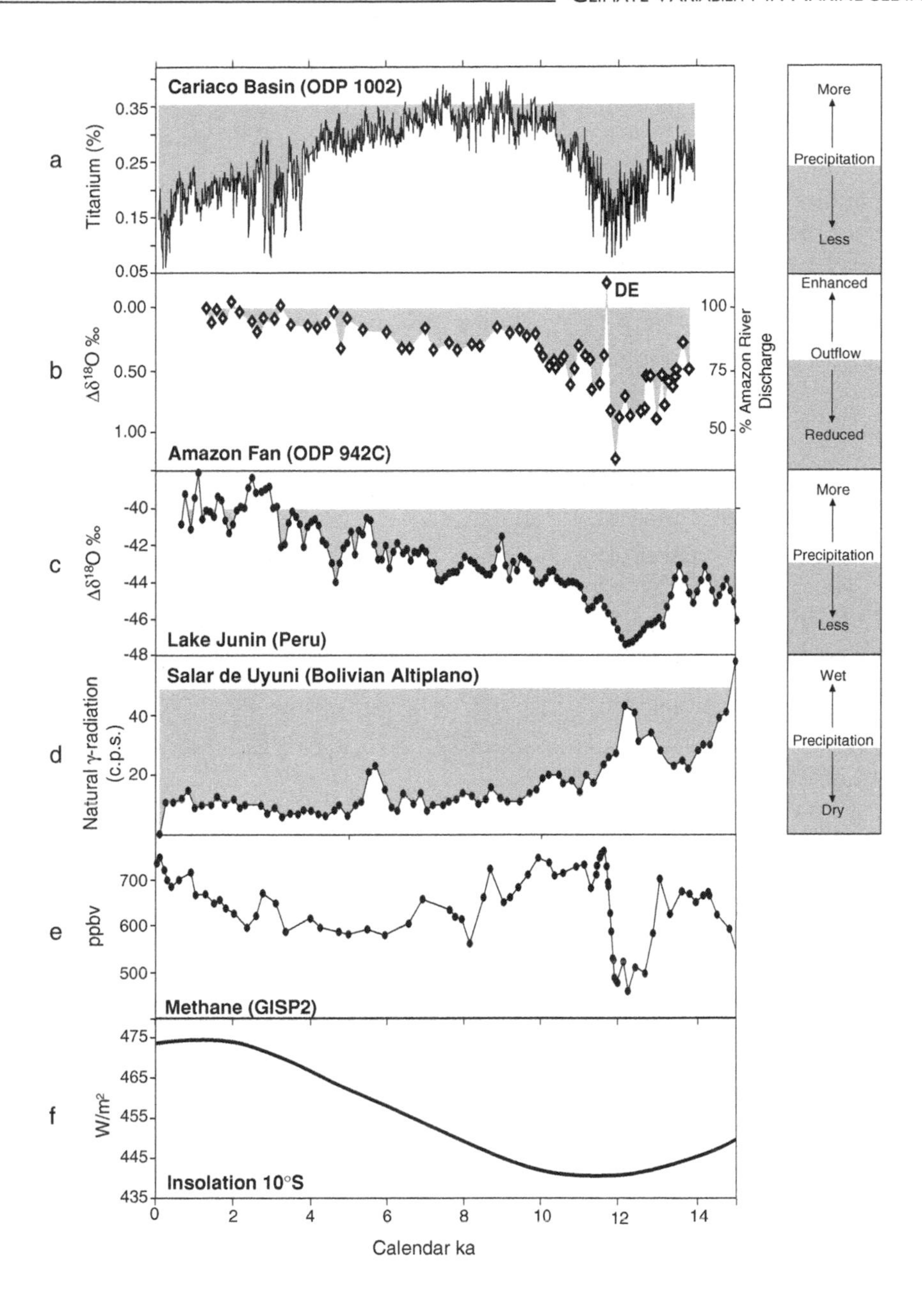

Figure 14.6 Amazon Basin: comparison of moisture reconstructions around the Amazon Basin. (a) Cariaco Basin ODP 1002, monsoon intensity record (Haug *et al.*, 2001); (b) Amazon Fan ODP 942C, Amazon River discharge record (Maslin and Burns, 2000); (c) Lake Junin, Peruvian high Andes, precipitation record (Seltzer *et al.*, 1999); (d) Salar de Uyuni, Bolivian Altiplano precipitation record (Baker *et al.*, 2001); (e) GISP2 methane record (Severinghaus and Brook, 1999); and (f) Solar Insolation at 10°S (Berger and Loutre, 1991). These records show the complexity of the moisture availability of the Amazon Basin and the need for multiple records from different sources, marine sediments, lakes and ice-cores.

found in the sediment, deMenocal *et al.* (2000b) found that during the Early Holocene, North Africa was relatively wet. This corresponds to continental records that show a marked reduction in the extent of the Sahara Desert from 14.8 ka to ~5.5 ka (e.g. Petit-Maire, 1994) which has been called the African Humid Period. After 5.5 ka there was a 300-year transition to much drier conditions in this area that have persisted to today. This demonstrates that the Holocene climate, far from being constant, contains major climatic thresholds, some of which still have to be fully explained. deMenocal *et al.* (2000b) also confirmed that the millennial-scale climate events recorded in both the North Atlantic and the Greenland ice-cores also occurred in the SST record off North Africa (Fig. 14.7), which leads on to the following discussion of Holocene climatic events and cycles recorded in marine sediments.

14.4 HOLOCENE CLIMATIC EVENTS?

Initial studies of the Greenland ice-core records concluded the absence of any major climate variation within the Holocene (e.g. Dansgaard *et al.*, 1993). This view is being progressively eroded (O'Brien *et al.*, 1995), particularly in light of new information being obtained from marine sediments (see Fig. 14.7). Long-term trends indicate an Early to Mid-Holocene climatic optimum, with a cooling trend in the Late Holocene (e.g. deMenocal *et al.*, 2000b). Superimposed on this trend are several distinct oscillations, or climatic cooling steps, which appear to be of widespread significance (see Fig. 14.7), the most dramatic of which occur at 8.2 ka, 5.5 ka, 4.2 ka and between AD 1200 and 1650.

The first cold event during the Holocene is at 9.9 ka, referred to as the Pre-Boreal cold event. However, much more dramatic is the second event at 8.2 ka which is the most striking and abrupt, leading to widespread cool and dry conditions lasting perhaps 200 years, before a rapid return to warmer and generally moister climates than at present. This event is noticeably present in the GISP2 Greenland ice-cores, from which it appears to have been about half as severe as the Younger Dryas to Holocene transition (Alley *et al.*, 1997; Mayewski *et al.*, 1997). Marine records of North African to Southern Asian climate suggest more arid conditions involving a failure of the summer monsoon rains (e.g. Sirocko *et al.*, 1993). Cold and/or arid conditions also seem to have occurred in northernmost South America, eastern North America and parts of northwest Europe (Alley *et al.*, 1997).

In the Mid-Holocene, at approximately 5.5–5.3 ka, there is a sudden and widespread shift in precipitation causing many regions to become either noticably drier or moister (e.g. deMenocal *et al.*, 2000b). The dust and SST records off northwest Africa show that the African Humid Period, when much of sub-tropical West Africa was vegetated, lasted from 14.8 to 5.5 ka (deMenocal *et al.*, 2000b). After 5.5 ka there was a 300-year transition to much drier conditions in this area. This shift also corresponds to a marked change in Arctic sea-ice variability (Jennings *et al.*, 2002). In addition there is a vegetation response with the decline of Elm (*Ulmus*) in Europe at about 5.7 ka, and of hemlock (*Tsuga*) in North America at about 5.3 ka. Both vegetation changes were initially attributed to specific pathogen attacks (Rackham, 1980; Peglar 1993); an alternative theory is that they may have been related to climate deterioration (Maslin and Tzedakis, 1996). The step to colder and drier conditions is analogous to a similar change that is observed in Marine Oxygen Isotope Stage 5e (Eemian) records (Maslin *et al.*, 1996).

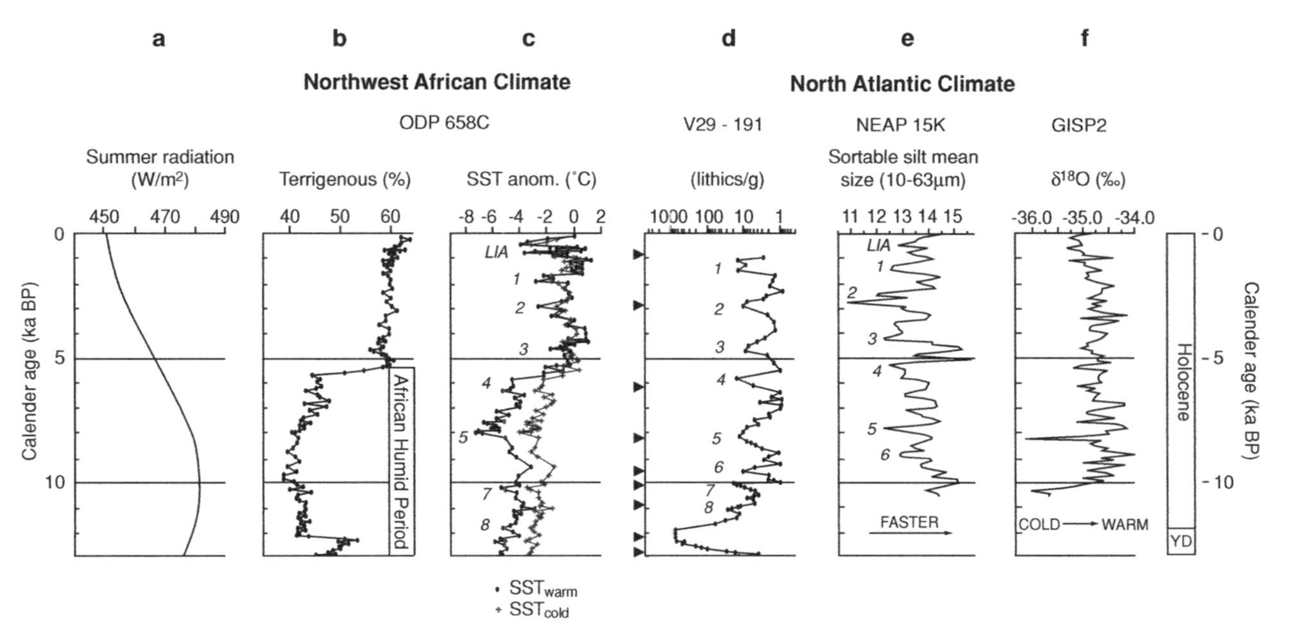

Figure 14.7 Comparison of summer insolation for 65°N (Berger and Loutre, 1991), with Northwest African climate (deMenocal *et al.*, 2000b) and North Atlantic climate (V29-191, Bond *et al.*, 1997; NEAP 15K, Bianchi and McCave, 1999; GISP2, O'Brien *et al.*, 1995). Note the climate threshold seen in the North West African dust record b and the similarity of events labelled 1 to 8 and the Little Ice Age (LIA).

There is also evidence for a strong, cold and arid event occurring at about 4.2 ka across the North Atlantic, northern Africa and southern Asia (Bradley and Jones 1993; O'Brien *et al.* 1995; Bond *et al.*, 1997; Bianchi and McCave, 1999; deMenocal *et al.*, 2000b; Cullen *et al.*, 2000). This cold, arid event coincides with the collapse of a large number of major urban civilizations including: the Old Kingdom in Egypt, the Akkadian Empire in Mesopotamia (at the head waters of the Tigris–Euphrates Rivers), the Early Bronze Age societies of Anatolia, Greece, Israel, the Indus Valley civilization in India, the Hilmand civilization in Afganistan and the Hongshan culture of China (Peiser, 1998; Cullen *et al.*, 2000; deMenocal, 2001). Detailed evidence of this event has come from marine sediments recovered from the Gulf of Oman. Cullen *et al.* (2000) show that at about 4025 yrs BP and for the subsequent 300 years there was a significant increase in the amount of dust reaching the ocean (Fig. 14.8), suggesting a period of extreme drought. Stable isotopes (Nd and Sr) confirm that this dust came from the Mesopotamian source area and thus this arid period may have led to the collapse of the highly developed Akkadian empire (Fig. 14.8).

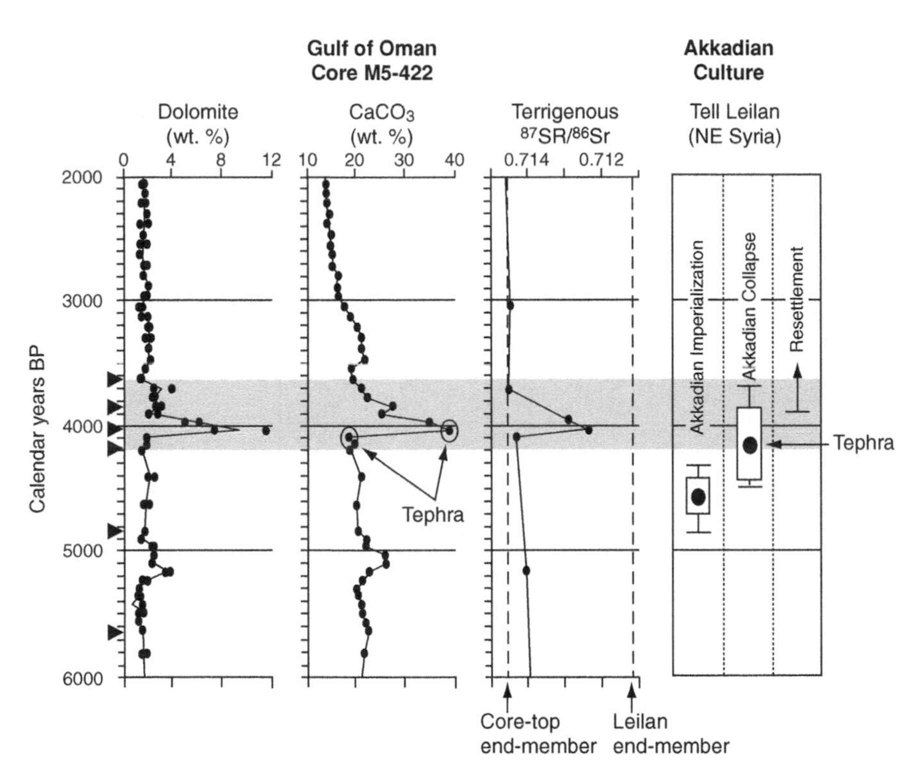

Figure 14.8 Mesopotamian palaeoclimate and the collapse of the Akkadian empire as shown by archeological remains at Tell Leilan in Syria and a deep-sea sediment core recovered from the Gulf of Oman (Cullen *et al.*, 2000; deMenocal, 2001). The massive increase in dolomite and calcium carbonate in the marine sediments, which occurred at 4025 BP and persisted for 300 years is excellent evidence for a huge prolonged drought. The strontium isotopes confirm that this additional dust came from Mesopotamia and suggests that this prolonged period of aridity lead to the collapse of the Akkadian empire. Adapted from deMenocal (2001).

The most recent Holocene cold event is the Little Ice Age (LIA) (see Figs 14.7 and 14.9). This event has been documented from regions as far apart as Greenland and the Ross Sea, Antarctica (Leventer *et al.*, 1996). It is an interesting aside to note that during the LIA in Antarctica, the majority of the continent was cold; however, the Antarctic Peninsula exhibited some warming (Leventer *et al.*, 1996). However, the LIA event is really two cold periods, the first follows the Medieval Warm Period (MWP) which ended a 1000 years ago and is often referred to as the Medieval Cold Period (MCP) or **LIA b** (e.g. deMenocal *et al.*, 2000b; Andrews *et al.*, 2001a, 2001b). The MCP played a role in extinguishing Norse colonies on Greenland and caused famine and mass migration in Europe (e.g. Barlow *et al.*, 1997). It started gradually before AD 1200 and ended at about AD 1650 (Bradley and Jones, 1993). The second cold period, more classically referred to as the Little Ice Age or **LIA a** (deMenocal *et al.*, 2000b; Andrews *et al.*, 2001a, 2001b) may have been the most rapid and largest change in the North Atlantic during the Holocene as suggested from ice-core and deep-sea sediment records (O'Brien *et al.*, 1995; Mayewski *et al.*, 1997; deMenocal *et al.*, 2000b). The LIA events are characterized by a fall of 0.5–1 °C in Greenland temperatures (Dahl-Jensen *et al.*, 1998), significant shift in the currents around Iceland (Jennings and Weiner, 1996; Andrews *et al.*, 2001a, 2001b) and a SST fall of 4 °C off the coast of West Africa (deMenocal *et al.*, 2000b) and 2 °C off the Bermuda Rise (Keigwin, 1996) (see Fig. 14.9).

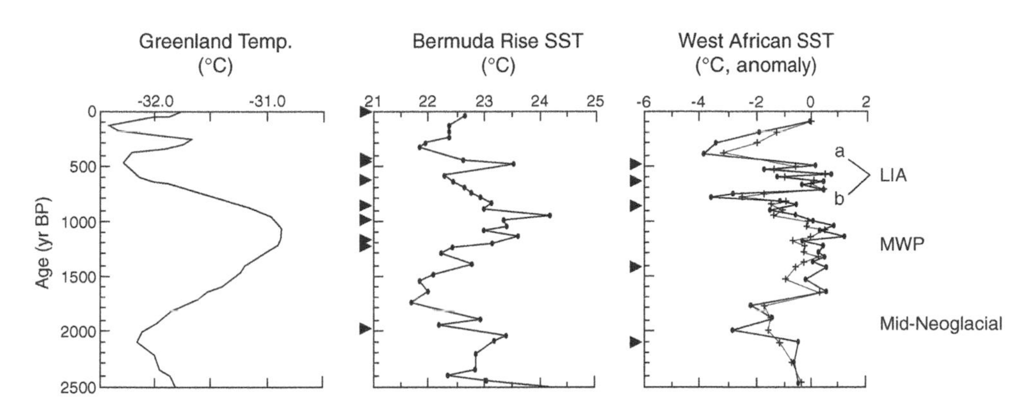

Figure 14.9 Comparison of Greenland temperature (Dahl-Jensen *et al.*, 1998), the Bermuda Rise sea-surface temperature (Keigwin, 1996) and West African sea-surface temperature (deMenocal *et al.*, 2000b) records for the last 2.5 ka. LIA = Little Ice Age, MWP = Medieval Warm Period. Solid triangles indicate radiocarbon dates.

14.5 HOLOCENE CLIMATIC 'DANSGAARD–OESCHGER' CYCLES

The above events are now regarded as part of the millennial-scale quasi-periodic climate changes characteristic of the Holocene (see Fig. 14.1). Similar cycles have also been documented for the last interglacial period (Bond *et al.*, 2001). These interglacial cycles are thought to be similar to glacial Dansgaard–Oeschger (D–O) cycles (Pisias *et al.*, 1973; O'Brien *et al.* 1995; Andrews *et al.*, 1997, 2001a, 2001b; Bond *et al.*, 1997; Bianchi and McCave, 1999; deMenocal *et al.*, 2000b). The periodicity of these Holocene D–O cycles is a subject of much debate. The first study to recognize the strong ~1500 cycles in both the Holocene and Late Pleistocene was by Pisias *et al.* (1973). This work was only rediscovered after the initial analysis of the GISP2

Greenland ice-core and North Atlantic sediment records revealed these cycles at approximately at the same 1500 (± 500)-year rhythm as that found within the last glacial period (O'Brien *et al.*, 1995; Bond *et al.*, 1997; Mayewski *et al.*, 1997; Campbell *et al.*, 1998; Bianchi and McCave, 1999; Chapman and Shackleton, 2000). In general, during the coldest point of each of the millennial-scale cycles (shown in Fig. 14.7c), surface-water temperatures of the North Atlantic were about 2–4 °C cooler than during the warmest part (Bond *et al.*, 1997; deMenocal *et al.*, 2000b).

14.5.1 Causes of Millennial Climate Cycles

Comparison of deep-sea sediment palaeoclimate records with other palaeoclimate data is providing us with essential insight to the possible cause of both glacial and interglacial Dansgaard–Oeschger cycles. Marine records have demonstrated that the deep-water system is involved. Bianchi and McCave (1999) analysed deep-sea sediment recovered from the North Atlantic and by measuring grain size they were able to make assumptions about the strength of the bottom-water currents. From this, they suggested that during the Holocene there have been regular reductions in the intensity of NADW (see Fig. 14.7e) linked to the 1500-year D–O cycles identified by O'Brien *et al.* (1995) and Bond *et al.* (1997). This makes sense as the deep ocean is the only candidate for driving and sustaining internal long-term climate change (of hundreds to thousands of years) because of its volume, heat-capacity and inertia (e.g. Broecker, 1995; Jones *et al.*, 1996; Rahmstorf *et al.*, 1996; Seidov and Maslin, 1999; Maslin *et al.*, 2001). The question is what causes the deep-ocean system to vary.

At the present time in the North Atlantic, the north–east trending Gulf Stream carries warm and relatively salty surface-water from the Gulf of Mexico up to the Nordic seas (Fig. 14.4b). Upon reaching this region, the surface-water has sufficiently cooled so that it becomes dense enough to sink forming the North Atlantic Deep Water (NADW). The 'pull' exerted by this dense sinking maintains the strength of the warm Gulf Stream, ensuring a current of warm tropical water into the North Atlantic that sends mild air masses across to the European continent (e.g. Schmitz, 1995; Rahmstorf *et al.*, 1996). Formation of the NADW can be weakened by two processes:

1. A shift in the position of the atmospheric polar front due to, for example, the presence of huge ice sheets over North America and Europe, preventing the Gulf Stream from travelling as far north. This reduces the amount of cooling that occurs and hence its capacity to sink which happened during the last glacial period (e.g. Broecker and Denton, 1989)
2. The input of freshwater, either from melting sea-ice or adjacent ice sheets or increased precipitation, which forms a lens of less-dense water preventing sinking (e.g. Rahmstorf *et al.*, 1996).

If NADW formation is reduced it weakens the warm Gulf Stream, causing colder and drier conditions particularly during winter throughout the entire North Atlantic region, and thus has a major impact on global climate (e.g. Broecker, 1995).

There are currently three suggested reasons for Holocene millennial-scale climate varability and the involvement of the deep-ocean system, all of which may play some part in the quasi-periodic cycles observed:

14.5.1.1 Internal Instability of the Greenland Ice Sheet or Arctic Sea-Ice

This theory is an extension of the explanations put forward for the cause of the glacial D–O and Heinrich cold events (e.g. MacAyeal, 1993). The suggestion is that large ice sheets and large areas of sea-ice such as the Arctic Ocean may have their own internal waxing and waning rates (Jennings *et al.*, 2002). These internal oscillations may be causing increased melt-water in the Nordic Seas which reduces deep-water formation leading to a Holocene D–O event (based on studies by Seidov and Maslin, 1999; van Kreveld *et al.*, 2000; Jennings *et al.*, 2002; Sarnthein *et al.*, 2001).

14.5.1.2 Ocean Oscillator and the Bipolar Climate Seesaw

The second possible cause is an extension of the suggested intrinsic glacial millennial-scale 'bipolar seesaw' (Broecker, 1998; Stocker, 1998; Seidov and Maslin, 2001) to the Holocene. One of the most important finds in the study of glacial millennial-scale events is the apparent out-of-phase climate of the two hemispheres, as observed in the ice-core climate records from Greenland and Antarctica (Blunier *et al.*, 1998). It has been suggested that this bipolar seesaw can be explained by variations in relative amount of deep-water formation in the two hemispheres and the resulting direction of the inter-hemispheric heat piracy (Seidov and Maslin, 2001; Seidov *et al.*, 2001) (see Fig. 14.10). The bipolar seesaw model suggested by Maslin *et al.* (2001) could be self-sustaining with melt-water events in either hemisphere triggering a chain of climate changes which eventually causes a melt-water event in the opposite hemisphere thus switching the direction of the heat piracy.

Figure 14.11 shows the feedback loops that could produce this self-sustaining bipolar climate seesaw system. For example, a D–O cold type event is typically accompanied by an increase amount of ice rafting or melt-water in the North Atlantic. This produces a low-density surface-water that acts as a lid, reducing NADW production (Fig. 14.10c). This in turn reduces and then reverses the heat exchange (or piracy) between the hemispheres: the northern hemisphere cools while the southern hemisphere warms. Over hundreds of years this additional southern warming could melt Antarctic continental- and sea-ice, adding freshwater to the surface-waters. Once a critical amount of freshwater is added, the system crosses a threshold and Antarctic Bottom Water (AABW) volume is dramatically reduced (Fig. 14.10d). This causes an increase in the formation of NADW; this re-invigoration of the NADW also reverses the heat exchange, and returns the dominance of the climate system to the North Atlantic. As heat builds up in the North Atlantic the probability of significant sea-ice melting and failure of a surrounding continental ice sheet increases, starting the cycle again. The century time-scale required for deep-water to warm or to cool the surface-waters in the respective hemisphere, could provide the delay necessary to produce the millennial time-scale cycles. This mechanism of altering the dominance of NADW and AABW could be applied to the Holocene as all that would be required are changes in the rate of formation or melting of sea-ice in the Arctic–Nordic Seas verses the Southern Ocean.

14.5.1.3 North Atlantic Changes Forced by Solar Variations

The latest twist to the tale is the Bond *et al.* (2001) study linking marine sediments from the North Atlantic and radionuclide records from the Greenland ice-cores. They show a correlation for the last 12,000 years between drift-ice-deposited material in the North Atlantic and the production of both ^{14}C and ^{10}Be in the atmosphere (see Fig. 14.12). Bond *et al.* (2001) suggest that millennial-

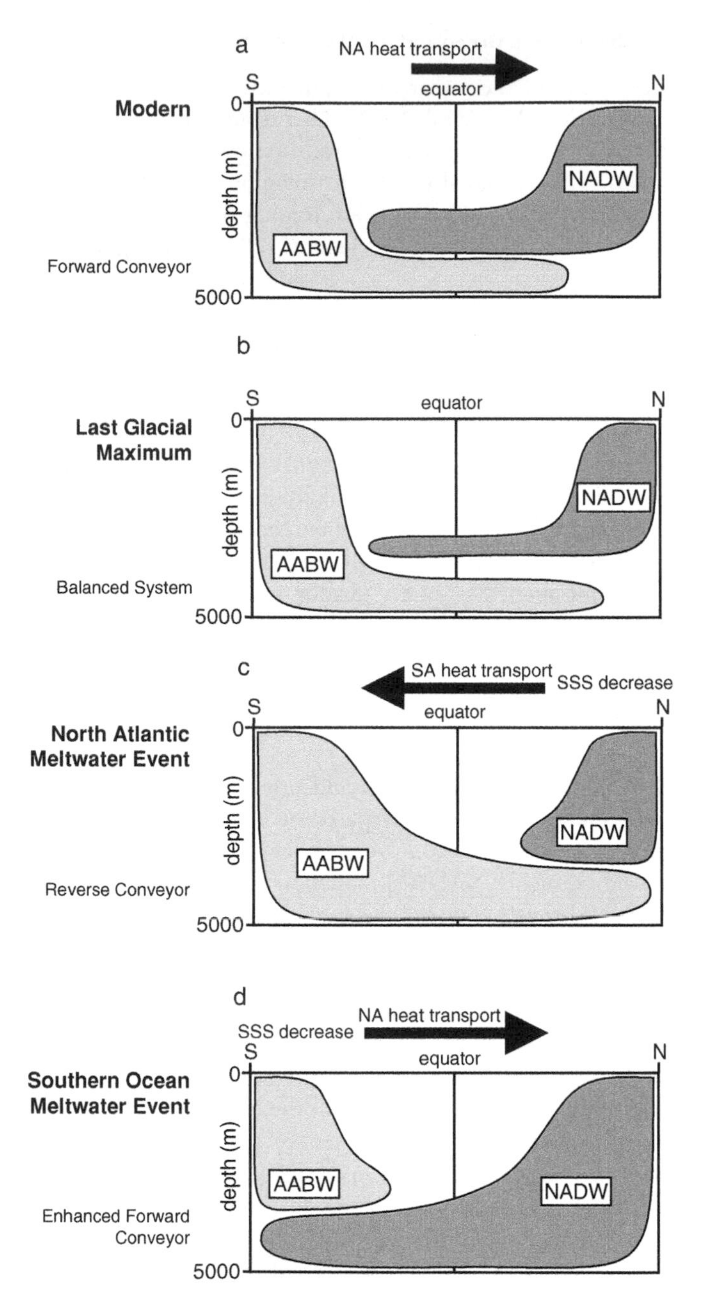

Figure 14.10 Illustration of the four different states of the deep ocean conveyor. (a) Modern circulation or *Forward* conveyor with the dominance of the NADW and heat transport to the northern hemisphere; (b) LGM circulation with a nearly balanced heat exchange between the two hemispheres; (c) melt-water perturbation in the North Atlantic, that results in a *Reverse* conveyor with the dominance of the AABW and heat transport to the southern hemisphere; (d) melt-water perturbation in the Southern Ocean, that results in an *Enhanced Forward* conveyor with the reduction of the AABW and enhancement of the NADW and strong heat transport to the northern hemisphere.

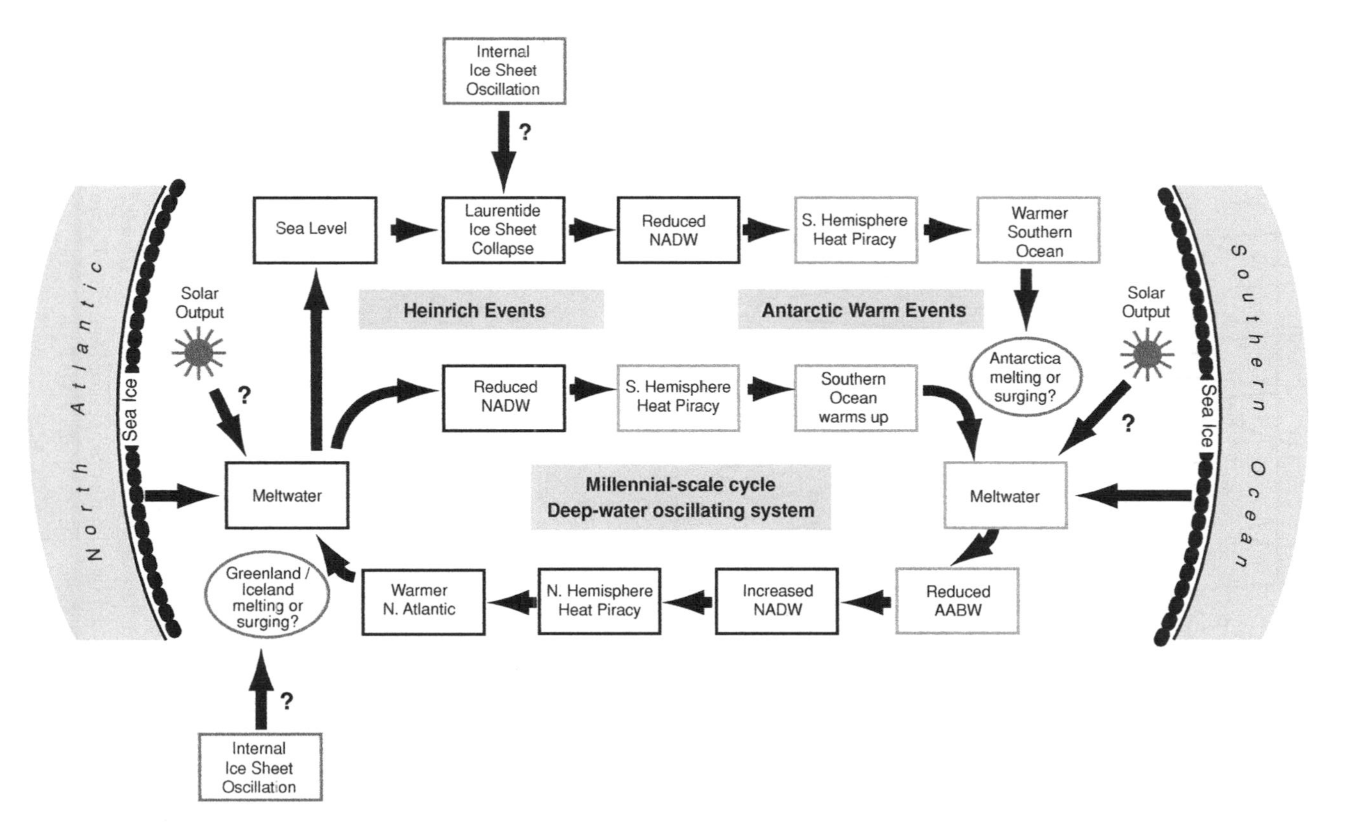

Figure 14.11 A possible millennial-scale deep-water oscillating system with heat transport to and from the different hemispheres and respective growth or decay of sea-ice controlling deep-water circulation. Note that there is separate route for the glacial Heinrich events which could be caused by (1) successive D-O cycles providing a critical sea-level that undermines and causes the Laurentide Ice Sheet to fail or (2) separate internal dynamics of the Laurentide Ice Sheet or (3) rerouting continental runoff from the Mississippi River to the Hudson or St Lawrence Rivers producing extra freshwater discharge to key areas of NADW formation. It is suggested that this same system could operate during interglacial periods forced by alternating melting of sea ice in the Southern Ocean and North Atlantic (Maslin *et al.*, 2001), possibly forced by solar changes (Bond *et al.*, 2001).

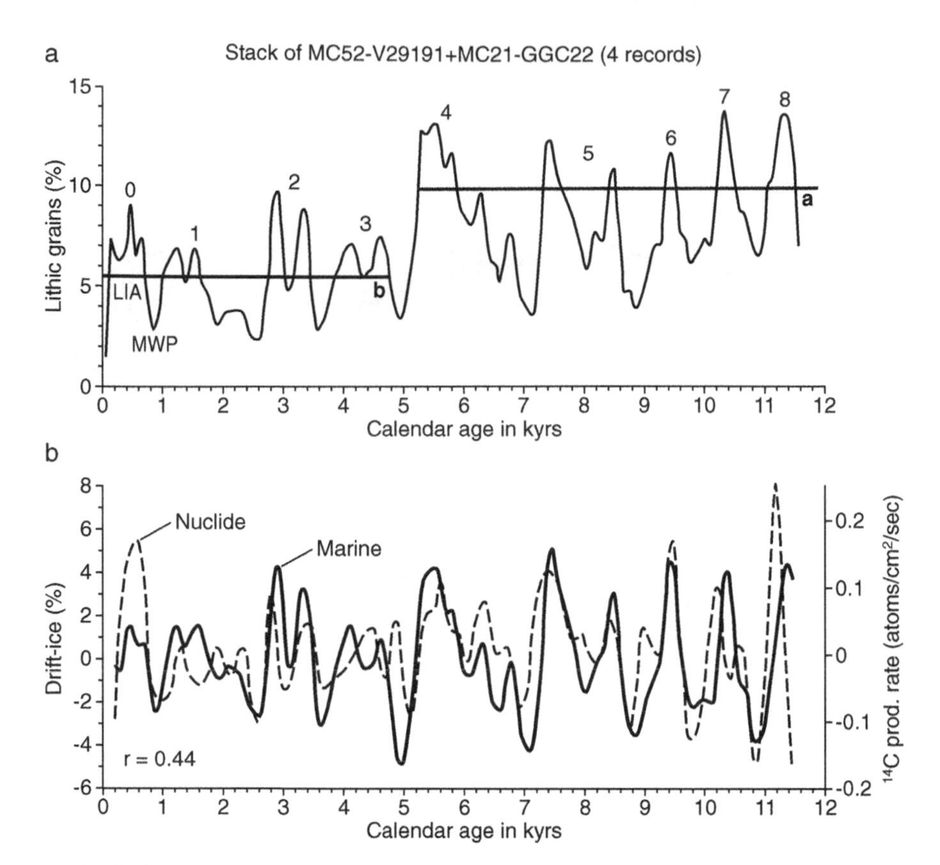

Figure 14.12 (a) Stacked record of petrologic tracers of drift-ice expressed as a percentage of lithic grains (ice rafted debris) in the 63- to 150-μm size range (Bond *et al.*, 2001). Numbers refer to the identified cycles; LIA = Little Ice Age, MWP = Medieval Warm Period. Notice the difference in the amount of ice rafting during the beginning and end of the Holocene; (b) comparsion between the normalized and detrended stacked 'drift-ice' lithic grain percentage with the smoothed detrended ^{14}C production rate over the last 12,000 years (Bond *et al.*, 2001).

scale variations in solar output have influenced the surface winds and surface hydrography of the subpolar North Atlantic region resulting in the D–O climate cycles. They also postulate that amplification of the relatively weak solar changes is due to the Arctic-Nordic Sea's sensitivity to changes in surface-water salinity. Hence increased freshening of this region would reduce the NADW and thus transmit the solar influence globally. There are, however, still a number of problems with the millennial-scale solar output theory. First, ^{14}C and ^{10}Be are used as indicators of solar output; however, both are influenced by climate change, for example ^{14}C can vary due to the switching on and off of deep-ocean circulation (Beck *et al.*, 2001) while ^{10}Be is influenced by the amount and type of precipitation over the recording site, both of which are known to occur during millennial-scale climate cycles. Second, it is not possible to determine a direct causal link between the solar output variations documented and climate change in the North Atlantic. This is mainly because the solar output variations are extremely small. Third, one the cores used by Bond *et al.* (2001), V28-14, is from the Denmark Strait and hence the IRD records may have been affected by the very high current strength. The record may, therefore, be influenced by winnowing and/or enhanced IRD transport. In addition, Andrews *et al.* (1997) IRD records from the suggested source area of Holocene IRD are dissimilar from those presented in Bond *et al.* (2001) producing a question of where the extra IRD could becoming from during these events. The debate about solar forcing on Holocene climate records will continue for a long time.

14.6 CONCLUSIONS

The Holocene or the last 10,000 years was once thought to be climatically stable. Recent evidence, including that from marine sediments, has altered this view showing that there are millennial-scale climate cycles throughout the Holocene. In fact we are still in a period of recovery from the last of the cycles, the Little Ice Age. It is still widely debated whether these cycles are quasi-periodic or have a regular cyclicity of 1500 years. In fact, Alley *et al.* (2001) have suggested that these cycles are due to stochastic resonance and hence require no cause. It is also still widely debated whether these Holocene Dansgaard–Oeschger cycles are similar in time and characteristic to those observed during the last glacial period. A number of different theories have been put forward to what causes these Holocene climate cycles, most suggesting variations in the deep-water circulation system. Hence we still have a gap in our understanding of climatic variability between the centennial- and millennial-scale. However, it is only by a combination of marine and other palaeoclimatic records can we develop these theories further. Future research is essential to understand the climate cycles so we can better predict the climate response to anthropogenic 'global warming'.

Acknowledgements

We would like to thank John Andrews for extensive comments on this manuscript. We would also like to thank C. Pyke, J. Quinn, E. McBay of the Department of Geography, Drawing Office for the help with the figures. This work was supported by ODP, Deutsche Forschungsgemeinschaft and NERC (grant GR9/03526).

HOLOCENE PALAEOCLIMATE RECORDS FROM PEATLANDS

Keith Barber and Dan Charman

Abstract: Peat formation is widespread across the globe but the majority of existing proxy climate records come from the ombrotrophic (rain-fed) raised and blanket peat bogs of temperate oceanic regions such as northwest Europe. In North America, peat initiation dates and the spread of peatlands have also been used to infer climate change during the Early to Mid-Holocene. Elsewhere, continuous records of change have been derived from profiles which may span most of the Holocene. Methods focus on the use of biological proxies, ideally used in a multi-proxy approach, to derive continuous bog surface wetness (BSW) curves which are an integrated record of effective precipitation, and on stable isotope measurements which may be directly influenced by temperature and precipitation. Both approaches have been used on recent peats in attempts to validate and calibrate the proxy records against documented climate records with some success, and the results suggest that the impact of summer temperatures on evapotranspiration from bog surfaces is the main factor in changes in BSW in northwest Europe. Within this region, changes in BSW are apparent in many sites at around 8200, 5900, 4400, 3500, 2700, 1700, 1400, 1100, 700 and 250 cal BP. These changes to a cooler/wetter climate are probably caused by circulation changes in the Atlantic Ocean and by solar variability. Ongoing research is providing decadal resolution records of proxy climate which may be used in testing climate models.

Keywords: Bog surface wetness, Multi-proxy records, Ombrotrophic bogs, Peat initiation

Peat forms in any area where decay of plant material is exceeded by production. Usually this means that conditions are sufficiently wet and/or cold to inhibit decay, yet growing seasons are adequately warm for plant growth. The dependence of peat growth on moisture and temperature conditions means that the occurrence, rate and nature of peat accumulation are always at least partly climatically determined. Other factors become important in particular situations, but it is abundantly clear, simply from the global distribution of peatlands, that climate is fundamental to peat formation (Fig. 15.1). Palaeoclimatic data from peatlands are derived in two main ways, which we consider below in some detail. First, the location and timing of peat initiation may be related to a change to colder and/or wetter climate. Second, variations in peat composition and characteristics through time reflect changes in precipitation and temperature in various ways. The potential value of the peat palaeoclimate record was recognized some time ago, most notably by Blytt (1876) and Sernander (1908), who developed a scheme that established the principal terminology of Holocene climate change used in Europe for over half a century. While this was an oversimplistic use of peat stratigraphic evidence, it demonstrates that interest in the peat record began very early. So it seems strange that research on peat palaeoclimate records was largely abandoned until the last 30 years of the 20th century.

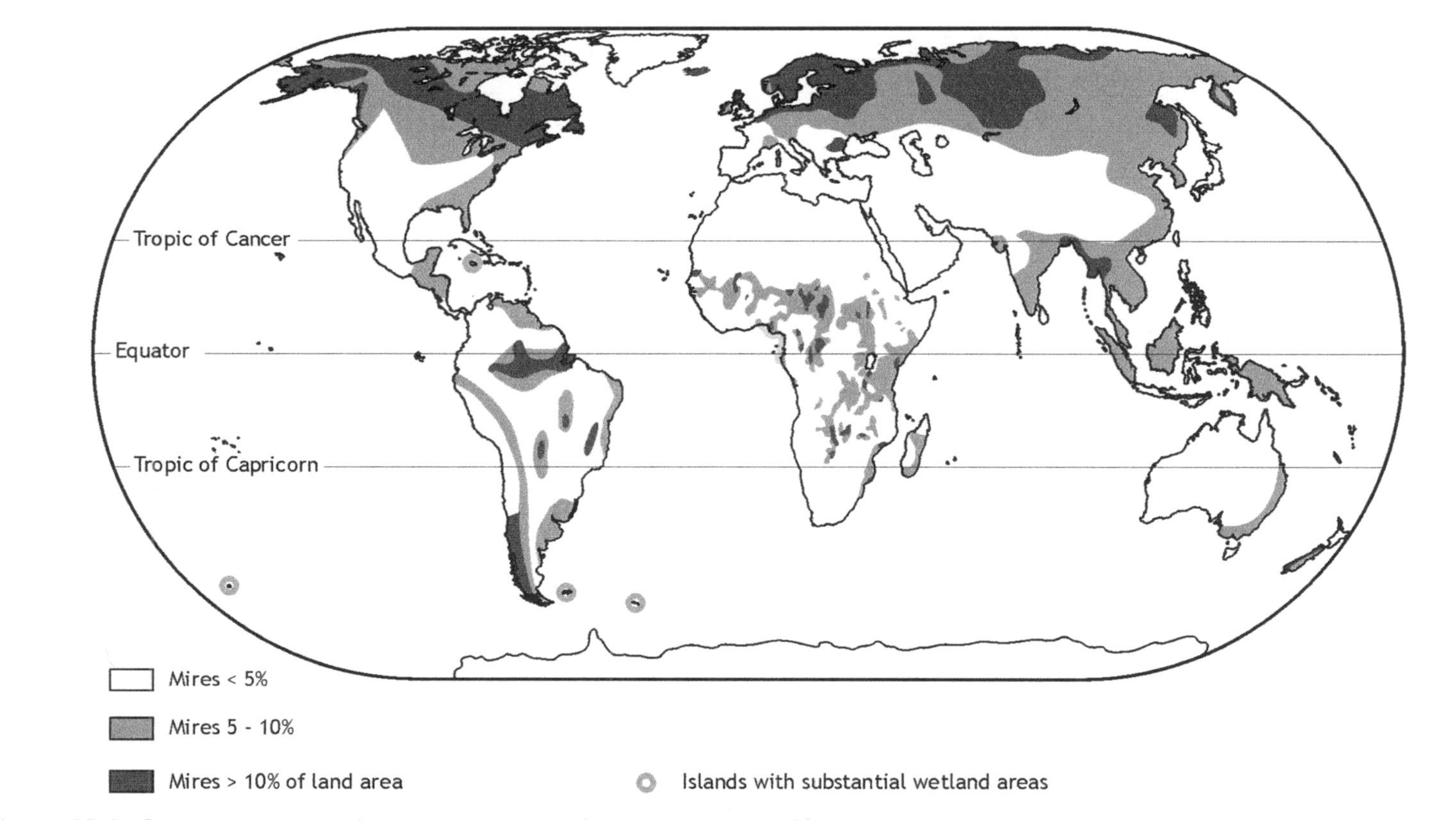

Figure 15.1 Global distribution of peatlands. Redrawn from Lappalainen (1996) with kind permission from the International Peat Society.

This partly reflects the increase in the number of other proxy climate methods (shown by the wide-ranging content of this book) but also a misplaced emphasis on autogenic peat development processes (Barber, 1994). Autogenic development and change provide the background 'noise' to the palaeoclimate record, but while it is important to be aware of these processes in locating suitable sites and peatland types, they do not mean high-quality palaeoclimate records are difficult to obtain (Charman, 2002). In this chapter we provide a review of key processes and characteristics of peat deposits relevant to palaeoclimate records. We also spend some time discussing chronological issues which are fundamental to correlation and interpretation of peat records and their links with other proxy data.

15.1 PEATLAND TYPES, HYDROLOGY AND ECOLOGICAL FUNCTIONING: IMPLICATIONS FOR PALAEOCLIMATE RECORDS

It is vital to understand something of the functioning of peatland systems if the palaeoenvironmental record is to be properly interpreted. Peatlands do not always act as simple repositories of environmental and climatic change but are subject to multiple influences from internal processes as well as external forcing. The challenge in deriving high-quality palaeoclimatic data from peatlands is to select sites, and locations within those sites, that represent the most sensitive records that are closely coupled to climate. The hydrology of a peatland is critical here. Peatlands are divisible into two main hydromorphological types, both of which are often referred to as '**mires**'. First, **ombrotrophic peatlands** ('bogs') depend only on water directly from precipitation. Thus their water balance is predominantly a function of climate determined by precipitation and evaporation, modified slightly depending on vegetation cover. If there are no large changes in vegetation physiognomy, the peat record can be interpreted as a record of changing 'effective precipitation'. Second, **minerotrophic peatlands** ('fens', sometimes, if nutrient-poor, popularly referred to as 'valley bogs') are also partly dependent on direct precipitation for their moisture supply but this is supplemented by surface runoff and/or groundwater. Thus the link between the peat record and climate is more tenuous in these sites. At best it is modified by the hydrological processes that determine runoff and groundwater supply. Such sites may hold a climatically related record, but much more care may be needed to determine and allow for other influences on the record such as anthropogenic landscape disturbance. Even in apparently ombrotrophic sites, groundwater may occasionally have been an influence in the past (e.g. Glaser *et al.*, 1997). To date, most peatland palaeoclimate records have used ombrotrophic sites that are unaffected by runoff or groundwater flow. These are the raised bogs and blanket bogs (Fig. 15.2) that occur particularly in the temperate oceanic areas of the world. However, dates of peat initiation at a much wider range of sites have been used to infer climate change with some success (see below) and some continuous climate records have been derived from minerotrophic sites within regions where ombrotrophic records are more conventionally used (e.g. Anderson, 1998).

15.2 SPACE AND TIME: DISTRIBUTION, RESOLUTION AND SENSITIVITY OF THE PEAT RECORD

The value of any palaeoclimate proxy record is determined not only by the strength of the climate–proxy link, but also by its distribution in space and time, as well as its temporal resolution. Peat deposits cover around 400 million hectares worldwide, with depths varying from 30 to 40 cm

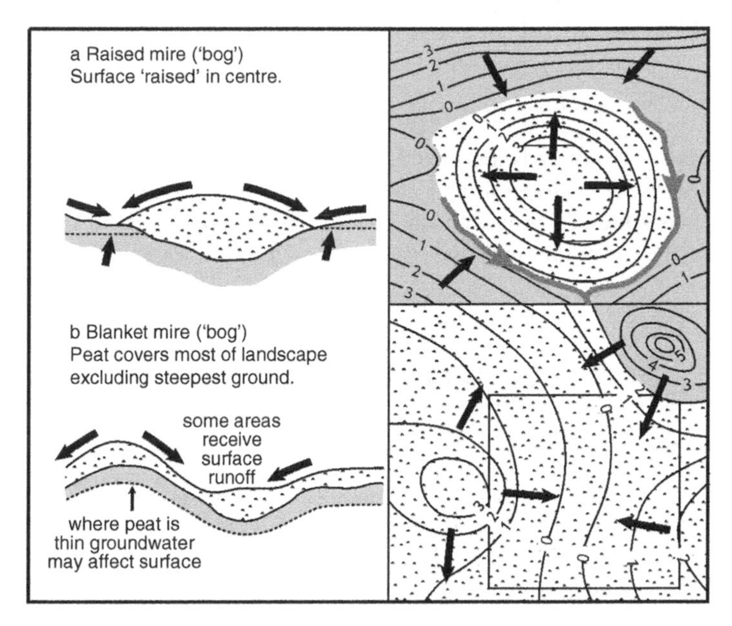

Figure 15.2 Cross-sections and plan views of typical ombrotrophic mires. (a) A raised mire, (b) a blanket mire. The arrows show the direction of water flow. In both cases the only water supply is from direct precipitation. Reproduced from Charman (2002).

up to almost 20 m. The exact extent of estimated peat cover varies between around 320 million hectares (Pfadenhauer *et al.*, 1993) and 421 million hectares (Kivinen and Pakarinen, 1981), the latter excluding Africa and parts of South America, or up to 1300 million hectares for 'wetlands' as a whole (Bouwmann, 1990). The total carbon stored in these deposits is somewhere in the region of 200–500 billion tonnes (10^{15} g) (Immirzi *et al.*, 1992). As dried peat is approximately 50 per cent carbon, these figures can be doubled for the total dry mass, giving a figure of up to 10^{18}g or 10^{12} tonnes. Peatlands are most abundant across the northern Boreal zones but extend southwards to about 40°N, especially at higher altitudes. In the tropics, many areas of peatlands are intermingled with other wetland systems, but there are large deep peat deposits in Southeast Asia, as well as smaller basin peatlands in mountainous areas (Gore, 1983). Peatland systems akin to the northern oceanic systems occur on the land masses in the southern hemisphere south of about 40°S, principally New Zealand, southern South America and the subantarctic islands (Lappalainen, 1996).

The vast majority of peatlands are of Holocene age, and many are limited to the middle to late parts of the epoch, and only a few records extend back to the earliest Holocene. Because accumulation rates are highly variable, the resolution of the record is also variable. Many of the raised mires of the oceanic margin of Europe where palaeoclimatic studies have traditionally been carried out accumulate approximately 1 cm every 10 years on average, so that decadal resolution is easily obtained. Blanket mires tend to accumulate more slowly and are often more variable with accumulation rates, typically 1 cm every 20–50 years. Narrower sample spacing may improve resolution here (e.g. Chambers *et al.*, 1997). Resolution of samples does not necessarily reflect the actual temporal resolution of the record, as bioturbation or other processes could affect it. However, as biological activity below the aerobic zone in peats is limited to microbial decay, it seems likely that sample resolution adequately reflects actual time, at least down to decadal levels.

In summary, given that peatlands are directly coupled to climate through their water balance relations, the deposits are widespread and abundant and their resolution is often decadal over 5–6 millennia, their value as a proxy climate record is promising. In the rest of this chapter we highlight some of the key approaches that have been used to unlock this potential in some areas of the world.

15.3 PEAT CHRONOLOGIES

15.3.1 Radiocarbon Dating

Chronological control is of course crucial to the successful use of any palaeoclimate proxy. Because peat when dried is approximately 50 per cent carbon and is entirely autogenic, it is undoubtedly one of the best Holocene materials for radiocarbon dating. Conventional radiocarbon dating has been used for many years with success. However, an increasing requirement for greater precision and accuracy, has led to a much more critical selection of dated materials and pre-treatments and latterly the use of accelerator mass spectrometry (AMS) ^{14}C analyses (see Pilcher, pp. 63–74 in this volume). Radiocarbon dating of peat is not without its problems, however. Different chemical and particulate fractions of some peats may yield different and unpredictable age estimates (Shore *et al.*, 1995), suggesting that the processes determining ^{14}C content may not be well understood. Basal peats are some of the most problematic materials to date, being strongly affected by mobile humic acids and root penetration, especially in shallow deposits (Smith and Cloutman, 1988). However, other peats provide near-perfect material for dating. *Sphagnum* moss peats are a good example, being entirely autogenic and not subject to contamination from root penetration from above or translocation of older carbon from below. Wiggle-matching of ^{14}C ages to achieve much greater precision is also now being used, especially to date peats from periods where there are plateaus in the calibration curve (van Geel *et al.*, 1998). Selection of pure *Sphagnum* material for AMS ^{14}C wiggle-matching perhaps represents the optimum strategy for dating of peat deposits (Kilian *et al.*, 2000).

15.3.2 Other Techniques

While radiocarbon dating remains the most widely applicable and useful dating method for many peat records, others are now being increasingly exploited. Tephrochronology is emerging as a useful technique in some parts of the world, especially with the discovery of microtephras in regions distant from volcanic sources (e.g. northern Britain) (Dugmore *et al.*, 1995). Icelandic tephras in northwest Europe have the potential to constrain radiocarbon chronologies and to make precise correlations between sites for particular time periods (Langdon and Barber, 2001 and in press). For peats covering the last few hundred years of climate change, other approaches in addition to ^{14}C analyses are also needed (Belyea and Warner, 1994). Age estimates from ^{210}Pb have produced variable results and are normally constrained with other short-lived isotopes (e.g. ^{241}Am) and other chronological markers such as changes in pollen spectra referable to known calendar ages (Oldfield *et al.*, 1995). Other lead isotopes (^{206}Pb, ^{207}Pb) related to changing fuel sources can also be used to refine ^{210}Pb chronologies (Shotyk *et al.*, 1998). Spheroidal carbonaceous particles (SCPs) have been used extensively for European lake sediments, but are only now gaining wider acceptance for dating peats (Chambers *et al.*, 1999; Yang *et al.*, 2001). It is essential to apply and further refine these techniques and wiggle-matched ^{14}C (Clymo *et al.*, 1990), if comparisons with instrumental climate data and other proxies are to be successful (see below).

15.4 PEAT INITIATION AND PALAEOCLIMATES

The causes of peat initiation at any single location are a function of vegetation, substrate, local hydrological position and climate. It is thus very difficult to use the age of peat initiation at a single point to infer any kind of climate deterioration. In addition, much of the early work on the age of peat initiation was carried out on blanket mires in northern Britain, where anthropogenic disturbance is associated with the start of peat growth, suggesting that local hydrological change as a result of forest clearance was the main cause of peat growth (Moore, 1975). Subsequently, research in other areas has also often shown that human activities have promoted peat growth or stratigraphic change in peatlands (Moore, 1993). However, it has now been shown that peat initiation and spread especially on large spatial scales is explicable only in terms of climate change (e.g. Halsey *et al.*, 1998). Only sites where peat growth proceeded via paludification rather than terrestrialization should be used for such inferences and areas where human impact, burning or non-climatic hydrological changes have occurred should be avoided. Because of site-specific influences and the difficulties of dating basal peats, interpretation of peat initiation data has to be limited to identifying broad-scale spatial and temporal patterns of climate change.

In western Canada, clear regional patterns of peat initiation can be identified. Initially a distinction between pre- and post-6000 yr BP conditions was identified, with earlier peat initiation north of latitude 54°30'N and in the foothills of the Rocky Mountains (Zoltai and Vitt, 1990). This was due to more arid conditions further south in the Early to Mid-Holocene. Progressively wetter climates after 6000 yr BP resulted in the present distribution of peatlands by around 2000–3500 years ago. Detailed patterns of initiation have been described more recently and these are related to climate change modified by local topographic and edaphic factors (Halsey *et al.*, 1998). Combining the influences of climate with local factors allowed the production of modelled peat initiation times over a wide area of western Canada (Fig. 15.3). Early peat initiation occurred in some zones in northern Alberta around 9000–8000 cal yr BP after an initial deglacial lag unrelated to climate. Further peatland initiation was at first limited by warmer summers but peat growth later expanded east as a result of decreasing summer insolation. The spread of peatland was more pronounced in the west (Alberta) as a result of incursions of moist Pacific air. However, after 6000 BP, peatlands spread south and east in response to the more general increasing climatic wetness identified from the earlier work. Time-series analysis of peat initiation data in western Canada appears to reflect the millennial-scale periodicity in climate detected from many other palaeoclimate records in ocean and ice-core records (Campbell *et al.*, 2000). This finding demonstrates that on large scales, peat initiation is driven by hemispheric or global climate changes. In addition, the pre- and post-6000 yr BP patterns of peatland initiation in the area are consistent with other palaeoclimatic proxies and with reconstructions generated from climate models (Gajewski *et al.*, 2000). In Siberia, patterns of peatland initiation show two main phases of acceleration (Smith *et al.*, 2000). The first followed post-glacial warming between 13,000 and 8000 yr BP, and the second occurred after 5000 yr BP. The two main phases of accelerated peat initiation were

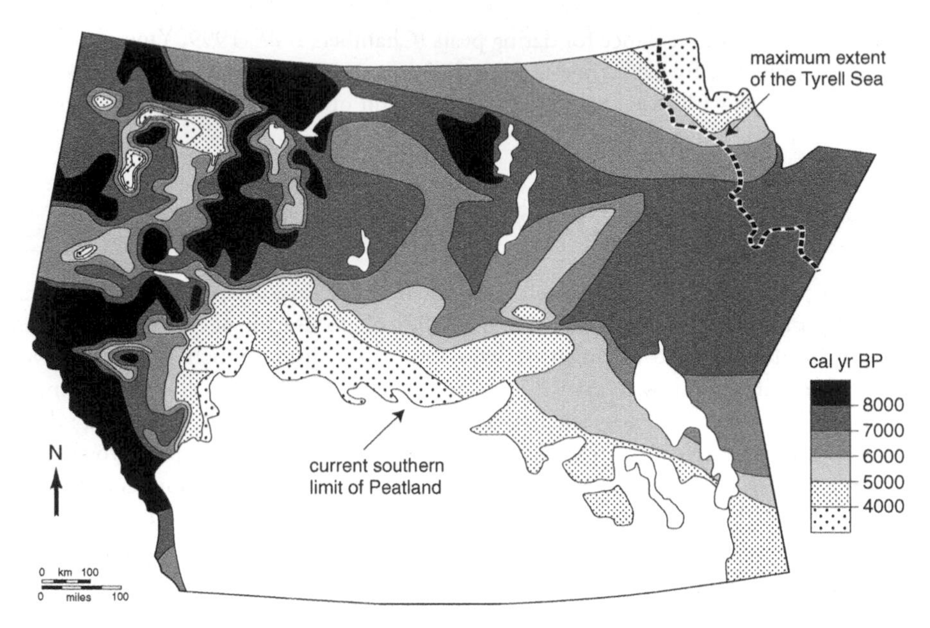

Figure 15.3 Large-scale peat initiation related to climate change: map showing the spread of peat initiation in western Canada. Redrawn from Halsey *et al.* (1998) from an original supplied by the authors, with kind permission of Kluwer Academic Publishers. See text for discussion.

interrupted by a period of reduced initiation between 8000 and 5000 yr BP, suggesting much warmer and drier conditions during this period. This is corroborated by the age of tree macrofossils which are mostly of Early and Mid-Holocene age (MacDonald *et al.*, 2000a). Furthermore, it suggests that the development of high-latitude peatlands in the northern hemisphere is linked to the variability in global atmospheric methane concentrations during the Holocene (Blunier *et al.*, 1995).

While it may be more difficult to use peat initiation data from smaller areas potentially subject to influences other than climate, there may be more potential for this than has often been realised. Given careful selection of samples and data sets, coherent patterns unlikely to be related to human impact may be found. Intensive studies within smaller areas can help identify regionally synchronous patterns of peatland initiation. For example, in southern Finland, two distinct phases of accelerated peat initiation and spread were identified between 8000–7300 and 4300–3000 cal yr BP. These coincided with wetter climate as indicated by lake levels in southern Sweden (Korhola, 1995). Even at individual sites it is often possible to find an association between climate and basal peat ages (e.g. Miller and Futyma, 1987; Korhola, 1996), although this does not necessarily imply a cause and effect relationship.

Clearly, the use of peat initiation data for palaeoclimatic reconstruction was underestimated until the last 10 years or so. The large-scale studies in particular demonstrate that broad-scale climate patterns are the main driver for continental-scale initiation and spread of peat growth. However, there will always be a number of problems and limitations of peat initiation data; the spatial and temporal precision are low, and climatic amelioration is more difficult to identify than climate deterioration. In addition, peat initiation does not necessarily show a linear relationship with climate over time, because as a larger area of the landscape becomes peat covered, the remaining area is less prone to peat formation (Payette, 1984).

15.5 CONTINUOUS RECORDS FROM OMBROTROPHIC PEATLANDS

15.5.1 Principal Approaches and Techniques

As mentioned above, ombrotrophic systems are most suited to palaeoclimate reconstruction because they receive all their moisture and nutrients directly from the atmosphere. There are two main ways in which the peat records changing climate over time. First, BSW changes can be reconstructed using various biological and chemical proxies and interpreted as a record of effective precipitation. These proxy records of surface wetness are the most widely used and have produced almost all of the palaeoclimate records from peatlands – these form the main focus for this chapter. Second, other proxies may record temperature and/or precipitation directly through processes independent of changing surface wetness. These proxies include isotopes of hydrogen (^{2}H), oxygen (^{18}O) and carbon (^{13}C) in bulk peat or in selected components. However, these records are strongly affected by changes in plant species composition in the peat (van Geel and Middeldorp, 1988). Changes in isotope chemistry may therefore be related to changing species composition rather than changing climate. For example, spatial variability of the ^{13}C content of *Sphagnum* is very high related to the local moisture conditions and species (Price *et al.*, 1997). The species-specific problem is known to affect ^{18}O and D as well as ^{13}C. To overcome this problem, a number of different approaches have been tried. Corrections can be applied based on the major peat components for each measurement (Dupont and Brenninkmeijer, 1984) and this seems to produce plausible results (Dupont, 1986). Alternatively, the species signal can be ignored if peat deposits composed predominantly of a single taxon can be found as in the site reported from China by Hong *et al.* (2000b, 2001). Finally, chemical compounds known to be derived from a single taxon can be isolated and used for the isotopic measurements (Xie *et al.*, 2000). A further potential direct measure of temperature has been proposed by Martinez-Cortizas *et al.* (1999). In this case the thermal lability of the accumulated mercury was used to reconstruct temperature. During periods of cooler climate more mercury with low thermal stability is retained, whereas during warmer periods, the mercury present in the peat has a higher thermal stability. All of these approaches show promise but they have not been extensively used to provide palaeoclimate records to date, and many of them need more investigation before the climate–proxy link is fully understood. The remainder of this section concentrates on the methods that aim to reconstruct BSW as a climate proxy record.

Surface wetness is the main control on the composition of communities of plants and other organisms on ombrotrophic peatlands. Biological proxies are therefore excellent indicators of past climate changes. Plant macrofossil analysis (Fig. 15.4) is the most widely used technique, particularly in *Sphagnum* peats as the main identifiable taxa show a strong gradient from 'wet' taxa such as *S. cuspidatum* to 'dry' taxa such as *Sphagnum* Sect *Acutifolia*. Such analyses have also recently been continued down through the ombrotrophic peat into the underlying fen peats to provide a record of change throughout the whole Holocene, identifying major events such as the fen/bog transition and the climatic character of the generally dry Early to Mid-Holocene period

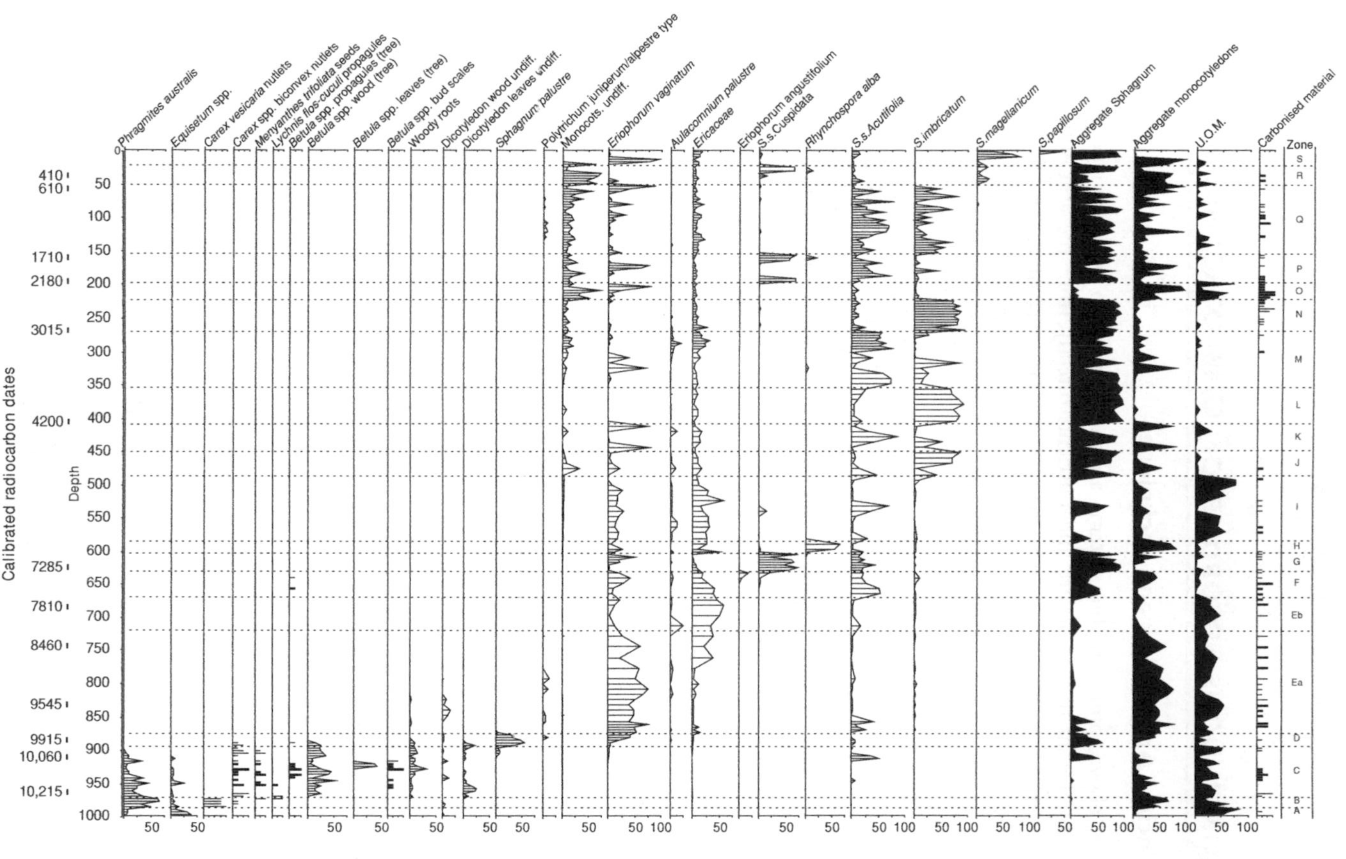

Figure 15.4 Plant macrofossil diagram of the whole Holocene from Walton Moss, Cumbria, core WLM-11. Linked histograms are quantitative estimates of the components of the peat derived using the Quadrat and Leaf Count method of Barber *et al*. (1994). Unlinked bar histograms are estimated abundances on a five-point scale. Modified from Hughes *et al*. (2000).

(Hughes *et al.*, 2000) (Fig. 15.4). Qualitative interpretation of changes in plant macrofossil data has been used to identify the main shifts in surface wetness with success (e.g. Barber, 1981). However, over the past 10 years it has become more usual to describe changes using ordination techniques to identify the latent wetness gradient in the data (e.g. Barber *et al.*, 1994). This approach has the advantage of being more objective and of providing a semi-quantitative measure of changing BSW (Fig. 15.5). Plant macrofossil analysis has been used most extensively in northwest Europe where the moisture gradient is reflected clearly in *Sphagnum* taxa and higher plants. It may not be so readily adapted for use where peat composition is less varied, such as in southern hemisphere or strongly continental peatlands. Other problems include the lack of a good modern analogue for *S. imbricatum*, a major peat former in many European peatlands up to approximately cal AD 1000 to cal. AD 1500 (Mauqoy and Barber, 1999b). Within northwest Europe the technique remains the main source of palaeoclimate data from peatlands from Ireland to Poland (Haslam, 1988; Barber *et al.*, 2000).

Testate amoebae are another group of organisms that respond primarily to hydrological conditions. Their small shells ('tests') are preserved after death in the peat and these can be extracted, identified and counted to reconstruct past surface wetness (Charman *et al.*, 2000). A number of ecological studies have provided data that can be used to model the relationship between testate amoebae assemblages and depth to water table numerically (e.g. Charman, 1992, 1997; Charman and Warner, 1997). This allows the reconstruction of mean annual water tables using a transfer function approach (Warner and Charman, 1994; Woodland *et al.*, 1998). While the use of these organisms has made good progress in the last few years, there remain problems with the approach. In particular, there are some taxa that occur abundantly in the past that are much less common in the modern analogue samples (e.g. *Difflugia pulex*). In humified peat, the concentrations of tests become too low to count with ease and in some cases they are absent (e.g. Charman *et al.*, 2001). Despite these difficulties, the geographical extent of studies using testate amoebae for surface wetness reconstructions is very wide, including North America (Warner and Charman, 1994), Switzerland (Mitchell *et al.*, 2001) and New Zealand (Wilmshurst *et al.*, 2002) as well as the UK (e.g. Hendon *et al.*, 2001).

Other biological indicators of surface wetness changes have been found such as some of the non-pollen microfossils described by Bas van Geel at the University of Amsterdam (van Geel, 1978, and subsequent work). On occasion, pollen analysis can also be used to show changes in major taxa associated with different hydrological conditions. However, these techniques seldom produce the main datasets for palaeoclimate reconstructions.

Surface wetness is also the major control on the amount of decay that takes place in the surface layer of peat. Drier peat surfaces are associated with deeper aerated horizons (the 'acrotelm') (Ingram, 1978) and increased decay of plant material, whilst when the surface is wetter, this zone is very shallow and decay is much reduced before the plant remains pass into the permanently saturated deeper horizons of the 'catotelm'. In peat profiles, the degree of decay ('humification') can be measured and variations in humification can be interpreted as records of changing BSW. Measurements are based on the absorbance of alkali extracts of dried peat measured in a spectrophotometer (Blackford and Chambers, 1993). The technique is especially useful in peats that are highly decayed where few plant fragments or other biological fossils are well preserved. There are some problems in accounting for species specific signals in the humification record and also uncertainty over exactly what the humification measurements are assessing (Caseldine *et al.*, 2000). However, because of its relative speed and

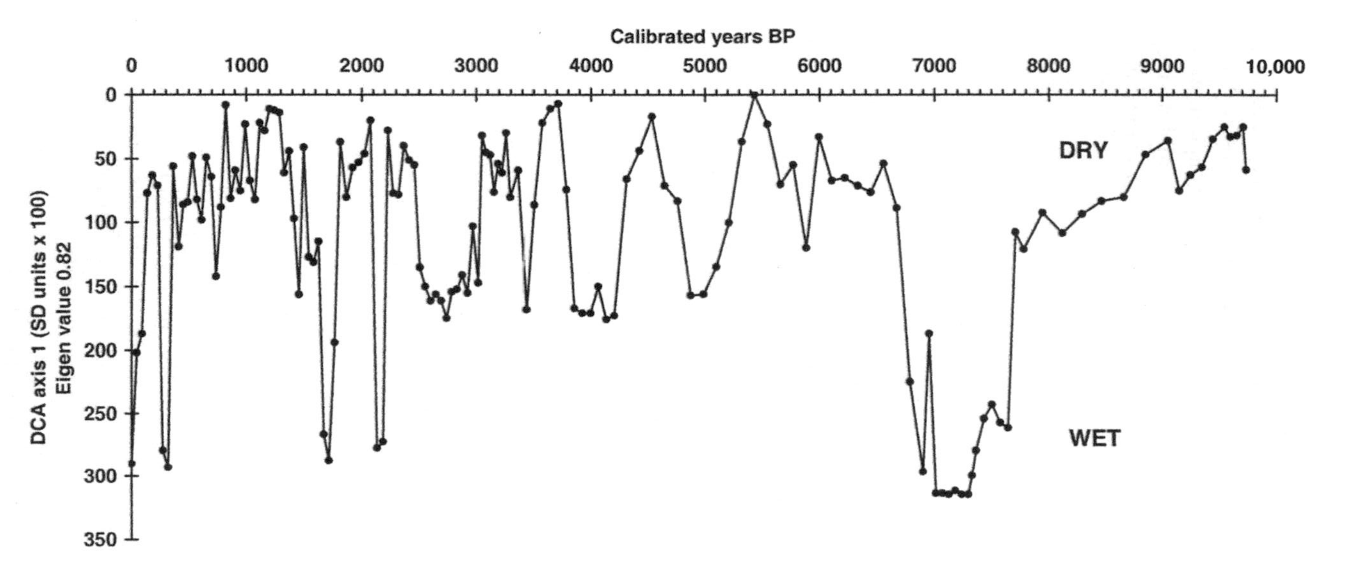

Figure 15.5 Proxy climate record of core WLM-11. The record is derived by plotting the axis 1 scores of a Detrended Correspondence Analysis of the raw data in Fig. 15.4. Modified from Hughes *et al.* (2000).

universal applicability it continues to provide data from a wide range of sites and locations. In fact, for many highly decayed peats, it is the only technique that is capable of yielding useful data. Most published data are from blanket peats in northwest Europe (e.g. Blackford and Chambers, 1991, 1995; Chambers *et al.*, 1997) but data have also been collected from peats in New Zealand (e.g. McGlone and Wilmshurst, 1999) demonstrating the wider applicability of the technique.

All of the techniques for recording BSW changes have their relative advantages and disadvantages, and it is becoming standard practice to derive multi-proxy records based on at least two of the methods where possible. There have been rather few comparative studies but the published data suggest that the main directional shifts in surface wetness are reflected in plant macrofossils, testate amoebae and humification (Mauquoy and Barber, 1999a; Charman *et al.*, 1999; Langdon *et al.*, 2003). The magnitude of change is more difficult to compare between proxies as the scales for each are different and may not always be linear. Normalization of plant macrofossil and testate amoebae reconstructions to a standard recent reference period appears to overcome some of these problems and demonstrates that these two proxies show the same magnitude of change as well as the same timing, for recent periods at least (Charman *et al.*, 1999). Much more data are required to test the sensitivity of the different proxies in more detail.

15.5.2 Methodological Issues

The relationship between the allogenic forcing factors, which produce the proxy climate record and the autogenic processes which occur within the peat body may be likened to 'signal' and 'noise' effects. Until recently many plant ecologists took the view that changes in surface vegetation and the accompanying changes in bog microtopography, such as hummock growth and pool expansion and contraction, were inbuilt autogenic processes, inherent in the life histories of the plants. These views, a hangover from the theory of cyclic bog regeneration (Osvald, 1923; Godwin, 1981) were first challenged by Walker and Walker (1961) and then formally falsified by Barber (1981). Nevertheless, there is still an issue as to the importance of the 'noise' in the record and this has been tackled by a number of recent studies. Barber *et al.* (1998) addressed the problem in two ways. First, they summarized the stratigraphic evidence for past bog surfaces being dominated by broad swathes of *Sphagnum*-rich vegetation, with a low proportion of hummocks – the microform which shows a low response to climatic forcing and may therefore be seen as the prime source of 'noise'. Second, they demonstrated that in a study of two adjacent raised bogs it was possible to assess the degree of 'noise' in core records by carefully recording field stratigraphy and by replicating macrofossil trends in 11 short surface cores. This approach, of replicating records between cores, has also been pursued by Woodland *et al.* (1998), by Mauquoy and Barber (1999a), by Charman *et al.* (1999) and by Hendon *et al.* (2001). As all these studies, and others, are from the same Anglo-Scottish Border region they are used as the basis for the compilation of major proxy climate changes in a later section.

The implications of these and other studies (Langdon, 1999; McMullen *et al.*, 2000) are that great care should be taken by way of site survey, including aerial photography, field stratigraphy and preliminary laboratory analyses to ensure that representative cores are taken for the time-consuming multiproxy analyses of plant macrofossils, testate amoebae and humification. If the protocols detailed in Barber *et al.* (1998) are followed then one may be confident of gaining proxy climate records with the maximum 'signal-to-noise' ratio.

15.5.3 Interpretation of Surface Wetness Changes: Temperature or Precipitation?

Disentangling the relative importance of the various climatic parameters to bog water tables and therefore to the proxy climate signal has been a major goal of research since the work of Aaby (1976), van Geel (1978) and Barber (1981). A clear correspondence between the summer wetness and winter severity indices of Lamb (1977) and a qualitative curve of BSW derived from macrofossil analyses at Bolton Fell Moss was found by Barber (1981). More recent research has significantly improved our understanding of the climatic parameters contributing to the peat palaeoclimate signal and our ability to estimate errors. Barber *et al.* (1994) compared a detailed macrofossil record from Bolton Fell Moss with Lamb's indices using a specific type of Generalized Linear Model and were able to show that the resulting Climate Response Model represented a very good fit – that is, the taxa responses matched the changes in the climatic indices.

Periods of increased rainfall may be thought *a priori* to be an obvious influence on BSW, but comparisons of macrofossil changes in two quite different bogs – a lowland raised bog at Fallahogy, Northern Ireland, and a montane blanket bog in the Cairngorms, Scotland – with long documentary climate records suggest that the temperature signal is more coherent and is dominant over the spatially and temporally incoherent precipitation signal (Barber *et al.*, 2000). The impact of summer temperatures on evapotranspiration from bog surfaces does therefore seem to be the main factor in changes in BSW, and Charman and Hendon (2000) and Charman *et al.* (2001) have also shown that changes in BSW may be correlated to the GISP2 ice-core record and indicators of sea-surface temperature and ocean circulation.

Calibration of BSW changes with instrumental climate data is an attractive approach to gauging the relative influence of seasonal and annual temperature and precipitation for BSW. However, precise correlations with instrumental climate data depend heavily on high-precision chronologies for peat deposited over the last 200 years (see above). Reconstruction of mean annual water tables from samples representing 2–7 years of peat growth in northern Europe suggest that BSW responds to precipitation and temperature but the precise nature of the response varies along an E–W gradient (Charman *et al.*, in press).

New ways of assessing the relative importance of temperature and rainfall are also coming from the utilization of molecular stratigraphic records based on lipid biomarkers of both Sphagna (Nott *et al.*, 2000) and sedges (Avsejs *et al.*, 2002). The strong correlation of compound-specific δD values of *n*-alkanes with recorded mean July temperatures for the past >200 years is an advance which opens the way for deriving temperature coefficients for use in palaeotemperature reconstructions (Xie *et al.*, 2000). Another approach has involved correlations between BSW changes at Walton Moss, Cumbria, and chironomid-based temperature reconstructions from a nearby lowland lake, Talkin Tarn. Comparison of the two records over the last 3000 years again supports the hypothesis that changes in BSW are driven primarily by summer temperature (Barber and Langdon, 2001).

15.5.4 Surface Wetness Changes in Northern Britain: Compilation of Evidence from an Intensively Studied Area

As research into the peat-based palaeoclimate record has progressed it has become apparent that some bogs in some climatic regions are more sensitive to climatic shifts than others. This was demonstrated by Haslam (1988) who analysed macrofossils during the period of the Main Humification Change (defined as the major change in each individual bog) in a transect of bogs from Ireland, through north Germany, Denmark and Sweden, to eastern Poland. He found that bogs in inland Germany and Poland displayed very dry stratigraphy overall, whereas those near the German coast and in Britain and Ireland showed frequent changes in humification and species assemblages. There are also differences in the proxy climate records from bogs within the same general climatic region, and besides the 'biological noise' factor mentioned above one must also consider differences that may arise due to local topography, the effects of sea-level changes on coastal bogs, or lake-level changes on adjacent bogs, and differences that may be linked to the size and shape of individual bogs. These last two factors, size and shape, will affect the front of efflux from a bog – a small, irregularly shaped bog will have a proportionally longer perimeter than a large, round bog, and this may affect the seepage rate of water from the edge of the bog as well as its hydrological stability. To obtain the best record of Holocene palaeoclimates it is clearly sensible to seek out sites that display sensitivity rather than complacency to climatic change, as advocated by Lowe (1993) in relation to isolating the climatic factors in Scottish woodland history. The large number of bogs in northern Cumbria and adjacent areas of Northumberland and Dumfries and Galloway, have been targeted as an ideal testing ground for the sensitivity of the bog record over the rainfall and temperature gradient from the Solway Firth coast to the inland hills of the Pennines. This area has been the scene of a number of studies over the last decade and a broadly coherent picture of Holocene palaeoclimates can be built up from the following publications outlined in Table 15.1.

From this considerable data set it is possible to pick out coherent phases of increased BSW and of drier phases, which are present in many of the cores. Table 15.2 presents a series of rounded dates for the beginning of significant wet shifts, as identified by the various authors above. The dates have been rounded to the nearest 50 years to avoid any suggestion of spurious precision, and it must also be noted that each author has calculated the wet shift dates in their own way – that is, some by means of an age/depth model, others by using single dates on particular stratigraphic events etc. Even so, the clustering of dates at around the following cal BP ages: 250, 600, 850, 1100, 1400, 1700, 2150, 2650, 2900, 3500, 4400 and 5300, with isolated older dates, does point to periods of high BSW which appear more or less consistently in bogs in this region. Many of these shifts are also found in Irish, Dutch and Swedish bogs (e.g. van Geel *et al.*, 1998) and are therefore significant at a sub-continental scale.

It must also be borne in mind that the fact that a particular bog does not show a wet shift at exactly the same time as another bog, or does not show a particular shift at all, does not invalidate either the method or the results. For example, core WLM-11 (Figs 15.4 and 15.5, Table 15.2) does not show a wet shift in the early Little Ice Age period, *c.* cal AD 1450 or 500 cal BP, but this is a widespread wet shift which is prominent in many other sites and in the field stratigraphic sections from Bolton Fell Moss examined by Barber (1981) and recently reaffirmed in two other cores from Walton Moss (Table 15.2), WLM-17 by Barber and Langdon (2001) and WLM-19 by Mauquoy *et al.* (2002), with a 'wiggle-matched' calendar age of AD 1464. Inspection of the macrofossil diagram from WLM-11 (Hughes *et al.*, 2000) (Fig. 15.4) reveals the reason for the lack of a wet shift at 500 cal BP – the peat at that time was dominated by Monocotyledonous remains, mostly cotton sedge, and only responded to increasing BSW by a later increase in *S. magellanicum* and then *Sphagnum* Sect. *Cuspidata* at around 350 cal BP. As Barber *et al.* (1998, p. 527) point out '. . . cores analysed in previously published work . . . have not yielded invalid data; they may simply not be telling the whole story'. In a full discussion of

1. Barber, 1981	Field stratigraphy from 14 sections and Abundance Estimate macrofossil analyses from 21 monoliths from Bolton Fell Moss (BFM), Cumbria; time range 0–2200 cal yr BP
2. Barber *et al.*, 1994	QLC macrofossil analyses (QLCMA) from core BFM-J; 0–6500 cal yr BP
3. Woodland *et al.*, 1998	Testate amoebae analyses (TAA) from BFM-J; 0–1500 cal yr BP.
4. Barber *et al.*, 1998	Replication study involving field stratigraphy from 23 cores and QLCMA from 11 short cores from BFM and Walton Moss (WLM); 0–850 cal yr BP
5. Mauquoy and Barber, 1999a	QLCMA and humification analyses (HA) from two cores from Coom Rigg Moss (CRG) and Felecia Moss (FLC), Northumberland; 0–3000 cal yr BP
6. Mauquoy and Barber, 1999b	QLCMA, TAA and HA from BFM core L, Walton Moss (WLM) core 11, CRG, FLC and in Scotland, Raeburn Flow (RBF) and Bell's Flow (BSF); time period focussed on decline of *Sphagnum imbricatum*; 0–1000 cal yr BP
7. Charman *et al.*, 1999	Replicate cores from Coom Rigg Moss analysed for TAA, QLCMA and HA; 0–4500 cal yr BP
8. Hughes *et al.*, 2000	QLCMA on Walton Moss core WLM-11; 0–9800 cal yr BP
9. Charman and Hendon, 2000	TAA composite record from Coom Rigg Moss and Butterburn Flow; 0–4500 cal yr BP
10. Xie *et al.*, 2000	Compound-specific δD values of *n* alkanes from a BFM monolith; 0–200 cal yr BP
11. Hendon *et al.*, 2001	TAA from four cores from Coom Rigg Moss, Butterburn Flow and The Wou; 0 – 5500 cal yr BP
12. Barber and Langdon, 2001 and in prep.	QLCMA from Walton Moss, WLM-17; 0–3000 cal yr BP
13. Mauquoy *et al.*, 2002	Volume abundance macrofossil analyses from WLM-19; 0–1050 cal yr BP

Table 15.1 Published research on peat-based palaeoclimate studies from southern Scotland and northern England, UK

their own and other data from this region, Hendon *et al.* (2001) also concluded that the main changes in the proxy climate records could be replicated within sites and between sites, but that one of the main limitations of the present dataset was the imprecision of the chronology at some sites. This is now being addressed by the use of AMS ^{14}C wiggle-matching (Mauquoy *et al.*, 2002) and tephrochronology (Langdon and Barber, 2001).

Bog and core no.	Reference no.										
Bolton Fell Moss	1 and 2	200	500	1000	1300		2900	3600	4400	5300	*
			650	1150						5700	
										6200	
BFM and Walton Moss (WLM)	4	250–300	600	*							
Coom Rigg Moss (CRG)	5	200	550	850	1400	1750	2150	*			
and Felecia Moss (FLC)				1050		1950	2550				
							2700				
Coom Rigg Moss (CRM)	7	100	450	850	1350		2150	3500	*		
			600	1000	1550		2700				
							2900				
WLM-11	8	100			1450	1750	2600	3500		5300	7800
		350								5900	
CRM and Butterburn Flow	9			950			2100	3700	4400	*	
(BBF) composite							2850				
CRM, BBF and The Wou	11		400	800	1500	1650	2800	3450	*		
			700	1100				3850			
WLM-17	12	240	500	850	1550	1750	2650	*			
			700	1100			2900				
WLM-19	13	350	500	*							
			750								

Notes:

1. Cores vary in depth and therefore age – for example at Walton Moss core WLM-11 covers the whole Holocene; WLM-17 the last 3000 years and WLM-19 the last 1000 years. The end of each core is denoted by an asterisk *.
2. Precise dates were not available for core analysed in reference 3, and reference 6 repeats most data from reference 5.
3. Reference 10 data only cover the last 200 years and are provisional results.
4. All dates have been rounded to the nearest 50 years, but note that WLM-19 dates are rounded from wiggle-matched AMS calendar dates of AD 1215, 1464 and 1601.

Table 15.2 Dates of wet-shifts, arranged in clusters according to age in calibrated years BP, from bogs in the Anglo-Scottish Border region

15.6 CONCLUSION AND FUTURE DIRECTIONS

Research on the peat-based proxy climate record has clearly advanced greatly over the last 20 years and is now at a most interesting and exciting stage. The techniques used in extracting the palaeoclimate signal have advanced markedly in the last decade and now include not just proven and tested methods – plant macrofossil, testate amoebae and humification analyses – but also novel biomarker and stable-isotope methods. Important methodological issues such as core replication have been confronted and resolved, and the spread of peat-based research outside of its European home is gathering pace. It is now possible to deliver, on decadal time-scales, detailed records of climatic changes from almost the entire Holocene time-scale. Furthermore, we can expect these records to be calibrated and validated against instrumental data for the last 300 years, and compared and tested against other palaeoclimatic proxy data in future studies.

The way forward is increasingly a matter of consensus amongst a small but growing community of peatland researchers who are coming together to derive independent high-resolution, calibrated multiproxy climate records from key sites around the world. Different approaches will clearly be needed in different regions. In North America, the peat initiation record is valuable but more precise climate reconstructions from work on continuous peat records are a possibility for the future. In the southern hemisphere, techniques are also being successfully developed to exploit the peat record from ombrotrophic mires (McGlone and Wilmshurst, 1999; Wilmshurst *et al.*, 2002). In the North Atlantic region changes in BSW are known to cluster around 8200, 5900, 4400, 3500, 2700, 1700, 1400, 1100 cal BP and the Little Ice Age. One may hypothesize that most, if not all, of these events resulted from thermohaline circulation (THC) changes in the Atlantic Ocean and future work is planned to test the linkages between oceans and land by providing high resolution records of terrestrial biosphere response to the oceanic changes. The link to solar variability is also assuming increasing importance (van Geel *et al.*, 1998; Chambers and Blackford, 2001). In the near future it should be possible to derive multiproxy data in a form suitable for testing climate models, perhaps focusing initially on the testing of natural variability and the use of downscaling techniques to link model output and site-based proxy data. The importance of good chronological control of the stratigraphic data cannot be overemphasized, and the use of wiggle-matched AMS ^{14}C dates, allied where possible to the precise 'pinning points' of tephra layers, will become the norm. As a scientific 'bonus' the derivation of palaeoclimate records from peat bogs gives us a much greater understanding of the ecological functioning of these fascinating ecosystems, whose archives of past conditions are unique amongst terrestrial plant formations.

Acknowledgements

The continuing support of the UK Natural Environment Research Council (NERC) in the provision of grants, studentships and radiocarbon dating facilities is gratefully acknowledged. We thank Professor Frank Chambers and Dr Pete Langdon for helpful review comments.

CHAPTER

LACUSTRINE PERSPECTIVES ON HOLOCENE CLIMATE

Sherilyn C. Fritz

Abstract: This chapter presents a selection of examples that illustrate how lake records have advanced our understanding of patterns and controls of Holocene climate. A hierarchy of temporal scales of climate variation influence lake behaviour: from orbital cycles, to millennial and centennial changes in sea-surface temperatures or solar radiation, to multi-annual variation in the linked ocean–atmosphere system associated with synoptic climate features, such as the El Niño Southern Oscillation or the North Atlantic Oscillation. Orbital forcing affects not only temperature of continental areas but also atmospheric circulation systems that drive winds and precipitation. Thus, there is considerable spatial variation in Holocene moisture patterns at orbital scales related to the positioning and strength of surface pressure systems that drive the penetration of moisture from the oceans onto the continents. In North America, the Laurentide ice sheet, which persisted until the Mid-Holocene, also had major impacts on circulation patterns and hence on moisture and temperature gradients in the Early and Mid-Holocene. In many cases, higher frequency oscillations are superimposed upon an overall envelope of change at orbital time-scales. Millennial-scale cooling events in the North Atlantic region, as well as solar variation at multiple temporal scales, appear to drive drought cycles, as evidenced in paleolimnological records throughout the Americas, Europe and Africa. How insolation forcing interacts with other modes of climate variation operating at millennial- to sub-annual-scales is not understood and will require long sequences of high-resolution data that span the entire Holocene.

Keywords: Lakes, Lake-level, Palaeoclimate, Palaeohydrology, Palaeolimnology

Lacustrine deposits are one of the best archives of continental climate and its environmental impacts, because lake deposits often span long periods of time and yield moderately high temporal resolution. In sediments that are varved, annual or even seasonal conditions can be resolved, although more commonly temporal resolution is on the scale of multiple years or decades. Lacustrine deposits are often continuous, and individual records commonly span thousands of years, yielding longer time series than other continental records of high temporal resolution, such as tree-rings or speleothems.

Climate reconstruction from palaeolimnological records is based on physical, biological and chemical proxies that are sensitive to climate-driven changes in a lake's energy balance (thermal structure), hydrologic balance, or catchment conditions affected by climate. The ability to discern climate from lacustrine records is limited by our knowledge of how various direct and indirect climatic forcings interact to influence the proxies that form the basis of reconstruction (Fritz, 1996; Fritz *et al.*, 1999). The use and limitations of various proxies in reconstructing climate is discussed in detail in other chapters in this text. An individual lake's response to regional climate is mediated by site-specific catchment and basin characteristics that affect energy and water budgets, including connection to local and regional groundwater flow systems; presence of inflows and outflows; soils, vegetation, and topography, which affect basin runoff and recharge; and lake morphometry and chemistry in terms of their influence on the penetration of solar radiation. Below a few of the factors that influence how the climatic signal is recorded in lake systems are highlighted. More detailed reviews can be found elsewhere (Fritz, 1996). The primary focus of this chapter is the insights about Holocene climate that have been generated from studying lake proxy records. Proxies that are derived from within the lake itself are primarily considered, rather than proxies that are derived from the lake catchment. Although pollen studies commonly have generated reconstructions of regional temperature, in many regions Holocene palaeolimnological studies have emphasized hydrologic rather than temperature change, because these influences are less ambiguous. In general, lacustrine records from tropical and temperate latitudes most strongly reflect hydrologic balance (precipitation minus evaporation), whereas those from higher latitudes may also record strong temperature influences. Please note that all **ages in this chapter refer to calendar years**, not radiocarbon years.

16.1 CONTROLS ON LAKE RESPONSE TO CLIMATE

Clearly, many characteristics of a lake and its watershed influence how climate impacts the system and thus the stratigraphic signal. Large lakes commonly have the advantage of large watersheds that integrate environmental dynamics of large areas, and even large lakes can respond quickly and sensitively to interannual climate variation. Lake Titicaca, for example, rose 4 m over its long-term average during several years of extreme precipitation in the late 1980s and flooded large areas of agricultural land (Pawley *et al.*, 2001). This sensitivity to precipitation variation (Baker *et al.*, 2001) is amplified on millennial time-scales, and seismic data show that the lake fell over 85 m during the Mid-Holocene lowstand (Seltzer *et al.*, 1998). Lakes and lake catchments that differ in surface area to volume can differ in the magnitudes and/or timing of change in response to the same climatic forcing (Mason *et al.*, 1994). Similarly, topographic and geologic settings can amplify or dampen the response of individual basins to a common forcing (Webster *et al.*, 2000). Geologic setting can also produce counter-intuitive situations that require a detailed understanding of the regional geomorphic response to climate change. For example, lakes in the Sandhills of Nebraska were formed and began accumulating sediment during dry intervals when migrating sand blocked drainage systems (Loope *et al.*, 1995), rather than during times of increased precipitation, as one might expect in other settings.

Several recent palaeolimnological studies have highlighted the influence of groundwater in mediating a lake's response to changes in moisture budget. In the semi-arid North American Great Plains, oxygen isotopic values became more depleted during the Mid-Holocene, when other proxies indicate evaporative enrichment of lakewater. Modelling studies suggested that changes in groundwater flow path during times of extreme drought (Smith *et al.*, 1997) are the likely explanation for what might, in other settings, be interpreted as increased moisture. Similarly, multi-proxy data from a crater lake in Ethiopia suggested that isotopic variation was likely influenced by changes in groundwater inflow that were not always tightly coupled to climate change (Lamb *et al.*, 2000). Chemical stability does not always imply climatic stability, as evidenced by two neighbouring basins in Turkey in which a groundwater fed lake showed little variation in lakewater salinity, whereas a nearby hydrologically closed basin underwent changes in salinity driven by changes in precipitation minus evaporation (Reed *et al.*, 1999). Such studies highlight the need for using multiple proxies (Rosén *et al.*, 2001) and multiple lakes (Fritz *et al.*, 2000) in developing regional palaeoclimatic inferences.

Several detailed multi-proxy studies from Scandinavia elucidate the influence of directional landscape evolution on lake response to climate change, particularly in the initial centuries and millennia following lake formation. Transfer functions for temperature reconstruction applied to diatoms, pollen and chironomids in a stratigraphic record from northern Sweden (Fig. 16.1) showed similar patterns of inferred climate after 6000 yr BP, but the inferences diverged considerably in prior millennia (Rosén *et al.*, 2001; Bigler *et al.*, 2002). Analyses suggested that the Early Holocene diatom response was largely governed by pH changes associated with long-term catchment acidification. The overriding influence of non-climatic influences, such as pH, on diatom composition is also suggested by modern studies of lake evolution (Engstrom *et al.*, 2000) and calibration studies comparing diatom temperature inferences with the instrumental record in lakes of stable versus fluctuating pH (Bigler and Hall, 2003). In regions or times with stable vegetation, however, pH change can be a sensitive indicator of climate change, as has been demonstrated in the Alps where climate change affects weathering and erosion of base cations and produces variation in lake pH (Sommaruga-Wögarth *et al.*, 1997).

16.2 ORBITAL FORCING OF CONTINENTAL CLIMATE

Changes in the Earth's orbit relative to the Sun drive climate by altering the amount and seasonality of radiation at the Earth's surface, particularly at high latitudes. The variation in incoming radiation changes the strength of the temperature gradient from the equator to the poles and the resultant atmospheric circulation. Insolation change can also impact regional climates via its influence on ocean circulation and glacier mass balance. The Holocene spans one half of the 22,000-year precessional cycle and hence spans major extremes in insolation. Reconstruction of the spatial pattern of climate change over this period provides a tool for understanding how the oceans, atmosphere and land surface interact in response to these radiation changes. Continental records also are crucial for unravelling the patterns and thresholds of ecosystem response to directional changes in forcing.

16.2.1 Tropical and Sub-Tropical Latitudes

In tropical latitudes, precipitation variation is governed by shifts in the latitude of tropical convection and associated change in the strength of the trade winds. During the annual cycle, as the intensity of solar insolation at the Earth's surface shifts from January to July, convection over the oceans (along the intertropical convergence zone, ITCZ) moves from south to north and with it the wind belts and thus the locus of continental precipitation. Interannual variation in radiation balance can alter the mean seasonal positions of the ITCZ, the position of subtropical high-pressure systems, and the intensity of winds bringing moisture into continental areas, and thus change the intensity and distribution of precipitation.

Based on lake records from the tropics, it has been suggested that orbital changes in insolation are the first-order control on tropical moisture budgets. The most extensive data come from the African tropics and sub-tropics, where several decades of research utilizing an array of palaeolimnological proxies indicate hydrological changes more drastic than those of recorded

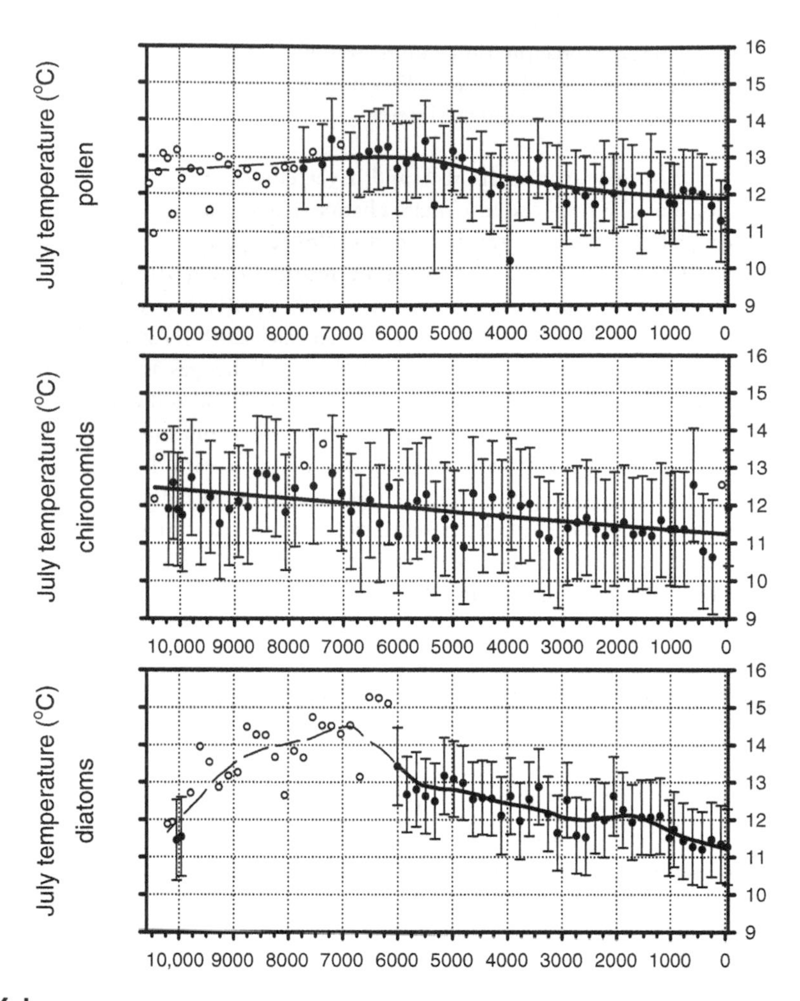

Figure 16.1

history and in some areas as large as those of glacial–interglacial cycles (reviewed in Gasse, 2000). Prominent among these is the greening of the now arid Sahara (*c.* 20°N) during the Early to Mid-Holocene, as evidenced by palaeo-lake deposits (Fig. 16.2), indicative of lakes maintained by intensified monsoon precipitation, associated, at least in part, with a northward shift of the ITCZ driven by increased northern hemisphere summer insolation. Similarly, lake-levels in the Early Holocene in equatorial East and West Africa and the Sahel were considerably higher than today and fell sometime around 4000 yr BP. In contrast, Lake Malawi (10°S) was low in the Early Holocene, suggestive of climate changes that are opposite in sign from north to south across the equator, at least during the Holocene (Baker, 2002).

The envelope of change in the tropical Americas suggests a strong influence of the precessional cycle on tropical moisture budgets. In the Caribbean (10–18°N), isotopic values in lacustrine ostracods became increasingly more depleted from the Early to Mid-Holocene, which suggests increased precipitation (Hodell *et al.*, 1991; Curtis *et al.*, 2001), followed by enriched values in the last few thousand years and an inferred reduction in moisture. Similarly, lake records from

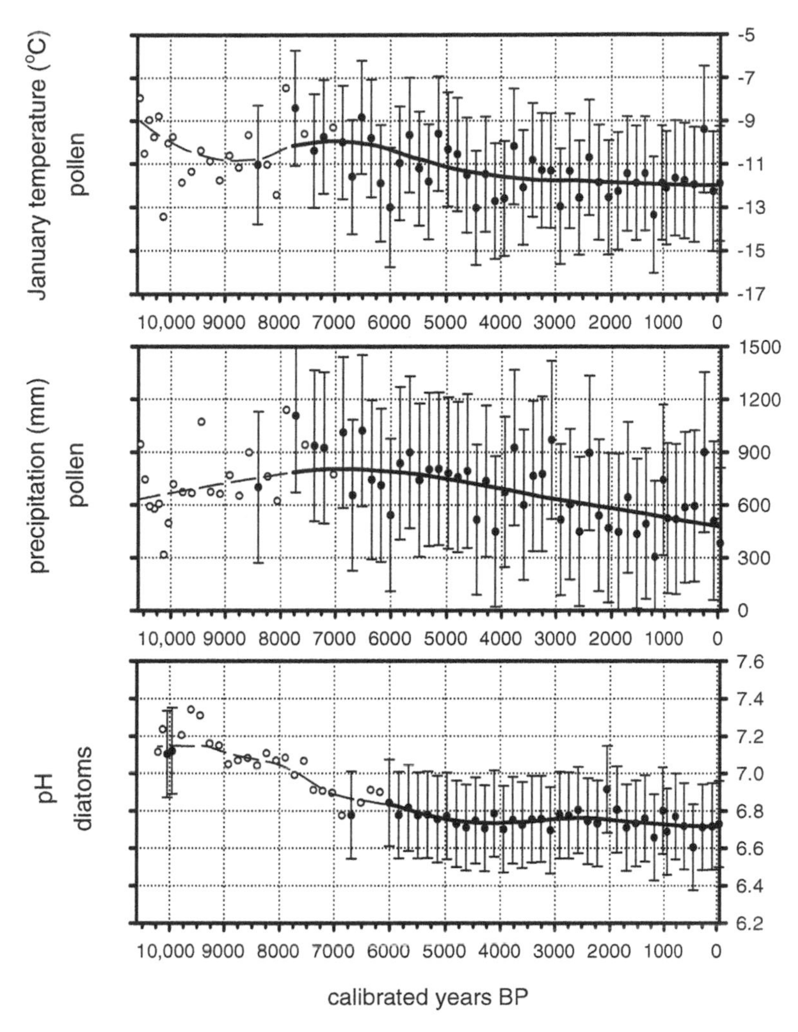

Figure 16.1 Multiproxy reconstructions of air temperature, precipitation, and pH from fossil diatoms, chironomids and pollen from a sediment core from a small lake in northern Sweden. Open circles indicate that the fossil samples have a poor fit to the modern calibration data set used to generate the transfer function. Smoothing of the data (solid line) is used to indicate overall trends. Note that poor fit of the diatom reconstruction in the Early Holocene corresponds with a period of significant pH decline associated with changes in vegetation and soils. Reprinted from Bigler *et al.* (2002), with kind permission from Arnold Publishers.

throughout tropical Mexico suggest that the Mid-Holocene was generally wet relative to the Earliest and the Late Holocene (Metcalfe *et al.*, 2000; Curtis *et al.*, 2001). The overall trend of a wet Mid-Holocene is correlated with an enhancement of the annual cycle (increased seasonality) in the northern hemisphere tropics, driven by changes in orbital forcing.

Overall patterns of lake-level fluctuation in the southern tropics of the Americas are broadly inverse to those in the northern hemisphere tropics, consistent with the hypothesis of orbital forcing of moisture budgets (Fig. 16.3). At Lake Titicaca (16°S) (Seltzer *et al.*, 1998; Baker *et al.*,

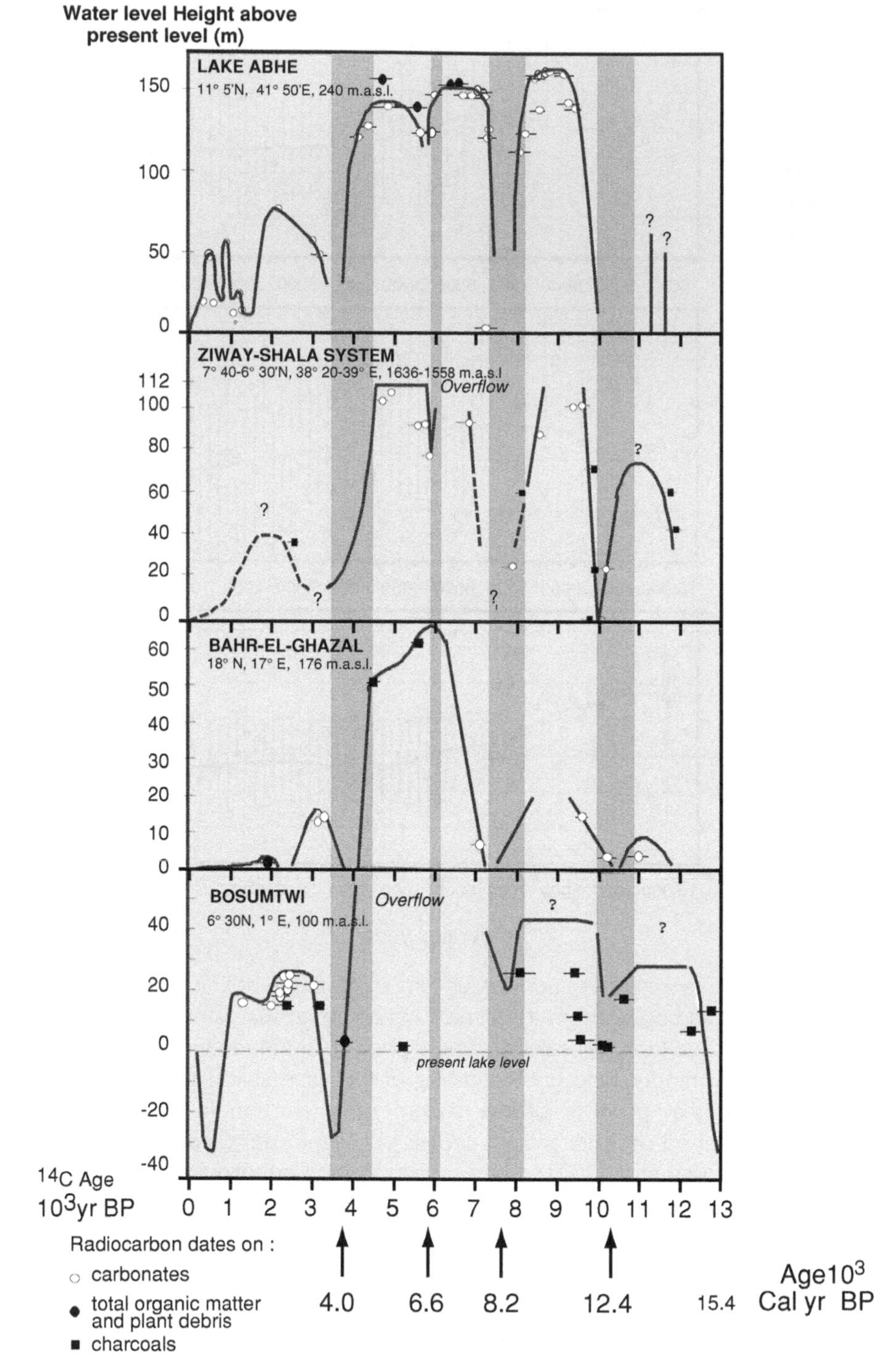

Figure 16.2 Lake-level reconstructions from Africa. Grey bands show times of regional lake-level lowering. Note that the continuous age scale (*x*-axis) is in ^{14}C years before present; calendar ages for select 'events' are indicated with the arrows. Modified from Gasse (2000) with kind permission from Elsevier Science.

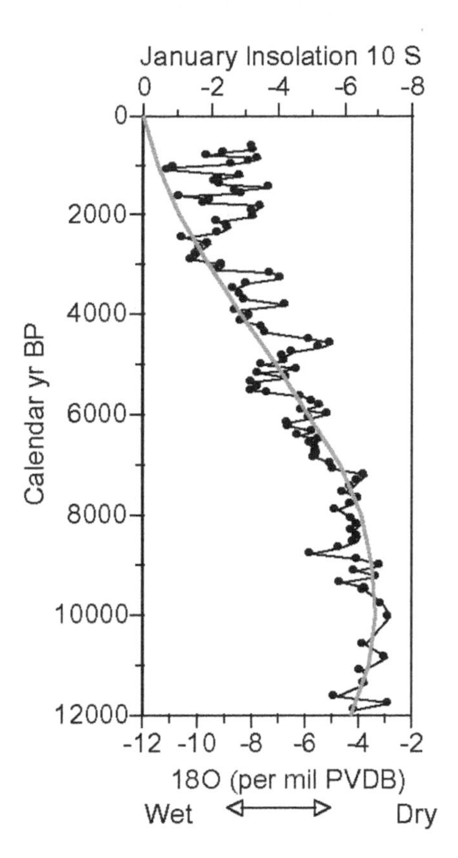

Figure 16.3 Comparison of the $\delta^{18}O$ record of authigenic calicite from Lake Junin, Peru (Seltzer *et al.*, 2000) (black line) with wet season (January) insolation for 10°S, expressed as per cent deviation from AD 1950 (grey line).

2001) and in smaller lakes in the Bolivian Andes (Abbott *et al.*, 1997), diatom and geochemical records suggest Mid-Holocene drying and enhanced precipitation in the last few thousand years. Similar patterns are inferred for adjoining sub-tropical (18–28°S) regions of Chile (Grosjean *et al.*, 1997; Schwalb *et al.*, 1999), suggesting that an arid Mid-Holocene was regionally characteristic. Further north in Peru (11°S), isotopic values of authigenic carbonates suggest reduced Early Holocene precipitation and wet conditions in the Late Holocene (Seltzer *et al.*, 2000). At these southern hemisphere latitudes, insolation during the summer rainy season (December to March) was reduced in the Early to Mid-Holocene and subsequently increased to near modern values about 3000 yr BP (Fig. 16.3). Thus, the overall pattern of inferred moisture change suggests that the intensity of convection and the resultant precipitation in the Andes followed changes in insolation driven by precession.

16.2.2 Northern Hemisphere Mid-Latitudes

At the onset of the Holocene, incoming January insolation in the northern hemisphere was reduced relative to today, whereas July radiation was enhanced. In North America, the Laurentide ice sheet was still present in central and eastern Canada and also exerted a major influence on atmospheric circulation across the continent (COHMAP Members, 1988). Lake records from mid-latitudes (40–50°N) across North America show differential patterns of lake-level change that suggest the influence of the ice sheet. In northwestern North America, to the west of the ice sheet, many lakes were dry or low in level at the onset of the Holocene, because of high summer temperature driven by enhanced insolation and intensification of the Pacific high-pressure system just off the West Coast, which suppressed moisture penetration into the northwest. These conditions generally persisted through the Mid-Holocene (Bennett *et al.*, 2001; Fritz *et al.*, 2001). In contrast, lake-levels in central North America were generally high, and waters were fresh (Fritz *et al.*, 2001), which suggests moderation of insolation forcing by the ice sheet, an interaction evident in modelling studies (Hostetler *et al.*, 2000). Lake-levels in parts of eastern North America were low at the onset of the Holocene (Shuman *et al.*, 2001), probably because of dry anticyclonic flow off the ice, which prevented the penetration of moisture. After about 8000 years ago, as the ice sheet waned, dramatic changes in lakes in central and eastern North America suggest altered circulation patterns across North America. In central North America, lake-levels dropped and, in the grassland regions, many lakes became saline because of evaporative concentration of salts (Frtiz *et al.*, 2001). In southern New England, lake-levels rose (Shuman *et al.*, 2001), whereas other parts of northeastern North America show low lake-levels (Almquist *et al.*, 2001; Fritz *et al.*, 2001). The overall pattern around 8000 yr BP suggests a strong ridge over central North America, which blocked the penetration of summer moisture from the Gulf of Mexico into the interior, yet allowed flow of moist subtropical air into more southerly areas in the east. As summer insolation was gradually reduced and winter insolation increased in the latter half of the Holocene, lake-levels generally rose in mid-latitudes from west to east across the continent.

Multi-core studies from mid-latitudes of Europe (Digerfeldt, 1988; Almquist-Jacobson, 1995) show complex patterns of lake-level change that do not mirror the directional and gradual changes in insolation (Harrison *et al.*, 1993). In southern Sweden, for example, lake-levels were high at the onset of the Holocene, as much as 7 m below modern in the Early Holocene, rose to modern levels about 8000 yr BP, subsequently fluctuated 3–6 m below modern, and then rose again to near modern levels about 3000 yr BP (Digerfeldt, 1988). Modelling studies show that insolation and temperature change cannot account for the lake-level patterns without precipitation variation. Precipitation changes, however, may have been largely insolation forced via changes in sub-tropical high pressure systems and thus the intensity and locus of winds and rain (Harrison *et al.*, 1993). One interesting result of simulation studies for southern Sweden is that precipitation had a significant impact on lake-levels only when winter temperatures were warmer than present and hence runoff was reduced (Vassiljev *et al.*, 1998). Similar modelling studies would be useful in other regions, as well, to better understand the extent to which lake-level is influenced by a series of non-linear interactions forced by orbital variation, including insolation, temperature, and precipitation amount and their seasonal distribution.

16.2.3 Boreal and Sub-Arctic Latitudes

In Boreal latitudes of North America (60–65°N), a strong west to east differential also existed in inferred Early and Mid-Holocene climate that suggests an interaction between insolation forcing and the Laurentide ice sheet. In northwestern Canada, treeline moved northward in the Early Holocene at the time of maximum summer insolation, and the chemistry and level of lakes in the interior of Alaska indicates reduced precipitation and increased summer evaporation relative to today (Hu *et al.*, 1998; Abbott *et al.*, 2000a; Anderson *et al.*, 2001). By about 6000 yr BP, northwestern North America began to cool and precipitation increased. In contrast, in central

Canada, changes in tree-line caused by warming and associated limnological shifts were delayed until 5700 yr BP, because of the influence of the remnant Laurentide ice sheet (MacDonald *et al.*, 1993; Pienitz *et al.*, 1999). Isotopic data from central Canadian lakes suggest that Mid-Holocene changes in lake productivity and transparency, driven by warming and treeline advance, were associated with increased precipitation (Wolfe *et al.*, 1996). Thus, the data suggest a strong temperature differential across high latitudes in North America from west to east in the Early to Mid-Holocene, although on average, both regions likely experienced increased Mid- to Late Holocene moisture and Late Holocene cooling.

Sites near modern treeline in north-central Russia (68–70°N) were not influenced in the Early Holocene by a remnant ice sheet and show a direct response to increased summer insolation, with warmer summer temperatures and the northward migration of treeline between 9000 and 4000 yr BP (MacDonald *et al.*, 2000a). $\delta^{18}O$ analyses of cellulose in lake sediments (Wolfe *et al.*, 2000) suggest that sites closer to the Nordic Seas were wet at 9000 yr BP, because of a strengthened Siberian Low and the penetration of warm maritime air, whereas sites inland had a warm, dry continental climate. These moisture gradients from coastal areas inland would have been enhanced by lowered sea level at the onset of the Holocene. As sea-surface temperatures (SSTs) declined and cyclonic activity weakened through the Mid- to Late Holocene, the strong moisture gradient weakened.

16.3 CLIMATE VARIATION AT SUB-ORBITAL SCALES

In many lake records, higher frequency oscillations are superimposed upon an overall envelope of change at orbital time-scales (Grosjean *et al.*, 1997; Hu *et al.*, 1998; Metcalfe *et al.*, 2000). In Greenland, for example, much of the millennial- and centennial-scale variation evident in ice-cores is also manifested in the limnologic record of temperature change (Willemse and Tornqvist, 1999), illustrating the sensitivity of lakes to a hierarchy of forcing mechanisms operating at multiple temporal scales. In Lake Titicaca, oscillations from low to high lake-level within the Late Glacial to Mid-Holocene can be related to SSTs in the North Atlantic (Baker *et al.*, 2001) and suggest an interaction of orbital forcing and Atlantic SSTs in controlling advection of moisture into the South American tropics. Similarly, in the Swiss Alps and the Jura of France, lake-level oscillations are evident in the Early to Mid-Holocene at millennial frequencies and appear linked to cooling cycles in the North Atlantic region (Magny, 1993b). Recent analyses of marine cores (Bond *et al.*, 2001) suggest that these environmental fluctuations may be part of recurrent oscillations in the climate system driven by solar variation (see below).

16.3.1 Abrupt Events

Many lacustrine records show times of abrupt change; some changes are very short-lived, whereas in other cases the abrupt change marks the onset of a new and persistent condition. A long-term shift in regime may be a threshold response to a gradually changing forcing, such as insolation, or alternatively a response to an abrupt climatic shift.

At the onset of the Holocene, a major cooling event centered at 11,300 yr BP, the Pre-Boreal Oscillation, lasted less than 200 years but is widespread throughout the North Atlantic region based on pollen and glacial records. It is correlated with drainage of the Baltic Ancylus Lake and IRD events in North Atlantic cores and also appears to have had dramatic effects on the continents, including decreased lacustrine productivity in Iceland and the Faeroe Islands based on biogenic silica and organic carbon analyses (Björck *et al.*, 2001).

Impacts of another cooling phase, the '8.2 event' – an abrupt and short-lived temperature depression seen in the Greenland ice-core about 8200 yr BP – are evident in lake records. A 200-year negative excursion of $\delta^{18}O$ in biogenic carbonate from Ammersee in southern Germany is coincident with the ice-core record and provides evidence that cooling occurred throughout the North Atlantic region (von Grafenstein *et al.*, 1998). Multi-proxy studies from lakes in northern Sweden (Rosén *et al.*, 2001) and Italy (Ramrath *et al.*, 2000) also suggest short-term cooling at this time. In the North American interior, distant from the immediate influence of the North Atlantic, abrupt changes occurred in lake records between 9000 and 8000 yr BP. In Deep Lake, Minnesota, a depletion in $\delta^{18}O$ of authigenic calcite beginning at 8900 yr BP (Fig. 16.4) suggests cooling (Hu *et al.*, 1999) or a change in hydrologic balance. It is followed by a sharp increase in varve thickness at 8200 yr BP. Immediately to the west in Moon Lake, North Dakota, declines in diatom-inferred salinity (Fig. 16.4) indicate brief periods of increased freshwater inflow (Laird *et al.*, 1998b) during the same time period. Hu *et al.* (1999) suggest that the changes originating around 8900 yr BP are distinct from the 8.2 event and probably stem from alternate climatic controls. It is also possible that the differences in timing resulted from differing thresholds of response of lacustrine and terrestrial systems, rather than separate climatic mechanisms. In any case, it is likely that the mid-continental response in North America resulted from rearrangement of atmospheric circulation following drainage of Glacial Lake Agassiz and Ojibway as the Hudson Bay ice dome collapsed. It also may be related to recurrent climate variation evident throughout the North Atlantic region and linked to solar variability (Bond *et al.*, 2001), as discussed later in this section.

About the same time, abrupt dry periods are widespread across the northern and equatorial tropics of Africa during the otherwise wet Early Holocene (Fig. 16.2), with lake-level depression sometimes as great as several tens of meters and large increases in lakewater salinity driven by weakening of the monsoon (Gasse, 2000). In some cases, the dry intervals are only several hundred years in duration and are centred around 8200 yr BP, whereas in other sites about a millennium of dry conditions persist with an onset sometime around 8000 yr BP. The latter may reflect a different environmental trigger or alternatively differences in the response of individual systems, such that some do not amplify the signal to such an extent or respond at a slower rate. Nonetheless, the correlation of dry intervals in the northern tropics with cold periods in the North Atlantic region suggests a common forcing – either solar (Bond *et al.*, 2001) and/or changes in Walker or Hadley circulation driven by altered ocean circulation (Alley *et al.*, 1997).

The period of 4500–4000 yr BP was also one of widespread abrupt change. Short-term dry periods inferred from lake-level lowering between 4500 and 4000 yr BP (Fig. 16.2) are evident in many equatorial and northern tropical sites in Africa (Gasse, 2000; Street-Perrott *et al.*, 2000) and are correlated with cooling in the North Atlantic. Drought is also prominent across Mesopotamia and has been implicated in cultural collapse and abandonment of dry-land farming throughout the eastern Mediterranean region (Weiss *et al.*, 1993). For example, in Lake Van, near the headwaters of the Tigris and Euphrates River, varve measurements show a fivefold increase in dust, and oxygen isotopic data suggest a drop in lake-level for a ~200-year period

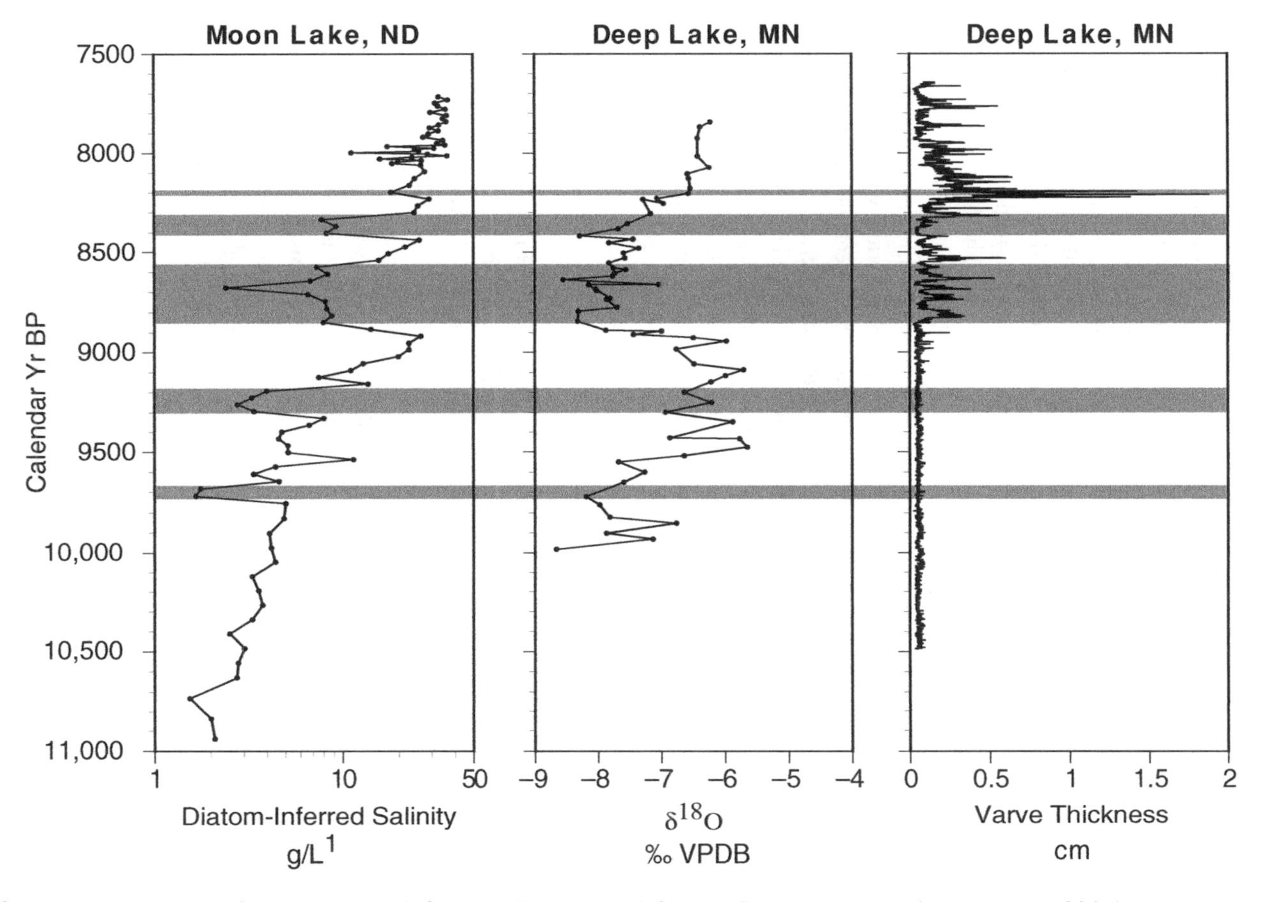

Figure 16.4 Moon Lake, North Dakota diatom-inferred salinity record for the Early Holocene (Laird *et al.*, 1998b) compared with the $\delta^{18}O$ record of authigenic calcite and varve thickness measurements from Deep Lake, Minnesota (Hu *et al.*, 1999). The grey bars show brief intervals of lowered salinity and increased moisture, as evident at Moon Lake, during the overall drying trend of the Early Holocene. Note that major freshwater excursions in Moon Lake are frequent during the period of depleted isotopic values from Deep Lake, suggesting that both sites are responding to regional climatic variation. The large increase in varve thickness at Deep Lake is correlative with the '8.2 event', but coincident signals in the isotopic record (Deep Lake) and diatom-inferred salinity (Moon Lake) are weak.

before 4000 yr BP (Lemcke and Sturm, 1997). The isotopic record from Lake Zeribar in Iran also suggests a short-term drying phase about 4000 years ago, although the chronology is not certain (Stevens *et al.*, 2001).

16.3.2 Late Holocene Climate Variation

The Late Holocene is of particular interest to palaeoclimatologists, because major boundary conditions (orbital relationships, continental ice, CO_2) are roughly equivalent to those of today. Hence, one can use the stratigraphic record to generate a long time-series of climate variation prior to major human-induced changes in greenhouse gas concentration. During the last 2000 years, the Little Ice Age (*c.* AD 1350–1850) and Medieval Period (*c.* AD 900–1300) have been identified as intervals of temperature anomalies in Europe, but additional sites are still needed to define whether these are intervals of anomalous climate on hemispheric or global scales and what the spatial manifestation of climate variation is during these times. The Medieval Period also provides an interesting benchmark against which to evaluate 20th century warming and drought variation. Modelling of hemispheric temperature trends from a variety of palaeoclimatic data suggests that recent warming is greater than any period within the last 1000 years, including Medieval times (Crowley, 2000). Although these hemispheric syntheses are compelling regarding the impact of human activities on global climate, our understanding of regional patterns of climate change over the last few millennia is limited. At high latitudes, lacustrine records have been used to reconstruct temperature change from lakes, because of temperature control of the duration of the ice-free season, which in turn controls many lacustrine processes (Joynt and Wolfe, 2001). Temperature reconstruction from a 1200-year sequence of glacio-lacustrine varves in the Canadian Arctic (Moore *et al.*, 2001) suggests that summer temperatures from AD 1200–1350 were as much as 0.5°C warmer than the long-term average of the last 1200 years and were followed by over 400 years of cooling. These data suggest that although late 20th century temperatures in this region are elevated over those of the last 400 years (Lamoureux and Bradley, 1996; Overpeck *et al.*, 1997; Hughen *et al.*, 2000), they do not exceed those of Medieval times. Clearly, additional sites are needed to determine the extent to which the apparent hemispheric trends are representative of the behaviour of individual regions.

Stine (1994) proposed that, in some parts of the world, moisture extremes were more prevalent than temperature anomalies during Medieval times. He defined two intervals (AD. 900–1100 and AD 1200–1350) when water levels and hence effective moisture in western North America were much lower than at any time during the 20th century. Subsequent palaeolimnological studies at decadal or sub-decadal resolution suggest that extreme drought was widespread throughout the Americas (Fig. 16.5) and Africa during Medieval times (Binford *et al.*, 1997; Fritz *et al.*, 2000; Li *et al.*, 2000; Verschuren *et al.*, 2000; Hodell *et al.*, 2001). These studies suggest that both the Medieval Period and Little Ice Age were hydrologically complex, with alternations between dry and wet climate. Spatial variation is also apparent, even on relatively small geographic scales (Fritz *et al.*, 2000; Johnson *et al.*, 2001), although presently it is unclear whether this spatial variation results from non-climatic influences or a spatially variable climate. In any case, in all regions, droughts more protracted and sometimes more extreme than those of the 20th century were recurrent, indicating that the 20th century does not display the full range of natural climate variation. In both Central and South America, major drought was coincident with cultural decline and thus may have played a contributing role (Binford *et al.*, 1997; Hodell *et al.*, 2001).

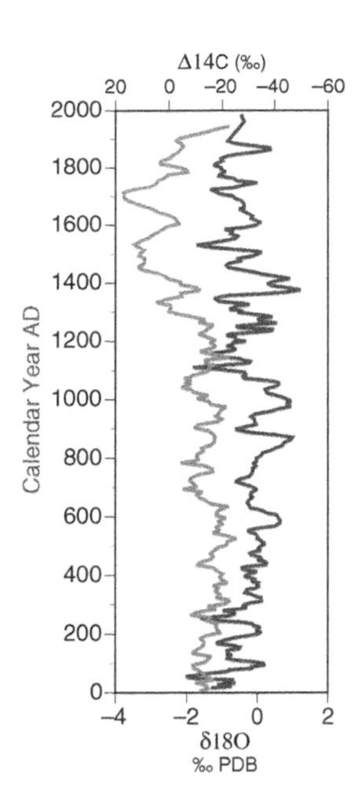

Figure 16.5 Comparison of the Late Holocene ostracod $\delta^{18}O$ record (black line) of Punta Laguna, Mexico (Curtis *et al.*, 1996) with atmospheric $\Delta^{14}C$ (grey line) (Hodell *et al.*, 2001). Note that periods of inferred drought (enriched $\delta^{18}O$) correspond with $\Delta^{14}C$ minima and hence times of increased solar activity. The prolonged drought between approximately AD 800 and 1000 is correlated with collapse of the Classic Maya civilization in the 9th century AD.

16.3.3 Sub-Orbital Cycles

It has been suggested for quite some time that lake-level oscillations may be cyclical in nature and linked to solar forcing at a variety of scales. In Holocene lake records from the Jura and Alps in Europe, alternating transgressive and regressive lake-level phases occur at roughly 2300-year intervals and have been correlated with North Atlantic temperature change and variation in atmospheric radiocarbon (Magny, 1993b). Time-series analysis of a diatom record from Africa (Stager *et al.*, 1997) also shows significant periodicities in the 2350-year band, as well as 1400- and 500-year cycles. A 1400-year cycle also occurs in some American lacustrine sequences (Campbell, 1998), and this frequency is equivalent to the Bond cycles noted in marine IRD records and linked to solar variability (Bond *et al.*, 2001).

Various lacustrine proxy records of drought have been linked to solar activity based on correlation with cycles evident in the ^{14}C and ^{10}Be records of cosmogenic nuclide production. Statistically significant cycles of 400 and 200 years are apparent in many Late Holocene records of inferred eolian activity and drought in the mid-continent of North America, as well as in the 130- and 84-year bands (Dean, 1997; Yu and Ito, 1999; Clark *et al.*, 2002). The dry periods at 200-year intervals are roughly equivalent to ^{14}C maxima and thus solar minima and to cold intervals in the Greenland ice-core record. Thus, the data suggest a link between drought in the North American mid-continent and cold climate in the North Atlantic region (Yu and Ito, 1999). Drought in the Yucatan of Mexico, as inferred from geochemical and isotopic proxies, is correlated with ^{14}C minima (high solar activity) (Fig. 16.5) and shows a significant 206-year cycle equivalent to solar variability (Hodell *et al.*, 2001). Major droughts inferred from the sedimentology of lakes in East Africa are coincident with the Wulf, Sporer and Maunder sunspot minima and thus also can be linked to solar activity (Verschuren *et al.*, 2000).

Higher frequency periodicities can be evaluated in sediments dated by varve counting. Significant periodicities at 40–50, 20–25 and 10–11 years have been noted in measures of varve thickness (Anderson, 1993; Zolitschka, 1996a; Livingstone and Hajdas, 2001). Solar forcing is suggested because of equivalent periodicity in ^{10}Be and ^{14}C measurements; the latter two periods may be related to the 22-year solar Hale cycle and the 11-year Schwabe sunspot cycle. The potential mechanisms linking solar forcing and lake records vary from system to system and may include temperature control on primary production and carbonate precipitation, wind influence on eolian inputs, and precipitation influence on overland flow. Non-varved sediments from lakes with high rates of sediment accumulation have been used to document fluctuations in climate variability at time-scales equivalent to El Niño Southern Oscillation (ENSO) frequencies (Rodbell *et al.*, 1999; Riedinger *et al.*, 2002). These studies suggest that the intensity of ENSO variation was weakened in the Early to Mid-Holocene. Although modelling studies can produce a similar pattern with known forcing (Clement *et al.*, 1999), additional data with good chronological control, particularly from regions strongly affected by ENSO, are clearly needed to evaluate the long-term nature of ENSO variability.

16.4 CONCLUSIONS

Although it is abundantly clear that orbital forcing drives climate variation on the continents, there is still much that is poorly understood about how insolation changes are propagated in various systems and across large geographic areas. Clearly, Early and Mid-Holocene climate differed from that of the Late Holocene, when the magnitude and seasonality of insolation was different from today. However, transition points from Early to Late Holocene climate differ from region to region, and it is not known whether the differences in timing are a function of the time-transgressive propagation of a single signal across large areas or because of differential responses of regions or types of systems to climate forcing. For example, the onset and duration of peak Holocene aridity differs across North America, and in many areas is still not clearly defined nor the spatial pattern understood. This is true for much of North America east of the Mississippi River (Fritz *et al.*, 2001) and north to south within the continental interior. Thus, despite years of effort, we still need better data coverage for many regions and critical syntheses of palaeohydrological data.

The interaction of insolation forcing with other modes of climate variation operating at millennial- to sub-annual-scales is not at all understood. Cycles that can be correlated with a variety of scales of solar variation (for example 1500, 400, 200, 22 years) or with ENSO activity have been observed in a number of lake records. However, it is not clear why certain periods of variation and not others are present in individual records, although much of this may be because of the small number of studies of sufficiently high resolution to detect centennial and sub-centennial cycles. Some of the high-frequency climate variation and its impact may be amplified or dampened under different large-scale boundary conditions (e.g. insolation), as has been suggested by modelling of lake-level (Vassiljev *et al.*, 1998) and in studies related to ENSO variation (Rodbell *et al.*, 1999). Nonetheless, there are far too few high-resolution palaeolimnological sequences, especially for time periods prior to the last few thousand years.

Presently, there is considerable interest in abrupt climate change and its ecosystem impacts, because of the possible analogies to Greenhouse warming (NAS, 2002). Understanding the environmental impacts of abrupt climate change, however, requires independent archives of local climate variation, and these data are not always available. Thus, it is often impossible in a single site to determine if an abrupt shift represents abrupt climate change or alternatively a threshold response to a gradual climatic forcing. Multiple sites, however, can corroborate the hypothesis that abrupt landscape change is driven by abrupt climate change. At sub-century-scale, however, it may be difficult to correlate confidently among sites because of errors associated with dating.

Many of the uncertainties about continental climate variation result from an incomplete understanding of the nature of climatic influences on organisms or lakes in varied regions and geomorphic settings. Climatic interpretations from multiple proxies in an individual lake are not always coherent (Fig. 16.1), and in some cases may result from differences in the seasonality of response (Stevens *et al.*, 2001; Rosenmeier *et al.*, 2002). Multiple sites within a region may show variable patterns of hydrologic change that can be caused by differences in microclimate, groundwater setting, or other aspects of the lake or lake catchment that affect energy and hydrologic budgets. These seeming inconsistencies can be resolved with continued development of spatial networks of sites to better define signal and noise, as well as process studies to refine our understanding of how organisms (Xia *et al.*, 1997a; Saros and Fritz, 2000) and lakes (Doran *et al.*, 2002) respond to and integrate climate variation at a suite of spatial and temporal scales.

CHAPTER

RECONSTRUCTING HOLOCENE CLIMATE RECORDS FROM SPELEOTHEMS

Stein-Erik Lauritzen

Abstract: Speleothems have advantages as Holocene palaeoclimatic archives because uranium-series dating yields very precise ages directly in calendar years. In spite of being subterranean deposits, speleothem formation is an integral part of the meteoric water cycle and therefore also governed by surface climatic change. Changes in the concentration of various trace components in speleothem calcite (isotopes, trace elements and organic matter) provide a link to external climate conditions. Each of these components displays individual sensitivity to climate conditions, some more directly than others, and these are discussed in this chapter. A multi-proxy approach to speleothem analysis can provide very precise climatic information.

Keywords: Dating, Holocene, Humic, Isotopes, Organic matter, Palaeoclimate, Speleothem, Trace elements, Uranium-series

Speleothems are karst cave deposits and serve as palaeoclimatic archives. Limestone karst is globally common, so that climatic signals can be studied in the same type of archive over a wide geographical range of terrestrial sites. Speleothem deposition occurred over long periods of the Quaternary, and the cave environment is sheltered against surface processes so that old deposits may survive extreme climatic changes like glaciations. Speleothems can be dated by extreme precision using uranium-series (U-series) methods and the time-resolution is very good (often annual). The climatic proxies recorded within them are linked to the meteoric water cycle and are relatively simple to obtain, although the underlying transfer processes are not completely understood. The analysis of speleothems is highly specialized and multi-disciplinary, including optical, mass spectrometric and chemical methods, by which reconstruction of long time-series and time slices of regional climate variation at key stages is possible. Caves often contain archaeological or palaeontological deposits, in themselves environmental archives, which in turn may be linked to the age of speleothems and to climatic change recorded within them. On the negative side of the recent success of using speleothems as palaeoclimatic archives is the growing tendency to destructive sampling of a limited resource.

A speleothem[1] is a secondary mineral deposit formed in caves. The term is connected to form and occurrence, not to mineralogy (Hill and Forti, 1997). We will basically deal with **calcite speleothems** deposited from dripping or flowing water, and we distinguish between three basic forms: **stalactites**, **stalagmites** and **flowstones**. They differ in their stratigraphic context and

[1] *Speleothem* (from Greek: σπηλαιον = cave and θεμα = content) is in fact a generic term meaning 'cave fill'.

inner structure. In palaeoclimatic research, stalagmites and flowstones are almost exclusively used.

17.1 FORMATION OF CARBONATE SPELEOTHEMS

Speleothems are formed by precipitation of calcium carbonate from supersaturated percolation water. The driving force of karstification and speleothem deposition is the meteoric water circulation system in combination with soil carbon dioxide production (Ford and Williams, 1989) (Fig. 17.1). Meteoric circulation is also the conveyor of the climatic signals that become preserved in the speleothem, although these signals may be modulated in the percolation zone above the cave (Fig. 17.1d). Rainwater at the cave site (Fig. 17.1c) originates from ocean water (Fig. 17.1a), by means of atmospheric transport (Fig. 17.1b). Part of the precipitation will penetrate plant cover and pass through the soil and epikarst zone where it takes up CO_2, produced by soil respiration. The CO_2 uptake produces carbonic acid, which in turn dissolve limestone:

$$CaCO_3(s) + CO_2(g) + H_2O(1) \underset{\text{precipitation}}{\overset{\text{dissolution}}{\rightleftharpoons}} Ca^{2+}(aq) + 2HCO_3^-(aq) \qquad 1$$

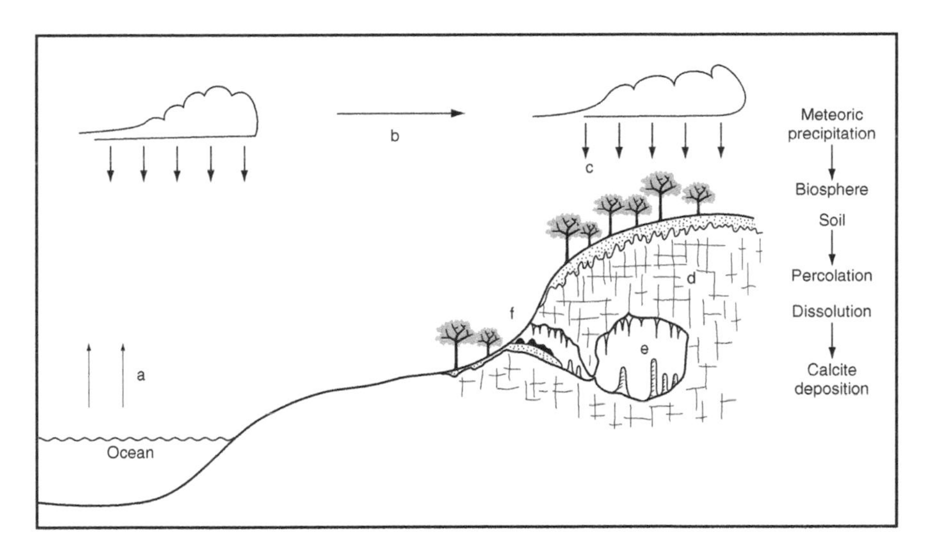

Figure 17.1 Speleothem deposition and the meteoric cycle. (a) Evaporation from ocean surface; (b) transport of moisture from source to cave site; (c) precipitation at cave site; (d) rainwater percolates through plant cover, epikarst soil and rock fractures, dissolving CO_2 and $CaCO_3$; (e) speleothem deposition in the deep cave environment by degassing excess CO_2 from dripwater; (f) speleothem deposition in the entrance environment, where evaporation may dominate. The right-hand column depicts the various components that transmit (and modulate) the surface climatic signals as they are finally recorded in the speleothem.

This is a three-phase reaction, involving solid (s), gaseous (g) and liquid (l, aq) phases. The overall reaction rates are dependent on molecular and ionic transport across phase boundaries, like uptake and degassing of CO_2 into and out of the aqueous phase. Reaction from left to right consumes CO_2 and dissolves bedrock. The opposite reaction precipitates calcium carbonate and liberates CO_2. Because gas phase is involved, the partial pressure of the CO_2 (P_{CO2}) in the soil microenvironment and in the cave – or rather the difference between them, $\Delta P_{CO2} = P_{CO2}$ (cave) – P_{CO2}(soil) – is the driving force of the reaction[2]. Typically, P_{CO2} of atmospheric air is $10^{-3.5}$ atm, while soil atmospheres can attain values up to $10^{-1.5}$–$10^{-2.5}$ atm. Due to ventilation effects, most caves have intermediate values ($P_{CO2} = 10^{-2.5}$–$10^{-3.5}$ atm). In the deep cave environment, where humidity is near 100 per cent and evaporation neglible, speleothem growth only occur when degassing can take place, i.e. when $\Delta P_{CO2} < 0$. Closer to entrances (Fig. 17.1f), or if the cave has draught between several entrances, evaporation may be sufficient to produce supersaturation and thereby speleothem growth.

The growth rate (R_{acc}) of a speleothem is dependent on:

- the water supply (i.e. drip rate)
- the chemical kinetics of calcite deposition
- the flow conditions, i.e. laminar or turbulent flow.

These factors translate to cave temperature, soil P_{CO2} (also dependent on temperature), and drip rate (i.e. precipitation rate). For laminar (non-turbulent water flow) this can be summarized as (Dreybrodt, 1999):

$$R_{acc} = \beta \cdot (c - 1.11 \cdot c_{eq}) \qquad 2$$

where c is the calcium concentration of the dripwater and c_{eq} is the concentration at thermodynamic equilibrium at the ambient P_{CO2} and temperature. The factor 1.11 expresses an inhibitory effect on calcite growth so that precipitation does not occur unless the solution is supersaturated somewhat (11 per cent) above the concentration given by its thermodynamic solubility product (Busenberg and Plummer, 1986; Dreybrodt, 1999). The expression within the brackets expresses the 'effective' supersaturation, $\Delta c = (c - 1.11 c_{eq})$. β is a kinetic factor that depends on the thickness of the water film and on temperature. Supersaturation is generally a function of soil P_{CO2} and temperature, but can also be attained independently through complex mineral dissolution in the percolation zone by the common ion effect (Atkinson, 1983). The thickness of the precipitating water film is proportional to the drip rate and thus to precipitation.

Stalagmites rarely extend more than 1 mm a^{-1} and mean rates as low as 0.001 mm a^{-1} are calculated from radiometric dating. Laboratory calibration of the kinetic parameter β and of Δc permit the growth rates of stalagmites to be predicted theoretically (Buhmann and Dreybrodt, 1985; Dreybrodt, 1988), which are in satisfactory agreement with corresponding observations of drip rates, water chemistry and growth rates of stalagmites in caves (Baker and Smart, 1995; Baker *et al.*, 1998).

[2] Here, ΔP_{CO2} is defined so that precipitation occurs when $\Delta P_{CO2} < 0$ and dissolution when $\Delta P_{CO2} > 0$

17.1.1 Speleothem Mineralogy

The most common speleothem mineral is calcium carbonate ($CaCO_3$: calcite or aragonite), but sulphate speleothems ($CaSO_4 + 2H_2O$: gypsum) may also be used in palaeoclimatic research. The precipitation of aragonite (which is the metastable form) instead of calcite in the cave environment is somewhat enigmatic (Hill and Forti, 1997), but it seems to be favoured by:

1. poisoning calcite growth by high Mg^{2+} levels at low supersaturation
2. drying
3. ambient cave temperatures above 10–15 °C (Moore, 1956).

Because aragonite can be formed in more than one way, its occurrence is not necessarily an indication of present or past temperatures.

17.1.2 Speleothem Morphology

17.1.2.1 Stalactites

Stalactites form in the cave ceiling where water emerges through a fracture opening, forming a drop (Fig. 17.2). Once in contact with the cave atmosphere, degassing cause supersaturation and precipitation of calcite. At first, this forms a tubular stalactite ('soda-straw') of

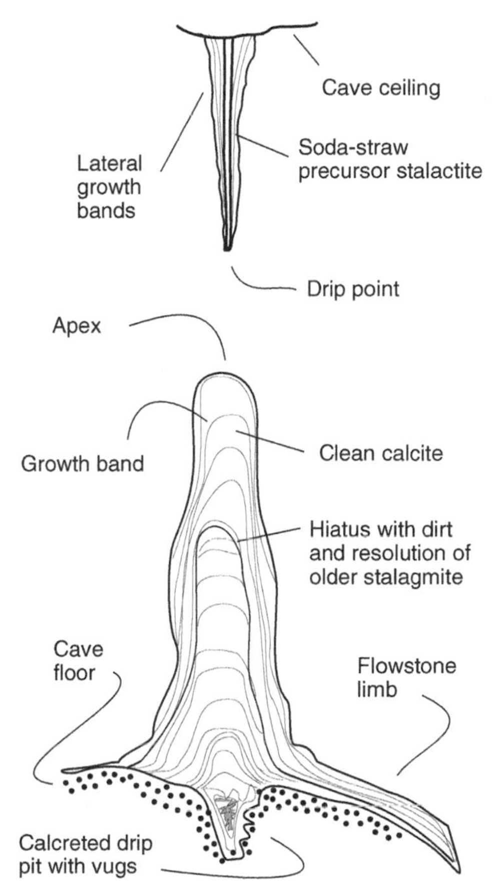

Figure 17.2 The internal structure of stalactite, stalagmite and flowstone.

approximately the same diameter as the drop emerging through a central canal. If this canal is clogged, water is forced on the outside of the straw, which is transformed into an icicle-shaped stalactite. Due to their complex cone-in-cone stratigraphy, stalactites are not preferred as palaeoclimatic archives.

17.1.2.2 Stalagmites

Stalagmites form on the cave floor at the impact point of dripwater, where further degassing occurs (Fig. 17.2). On unconsolidated sediments, the dripwater will first erode a drip-pit, before the surrounding sediment becomes cemented and it gradually transforms into a tiny pool. Subsequently, the pool fills up with speleothem material and a normal stalagmite starts to grow. Thus, stalagmites on a sediment floor have vuggy 'roots', into which a considerable part of the early growth history is hidden. Under very uniform and slow-flow conditions, columnar stalagmites of uniform thickness ('broomsticks') will form. Here, water flow is slow enough to exhaust all supersaturation before the dripwater reaches the side of the stalagmite. The diameter is roughly proportional to drip rate, and at very low rates minimum diameters down to some 3 cm occur. At higher flow rates, accretion will also occur down the sides of the stalagmite, and conical or tapered shapes will form. Along the central growth axis, stalagmites display parallel growth layers, which are easy to sample for palaeoclimatic studies.

17.1.2.3 Flowstones or Stalagmite Plates

These are deposited from a rather uniform flow and accumulate parallel to the host surface, which is either the cave floor or walls (Fig. 17.2). Their internal stratigraphy usually display very well developed parallel banding. Flowstones may, along with stalagmites, occur deeply buried under sediments, testifying earlier periods of growth.

17.1.2.4 Crystal Fabric

Speleothem calcite occurs either as macrocrystalline length-fast, or microcrystalline, length-slow aggregates, depending on nucleation rate and competition between individual crystallites (Folk and Assereto, 1976; Onac, 1997). Massive, sparry crystals suggest slow growth or recrystallization, microclusters often occur above hiatuses and layers where growth was impeded, for instance by absorbed humics (Lauritzen *et al.*, 1986; Ramseyer *et al.*, 1997). Microscopic study of crystal habit and fabric yield information on nucleation, crystallization rates and supersaturation conditions (Dickson, 1978, 1993; Gonzàles *et al.*, 1992; Frisia, 1996; Frisia *et al.*, 2000).

17.1.2.5 Growth Hiatuses

These cannot always be detected by the naked eye, and are commonly either corrosive and/or non-corrosive. For dissolution of calcite, there is no inhibitory effect of the same magnitude as for precipitation, see equation 2. There is a delicate balance between deposition and dissolution of speleothem, which may be seasonal; every growth layer may contain a minute seasonal hiatus. A **corrosive hiatus** occurs when the dripwater, due to chemical changes, becomes aggressive and re-dissolves previously deposited calcite. This may be caused by increasing ΔP_{CO2}, increasing flow rate, or by flooding of the cave passage. Flooding often causes bulk resolution, giving a discordant hiatus, which may contain organics and cave mud. **Non-corrosive hiatuses** occur during periods of draught and may be accompanied with dust layers or whitish 'drying horizons' where water is lost from fluid inclusions. Yet other hiatuses may not be visible, but are revealed by distinct time-jumps when the sample is dated.

17.1.3 Sampling of Speleothems

The almost explosive interest and success in speleothem research seen lately is a two-edged sword. Speleothems yield unique terrestrial palaeoclimatic information, but are at the same time the most spectacular and vulnerable elements of cave interiors. Sampling any speleothem **inevitably destroys the cave environment**, a very limited, non-renewable resource. The amateur (caving) community is well disciplined in non-destructive action, in contrast to increasingly destructive sampling from scientists (Forti, 2001). The codes of behaviour in caves must be respected, even in remote places: the cave environment is in every way more valuable on a longer time-scale than anybody's transient career! Therefore, new samples must be selected with the greatest care and modesty, and the collected samples must be analysed as completely as possible. Preferably, samples should be taken from caves in quarries or from show caves where speleothems would be removed or destroyed in any case. In a show cave, the results of the speleothem study may be exhibited for the public on site, to the benefit of both cave-owner and scientist.

Because the interior of a speleothem is not easily judged from its outside, finding a good sample is tricky. Sampling is a compromise between judgement of stratigraphic context in the cave, the specimen's stratigraphic resolution, mineralogical quality, hydrological growth conditions on the site and cave conservation issues. Drilling cores is a preferred and less destructive alternative to bulk sampling. In many cases, it is necessary to monitor dripwater behaviour and ambient temperature over some time before a specimen is selected. Near the cave entrance, speleothems may be in contact with important archaeological or palaeontological remains, but this is the zone of evaporation and samples may be extremely porous and full of detrital dust, rendering them less suitable for palaeoclimatic work. Much effort has been put into dating such enigmatic specimens (Schwarcz and Blackwell, 1992). The **ideal stalagmite** consists of massive, translucent calcite forming a closed system for radionuclides, stable-isotopes and luminescent organics. According to Murphy's Law, deceivingly good-looking specimens (in particular, broomsticks) may show up to be porous, full of solutional vugs and discordances or display signs of severe recrystallization. Smaller stalagmites and flowstones are often more massive and may yield better information than larger specimens.

17.2 GEOCHRONOLOGY OF SPELEOTHEMS

It is of paramount importance to obtain an **independent chronology** of the speleothem record. The growth rate cannot be assumed as constant, nor is it acceptable to assume growth bands to be strictly periodic (i.e. annual) unless proven by independent dating techniques. Therefore, radiometric dating is at present the only acceptable tool for converting the stratigraphic scale into a time-scale.

^{14}C **dating** of speleothem calcite has been used to some extent in the past (Geyh and Franke, 1970; Bluszcz *et al.*, 1988), but its application rests on assumptions of the initial ^{14}C concentration. It is seen from equation 1 that atmospheric CO_2, which is the carrier of ^{14}C, is diluted by carbonate stemming from the bedrock, so that the precipitated calcite contain an unknown proportion of ^{14}C. This is called the **dead carbon fraction** (DCF) and introduces an age offset analogous to the marine reservoir age corrections. The DCF can be calibrated by dating either recent material, or old material of known age (Genty and Massault, 1997). As the atmospheric production of radiocarbon varies over time, radiometric dates also need to be converted to calendar years to be comparable with U-series dates. Therefore, ^{14}C in speleothem is rather used as an environmental isotope (see below).

The most widely used and best dating methods for speleothems are based on **uranium-series disequilibria**, which includes dating methods like $^{234}U/^{238}U$ (up to 1.2 million years), $^{230}Th/^{234}U$ (up to *c.* 500 ka), $^{231}Pa/^{235}U$ (up to 200 ka), $^{206}Pb/^{238}U$ (up to the age of the Earth), and ^{210}Pb (up to 200 years) (Ivanovich and Harmon, 1992). Of these, the $^{230}Th/^{234}U$ method is the single most important method applied to speleothems.

17.2.1 The $^{230}Th/^{234}U$ Dating System

Part of the ^{238}U decay series (Fig. 17.3) involves its daughter nuclide ^{234}U and the granddaughter ^{230}Th. Geochemically, uranium is highly mobile in an oxidizing, near-surface environment, whilst thorium tend to be immobilized under the same conditions (Gascoyne, 1992a). In this way, almost pure uranium is incorporated into the crystals of a growing speleothem, forming a closed system where authigenic ^{230}Th will grow into equilibrium with its mother nuclide, ^{234}U. If there is no initial ^{230}Th, or if this amount is known, then the calcite crystal lattice acts as an ideal radiometric clock. According to the half-lives of these two nuclides, radioactive equilibrium is attained after 450–750 ka, depending on the initial activity ratio[3] of $^{238}U/^{234}U$. Any time after deposition, the $(^{230}Th/^{234}U)_t$ ratio is:

$$\left(\frac{^{230}Th}{^{234}U}\right)_t = \frac{^{238}U}{^{234}U}\cdot\left(1-e^{-\lambda_{230}\cdot t}\right)+\left(1+\frac{^{238}U}{^{234}U}\right)\frac{\lambda_{230}}{\lambda_{230}-\lambda_{234}}\left(1-e^{-(\lambda_{230}-\lambda_{234})\cdot t}\right) \quad 3$$

where λ_x is the decay constant[4] of the respective nuclide (^{230}Th and ^{234}U). In Fig. 17.4a, equation 3 is solved graphically for all realistic cases of $^{230}Th/^{234}U$ and $^{238}U/^{234}U$. The equation is valid only if:

1. the sample acted as a closed system since deposition
2. the system contained no ^{230}Th initially
3. there is enough uranium in the sample for isotopes to be measured by sufficient accuracy.

Measurements are either done radiometric (α-particle spectrometry) or by thermal ionization mass spectrometry (TIMS). The two methods have vastly different precision[5] (Fig. 17.4b), and are used for different types of sample. α-Particle counting will yield dates of about 5 per cent (1σ) error at 100 ka and is mainly used for geomorphologic studies, explorational (testing U-content and approximate age) and for isochron dating of 'dirty calcites', see below. TIMS will

[3] $(^{238}U/^{234}U)_t$ is substituted for $1/(^{234}U/^{238}U)_t$, which makes equation 3 simpler but slightly different from the form encountered in most other textbooks.

[4] The decay constant, λ, of a nuclide has the dimension yr^{-1} and is the probability for one atom of that nuclide to decay. For ^{234}U, this probability is $2.835 \cdot 10^{-6}$ year^{-1}, i.e. about 2.8 million atoms are needed to record an average of 1 decay per year. Therefore, the total radioactivity (decays per year) is the number of atoms, N, multiplied with the decay constant: $A=N\lambda$. The *half-life* ($t_{1/2}$) of a radionuclide is the time it takes for half of the amount of atoms given at any time to decay away. It is related to the decay constant as: $t_{1/2} = \ln(2)/\lambda$.

[5] The error of radiometric counts (n) is $\pm\sqrt{n}$. The decay constant of ^{230}Th is $9.15\cdot10^{-6}$ yr^{-1}. In order to count one α decay of this isotope per minute (with 50 per cent yield), at least $2.9.10^{10}$ atoms is needed; counting this activity for 2 days will yield 2800 counts which have a 1σ error of $\pm\sqrt{2800} = 53.6$ or about 2 per cent. If the *same number of atoms* are counted by mass spectrometry (with a typical ionization yield of 10^{-4}), then still 2.9 *million* counts can theoretically be measured, this time with an error of 0.06 per cent.

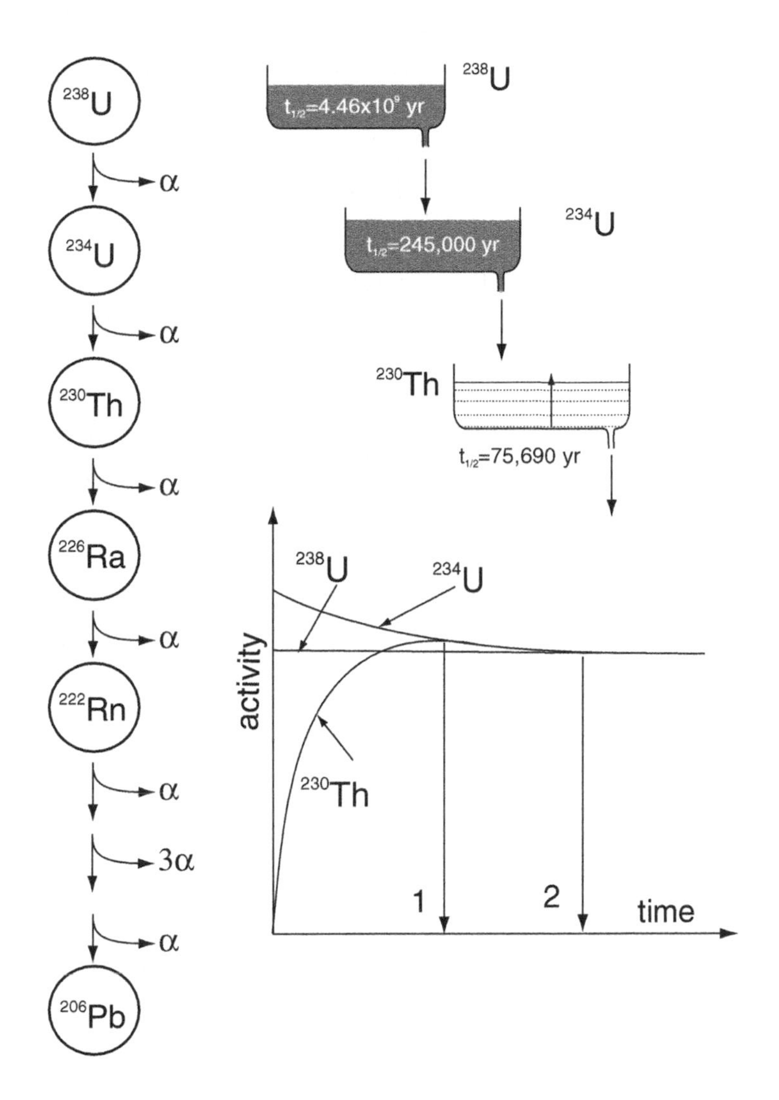

Figure 17.3 Uranium-series disequilibrium dating. *Left column*: The ^{238}U decay series, where uranium-238 breaks down to other daughter nuclides by emission of α-particles. α-Particles are helium nuclei ($^{4}He^{2+}$), so that the mass number decreases with 4 units for each decay step. After emission of a total of nine α-particles, ^{238}U is transformed into a stable-isotope, ^{206}Pb. *Top right*: A radioactive decay series may be viewed as a cascade of leaking buckets, each of which represent a nuclide, having an aperture according to its half-life. When the flow into and out of a given bucket is the same, the nuclide is broken down at the same rate as it is formed by its mother nuclide, and the system is in a steady state, radioactive equilibrium. In very old uranium minerals the whole series is in equilibrium and there is a net flux from the ^{238}U- bucket into the ^{206}Pb-bucket. Chemical weathering and transport (e.g. formation of a speleothem) may leave only uranium isotopes in the calcite, i.e. with an empty ^{230}Th-bucket. The amount of ^{230}Th will then slowly grow until it reaches equilibrium. *Lower right diagram*: Activity as a function of time for the ^{230}Th/^{234}U dating system. At $t = 0$, there is no ^{230}Th present, which grows in to equilibrium at arrow (1). The two uranium isotopes will have attained equilibrium at (2).

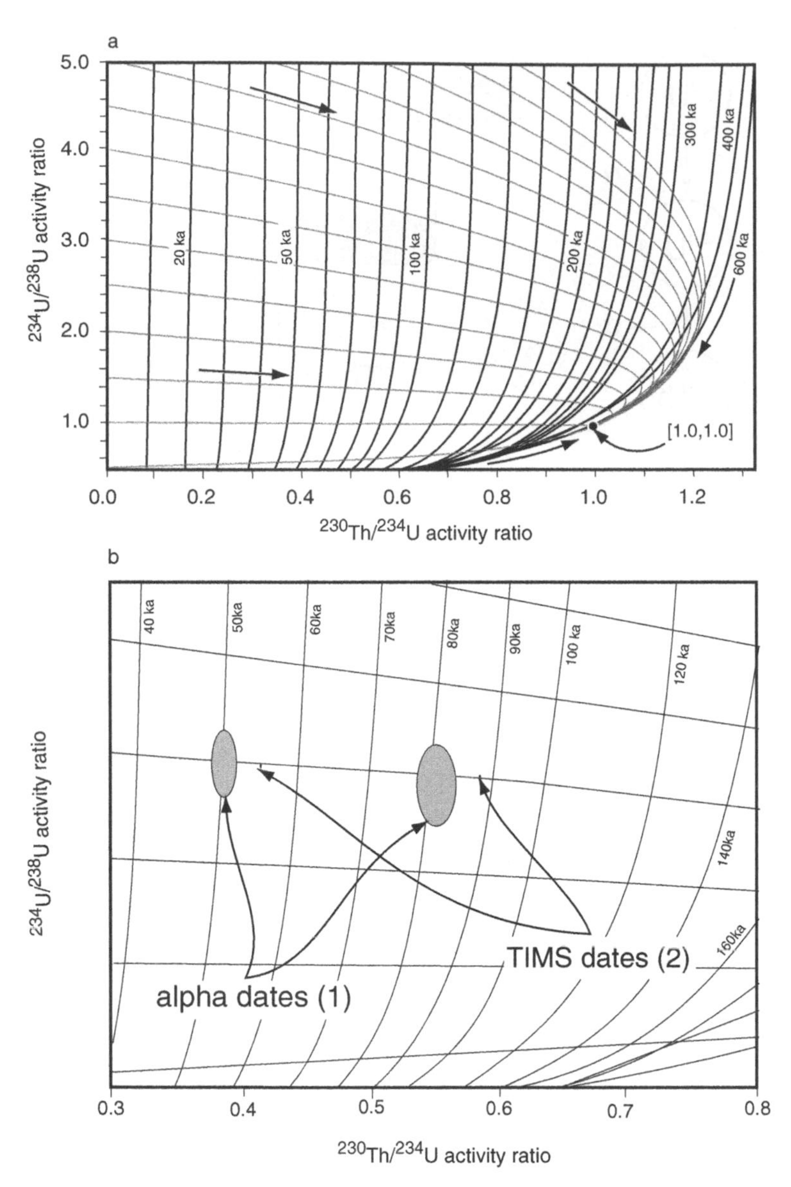

Figure 17.4 (a) *Isochron diagram*, or graphic solution of equation 3 for all possible values of $(^{230}Th/^{234}U)_t$ and most realistic values of $(^{234}U/^{238}U)_t$. Horizontal lines are evolution paths for closed systems with various initial $(^{234}U/^{238}U)_0$ as a function of time, i.e. ^{230}Th ingrowth (increasing $(^{230}Th/^{234}U)_t$). The arrows depict this process. At t > 1.25 million years, or 'infinite time', all lines will meet in one point (1,1) in the diagram, which is the *ultimate* limit of the dating method. The *practical* limit is much less and dependent on analytical precision. Vertical lines represent the state of systems of the same age, and are therefore named *isochrons*. For $(^{230}Th/^{234}U)_t < 0.6$, the time-dependence is approximately linear, which is the region of highest resolution and sensitivity of the method. (b) *Enlarged section* of the isochron diagram, depicting the tremendous difference in precision between α-dating and TIMS, represented as error ellipses for the two methods. α-counted error ellipses (grey) represent 1σ, TIMS ellipses (black) 2σ error.

typically produce ages at 100 ka with (2σ) error of 0.5 per cent and accept much smaller samples. TIMS is much more labour-intensive and is used for making precise age models with great detail.

17.2.1.1 Crystal Fabric

Non-porous, macrocrystalline fabric showing intact growth bands generally yield ages in perfect stratigraphic order, which is one of the best tests available for closed-system conditions. TIMS dating close to corrosional hiatuses may yield stratigraphically inverted dates due to leaching.

17.2.1.2 Initial ^{230}Th

The presence of non-authigenic ^{230}Th at the time of deposition will result in an age that is too high. However, any non-authigenic ^{230}Th must be accompanied with ^{232}Th, the common, very long-lived Th isotope in nature. Hence, significant concentration of ^{232}Th in the sample implies that some ^{230}Th also was present, and the ratio ^{232}Th/^{230}Th may serve as an index of contamination (for very young samples, with little authigenic ^{230}Th, the ratio ^{232}Th/^{238}U is a more robust measure). Thus the present $(^{230}\text{Th}/^{234}\text{U})_t$ consists of a mixture of authigenic and non-authigenic ^{230}Th, of which the latter is unsupported and decays away with time. Hence, contamination is less serious for old samples. If the initial $(^{230}\text{Th}/^{232}\text{Th})_0 = B_0$, then the left-hand side of equation (3) becomes (Schwarcz, 1980):

$$\left(\frac{^{230}\text{Th}}{^{234}\text{U}}\right)_t - \left(\frac{^{232}\text{Th}}{^{234}\text{U}}\right)_t B_0 \cdot e^{\lambda_{230} \cdot t} \qquad 4$$

B_0 may be estimated independently from detrital matter at the site, or by dating very young (historical) material of known age.

Non-authigenic Th is generally associated with detrital (dust) components, so that a given carbonate sample may be viewed as a microscopic, inhomogeneous mixture of 'clean' and 'dirty' calcite. So-called **isochron dating** corrects for detrital Th in several coeval samples of the same age but with variable contamination, or by successive acid leaching of a single sample (Ivanovich and Harmon, 1992). Successive leachates have increasing amounts of contamination, as the purest calcite grains will dissolve faster than the detrital components (Schwarcz and Latham, 1989; Przbylowicz *et al.*, 1991). In both cases, a regression technique is used to calculate uncontaminated $(^{230}\text{Th}/^{234}\text{U})_t$ and $(^{238}\text{U}/^{234}\text{U})_t$ for input into equation 3.

17.2.1.3 Uranium Concentration

Speleothems contain between <0.01 and >120 ppm U; lowest concentration is found in areas of thick, homogeneous limestones, and highest where runoff comes from black shales, coal seams or granites. A value of 0.01 ppm is the practical detection limit for conventional, α-particle dating. For mass spectrometry, a minimum (empirical) amount of some 0.6 μg ^{238}U is required for dating a sample of 100 ka, corresponding to 0.06 g calcite of 1 ppm U. Correspondingly greater amounts are needed for lower U concentrations or younger samples.

17.2.2 Age Model Precision

The error of age interpolation from the resulting stratigraphic age model is not only dependent on the analytical precision of the dates, but also on the stratigraphic thickness of the samples. As a TIMS date may have an error corresponding to only a few tens of years, it is only an average of a much longer time interval in the sample. The optimum sample thickness, which is dictated by the uranium content, the age of the sample, and to some extent on operator skills, must be carefully balanced against the analytical precision. Ideally, the time interval integrated in the sample thickness should match the analytical error (Fig. 17.5).

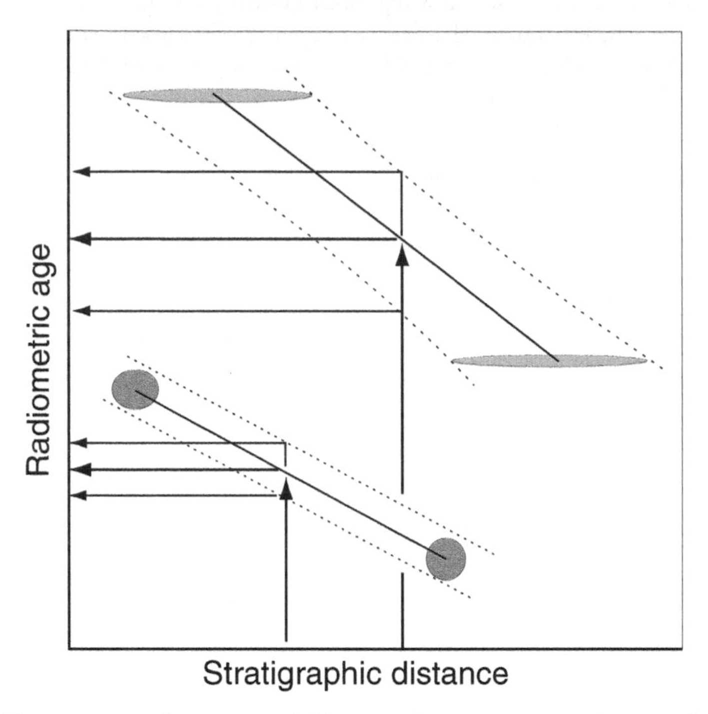

Figure 17.5 The purpose of an age model is to perform accurate estimates of the age of any stratigraphic position outside dated horizons. The thickness of these horizons relative to the analytical precision is critical for this estimate. In spite of accurate dating, too thick sub-samples obscure the accuracy of the date (upper diagram), whilst a better match between the two measures actually increases the accuracy of the age model (lower diagram).

17.3 CLIMATE SIGNALS IN SPELEOTHEMS

As depicted in Fig. 17.1, various products of the surface environment may be transformed and transported into the cave by the dripwater and recorded within the speleothem stratigraphy. Practically all these environmental proxies are **trace concentrations** of isotopes, organic matter and inorganic elements. Also the crystallographic fabric and the growth rate of the calcite carry environmental information. If the mechanisms involved in these processes are fully understood and quantified, or environmental dependency can be calibrated empirically, then the speleothem becomes a climatic record of very high chronological precision. Climatic information can be qualitative, quantitative or both. **Qualitative** information is linked to occurrence of a specific type of product, e.g. a single fossil, molecule or element, which in turn can be linked to a specific environmental condition. The stratigraphic variation of the **quantity** of a variable (which is not necessary specific) gives information about magnitude and frequency of environmental change.

17.3.1 Collective Properties of Speleothem Growth

Growth frequency of speleothems in a region is discontinuous; there are periods of enhanced and retarded speleothem growth, which corresponds roughly to known warm and cold periods during the Upper Pleistocene. The main controls of speleothem growth are soil CO_2 production, rainfall and temperature. At high latitudes, sub-zero temperatures would hamper percolation as well as attenuate plant growth. During periods of glacier cover, sub-glacial flooding provides the ultimate quenching of speleothem growth (Lauritzen, 1995). Growth frequency has been explored in great detail at various latitudes (Gordon *et al.*, 1989; Baker *et al.*, 1993b; Lauritzen, 1993, 1995; Lauritzen and Onac, 1996; Lauritzen *et al.*, 1996), confirming that speleothem growth tends to be more continuous at lower than at higher latitudes, reflecting palaeoclimatic gradients (Fig. 17.6a). The lesson to be learnt from this is that speleothem records are available from all warm periods of the Upper Pleistocene, so that Holocene records can be compared directly with older interglacial records, using the same proxies. Also, for temperate regions, speleothem records are also available for studying climatic changes during cold periods as well as Late Glacial/Holocene transitions (Bar-Matthews and Ayalon, 1997).

17.3.2 Growth Rate

Growth rate may be determined through dating and through growth band analysis. The latter is in turn calibrated by radiometric dating.

Differentiation of the age model yields change in growth rate as a function of time. The precision of this determination is dependent on the density of dated horizons, the stratigraphic thickness of each dated sample, and the precision of the dating itself. Sufficiently frequent dating along a stratigraphic sequence can reveal long-term climatic change (Linge *et al.*, 2001b) (Fig. 17.6b), and may also reveal hidden hiatuses caused by dripwater cut-offs (Baker *et al.*, 1995). Growth bandwidth analyses can give annual resolution and thus high-frequency growth rate variation.

17.3.3 Growth Bands with Annual Resolution

Although growth bands have long been recognized in speleothem (Hill and Forti, 1997), attention to the fact that these bands may be annual like tree-rings (Broecker and Olson, 1960) has recently been revived (Shopov *et al.*, 1989, 1994; Baker *et al.*, 1993a; Genty and Quinif, 1996). Growth bands are not always visible, as they may be obscured by recrystallization or by chaotic crystal fabric. One may distinguish between millimetre-sized **visible growth bands** that can be detected in normal light, and microscopic **luminescent or fluorescent bands**[6], that are only visible under ultraviolet excitation. Both types of banding can have annual resolution, but requires confirmation in every new specimen.

17.3.3.1 Visible (Fabric-Dependent) Bands

Variations in crystal fabric and content of fluid inclusions, detrital or coloured, organic (humic) matter display a rhythmic pattern of growth bands which may attain bandwidths up to ≈1 mm. Series of dark compact laminae (DCL) and white porous laminae (WPL) corresponds to

[6]Also named *Shopov Bands*, after their discoverer (Shopov *et al.*, 1989).

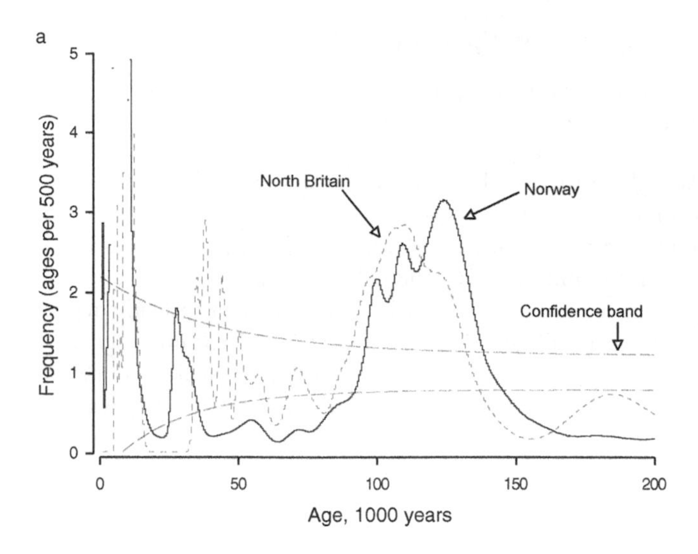

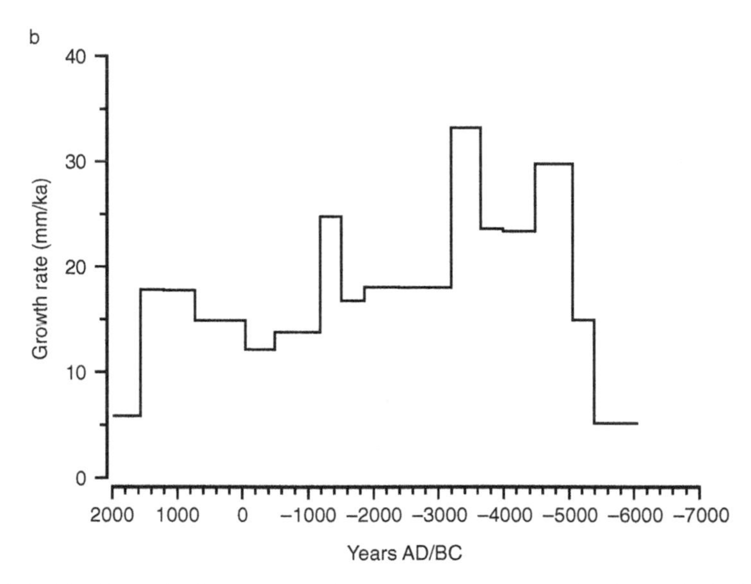

Figure 17.6 (a) Growth frequency of speleothems through time, depicting latitudinal contrasts between Norway (solid curve) and Northern England (dashed curve). Further south (e.g. Poland, Romania and Spain) speleothem growth tend to continue through the glacial. From Lauritzen *et al.* (in prep.) (b) Growth rate variation of a single speleothem through the Holocene measured from TIMS U-series dates. Adapted from Linge *et al.* (2001).

variation of water excess (Genty *et al.*, 1996). The morphological difference between the two types is the density of intercrystalline pores and organic content. They occur as annual bands in fast-growing samples, in which 11-year (solar) cycles have been suggested (Genty and Quinif, 1996), but also in slow-growing specimens where such bands necessarily represent much longer intervals.

17.3.3.2 Luminescent Bands

Many speleothems display **luminescence**, or more specifically, **fluorescence** – light emission on excitation in ultraviolet light[7] – the intensity and wavelength of the emitted light is strongly sensitive to depositional conditions (Shopov, 1997). Some, but not all, specimens display distinct growth bands that emit stronger light than the surrounding calcite (Shopov *et al.*, 1994). These bands are typically a few microns wide and regularly spaced at 10–100 microns (Plate 9). They are different from the larger, visible (structural) bands although both can occur in the same specimen with similar spacing (Genty *et al.*, 1997). Spectroscopic measurements of ultraviolet-fluorescent organics in speleothem (Baker *et al.*, 1996b; Genty *et al.*, 1997; White, 1997) and in dripwater (Baker *et al.*, 1996a, 1997a) have demonstrated that the luminescence is caused by humic matter. Chemically, humics contain conjugate π-electron systems (aromatic rings and carbonyl groups) as chromophores (see below and Fig. 17.10). Fluorescent yield increases with increasing concentration of chromophores and can be used as an assay of humic concentration, fingerprinting, and as an environmental proxy. However, at higher concentrations (dark specimens), self-absorption occurs, so that the fluorescence yield is optimal at moderate humic concentrations.

Variations in luminescence intensity may reflect variations in plant cover and productivity (Baker *et al.*, 1996b). Numerous studies demonstrate that growth rate variations are controlled by precipitation (Baker *et al.*, 1993a, 1999; Brook *et al.*, 1999a, 1999b; Qin *et al.*, 1999). In a recently deposited Belgian speleothem, where visible and luminescent laminae of similar bandwidth was studied simultaneously and compared with climatic records, luminescence was strongest in the WPL (Genty *et al.*, 1997). This corresponds to increased water flux in the winter when the concentration of humic matter is highest, although no simple connection was observed between laminae thickness and contemporary records of annual rainfall, water excess or temperature. Therefore, the connection between bandwidth and rainfall requires verification in each case.

In some cases, there is good correlation between speleothem proxies and other climatic variables, like lamina-derived growth rates and tree-ring growth (Holmgren *et al.*, 1999). Another example is reasonable correlation during the last 1200 years between mean annual rainfall in northwest Scotland and temperature in north Norway reconstructed from luminescent laminae and stable-isotopes (Fig. 17.7), suggesting a common linkage with North Atlantic Oscillation (NAO) strength (Baker *et al.*, 2000).

17.3.4 Isotopes

Isotopes of hydrogen (^{1}H, ^{2}H), oxygen (^{18}O, ^{16}O), carbon (^{12}C, ^{13}C, ^{14}C) and possibly also nitrogen (^{14}N, ^{15}N) are all environmental isotopes and potential carriers of surface information into speleothem carbonate, fluid inclusion water and organic matter. Strontium isotopes are discussed below, under trace elements.

[7] *Luminescence* is a generic term meaning emission of light; *fluorescence* means (short-time) emission of light upon excitation with photons at specific wavelength, the emitted light is of lower energy and longer wavelength than the exciting photons. Many chemical structures (also humics) display a characteristic *absorbtion spectrum* and a corresponding *emission spectrum*, allowing us to distinguish between humic and fulvic acids. Light emission which persists for some time after excitation (afterglow), is called *phosphorescence*. Calcite display both fluorescence and phosphorescence which is responsible for occasional discoloration (blue or green) of speleothems on flashgun-based photographs.

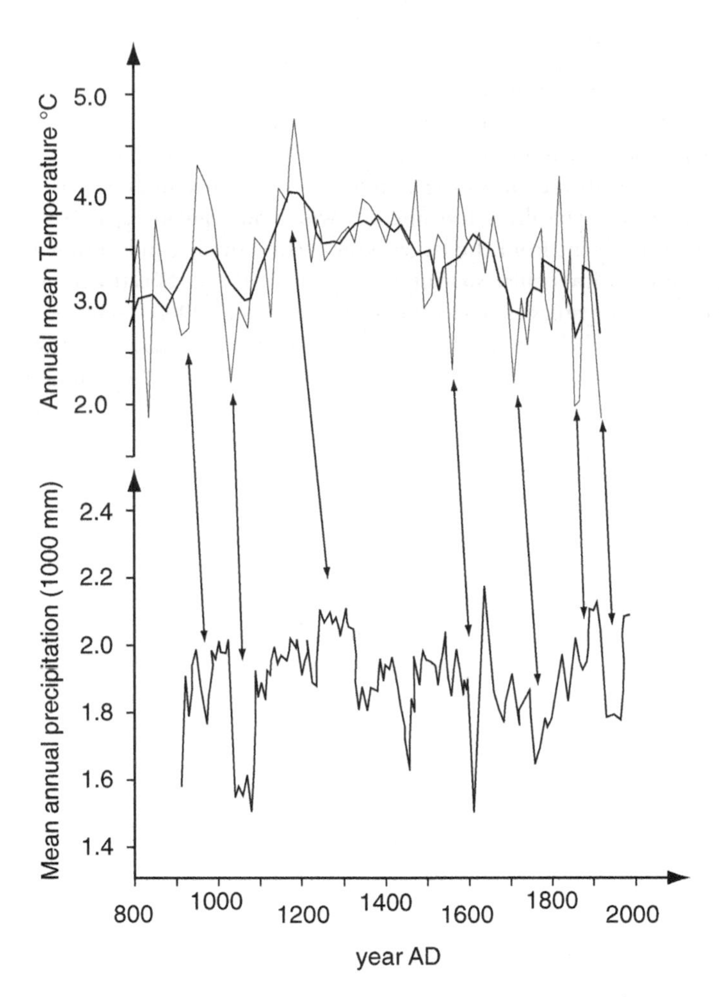

Figure 17.7 Comparison of Scottish (lamina) and Norwegian (isotope) speleothem records during the last two millennia. After Baker *et al.* (2000).

17.3.4.1 Oxygen and Hydrogen Isotopes

The degree of isotopic distribution between calcite and water is temperature-dependent. As illustrated in Fig. 17.1, the isotopic composition of the dripwater is dependent on rainwater composition, which in turn is a function of the atmospheric circulation pattern and the temperature at the site of rainfall. Provided that the calcite is precipitated in isotopic equilibrium with the dripwater (Hendy, 1971; Gascoyne, 1992b), these effects may be expressed together as the '**Speleothem Delta ^{18}O Function**' (Lauritzen, 1996):

$$\delta^{18} O_c = e^{\left[\frac{a}{T_1^2} - b\right]} [F(T_2, t, g) + 10^3] - 10^3 \quad 5$$

where $\delta^{18}O_c$ is the oxygen isotopic composition of speleothem carbonate, T_1 is (°K) temperature inside the cave, T_2 is the surface temperature above the cave, t is time and g is geographical position of the site. In well-ventilated caves, $T_1 = T_2$ (Wigley and Brown, 1976). The two constants, *a*, and *b*, govern the thermodynamic fractionation factor between the carbonate mineral and water $(\alpha_{c\text{-}w})^8$.

The right-hand side of equation 5 has two terms: the first (exponential) term represents the thermodynamic fractionation (T_{Fr}) between calcite and water, and the second term contains the **dripwater function**, *F(T,t,g)*. Here, the T-dependence relates to the atmospheric precipitation at a site, whilst the t- and g-dependence represent the transport history of water and various properties of the weather system producing the rainfall. Depending on geographic position, these may be controlled by **temperature**, by the **amount of rainfall**, or **both**. This sensitivity is further modulated by storage and mixing of the percolation water in epikarst and the fissure aquifer above the cave. There may also be a seasonal interception bias in aquifer recharge above the cave, dependent on the fraction of precipitation that actually enters the epikarst (Lauritzen, 1995; Lauritzen and Lundberg, 1999b).

The components T_{Fr} and F(T,t,g) have different temperature sensitivities. T_{Fr} always has a negative response to temperature (i.e. heavier $\delta^{18}O_c$ values imply decreasing temperature), whilst F(T,t,g) may respond either positively or negatively (or not at all), depending on regional meteorology, like rainout effects. Consequently, the temperature response of speleothem carbonate is entirely dependent on the relative magnitude of T_{Fr} and F(T,t,g). The **temperature response** (μ) of equation 5 is, in mathematical terms, its T-derivative (Lauritzen, 1995) defined as:

$$\mu = \frac{\partial}{\partial T}(\delta^{18}O_c) \qquad 6$$

which can, in principle, be negative, zero or positive, depending on the relative magnitudes of the T-responses of (T_{Fr}) and of F(T,t,g). However, this ambiguity may be overcome in several ways that either aim at estimating F(T,t,g), or the sign of μ:

17.3.4.2 Palaeoprecipitation from Fluid Inclusions

F(T,t,g) at a given point in time and space can be estimated from fluid inclusions within the calcite, which are actual samples of the original dripwater at the time of precipitation. Due to possible post-depositional exchange with the surrounding calcite, it is believed that $\delta^{18}O_w$ = F(T,t,g) cannot be measured directly in the water, but must be estimated from δ^2H_w, via the so-called 'meteoric water line' (Craig, 1961; Gat, 1980):

$$\delta^{18}O_w = \tfrac{1}{8}\delta^2H_w - \tfrac{10}{8} \qquad 7$$

Recent experiments do, however, suggest that a robust $\delta^{18}O_w$ can be recovered from some speleothem fluid inclusions, and deserve further investigation (Dennis *et al.*, 1998, 2001). Then, temperature can be calculated directly from the thermodynamic term of equation 1 (Schwarcz and Yonge, 1983).

17.3.4.3 Calibration of the Speleothem Delta Function

Given that some temperature estimates exist together with calcite $\delta^{18}O_c$-values for both present and past conditions, such as present-day and well-defined historic thermal events, (e.g. the Little Ice Age), the dripwater term F(T,t,g) in equation 5 may, in principle, be determined and $\delta^{18}O_c$ transformed to absolute temperature (Lauritzen, 1996; Lauritzen and Lundberg, 1998).

[8] Isotopic fractionation means that heavier or lighter isotopes become enriched in one of the phases during the process, and the degree of fractionation is temperature-dependent.

17.3.4.4 Qualitative Comparison with Recent Calcite

Finally, by comparing trends in the $\delta^{18}O_c$ time-series with present-day $\delta^{18}O_c$ of stalactite tips and known climatic changes in the past (warmer-colder or wetter-drier), the sign of (μ) may be judged and assumed valid for the rest of the time-series (Schwarcz, 1986). In this way an isotope record can be interpreted accordingly in terms of 'warmer' and 'colder', or 'wetter' and 'drier'.

17.3.4.5 Carbon Isotopes

Carbon isotopes in speleothem stem in part from bedrock carbonate, part from soil and atmospheric CO_2. This affects both $\delta^{13}C$ and $\delta^{14}C$. In contrast to oxygen isotopes, which are dominated by exchange from a huge amount of meteoric water relative to the amount of oxygen precipitated as carbonate, carbon isotopes are exchanged between a gas and a solid phase which are in approximate equimolecular proportion and are much more sensitive to the composition and proportion of the two phases, and to the progress of the precipitation process. This in turn makes $\delta^{13}C$ dependent on more variables than those we can measure ($\delta^{13}C_c$), and the isotope signal becomes difficult to interpret. For instance, the total amount of carbon species changes significantly with the amount of precipitated calcite, so that $\delta^{13}C_c$ changes as precipitation proceeds from the apex and down the sides of a stalagmite (Hendy, 1971). Stratigraphic variations of $\delta^{13}C$ in speleothem carbonate is governed by the metabolic processes that control the composition of the soil CO_2, the drip rate of the cave water, the amount of bedrock carbonate that goes into solution and the rate of outgassing and precipitation (Schwarcz, 1986; Dulinski and Rozanski, 1990). Often, $\delta^{13}C$ and $\delta^{18}O$ display some co-variation along the growth axis of a speleothem, suggesting that they both may depend on the same external parameter, e.g. temperature and/or precipitation.

In regions where a change between C3 and C4 vegetation is possible, speleothem carbonate with $\delta^{13}C$ values of around –13‰ (PDB) reflects an environment dominated by C3 vegetation, whereas carbonates with $\delta^{13}C$ values around +1.2‰ reflect a pure C4 biomass. For instance, C4 grasses are adapted to drought stress and grow preferentially where temperatures in the growing season are above 22.5 °C and minimum temperatures never go below 8 °C, e.g. South Africa and Israel (Talma *et al.*, 1974; Holmgren *et al.*, 1995; Bar-Matthews and Ayalon, 1997). In northern latitudes, C4 vegetation is lacking, so that shifts in $\delta^{13}C$ may be interpreted as changes in soil productivity and amount of bedrock interaction, which in part is controlled by temperature (Linge *et al.*, 2001a).

As mentioned, ^{14}C is not recommended for dating speleothems, but may on the contrary be used as an environmental isotope (Gascoyne and Nelson, 1983; Shopov *et al.*, 1997). $\delta^{14}C$-variations reflect cosmogenic production rates of ^{14}C, and these rates can be assayed through combination with independent U-series dates. This can greatly extend the tree-ring-based correction curves for carbon dating and reveal climate-related variations in cosmic ray flux (Bard *et al.*, 1990; Holmgren *et al.*, 1994; Hercman and Lauritzen, 1996; Genty *et al.*, 1999, 2001; Beck *et al.*, 2001).

17.3.4.6 Published Isotope Records

Stable-isotopes were the first climatic proxies to be studied in speleothems, e.g. Gascoyne (1992b), and the literature on the topic is substantial. Recently, impressively detailed time-series have been produced: from Oman (Neff *et al.*, 2001), South Africa (Holmgren *et al.*, 1994, 1999; Lee-Thorp *et al.*, 2002), African deserts (Brook *et al.*, 1990), Eastern Mediterranean (Bar-Matthews and

Ayalon, 1997; Frumkin *et al.*, 1999; Bar-Matthews *et al.*, 2000), France (Perrette *et al.*, 2000), northern Italy, Belgium and western Ireland (McDermott *et al.*, 1999, 2001) and northern Norway (Lauritzen and Lundberg, 1999a; Linge *et al.*, 2001a). From North America, recent studies have been done on samples from New Mexico, the Ozarks and Iowa (Dorale *et al.*, 1992; Denniston *et al.*, 1999, 2000; Polyak and Asmerom, 2001). From Australasia, records are published from New Zealand (Williams *et al.*, 1999), Tasmania (Goede *et al.*, 1990) and China (Yan *et al.*, 1984; Wang, 1989; Liu and He, 1990; Bin *et al.*, 1997). An example of a palaeoprecipitation record is shown in Fig. 17.8.

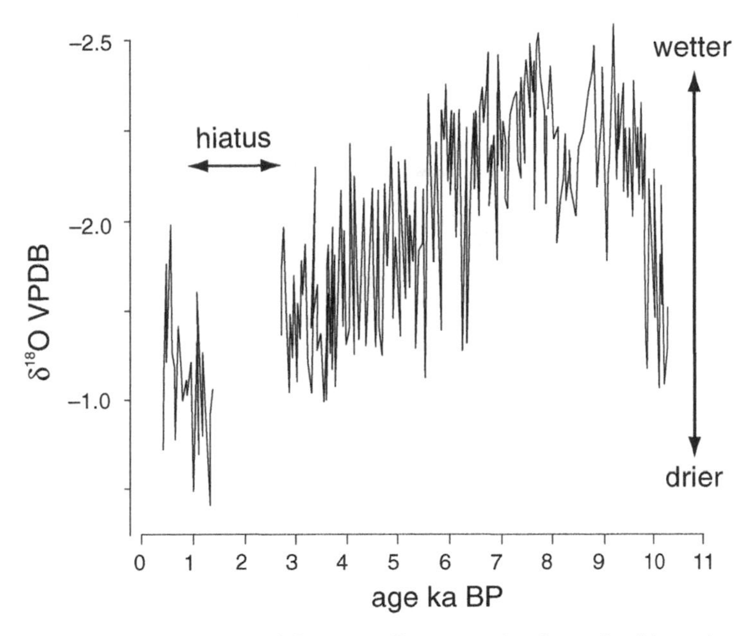

Figure 17.8 Oxygen isotope record from an Oman speleothem. In this region, rainout and amount effects dominate the $\delta^{18}O$ signal, so that shifts are interpreted in terms of 'wetter' and 'drier'. After Fleitmann *et al.* (2002).

17.3.5 Organic Matter

Speleothems contain various types of organic matter, which range from almost intact fossils like bones, insects, mites, pollen and bacteria to humic macromolecules and simple, low molecular weight organic compounds. The transport routes from the surface into the speleothem is depicted in Fig. 17.9. The information gained from microfossils, like pollen (Bastin, 1978; Lauritzen *et al.*, 1990; Burney *et al.*, 1994; Baker *et al.*, 1997b), or mites (Polyak *et al.*, 2001) is often specific and direct interpretation of surface climate is possible. Pollen transport into and through caves takes place mainly by air currents, so that entrance fascies speleothems are expected to be richer in pollen than deep-cave specimens. Tests should be done on possible bias in pollen composition spectra at various locations in the cave compared to the outside.

Organic colloids and solutes are formed by microbiological fragmentation and transformation of organic matter in the soil and bedrock, and information on the surface environment is less specific, unless unique 'biomarkers' can be found. Organic matter of natural origin transported as colloids or as solutes is called **natural organic matter** (NOM) – in contrast to organic

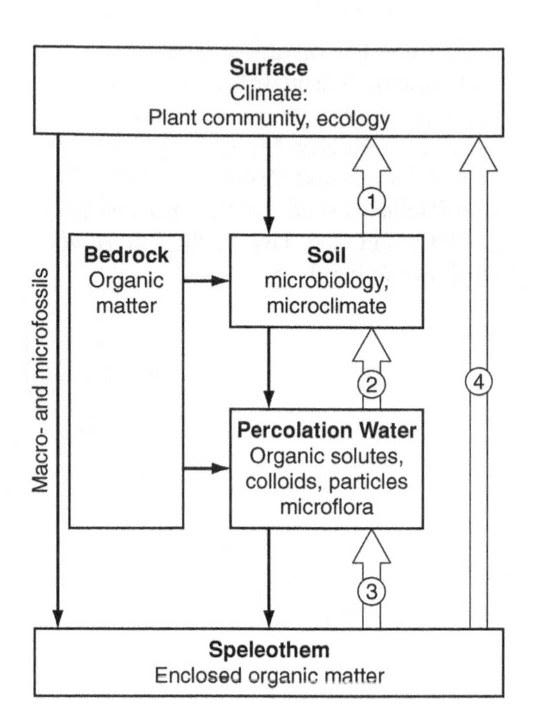

Figure 17.9 Transport routes (single arrows) for various types of organic matter from the surface into speleothem. Inference about the *source* of organic matter (double, implication arrows 1..4) of which the preferred implication (4) is valid only if the modulation effects of soil, bedrock and microflora is understood.

pollutants of anthropogenic origin. NOM may be divided into a macromolecular or polymeric fraction (humic matter) and a low molecular weight fraction consisting of simple organic molecules.

17.3.5.1 Humic Matter

Biodegradation of organic litter and soil produce yellow- to brown-coloured macromolecular, complex mixtures, named **humic matter**. This kind of organic matter is found in speleothems in concentrations up to 1000 ppm (Lauritzen *et al.*, 1986; Ramseyer *et al.*, 1997). An operational classification of humics is based on average molecular weight, acidity and solubility:

- *Humin*: macromolecular, insoluble in both alkali and acids
- *Humic acid*: macromolecular, soluble in alkali, but insoluble at pH <2
- *Fulvic acids*: lower molecular weight than humic acids, and more acidic groups; soluble in alkali and at pH <2.

The solubility reflects molecular size; humic acids range from about 50,000 to 100,000 u, whilst fulvic acids have molecular sizes of 500–2000 u (Stevenson, 1982). Their acidic property is ascribed to carboxylic and phenolic groups, giving affinity to the calcite surface. **Aquatic humus** is the naturally water-soluble and colloidal fraction consisting of fulvic acids and some humic acid. Humics are found within the calcite crystals (probably as calcium salts) and as thicker aggregates within intercrystalline voids (Lauritzen *et al.*, 1986; Ramseyer *et al.*, 1997).

Due to aromatic components and conjugated π-electron systems (chromophores and fluorophores), humic matter display fluorescence when irradiated in ultraviolet light (Miano *et al.*, 1988), and can therefore easily be assayed in speleothem and dripwater by non-destructive methods as is discussed in connection with luminescent bands.

There is an established connection between the composition of lignin in the plant cover and the composition of the derived humic/fulvic acids (Stevenson, 1982; Standley and Kaplan, 1998). Lignin is a phenolic macromolecule found uniquely in vascular plants (Sarkanen and Ludwig, 1971). The building blocks of lignin are polyphenol units. Lignin from deciduous trees contain the **syringyl unit** (3,4,5-trioxy-phenyl), conifers the **vanillyl unit** (3,4-dioxy-phenyl), and grasses the **para-hydroxyphenyl** (4-oxyphenyl) unit, respectively (Fig. 17.10). These structures, or their dimers (Goñi *et al.*, 1993), can be recognized upon analytical fragmentation of humic matter and serve as biomarkers. Analysis of humic and fulvic acid extracted from speleothems also yield these structures (Lauritzen *et al.*, 1986).

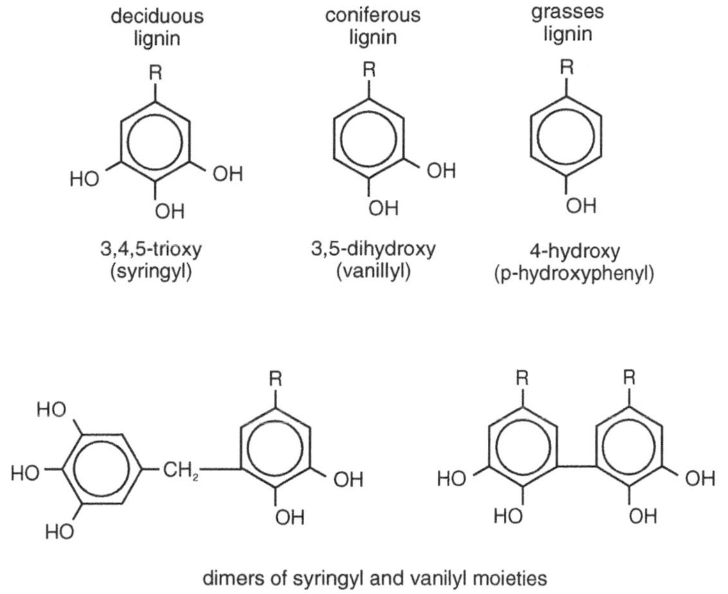

deciduous lignin — 3,4,5-trioxy (syringyl)

coniferous lignin — 3,5-dihydroxy (vanillyl)

grasses lignin — 4-hydroxy (p-hydroxyphenyl)

dimers of syringyl and vanilyl moieties

Figure 17.10 Lignin-derived polyphenols found in humic matter. Syringyl, vanillyl and p-hydroxyphenyl units occur in different proportion in different types of trees. These simple structures, or their dimers, are also found in soil humics, permitting an estimate of the type of vegetation it was derived from. The vanillyl unit was found in speleothem humic acid from boreal latitudes (Lauritzen et *al.*, 1986).

17.3.5.2 Low Molecular Weight NOM

This includes smaller molecules like fatty acids, waxes, terpenes, amino acids, etc. Some of them may have biomarker specificity to certain plants, but the field is essentially unexplored. Dark-coloured speleothems may contain up to about 1 μmol/g amino acids, which has been used for racemization geochronology (Lauritzen *et al.*, 1994). Incorporation of non-racemized (i.e. young) amino acids in growing speleothems suggests that the source may be bacteria living on the speleothem surface.

17.3.6 Trace Elements

During the process of calcite crystal growth, trace components in the solution may become incorporated into the solid phase, either by surface adsorption or by acceptance within the crystal lattice. The ratio of the trace component concentration (M), normalized to calcium (Ca) in the liquid phase is related to the same ratio in a homogeneous solid phase by the partition coefficient (D_M) between the two phases:

$$D_M = \frac{([M]/[Ca])_s}{([M]/[Ca])_t} \qquad 8$$

Under ideal, thermodynamic equilibrium conditions, D_M is constant at constant temperature and at low total concentration of M. D_M is temperature-sensitive, and provided that the concentration in the water $[(M)/(Ca)]_l$ is known, trace element concentration in speleothem minerals is a potential geothermometer as suggested for Mg in early studies (Gascoyne, 1983). In practise, various competing processes disturb the equilibrium so that the effective partition coefficients cannot be strictly regarded as a thermodynamic quantity. Finally, the temperature effect may be much less than the effect of variations in the dripwater $[(M)/(Ca)]_l$. Variation of trace element concentration in dripwater is controlled by water flow and dissolution–precipitation processes in the overlying karst system.

Proper interpretation of a trace element record in speleothem requires site calibration (Fairchild *et al.*, 2000) as well as calibration of the partition process under laboratory conditions that mimic the conditions in the cave as closely as possible. This is discussed in an elegant experimental study of Huang and Fairchild (2001), who determined D_{Mg} and D_{Sr} at various temperatures and studied the effect of crystal growth on the partition.

Earlier studies interpreted Mg/Sr ratios in speleothems reflecting palaeotemperature (Gascoyne, 1983; Goede and Vogel, 1991). Ayalon *et al.* (1999) and Goede *et al.* (1998) interpreted changes in Sr/Ca and $^{87}Sr/^{86}Sr$ as changes in rainfall intensity, temperature and input of airborne dust and sea spray. Verheyden *et al.* (2000) interpreted Mg/Ca, Sr/Ca and $^{87}Sr/^{86}Sr$ variations in a Belgian speleothem in terms of changes in water residence times and changes in weathering processes, probably induced by changes in West European climate. Similar arguments are valid for $^{234}U/^{238}U$ systematics in speleothem (Kaufman *et al.*, 1998; Ayalon *et al.*, 1999).

Trace element records seem promising for studies at very high resolution, although between-sample reproducibility needs thorough control. High-resolution microprobe ablation combined with mass spectrometry (SIMS) revealed annual cyclicity in Sr distribution in speleothem calcite (Roberts *et al.*, 1998), but Roberts *et al.* (2001) observed no similarities between three parallel Sr/Ca records taken from the same cave.

17.4 FUTURE PROSPECTS

For 25 years, speleothems held 'great promise' as palaeoclimatic archives and only a few stubborn enthusiasts kept working with them. The number of speleothem records of isotopes, trace elements and laminae has virtually exploded during the last couple of years, in hand with fundamental, *in situ* studies of hydrology, water chemistry and luminescent properties of the groundwater, providing calibration of the underlying processes. Temporal resolution has also increased, as 'annual' and 'decadal' records exist for isotopes, laminae and trace elements.

Future studies should focus on robustness of these signals. The fundamental climatic parameters in palaeoclimatology that can easily be tested in global circulation models, is temperature, amount of precipitation and its isotopic composition. None of the proxies so far known from speleothems yield a clean 'temperature' or 'rainfall' signal; each of them are affected by both, but to various degrees. Multivariate calibration is therefore an obvious way to go, along with experimental studies of the fundamental mechanisms, (e.g. Huang and Fairchild, 2001). Apart from Group II elements, which have been the obvious focus of 'trace elements' in speleothems, there are, in principle, some 80 other elements available for study at trace levels in calcite! The biomarker approach to organic matter is virtually unexplored, possibly because of the low level of suitable organics in most speleothems. More serious is the paucity of studies focussing on the problem of initial ^{230}Th in dripwater which will produce a severe bias to young material; a particularly serious problem for Holocene studies. It is also necessary to establish interlaboratory standards in order to make TIMS dates from different laboratories comparable. This is crucial for correlation of rapid events.

CHAPTER

18

GLACIERS AS INDICATORS OF HOLOCENE CLIMATE CHANGE

Atle Nesje and Svein Olaf Dahl

Abstract: Records of Holocene glacier fluctuations contribute important information about the range of natural climate variability and rates of climate change. Fluctuations in the equilibrium-line altitude (ELA) or snowline, variations in glacier mass balance and frontal fluctuations are important indicators of glacier response to climate change which may allow reconstructions of both past, present and future climate change. Glaciers, glacial landforms and proglacial lacustrine sediments are well-suited archives to record past climate change, as demonstrated through the history of glacier fluctuations obtained from different regions of the world. Based on an exponential relationship between winter accumulation and ablation-season temperature at the ELA of Norwegian glaciers, mean winter precipitation can be quantified by combining reconstructed ELA variations and mean ablation-season temperature reconstructed from independent biological proxies.

Keywords: Equilibrium-line altitude (ELA), Frontal fluctuations, Glaciers, Holocene, Marginal moraines, Mass balance

Glaciers and associated marginal moraines and proglacial lake sediments (Fig. 18.1) provide information about past climate change, as demonstrated through the history of glacier fluctuations obtained from the glaciated regions of the world. Records of glacier fluctuations contribute important information about the range of natural climate variability and rates of environmental change (e.g. Bradley, 1999). Reconstructions indicate that the glacier extent in many mountain ranges has varied considerably during the Holocene, exemplified by the Early Holocene glacier demise, the 'Little Ice Age' (LIA) glacier expansion and 20th century general frontal retreat. The general shrinkage of alpine glaciers during the 20th century (data from World Glacier Monitoring Service (WGMS)) is a major reflection of rapid change in the energy balance at the Earth's surface. As an example, the extent of ice cover on Mt Kilimanjaro decreased by *c.* 80 per cent between 1912 and 2000 (Alverson *et al.*, 2001). Hence, glacier mass balance variations and glacier front fluctuations are key indicators for evaluating past and future climatic trends.

Over the past decades the techniques of studying glaciers and glacier variations have greatly improved. For example, satellite images have improved the accuracy measuring ice movement and mass balance. The ice-cores retrieved from the Antarctic and Greenland ice sheets have greatly improved our knowledge of past environmental changes (see Fisher and Koerner, pp. 281–293 in this volume). Computer models have increased our understanding of glacier

Figure 18.1 The Pattullo Glacier, British Columbia, showing a proglacial lake, trimlines and marginal moraines. With kind permission from A. Post and E.R. Lachappelle, and the University of Washington Press.

growth and potential stability/instability as a result of climate variations. In addition, there has been a growing knowledge of the spatial and temporal development of glaciers (e.g. Lowe and Walker, 1997; Benn and Evans, 1998; Bradley, 1999; Nesje and Dahl, 2000).

International monitoring of glacier variations started in 1894. At present, the WGMS of the International Commission on Snow and Ice (ICSI) collects standardized glacier information as a contribution to the Global Environment Monitoring System (GEMS) of the United Nations Environment Programme (UNEP) and to the International Hydrological Programme (IHP) of the United Nations Educational, Scientific and Cultural Organization (UNESCO). The database includes observations on changes in length and, since 1945, also mass balance. Most of the data are from the Alps and Scandinavia.

In this chapter, we will focus on glacier mass balance records, response time, glacier front variations and how glacier variations can be reconstructed.

18.1 GLACIATION THRESHOLD AND EQUILIBRIUM-LINE ALTITUDE

The **glaciation threshold** is defined to be in the altitudinal range between the lowest topographically suited mountain hosting a glacier and the highest mountain not carrying a glacier (e.g. Paterson, 1994). The **equilibrium-line altitude** (ELA) marks the area or zone on the glacier where accumulation equals ablation. The ELA is sensitive to variations in winter precipitation, summer temperature and wind transport of dry snow. When the annual net mass balance is negative, the ELA rises, and when the annual net balance is positive, the ELA drops. The adjustment of a glacier to a change in its mass balance takes place over several years. The larger the glacier, the larger the period will have to be before the dimensions of the glacier will remain constant. The glacier is then in a **steady state** (Paterson, 1994). The **climatic ELA** is the average ELA over a number of years, commonly a 30-year period (corresponding to a climate 'normal' period).

On plateau glaciers, snow deflation and drifting dominate on the windward side, whereas snow accumulates on the leeward side. By calculating the mean ELA in all glacier quadrants, the influence of wind on plateau glaciers may be neglected. The resulting ELA is therefore defined as the temperature/precipitation ELA (TP-ELA) (Fig. 18.2). The TP-ELA reflects the combined influence of the regional ablation-season temperature and accumulation-season precipitation (Dahl and Nesje, 1992, Dahl *et al.*, 1997).

In addition, wind transport of dry snow is an important factor for the glacier mass balance. In deeply incised cirques and on valley glaciers surrounded by large and wind-exposed mountain plateaux, the snow may deflate from the plateaux and accumulate in the cirques and valleys, either by direct accumulation on the cirque/valley glaciers, or by avalanching from the mountain slopes. This may thereby increase significantly the accumulation on the cirque/valley glaciers (Dahl and Nesje, 1992; Tvede and Laumann, 1997). Consequently, the mean ELA on a plateau glacier (average for all quadrants) defines the TP-ELA, while the ELA on a cirque glacier, commonly influenced by wind-transported snow, gives the temperature/precipitation/wind ELA (TPW-ELA). Therefore, the TPW-ELA is commonly lower than the TP-ELA (Fig. 18.2).

18.1.1 Reconstruction of the Equilibrium-Line Altitude

Fluctuations in the ELA provide an important indicator of glacier response to climate change which may allow reconstructions of palaeoclimate, but also of future glacier response to a given climate change. The most common approaches in reconstructing palaeo-ELAs are to use the maximum elevation of lateral moraines (MELM), the median elevation of glaciers (MEG), the toe-to-headwall altitude ratio (THAR), the ratio of the accumulation area to the total area (AAR) and the balance ratio method (e.g. Torsnes *et al.*, 1993). Due to the nature of glacier flow towards the centre and the margin of the glacier above and below the ELA, respectively, lateral moraines are theoretically only deposited in the ablation zone below the ELA. As a result, the maximum elevation of lateral moraines reflects the corresponding ELA (Fig. 18.2).

It is commonly difficult to assess whether or not lateral moraines are preserved entirely in the upper part and whether moraine deposition started immediately down-glacier of the ELA. ELA estimates derived from eroded and/or non-deposited lateral moraines may therefore be too low. On the other hand, the assumption that the maximum altitude of lateral moraines is obtained

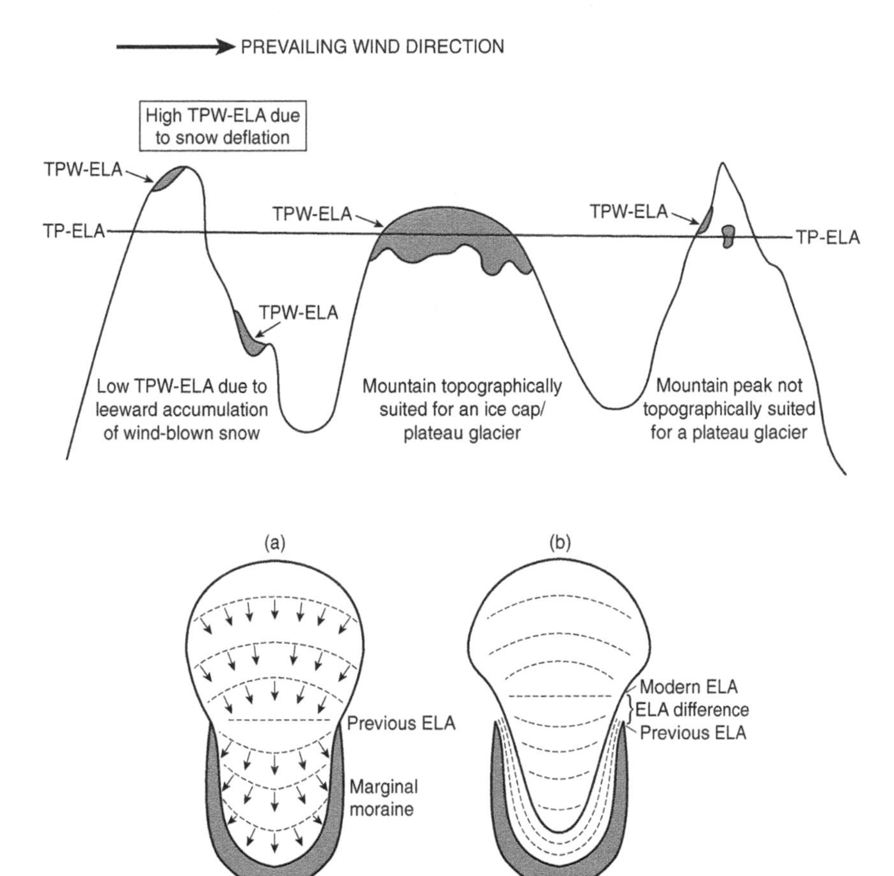

Figure 18.2 Upper panel: Schematic examples showing the difference between the TP-ELA (temperature–precipitation equilibrium-line altitude) at plateau glaciers and the TPW-ELA (temperature–precipitation-wind equilibrium-line altitude) at cirque glaciers. Modified from Dahl and Nesje (1992). Lower panel: The principle of calculating the depression of equilibrium-line altitude on a glacier based on the maximum elevation of lateral moraines. The (a) previous extent is compared with the (b) modern extent of an idealized glacier. Dashed lines indicate surface contours and arrows indicate ice flow direction. Adapted from Nesje (1992).

during steady-state conditions may overestimate the ELA. If initial glacier retreat is slow, additional moraine material may be deposited in the prolongation of the former steady-state lateral moraine. A continuous supply of debris from the valley or cirque walls may lead to the same source of error in the ELA calculations.

The ratio of the accumulation area to the total area (AAR) is based on the assumption that the steady-state AAR of former glaciers is 0.6 ± 0.05, or 2/3 = 0.65 (Porter, 1975). This value has been derived from temperate cirque and valley glaciers from different regions of the world, mostly northwest North America. The AAR of a glacier varies mainly as a function of its mass balance; ratios below 0.5 indicating negative mass balance, 0.5–0.8 corresponding to steady-state conditions and values above 0.8 reflecting positive mass balance regimes (Andrews, 1975).

An AAR of 0.6 ± 0.05 is generally considered to characterize steady-state conditions of valley/cirque glaciers. Ice caps and piedmont glaciers may, however, differ significantly from this ratio.

The largest source of inaccuracy related to the AAR method of determining the ELA on former glaciers is the reconstruction of the surface contours, especially if the glacier margins intersect valley-side topographic contours at small angles or coincide with them for some distance. However, this source of error is considered to be randomly distributed and is not considered to introduce major deviations from representative conditions. In addition, this method only requires glacier reconstruction as high as the former ELA. A theoretical evaluation of the AAR approach, using changing slope angles and valley morphology on idealized glaciers shows that glaciers advancing into flat areas underestimate the ELA depression, whereas glaciers moving into areas of increasing slope angle, overestimate the climatic ELA difference (Nesje, 1992). Consequently, topographical and morphological effects on calculated ELA depressions on glaciers must be carefully evaluated. As demonstrated above, one shortcoming of the AAR method, and also the MEG approach, is that they do not fully account for variations in the hypsometry (distribution of glacier area over its altitudinal range). To overcome this problem, a balance ratio method was developed by Furbish and Andrews (1984). This approach takes into account both glacier hypsometry and the shape of the mass balance curve and is based on the fact that, for glaciers in equilibrium, the total annual accumulation above the ELA must balance the total annual ablation below the ELA. This can be expressed as the areas above and below the ELA multiplied by the average accumulation and ablation, respectively (for further details, see Furbish and Andrews, 1984; Benn and Evans, 1998).

18.2 GLACIER MASS BALANCE

Glaciers and ice sheets grow by input of snow and ice accumulation, and lose mass by different ablation processes. The difference between accumulation and ablation over a given time span is the net mass balance, which can be either positive or negative. Mass balance measurements can therefore give information on the causes of retreat or advance of glaciers.

The most important **accumulation** factor on glaciers is snowfall. The amount and distribution of snow vary considerably, however, geographically and seasonally. The highest accumulation rates are observed in maritime, mountainous regions with prevailing winds blowing from the sea on land, as for example in western North America, west coast of New Zealand, western Patagonia, southern Iceland and western Scandinavia. Conversely, snowfall is lowest far away from oceanic sources and in precipitation 'shadows' in downwind position of high mountains. Locally, accumulation may be heavily influenced by wind transport of dry snow and by snow avalanches. Ice and snow crystals, or rime ice, can also form on glacier surfaces by freezing of wind-transported, super-cooled vapour or water droplets. This process is most common on maritime glaciers.

Ablation refers to the processes causing mass loss from the glacier, including wind deflation, avalanching from the front, calving of icebergs, melting from runoff, evaporation and sublimation. Wind deflation is wind scouring of snow resulting in the removal of snow and ice from the glacier surface. The process is most efficient in areas of strong katabatic winds and on narrow valley glaciers. Avalanching may be an important ablation factor, especially where the ice front terminates above steep rock cliffs. Ice, which breaks off from the glacier front, falls down and if the avalanching rate is greater than the melt rate, regenerated glaciers may form below. Iceberg calving is the mass loss at the margins of glaciers and ice sheets terminating in water (lake or sea). Calving events may vary considerably in scale, from small blocks to enormous icebergs. Melting, evaporation and sublimation are processes causing transformation of ice to water, water to vapour, and ice to vapour, respectively. These processes take place if there is extra energy available at the glacier surface when the temperature has been raised to the melting point. A net deficit of energy, on the other hand, can lower the ice temperature or cause ice accumulation by condensation of vapour or freezing. The energy balance is the surplus or deficit of energy over time, and is an important factor for ablation rates (Paterson, 1994). Energy balance factors on a glacier surface are solar radiation, long-wave radiation, sensible and latent heat, freezing, condensation, evaporation and sublimation.

The relative importance of each of the energy balance components varies both temporarily and spatially. Commonly, the net radiation (both short- and longwave) is the most important component, the highest proportions being associated with clear sky. In areas with continental climate, net radiation has been calculated to amount for more than 60 per cent of the ablation energy. In more humid, maritime climates, this value may be reduced to 10–50 per cent. Debris on the surface of snow and glacier ice influence ablation rates in two ways. Rock surfaces can heat up and re-emit longwave radiation, causing melting of adjacent ice and snow. If the debris layer, on the other hand, is thicker than 1–2 cm, the debris will protect the ice and snow from ablation. On glaciers with a thick debris cover, the annual ablation may be negligible.

The amount of snow and ice stored in glaciers is subject to systematic changes during a year, due to cycles of accumulation and ablation. Several types of cycles occur, depending on the timing of warm and cold seasons, maximum precipitation and variations in the proportion of precipitation falling as snow. The most common cycles are:

(a) winter accumulation type, with a well-defined winter accumulation season and summer ablation season
(b) summer accumulation type, with maxima in accumulation and ablation taking place at the same time during the summer season
(c) year-around ablation type, with one or more accumulation maxima occurring simultaneously with wet seasons.

The mass balance of a glacier is preferentially measured at representative points on its surface. The results of the mass balance measurements are integrated and reported as a value averaged for the whole glacier surface so that comparisons may be made between different glaciers. The mass balance components are expressed in metres water equivalents. The methodology and technique used to measure glacier mass balance commonly follow guidelines from the Commission on Snow and Ice of the International Association of Scientific Hydrology (UNESCO, 1970). The different mass balance terms used are illustrated in Fig. 18.3.

The **winter balance** is commonly measured in April and May by sounding the snow depth at several points on the glacier surface. The soundings always refer to the last summer surface, which may consist either of glacier ice or firn, depending on where you are on the glacier. The density of the snow is measured at a few sites, preferentially at different elevations. The water equivalents are thereafter calculated on the glacier. The points are plotted on a map and isolines

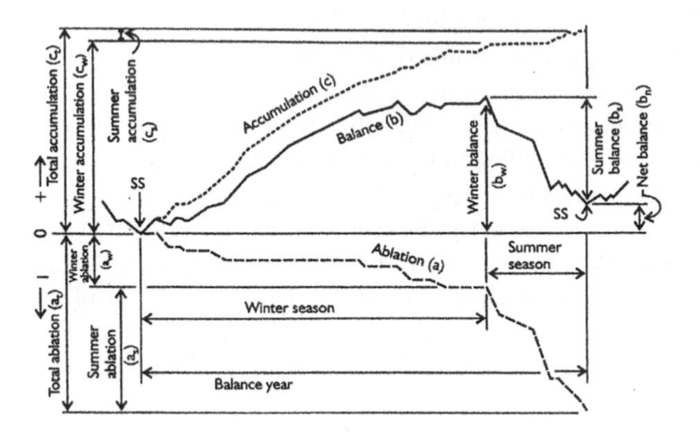

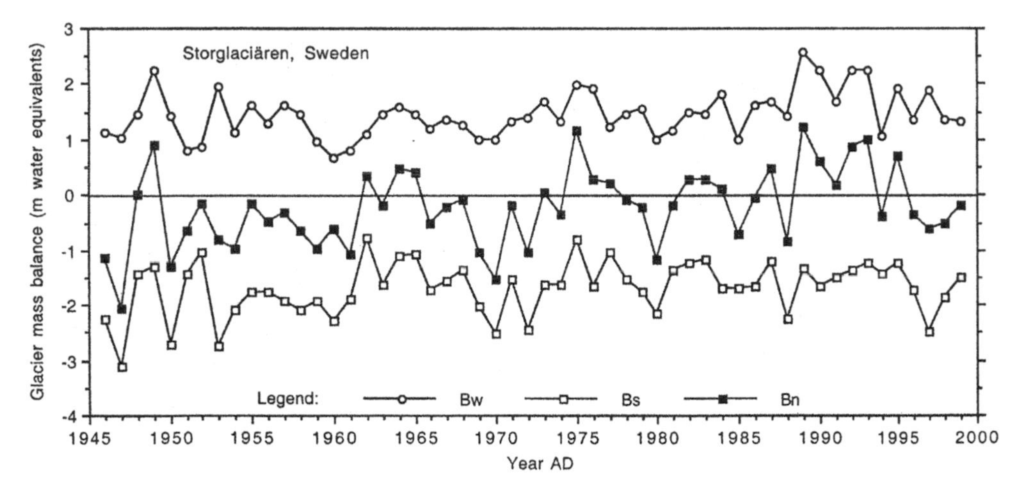

Figure 18.3 Upper panel: Terms used in mass-balance studies for one balance year. Adapted from UNESCO (1970). Lower panel: Annual winter (Bw), summer (Bs) and net (Bn) balance of Storglaciären, northern Sweden in the period 1946–1999. Data: Holmlund *et al.* (1996) with later updates.

of winter accumulation are drawn. Usually, some snow falls on the glacier after the measurements of the winter accumulation are finished. This additional accumulation may be measured, but the most common approach is to calculate it from precipitation and temperature measurements at meteorological stations close to the glacier.

The **summer balance** is calculated at several stakes drilled into the glacier surface by measuring the lowering of the snow/ice surface during the ablation season. The summer balance measured at the stakes is then transferred to a glacier map and isolines of the summer balance can be drawn. The summer balance is commonly more evenly distributed than the winter balance, because in most cases it decreases with rising elevation. The net balance is calculated as the algebraic sum of the winter balance minus the summer balance ($b_n = b_w - b_s$). The net balance is positive if the winter balance is greater than the summer balance, and negative if the summer balance is greater than the winter balance. The world's longest glacier mass balance series is from Storglaciären, northern Sweden (Fig. 18.3), dating back to 1946 (e.g. Holmlund *et al.*, 1996).

Glacier mass balance can also be calculated by measuring other parameters, such as precipitation, runoff and evaporation. Thus, the net balance (Bn) of a glacier can be expressed as: Bn = P – R – E, where P is precipitation, R is runoff and E is evaporation. This approach to calculate glacier mass balance is termed the hydrological method.

Where detailed mass balance data are not available, a statistical approach can be adopted to estimate ablation rates using mean annual or monthly temperatures or **positive degree-days**, defined as the sum of the mean daily temperature for all days with temperatures above 0 °C. For some Norwegian glaciers, Laumann and Reeh (1993) found melt rates of 3.5–5.6 mm of water equivalents per positive degree-day for snow and 5.5–7.5 mm of water equivalents per positive degree-day for ice. The difference is due to higher albedo of snow. Melt rates per degree-day are higher for maritime glaciers, because higher wind speeds and humidity cause more melting due to transfer of sensible heat and the latent heat of condensation.

Glacier mass balance can also be calculated from aerial photographs and satellite images obtained from successive years or over longer periods. Changes in glacier volume can be measured by changes in the altitude of the glacier surface. This can be converted into mass of water by estimating or measuring the density of snow, firn and ice on different parts of the glaciers. High-quality air photographs and satellite images are quite expensive to obtain, but can be used to study mass balance variations in very remote areas. So far, the radar altimetry used for studies of altitudinal variations of glacier surfaces has not been accurate enough for precise estimates. The use of laser altimetry may, however, give sufficient precision for such investigations.

On most glaciers, the amounts of annual ablation and accumulation vary quite systematically with altitude. The rates of which annual accumulation and ablation change with altitude are termed the accumulation and ablation gradient, respectively. Together, they are defined as the mass balance gradient. Steep mass balance gradients are the result of heavy snowfall in the accumulation area and high ablation rates near the front, characteristic for maritime glaciers. Low mass balance gradients, on the other hand, indicate small differences in mass balance with altitude, characteristic for slow-moving, low-gradient, continental glaciers.

On valley and cirque glaciers, the net annual accumulation increases in general with increasing altitude. In southern Norway, the precipitation–elevation gradient is in general *c.* 8 per cent 100 m^{-1} (Haakensen, 1989; Dahl and Nesje, 1992; Laumann and Reeh, 1993). If, however, high mountains stand above snow-bearing weather systems, the accumulation may decrease with altitude. The accumulation gradient can also be influenced by topography and by snow avalanching from adjacent valley sides. The amount of mass gained or lost by a glacier in response to a change in the ELA depends on the hypsometry of the glacier. If, for example, a glacier has a large part of its area close to the ELA, a rising or lowering of the ELA will cause significant variations in mass. If, on the other hand, a minor proportion of the glacier is close to the ELA, ELA variations will have little effect.

Accumulation and ablation gradients have in general different values, because they are controlled by different climatic variables. The ablation gradient is normally steeper than the accumulation gradient, showing an inflection at the ELA. The ratio between the two gradients is termed the balance ratio (BR), given as: $BR = b_{nb}/b_{nc}$, where b_{nb} and b_{nc} are the mass balance gradients in the ablation and accumulation areas, respectively. The balance ratio ignores any non-linearity, which may exist in the respective mass balance gradients, but it is a useful parameter that summarizes the balance curve of a glacier. For 22 Alaskan glaciers, Furbish and Andrews (1984) found mean balance ratios of 1.8. A value around 2 may be representative for mid-latitude maritime glaciers, while balance ratios for tropical glaciers may exceed 20 (Benn and Evans, 1998).

There is an exponential relationship between mean ablation-season temperature t (1 May–30 September) and winter accumulation A (1 October–30 April) at the ELA of modern Norwegian glaciers (Liestøl in Sissons, 1979; Sutherland, 1984), and expressed by the regression equation (Ballantyne, 1989):

$$A = 0.915\ e^{0.339t}\ (r^2 = 0.989,\ P < 0.0001) \qquad 1$$

where A is in metres water equivalent and t is in °C. The positive correlation between these two variables for different glaciers reflects that higher levels of mass turnover at the ELA require higher ablation and thus higher summer temperatures to balance the annual mass budget. Loewe (1971) and Ohmura *et al.* (1992) have also demonstrated this relationship. The scattering of the data points in the compilations by Loewe (1971) and Ohmura *et al.* (1992), are due to the fact that they have included glaciers strongly influenced by non-climatic factors (e.g. calving) and glaciers with short mass balance records.

A similar approach was used to expand the range of summer temperature and winter precipitation of this glacier/climate relationship by using annual winter (1 October–30 April) accumulation measurements and summer (1 May–30 October) temperature at the ELA in the corresponding years calculated from adjacent meteorological stations (Nesje and Dahl, 2000). The four glaciers, Ålfotbreen, Hardangerjøkulen, Hellstugubreen (all three in southern Norway) and Brøggerbreen (Svalbard) were used together with summer temperature data from the adjacent meteorological stations: Sandane, Finse, Øvre Tessa and Isfjord Radio, respectively. The correlation (r^2) using this approach is 0.84.

18.2.1 Calculation of Winter Precipitation from Fluctuations in the Equilibrium-Line Altitude and Summer Temperature Variations

Based on equation 1, mean winter precipitation can be quantified when mean ablation-season temperature has been measured or reconstructed from an **independent**, in most cases biological, proxy (see Dahl and Nesje, 1996; Nesje and Dahl, 2000). The procedure calculates what mean winter precipitation is or has been at the present ELA of a glacier in steady state. For example, the method used to reconstruct variations in Holocene winter precipitation at Hardangerjøkulen is illustrated in Fig. 18.4 (see Dahl and Nesje, 1996). The Holocene variations in glacier extent (Fig. 18.4a) must be converted into an ELA curve adjusted for land uplift (Fig. 18.4b). Holocene summer (ablation-season) temperature variations used to reconstruct the Holocene winter precipitation curve (Fig. 18.4d) are based on a Holocene July temperature curve at Finse (just north of Hardangerjøkulen) reconstructed from chironomids (non-biting midges) from lake sediments (Fig. 18.4c) (Velle, 1998 and in preparation). The reconstructed winter precipitation curve is in close agreement with a winter precipitation curve constructed for the Jostedalsbreen area (Nesje *et al.*, 2001).

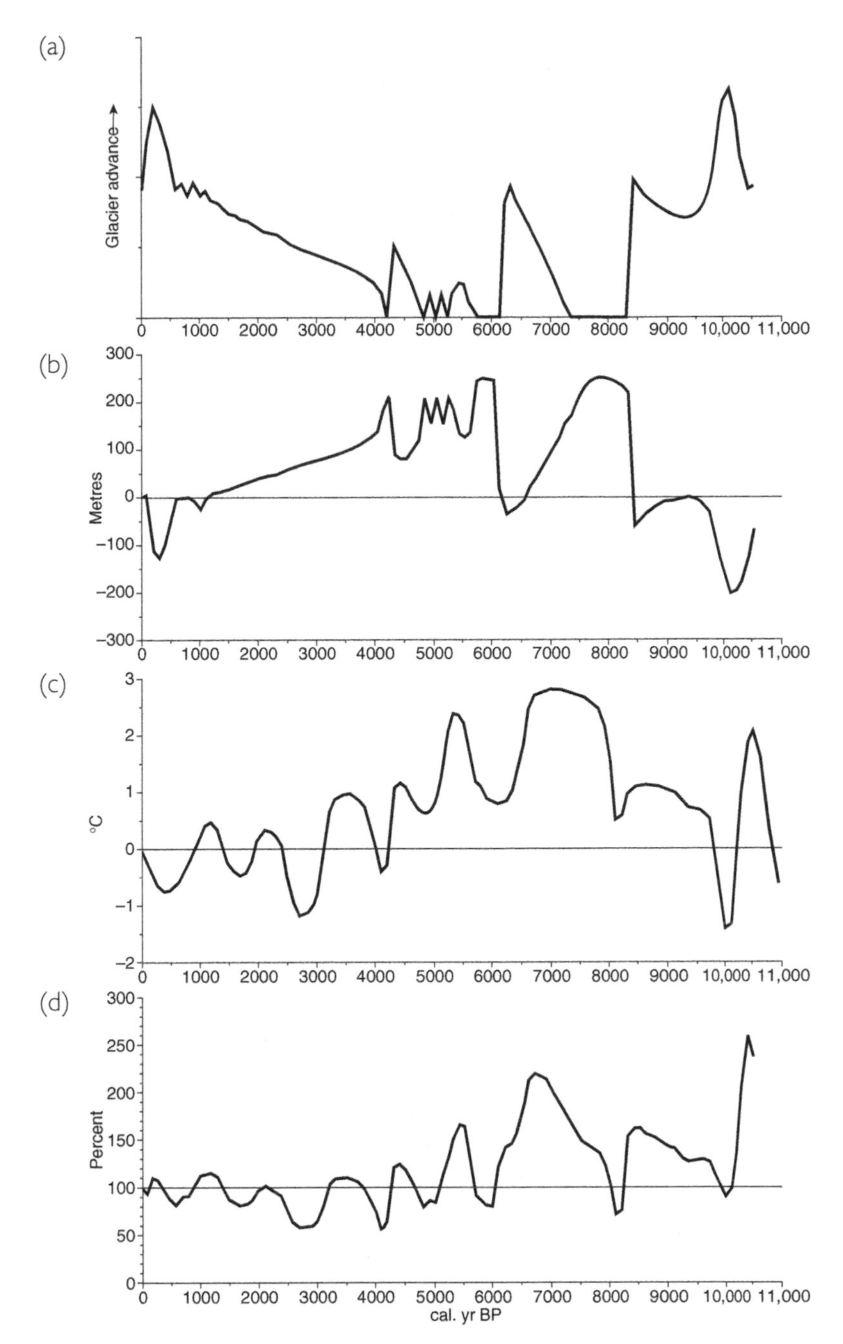

Figure 18.4 (a) Holocene glacier variations of Hardangerjøkulen (Dahl and Nesje, 1994); (b) Holocene equilibrium-line altitude (ELA) variations at Hardangerjøkulen (Dahl and Nesje, 1996); (c) Holocene July temperature variations based on chironomids (non-biting midges) from lake sediments at Finse north of Hardangerjøkulen (Velle, 1998). Due to poor age control in the upper part, the curve has been tuned towards a pine-tree curve for southern Norway (Lie *et al.*, unpublished); (d) Holocene variations in winter precipitation (in per cent, 100 per cent = 1961–1990 normal) in the Hardangerjøkulen area (Dahl *et al.*, unpublished data).

18.3 THE NORTH ATLANTIC OSCILLATION AND GLACIER MASS BALANCE

The North Atlantic Oscillation (NAO) is one of the major modes of climate variability in the North Atlantic region (e.g., Hurrell, 1995). Atmospheric circulation during winter commonly displays a strong meridional (north–south) pressure contrast, with low pressure (cyclone) centred close to Iceland and high pressure (anticyclone) near the Azores. This pressure gradient drives mean surface winds and mid-latitude winter storms from west to east across the North Atlantic, bringing mild, moist air to northwest Europe. Interannual atmospheric climate variability in northwest Europe, especially over Great Britain and western Scandinavia, has mainly been attributed to the NAO, causing variations in winter weather over the northeast North Atlantic and the adjacent land areas. A considerable impact of the NAO on regional winter precipitation has been observed. Positive NAO-index winters are related to above-normal precipitation over Iceland, Great Britain and Scandinavia, and below normal precipitation over central and southern Europe, the Mediterranean region and northwest Africa (van Loon and Rogers, 1978) (important for the winter mass balance on maritime glaciers in Scandinavia and for glaciers in the European Alps) (Fig. 18.5). A comparison between NAO and winter precipitation between AD 1864 and 1995 in western Norway shows that these are highly correlated (r = 0.77; Hurrell, 1995). Variations in NAO are also reflected in the mass balance records of Scandinavian glaciers (Nesje *et al.*, 2000), the highest correlation is with mass balance records from maritime glaciers in southern Norway (e.g. Ålfotbreen r = 0.71). Reichert *et al.* (2001) inferred that precipitation is the dominant factor (1.6 times higher than the impact of temperature) for the close relationship ($r = 0.73$) between net mass balance on Nigardsbreen and the NAO (observation period AD 1962–2000). For Nigardsbreen and Rhonegletscher (Switzerland) they also found a high positive correlation ($r = 0.55$) and a high negative

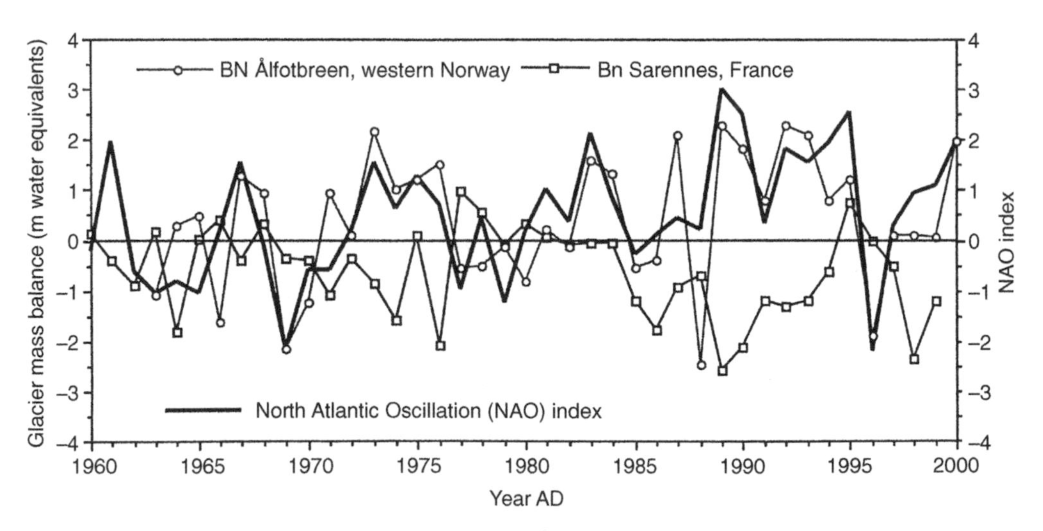

Figure 18.5 The annual net mass balance of Ålfotbreen, western Norway (data: Kjøllmoen 1998 with later updates), annual net mass balance of Sarennes Glacier in SE France (data: World Glacier Monitoring Service) and the North Atlantic Oscillation index (Jones *et al.*, 1996 with later updates).

correlation ($r = -0.64$), respectively, between decadal variations in the NAO and in glacier mass balance model experiments.

18.4 RESPONSE TIME

Advance and retreat of the glacier front normally lag behind the climate forcing because the signal must be transferred from the accumulation area to the snout. This is referred to as the **time lag** or preferentially the **response time**, which is longest for long, low-gradient and slowly moving glaciers, and shortest on short, steep and fast-flowing glaciers (e.g. Johannesson *et al.*, 1989; Paterson, 1994). Kinematic wave theory has been applied on calculating response times (Nye, 1960; Paterson, 1994). However, physically-based flow models may help determine the response times more precisely (Kruss, 1983; Oerlemans, 1988; Greuell, 1992; Oerlemans and Fortuin, 1992; McClung and Armstrong, 1993; van de Wal and Oerlemans, 1995; Raper *et al.*, 1996).

Theoretically, if the mass balance was constant for several years, the glacier would reach a steady state and then the glacier size would remain the same, termed the **datum state** (Paterson, 1994). An increase in mass balance maintained for several years would lead to a new steady state. The altitudinal difference between two glacier surface profiles increases steadily from the upper part and reaches a maximum at the position of the datum terminal position. Consequently, the head of the glacier does not change significantly, whereas the frontal part does, because the change in ice flux produced by the change in mass balance accumulates down-glacier. The response time is defined as the time a glacier takes to adjust to a change in mass balance (Paterson, 1994), or as the time the mass balance perturbation takes to remove the difference between the steady-state volumes of the glacier before and after the change in mass balance (Johannesson *et al.*, 1989). Glaciers in a temperate maritime climate with a thickness of 150–300 m and an annual ablation at the terminus of 5–10 m, have estimated response times of 15–60 years. On the other hand, ice caps in Arctic Cascade with thickness of 500–1000 m and an annual ablation of 1–2 m, have estimated response times of 250–1000 years. The response time of the Greenland Ice Sheet is estimated to be *c.* 3000 years (Paterson, 1994). It is, however, difficult to test these estimates, because glaciers are constantly adjusting to a complex series of mass balance changes. Changes in mass balance are propagated down the glacier as kinematic waves, or more accurate as a point moving with a velocity different from the ice velocity (normally three to four times faster).

18.5 RECONSTRUCTING AND DATING GLACIER FRONT VARIATIONS

The retreat and advance of glacier fronts has been used as a measure of climatic variations as long as humans have lived close to glaciated environments. Climate is constantly changing, with annual fluctuations superimposed on long-term trends. Such climatic changes are reflected as variations in the glacier extent, e.g. glacier advances during the Little Ice Age (LIA) and the subsequent frontal retreat. On a large scale, advances and retreats may be broadly synchronous. On a more detailed scale, however, the picture is more complex. In the same region, some glaciers may be advancing, while others may be retreating (Fig. 18.6). Differences in local climates, aspect, size, steepness and speed of individual glaciers may explain the different behaviour. In addition, the effect of a given climatic fluctuation on the glacier mass balance depends on the area–altitude distribution of the glacier. As a result, glaciers in the same area are likely to react differently, or at different rates, to the same mass balance variation (Paterson, 1994).

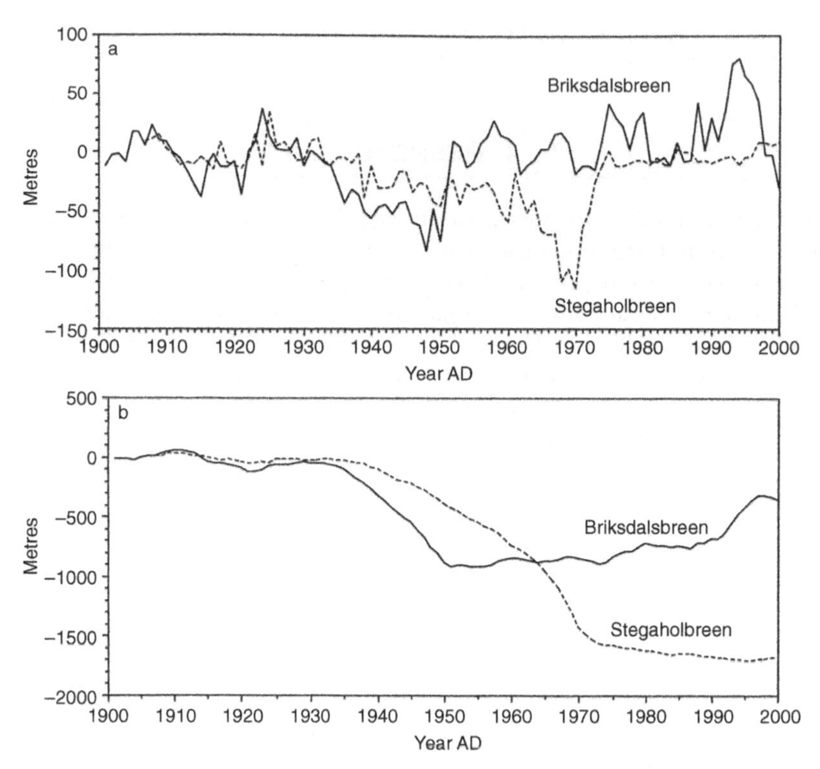

Figure 18.6 (a) Annual and (b) cumulative front variations of two outlet glaciers from Jostedalsbreen, located on the opposite side of the ice cap.

Glacier fluctuations contribute information about natural climate variability and rates of change with respect to short- and long-term energy fluxes at the glacier surface. Historical and Holocene glacier fluctuations reconstructed from direct measurements, paintings, written sources and moraines indicate that the glacier extent in many mountain regions have fluctuated considerably (Fig. 18.7). The compilation by Grove (1988) demonstrates the complexity of the records and the difficulty of discerning worldwide synchronous periods of glacier expansion. The range of variability is defined by the Early Holocene climate optimum and today's reduced stages and the maximum LIA glacier extent. On longer time-scales, glacier fronts are subject to advance and retreat as a result of climate change or **internal instabilities.** Climatic influences on the frontal response can be divided into factors causing changes in ablation and accumulation. Debris-covered glacier fronts are, however, rather insensitive to changes in mass balance.

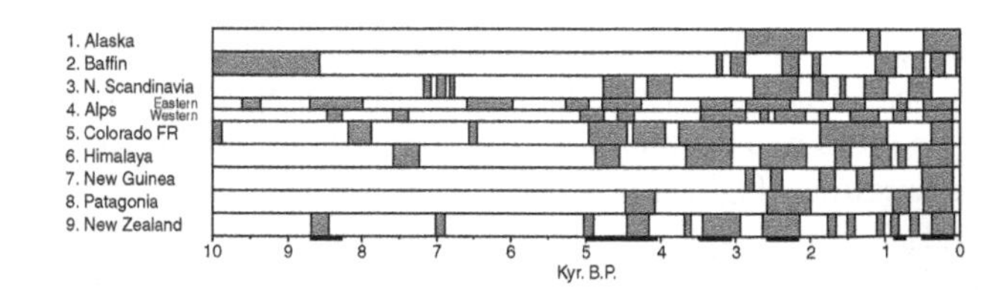

Figure 18.7 Summary diagram of Holocene glacier expansion phases in different areas of the world. Adapted from Grove (1988).

18.5.1 Direct Measurements

Long-term glacier observations, which were coordinated internationally, started in 1894 with the establishment of the International Glacier Commission in Zurich, Switzerland. The goal of this worldwide monitoring programme was to provide information on mechanisms of modern climate and glacier variations. At present, the record of glacier variations are recognized as summer temperature and winter precipitation indicators used in early detection of possible man-induced climate change given in the IPCC reports. Figure 18.6 shows the annual glacier front variations of Briksdalsbreen, a western outlet glacier from the Jostedalsbreen ice cap in western Norway. The record shows the extensive glacier retreat in the 1930s and 1940s mainly due to higher than normal summer temperatures and the significant glacier advance in the 1990s due to increased winter precipitation (Nesje *et al.*, 1995 with later updates).

18.5.2 Historical Data

Historical documents have been considered to be one of the most accurate sources of reconstructing recent glacier variations (e.g. Pfister and Brazdil, 1999). In addition, this information has been used to calibrate data to extend information on glacier variations further back in time. The Icelandic Sagas (AD 870–1264) seem to be the oldest documents related to glacier variations. Some of the oldest historical data do not, however, fulfil modern scientific standards and must be carefully evaluated (see Brimblecombe, pp. 159–167 in this volume).

In Iceland, Norway and the Alps, agricultural land was abandoned due to expanding glacier fronts during the LIA (e.g. Grove, 1988). Around Jostedalsbreen in western Norway, historical evidence shows that the advances of Nigardsbreen in Jostedalen and Brenndalsbreen in Olden caused the most severe damage. Information about the LIA glacier damage in Norway has been obtained through records of tax reductions (e.g. Grove and Battagel, 1983). From the 17th century onwards several persons visiting the glaciers left paintings, drawings and photographs, providing material for reconstruction of glacier positions and later fluctuations. In the Swiss Alps, as an example, the Lower Grindelwald Glacier has 323 illustrations to document its former extent which, together with written evidence, form the basis for a detailed reconstruction of the glacier back to AD 1590 (e.g. Grove, 1988).

18.5.3 Marginal Moraines

Present and former marginal positions of glaciers are marked by different moraine types formed by the deposition of sediment at the margin of glaciers, or by stresses induced by the glaciers. Such deposits exhibit a number of features, like glacitectonic landforms, push-and-squeeze moraines, dump moraines and later/frontal fans and ramps. In nature, it is commonly difficult to classify ice marginal deposits (e.g. Benn and Evans, 1998). Moraines of supraglacial and englacial origin are difficult to recognize because the material is flowing to the ground during the lowering and retreat of the glacier margin. The outer moraine ridge formed at the limit of the glacier advance is commonly termed the **terminal moraine**, while younger moraines within the terminal moraine are called **recessional moraines.** Recessional moraines form during minor advances or still stand during general retreat. Terminal and recessional moraines may be subdivided into frontal and lateral parts, or latero-frontal moraines. Recessional moraines formed on a yearly basis are termed **annual moraines**.

Rock surface colour (Mahaney, 1987), rock disintegration (Innes, 1984), rock surface hardness and roughness (Matthews and Shakesby, 1984; McCarroll, 1989; McCarroll and Nesje, 1993) and weathering-rind thickness have been used to obtain relative ages of moraine systems. Theoretically, weathering rates or degree of rock surface weathering can be calibrated with other dating techniques, for example the radiocarbon method, to get absolute dates (e.g. Colman, 1981). Another technique uses tephra layers form stratigraphic marker horizons, which may indicate the relative age (absolute age when the age of the actual tephra layer is known) of the overlying or underlying deposits. Tephra layers have been used to date moraines in areas subject to volcanic eruptions, particularly in Iceland (e.g. Dugmore, 1989) and North America (e.g. Porter, 1981).

Two biological dating techniques have proved useful for dating glacier forelands, dendrochronology and lichenometry. In regions where glaciers descend into areas with trees, the annual pattern of tree growth may be affected by the proximity to the glacier. In recently deglaciated glacier forelands it is important to establish the age of living trees by counting annual rings and the age of abnormal (normally reduced) growth rate both in living and dead trees (e.g. Schweingruber, 1988). The age of the oldest living tree provides a minimum age for deglaciation. This technique has been used with success, especially in western North America (e.g. Luckman, 1986). A dendrochronological technique tries to date glacier-induced growth rates from trees partly broken or tilted but not overrun or killed by the glacier (e.g. Luckman, 1986). Trees killed by the glacier and later exposed by glacier retreat may be cross dated with living trees or dated by the radiocarbon method, an approach widely used in the European Alps (e.g. Holzhauser, 1984).

Lichenometric dating, developed in the context of recently deglaciated terrain by Beschel (1950, 1957, 1961), has been widely used, in particular the yellow-green *Rhizocarpon geographicum* (e.g. Innes, 1985a, 1985b, 1986a, 1986b). There are two main applications of the method. The 'indirect' approach is based on the assumption that there is a relation between lichen size and terrain age, and besides, several substrate sites of known age. Interpolation between points of known age can be used to date other surfaces by using lichen size. The greatest limitation of the indirect lichenometric approach is the need for several surfaces of known age (control points). Commonly, the age estimates obtained by this method are given with an accuracy of ± 10 per cent (Bickerton and Matthews, 1993). McCarroll (1994) developed a new approach to lichenometric dating, based on large samples of the single largest lichen on each boulder. On surfaces of uniform age, lichen sizes are close to normally distributed, and mean values can therefore be used to construct lichenometric-dating curves. Direct lichenometric dating means that a growth curve is established by direct measurements of lichen growth rates. This dating technique, however, has had little success, mainly due to slow lichen growth, great variability of lichen growth rates and problems of linking growth curves to site age (Matthews, 1992).

18.5.4 Buried Soils

Radiocarbon dating of different soil fractions from palaeosols has provided information on glacier advance and retreat. This approach has been used with great success in New Zealand, the European Alps and in Scandinavia. Dating of thicker soils developed over a considerable time span is somewhat problematic, because of problems of isolating organic layers/fractions of different age due to greater mean residence time (Matthews, 1985). In southern Norway, Matthews and Dresser (1983) and Matthews and Caseldine (1987) reported of steep age/depth gradients in soils beneath LIA moraines, ranging from about 4000 ^{14}C yr BP at the bottom and several hundred years at the top. The radiocarbon method itself poses a imprecision problem.

Age limits given with two standard deviations (95 per cent certainty) give normally an age uncertainty of ± 100 years. When calibrating radiocarbon dates into calendar ages, dates younger than about 400 years give equivocal dates (see Pilcher, pp. 63–74 in this volume). As an example, a ^{14}C age of 220 ± 50 (1 sigma) ^{14}C yr BP is equivalent to calendar age ranges of 150–210, 280–320 and 410–420 calibrated yr BP (Porter, 1981).

18.5.5 Proglacial Peat Sections and Lake Sediments

Sediments accumulating in proglacial lakes contain information about glacier fluctuations in the form of variations in particle size, sediment thickness and organic minerogenic content (e.g. Karlén, 1976, 1981, 1988; Leonard, 1986; Nesje *et al.*, 1991, 2001; Karlén and Matthews, 1992; Matthews and Karlén, 1992; Dahl and Nesje, 1994, 1996; Leeman and Niessen, 1994b; Matthews *et al.*, 2000). Such sediments commonly form continuous records reflecting climate and glacier fluctuations, because the relative amount of minerogenic silt and clay eroded by the glacier varies with glacier activity. In lakes with rhythmic sedimentation, precise dating may provide information about climatic/environmental changes and different response times between physical and biological systems.

Several factors including the size, depth and bathymetry of the lake, the altitude of the lake, its distance from the glacier, the proportion of the catchment glacierized, whether coarse sediments are trapped in upstream lakes, and the form of the surrounding hill slope and especially their exposure to avalanche activity must be seriously evaluated in lacustrine sediment studies. Especially, the possibility that the minerogenic fraction and the thickest varves may reflect warming, rapid glacier retreat and paraglacial reworking of exposed glacier forelands must be seriously evaluated (Ballantyne, 2002). Proglacial lacustrine sediment studies commonly include analyses of loss-on-ignition, water content, dry weight, bulk density, palaeomagnetic properties (preferentially magnetic susceptibility), grain-size distribution (sedigraph), X-radiography of sediment cores and if present, visual counting of laminae/varves (see Zolitschka, pp. 92–106 in this volume). The sediments may be dated by ^{210}Pb and 'conventional' or accelerator mass spectrometry (AMS) radiocarbon dating.

18.6 CONCLUDING REMARKS

Observed or reconstructed glacier fluctuations provide important information on natural climate variations as a result of changes in the mass and energy balance at the Earth's surface. Variations in glacier mass balance are the direct reaction of a glacier to climatic variations. Variations in glacier length, on the other hand, are the indirect, filtered and commonly enhanced response.

Glaciers and ice sheets are some of the best archives of past environmental change, as demonstrated by ice-cores and through the history of glacier fluctuations obtained from the glaciated regions of the world. Records of glacier fluctuations contribute important information about the range of natural climate variability and rates of change with respect to energy fluxes at the Earth's surface on long time-scales. Reconstructed Holocene and historical glacier fluctuations indicate that the glacier extent in many mountain ranges have varied considerably during the recent millennia and centuries, exemplified by the LIA and late-20th century weather extremes. The general worldwide shrinkage of alpine glaciers during the 20th century is a major reflection of rapid change in the energy balance at the Earth's surface.

Acknowledgements

We want to express our gratitude to John A. Matthews and an anonymous referee, whose comments helped to improve and clarify the manuscript.

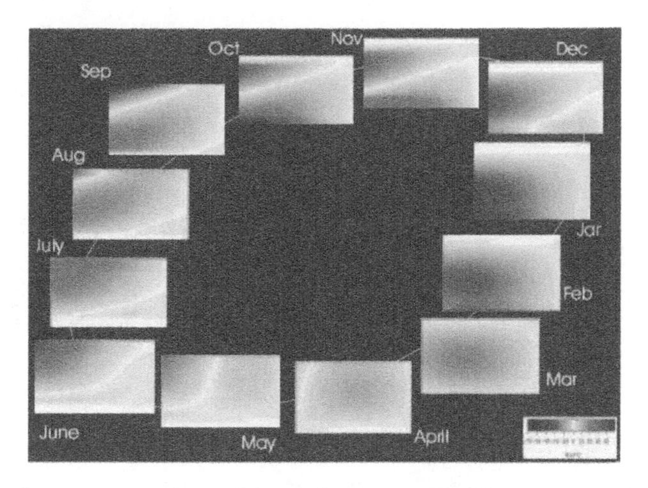

Plate 1 Schematic representation of insolation anomalies at the top of the atmosphere, relative to 1950 levels. Anomalies are colour-coded in W m^{-2}. Each **monthly panel** shows (on the *x*-axis): changes from 10,000 calendar years BP (left) to today (right), and on the *y*-axis: latitude, from 90°N at the top, to 90°S at the bottom (figure prepared by A. Waple).

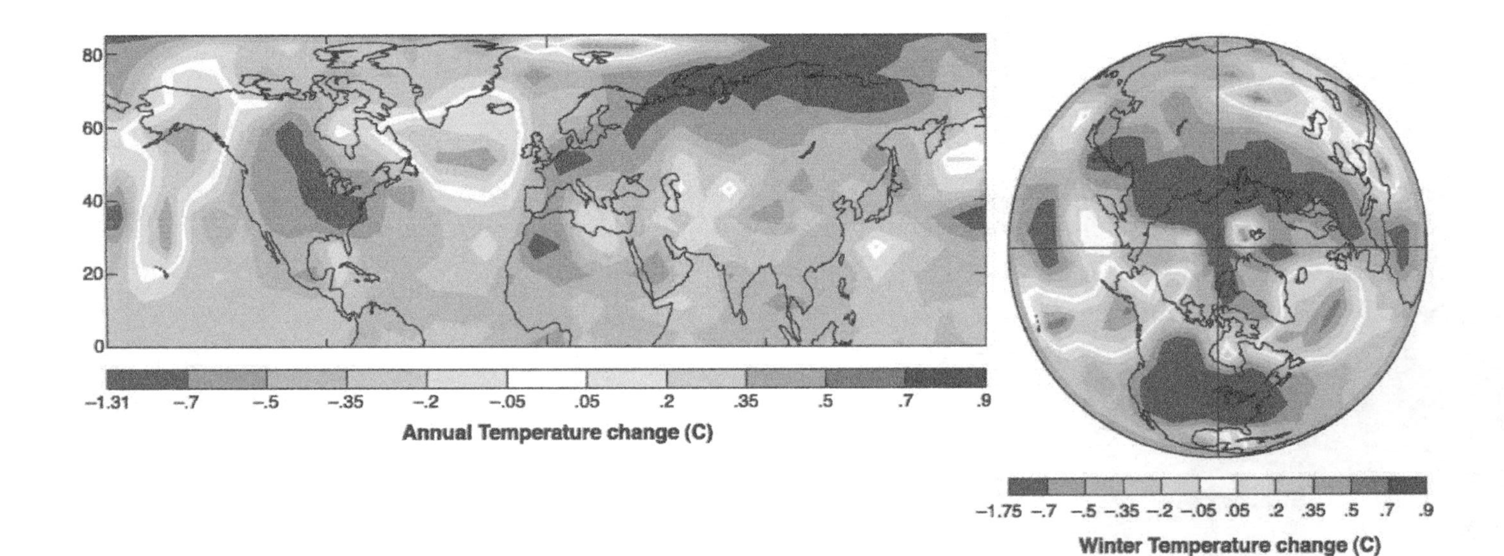

Plate 2 Modelled winter (November to April) temperature difference between AD 1780 and 1680, showing the temperature 'anomalies' associated with the late Maunder Minimum of solar activity, relative to a period (1780) with higher levels of activity (cf. Fig. 2.4). Nearly all points are statistically significant (not shown) because of the large number of model years. Reprinted from Shindell *et al.* (2001), with kind permission from the American Association for the Advancement of Science.

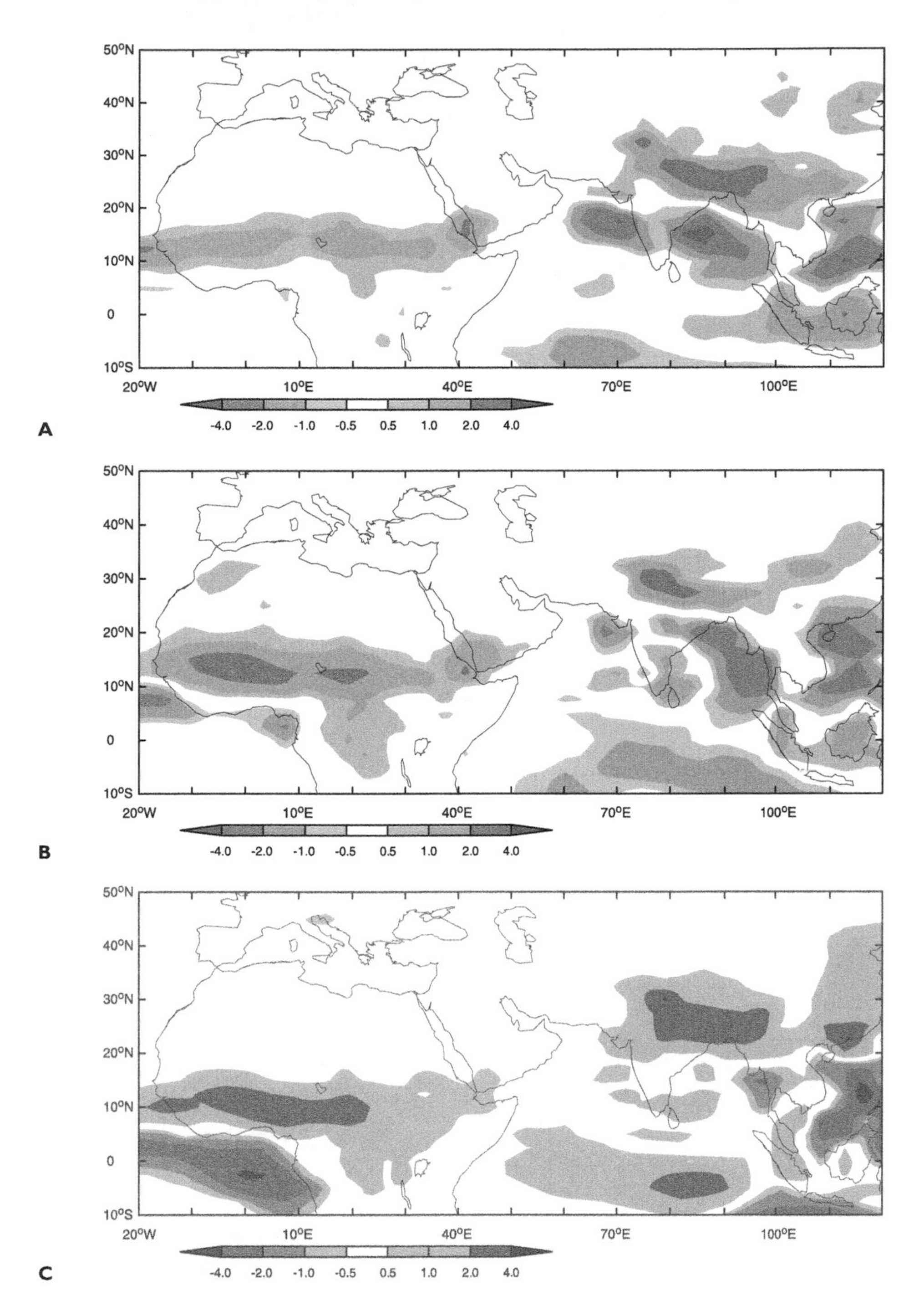

Plate 3 A–C The change in JJA precipitation (Mid-Holocene – present day) for the North African and South East Asian region for three different Hadley Centre model configurations: (a) is for no change in sea-surface temperature (as in PMIP), (b) using a simple slab ocean model to predict sea-surface temperature and (c) a fully dynamic ocean model to predict sea-surface temperature. All other aspects of the simulation are identical. The shading corresponds to –4, –2, –1, –0.5, 0.5, 1, 2, 4 mm/day and blue colours indicate that the Mid-Holocene was wetter than present day.

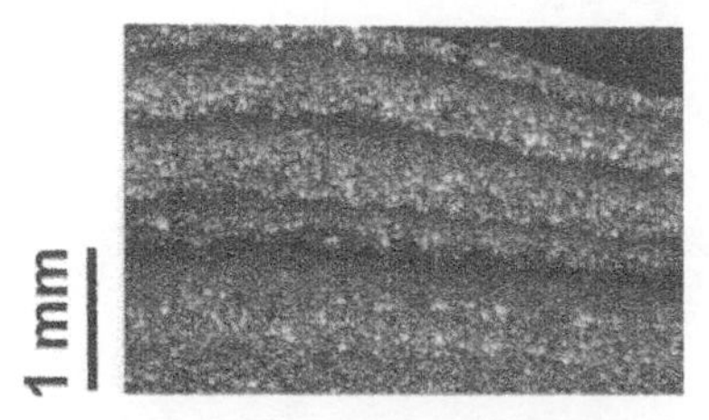

Plate 4 Non-glacial clastic varves from the sediment surface of Lake C2 (core C2-19, 0–3 mm), northern Ellesmere Island, Northwest Territories, Canada (Zolitschka, 1996b). Brownish fine-grained silt/clay caps top pale coarse-grained sub-laminae. The water–sediment interface is undulated with the water on top displayed in dark under polarized light. Microphotograph of a thin section by B. Zolitschka.

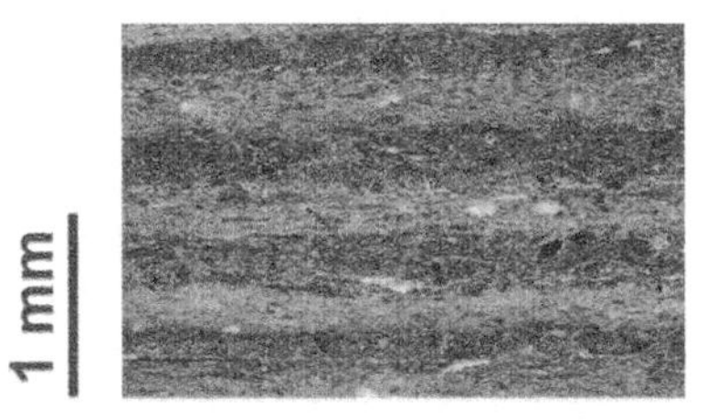

Plate 5 Organic varves from Lago Grande di Monticchio, southern Italy (core LGM-D, 5.4 m, 7200 varve yr BP) with pale sub-laminae consisting of diatom frustules and dark sub-laminae consisting of organic detritus with varying origin (Zolitschka and Negendank, 1996; Allen *et al.*, 1999). Mean varve thickness of this section is 5 mm. Microphotograph of a thin section by B. Zolitschka.

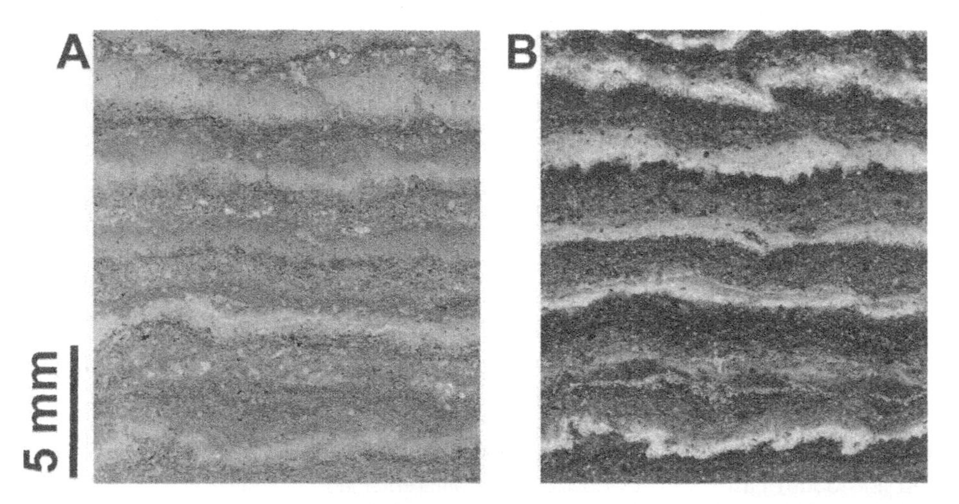

Plate 6 Carbonaceous organic varves from the 20th century of Lake Sacrow (core SAC 98-3, 11 cm), northeastern Germany (Lüder and Zolitschka, 2001) scanned with (a) normal and (b) polarized light. With normal light pure diatom laminae are pale becoming very dark under polarized light. Calcite laminae are faintly visible under normal light but show a clear white under polarized light. Two flatbed scans of the same thin section by D. Enters.

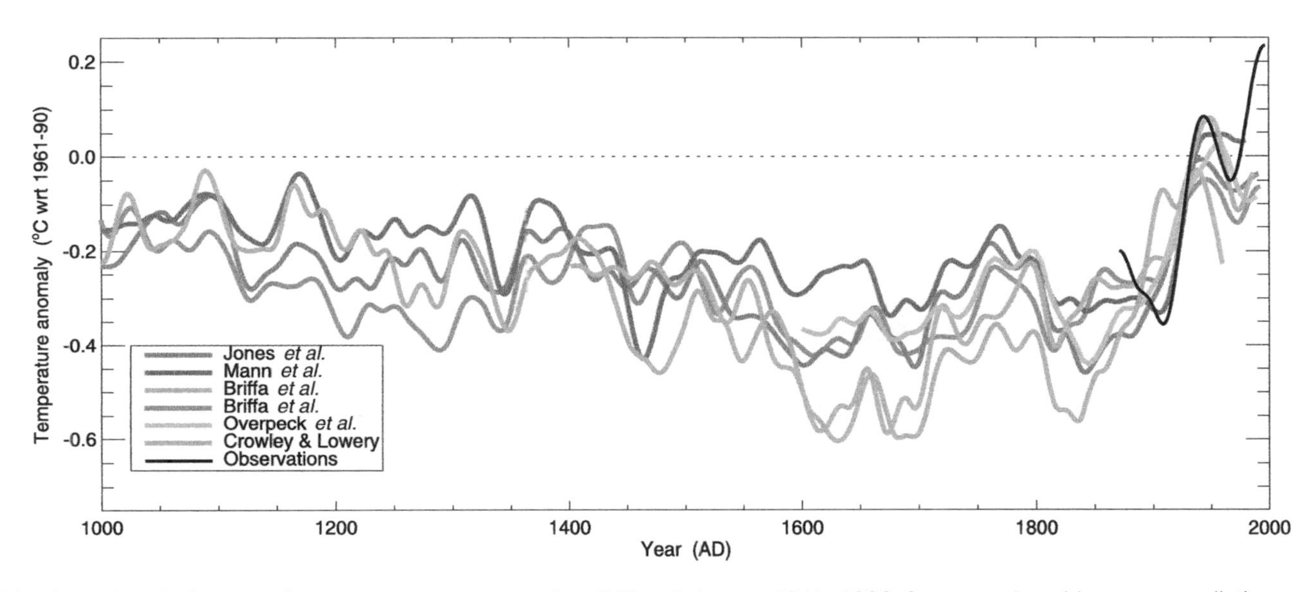

Plate 7 Northern hemisphere surface temperature anomalies (°C), relative to 1961–1990 for several multiproxy compilations of millennial temperatures. All series are discussed in the text and each is smoothed with a 30-year Gaussian weighted filter.

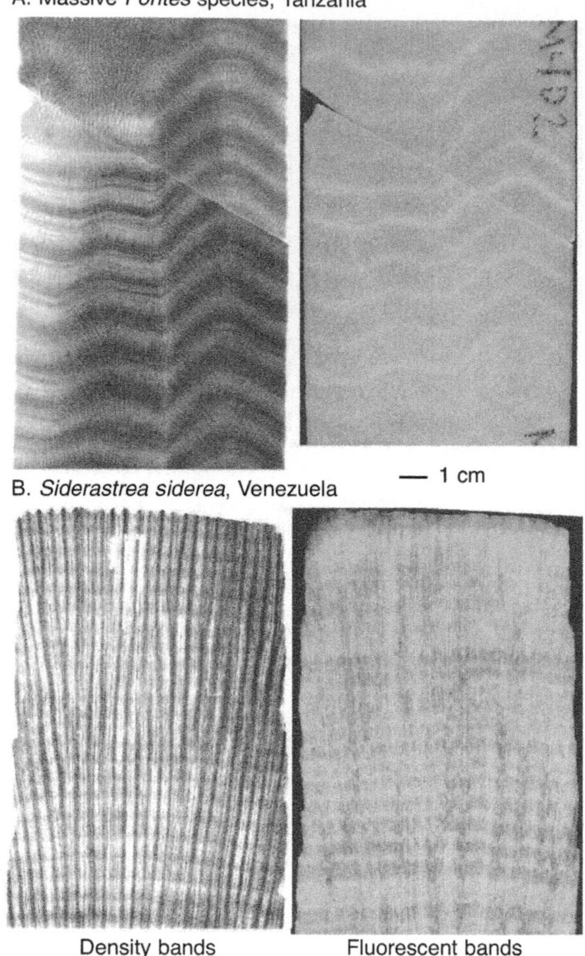

Plate 8 Images of annual density and fluorescent banding from two different coral taxa, a massive *Porites* species (a) and *Siderastrea siderea* (b). Fluorescent bands are visible under ultraviolet light (blue-tinged images) and in these cases, probably result from the incorporation of river-borne organics which influence the coral seasonally. The Venezuela site is influenced by the Orinoco River, and the Tanzania site lies off the mouth of the Rufiji River. Density bands result from seasonal alternation in extension and calcification by the coral organism. Although both images are shown at the same scale, the *S. siderea* reveals a slower and more complex growth pattern than does the *Porites*. Such growth complexity needs to be taken into account when sampling the coral skeleton at the millimeter or smaller scales typical of most paleoclimate studies.

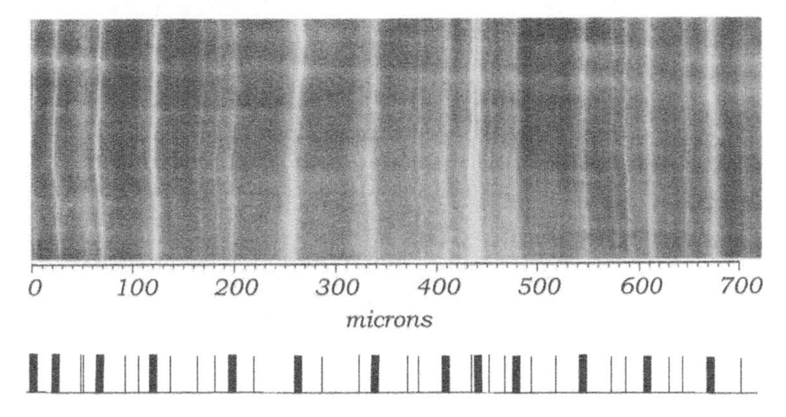

Plate 9 Microphotograph of fluorescent bands in speleothem as they appear under UV excitation. Major and minor bands are indicated on the diagram below.

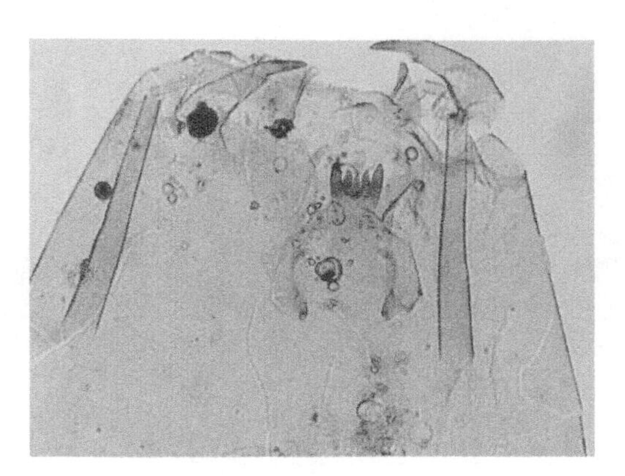
A

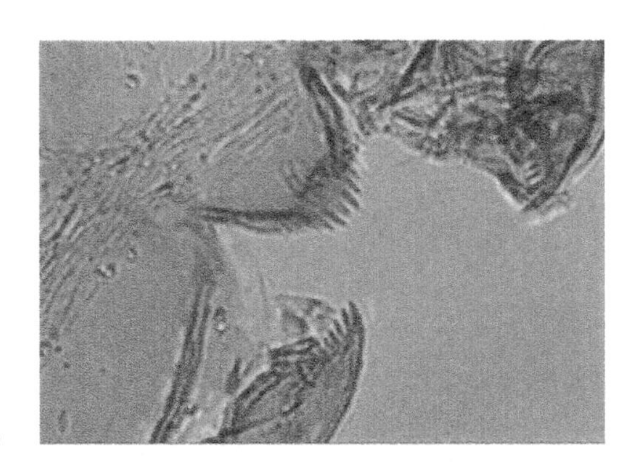
B

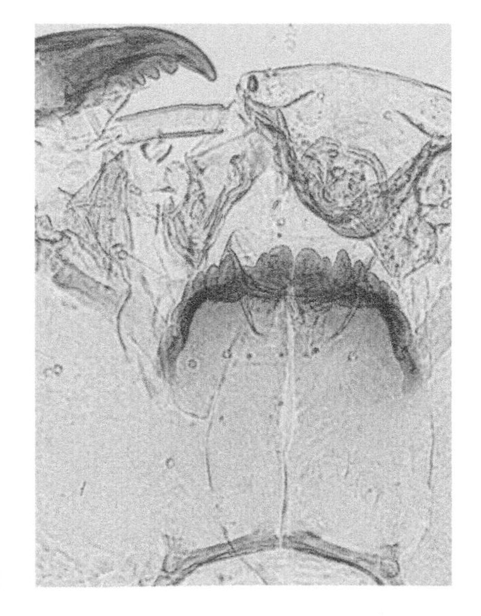
C

D

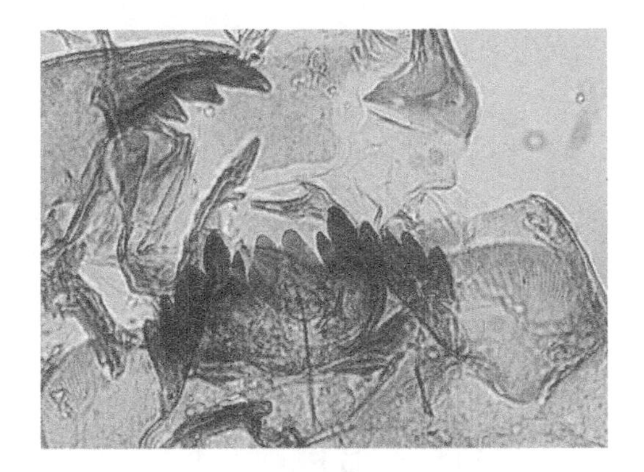

E

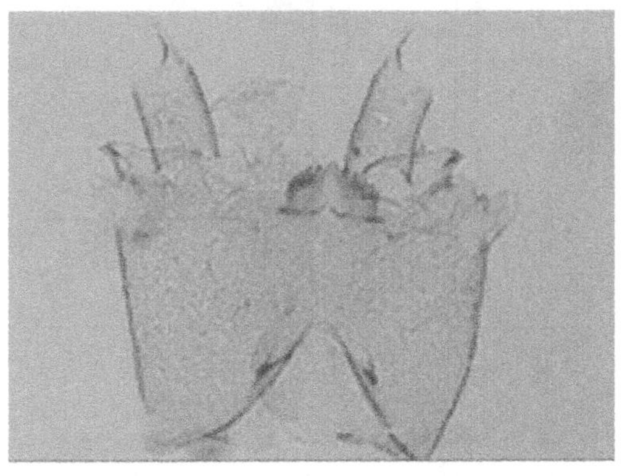

Plate 10 A–E Sub-fossil chironomid larval head capsules as examples of the major sub-families and tribes: (a) *Ablabesmyia* (Tanypodinae); (b) *Diamesa* (Diamesinae); (c) *Heterotrissocladius subpilosus*-type (Orthocladiinae); (d) *Microtendipes* (Chironomini); (e) *Tanytarsus chinyensis-type* (Tanytarsini).

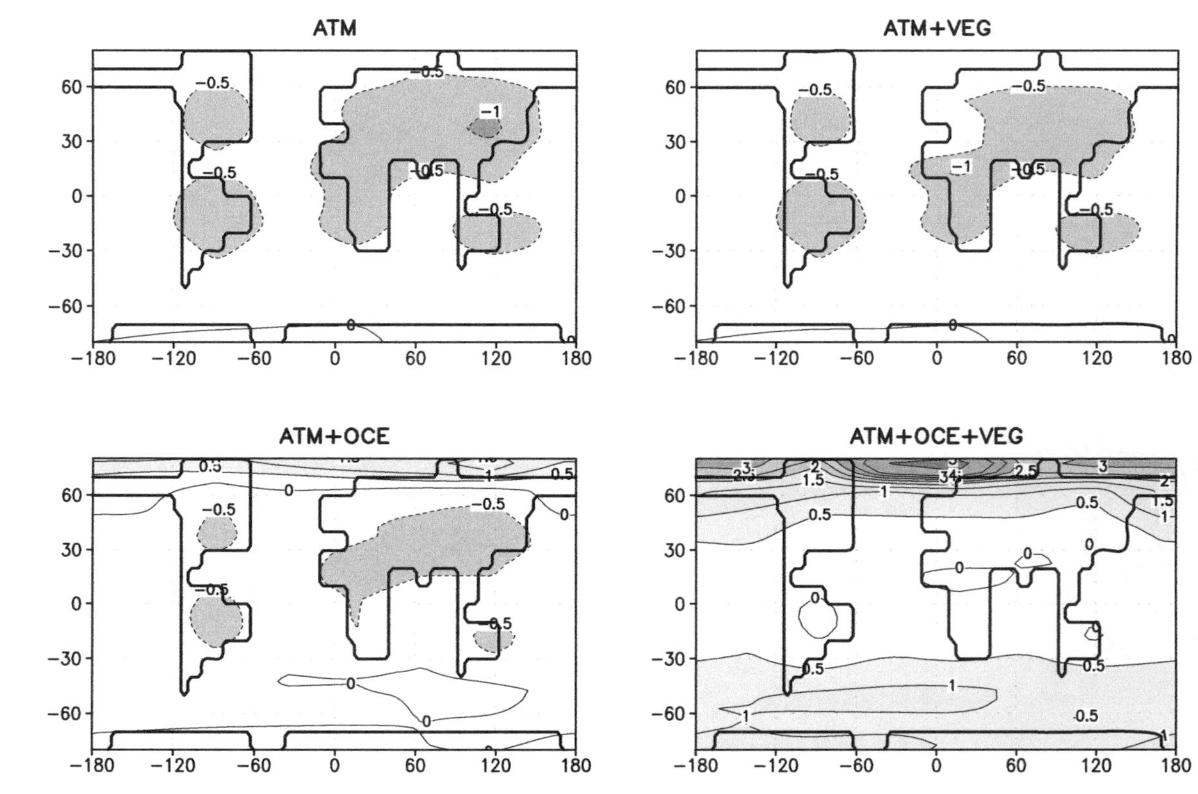

Plate 11 Differences in near-surface temperatures during northern hemisphere winter (December, January, February) between Mid-Holocene climate and present-day climate simulated by Ganopolski *et al.* (1998a). The authors used different model configurations: the atmosphere-only model (labelled ATM), the atmosphere–ocean model (ATM + OCE), the atmosphere–vegetation model (ATM + VEG) and the fully coupled model (ATM + OCE + VEG). In ATM, ATM + OCE, and ATM + VEG, present-day land-surface and ocean-surface conditions – depending on the model configuration – are used. Reprinted from Wasson and Claussen (2002) with kind permission from Elsevier Science.

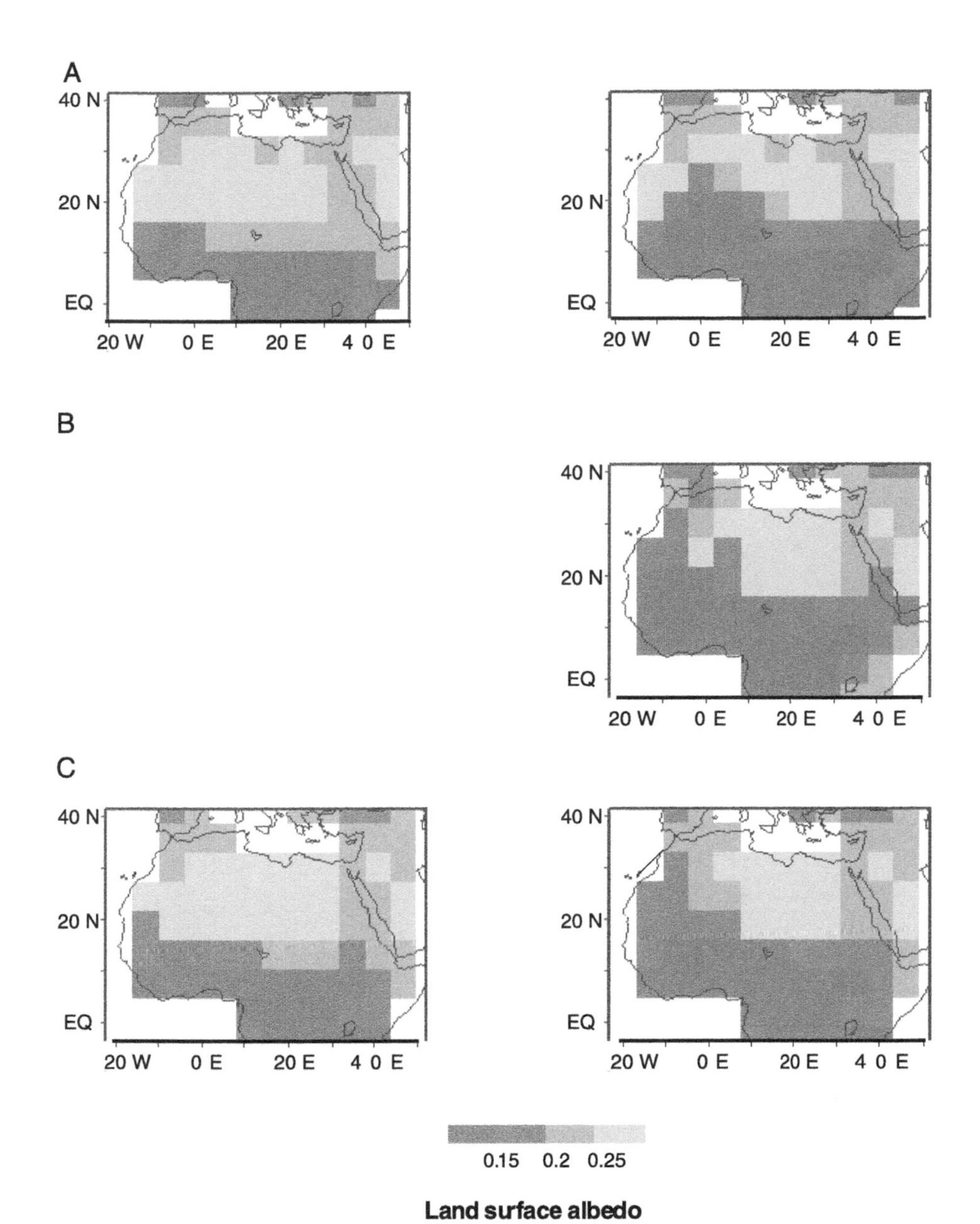

Plate 12 Multiple equilibria computed for present-day climate (a), and for the climate of the Last Glacial Maximum (c). For Mid-Holocene conditions, only one solution is obtained (b). This figure summarizes the results of Claussen (1997) (a), Claussen and Gayler (1997) (b) and Kubatzki and Claussen (1998) (c). Reprinted from Claussen (2002) with kind permission from Springer-Verlag.

CHAPTER

19

HOLOCENE ICE-CORE CLIMATE HISTORY – A MULTI-VARIABLE APPROACH

David A. Fisher and Roy M. Koerner

Abstract: Holocene paleoclimate history derived from multi-variable ice-core time-series in both hemispheres is presented and interpreted. The major features discussed are the early Holocene Warm Period and the long, cooling trend culminating in the Little Ice Age and interrupted by warm intervals like the Medieval Warm Period. The Early Holocene cold event at 8200 BP is examined as are the last few centuries of high-resolution data that reflect the modern warming. The effects of the warming on glacier melting are summarized. The advantages of merging ice-core data with other palaeo-data types, like tree-rings, are highlighted and discussed.

Keywords: Global, Holocene, Ice-cores, Multi-proxy, Palaeoclimate

One of the strengths of ice-cores is that many variables, or proxies, can be measured on a single sample and their production processes related. Some relate to causative climate processes (e.g. volcanic acidity), others to responsive processes (accumulation or stable-isotopes) and still others are passive (e.g. the record of extra-terrestrial particles). The marching orders to the glaciological ice-core community were given by A.P. Crary at the September 1968 ISAGE symposium in New Hampshire (Crary, 1968): 'So my suggestion for future glaciological studies is simple: add the thin dimension. **Drill, drill and drill some more:** know the ice–rock interface as well as the surface is presently known. Study the internal ice so that we can learn and understand the history of accumulated snow and other material that is available to us as far back as the cores take us; drill on the continental divides, on the slopes and on the shelves.' Crary's slogan has driven the international programme since then and the results have been worth it.

Key variables are discussed here from a wide range of sites. The site properties are listed in Table 19.1 and the variables and their meaning are listed in Table 19.2. A good textbook chapter on ice-core methods and glaciology can be found in Paterson (1994).

Many ice-core studies, particularly those from the great ice sheets of Greenland and Antarctica, have concentrated on the pre-Holocene record (e.g. Lorius *et al.*, 1979; Watanabe *et al.*, 1999; Dansgaard *et al.*, 1993). Ice-cores from elsewhere, such as those from Arctic Canada (e.g. Fisher *et al.*, 1998), the Russian Arctic islands (e.g. Kotlyakov *et al.*, 1991), Svalbard (e.g. Fujii *et al.*, 1990) and the tropical ice caps (e.g. Thompson and Mosely-Thompson, 1992; Thompson *et al.*, 1998) have by necessity devoted more attention to the Holocene record, partly because the pre-Holocene ice is either absent or present at a very low resolution. Records from the pre-

Name of site	Latitude	Longitude	Elevation (m a.s.l.)	Melt (per cent)	Time–scale error (per cent)	Accumulation (cm ice/year)	Average temp (°C)	Distance from coast (km)
Aggasiz, Ellesmere Island	80.7°N	73.1°W	1730	5	5–10	9.8	–24.52	80
Devon Ice Cap	75.42°N	82.50°W	1800	8	5–10	24	–23	50
Penny Ice Cap, Baffin Is.	67.25°N	65.75°W	1900	40	5–15	36	–15	100
Renland, E. Greenland	71.30°N	26.72°W	2340		5–10	50	–18	150
Camp Century, Greenland	77.18°N	61.02°W	1890	<5	<5	38	–24.35	200
Dye–3, Greenland	65.18°N	43.83°W	2490	17	<5	56	–20	150
Summit, Greenland, GRIP	72.57°N	37.62°W	3240	~0	~0.5	23	–32	550
North GRIP, Greenland	75.1°N	42.3°W	2921	~0	~0.5	21.5	–32	500
Academii Nauk, Russia	80.5°N	94.8°E	810	42	>10	35	n/a	
Guliya, W. China	35.28°N	81.48°E	6710	N/a	?	22	–15.6	
Dunde, W. China	38.1°N	96.4°E	5325	<5	?	47	–7.3	–
Sajama, Bolivia	18.1°S	68.88°W	6542	N/a	?	49	–10.3	
Huascaran, Peru	9.1°S	77.6°W	6048	N/a	?	144	–9.8*	–
Quelcaya, Peru	13.9°S	70.8°W	5670	Temp	<5	122	–3.0	
Vostok, Antarctica	78°S	106°E	3488	0	<5	2.6	–55.5	
Taylor Dome, Antarctica	77.8°S	158.72°E	2365	0	<5	7.2	–42	
Dome Fujii, Antarctica	77.32°S	39.7°E	3810	0	<5	3.6	–58	

*Air temp. nearby station; temp: firn is temperate (0 °C).

Table 19.1 Site statistics

Variable	Units	Proxy for...	References
Stable-isotopes $\delta(D)$ and $\delta(^{18}O)$	‰	accumulation weighted temperature and elevation at the deposition site, source water, storminess. Noise from local re-working of snow and sastrugi	Dansgard *et al.* (1973) Johnsen *et al.* (2001) Fisher (1990, 1992) Jouzel *et al.* (1997)
Melt per cent or amount	Per cent or $g/cm^2/year$	peak summer temperature at the site	Koerner and Fisher (1990) Koerner (1997)
Borehole temperatures	°C	mean annual air temperature at the site, so elevation of deposition point is a factor	Johnsen *et al.* (2001) Cuffey and Clow (1997) Dahl-Jenssen *et al.* (1993)
Pollen	num/L	pollen productivity at source, storm intensity and directions	Bourgeois *et al.* (2000)
Salts	ppb	sea–ice extent, marine storminess, and atmospheric water vapor content over the whole cycle	Mayewski *et al.* (1997) Grumet *et al.* (2001)
Acids	$[H^+]$	volcanic activity, marine biological productivity and recently, anthropogenic acid pollution. The water vapor content over the delivery cycle is important	Hammer *et al.* (1985) Zielinski *et al.* (1997) Clausen *et al.* (1995)
Mineral dust	ppb	distance to continental source areas, windiness and water vapor content over the whole cycle	Hammer (1980) Zdanowicz *et al.* (2000)
Gases	ppb–v cm^3/g	Because gases mix so quickly and homogeneously they should be the same in all ice-cores. As such they are used in dry snow areas and times to cross–date distant ice-cores	Chappellaz *et al.* (1997) Sowers *et al.* (1997)
MSA	ppb	CCN: MSA (a daughter product of DMS) is related to marine surface water productivity of living planktonic algae, sensitive to salinity, temperature. DMS is a precursor for cloud condensation nuclei (CCN), so MSA could be an indicator for CCNs	Legrand (1995) Legrand *et al.* (1997, 1991) Saltzman (1995) Saltzman *et al.* (1997)

Table 19.2 Ice–core variables

Holocene part of the Greenland and Antarctic ice-cores and ice sheets have proved particularly useful in establishing the parts played by greenhouse gases, mineral particulates, salt and acid aerosols in climatic change (Lorius *et al.*, 1979). However, inclusion of the Pleistocene records from the smaller ice caps allows for comparisons among all the world's ice stratigraphies. The comparisons led to the conclusion that the last interglacial period saw the demise of those same ice caps and caused a substantial retreat of the Greenland ice sheet which effected the larger part of the 6 m sea level rise during that interglacial (Koerner, 1989). It will be shown below that a similar approach has led to a history of ice cap retreat and regrowth in the circumpolar Arctic region through the Holocene period.

19.1 THE HOLOCENE: A DOMINATING LONG-TERM COOLING TREND?

The stable-isotopes of oxygen and hydrogen tend to be the primary variables analysed in any ice-core and they are typically related to site temperature weighted to the accumulation rate. It is known now that many other factors (see Table 19.2), influence the 'isotope' palaeo-thermometer and these will touched on. The listed time-scale error estimates should be kept in mind especially for the smaller ice caps.

Ice-core records in the northern hemisphere (Table 19.1) show that the last Ice Age ends abruptly (Dansgaard *et al.*, 1989; Alley *et al.*, 1993) at 11550 ±70 BP (calendar years before present i.e. AD 2000) (Fig 19.1a), although this is not the case in Antarctica, where the transition is more gradual (Fig. 19.4). There is some variation among the various non-polar ice-core $\delta(^{18}O)$ records (Fig. 19.5). For example, the Huascaran (Peru) and Guliya cores show a more Antarctic-like warming but the Dunde ice-core suggests that the area it represents in Western China experienced only a remarkably slow and small change. From the Early Holocene (defined here as 7–11.5 ka BP and lightly shaded in Fig. 19.2) many, but not all, of the $\delta(^{18}O)$ records show there was a general and gradual cooling that culminated in the minimum of the Little Ice Age (LIA) about AD 1500–1900 (Dansgaard *et al.*, 1975; Fisher *et al.*, 1996; Koerner, 1997). The Agassiz (melt and $\delta(^{18}O)$), Penny-95 ($\delta(^{18}O)$) (Fig. 19.1, #9, #4, #3, respectively) and Academii Nauk (melt and $\delta(^{18}O)$) (Fig. 19.6), records all show the persistent cooling trend, where that for melt begins earlier than that for $\delta(^{18}O)$. The persistent cooling is also clear in all the Antarctic stable-isotope records (Fig. 19.4) and in that for Huascaran (Fig. 19.5). This trend is present for only some of the Greenland $\delta(^{18}O)$ records (Fig. 19.1a) and not for the other non-polar records (Fig. 19.5).

The reason for some of the disagreement with respect to the Holocene cooling trend between the ice-core records is that their interpretation must be linked to the site particulars (Table 19.1). The records are affected, first, by local stratigraphic noise (Fisher *et al.*, 1985) and, second, by systematic biases related to long-term thickness changes and flow of the ice (Cuffey and Clow 1997; Fisher *et al.*, 1998). For example, the Camp Century and Dye-3 sites have been slowly moving down their respective flow-lines so that their records include an elevation effect. In their case, warming by lowering of the drill site elevation cancels out climatic cooling. However, the Summit site is at the top of the flow-line and yet Fig. 19.2f shows a change in the air content of the ice from the GRIP core in central Greenland (Raynaud *et al.*, 1997), which must largely be due to changing (decreasing) elevation. Johnsen *et al.* (2001) have therefore corrected the original record (Fig. 19.1a, #7, light grey) for the effect of changing elevation. The corrections are based mainly on the record of margin retreat of the Greenland Ice Sheet and the accumulation record (Fig 19.1a, #8) (Dahl-Jensen *et al.*, 1993, Johnsen *et al.*, 1995; Kapsner *et al*, 1995; Cuffey and Clow, 1997). The corrected GRIP record (Fig. 19.1a, #7, black line) then shows the persistent cooling trend which was already evident in the records from Antarctica, Huascaran and the Arctic ice caps where, at least in the Canadian Arctic, elevation changes do not affect the $\delta(^{18}O)$ values (Koerner and Russell, 1979).

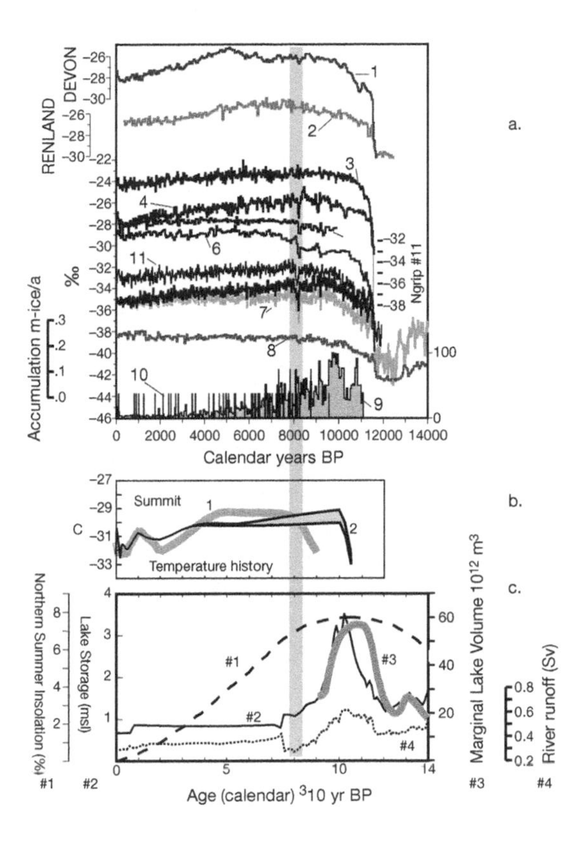

Figure 19.1 Holocene records for Greenland and North America. Years are calendar and with respect to AD 2000. The shaded strip highlights the 8200 BP cold event. The end of the Laurentide Ice Age is nominally taken as the end of the Younger Dryas as given by the sudden changes in ice-core properties at 11,550 BP. All the ice-core records have some time-scale error, listed in Table 19.1. Furthermore, they have all been 'forced' to have the Summit transition age, and when obvious and within the errors, to place the 8200 event at that age. (a) The numbered curves #1 to #11 are from sites described in Table 1 and the variables are described in Table 19.2. #1 Devon Island Ice Cap $\delta(^{18}O)$ (Paterson *et al.*, 1977); #2 Renland $\delta(^{18}O)$ (Hansson, 1994); #3 Penny 1995 Ice-Core, Baffin Island $\delta(^{18}O)$ (Fisher *et al.*, 1998); #4 Agassiz Ice-Core (1984 and 1987), Ellesmere Island $\delta(^{18}O)$ (Fisher *et al.*, 1995); #5 Dye-3, South Greenland $\delta(^{18}O)$ (Dansgaard *et al.*, 1982); #6 Camp Century northwest Greenland $\delta(^{18}O)$ (Dansgaard *et al.*,1971); #7 Summit (GRIP) $\delta(^{18}O)$ (Johnsen *et al.*, 2001) grey, uncorrected and black corrected for Holocene elevation changes; #8 Summit (GISP2) accumulation rate inferred from annual layer thickness (Cuffey and Clow, 1997); #9 Agassiz ice-core (1984 and 1987) melt per cent (Koerner and Fisher, 1990); #10 Summit (GISP2) melt layers (Alley and Anandakirshnan,1995), note that as the air bubbles disappear at 1200 m depth this record terminates at the equivalent age of about 8000 BP ; #11 North GRIP (NGRIP) $\delta(^{18}O)$ (Johnsen *et al.*, 2001). (b) Reconstructed air temperature records using measured borehole temperatures. Allowance has been made for the Early Holocene having higher elevations and the records are 'normalized to a fixed modern elevation. Curve #1 (Dahl-Jensen, 1997) assumes nothing about the form of the air temperature history, but finds the solution using a Monte-Carlo statistical procedure. Curve #2 (Clow and Waddington, 2003) assumes a range of marginal histories (shaded zone) and consequent site surface elevation histories. Furthermore, they factor in a sharp temperature change at the end of the Younger Dryas, which is not unrealistic. The differences between curve #1 and theirs (#2) have to do with their using this 'extra bit' of information. (c) Curve #1 shows the northern Holocene insolation (Berger, 1978) has a maximum just around the transition. Curves #2 and #4 are model calculated marginal lake volumes (in equivalent meters of sea-level change; msl) and melt-river flow rates from the Laurentide Ice Sheet (Marshall and Clarke, 1999). Curve #3 shows estimated Laurentide marginal lake volumes (Teller, 1987).

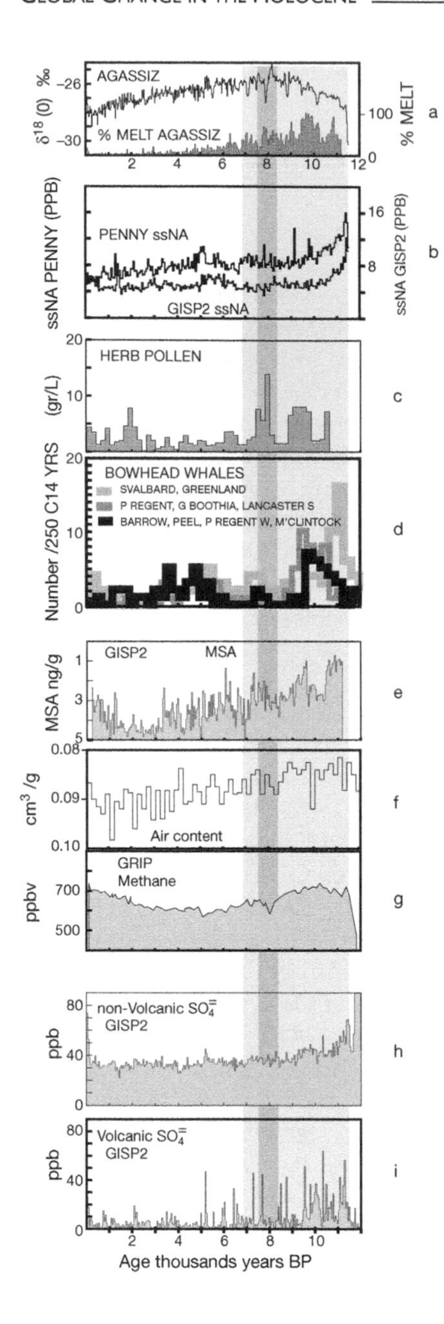

Figure 19.2 Holocene multivariables. The lighter shading marks what we refer to as the Early Holocene and the 8200 BP event is marked with the darker strip. (a) Agassiz Ice Cap $\delta(^{18}O)$ and melt percent (Koerner and Fisher, 1990; Fisher *et al.*, 1995). (b) Sea salt sodium concentration, ssNa, from the Penny Ice Cap (Fisher *et al.*, 1998) and the GISP2 cores (Mayewski *et al.*, 1997). (c) Herb pollen concentration in the Agassiz (1987) core (Bourgeois *et al.*, 2000). (d) Bowhead (mostly) whale bone counts by region (Dyke *et al.*, 1996; Dyke and Morris, 1997). Note the ages are in calendar years but the counting bins are originally 250 ^{14}C years. (e) MSA (methanesulfonate) concentration in the Summit cores (Legrand *et al.*, 1997; Saltzman *et al.*, 1997). (f) Total air content in Summit ice relates mainly to site elevation history (Paterson, 1994; Raynaud *et al.*, 1997) though other influences have secondary effects. (g) Methane concentration in Summit cores (Chappellaz *et al.*, 1993, 1997). (h and i) Fifty-year averages of volcanic and non-volcanic sulphate ion concentration. The method for partitioning between the types is given in (Zielinski *et al.*, 1997). The Early Holocene seems to be more volcanically active.

However, there are differences among the circum-polar Arctic records in the pattern of climatic change, particularly between the (only) melt record going back as far as 11.5 ka (Fig 19.1a, #9), and all the other $\delta(^{18}O)$ records. The summer-melt maximum in the Agassiz cores is between 9 and 11 ka, whereas the $\delta(^{18}O)$ max is between 8.5 and 10 ka (Fig 19.1a, #4). The melt record peaks at 10 ka but $\delta(^{18}O)$ peaks 1500 years later (Fig. 19.1a, #9). The GRIP and Penny 96 $\delta(^{18}O)$s peak at about 9.5 ka, which is somewhat similar to the melt record on Agassiz. However, whereas the melt record suggests summers between 10.5 and 11.5 ka were

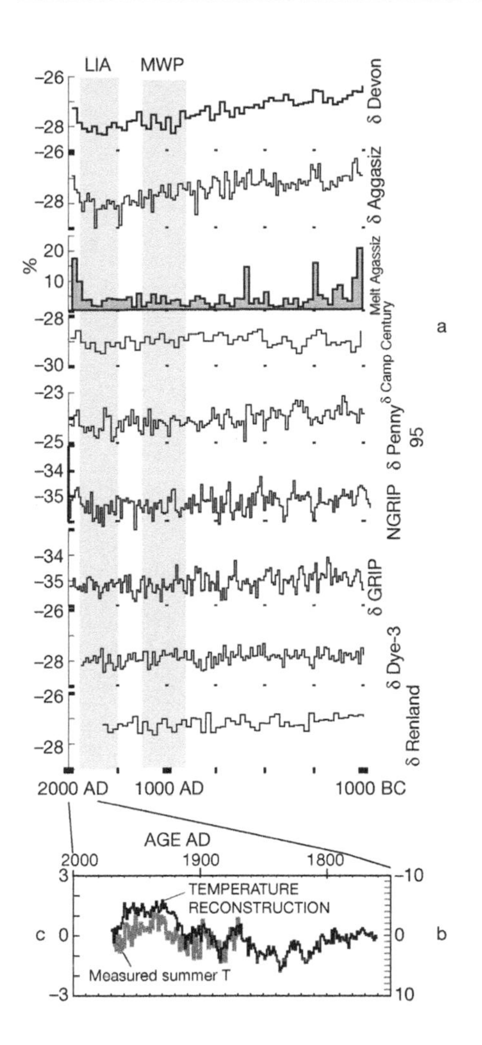

Figure 19.3 (a) The last 3000 years of Greenland and Canadian δ(^{18}O) records and the Agassiz melt record. All these records have time-scales that are good to ±5 per cent or better in this time range. Some of the records have LIAs (Little Ice Ages), shaded between AD 1500 and 1900 and some possibly have a MWP (Medieval Warm Period) also shaded between AD 900 and 1250. The differences are real and give information about regional differences (Fisher *et al.*, 1996). (b) High-resolution north of latitude 40° reconstruction of summer temperatures using ice-core records, tree-rings, documentary and other sources. The various proxy variables can be used together, and in spite of, or possibly because of their different spectral sensitivities, time-scale errors and noise contents compliment each other and help build up a very accurate picture of the temperature history (Bradley and Jones, 1993; Overpeck *et al.*, 1997; Mann *et al.*, 1998; Fisher, 2002). The correlation coefficient with the measured record (in grey) is 0.70 over the common period.

approx. 1.0–2.0°C warmer than today, the δ(^{18}O) records show much cooler conditions in the immediate post-glacial period.

To investigate this problem further one needs to examine other climate proxies. Grumet *et al.* (2001) showed that the sea-salt record from the Penny Ice Cap (1995) core was related to sea-ice extent in nearby Baffin Bay, whereby higher Na^+ concentrations were associated with more open water. The 12,000-year Na^+ record shown in Fig. 19.2b (Fisher *et al.*, 1998) suggests that, in Baffin Bay, the sea-ice extent has slowly increased, through the Holocene, from a remarkably low minimum at 10.5–11.5 ka. The GRIP record of Na^+, shown alongside the Penny one, does not show such a strong trend through the Holocene. However, it does show a very high concentration in ice deposited in the very Early Holocene. If the relationship between Na^+ and sea-ice holds for that record too, it means there was a maximum area of open water in the Na-source region for Greenland at that time, i.e. conditions were very warm. Salt content of the ice is also significantly influenced by storm-driven advection and not just open water although the two are related (Fischer, 2001).

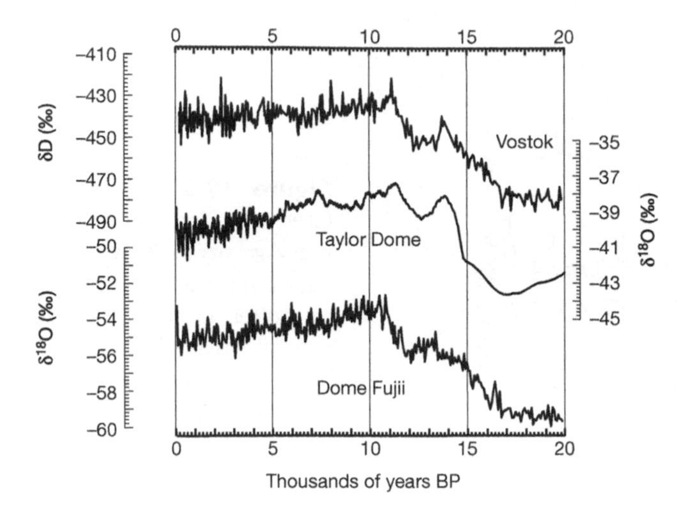

Figure 19.4 Antarctica: Vostok, deuterium – δD (Petit *et al.*, 1999), Taylor Dome, $\delta^{18}O$ (Steig *et al.*, 2000), and Dome Fujii, $\delta^{18}O$ (Watanabe *et al.*, 1999).

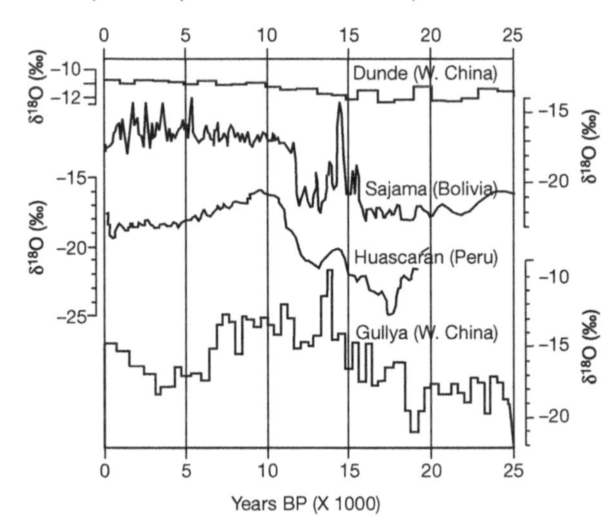

Figure 19.5 Tropical ice-core $\delta^{18}O$'s: Dunde Ice Cap (Thompson *et al.*, 1989), Huascaran, Guliya and Sajama (Thompson *et al.*, 1998).

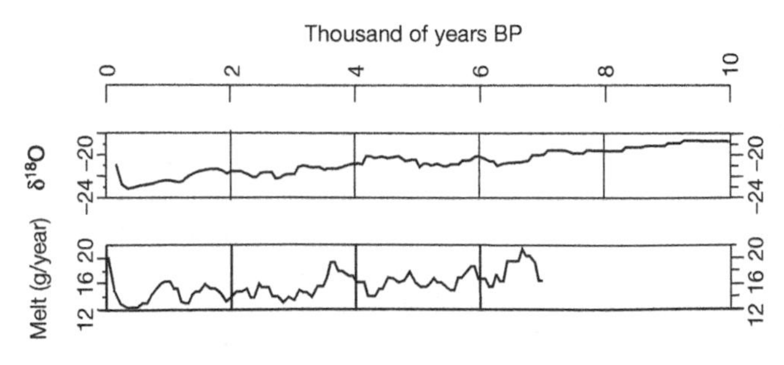

Figure 19.6 Academii Nauk (Russian Arctic Islands), $\delta^{18}O$ and melt ($g/cm^2/y$), Kotlyakov *et al.*, 1991).

Legrand *et al.* (1997) provide evidence of a positive relationship between methane sulphonate (MSA) and salinity. They attribute the Early Holocene MSA minimum to lower salinity in the Atlantic, caused by melt-water, as the algae that produce dimethylsuphide (DMS) prefer saltier water. Salt rejection under sea-ice increases the salinity promoting a parallel positive relationship between MSA and sea-ice coverage. The Holocene record from GISP2 (plotted inversely in Fig. 19.2e) shows low concentrations of MSA (i.e. low concentrations of sea-ice) in central Greenland between 10.5 and 11.5 ka, increasing to high concentrations at approximately 2 ka. Again, the implications are that sea-ice concentrations around Greenland were at a minimum at the beginning of the Holocene but increased thereafter for 8000 years (Legrand *et al.*, 1997). While the MSA record suggests that salinity in the Early Holocene was lower, MSA in ice-cores and its relation to climate is still only partly understood (Saltzman, 1995; Legrand, 1995).

The record of Bowhead Whales in the Canadian Arctic archipelago waters (Dyke *et al.*, 1996), with maximum numbers between 10 and 11.5 ka (Fig. 19.2d), suggests a minimum sea-ice coverage at that time. These records **all** suggest much more open water, indicative of warmer conditions than today, in the seas and oceans affecting the ice caps and the Greenland Ice Sheet at that time.

The temperatures measured in the GRIP borehole have been used to construct a Holocene temperature record. The reconstructions (shown in Fig. 19.1b) have been corrected for changes in thickness of the ice sheet at the drill site, where increased ice thickness was primarily due to a rapid increase in the accumulation rate in the Early Holocene. Reconstruction #1 (Dahl-Jensen, 1997) uses a 'Monte Carlo' procedure for reconstructing the air temperature and does not assume an Ice Age transition at 11.5 ka BP. The more recent reconstruction (#2, Clow and Waddington, 2003) assumes, first, a thickness change and, second, a sudden temperature shift at 11.55 ka. The refinement includes the effect on drill site ice thickness of a range of possible Early Holocene ice margins from 200 to 100 km beyond present (shown by the grey region in Fig. 19.1b, #2). The more recent reconstruction (#2) shows an Early Holocene thermal maximum at 10 ka and warmer than present temperatures until approx. 3–4 ka. Johnsen *et al.* (1995) with similar assumptions produced a similar temperature history and have shown that in the Holocene the $\Delta\ \delta(^{18}O)/\Delta T$ transfer ratio is close to 0.67‰/°C, provided the elevation and source water is constant. This ratio varies in the pre-Holocene.

Pollen concentrations in the Canadian cores (Agassiz record, Fig. 19.2c), show an Early Holocene maximum even though some of the source regions were still covered by ice (Bourgeois *et al.*, 2000). There are *higher* concentrations between 11 and 7 ka, with decreasing concentrations until 2 ka. The increase at that time has not yet been adequately explained.

In summary then, the $\delta(^{18}O)$ records from Summit, Academii Nauk, Agassiz and Penny 95, the melt record from Agassiz, the MSA record from Summit and the Na^{+} records from Summit and Penny all support an Early Holocene thermal maximum between about 8 and 11.5 ka and a cooling between 8 ka and about 0.3 ka. The Antarctic and Huascaran records show the same trends. However, there is a wide divergence among the other Holocene non-polar records (Fig. 19.5, Dunde, Sajama and Guliya), suggesting lower latitudes have shown greater regional variations in the Holocene period than the polar regions.

There are still two $\delta(^{18}O)$ records (Camp Century and Dye-3) that show very little change throughout the Holocene and two (Devon Ice cap and Renland) that show very delayed thermal maximums. The Camp Century and Dye-3 sites lie some distance down their respective flow-lines. Down-slope movement means their records (Fig 19.1a, #6, #5) include an elevation effect. Thus, warming due to lowering of the drill site by down-slope movement has most likely cancelled out any climatic cooling effect. The Renland δ (^{18}O) record (Fig. 19.1a, #2) (Hansson, 1994) is from an independent and isolated ice cap on the east coast of Greenland and shows a Mid-Holocene maximum (about 7 ka). Uplift of the ice cap, mainly in the Early Holocene due to a shrinking Greenland Ice Sheet, may have affected Renland's record. Two recent (unpublished) ice-cores from Devon Ice Cap show that the ice cap has a more complicated dynamic history than that shown in Fig. 2(#1) from Paterson *et al.* (1977). This may be attributable to its position in the Early Holocene at the southern edge of the Pleistocene Innuitian Ice Cap (Blake, 1993), and the northern edge of the Laurentide Ice Sheet.

The divergence of the two Penny δ(^{18}O) records (Fig. 19.1a, #3 shows the Penny 95 record) after 8 ka is an indication that the flow-lines have changed from a common one when the two sites were close and part of a larger (Laurentide) ice sheet, to independent flow-lines on a smaller ice cap (Fisher *et al.*, 1998).

The question is why the Early Holocene δ(^{18}O) records show cooler conditions than the melt and chemistry proxies and borehole temperature reconstructions. The answer may rest in the effects on δ(^{18}O) of melt water from the shrinking Laurentide Ice Sheet. There is a maximum in marginal melt water lakes and river (melt) flow between 9 and 12 ka BP (Fig. 19.1c, #2, #3, #4) (Teller, 1987; Marshall and Clarke, 1999). Source water for Greenland and Arctic Canada comes from many places during the Late Glacial and Early Holocene. During the summers 13,000 years ago, it has been estimated that the area covered by marginal lakes in North America was about 35 per cent (Marshall and Clarke, 1999) of the area of the Atlantic source area between 30 and 60°N. The wet ablation area of the ice sheet would increase this source area. In this case the initial δ(^{18}O) for all this melt water would be about −35‰ (Fisher, 1992). The Atlantic surface waters themselves would have a δ(^{18}O) somewhere between the bulk mixed ocean value of +1.2‰ and the ice sheet value of about −35‰ (Fisher, 1992). For example, if the mixing ratio in the surface layer between bulk ocean water and melt water was 9 : 1 then the surface layer δ(^{18}O) could be as low as −2.4‰. In the summer melt season then, modelling suggests that the snows falling on Greenland and Arctic Canada could be biased toward too negative values by about 2.7‰ (Fisher, 1992). This biasing would be expected to be a maximum when the marginal lakes and river runoff are also at a maximum (Fig. 19.1c) and could possibly explain the offset between the δ(^{18}O) and melt layer records for Agassiz (Fig. 19.1a, #4 and #9, respectively).

There is some corroborative evidence in the Summit MSA record. The MSA (Legrand *et al.*, 1997; Saltzman *et al.*, 1997) record from Summit has been interpreted as being a measure for salinity of the surface source water. As noted above, the living planktonic algae that emits the DMS (that becomes sulphate particles and MSA), do not thrive in fresh water (Legrand *et al.*, 1997). These products of DMS are thought to be key source of cloud condensation nuclei (CCNs) (Charlson *et al.*, 1987). The MSA does have its minimum in the Early Holocene and this is in anti-phase with the Agassiz melt record and offset with the δ(^{18}O), suggesting the melt water bias explanation of the offset is at least partly correct. The same of course would affect the other δ(^{18}O) records. There must have been considerable melt-water in the Atlantic surface waters to cause the Younger Dryas shutdown of the Thermohaline circulation and the possible near shut-down at 8200 BP (Alley *et al.*, 1997).

The following is a speculative positive feedback loop that would enhance melting during the Early Holocene: [solar isolation increase] → [increased fresh water from summer melting] → [decreased salinity in the ocean (Atlantic)] → [decrease in summer DMS and MSA] → [decrease in CCN and increase in clear summer skies] → [increased effectiveness of the higher summer insolation] → [increased melting of ice sheets].

Over the last glacial cycle the MSA record in the Antarctic cores is different to that from the Greenland cores (Legrand *et al.*, 1991), perhaps due to the availability of fresh melt-water in the northern interglacial.

19.2 THE 8200 BP COOLING EVENT

The 8.2 ka event (Figs 19.1 and 19.2, darker shaded strip) shows as a (cold) minimum in $\delta(^{18}O)$ in all the Greenland sites, except Renland, and in the Agassiz records from Ellesmere Island, Canada. It shows up in the middle of a minimum in Arctic whale numbers (Fig. 19.2d) (Dyke and Morris, 1997), and in a Summit methane minimum (Fig. 19.2g) (Chappellaz *et al.*, 1993). There is an accumulation rate dip at 8.2 ka in the Summit accumulation rate history (Fig. 19.1a, #8) (Cuffey and Clow, 1997). However, at 8.2 ka there is a secondary maximum in summer melt in the Agassiz record (Fig. 19.2a) and a trend to lower MSA concentrations at Summit (Fig. 19.2e), which is an indication of more fresh water in the Atlantic. The volcanic $SO_4^=$ record from Summit (Zielinski *et al.*, 1997) shows no unusual clustering of eruptions prior to 8.2 ka (Fig. 19.2i). The event is not apparent in the Antarctic or tropical ice-cores.

The 8.2 event has been convincingly related by Barber *et al.* (1999a) to the sudden collapse of the Laurentide Ice Sheet over Hudson Bay and the release of fresh water and ice into the Labrador Sea and Atlantic at flow rates somewhere between 6 and 0.06 Sv (Sverdrups; $1\ Sv = 10^6\ m^3/s$) . The Labrador Sea Intermediate Water (LSW) and North Atlantic Deep Water (NADW) formation can be threatened by freshwater inflow rates in this range (Alley *et al.*, 1997; Barber *et al.*, 1999a). So the hypothesis is that the fresh water added to the marine surface waters caused a serious slowdown of the thermohaline circulation (i.e. a failed Younger Dryas event). There is still some dispute about the exact timing, extent and cause of this event (Hu *et al.*, 1999). The cold excursion in ice-core $\delta(^{18}O)$ possibly exaggerates the cooling by the presence of the low δ fresh water needed for the shutdown. Surface water $\delta(^{18}O)$ changes of 0.5‰ at 8.2 ka have been reported from ocean cores just east of Newfoundland (Barber *et al.*, 1999a). The MSA minimum and high summer melt discussed above also point to lower salinity and higher melt-water for the 8.2 ka event. Thus an increase in summer warmth can trigger a fresh water break out and result in a cold event in the ice-core stable-isotopes.

However, it should be stressed that the 8.2 ka event, although it occurred during the present interglacial period, is related to the presence of the Pleistocene ice sheets (like the Younger Dryas and the Dansgaard–Oeschger events of the Glacial Period). Melt-water appears to have played an important part, but the availability of similarly large outbursts of melt-water are not possible in the modern world where ice masses are either substantially smaller or in very high latitudes (Koerner, 2001).

19.3 MID-HOLOCENE SECONDARY WARM PERIOD: 3-5 KA BP

There is a small, secondary maximum or inflection indicative of warmer temperatures between 3 and 5 ka, in the Summit borehole temperature reconstruction (Fig. 19.1b, #2), whale numbers (Fig. 19.2d) and $ssNa^+$ (Summit and Penny, Fig. 19.2b) which corresponds to the 'classical' Mid-Holocene optimum (MHO). O'Brien *et al.* (1995) attribute it to either a contraction of the North Polar vortex or a decrease in meridional air flow and suggest there are several such secondary warmings interspersed with cold excursions, the most extreme cold period being the LIA. This MHO warming is not present in any of the $\delta(^{18}O)$ records except that from Devon Ice Cap (Fig. 19.1a, #1). Neither is it present in the melt layer record from Agassiz (Fig. 19.1a, #9) although there is a small shift to cooler summers at about 4 ka. There is a shift in MSA toward higher values at 3 ka (Fig. 19.2e). Lower levels of volcanic activity follow after 5 ka (Fig. 19.2i), but as volcanic veils generally result in cooler temperatures, this last observation is counter intuitive. Johnsen *et al.* (2001) have noted that during the MHO interval that there is a small but consistent divergence between the $\delta(^{18}O)$ records from Summit and North GRIP (from about 4 to 8 ka BP) and suggest similar explanations to those of O'Brien *et al.* (1995). The abrupt Early to Mid-Holocene climatic transition to full post-glacial conditions discussed by Stager and Mayewski (1997) is not universally apparent in the ice-core records of either polar regions or the non-polar sites. However, it should be stressed that the $\delta(^{18}O)$-records from the Arctic regions show a strong cooling in the second half of the Holocene, which includes what is often termed the neoglacial period. This cooling continues right up to the LIA. The same cooling is not apparent in the Dunde and Sajama cores but ends earlier in the Guliya core (Fig. 19.5).

19.4 THE LAST MILLENNIUM

The culminating cold period of the Holocene cooling trend has been the LIA (about AD 1500–1900). This period has perhaps been somewhat overemphasized and searched for globally, as it occurs well within the historic period and has a cultural significance in both Europe and North America. It stands out largely (but not in all the records) because it is followed by the modern period of warming. The modern 'global warming' is more apparent in the $\delta(^{18}O)$- and melt-records from the ice caps in the Canadian Arctic, the Russian Arctic Islands (Fig. 19.6) and Svalbard (Koerner, 1997) than in those from Greenland. Recent warming has also effected an increase in herb pollen concentrations in the Agassiz core (Fig. 19.2c) (Bourgeois *et al.*, 2000).

To some extent the Medieval Warm Period is culturally defined in Europe by the LIA. The stable-isotopic and melt layer record for many ice-cores shows an LIA and an older warmer period that could be called an MWP. For example, Fig. 19.3a demonstrates that the Devon, Agassiz, Camp Century and North GRIP have the LIA-MWP couple, but that Summit and Dye-3 do not. These regional differences have been shown to be real and not simply noise (Fisher *et al.*, 1996).

19.4.1 Local Noise and the Multi-Proxy Approach

There is a lot of local noise (Fisher *et al.*, 1985) and regional differences between cores. The local noise can be reduced by averaging or stacking several cores and the regional differences extracted by using eigenvector methods (Hibler and Johnsen, 1979; Fisher *et al.*, 1996; Fisher, 2002). Once it is admitted that regional differences are important, then using ice-core records in conjunction with other proxy climate records immediately suggests itself as a means of delineating the spatial–temporal climate story.

Multi-proxy reconstructions have been based on ice-cores, tree-rings, coral growth, varves, documentary evidence and any other dateable records. Each type of record, in spite of its varying levels of error and sensitivity, can be used (Bradley and Jones,1993; Overpeck *et al.*, 1997; Mann *et al.*, 1998; Fisher, 2002) (see Lotter, pp. 373–383 in this volume, of an assessment of multiproxy studies). The reconstructed summer temperature series for continental regions north of 40°N are presented in Fig. 19.3b, dark curve (Fisher, 2002). Over 50 sites are used and hundreds of individual records from trees, ice-cores and documentary sources. The average measured summer temperature is given by the grey curve and its correlation coefficient with the reconstruction is 0.70. Model studies (Fisher, 2002) suggest that the multi-proxy reconstructions of temperature can capture most of the actual temperature history.

19.5 GLACIER BALANCE IN THE HOLOCENE

In general, the $\delta(^{18}O)$- and melt-records indicate that present-day temperatures (with the exception of the Academii Nauk melt record; Fig. 19.6) are much lower than temperatures in most of the first half of the Holocene. Glacier mass balance measurements made in the circumpolar Arctic over the past 40 years have shown that the glaciers have negative balances; i.e. they are losing mass (Dowdeswell *et al.*, 1997). It follows, therefore, that the first few thousand years of the Holocene must have been a time of rapid glacier retreat, not only of the Laurentide and FennoScandian Ice Sheets, but also of all the circumpolar ice caps and icefields, and the Greenland Ice Sheet. In fact, some of the ice-cores from ice caps in Svalbard and Vavilov Ice Cap (Kotlyakov *et al.*, 1991; Koerner, 1997; Koerner and Fisher, 2002) show no evidence of pre-Holocene ice. This means the ice caps they were drilled from completely melted away, as the sites are close to, or at the tops of those ice caps. By the second half of the Holocene, the cooling trend discussed here began promoting ice cap/ice sheet regrowth. Firn/equilibrium lines (an elevation dividing the area of positive balance (above) from that of negative balance (below)) dropped below the elevation of the ice cap margins (Koerner and Fisher, 2002; see also Nesje and Dahl, pp. 264–280 in this volume for further explanation). The ice caps reached maximum dimensions probably in the LIA but have retreated again due to the modern warming which has produced the negative balances mentioned above. Thompson (1996) has stressed that some of the tropical ice caps may not survive modern warming. The same is also true of the smaller Arctic ice caps and glaciers such as those in Svalbard and the Canadian Arctic islands (e.g. Meighen and Melville ice caps). In fact, the very small Arctic ice caps that appear on aerial photography taken in the 1950s have already disappeared. However, that retreat is not yet as extensive as that of the Early Holocene period (Koerner and Fisher, 2002).

CHAPTER

20

APPROACHES TO HOLOCENE CLIMATE RECONSTRUCTION USING DIATOMS

Anson W. Mackay, Vivienne J. Jones and Richard W. Battarbee

Abstract: Diatom analysis has been used extensively to reconstruct past environments, and increasingly attention is being given to developing the technique to model Holocene climate variability. This chapter reviews progress in the field, including both qualitative and quantitative interpretations: (i) the relationships between diatoms and climate indicators such as solar insolation (Elk Lake, Minnesota), snow cover (Lake Baikal) and ice-cover (Elison Lake, Ellesmere Island); (ii) the development of models to reconstruct diatom inferred climates, either directly (e.g. surface water temperature, Scandanavia) or indirectly (e.g. pH in the Austrian Alps and on Baffin Island; e.g. salinity in lakes in the Northern Great Plains region, North America); and (iii) the development of high resolution studies in coastal and marine environments (including the Icelandic Shelf, the Antarctic Peninsula and the eastern Norwegian Sea). The importance of autecological and taphonomic studies is highlighted, although they still receive too little attention when attempts are made to interpret past climates using diatom analysis.

Keywords: Closed lakes, Diatoms, Marine diatoms, Open lakes, Qualitative reconstructions, Quantitative reconstructions, Transfer functions

Diatoms are unicellular algae and are used extensively in palaeoecological studies because they are excellent indicators of past environmental conditions. They are particularly useful palaeoecological proxies because they can be identified to the species level using light and scanning electron microscopy. Thus by inspection of assemblages in sedimentary records, we can make direct and indirect inferences about past environmental conditions. A detailed recent review on diatom analysis and applications can be found in Battarbee *et al.* (2001b).

Diatoms have been used as proxy indicators to reconstruct Holocene climate variability in every continent, and work on continental aquatic ecosystems has been much more common than marine or coastal studies. The majority of recent studies use quantitative multivariate techniques to reconstruct past climatic variables either directly, such as surface-water temperature (e.g. Vyvermann and Sabbe, 1995; Rosén *et al.*, 2000; Bigler and Hall, 2002) and air temperature (e.g. Korhola *et al.*, 2000), or indirectly by reconstructing, for example, salinity (e.g. Fritz *et al.*, 1991; Laird *et al.*, 1996; Gasse *et al.*, 1997; Verschuren *et al.*, 2000), DOC (Pientiz *et al.*, 1999), conductivity (e.g. Davies *et al.*, 2002) and pH (e.g. Psenner and Schmidt, 1992; Koinig *et al.*, 1998a). However, qualitative information provided by assessing changes in diatom species themselves, especially with respect to their habitat, survival strategies and autecologies in both freshwater (e.g. Smol, 1988; Bradbury *et al.*, 2002) and marine environments (e.g. Rathburn *et al.*, 1997; Gersonde and Zielinski, 2000) should not be ignored.

In this chapter we review some recent developments in the use of diatom analysis to reconstruct climate variability during the Holocene. We describe some recently observed relationships between planktonic taxa in freshwater lakes and possible climate forcing factors, such as solar insolation. We then outline two studies that make use of diatom assemblages in the sediments of open-basin systems to (i) reflect past ice-cover in North American polar regions and (ii) reconstruct past climates in Northern Europe, both directly through reconstructing surface water temperatures, and indirectly by reconstructing pH and establishing links with changing air temperatures. We then outline the potential for using diatom-inference models for reconstructing past salinity in closed-basin systems in agriculturally important sub-humid regions of North America (where water quantity and quality are vulnerable) and, finally, we summarize recent progress in using diatom analysis in marine sequences to reconstruct Holocene climate variability.

Interpretation of the climate record held by diatom proxies in lake and marine sediments is enhanced in cases where there is knowledge of population growth and succession in the water column, and an understanding of how diatoms are transported to the sediment surface, and finally incorporated into the sedimentary record (e.g. through what we will call here integrated studies). In this way the quality of the diatom preservation can be assessed and key, controlling climatic variables evaluated. Diatom-inferred climatic interpretations therefore require knowledge of taphonomic processes, most notably dissolution (Flower, 1993). In saline systems, it is well known that diatom valve preservation is affected by the dissolution of biogenic silica; valves may either be partly dissolved so that identification becomes more difficult, if not impossible, or whole assemblages may be simply removed from both the training set and the stratigraphic record. Accordingly, recent attempts to improve diatom-inferred salinities from saline lake training sets have employed a very different approach of quantifying the relationships between the resistance of specific taxa to dissolution through the establishment of dissolution indices to help validate transfer functions (Ryves *et al.*, 2001). Dissolution is also a problem in many freshwater ecosystems, such as Lake Baikal in central Asia. A recent study found that approximately 50 per cent of diatom frustules in surface sediments are affected to some degree by dissolution, and provisional results suggest that only a small proportion of the cells produced in the water column are ultimately preserved in the sediments (Mackay *et al.*, 2000). Nevertheless, transfer functions from dissolved assemblages can still be effective (e.g. Pichon *et al.*, 1992), and although there is not the scope in this chapter to give a thorough representation of the problems of dissolution in lacustrine ecosystems, readers are instead encouraged to read Fritz *et al.* (1999) and Ryves *et al.* (2001), and references contained therein.

20.1 PLANKTON RESPONSES TO CLIMATE VARIABLES

Integrated studies provide an effective way of maximizing our knowledge of the processes that affect diatoms, which can subsequently be used to aid interpretation of past climates. Such studies ideally incorporate:

1. monitoring of the diatom populations, and sometimes also of the hydrophysical and hydochemical properties of the water body through space and time
2. an estimation of fluxes of diatoms down through the water column using open and/or sequencing sediment traps, followed by
3. an assessment of the rate of diatom incorporation into the surface sediments.

Here we present two examples of approaches taken towards integrated studies, both of which make use of sediment traps and accompanying environmental data. The first example is of Elk Lake, in Minnesota (e.g. Bradbury, 1988; Bradbury *et al.*, 2002) and the second is of research currently being carried out on Lake Baikal (Jewson and Granin, 2000; Ryves *et al.*, 2003).

Elk Lake is a dimictic lake with annually laminated sediments that began forming approximately 11,000 years ago. Between 1979 and 1984, phytoplankton populations and succession were monitored using sediment traps (Bradbury, 1988), and population blooms of specific diatoms were linked to ice-cover and lake circulation patterns, especially the spring and autumn overturn. In this study, the importance of air temperature in controlling ice-cover, and circulation patterns in controlling nutrient availability, were highlighted especially with regard to two diatom species, *Fragilaria crotonensis* and *Stephanodiscus minutulus*. The study concluded that *S. minutulus* is indicative of a long, and 'vigorous' spring overturn period, which itself could be brought about by thin ice-covering over the lake, a warm, dry, early spring (to help prevent ice-cover becoming thick) and a cool late spring to help promote wind-driven circulation and postpone summer stratification. Bradbury suggests that after long, cold springs, when ice-cover is thick, the water column in lakes can quickly become stratified once the ice-cover has disappeared due to prevailing warm weather conditions. Water stratification prevents the resuspension of nutrients such as phosphorus from the bottom sediments; consequently, only diatoms that require small amounts of phosphorus in relation to silica (derived from groundwater and other allochthonous inputs) can bloom, e.g. *F. crotonensis*. Taxa such as *S. minutulus*, on the other hand, bloom when the Si : P ratio is low (e.g. < 1).

Information obtained from the trapping experiments was used in a subsequent study to help interpret changes in the diatom assemblage composition throughout the Holocene in the context of climate variability (Bradbury and Dieterich-Rurup, 1993) (Fig. 20.1). Changing ratios between *S. minutulus* and *F. crotonensis* suggest shifts between long, dry springs, when the ratio between the two species is high, and warm stormy summers, when the ratio between the two species is low. Knowledge of the autecology of a further species, *Aulacoseira ambigua*, is used to suggest that during the Mid-Holocene at Elk Lake (between *c.* 6 ka yr BP and 4 ka yr BP) (Fig 20.1) the region experienced a marked increase in stormy weather during the summer and late autumn: *A. ambigua* is a summer species which blooms when silica is abundant, but also requires an increased degree of turbulence in the water column to keep the species in suspension.

The relationship between *S. minutulus* and *F. crotonensis* and other taxa, including benthic diatoms (e.g. *Cymbella* spp.) and chrysophyte cysts, has been further studied at Elk Lake over the last 1500 years (Bradbury *et al.*, 2002), using a combination of wavelet transformation analysis and transfer function methodologies. Wavelet transformation analysis differs from Fourier analysis as it can be used to analyse data that exhibit periodicity (i.e. variation with both amplitude and frequency in time) (Torrence and Campo, 1998). In this study, wavelet analysis of, for example *Cymbella* spp., suggests strong periodic correlations between individual diatoms species and various sunspot cycles, including the 11-year Schwabe cycle, the 22-year Hale sunspot cycle, together with other multi-decadal- and multi-centennial-cycles (Fig. 20.2). These results suggest that solar insolation is an important explanatory variable for diatom population succession, whether directly, for example through ultraviolet penetration, or indirectly through changes in air masses. However, the results also clearly indicate that controls on diatom populations must come from a combination of factors, confirming that knowledge of the

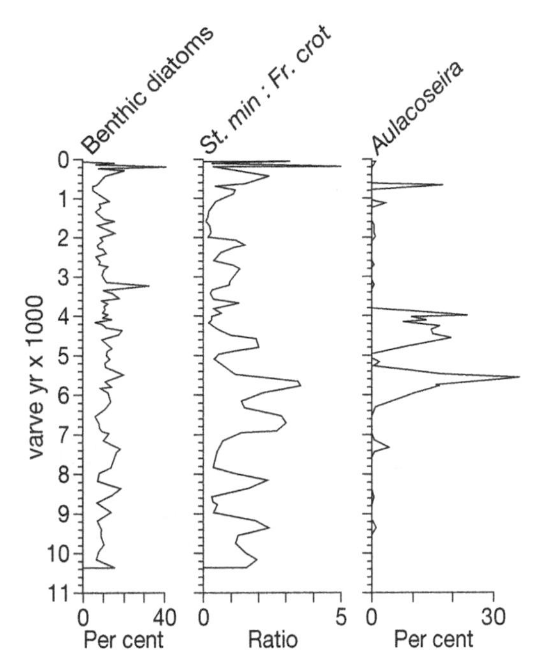

Figure 20.1 Relative proportion of benthic diatoms and *Aulacoseira* species, and the ratio between *S. minutulus* and *F. crotonensis* in varved sediments from Elk Lake. Adapted from Bradbury and Dieterich-Rurup (1993).

autecologies of diatom species greatly helps in palaeolimnological interpretations of Holocene climate.

Climate predictions for southern Siberia suggest that temperatures in winter will increase by 2–5 °C in the next 50 years (IPCC, 2001). Lake Baikal in southeast Siberia is therefore a key site for Holocene palaeoclimate research as it is positioned close to the boundaries of the Siberian high-pressure and the Asian monsoon weather zones, far from oceanic influences. Lake Baikal is extremely continental and its mid-latitude position makes it sensitive to insolation changes: winters are long, cold and dry, and summers (although short) are relatively warm and wet. Studies of the phytoplankton in Lake Baikal were pioneered by Skabitchevsky (1929), and have since been followed up by many other studies (for more details, see Jewson and Granin, 2000). The diatom phytoplankton of Lake Baikal is largely affected by physical parameters of the lake, e.g. ice-cover, seasonal overturn and other mixing properties, and recent studies have used diatoms to reconstruct Holocene climate variability in this region (e.g. Bradbury *et al.*, 1994; Edlund *et al.*, 1995; Mackay *et al.*, 1998; Bangs *et al.*, 2000).

Seasonal ice formation is an important feature of Lake Baikal, lasting for about 4–6 months of the year, although ice persists longer in the north basin than in the south. Ice formation and eventual duration is complex (Verbolov *et al.*, 1965) and recent studies have shown ice-cover to be linked to large-scale atmospheric circulation patterns, including the Scandinavian and Arctic Oscillation patterns, the position of the Siberian High, and sea surface temperature anomalies in the North Atlantic Oscillation (NAO) during the autumn winter period (Livingstone, 1999;

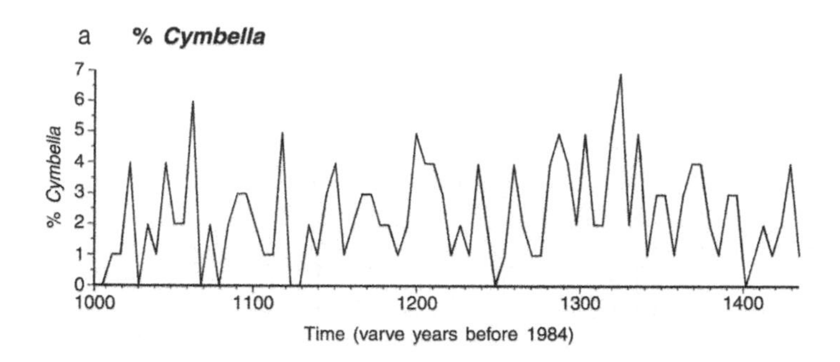

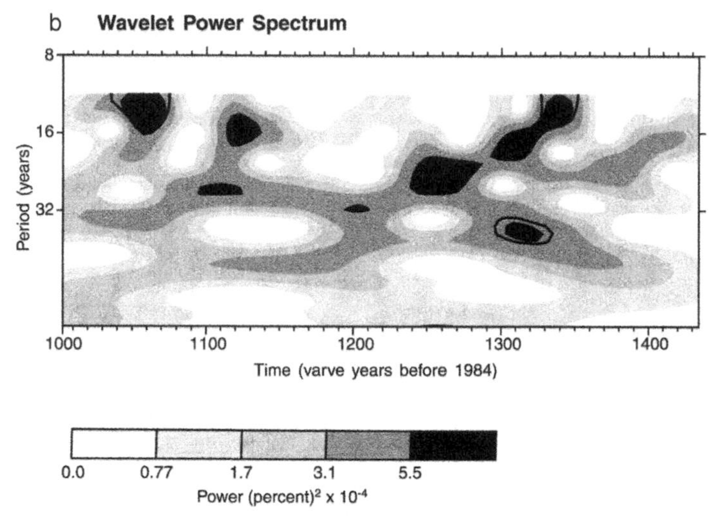

Figure 20.2 (a) Per cent *Cymbella* species (detrended and standardized) in the varved Elk Lake core between 1535 and 1000 years ago. (b) Wavelet power spectrum of *Cymbella* per cent. Shaded contour levels were chosen so that 75, 50, 25 and 5 per cent of the wavelet power is above each level, respectively. Black contour lines represent the 10 per cent significance level relative to the global wavelet background spectrum. Redrawn with kind permission from Bradbury *et al.* (2002) and Kluwer Academic Publishers.

Todd and Mackay, 2003). The water column under clear ice supports a dynamic and extensive diatom assemblage, including the heavily silicified endemic species *Aulacoseira baicalensis*; thermal heating and convective mixing keeps these large cells in suspension (Jewson and Granin, 2000). Kelly (1997) estimates that between 4 and 11 per cent of solar radiation is able to penetrate through clear ice, providing enough energy for algal growth in spring. Light penetration, however, is also dependent on snow cover on top of the ice: as little as 5 cm of snow cover on the ice can reduce solar transmission by a factor of 50 (Kelly, 1997), and when snow cover on Lake Baikal exceeds 10 cm, population growth of *A. baicalensis* is inhibited (Jewson and Granin, 2000). Over the last 150 years, populations of *A. baicalensis* have dominated the sediment record, but before then concomitant with the period commonly known as the Little Ice Age (LIA) the autumnal blooming endemic *Cyclotella minuta* is the dominant species (Mackay *et al.*, 1998), suggesting that ice and/or snow conditions on the lake were too extreme at this time for spring diatom populations to flourish. Work is ongoing to further establish relationships between both endemic and cosmopolitan diatom species in the lake through culturing studies (Jewson *et al.*, unpublished data) and integrated monitoring of diatom populations and hydrophysical properties of the water column (Ryves *et al.*, 2003). A recent study by Mackay *et al.* (2003) explored the relationships between diatoms and environmental variables using multivariate techniques, including the direct gradient technique of canonical correspondence analysis (CCA), together with forward selection, a form of step-wise regression. In order of importance it was found that snow cover on the lake, water depth, suspended matter, annual solar radiation and mean July temperature of the surface water were significant in independently explaining diatom distribution across the lake, opening up the possibility of reconstructing Holocene climate variability using diatom analysis of the bottom sediments.

20.2 ICE-COVER RECONSTRUCTION USING DIATOM ANALYSIS

The IPCC (2001) also reported that the effects of global warming are likely to be more keenly experienced in high latitude regions, such as the Arctic. However, these regions are remote and the availability of long-term monitoring records is poor. In these situations, palaeoecological records with robust chronologies can provide important information on when the most recent trend in global warming first started to have significant ecological impacts. In other regions, pollen and macrofossil studies are important tools used to reconstruct past climates (see Birks and Birks, pp. 342–357 in this volume) but in the Arctic trees are sparse and so other techniques need to be employed. Consequently, the use of proxy data from lake ecosystems, such as diatoms, has allowed researchers to extend climate records where monitoring data are otherwise absent (e.g. Overpeck *et al.*, 1997).

In polar regions, ice-cover is an important variable in determining functioning of lake ecosystems. Ice-cover influences light penetration into surface waters, which has a direct influence on diatom photosynthesis, and consequently population growth. Ice-cover also plays a major role in the stratification of the water column, which in turn has an influence on lake mixing and nutrient cycling processes (e.g. as determined by Bradbury, 1988, summarized above). Together, these influences affect the extent and types of habitat available to diatom species, which Smol (1988) suggests can be used to reconstruct palaeoclimates in polar environments. For example, in extreme polar environments, lakes (such as Lake Vanda in Antarctica) are permanently covered by ice and biological productivity is very low. However, ice-cover in many other polar lakes at least partially thaws around the margins every summer, resulting in an increased area of littoral habitat. In warmer years this area increases, sometimes even to the extent that the whole of the lake is ice-free (Smol, 1988). It follows that as a greater area of littoral habitat is opened up, then diatom responses change accordingly (see Table 20.1). If the whole of the lake does become ice-free during summer months (and providing that the lake is deep enough) then planktonic taxa will become more abundant in the lake. Several studies have investigated the interactions between diatom assemblages and ice and snow cover in lakes in extreme environments, including the Arctic (e.g. Smol, 1988; Douglas *et al.*, 1994; Sorvari *et al.*, 2002), high-altitude regions such as the Alps (e.g. Lotter and Bigler, 2000; Ohlendorf *et al.*, 2000) and in very continental regions such as Lake Baikal in central Asia (e.g. Mackay *et al.*, 2003; Ryves *et al.*, 2003). Here we review the case study of Elison Lake in the arctic region of North America.

Prevailing climate	Lake conditions	Diatom responses
Cold	Extended ice and snow cover Very small moat	Aerophillic diatoms Shallow water taxa Low production
Moderate	Moderate sized moat	Deeper water taxa Moderate production
Warm	Pelagic region of the lake is ice-free	Progressively more deep water and planktonic taxa High production

Table 20.1 Climatic changes and the corresponding responses in the diatom community. Adapted from Smol (1988)

Elison Lake is located in Cape Herschel on Ellesmere Island, the northernmost island in the Canadian Arctic archipelago (Douglas *et al.*, 1994). It is one of a number of small freshwater ponds and lakes abundant in the region, which for most of the year are covered by ice, except for a short period of about 1–2 months between June and August. Douglas *et al.* (1994) provide evidence for marked increased ice-free periods on Elison Lake since the beginning of the 19th century, which were not evident in recent millennia, and which they attribute to recent increases in global warming.

The sediment records were dated using a combination of ^{210}Pb analysis for the uppermost samples, and ^{14}C for the basal samples, indicating that the base of their core was just over 4000 calibrated years BP. Diatom analysis revealed (Fig. 20.3) that for the majority of this time,

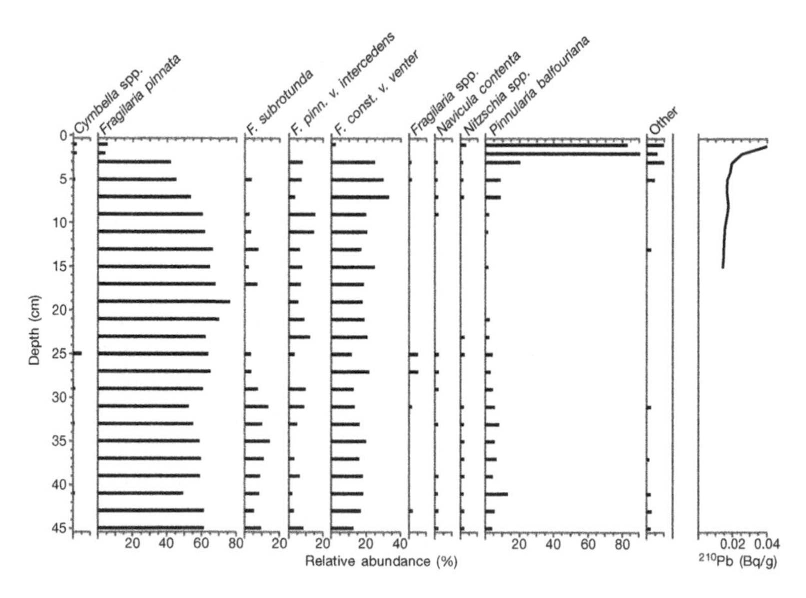

Figure 20.3 Relative percentages of major diatom taxa in a core taken from Elison Lake, Ellesmere Island. ^{210}Pb activity is shown on the right. Redrawn from Douglas *et al.* (1994) with kind permission from the American Association of the Advancement of Science.

assemblages were dominated by benthic *Fragilaria* taxa (especially *F. pinnata* and *F. construens* v. *venter*) which are characteristic of species found growing in the littoral zones of lakes. Lotter and Bigler (2000) suggest that *Fragilaria* species are r-strategists, and are thus able to colonize habitats that undergo frequent environmental changes, such as the littoral zones of lakes that have seasonal ice-cover. However, since the early 19th century up to the present day, the diatom assemblage at Elison Lake changed dramatically, with *Pinnularia balfouriana* becoming the dominant species. In polar regions, *P. balfouriana* is found growing on arctic mosses, and the Douglas *et al.* (1994) study indicates, therefore, that the lake has opened up sufficiently to allow aquatic mosses to colonize. Douglas *et al.* (1994) also convincingly argue that other causal agents, such as anthropogenic pollution or increasing ultraviolet radiation could not have been the determining factors. These findings are supported by a more recent study of contemporary processes on the effects of ice-cover on diatom habitats in the high altitude Alpine lake, Hagelseewli (Lotter and Bigler, 2000), confirming that ice-cover acts to inhibit plankton development, while prolonged periods of ice-cover tended to favour the growth of *Fragilaria* species.

20.3 QUANTITATIVE CLIMATE RECONSTRUCTIONS IN OPEN-BASIN SYSTEMS

20.3.1 Diatom-Inferred Surface Water Temperatures

An increasing number of studies have employed numerical techniques to reconstruct climate directly (e.g. temperature) and indirectly (e.g. pH, DOC and salinity) from both open- and closed-lake systems. Perhaps the more contentious palaeoenvironmental reconstructions using diatoms in recent years have been those for summer surface-water temperatures (Pienitz *et al.*, 1995; Vyvermann and Sabbe, 1995; Joynt and Wolfe, 2001; Wolfe, 2003) and mean July air temperatures (Korhola *et al.*, 2000; Rosén *et al.*, 2000; Bigler and Hall, 2002) from diatoms in lake sediments (see Anderson, 2000 for a critical review).

There is some controversy as to whether diatoms can be effectively used as proxies to reconstruct past temperatures. On the one hand, (i) water temperature plays a crucial role in regulating algal photosynthetic processes and metabolic activity (e.g. see appropriate references in Pientiz *et al.*, 1995); (ii) optima and tolerances have been determined for many species using culturing studies (e.g. Dauta *et al.*, 1990; Richardson *et al.*, 2000; Jewson *et al.*, unpublished data); and (iii) many taxa exhibit distinct relationships with increasing latitude and altitude (e.g. Foged, 1964; Vyvermann and Sabbe, 1995). On the other hand, however, (i) it is argued that over extended time periods such as the Holocene, indirect climatic effects on lake catchment processes (which in turn affect e.g. DOC, pH and nutrients in lakes), are likely to have a much greater influence on diatom species composition than changes in temperature alone (see Anderson 2000) as the amplitude of temperature change during the Holocene is often close to the prediction errors for diatom-based temperature models (Battarbee, 2000); and (ii) throughout the Holocene, changes in lake depth due to e.g. hydroseral succession, result in changing habitat availability to diatoms and alter the thermal regime of the water column.

Recent studies have tried to take account of some of these concerns, especially with regard to initial project design. One such development includes the use of multi-proxy studies, so that models are constructed using other proxies as well as diatoms from the same stratigraphical samples. For example, Lotter (pp. 373–383 in this volume), highlights the study by Rosén *et al.* (2001), who demonstrate that modelled temperature reconstructions from diatoms, pollen, chironomids and near-infrared spectroscopy of organic sediments can be used to confirm mean July air temperature oscillations over the last 7300 cal yr BP in northern Sweden. Another important development has been the use of independent cross validation techniques for diatom-inferred temperature models (Bigler and Hall, 2002). This study, together with that by Rosén *et al.* (2000), have modelled temperature optima for species in both datasets, and comparison between studies demonstrate an encouraging level of agreement for many taxa, although some important differences are apparent. An alternative approach has been taken by a joint consortium of European and Russian scientists working on Lake Baikal: interestingly, many of the initial objections raised by Anderson (2000) in terms of using diatoms to reconstruct climate parameters are not valid for this lake. For instance, the lake is so deep, that infilling is not an issue, and even though there is concern about pollution, the recent sediment record of diatom assemblages does not indicate any significant acidification or eutrophication trends in the lake (Mackay *et al.*, 1998). Anderson further highlighted that many of the diatom taxa currently used in training sets to reconstruct temperatures are very cosmopolitan, and so are unlikely to have distinct optima and narrow tolerances for surface-water temperature. However, in Lake Baikal, the majority of the dominant taxa are endemic, and preliminary investigations suggest that many of them have very clearly defined responses to temperatures (Richardson *et al.*, 2000; Jewson *et al.*, unpublished data).

Much research has still to be done, therefore, to fully understand the complexity of diatom–temperature interactions, and although the criticisms by Anderson (2000) need to be taken into account, recent studies continue to demonstrate the potential of diatom-inferred temperature models.

20.3.2 Diatom-Inferred pH and its Relationship to Prevailing Air Temperatures

Because of their remote, high-altitude location, alpine lakes are sensitive to changes in climate that influence ice and snow cover on the lake and in the catchment: for example, as climate ameliorates, both weathering of the catchment and the growing period in the lake, increase. The impact of anthropogenic contamination on alpine lakes is generally lower than that for lowland lakes, although they are still prone to acidification from industrial pollutants (Jones *et al.*, 1993). Given the above uncertainties of reconstructing past temperatures directly using diatom-based transfer functions, indirect climate relationships via changing diatom-inferred pH (e.g. Psenner and Schmidt, 1992; Sommaruga-Wögrath *et al.*, 1997; Koinig *et al.*, 1998a, 1998b) and DOC (Pienitz *et al.*, 1999) in freshwater lakes have proved fruitful in recent years. This approach has been adopted in the Austrian Alps, and a strong relationship between diatom-inferred pH values with changing Austrian mean air temperatures for the period 1778–1991 demonstrated (Koinig *et al.*, 1998a).

Using a previously developed training set for diatoms and pH, initial pH reconstructions at Schwarzsee ob Sölden and Brechsee (two remote, soft-water lakes in the study region) reveal interesting relationships between diatom-inferred pH and air temperatures and glacier mass trends respectively (used as a climate proxy prior to 1780) (Fig. 20.4). At Brechsee, the record extends the pre-industrial period, and results show that during periods of warmer climate, pH values increased, and vice versa (i.e. during periods of colder climate, lake water became more acidic). Even during industrialization, and in spite of increasing deposition of acidifying

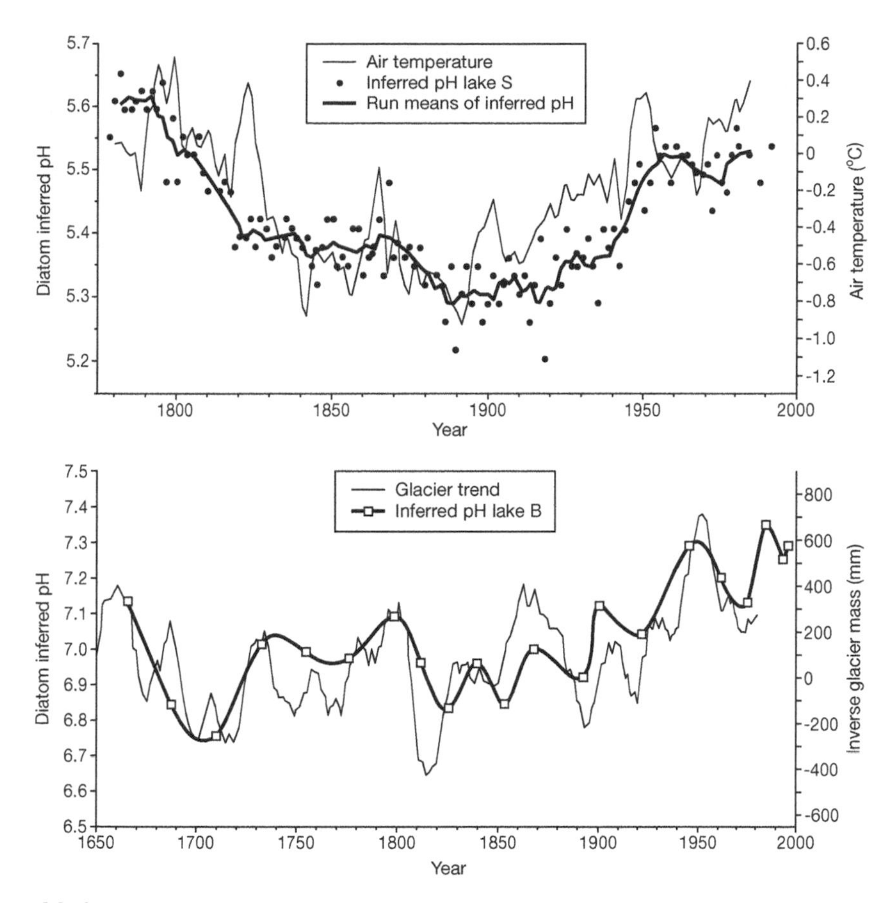

Figure 20.4 The upper graph shows the comparison between reconstructed pH and mean air temperature for Schwarzsee ob Sölden (lake S). The lower graph shows the comparison between reconstructed pH for Brechsee (lake B) with glacier trends. (Note different time-scales). Redrawn from Koinig *et al.* (1998a) with kind permission from AA Balkema Publishers.

compounds at these two sites, diatom-inferred pH continued to follow closely mean air temperatures (and glacier trends). This is in contrast to other sites in the region where pH reconstructions often showed a decoupling of inferred pH and temperature as acid deposition increased, i.e. as levels of acid rain increased during the period of recent warming, diatom-inferred pH also declined (Koinig *et al.*, 1998a).

There are several mechanisms that can account for increasing values of diatom-inferred pH during warmer climates, linked to prevailing snow and ice conditions in the catchment and on the lake. Periods of warmer weather lead to decreased amounts of snow and ice-cover, which in turn allows for an increase in catchment weathering rates and increased primary productivity in the lake itself, due to the concomitant lengthening of the growing-season and light penetration into the lake (Psenner and Schmidt, 1992; Sommaruga-Wögrath *et al.*, 1997). Sommaruga-Wögrath *et al.* (1997) analysed monitoring data for 57 low-alkalinity high mountain lakes (all between 2000 and 2900 m a.s.l.). Each lake was sampled annually (during the autumn overturn between 1985 and 1995 for a range of chemical variables, including pH, conductivity, alkalinity, silica, dissolved inorganic nitrogen (DIN), major anions and cations, amongst others. Rather surprisingly, despite falling levels of sulphur deposition and small increases in nitrogen deposition, linked to prevailing deposition trends, concentrations of base cations and SO_4^{2-} increased in most lakes, whilst for DIN they declined. These changes were also accompanied by small increases in overall lake water pH and silica concentrations. Sommaruga-Wögrath *et al.* (1997) linked these changes to a marked increase in mean annual Austrian air temperatures measured over the same period of *c.* 1 °C, confirming distinct temperature- and pH-related processes within the lakes.

While these alpine studies are confined to records going back no more than 200 years, Wolfe (2002) investigated the relationship between pH and climate in two ultra-oligotrophic Arctic lakes (Kekerturnak Lake and Fog Lake) on Baffin Island over the last 5000 years. His study shows a distinct fall in diatom-inferred pH, coincident with regional neoglacial cooling over the last *c.* 4000 years. Whereas Sommaruga-Wögrath *et al.* (1997) attribute a decline in alpine lake pH during colder years to reduced catchment weathering, aeolian dust transport and reduced primary production in the lake, Wolfe (2002) attributes the fall in pH in the Arctic lakes mainly to a decline in photosynthesis and primary productivity due to enhanced snow and ice-cover. Wolfe (2002) does not consider weathering of the catchment and aeolian transport of minerals as important pH regulators in these lakes, as the bedrock consists of hard Archaen granite and gneiss, and the lakes are ice-bound for about 10 months of the year, which precludes increased input of base cations. Furthermore, increased snow and ice-cover also tend to inhibit photosynthesis and primary production, which in turns reduces the uptake of CO_2 by biological processes, leaving excess CO_2 in the water. Speciation of dissolved inorganic carbon (DIC) in these lakes is very important, as the lakes are covered by ice for such a long period of the year, resulting in supersaturation of lake water CO_2 as water–air exchange of CO_2 is inhibited and photosynthesis reduced (Wolfe, 2002). During warmer periods it is hypothesized that ice-cover is reduced, which in turn leads to reduced CO_2 in the water and a concomitant increase in pH.

20.4 QUANTITATIVE CLIMATE RECONSTRUCTIONS IN CLOSED-BASIN SYSTEMS

10.4.1 Diatom-Inferred Salinity Relationships

Closed-basin systems are extensive throughout arid and semi-arid regions of the world, where the balance between precipitation and evaporation (P–E) and the influence of groundwater principally control lake water levels. These lakes are therefore found in regions of the world where pressures on water resources are high, both from natural (e.g. drought) and anthropogenic (e.g. abstraction) processes. As in open-basin systems, diatoms are an important component of closed-basin lakes (thereafter referred to as saline lakes), and although the comprehensive nature of the relationship between diatoms and changing ionic composition of lake water is still poorly understood (Fritz *et al.*, 1999), diatoms have been extensively used in recent years to reconstruct past climates indirectly.

The relationships between diatoms and lake water salinity have been explored in detail (e.g. Bradbury *et al.*, 1981; Gasse *et al.*, 1987), culminating in an extensive range of projects since the 1990s, including the seminal study by Fritz *et al.* (1991) on reconstructing Holocene climate in the Northern Great Plains (NGP), North America. There have been many subsequent studies establishing the relationship between diatoms, salinity and brine composition and conductivity in closed-lake basin systems to Holocene climates from around the world, e.g. North America (Cumming and Smol, 1993; Wilson *et al.*, 1994; Laird *et al.*, 1996, 1998a, 1998b; Wilson *et al.*, 1997; Last *et al.*, 1998), Central America (Metcalfe, 1995; Davies *et al.*, 2002), Africa (Gasse *et al.*, 1997), Australia (Gell, 1997), Europe (Reed, 1998; Reed *et al.*, 1999), Antarctica (Roberts and McMinn, 1998) and Greenland (Ryves *et al.*, 2002).

Here we focus on one of the more comprehensive studies carried out at Moon Lake, situated within the NGP region (Laird *et al.*, 1996, 1998a, 1998b), which provides an example of how diatom-inferred salinity reconstructions can be used to reconstruct past climates at high resolution, and of the potential problems associated with these types of analyses. The NGP region lies in continental North America, where winter temperatures are very low, and summer temperatures hot: mean annual temperature at Moon Lake is *c.* 6 °C (range –29 °C to +38 °C) (Laird *et al.*, 1996). The negative effective moisture gradient is responsible for prevailing prairie and steppe vegetation, and for the hundreds of saline lakes found in the NGP region. Overall, the region is especially important for agriculture but it is also prone to severe drought, for example the dust-bowl event during the 1930s and 1940s.

The training set used to develop the diatom-inferred salinity model was adapted from Fritz *et al.* (1991), but its suitability was first validated by comparing inferred values to historical records of mean annual P–E, determined from nearby climate stations recorded since the late 19th century. An initial diatom-inferred salinity reconstruction for the Early Holocene at Moon Lake (see Fig. 16.4 in Fritz, pp. 227–241 in this volume) suggested that during the Early Holocene, Moon Lake was actually a freshwater lake. However, concomitant with the development of a prairie landscape from a forested one, lake levels fell, turning it into a closed basin: these features taken together point to a strong shift from cool moist conditions to a warmer, drier climate. In a subsequent study, climate variability at Moon Lake was reconstructed at a higher resolution over the last two millennia to establish drought patterns in the context of a longer-term perspective (Laird *et al.*, 1998a). This study is important both from a methodological point of view (model validation), and in its societal contribution in terms of adding to the debate on water resources in this vulnerable, but agriculturally important region. Initially, diatom-inferred salinity values were compared to instrumental records of salinity over the last 100 years (Bhalme and Mooley drought index (BMDI) – see Laird *et al.*, 1998a for full details). Overall, a good correspondence was found between modelled and historical values (Fig 20.5), resulting in increased confidence that modelled values for the remaining Holocene period are a good reflection of past climates (Laird *et al.*, 1996). The importance of validating such climate models cannot be overestimated (e.g. see Fritz *et al.*, 1994 for another case study). Diatom-inferred salinity changes over the last 2300 years (Fig. 20.6) suggest that droughts were more prevalent prior to 1200 yr BP, and that these droughts were more extreme than those witnessed during the dust bowl years of the 1930s and currently have no modern analogues (Laird *et al.*, 1998a). Their analysis suggests that forcing mechanisms causing climate change during the last 2000 years in the NGP region have shifted more frequently than in the Early Holocene, and may be linked to changing patterns in El Niño Southern Oscillation (ENSO) development, culminating in a general shift to wetter conditions, coincident with the Little Ice Age (LIA) (Laird *et al.*, 1998a).

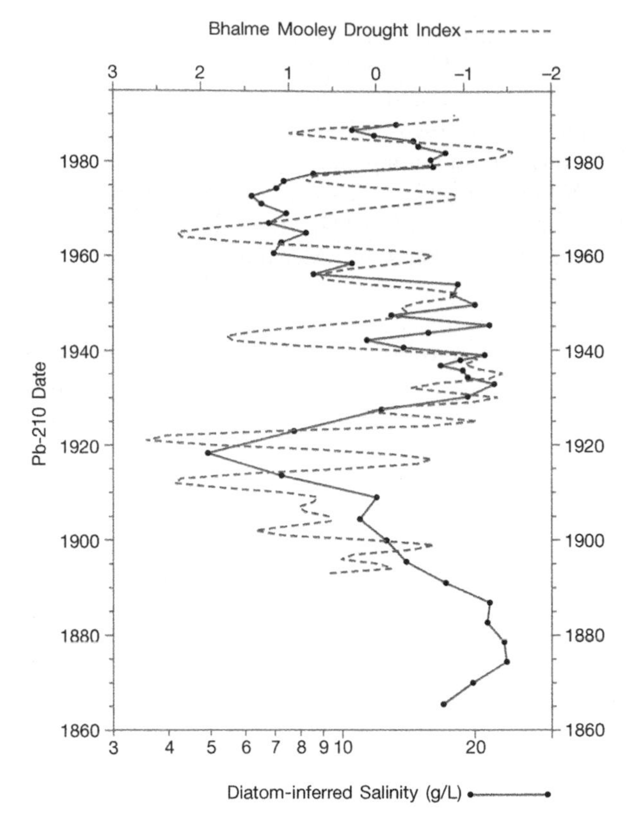

Figure 20.5 Comparison between diatom-inferred salinity at Moon Lake (solid line), with the Bhalme Mooley Drought Index (BMDI) (dashed line), developed for the NGP region from monthly summer precipitation records from nearby climate stations. Redrawn from Laird *et al.* (1998a) with kind permission from Kluwer Academic Publishers.

Diatom-inferred climate change studies from saline lakes are affected by both diatom dissolution (as outlined in the Introduction), and problems of 'no analogue' (see Birks, pp. 107–123 in this volume). It is appropriate therefore to outline some of the problems encountered in the development of the salinity model for the NGP region. In the Great Plains, dissolution has resulted in poor diatom preservation in approximately 20 per cent of the samples used in the training set, and there is a direct relationship between the number of samples with poor preservation and increasing salinities (Fritz *et al.*, 1993). Dissolution therefore results in selective preservation of the more robust valves, such as *Cyclotella quillensis*. This in turn results in samples with low species diversity, which leads to bias in reconstructed salinity values. A second major weakness found during development of diatom-inferred salinities in North America is the problem of 'no analogue' situations between the training set used, and the core selected for reconstruction. Both Fritz *et al.* (1993) and Laird *et al.* (1996) acknowledge that the original training set suffers from a lack of sites at the lower end of the salinity gradient, i.e. freshwater lakes. At Moon Lake, samples with no analogues are further exacerbated by the presence of high proportions of *C. choctawatcheeana* in the core, which are not present in the training set.

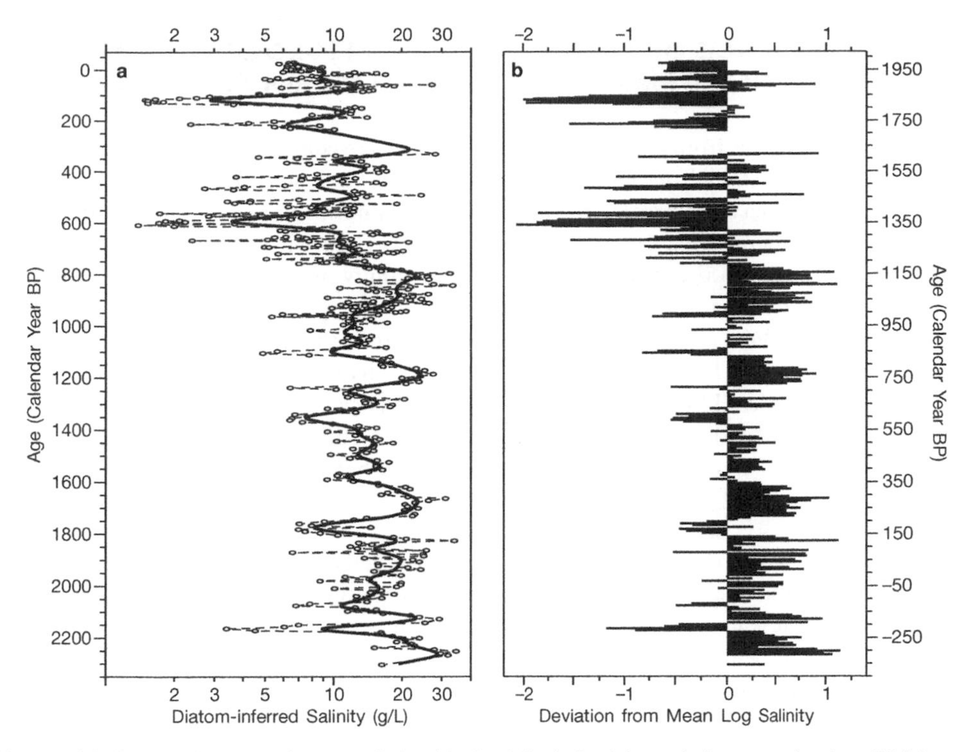

Figure 20.6 (a) Diatom inferred salinity (dashed line) for Moon Lake over the last 2300 years plotted alongside a Fast Fourier transformation (see Laird *et al.*, 1998a for full details). (b) Deviations from the mean log salinity. Redrawn from Laird *et al.* (1998a) with kind permission from Kluwer Academic Publishers.

To overcome these difficulties and to improve estimates of diatom-inferred salinities, Juggins *et al.* (1994) recommend the development of regional training sets, which serves to increase the number of sites used at the lower concentrations of salinity for example, and to increase the number of species present for model development. The CASPIA project was set up specifically to harmonize diatom collections from saline lakes from the NGP, East and North African and the southeastern Australian datasets (Juggins *et al.*, 1994). More recently, the EDDI project was designed to harmonize diatom-inferred pH, nutrient and salinity training sets, the latter collected from Africa, Europe and western Asia (Battarbee *et al.*, 2000).

In summary, saline lakes in arid and semi-arid regions of the world provide unparalleled opportunities for palaeoclimate reconstructions. However, these studies also need to take into account problems of taphonomy, such as dissolution, and no analogue situations, where the range of sites within any one training set is too small. Major steps are being taken to compensate for these problems, with an increasing amount of modern process-based studies (e.g. Lent and Lyons, 2001; Ryves *et al.*, 2001) and the merging of training sets, where appropriate, from around the world.

20.5 EVIDENCE OF DIATOM-INFERRED HOLOCENE CLIMATE VARIABILITY FROM COASTAL AND MARINE REGIONS

In comparison to studies from lacustrine ecosystems, there are comparatively few studies that use diatoms to reconstruct high-resolution Holocene climate variability from marine sequences, although there has been an increase in the number of such studies in recent years, e.g. off the coast of Antarctica (Rathburn *et al.*, 1997; Cunningham *et al.*, 1999; Taylor *et al.*, 2001; Taylor and McMinn 2001, 2002); in the shallow waters in the proximity of the Baltic Sea (e.g. Jiang *et al.*, 1998; Westman and Sohlenius 1999), off the coast of Iceland (e.g. Jiang *et al.*, 2001, 2002) and in the Norwegian Sea (Birks and Koç, 2002). Many of these studies tend to use information provided by diatoms in the form of habitat availability. For example, diatoms are usually the most abundant organisms under sea-ice (Horner, 1985), and their abundance can be used to indicate changing patterns in sea-ice in relation to volume of open water, and ultimately to climate (Gersonde and Zielinski, 2000), e.g. *Nitzschia curta* forms a large biomass when cells are released from melting ice (Wilson *et al.*, 1986).

Other studies have taken a multivariate approach, in relating diatoms in marine surface sediments to environmental variables, including sea-surface temperatures (SST). Jiang *et al.* (2001) used CCA to investigate the relationships between diatom species and environmental variables (including SST and winter SST) off the north Icelandic shelf. They recognized five diatom assemblages characteristic of different oceanic habitats: sea-ice, cold water, warm water, mixed water and coastal water. A diatom-inferred SST model was subsequently developed and applied to a marine core, located 50 km offshore on the north Icelandic shelf (Jiang *et al.*, 2002). Their diatom model suggests a distinct period of cooling from around 2200 cal yrs BP up to the present day, coincident with neoglacial cooling in other regions. However, as with temperature reconstructions from freshwater lakes, the magnitude of reconstructed SSTs are relatively small, and within the range captured by error analysis.

Perhaps a more conclusive study of marine diatom evidence for Holocene climate variability comes from a multi-proxy study by Taylor *et al.* (2001) from the Antarctic Peninsula. They make use of known ecological characteristics of key diatoms, together with multiple regression analysis between sedimentary environmental variables and diatom data. Their results indicate three distinct climatic periods during the Holocene in this region: an early deglaciation phase between 10,600 and *c.* 7800 yr BP, a mid Holocene climatic optimum between 7800 and 4000 yr BP, ending in a neoglacial period from c. 4000 yr BP up to the present day. The diatom assemblages during the neoglacial are different from those present during the Early Holocene, and so may represent a response to an increase in ice extent, and or an advance of the Müller ice shelf (Taylor *et al.*, 2001). Further during this period, as summer insolation in this region was at a maximum at *c.* 2000 yr BP, the neoglacial cooling prior to then is likely to be due to other factors such as changes in deep water circulation. Diatom-inferred climate studies have therefore opened up some interesting questions on climate forcing mechanisms in this region of the world and the concomitant biological and physical responses.

The last case study presented in this chapter combines traditional palaeoceanographic approaches to reconstructing SST (i.e. Imbrie and Kipp, 1971; factor analysis methodology) with the now commonly used palaeolimnological technique of weighted averaging partial least squares (WA-PLS) (see Birks, pp. 107–123 in this volume). Birks and Koç (2002) used high-resolution diatom analysis to reconstruct Late Glacial to Holocene, February and August SSTs from the Vøring Plateau in the eastern Norwegian Sea. Both models were robust, i.e. had high r^2 values, low error and a maximum bias of *c.* 1 °C. Throughout the Holocene the models show similar trends, although differences between the two were apparent. Both models, for example, record maximum SSTs between *c.* 9700 and 6700 cal yr BP. However, February SSTs reconstructed using WA-PLS fluctuated between 9.5 °C and 11 °C from *c.* 8750 to 7250 cal yr BP before falling to between 8.5 °C and 9.5 °C from 7250 to 6700 cal yr BP. Whereas factor analysis recorded higher temperatures fluctuating around 11 °C throughout the whole of this period. With regard to the varimax factors associated with the time period, the dominant diatom assemblages suggest a strong influence from North Atlantic Water. Interestingly, any response of the diatoms to the 8.2 ka event remains inconclusive: whilst there is a drop in concentrations at this time, there is no concomitant reconstructed cooling of SSTs. The role of sea-ice in influencing diatoms is also highlighted, and the authors conclude that diatom-free levels, e.g. during the period coincident with the Younger Dryas, can be attributed to extensive ice-cover prohibiting photosynthesis thereby reducing production, although the authors, rightly, cannot and do not preclude the influence of dissolution processes affecting the sedimentary diatom record.

20.6 CONCLUSIONS

In recent years, the number of diatom-based Holocene climate reconstructions has increased dramatically, although most studies are still confined to terrestrial ecosystems, rather than marine or coastal environments. As with other chapters in this book, it has not been possible to cover the full range of studies that have used diatom analysis; rather the intention has been to outline selected highlights in diatom research (including both qualitative and quantitative approaches), and recent developments that take account of the importance of site selection, sampling resolution, application of numerical methods and taphonomical processes, such as dissolution.

Implicit in each of the case studies is knowledge of the autecology of individual diatom species. Both direct and indirect climate reconstructions have furthered our understanding between diatoms and their environment, although the results are often contentious, and while quantitative studies are increasingly being improved, interpretations from more qualitative work should not be ignored. Many challenges to the discipline remain, which include increasing our understanding of the ecological requirements of diatom species (from *in vivo* and culturing studies) and diatom life-cycle strategies, while databases should be used more widely to explore the relationships between diatom biogeography with climate, habitat and environmental chemistry, thereby generating new hypotheses for future study.

Acknowledgements

We thank Patrick Rioual for constructive comments on an earlier version of the manuscript.

CHAPTER

NON-MARINE OSTRACOD RECORDS OF HOLOCENE ENVIRONMENTAL CHANGE

Jonathan A. Holmes and Daniel R. Engstrom

Abstract: Ostracods are small bivalved crustaceans. They secrete shells made of low-Mg calcite, which are commonly preserved in Holocene sediments. This chapter examines the contribution of ostracods to Holocene environmental reconstruction, with particular emphasis on non-marine and marginal-marine environments. Three specific areas in which ostracods have contributed to Holocene research are examined, namely palaeosalinity and palaeo-solute reconstruction, hydrological habitat type and lake-level change, and finally palaeotemperature reconstruction. Salinity and solute composition are important aquatic variables. In hydrologically-closed lakes, salinity and solute composition change as a result of variations in effective precipitation. In coastal regions, salinity and solute composition of waterbodies may change with varying inputs of marine and non-marine water, such as may occur as a result of sea-level change. Ostracods respond to such variations in water composition by changes in the abundance and presence of different species, through ecophenotypic changes and by changes in shell chemistry. Many ostracod species are closely associated with a specific habitat type and this association can be used in environmental reconstruction. For example in lakes, individual species are often found in specific water depths, although this is often a response to the presence or absence of a particular substrate type or aquatic plants, rather than a depth control *per se*. The ostracod–depth relationship can be used to reconstruct lake-level changes, which may be a response to changing climate or hydrology. Ostracods may also be used to track variations in temperature. Individual species may have specific water temperature preferences; moreover, the uptake of the trace element magnesium into the ostracod shell is positively correlated with water temperature. The oxygen isotope composition of benthic ostracods in deep, hardwater lakes in humid-temperate regions may also provide information about past air temperature.

Keywords: Lake, Ostracods, Palaeolimnology, Stable-isotopes, Trace elements

Ostracods are small, bivalved aquatic crustaceans that produce valves, or shells, composed of low-magnesium (Mg) calcite. They are common in most types of water from small, temporary ponds to the deep oceans. Although ostracods may be found in slightly acidic waters, they are more abundant at neutral to alkaline pH; moreover, their shells are only normally preserved in sediments with non-acidic porewaters.

Ostracods are now used quite commonly in Quaternary research and have provided important information about Holocene environments. Their contribution to Holocene environmental reconstruction has been through both classical palaeoecological approaches and through the use of their calcitic shells as sources of material for geochemical analysis. The aim of this chapter is to review the contribution of ostracods to Holocene environmental reconstruction. We examine both palaeoecological and geochemical approaches. We first provide a brief outline of ostracods as organisms and direct readers to further literature aimed at the understanding of ostracod biology, ecology, taxonomy and identification. We then outline the fundamental aspects of both palaeoecological and geochemical analysis and then finally illustrate important points using examples. We focus our discussion on lacustrine and marginal-marine environments.

21.1 OSTRACODS AS ORGANISMS

In this section we provide an outline of ostracod biology, ecology and taxonomy. Whilst it is impossible to be comprehensive, we argue that some understanding of the living organism is vital if ostracods are to be used successfully in palaeoenvironmental reconstruction. A detailed reference list will allow individual topics to be explored in more detail.

21.1.1 Biology

Ostracods are crustaceans and hence arthropods. They have a bivalved carapace that is hinged dorsally and completely encloses the body parts and limbs: the latter, which comprise five to eight pairs depending on group, protrude through the ventral gape in the carapace. The two components of the carapace are commonly termed 'valves' although the less specific term 'shell' is sometimes used to describe all of these components. The carapace, which is secreted by the epidermis, is composed of low-Mg calcite and is commonly preserved in lacustrine and marine sediments. In addition to Ca, the shell material contains trace elements such as Mg, Sr, Ba and others. As we will show, the composition of ostracod shells can provide valuable palaeoenvironmental information. Adult ostracods are typically somewhere between about 0.5 and 2.0 mm long (Horne *et al.*, 2002) although larger and smaller forms are known. Because their study requires a microscope, ostracod shells are classed as microfossils. The structure and biology of ostracods are reviewed in Horne *et al.* (2002).

Like other arthropods, ostracods grow by moulting, typically eight times following hatching from eggs. No further moulting occurs once the adult stage is reached, and the valves appear to be formed solely from dissolved ions taken from the ambient water at the time of shell secretion. That said, the details of the secretion process are not fully understood. During each moult, the carapace is discarded and a new one formed, typically within a few hours; full calcification may take up to several days. Each moult stage, or instar, is progressively larger and better calcified, and closer in features to the adult form. The timing of moulting and shell formation can have a large influence on shell chemistry, especially in temperate latitudes where water temperature (and sometimes chemistry) has large seasonal amplitude (Xia *et al.*, 1997a).

Ostracods reproduce either sexually or by parthenogenesis, in which case only female specimens are present. Sexual reproduction is the norm in marine species, whereas parthenogenesis is quite common in some non-marine taxa. Some non-marine parthenogens show geographical parthenogenesis; that is they reproduce sexually in some regions and parthenogenetically in others. Horne *et al.* (1998) and Griffiths and Horne (1998) review reproductive modes and strategies in non-marine ostracods. Sexual dimorphism is precocious in the shells of some, but not all, species and so past reproductive strategy can sometimes be determined from the fossil record.

21.1.2 Ecology and Life History

The distribution and abundance of ostracods is a function of a range of factors, which can be grouped broadly under hydrological habitat type and water characteristics and composition. Smith and Horne's (2002) review of the ecology of ostracods includes a detailed examination of non-marine environments.

At a gross level, the general habitat type may often control ostracod assemblages, as many species show affinity for a particular type of habitat. In particular, different species may occupy springs, small ponds, lakes and streams (e.g. Absolon, 1973) (Fig. 21.1). Moreover, some species can be found in groundwater habitats, although their shells can often be transported into an adjacent lake upon death. Some ostracod species can tolerate temporary (e.g. seasonal) desiccation and their occurrence in the stratigraphic record may point to seasonal drying out of a waterbody.

Within lakes, water depth appears to exert control on species distribution, although this appears to be through the influence of macrophytes zonation, substrate composition, or oxygen levels, rather than water depth *per se*. Although some non-marine ostracods are strong swimmers, there are no truly planktonic non-marine species and even the swimming forms tend to stay within the shelter of aquatic plants. Non-swimming species tend to crawl on their substrate or may be infaunal (live within the sediments).

Knowledge about habitat preferences (e.g. Absolon, 1973), depth preferences (e.g. Mourguiart and Montenegro, 2002), energy-level tolerance and desiccation resistance (e.g. Evans *et al.*, 1993) has been used to good effect in Holocene environmental reconstruction using non-marine ostracods.

Water characteristics, in many ways, are the prime determinant of ostracod species occurrence. Important characteristics include water temperature, salinity, solute composition and dissolved oxygen content. Information about the tolerances and the sensitivity of ostracod species to these variables is of great value in palaeoenvironmental reconstruction.

Over Holocene time-scales, the salinity of waterbodies may change as a result of variations in effective precipitation, especially in hydrologically-closed lakes or, in coastal localities, through varying marine intrusion. In both cases, salinity change is generally accompanied by changes in solute composition. Ostracods appear to be sensitive to salinity, although there are numerous euryhaline taxa for which there seems to be greater sensitivity to solute composition – especially the ratio of alkalinity to Ca (e.g. Curry, 1999) – than salinty *per se*. Salinity and ionic tolerance for individual ostracod species are given in Keyser (1977), Neale (1988) and Holmes (2001), amongst others.

Water temperature is known to control ostracod distribution, a fact that is confirmed by the broad latitudinal distribution of some ostracod species. Laboratory studies have shown that temperature may influence the reproduction, growth rate, size and lifespan of ostracods (e.g. Martens *et al.*, 1985). Whilst of potential value in palaeoenvironmental work, information on temperature tolerances of individual species is lacking. Moreover, many species have quite broad tolerances, even if their optimum tolerances, at which productivity is maximised, are much narrower. Temperature preference data for selected species are given in Holmes (2001).

Palaeoecological Group	Species	Groundwater	Spring	Stream	Pond	Lake	Temporary Water
	Psuedocandona eremita	0		•	•		
a	Cryptocandona kieferi	0			•		
	Cavernocypris subterranea	0	•	•			
	Potamocypris zschokkei		•	•			
b	Fabaeformiscandona brevicomis		•	•			
	Psychrodromus olivaceous		•	•			
	Psychrodromus fontinalis		•				
c	Ilyocypris bradyi		•	•	•		
	Eucypris pigra		•	•	•		
	Scottia pseudobrowniana		•	•	•		
	Cryptocandona vavral			•	•		
	Cyclocypris diebeli			•	•		
	Psuedocandona albicans			•	•		
	Fabaeformiscandona fabelia				•		
d	Nannocandona faba				•		
	Candonopsis kingsteii				•		
	Paracandona euplectella				•		
	Candona rostrata			•	•		
	Dolerocypris fasciata				•		
	Physocypria kraepelini				0		
	Cypridopsis elongata				0		
	Cyclocypris ovum			•	•	•	
	Cyclocypris laevis			•	•	•	
	Psuedocandona marchica			•	•	•	
	Candona candida			•	•	•	
	Cypridopsis vidua				•	•	
	Heterocypris salina		•		•	•	
	Potamocypris unicauata				•	0	
	Fabaeformiscandona lozeki					•	
	Candona protzi					•	
	Psuedocandona compressa				•	•	
	Fabaeformiscandona fabaeformis				•	•	
	Cypria ophtalmica					•	
e	Cypria exsculpta					•	
	Notodromas monacha					•	
	Herpetocypris reptans					•	
	Potamocypris villosa			•	•	•	
	Darwinula stevensoni			•	•	•	
	Limnocythere inopinata				•	•	
	Limnocythere sanctipatricii				•	•	
	Metacypris cordata					•	
	Candona neglecta				•	•	
	Candona weltneri obtusa					•	
	Cytherissa lacustris					•	
	Fabaeformiscandona hyalina					0	
	Fabaeformiscandona levanderi					0	
	Cyclocypris serena					0	
	Heterocypris incongruens						0
	Cypris pubera						0
f	Paralimnocythere relicta						0
	Cyprois marginata						0
	Bradleystrandesia hirsutus						0

Figure 21.1 Hydrological habitat affinities of selected European non-marine ostracod species. Adapted from Absolon (1973) with revised taxonomic nomenclature.

Although some non-marine species are found all year round, many taxa are markedly seasonal in their distribution, hatching and reaching maturity at specific times in the year. This is especially so for desiccation-resistant taxa, whose life-cycle must be completed prior to the drying up of the host waterbody. The total lifespan of an ostracod varies markedly between species, from several weeks to several years (Delorme, 1978). Other important ecological factors include food supply, predation and dissolved oxygen levels

21.1.3 Taxonomy and Identification

All non-marine and most marginal-marine ostracods belong to the order Podocopida, within which there are three superfamilies, namely Darwinuloidea, Cytheroidea and Cypridoidea (Martens *et al.*, 1998). Further discussion of taxonomy is beyond the scope of this chapter: Horne *et al.* (2002) include a detailed discussion of the taxonomy of the Ostracoda as a whole and present a summary of diagnostic characteristics of both limbs and carapaces for each of the higher taxonomic categories.

21.2 OSTRACOD SHELLS AS FOSSILS

Ostracod shells are often well preserved in Quaternary sediments, but only when sediment porewaters are neutral to alkaline. Hence, lakes that contain diverse ostracod faunas may sometimes fail to record fossil assemblages within the sediments as a result of dissolution. Absence from the sedimentary records, therefore does not automatically mean that ostracods were absent from the waterbody. Typically, only shells are preserved: appendages and other body parts tend to decompose quite rapidly after death of the organism (although see Matzke-Karasz *et al.*, 2001, for examples of soft-part preservation).

The ostracod shell material may be preserved as intact carapaces (valves articulated) or as disarticulated valves. Both adults and juvenile material may be found within sediments, although the earliest moult stages are generally not preserved. Shells may be present in varying degrees of preservation from a pristine state to a high degree of breakage and/or chemical alteration (dissolution or mineralized overgrowths). Information relating to all of the above can be of great palaeoenvironmental value.

Because ostracod carapaces are generally disarticulated upon moulting, the presence of large numbers of intact juvenile carapaces in the sediment record suggests large-scale juvenile mortality, probably brought on by the rapid onset of unfavourable conditions. However, the preservation of intact carapaces also requires low energy conditions and rapid burial following death, implying rapid sedimentation (e.g. De Deckker, 2002). The presence of large numbers of adult carapaces implies quiet conditions and rapid burial, although the tendency for valves to remain articulated also varies between species.

The age structure, that is the number of different moult stages of a fossil assemblage, can be used to assess the presence and degree of post-mortem reworking of an ostracod sample (e.g. Whatley, 1983, 1988; De Deckker, 2002). Because the different moult stages of a species have different sizes and weights, and hence differing hydrodynamic properties, the presence of adults and most juvenile moults within a sample suggests that it has been subjected to minimal reworking. In contrast, samples lacking adults or earlier juvenile moults (after allowance for any bias introduced by sample processing) have probably been reworked. Information about assemblage reworking is invaluable in palaeoecology (Whatley, 1988).

The preservation state of ostracod shell material may also yield palaeoenvironmental information, or at least serve to flag badly altered fossil assemblages. The degree of breakage may reflect energy conditions related to changes in water depth. On the other hand, badly eroded valves or those with visual evidence of chemical alteration (staining, overgrowths) should be avoided in analyses of shell chemistry. However, breakage and, to a lesser degree, chemical alteration can arise during sample collection (especially coring) and processing (e.g. Hodgkinson, 1991).

Most palaeoecological work is based on the identification and enumeration of individual ostracod taxa. Information about the ecological preferences of the species encountered within an assemblage is then used to reconstruct past environmental state, assuming uniformitarian conditions. Ostracod workers have tended to use descriptive approaches to the analysis of fossil assemblages, although there are notable examples of quantification (e.g. Smith *et al.*, 1992; Mezquita *et al.*, 1999; Mourguiart and Montenegro, 2002).

Some ostracod species show ecophenotypic responses in shell morphology to environmental conditions. The best known case is that of the euryhaline ostracod *Cyrpideis torosa*, the shells of which are noded in low salinity water but smooth at high salinity (see Van Harten, 2000 for further information).Whilst such variations have great potential for palaeoenvironmental reconstruction, their exact causes are often poorly understood (e.g. Roberts *et al.*, 2002) and, in many cases, ecophenotypy should be used as a palaeoenvironmental indicator only with caution.

Ostracod shells also provide material for geochemical analyses, especially of the trace element and stable-isotope composition. For trace elements, Mg and Sr have attracted most attention. Both Mg and Sr are partitioned from the host water into the ostracod shell at the time of shell secretion. Although adult shells discriminate against the trace elements, the uptake is proportional to the ionic activity of the trace element in the water and can be described by partition coefficients, which tend to be genus-specific. Strictly speaking, shell calcification is a physiological process that imparts non-thermodynamic behaviour (vital effects) to trace element uptake. Most partitioning coefficients depart significantly from those for inorganic calcite, with decidedly non-linear effects notable at high trace element concentrations or with lightly calcified shells (Chivas *et al.*, 1986a, 1986b; Engstrom and Nelson, 1991; Xia *et al.*, 1997b).

A knowledge of partition coefficients, coupled with trace element analyses of fossil shells allows reconstruction of past trace element levels within the host water. In some circumstances, there is a positive correlation between trace element levels within a waterbody and salinity. Such a correlation may arise either through evaporative concentration of a waterbody under low effective precipitation, or through mixing of fresh and marine waters in a coastal region. Mg and Sr are generally more soluble than calcium and so increase relative to Ca with increasing salinity. Notable exceptions include the preferential uptake of Sr in aragonite, which can lower Sr/Ca_{water} at higher salinities (Haskell *et al.*, 1996). In addition, Mg partitioning is quite strongly dependent on water temperature. Coupled Mg and Sr determinations of ostracod shells thus have the potential to yield information about past water temperature and salinity (see Holmes and Chivas, 2002 for a review).

Although oxygen and carbon isotope analyses are applied less routinely in palaeolimnology than in palaeoceanography, they are now used quite commonly in lake-based studies (see Leng, pp. 124–139 in this volume), with ostracod shells being the isotopic archive of choice. The oxygen isotope composition of ostracod shells is controlled by water temperature and the isotopic composition of the water, together with any vital effects. In many lakes, variations in the isotopic composition of the water have the most significant effect on the isotopic composition of the ostracod shell. The oxygen isotope composition of waterbodies is controlled by a number of factors, including the $^{18}O/^{16}O$ ratio of precipitation, and the ratio of evaporation to precipitation (or effective precipitation), together with the inputs from, and composition of, catchment runoff and groundwater (Smith *et al.*, 1997, 2002). The carbon isotope composition of ostracod shells is controlled by the carbon isotope composition of the dissolved inorganic carbon (DIC) within the host water. DIC composition depends on the balance of aquatic photosynthesis to decay, inputs of allochthonous organic and inorganic carbon from the catchment, certain microbial processes, such as methanogenesis, and exchange of CO_2 with the atmosphere (for further information, see Leng, pp. 124–139 in this volume). For both oxygen and carbon isotopes, determining the dominant controlling factors is vital and best achieved through studying the modern isotope systematics of the lake. For oxygen isotopes, ostracod shells show positive offsets from equilibrium values predicted for synthetic calcite (Xia *et al.*, 1997a), although there is some evidence that the ostracod may manipulate water chemistry at the precise site of calcite precipitation (Keatings *et al.*, 2002). In contrast, carbon isotope values appear to be in 'isotopic equilibrium'.

21.3 OSTRACODS AND HOLOCENE ENVIRONMENTAL RECONSTRUCTION

Non-marine ostracods have been used extensively to reconstruct past Holocene environments and there is a correspondingly large literature on the subject that is impossible to review comprehensively here. Instead, we examine three areas in which ostracods have specific value, namely hydrological habitat type and lake-level change; palaeosalinity and palaeo-solute composition; and temperature reconstruction. In each of these areas, we examine basic principles and specific studies, focusing both on palaeoecological and geochemical approaches where possible.

21.3.1 Palaeosalinity and Palaeo-Solute Composition

Salinity change and accompanying solute evolution may occur in hydrologically-closed lakes in semi-arid regions in response to changing effective precipitation or in coastal waterbodies as a result of varying marine intrusion. Palaeosalinity and palaeosolute reconstruction may thus provide information on Holocene climate, hydrology, sea-level change and coastal dynamics. Given the sensitivity of ostracods to water salinity and composition, ostracod records have excellent potential for the reconstruction of past salinity in both lacustrine and marginal-marine environments. In some instances, individual ostracod species are quite euryhaline, and provide a better indication of ionic composition rather than salinity *per se*. However, this relationship is still valuable in environmental reconstruction, since changes in ionic composition of water accompany both evaporative evolution and the mixing of marine saline and non-marine freshwater.

Figure 21.2 shows how non-marine water may evolve through evaporative evolution, which results in an increase in salinity and a change in the balance of alkalinity (principally HCO_3^- and

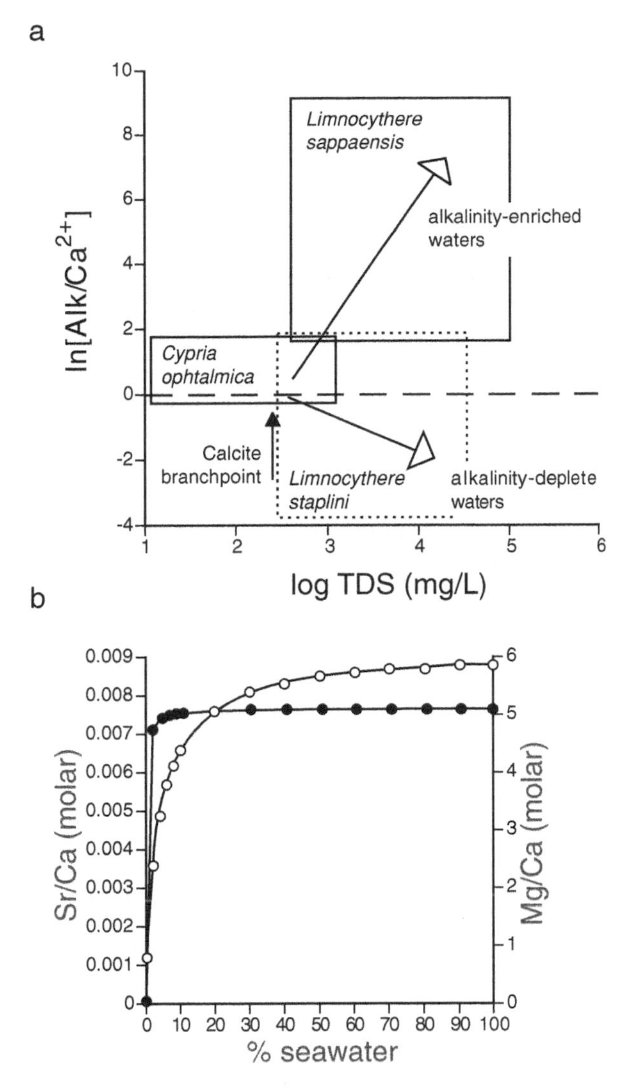

Figure 21.2 (a) Occurrence of three ostracod species in relation to evaporative evolution of non-marine waters in the American Midwest. Adapted from Curry (1999). (b) The relationship between molar M/Ca ratios and salinity arising from the mixing of fresh and marine saline waters, for example in an estuary. The form of the curve reflects the fact that trace elements concentrations are generally greater in marine water than fresh water. In this example, the Sr/Ca (open circles) and Mg/Ca (filled circles) molar ratios of the freshwater end member of the mixture are 0.0012 and 0.028 respectively: typical seawater values, of 0.0088 and 5.1 respectively, are used.

CO_3^{2-}) to calcium, the latter dependent on the initial composition of the unevaporated water (Eugster and Jones, 1979). Waters with similar salinity values can have contrasting composition, depending on the evaporative pathway. Some ostracod species are highly sensitive to water composition, and will only be found in saline water that is either enriched or depleted in alkalinity. Figure 21.2 shows the affinities of two species of the non-marine genus *Limnocythere* in the American Midwest. Although both species are found in saline waters, *L. sappaensis* is restricted to waters that are enriched in alkalinity compared to calcium, whereas *L. staplini* is found only in water that are alkalinity deplete. A third group of ostracods is found in waters that are relatively un-evaporated and of low salinity: for example *Cypria ophtalmica* on Fig. 21.2.

The relationships between water composition and ostracod species has been used to reconstruct Holocene lake water composition, both qualitatively and quantitatively. Most studies have been undertaken on lakes in continental North America (Smith, 1993; Forester and Smith, 1994; Smith and Forester, 1994), where there are clear gradients in water composition that relate broadly to evaporative concentration and, ultimately, to effective precipitation. However, there is scope to undertake similar studies in other parts of the world's arid regions and varying amounts of work have been undertaken in Australasia (e.g. De Deckker, 1982a, 1982b), Asia (Sun *et al.*, 1999), Africa (e.g. Holmes *et al.*, 1998: see also below) and semi-arid Europe (e.g. Anadón *et al.*, 1986).

The chemical and isotopic composition of ostracod shells may provide an additional line of evidence for variations in salinity and evaporative concentration in some cases. This is especially important if a sequence is dominated by euryhaline taxa, which are by definition insensitive to salinity change and so yield limited information when species composition alone is considered. Moreover, salinity-insensitive taxa are less likely to vary in abundance with changing lake hydrochemistry; a taxon that is present throughout a sedimentary sequence, regardless of salinity, is an ideal candidate for solute reconstructions based on shell chemistry. Hydrologically-closed lakes often respond to variations in effective precipitation by changes in salinity. Such changes may also be accompanied by variations in the oxygen isotope composition of the water and carbon isotope composition of the DIC. In some (but not all) lakes, changes in salinity are accompanied by changes in the Mg/Ca and Sr/Ca ratio of the lake water. In such cases, determinations of Mg/Ca and Sr/Ca ratios of fossil ostracod shells can be used in conjunction with knowledge of trace element partition coefficients (K_D values) to reconstruct past salinity. Oxygen isotope values may provide supporting evidence for evaporative evolution. It must be stressed, however, that the correlation between M/Ca_{water} (where 'M' is either Mg or Sr) and salinity is not universal (Haskell *et al.*, 1996; Xia *et al.*, 1997a; De Deckker *et al.*, 1999) and should be confirmed through modern limnological investigations. Similarly, lake-water oxygen isotope variations are not solely a response to changes in evaporative concentration: other factors may be involved. For example, Smith *et al.* (1997) demonstrate how pre-concentration of oxygen isotopes in upstream wetlands and lakes can lead to heavier (rather than lighter) isotopic ratios in ostracods during wet periods – and the opposite during droughts. Ideally tandem analyses of both trace elements and isotopes from the same shells along with independent proxies of salinity or climate (e.g. diatoms) should be used to help resolve competing interpretations and complex environmental forcing.

The study of the Holocene evolution of Kajemarum Oasis in north-eastern Nigeria provides an example of the combined use of ostracod palaeoecology and shell chemistry for palaeoenvironmental reconstruction (Fig. 21.3). The oasis lies in a depression within the relict dunefields in the semi-arid Manga Grasslands of the Nigeria – Niger border region. Its recent hydrology has been controlled primarily by effective precipitation. The lake water of this oasis and other lakes in the Manga Grasslands evolves by evaporation giving rise to increases in the salinity and oxygen isotope composition of the water. There is also a strong positive correlation between the salinity and Sr/Ca ratio of the extant waterbodies, suggesting that Sr/Ca ratios in

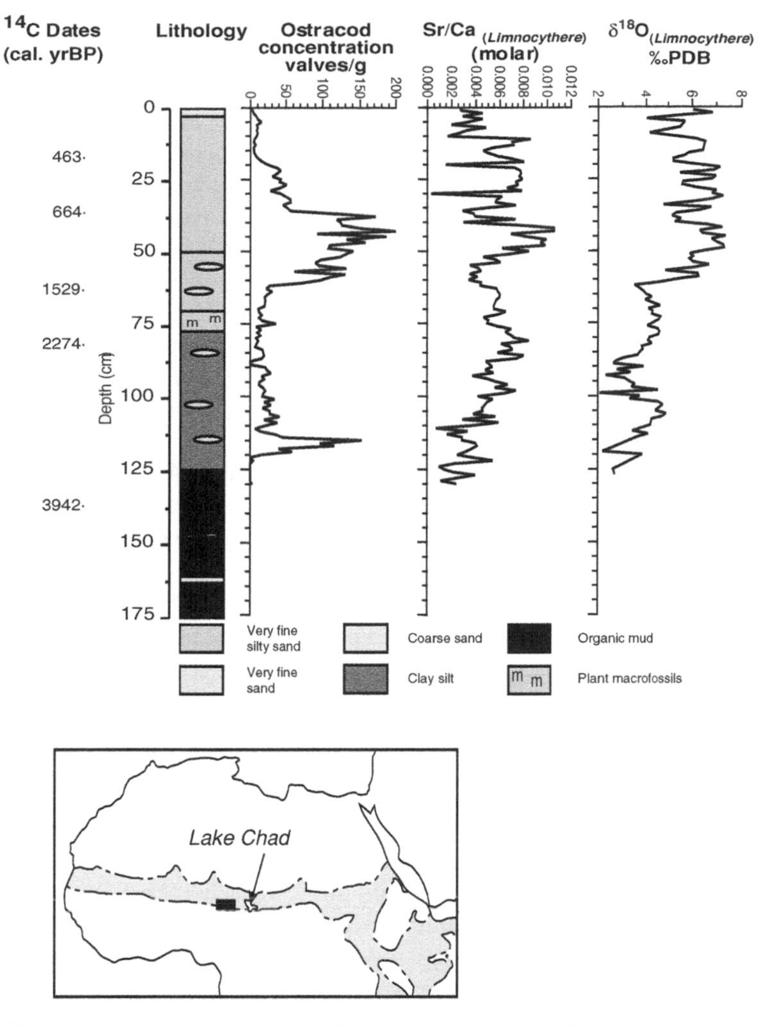

Figure 21.3 Holocene ostracod record from Kajemarum Oasis, northeastern Nigeria. Modified from Holmes *et al.* (1997).

ostracod shells can be used to reconstruct past salinity. The lack of a correlation for Mg/Ca, however, which is thought to result from Mg sorption by clays in the lake basin, precludes the use of Mg/Ca in ostracod shells as an indicator of past water composition.

The stratigraphic record from Kajemarum Oasis begins in the mid Holocene. Although evidence from pollen and other sources point to the existence of a deep, dilute lake at the time, ostracods are absent, probably as a result of dissolution in the organic-rich sediments. The ostracod record begins at 130 cm within the sequence, around 3800 cal. yr BP. The ostracod faunas in the lowest part of the sequence are dominated by *Candonopsis* sp., which probably reflects fresh or slightly saline water. The low salinity and low degree of evaporative enrichment is also indicated by the low Sr/Ca ratios and $\delta^{18}O$ values in ostracod shells. A switch to lower abundance ostracod assemblages dominated by the more-euryhaline species *Limnocythere inopinata* and, to a lesser degree, *Heterocypris giesbrechti* around 3000 cal. yr BP (112 cm) is accompanied by an increase in Sr/Ca ratios and $\delta^{18}O$ values in ostracod shells, indicating a rise in salinity, most likely caused by a reduction in effective precipitation. Low frequency and amplitude variations in Sr/Ca ratios and $\delta^{18}O$ values characterised the record until about 1500 cal. yr BP, after which time there was a switch to greater variability accompanied by an increase in *H. giesbrechti*: significantly, members of the genus *Heterocypris* tend to favour temporary or seasonably-variable waterbodies. The divergence of the Sr/Ca ratios and $\delta^{18}O$ values between 1500 and 1300 cal. yr BP probably reflects a change in the isotopic composition of precipitation at that time, which gave rise to increased $\delta^{18}O$ values at a time of high effective precipitation (Street-Perrott *et al.*, 2000). At around 1200 cal. yr BP, the parallel rise in both Sr/Ca ratios and $\delta^{18}O$ values points to a reduction in effective moisture leading to a rise in salinity. Thereafter, both the trace element and stable-isotope values showed high-frequency and high-amplitude variation in response to marked climatic variation.

Ostracod studies from the northern Great Plains (NGP) of the USA also demonstrate the utility of shell geochemistry in hydrological reconstructions in semi-arid landscapes. However, in this region, ostracod Mg/Ca, rather than Sr/Ca or oxygen isotopes, has proven the more robust and internally-consistent proxy. In the first such study from the NGP – and one of the rare examples in which ostracod records are compared to instrumental data – trace element ratios from *Candona rawsoni* were analysed from a ^{210}Pb-dated sediment sequence from Devils Lake, North Dakota, USA (Engstrom and Nelson, 1991). The Mg/Ca profile tracked closely water chemistry and lake-level records going back nearly a century, including the infamous 'dust-bowl' drought of the 1930s. Sr/Ca on the other hand, showed little correspondence with records of measured salinity, and in fact stayed relatively constant, a result consistent with water-chemistry data of the last half-century, which showed little change in Sr/Ca despite large excursions in salinity. The subsequent analysis of Late- and full-Holocene records from Devils Lake confirmed the validity of the Mg/Ca story – relative to diatom-reconstructed salinity – and further showed that Sr/Ca actually declined during major salinity/drought events (Fritz *et al.*, 1994; Haskell *et al.*, 1996). The mechanism for this antipathetic relationship between Mg and Sr is the preferential uptake of Sr over Ca during the abiotic precipitation of aragonite ($K_D > 1$); aragonite is the dominant inorganic carbonate phase in the high Mg/Ca waters of Devils Lake, and in most other saline lakes in the NGP.

The results from Devils Lake have since been corroborated by several high-resolution studies of Late Holocene records from other NGP lakes (Fritz *et al.*, 2000; Ito, 2002; Yu *et al.*, 2002). In the example from Coldwater Lake, North Dakota, ostracod Mg/Ca and diatom-inferred salinity (based on a weighted averaging regression model) showed extremely close correspondence, demonstrating conclusively that Mg/Ca is a robust salinity proxy in NGP waters (Fig. 21.4). The Sr/Ca record (previously unpublished) is also instructive in that it exhibits an inverse relationship with Mg/Ca (and diatom-salinity), showing once again the modulation of Sr chemistry by precipitation of inorganic aragonite during high-salinity events. Oxygen isotopes, on the other hand show episodes of both positive and negative covariance with salinity, a consequence of the complex relationship between lake-water $\delta^{18}O$ and climate. Synoptic changes in the isotopic composition of precipitation, the seasonality of aquifer recharge, and pre-evolution of isotopic ratios in upstream wetlands, as well as evaporative enrichment within the lake itself are all thought to play a role in this complex story. These studies indicate clearly the need to understand modern lake hydrochemistry before proceeding with environmental reconstructions based on ostracod or other geochemical archives.

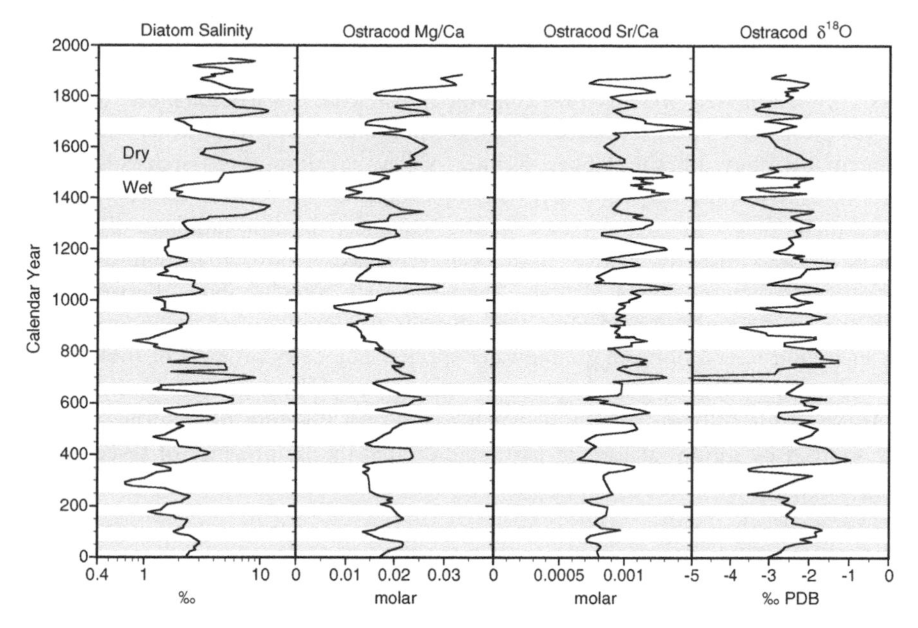

Figure 21.4 Comparison between ostracod geochemistry and diatom-inferred salinity from Coldwater Lake, North Dakota. The ostracod Mg/Ca and salinity data are from Fritz *et al.* (2000). Shaded intervals represent periods of high salinity and climatic aridity.

The occurrence of marginal-marine ostracods is governed by the same general factors as that of non-marine taxa in saline waters, i.e. by salinity and solute composition. Forester and Brouwers (1985) note that estuarine ostracods are usually closely related to marine taxa, although there are a few truly estuarine species, such as *Cyprideis torosa*. They further note that freshwater ostracods are rarely found living in estuaries, except at the freshwater end and conclude that this represents the influence of solute composition because whereas freshwaters are dominated by Ca–HCO_3, estuaries are dominated by Na–Cl. Once the Na–Cl composition of estuarine water is reached (usually around 3000 ppm total dissolved solids (TDS)), the ionic composition remains near constant even with further salinity change. It therefore follows that whilst ostracods can be used effectively to distinguish between fresh and marine saline water composition, they are unlikely to yield precise information about salinity. However, the mixing of fresh and marine saline waters, which generally have contrasting Ca, Mg and Sr contents, gives rise to an estuarine water with M/Ca ratios that reflect the proportions of the fresh and marine component (Fig. 21.2). Hence, the Mg and Sr content of fossil marginal marine ostracods can be used to reconstruct palaeosalinity (e.g. Anadón *et al.*, 2002; Boomer and Eisenhauer, 2002).

21.3.2 Hydrological Habitat

The association of many non-marine ostracod species to specific habitats, as noted earlier, means that faunal assemblages can be used to reconstruct hydrological environment. This capability has been used quite extensively in archaeological investigations (e.g. Bradbury *et al.*, 1990; Evans *et al.*,

1993; Palacios-Fest, 1997; Griffiths, 1998; Holmes and Griffiths, 1998) and also in the reconstruction of environments associated with tufas and travertines (e.g. Taylor *et al.*, 1994; Griffiths *et al.*, 1996).

Griffiths (in Evans *et al.*, 1993) combined ostracod palaeoecological analyses with sedimentological and other palaeontological indicators to reconstruct the Holocene environmental history of the Upper Kennet Valley, an important archaeological site in Wiltshire, southern England. Figure 21.5 shows a sequence from West Overton, a site in the Upper Kennett Valley. At the very base of the sequence, the low numbers of adults in the assemblage suggest that the faunas have been reworked. Above that point, however, the population age structures suggest that the assemblages are autochthonous. Between about 175 and 138 cm, in the Early to Mid-Holocene, there is an abundance of groundwater-indicating species, including several obligate groundwater species. The co-occurrence of these species with pond-dwelling taxa suggests that the environment was one of seasonal groundwater feeding a pond. Above about 138 cm, there is a switch to seasonal flooding from surface-water runoff, possibly sourced by springs at least in part, as indicated by the presence of several crenophilic taxa. Above 115 cm there is evidence of gradual drying.

21.3.3 Past Water and Air Temperature

The biogeographical zoning of many ostracod species indicates that water temperature, in addition to hydrochemistry, is an important control on species occurrence. Most lakes are sensitive to climate through the link between water temperature and air temperature. Ostracod-based reconstructions of water temperature in such lakes are therefore potentially valuable palaeoclimatic indicators. There is, however, a dearth of information about the temperature preferences of many ostracod species. A good exception is provided by the study of Forester *et al.* (1987) on Mid-Holocene climate of the American Midwest reconstructed using subfossil ostracods from Elk Lake, Minnesota. In this study, quantitative palaeoecological reconstructions of climate, including temperature, were undertaken by applying a quantitative analogue approach to the ostracod assemblages using a modern database of ostracods and environmental variables for around 5500 Canadian lakes. Confidence in the results was bolstered by the fact that the modern reconstructed temperature was near-identical to present-day observed values.

The thermodependence of Mg uptake into ostracod shells provides a second means of reconstructing water temperature. However, the method only works under clearly defined conditions, specifically when there is minimal long-term change in the Mg/Ca of the lake water (as Mg uptake is also affected by water composition) and the Mg/Ca of the water exceeds a critical minimum threshold of around 0.5–1.0. Wansard (1996) was able to use the Mg content of shells of *C. torosa* from a lake sediment sequence from Lake Banyoles in Spain to reconstruct changes in water temperature during marine isotope stage 2. In this case, it was shown that Mg/Ca_{water} had remained constant during the period covered by the record. In many lakes, however Mg/Ca_{water} rarely remains constant over long periods. Moreover, in lakes with very low Mg/Ca_{water} values, Mg partitioning into the shell is highly sensitive to Mg/Ca_{water}, such that even small variations in water composition will tend to mask any temperature effect.

A third approach to temperature reconstruction using ostracods involves oxygen isotope analyses of benthic specimens from deep, hardwater lakes in non-arid, temperate regions. In such lakes,

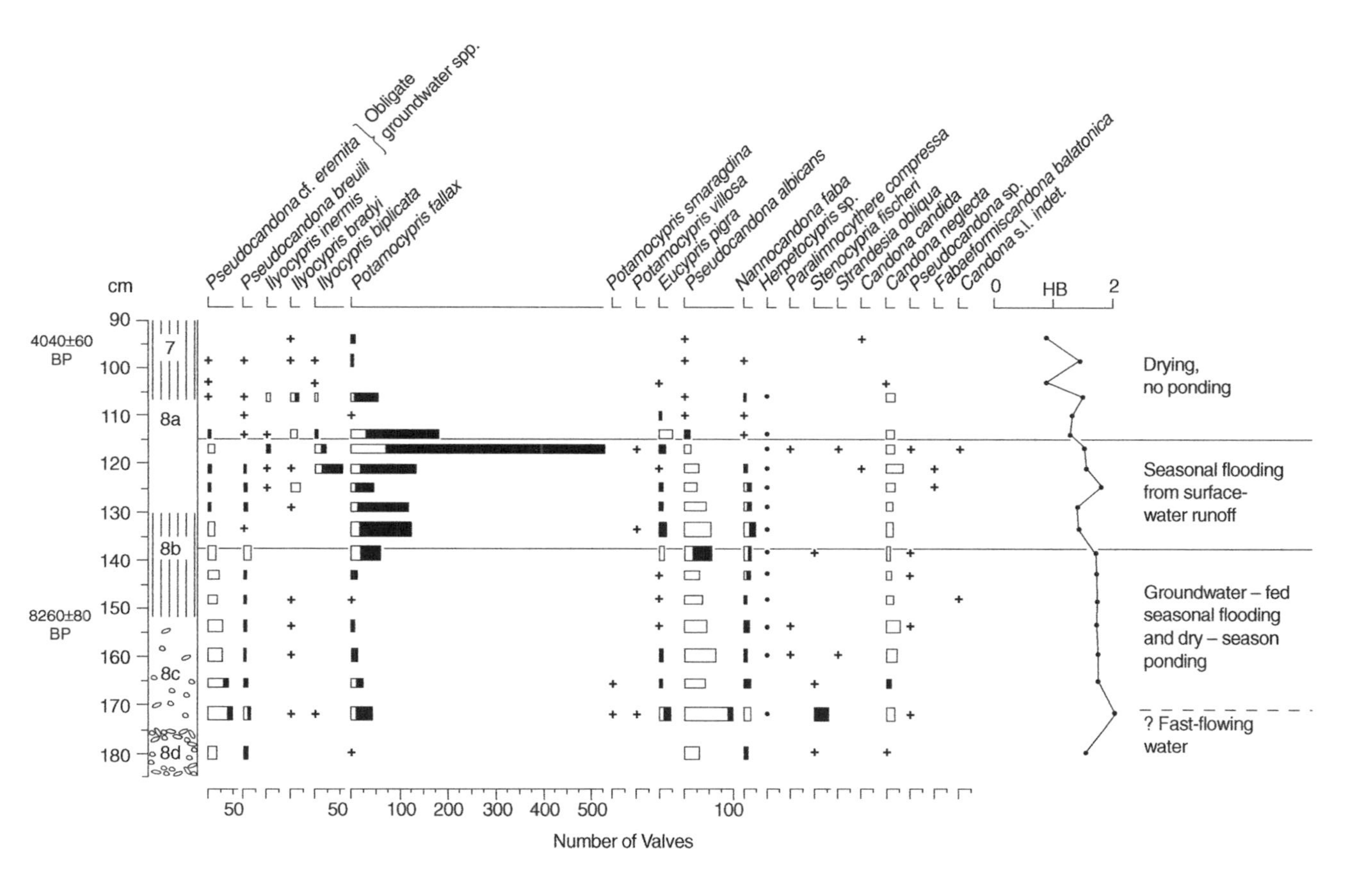

Figure 21.5 Holocene ostracod fauna from West Overton (Wiltshire, UK) and inferred hydrological habitat type. Redrawn from Evans et *al.* (1993).

bottom-water temperatures typically remain around 4 °C and, in the absence of significant evaporation, the oxygen isotope composition of the lake water is closely related to the composition of precipitation, which is in turn controlled by air temperature (e.g. von Grafenstein, 2002). An example of this approach can be found in von Grafenstein *et al.* (1998) and is illustrated in Fig. 21.6. The authors used the relationship between the isotopic composition of lake water and ostracod shells to reconstruct the long-term air-temperature history for southern Germany during the Late Glacial and Holocene. Moreover, such records can be used to reconstruct the oxygen isotope composition of palaeo-precipitation, which is also a valuable indicator of past atmospheric circulation (von Grafenstein, 2002).

21.3.4 Water Depth and Lake-Level Change

Lakes in many regions undergo changes in level and volume, often in response to variations in climate, for example, in effective precipitation. Ostracods provide a means of reconstructing

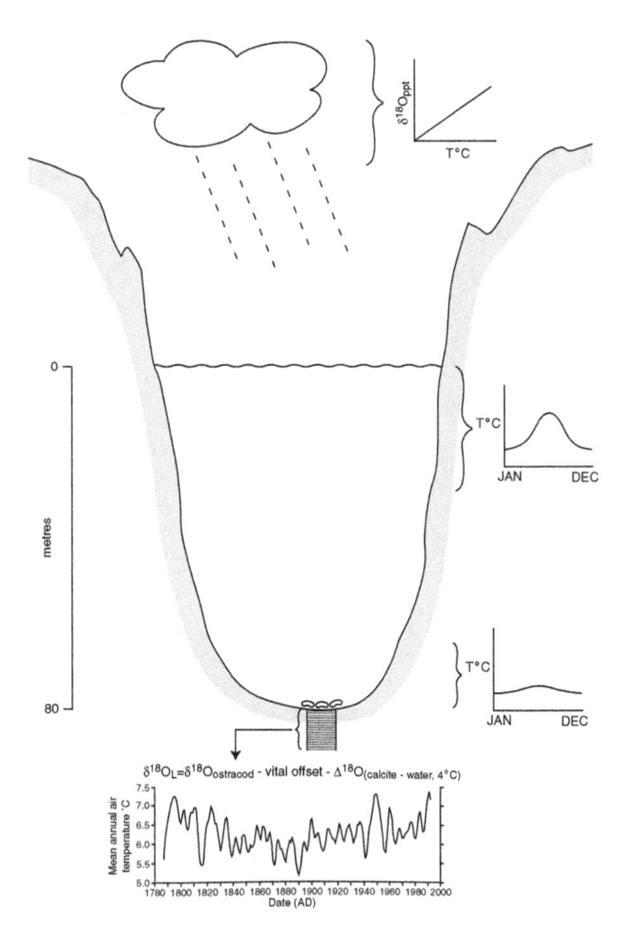

Figure 21.6 Schematic diagram to illustrate the use of the oxygen isotope composition of benthic ostracods from deep hardwater lakes to reconstruct past air temperature. Time-series of reconstructed air temperature is from von Grafenstein et *al.* (1996).

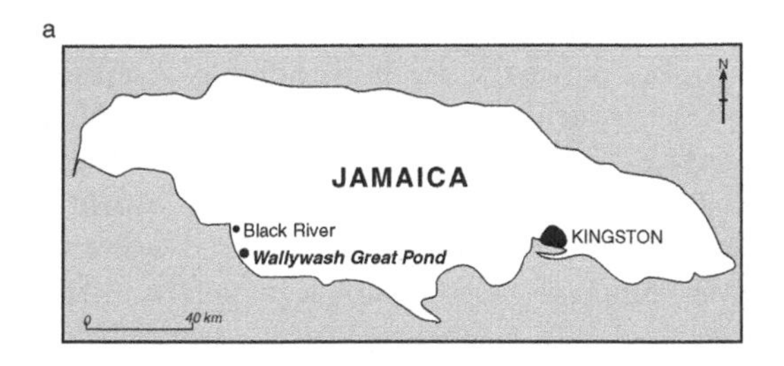

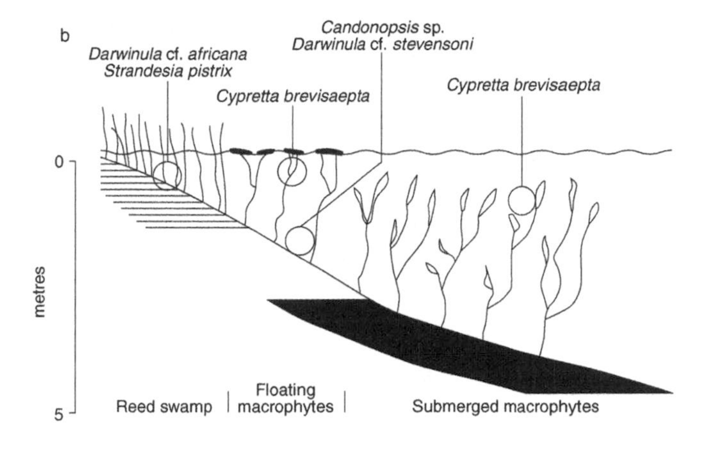

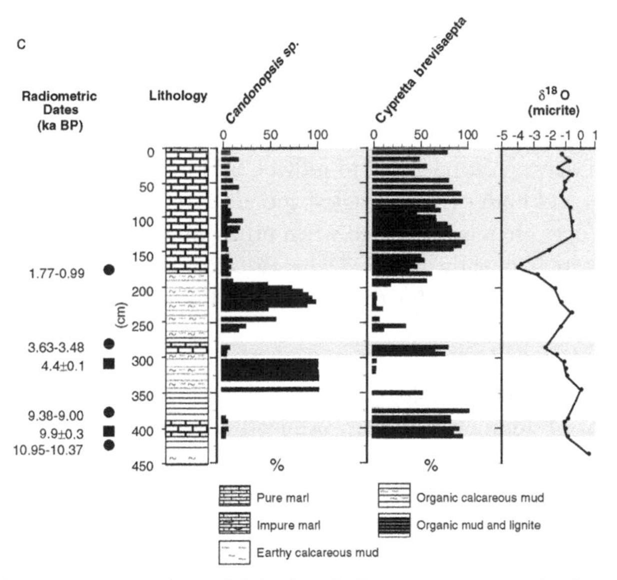

Figure 21.7 Holocene ostracods and lake-level change: an example from Wallywash Great Pond, Jamaica. (a) Location of Wallywash Great Pond; (b) summary of modern ostracod distribution in the lake and its relation to water depth and aquatic macrophytes; (c) variations in the percentage of *Candonopsis* and *Cypretta* in the Holocene sediments of the Great Pond together with oxygen isotope variations in fine-grained carbonate, which provide an indication of effective precipitation. Shaded intervals represent deep-water phases. For further information, see text. Adapted from Holmes (1998).

lake-level, since many species have specific depth preferences. It is not depth *per se* that controls species distribution, but rather the depth distribution of aquatic macrophytes together with variations in energy level (which affects substrate), food supply and dissolved oxygen at the sediment–water interface (Mourguiart *et al.*, 1998). Mourguiart and colleagues (Mourguiart *et al.*, 1992; Mourguiart and Carbonel, 1994; Wirrman and Mourguiart, 1995; Mourguiart and Montenegro, 2002) have developed ostracod–water depth transfer functions for lakes on the Bolivian Altiplano and used these to reconstruct Holocene lake-level variations.

Holmes (1997) used a study of the modern ostracod distribution in Wallywash Great Pond, Jamaica, to reconstruct Holocene variations in the water level of the lake. Wallywash Great Pond is a small, shallow karstic lake in the limestone region of the island. It has well-defined aquatic vegetation belts from reed swamp in the shallowest water, through floating macrophytes to submerged macrophytes at depth. Each vegetation belt is associated with a distinctive sedimentary environment and ostracod fauna (Fig. 21.7). A Holocene sediment sequence raised from the deepest part of the lake shows cyclical variations in sediment type, which reflect changing water depth at the core site. Oxygen isotope values in authigenic carbonate suggest that the variations in lake-level were probably caused mainly by changes in effective precipitation. The inferred lake-level variations are accompanied by changes in ostracod faunas. These changes are best described by the relative proportions of two taxa, namely *Cypretta*, which is most closely associated with deep, open water, and *Candonopsis* sp., which seems to prefer the zone of floating macrophytes in the shallower parts of the lake.

21.4 CONCLUSIONS AND FUTURE DIRECTIONS

Non-marine ostracods have great potential to provide information about Holocene climatic and environmental change and both palaeoecological and geochemical approaches may yield valuable data. Ostracod shells may often be preserved when other microfossils are absent: for example, in alkaline lakes diatoms are frequently dissolved whereas ostracods are generally very well preserved. However, additional work is needed for the full potential of ostracods to be realised. This work includes:

1. The collection of quantitative ecological information for more ostracod species.
2. The development of additional modern ostracod training sets for variables such as water temperature and chemistry.
3. Further modern limnological studies to understand ostracod life history.
4. Additional field and culture studies to understand and quantify trace element partitioning and stable-isotope fractionation.
5. Additional modern calibration studies, whereby ostracod-based palaeoenvironmental reconstructions are calibrated using instrumental and observational records.

As with all approaches to environmental reconstruction, ostracods are best used in concert with other independent proxies, both to evaluate competing environmental interpretations, and to refine our understanding of the ecological, geochemical and taphonomic processes that influence the sedimentary record.

Acknowledgements

We thank Emi Ito for the use of unpublished isotopic data from Coldwater Lake and Ian Boomer for constructive comments on an earlier version of the manuscript. This chapter is dedicated to the memory of Huw Griffiths, whose untimely death robbed the ostracod world of a much-valued colleague.

CHAPTER

22

CHIRONOMID ANALYSIS TO INTERPRET AND QUANTIFY HOLOCENE CLIMATE CHANGE

Stephen J. Brooks

Abstract: Chironomids are useful quantitative indicators of climatic change. The larval head capsules are well-preserved, abundant and diverse in lake sediments; they are readily identifiable, usually to genus- or species-group level, using identification keys to modern larvae; chironomids respond rapidly and characteristically to climate change; they occur in a wide variety of environmental conditions; and many taxa are stenothermic. Temperature has an important influence on many aspects of chironomid biology and ecology including rate and success of egg-hatching, larval development time, time of adult emergence and swarming behaviour of adults. Several low-error chironomid-air temperature inference models have now been developed and these have produced good climatic reconstructions for the Late Glacial. However, reconstructing Holocene temperatures is more problematic because expected temperature changes are close to the error margins of the models. In addition, the influence of the relatively small-scale temperature changes during the Holocene on chironomid assemblages can be overwhelmed by the impact of changes in other environmental variables, especially pH, water-level and dissolved oxygen, and this may compromise chironomid-inferred temperature reconstructions. Despite these problems, responses to many of the significant climatic fluctuations during the Holocene have been detected in many chironomid stratigraphies, and chironomid-inferred temperatures have been estimated for these events. Chironomids have also been used successfully to reconstruct quantitatively changes in salinity during the Holocene, which is closely linked to climate change, especially in subtropical regions.

Keywords: Chironomid biology, Chironomid ecology, Chironomid-salinity inference model, Chironomid taxonomy, Chironomid-temperature inference model, Holocene

Chironomidae, non-biting midges, are one of the larger families of two-winged flies (Diptera) and include about 5000 species (Cranston and Martin, 1989). The aquatic larval stage is the focus of palaeoecological studies because the head capsules are abundant and well-preserved in lake sediments (examples of some head capsules are shown in Plate 10). A comprehensive review of chironomid biology and ecology can be found in Armitage *et al.* (1995), and an account of palaeoecological chironomid analysis and methodology appears in Walker (2001). Most Holarctic chironomid larvae can be identified, at least to generic- or species-group level, using Wiederholm (1983) and Rieradevall and Brooks (2001). These identification guides are also useful for specimens derived from outside this region; in addition, Cranston (1997) and Brundin (1966) provide keys to elements of the Australian and Patagonian fauna, respectively. Identification to species-level is more problematic because of fragmentation of subfossil material, a lack of reliable taxonomic characters, and insufficient knowledge of the larval stages.

22.1 CHIRONOMIDAE AS ENVIRONMENTAL INDICATORS

22.1.1 Strengths

Chironomid larvae have several attributes that make them useful as environmental indicators.

22.1.1.1 Stenotopic

A large proportion of chironomid taxa are stenotopic; they have narrow ecological optima. Different elements of the chironomid assemblage respond in a characteristic way to particular environmental perturbations.

22.1.1.2 Ubiquitous

The larvae occur in practically all aquatic biotopes, but in lakes the profundal element of the fauna may be absent during times of anoxia.

22.1.1.3 Abundant

Among benthic macroinvertebrates, only oligochaetes are more abundant than chironomids but oligochaetes are poorly preserved in lake sediments. The high abundance of chironomid larval head capsules usually means that only small samples (2–4 g) of sediment are required to obtain sufficient capsules (about 100) for analysis so that sediments can be sampled at high temporal resolution (e.g. 2-mm intervals). The numbers of head capsules recovered from marl sediments is frequently low but yields can be improved by treating the sediments in a sonic bath (Lang *et al.*, accepted).

22.1.1.4 Identifiable

The heavily chitinized head capsules are preserved in good condition in most sediments. The taxonomic literature, at least for Holarctic taxa, is good and permits identification of most specimens at least to generic level. This level of taxonomic resolution is sufficient to make ecological inferences from a subfossil chironomid assemblage.

22.1.1.5 Species-Rich

Fifty or more taxa are often present in a north temperate Holocene sequence. This makes the assemblage extremely sensitive to environmental change as different species are likely to be affected by different environmental conditions.

22.1.1.6 Complementary

Inferences drawn from the response of the chironomid assemblage can complement the information inferred from other proxy indicators. For example, diatoms are good indicators of the pelagic environment whereas chironomids are indicative of benthic and profundal conditions. Diatoms can give an indication of changes in pH or trophic conditions against which the response of the chironomid fauna can be compared.

22.1.1.7 Sensitive

Because of the rapid generation time and ability of the winged adults to move readily from site to site, the chironomid response to environmental change is effectively instantaneous within the sampling resolution of most palaeoecological studies. Chironomid larvae develop *in situ* which means the analyst can be confident that larvae in their sediment samples are responding to local environmental change, and are not derived from locations remote from the sampling site.

22.1.2 Weaknesses

22.1.2.1 Air Temperature Versus Water Temperature

Air temperature rather than water temperature is often inferred from subfossil chironomid assemblages because of the superior performance of air temperature inference models (discussed below). However, the influence of water temperature on the distribution and abundance of chironomid larval assemblages is probably greater than air temperature. This becomes a problem if the relationship between water and air temperature was different earlier in the Holocene (perhaps because of the greater influence of snow beds and glacier-fed streams on lake water temperature) than it is now.

22.1.2.2 Group Level Identification

Most chironomid larvae cannot be identified beyond genus or species-group level. This limits the applicability of a calibration set: (i) to the region in which it was assembled because species within the same genus are likely to have different temperature optima in different climatic regions; (ii) to late Quaternary deposits, because species migrating from other biogeographic regions may not be distinguishable from local species; (iii) temperature optima of speciose genera distributed over a broad geographic range are less informative and possibly unreliable.

22.1.2.3 Environmental Factors

In some circumstances the distribution and abundance of chironomid assemblages may be influenced by certain environmental variables, especially pH, total phosphorus (TP), dissolved oxygen (DO) and lake depth, more strongly than temperature, and this may compromise the reliability a of chironomid-inferred temperature reconstruction.

22.1.2.4 Preparation Time

Compared with the preparation time needed for many other proxies, chironomid analysis is relatively time-consuming and typically 1–1.5 days may be required to sort, prepare and identify one sample.

22.2 INFLUENCE OF TEMPERATURE ON CHIRONOMID DISTRIBUTION AND ABUNDANCE

Temperature is of over-riding importance to the functioning of aquatic ecosystems (Wiederholm, 1984) and has a direct or indirect effect on chironomid pupation and emergence (e.g. Kurek, 1979), rates of egg development (Iwakuma, 1986), larval growth and feeding rates (e.g. Oliver, 1971) as well as flight and swarming behaviour (Kon, 1984). Cold-water and warm-water lakes each have their own characteristic assemblages of chironomid species. Typically, the large

Chironomini and Tanypodinae larvae are warm-adapted, whereas the smaller Orthocladiinae, Tanytarsini, Diamesinae and Podonominae are cold-adapted. Chironomini begin to dominate the fauna as temperature and trophic-level rises and oxygen levels fall. Conversely, low temperatures adversely effect the ability of large, warm-adapted taxa to complete their development. The metabolic rate is reduced but, in addition, lack of food in cold, oligotrophic lakes may also be a limiting factor. In cold lakes the littoral becomes dominated by cold stenothermic taxa, including taxa which typically occur in the profundal of warmer lakes and taxa typical of stream habitats. However, profundal taxa may be eliminated following prolonged ice-cover and winter anoxia. As lake temperatures rise, cold-stenothermic taxa may migrate to the profundal zone (Hofmann, 1971).

22.3 CHIRONOMIDS AS PALAEOCLIMATE INDICATORS

Chironomids have long been recognised as useful indicators of lake trophic status and several lake classification schemes have been devised based on characteristic suites of chironomid taxa (reviewed by Lindegaard, 1995). Other workers have used this relationship to look at the trophic development of lakes in the Holocene (e.g. Warwick, 1980; Brodin, 1986; Itkonen *et al.*, 1999). However, Walker and Mathewes (1987) were the first to suggest that post-glacial changes in chironomid distribution and abundance could be driven primarily by climate rather than trophic change. This provoked criticism from Warner and Hann (1987), Warwick (1989) and Hann *et al.* (1992) who argued that chironomid distribution and abundance was most affected by in-lake variables, such as lake depth, pH, dissolved oxygen, trophic status and substrate type and, as a consequence, chironomids had little potential as palaeoclimate indicators. However, Walker *et al.* (1992) countered with evidence that climate has broad-scale, regional control over midge distribution and abundance, often over-riding the influence of local in-lake variables. Since the early 1990s, the use of chironomids as palaeoenvironmental indicators has burgeoned and there is now overwhelming evidence of the significance of temperature in controlling the distribution and abundance of midges during periods of high amplitude climate change such as the Late Glacial. Because of the attributes of chironomids discussed above they are probably among the best palaeoclimate indicators and have great potential for quantifying Late Glacial climate change (Battarbee, 2000).

Chironomids are best used as part of multi-proxy studies (see Lotter, pp. 373–383 in this volume) in which the response of one proxy to environmental change can be compared with another. Each proxy has its own strengths and weaknesses. Chironomid-inferred temperatures can be validated by comparing them with data derived from plant macrofossils. For example, at Kråkenes Lake, western Norway, plant macrofossils suggested that birch woodland was present during the Early Holocene, indicating that the mean July air temperature must have been at least 11 °C. This inference was used to corroborate the chironomid-inferred mean July air temperature (Brooks and Birks, 2001). Similarly, the general trend of a chironomid-inferred temperature curve can be validated by comparison with a curve derived from ice-core oxygen isotope ratios. For example, Brooks and Birks (2000) demonstrated a close similarity between a chironomid-inferred temperature curve for the Late Glacial at Whitrig Bog, southeast Scotland and an oxygen isotope curve from the GRIP ice-core covering the same period. Alternatively, diatom-inferred pH values can be used to assess the degree of influence of pH change on the chironomid assemblage (e.g. Birks *et al.*, 2000).

22.4 CHIRONOMID-TEMPERATURE CALIBRATION SETS AND TRANSFER FUNCTIONS

The relationship between temperature and the distribution and abundance of chironomids has been exploited to develop chironomid-temperature inference models (see Birks, 1998 for a review of methods). Such work was pioneered by Ian Walker and co-workers (Walker *et al.*, 1991, 1997) who developed a modern calibration set from the distribution and abundance of taxa in surface sediment samples along a surface-water temperature gradient in eastern Canada. The use of water temperature as an explanatory variable for midge distribution and abundance makes good biological sense but the performance of models based on surface-water temperatures is compromised by the quality of the environmental data. In all the models published to date, the temperature data for each lake is based on one spot measurement taken in July at the time of the surface sediment sampling. When the amount of fluctuation in water temperature over the period of one month is considered (Brodersen and Anderson, 2000) it is surprising that the models work as well as they do. Table 22.1 gives the summary performance statistics for four chironomid-water temperature models recently developed in Canada (Walker *et al.*, 1997), Finland (Olander *et al.*, 1999), Sweden (Larocque *et al.*, 2001) and Norway (Brooks and Birks, 2000).

In order to improve the performance of chironomid-temperature inference models there have been three major developments. More lakes have been added to the models, evenly distributed over the temperature gradient and covering a wider range of altitudes and latitudes to maximize the temperature gradient length. This increases the chance that the full range of species responses to temperature change are covered in the model. The taxonomic resolution of the larval identifications has been improved, especially in Tanypodinae (Rieradevall and Brooks, 2001) and Tanytarsini (Brooks, unpublished data). The temperature data is based on mean July air temperatures, derived from local meteorological stations, averaged over a 30-year period, and corrected for altitude and distance from the coast (Brooks and Birks, 2001). The performance of these second generation inference models is a distinct improvement over the earlier models (Table 22.2). The strong statistical relationship between air temperature and midge distribution

	Canada	**Finland**	**Sweden**	**Norway**
No. of lakes	39	53	100	111
No. of taxa	34	38	48	119
Range (°C)	6.0–27.0	6.1–15.4	7.0–14.7	0.3–23.0
RMSEP (°C)	2.26	1.53	2.03	2.13
r^2_{jack}	0.88	0.37	0.32	0.86
Maximum bias (°C)	2.40	3.88	6.66	2.84
WA-PLS components	2	1	1	3

Table 22.1 Performance statistics based on leave-one-out (jack-knifed) cross-validation of chironomid-mean July surface-water temperature inference models, using weighted-averaging partial least squares (WA-PLS), from Canada (Walker *et al.*, 1997), Finland (Olander *et al.*, 1999), Sweden (Larocque *et al.*, 2001) and Norway (Brooks and Birks, 2001)

	Switzerland	Canada	Finland	Sweden	Norway
No. of lakes	50	39	53	100	153
No. of taxa	58	34	38	48	145
Range (°C)	6.6–17.3	5.0–19.0	8.5–14.9	7.0–14.7	3.5–16.0
RMSEP (°C)	1.37	1.54	0.87	1.13	1.01
r^2_{jack}	0.85	0.85	0.50	0.65	0.91
Maximum bias (°C)	1.67	1.71	2.93	2.10	0.93
WA-PLS components	2	2	2	2	3

Table 22.2 Performance statistics based on leave-one-out (jack-knifed) cross validation of WA-PLS chironomid-mean July air temperature inference models from Switzerland (Lotter *et al.*, 1997), Canada (Walker *et al.*, 1997), Finland (Olander *et al.*, 1999), Sweden (Larocque *et al.*, 2001) and Norway (Brooks and Birks, unpublished).

and abundance also has a sound biological basis as air temperature impinges directly on the adult stage, and also on the larval stage because of the relationship which generally holds between air temperature and water temperature (Livingstone and Lotter, 1998; Livingstone *et al.*, 1999).

22.5 RESPONSE OF CHIRONOMID ASSEMBLAGES TO CLIMATE CHANGE

There are now many examples in the literature documenting the response of chironomids to climate change and using chironomid-temperature inference models to quantify this change. However, at present, most of these studies have focussed on the Late Glacial and relatively few have dealt with the Holocene. Significant examples from the Late Glacial include the discovery and quantification of the Killarney Oscillation (Amphi-Atlantic Oscillation) (Levesque *et al.*, 1993b), of steep thermal gradients in the proximity of the Laurentide Ice Sheet (Levesque *et al.*, 1997) and of a series of cool oscillations during the Interstadial that closely parallel oxygen isotope records from Greenland (Brooks and Birks, 2000).

With the success of these Late Glacial temperature reconstructions, attention is now turning to the more difficult problem of quantifying Holocene climate change. Temperature fluctuations during the Holocene are thought to have been of the order of ±1–2 °C from present-day temperatures, much less than those during the Late Glacial, and close to the error of current chironomid-air temperature inference models. In some studies (e.g. Walker and Paterson, 1983; Schakau, 1986) the chironomid fauna has remained rather complacent throughout the Holocene, but at other sites the chironomid assemblage appears to be more sensitive to environmental change. The key is to find sites which support taxa close to their thermal thresholds. Some studies (e.g. Walker and MacDonald, 1995; Porinchu and Cwynar, 2000) have shown that sites situated at the modern tree-line show a high level of chironomid taxon turnover. But chironomid assemblages at such sites may also be influenced by pH changes, as a result of successional changes in vegetation, so chironomid assemblages in well-buffered sites are likely to be most sensitive to temperature change. Indeed, in an alkaline alpine lake in Switzerland, Heiri *et al.* (submitted) found that there was little relationship between changes in the chironomid assemblage and vegetational changes in the lake catchment. Periods of anoxia, often resulting from trophic change brought about by human influences in the catchment (Warwick, 1980; Heiri and Lotter, 2003), may completely eliminate the profundal chironomid fauna and thus influence any chironomid-inferred temperature reconstruction. In order to minimize the effects of human-induced change and to maximize the chironomid response to climate change, lakes situated at high altitude or high latitude should be sampled. However, in the past 200 years chironomid assemblages in even remote lakes have been influenced by the impact of post-industrial air-borne pollution, especially acid-rain, which may mask the response to recent climate change (e.g. Brodin and Gransberg, 1993; Schnell and Willassen, 1996). There may also be problems if cores are taken from shallow lakes. For example, Korhola *et al.* (2002) found that as Lake Tsuolbmajavri in northern Finland became more shallow during the Late Holocene, the chironomid fauna became dominated by littoral taxa with relatively high thermal optima and this may have influenced the chironomid-inferred rise in mean July air temperature.

What follows is a brief review of some of the studies into the response of midge assemblages to climatic change during the Holocene in north temperate regions. These results will then be compared with studies of Holocene midge faunas from the Afrotropics and southern hemisphere.

22.5.1 Early Holocene

At least two temperature reversals have been recognized in the North Atlantic region during the period of rapid warming at the beginning of the Holocene. The first, the Pre-Boreal Oscillation, is dated to about 11,150–11,300 yr BP (Björck *et al.*, 1997); the second, the Erdalen event, is estimated to have occurred between 10,750 and 10,200 yr BP (Dahl and Nesje, 1994; Matthews *et al.*, 2000). A response to two cold oscillations is indicated by changes in the chironomid assemblage at Kråkenes Lake, in western Norway (Brooks and Birks, 2001). During the first cool oscillation, dated to about 11,100–11,300 yr BP and presumably corresponding to the Pre-Boreal Oscillation, the chironomid-inferred mean July air temperature (CI-T) fell from about 10.8 °C to 10.2 °C. This was followed by a second cool oscillation when CI-T fell from about 11.9 °C to 10.8 °C. This oscillation is dated between about 10,800–10,500 yr BP and may be referable to the Erdalen event.

A response to the Pre-Boreal Oscillation was also detected in chironomid assemblages at two Swiss lakes (Brooks, 2000). Unfortunately, it was not possible to make a chironomid-inferred temperature reconstruction from these stratigraphies because several key cold-indicator taxa, which were present in the fossil assemblages, were absent from the modern Swiss calibration set, notably *Parakiefferiella* sp. A and *Einfeldia*. At Leysin (1230 m a.s.l., northern Swiss Alps) the early part of the Pre-Boreal Oscillation (11,371–11,320 yr BP) was characterized by the presence of *Einfeldia*. This taxon was otherwise present at this site only during cold climatic episodes in the Late Glacial (the Gerzensee Oscillation and the Younger Dryas). After 11,320 yr BP, *Einfeldia* disappeared from the fauna together with *Pagastiella*, a taxon favouring oligotrophic conditions (Brundin, 1956), which suggests conditions in the lake may have become more nutrient-rich following climatic warming. Thermophilic taxa increased rapidly after 11,320 yr BP, suggesting that the later stages of the Pre-Boreal Oscillation were considerably warmer than the early part at this site. At Gerzensee (603 m a.s.l., Swiss Plateau) the Pre-Boreal Oscillation was characterized not by increases in cold-tolerant taxa but rather by fluctuations in thermophilic taxa with the abundance of *Chironomus* spp. alternating with *Dicrotendipes*, *Microtendipes* and *Cladopelma*. This is consistent with a response to water-level fluctuations. *Chironomus* spp. are typically profundal taxa whereas *Microtendipes*, *Dicrotendipes* and *Cladopelma* dominate in shallow water. The core was sampled from the littoral of Lake Gerzensee making it especially sensitive to fluctuations in lake-level. Other evidence for fluctuations in precipitation during the Pre-Boreal Oscillation are widespread across Europe (e.g. Gaillard, 1985; Magny, 1995). At another site in Switzerland, Heiri *et al.* (2003) record a response by a chironomid assemblage to a cold period between 10,500 and 10,400 yr BP, when CI-T fell by at least 0.8 °C.

Oxygen isotope records from the Greenland ice-core suggest that a major cooling event occurred in the North Atlantic region at about 8200 yr BP (e.g. Dansgaard *et al.*, 1993). Evidence of a response to this event by chironomid assemblages was detected by Velle (1998) at Finse in northern Norway where CI-T declined by 1.2 °C, a decline of a similar magnitude was inferred by Korhola *et al.* (2002) in northern Finland and by Rosén *et al.* (2001) in northern Sweden. A 1.0 °C decline in CI-T was also recorded at this time at Holebudalen, a lake situated at the tree-line in southern Norway, when *Chironomus anthracinus*-type, a thermophilic taxon, was temporarily eliminated from the chironomid fauna (Fig. 22.1). A response to cooling by chironomid assemblages was also recorded at Hinterburgsee, Switzerland, by Heiri *et al.* (2003) between 8200 and 7700 yr BP when CI-T fell by over 1.0 °C.

Studies in British Colombia, Canada, suggest that thermophilic taxa were already dominating midge assemblages in the Early Holocene. Pellatt *et al.* (2000) found that from about 10,000 to 7,000 ^{14}C yr BP, a dry, sparsely vegetated spruce parkland co-existed with a diverse, warm-adapted chironomid fauna and the CI-T remained at about 13 °C, up to 4 °C higher than today. Similarly, Walker and Mathewes (1987) found that the cold stenothermic *Heterotrissocladius* was virtually eliminated from the chironomid fauna at Marion Lake, British Colombia, during the inferred xerothermic interval between 10,000 and 7000 ^{14}C yr BP.

Corynocera ambigua Zetterstedt is often regarded as a cold stenotherm favouring oligotrophic conditions (Walker and Mathewes, 1988) and may be common in Early Holocene sequences (Brodin, 1986; Sadler and Jones, 1997; Brooks, unpublished data from Lochnagar, Scotland). In the 153-lake modern Norwegian calibration set, the species has a mean July air temperature optimum of about 9.7 °C (Brooks and Birks, unpublished data) and it is largely restricted to lakes with a mean July air temperature between 8 and 12 °C. Therefore, the decline of this species in the Early Holocene has often been regarded as a response to rising temperatures towards the Holocene thermal optimum. However, although Brodersen and Lindegaard (1999) found *C. ambigua* in samples from the Younger Dryas and the last Interstadial, they also found it was abundant in modern warm (summer water temperature 18–25 °C), shallow, eutrophic lakes in Denmark. In fact, they found no significant distinguishing characteristics of lakes supporting *C. ambigua* in Denmark, although an association with *Chara* was often recorded. In their study of the influence of tree-line of chironomid distributions in northeast Siberia, Porinchu and Cwynar (2000) found that *C. ambigua* was associated with relatively warm, forested sites and did not occur at sites above tree-line.

22.5.2 Mid-Holocene

In northern Europe, many chironomid assemblages during the Holocene thermal optimum, between about 8000 and 5000 yr BP, were dominated by thermophilic Chironomini such as *Pseudochironomus*, *Microtendipes*, *Dicrotendipes* and *Chironomus* (Hofmann, 1991; Brooks, 1996; Brooks, unpublished data from Norway and Scotland). During the Mid-Holocene, there was often also an increase in the relative abundance of acidophilic taxa, such as *Psectrocladius*, *Heterotanytarsus* and *Zalutschia* (Brodin, 1986; Brooks, 1996; Porinchu and Cwynar, 2002; Brooks, unpublished data from Norway and Scotland), in response to a depression of lake water pH, possibly resulting from vegetational developments within the lake catchment. The chironomid response to pH change can lead to problems when inferring air temperatures from a midge assemblage as the response of a taxon to acidification may be stronger than its response to temperature change. For example, the high abundance of a particular taxon may be in response to conditions of optimal pH rather than to a temperature optimum, but nevertheless this taxon may make a significant contribution to the chironomid-inferred temperature reconstruction. This problem was discussed by Anderson (2000) in relation to the response of diatoms to pH change and their potential as temperatures proxies. A further problem may result from the presence of melting snow beds close to the lake. The input of cold water may cool the lake sufficiently to support a suite of cold-water taxa that would not otherwise be present if the lake water temperature was closer to the air temperature. The effect would be to produce chironomid-inferred air temperatures that were cooler than those derived from terrestrial temperature proxies such as plant macrofossils. In order to alleviate this problem, well-buffered sites which do not accumulate snow beds should ideally be chosen for Early to Mid-Holocene chironomid-inferred temperature reconstructions.

In western Canada, the period between about 7000 and 3500 ^{14}C yr BP appears to have become wetter and slightly cooler than in the Early Holocene with the CI-T at 3M Pond falling from 13 °C to 11–12 °C (Pellatt *et al.*, 2000). Engelmann spruce and subalpine fir forest became established and the chironomid fauna became dominated by *Dicrotendipes* (up to 80 per cent of the fauna). This domination of the fauna by one taxon for over 3000 years is very unusual and suggests a serious perturbation of the ecosystem. Pellatt *et al.* (2000) note the change occurred shortly after the eruption of Mount Mazama and suggest that this may have played a role in triggering the shift in the chironomid community.

22.5.3 Late Holocene

In northern Europe, following the Mid-Holocene thermal maximum, the climate is thought to have cooled during the Sub-Boreal (5700–2600 yr BP) and again at the onset of the colder, wetter Sub-Atlantic period (2600 yr BP to present; van Geel *et al.*, 1996). A marked and sustained decline in chironomid-inferred temperatures of up to 2 °C is indicated at Tsuolbmajavri in northern Finland (Korhola *et al.*, 2002) between 5800 and 4200 cal yr BP. The Sub-Atlantic was marked in many European chironomid stratigraphies by the decline or elimination of certain temperate Chironomini, notably *Microtendipes*, *Chironomus* and *Dicrotendipes*. These taxa were often replaced by cool stenotherms, especially Tanytarsini such as *Micropsectra*, *Stempellinella* and *Tanytarsus lugens*-group or Orthocladiinae such as *Heterotrissocladius*. Examples can be found from Lochan Uaine and Lochnagar, Scotland (Brooks, 1996; Brooks, unpublished data), Holebudalen (Fig. 22.1) and Bjornfjelltjørn, Norway (Brooks, unpublished data), Finse, Norway (Velle, 1998), Toskal, Finland (Nyman, unpublished data), Sjuodjijaure, Sweden (Rosén *et al.*, 2001), Vuoskkujávri, Sweden (Bigler *et al.*, 2002), Hinterburgsee, Switzerland (Heiri *et al.*, submitted)

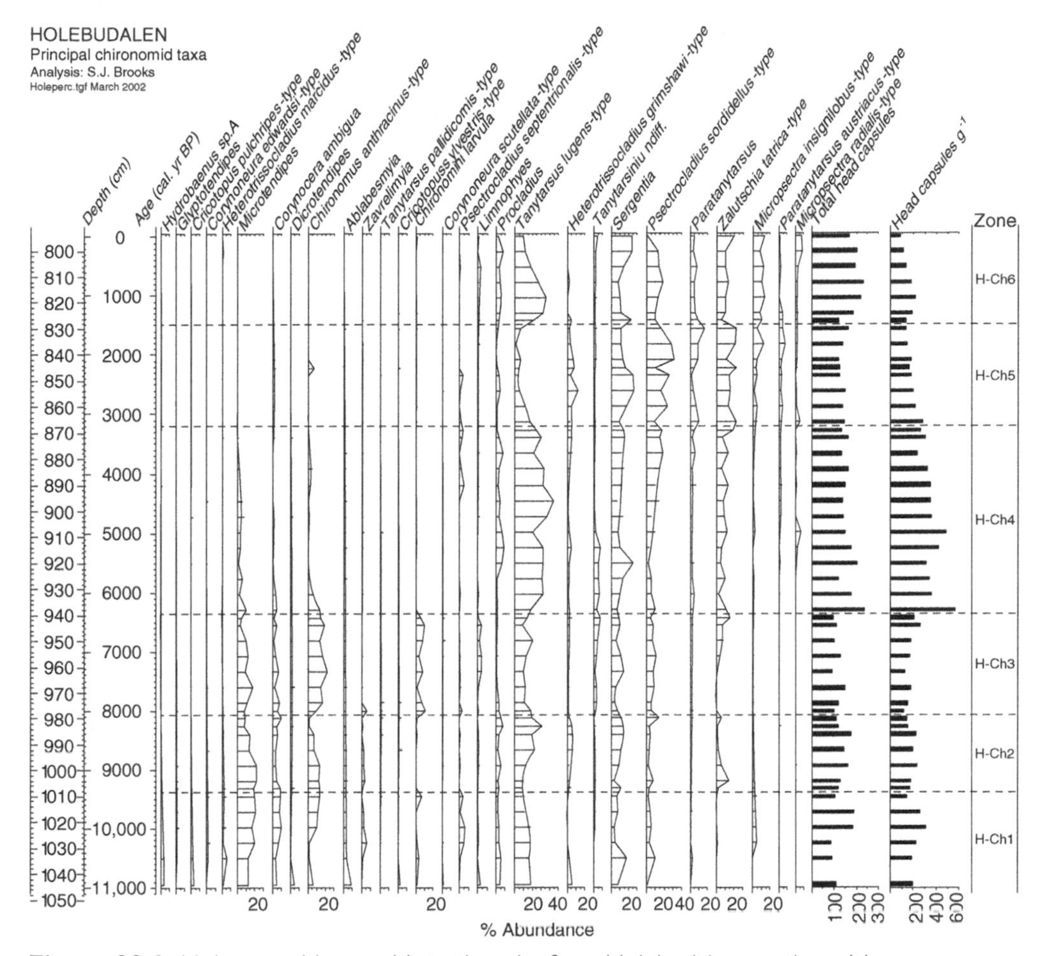

Figure 22.1 Holocene chironomid stratigraphy from Holebudalen, southern Norway.

and Rédo, Pyrenees (Rieradevall, unpublished data). The chironomid-inferred mean July air temperature fell by about 1.5 °C at Holebudalen at the transition to the Sub-Atlantic, about 2500 yr BP (Fig. 22.2).

In British Columbia, Canada, a response to cooling is apparent after about 3500 ^{14}C yr BP when Engelmann spruce and subalpine fir forest was replaced by open subalpine parkland. At the same time chironomid faunas became dominated by cold tolerant Tanytarsini (Walker and Mathewes, 1988; Pellatt *et al.*, 2000). At 3M Pond, the chironomid-inferred mean July air temperatures fell from 11–12 °C to 7–8 °C. These chironomid-inferred temperature data are consistent with temperatures inferred from glacial evidence in the region by Clague and Mathewes (1996).

At Lochan Uaine, in the Cairngorms, Scotland, the concentration of chironomid head capsules was found to oscillate in phase with loss-on-ignition (LOI) values with a periodicity of about 200–250 years throughout the second half of the Holocene (Battarbee *et al.*, 2001a), such that high LOI values coincided with high concentrations of chironomid head capsules. These results

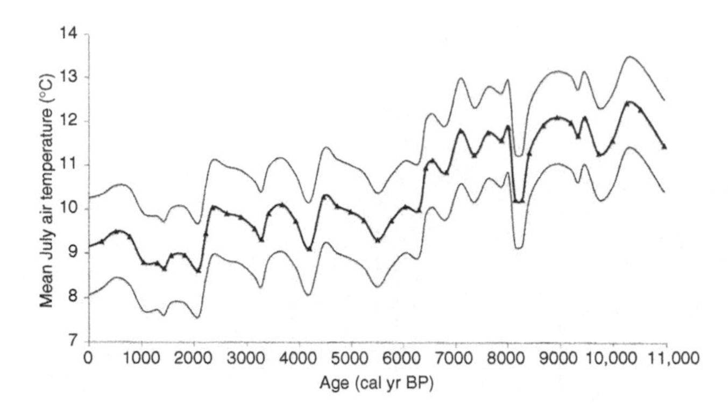

Figure 22.2 Holocene chironomid-inferred mean July air temperature reconstruction for Holebudalen, southern Norway. The upper and lower trend lines show sample specific errors calculated by boot-strapping.

indicate that higher numbers of chironomids could be supported during periods of high lake productivity, as signified by high LOI values, than at other times.

In northern Europe, glaciers reached their Holocene maximum extent during the so-called Little Ice Age (LIA) (Grove, 1988), which began 500–600 years ago. On Svalbard, the LIA lasted until the mid-19th or early 20th century (Svendsen and Mangerud, 1997) and this coincided with marked changes in the chironomid fauna of two lakes studied by Brooks and Birks (2004). One surprising result of this study was that although changes in the chironomid assemblages of both lakes were contemporaneous (about AD 1880), and therefore presumably in response to the same climatic signal, the changes were in opposite directions. In one lake the fauna became dominated by taxa associated with relatively eutrophic conditions, whereas in the other lake a more oligotrophic fauna became dominant. A possible explanation is that sea birds may have used the lake, which is situated on a coastal strand-flat, in increasing numbers during the LIA as one of the few open freshwater bodies in the area, leading to increases in nutrient input from bird guano. The other lake is situated further inland and had glaciers in its catchment during the LIA which would have cooled the lake waters as they began to melt, leading to a decline in lake productivity.

Chironomid assemblages from many sites in northern Europe show a marked response to post-industrial acidification (e.g. Henrikson and Oscarson, 1985; Brodin and Gransberg, 1993; Brooks, 1996; Schnell and Willassen, 1996) and in most cases this appears to have masked any response to recent climate change. However, a study by Granados and Toro (2000) on a lake in central Spain revealed that the abundance of *Chironomus* declined sharply after AD 1990. The authors interpreted this as a response to a reduction in the length of ice-cover and periods of winter anoxia. This also coincided with declines in cold stenothermic taxa including *Heterotrissocladius* and *Diamesa*. Using a chironomid calibration set from the Swiss Alps Lotter *et al.* (1997a) suggested that mean July air temperatures may have increased by about 1.5 °C since AD 1820, with a steep increase after AD 1990.

Battarbee *et al.* (2002) describe a study, which includes an investigation into the response of chironomids to climate change in the last 200 years in seven arctic and alpine lakes. Although a response to climatic warming was detectable by instrumental records at all seven sites, only a weak response could be demonstrated at two of the sites by the chironomid assemblages.

An example of a Holocene midge stratigraphy and the accompanying chironomid-inferred summer air temperature reconstruction is shown in Figs 22.1 and 22.2. Holebudalen Lake is situated in southern Norway, above the present-day tree-line, at an altitude of 1050 m a.s.l. The present-day mean July air temperature is 8.2 °C. The lake has a surface area of about 700 m^2, a maximum depth of 7.9 m and an average pH of 5.5. The chironomid-inferred mean July air temperatures were estimated using a modern Norwegian calibration set of 153 lakes with a mean July air temperature range of 3.5–16.0 °C (Brooks and Birks, unpublished). The performance statistics of this inference model are shown in Table 22.2.

A response to a brief cool episode centred on 8200 cal yr BP is indicated by the decline in the thermophilic *Chironomus anthracinus*-type and an increase in the cold stenotherm *Tanytarsus lugens*-type. This is reflected by a sharp fall in the chironomid-inferred mean July temperature estimates at this time. A response to a long-term cooling trend after the Mid-Holocene is shown by the general decline in thermophilic taxa such as *Chironomus anthracinus*-type, *Microtendipes* and *Dicrotendipes* and increases in cold stenotherms including *Micropsectra* spp. . This response is reflected as a downward trend in the chironomid-inferred temperature estimates, with a particularly marked decline at about 2500 cal yr BP. The increase in *Zalutschia lingulata*-type and *Psectrocladius sordidellus*-type after the Mid-Holocene is probably in response to falling pH. Evidence from plant macrofossils at the site suggest that the midge-inferred temperature estimates between 6500 and 2500 cal yr BP may be 1.0–1.5 °C too low but full details and analysis of this site will be published elsewhere (Brooks, in preparation).

22.6 CHIRONOMIDS AS INDICATORS OF CHANGES IN SALINITY

Salinity changes and water-level fluctuations in lakes are important indicators of climatic change in semi-arid regions, such as tropical East Africa, western North America, the central Andes and parts of Australia. Diatoms and ostracods have been used to infer changes in salinity (see Mackay *et al.*, pp. 294–309 in this volume; Holmes and Engstrom, pp. 310–327 in this volume) but in some saline lake sediments chironomids are also abundant and therefore are potentially valuable indicators.

Work by Paterson and Walker (1974) on sediment cores from an Australian saline lake revealed that chironomid assemblages changed in response to fluctuations in salinity. *Chironomus duplex* was present in freshwater conditions but was replaced by *Tanytarsus barbitarsus* and *Procladius paludicola* as salinity increased. *T. barbitarsus* was unable to compete with the chironomids that occurred in water with a low salinity but dominated once these taxa were eliminated. Similarly, Wiederholm (1980) studied Lake Lenore, Washington, USA, over a 24-year period during which time salinity gradually decreased, resulting in a succession of chironomid species. Ionic composition as well as salinity was found to be an important factor in effecting chironomid distribution. The potential of chironomids as indicators of changes in salinity during the Holocene was also demonstrated by Pienitz *et al.* (1992) who showed that changes in the chironomid fauna in a Canadian subarctic lake coincided with changes in diatom-inferred salinity.

From a calibration set of 86 lakes in western Canada, Walker *et al.* (1995) developed a chironomid-salinity transfer function. This can usefully infer the transition from fresh to moderately saline waters, but at salinities above 10 g/L the chironomid fauna is overwhelmingly dominated by a single taxon (*Cricotopus/Orthocladius*), and so the transfer function cannot track salinity changes in highly saline waters. This calibration set was used in a study to compare changes in salinity at a site in British Columbia, western Canada, throughout the Holocene using chironomid- and diatom-based inference models (Heinrichs *et al.*, 1997). The midge assemblage changed from a diverse freshwater community in the early post-glacial to one indicative of saline conditions (2.4–55.2 g/L), dominated by *Cricotopus/Orthocladius* and *Tanypus*, which remained in place throughout most of the Holocene. There was no indication of a Mid-Holocene mesothermal period. A cooler, wetter and less saline period occurred after about 1000 yr BP, indicating increased precipitation and/or decreasing temperatures. The midge and diatom-inferred salinities compared favourably with each other and with climatic trends inferred from palynological evidence.

Verschuren (1994, 1997) has demonstrated the relationship between the distribution and abundance of chironomids and precipitation and salinity in East Africa. Verschuren *et al.* (2000) presented a decadal-scale reconstruction of rainfall over the last millennium at Lake Naivasha, Kenya. These authors showed that the climate was significantly drier between AD 1000 and 1270 (the Medieval Warm Period) and was relatively wet between AD 1270 and 1850 (the Little Ice Age), interrupted by three dry episodes. There were strong chronological links between the inferred changes in precipitation and the pre-colonial cultural history of East Africa. Similarly, in northwest Sudan, East Africa, Mees *et al.* (1991) used chironomids, among other indicators, to reconstruct changes in precipitation during the Holocene. They demonstrated that a humid episode during the Early to Mid-Holocene was followed by a progressive increase in aridity.

In Arctic and north temperate regions the response of chironomids to salinity changes can also be helpful in understanding the influence of past climatic change. For example, Hofmann (1987) used chironomids to trace changes in the salinity of the Baltic Sea, which was linked to the retreat of the Scandinavian ice sheet. A freshwater fauna dominated by cold stenothermic, oligotrophic taxa including *Paracladopelma nigritula*, *Monodiamesa* and *Heterotrissocladius subpilosus*, was present in the Early Holocene but by 5000 yr BP it was replaced by taxa tolerant of brackish and fully marine water, namely *Chironomus salinarius*, *Halocladius* and *Clunio marinus*.

22.7 SOUTHERN HEMISPHERE

Relatively few palaeoecological studies have been carried out on chironomids in the southern hemisphere, despite the importance of the southern oceans in regulating global climate dynamics and the potential of chironomids to provide land-based palaeotemperature reconstructions. Massaferro and Brooks (2002) have completed the only detailed study of midge assemblages in the Holocene for Patagonia at Laguna Stibnite (46°S, on the Taitao Peninsula). Cyclical changes in the chironomid assemblage were interpreted as a response to fluctuations in precipitation, brought about by north–south oscillations in the westerly winds. This climatic interpretation was supported by changes in the terrestrial vegetation (Lumley and Switsur, 1993).

Schakau (1986, 1990) studied midges from Holocene sequences in two New Zealand lakes. Climate change was not considered as a significant factor driving the observed changes in the midge assemblages. Instead, changes in the rate and type of sedimentation and changes in lake productivity were regarded as the most likely causes of change in the midge assemblages. In another study from New Zealand, Boubée (1983) also interpreted changes in the chironomid assemblage at Lake Maratoto as a response to changes in lake productivity. A gradual change from an oligotrophic system in the early Holocene to a dystrophic system was indicated by an increase in *Polypedilum* and Orthocladiinae. At the same time, short-lived sporadic peaks in *Corynocera* and *Cladopelma curtivalva* were thought to be in response to water-level fluctuations.

22.8 CONCLUSIONS

Chironomids are useful proxy indicators of climate change in the Holocene. Inference models are now available that can provide quantitative air-temperature and salinity reconstructions. However, temperature change during the Holocene is close to or within the errors of current models (about 1 °C). Another problem is that changes in pH and water temperature may sometimes have a greater influence on the distribution and abundance of midges than air temperature and this may compromise the reliability of chironomid-inferred air temperature reconstructions. One challenge for the future is to further improve the performance of chironomid-temperature inference models by improving taxonomic resolution. The reliability of chironomid-temperature inferences will also be improved by carefully selecting lakes for palaeoclimatic investigations in which past changes in environmental variables other than temperature are at a minimum.

Acknowledgements

I am indebted to many colleagues over the years for stimulating discussions concerning the use of Chironomidae as palaeoenvironmental indicators but I would particularly like to thank John Birks for his encouragement. I would also like to thank my friends in the chironomid world who have freely exchanged ideas especially Klaus Brodersen, Peter Cranston, Evastina Grahn, Pat Haynes, Vanessa Heider, Oliver Heiri, Barbara Lang, Pete Langdon, Isabelle Larocque, Julieta Massaferro, Marjut Nyman, Dave Porinchu, Maria Rieradevall, Gaute Velle, Ian Walker and Endre Willassen.

CHAPTER

RECONSTRUCTING HOLOCENE CLIMATES FROM POLLEN AND PLANT MACROFOSSILS

Hilary H. Birks and H. John B. Birks

Abstract: This chapter discusses the combined use of pollen grains and spores and of plant macrofossils to reconstruct Holocene floras, vegetation and environments, including climates. The basic approaches to climate reconstruction are outlined, and the complementary nature of pollen-analytical and plant-macrofossil data is discussed in terms of dispersal, identification, improvement of interpretation and climatic value. Some studies that have successfully combined pollen and plant-macrofossil data to enhance understanding of Holocene climate history are reviewed. They include examples from North America and Europe on tree migration and on altitudinal tree-limit changes, on latitudinal tree-limit shifts in arctic Canada and Arctic Eurasia, and on tundra and prairie history in Spitsbergen and the Great Plains region of North America. The use of plant macrofossils to provide an independent validation of a pollen-based climate reconstruction is illustrated by recent work in northern Norway.

Keywords: Climate reconstruction, Pollen analysis, Plant macrofossils, Treeless environments, Tree-lines, Tree migration

Before the development of pollen analysis in the early 20th century, considerable knowledge had already been acquired about Holocene climate change from plant macrofossil evidence alone. Iversen (1973) outlines the history of peat-bog investigations in Scandinavia, which included the discovery of tree-remains buried in the peat. The interpretation of these and of peat-bog stratigraphy in Scandinavia by Blytt and Sernander as evidence of distinct climate periods (Pre-Boreal, Boreal, Atlantic, Sub-Boreal, Sub-Atlantic) became well established and the scheme was subsequently transferred to Scotland by correlation with the discoveries of Geike, Lewis and Samuelsson (see Birks, 1975). These 'climatic periods' were subsequently correlated with pollen zones established by von Post and Jessen. Although their climatic significance has largely been discarded, the scheme provided a useful subdivision of the Holocene, particularly for the correlation of pollen diagrams in northwestern Europe (e.g. Mangerud *et al.*, 1974) but its general use is now fading in light of more precise chronologies provided by ^{14}C dating. In contrast to the alternating wet and dry periods of Blytt and Sernander, the early work on macrofossil remains in peat bogs by Andersson (1902) led to the conclusion of a pattern of increasing warmth during the Holocene to a period with temperatures warmer than today [evidence from the wider and more northerly distribution of thermophilous plants such as *Corylus avellana* (hazel)] followed by a steady decrease in temperature to the present. Pollen analysis has subsequently documented Holocene vegetation and climate successions more effectively than plant macrofossil records. In northwestern Europe and elsewhere, the general Holocene climate pattern inferred from many diverse proxies on land, in the sea and from ice-cores conforms to that originally described by Andersson. That the temperature curve was not smooth, but contained small, short-term oscillations, was recognized for example by Behre (1967), Iversen (1973) and Odgaard (1994). The broad-scale changes, particularly in forest vegetation, can be interpreted as responses to climate change, that are modified by differential migration rates from glacial refugia, soil development and competitive interactions, particularly for nutrients and light (e.g. Iversen, 1958, 1973; Andersen, 1966, 1969; Berglund, 1966; Birks, 1986; Odgaard, 1994).

In this chapter we discuss approaches to reconstructing Holocene climate from palaeobotanical sources. We compare pollen and plant macrofossils and discuss how they complement each other in the reconstruction of past flora, vegetation and environment. This complementarity is illustrated with reference to integrated pollen and plant macrofossil studies on Holocene tree migrations and arrivals, changes in alpine and arctic tree-lines, and the history of treeless tundra and prairie areas. The use of plant macrofossils in validating pollen-based Holocene climate reconstructions is discussed for a site near the tree-line in northern Norway.

23.1 APPROACHES TO CLIMATE RECONSTRUCTION

Because they derive from plants, pollen and plant macrofossils reflect the flora and vegetation in which they originated, and the ecological and climatic conditions in which the plants grew. Because neither pollen nor macrofossils represent the complete vegetation or are linearly related to the numbers of plants on the surrounding landscape, both pollen and plant macrofossils are indirect proxies for climate reconstruction, and the records of both should be interpreted in terms of past flora and vegetation before climate reconstructions are made.

As described by Birks and Birks (1980), it is necessary to decide which plants are locally present and which fossil remains (usually pollen) have been dispersed from remote regions (long-distance dispersal). Two general interpretive approaches to environmental reconstruction may then be followed.

1. The **individualistic approach**: from a knowledge of the present-day ecology and environmental tolerances and optima of a plant species, the past conditions that allowed its growth can be inferred (see Birks, 1981; Birks, pp. 107–123 in this volume). This is a very important approach for all fossil organisms, and it applies particularly to plant macrofossils that can be identified to the species level.
2. The **assemblage approach**: assemblages of fossils can be compared with modern assemblages (e.g. modern analogues of vegetation types in certain climate/biogeographical regions) and knowledge of the conditions for the occurrence of the assemblages can be transferred to the past. This approach is particularly useful for palaeoecological interpretations (e.g. inferring lake-level changes; Watts and Winter, 1966; Hannon and Gaillard, 1997).

A consistent interpretation from an assemblage has more strength than an interpretation from the occurrence of an individual species. Various numerical methods have been developed for inferring the number of plants from the number of fossil pollen grains (e.g. representation factors; Bennett and Willis, 2001) and for comparing fossil assemblages with modern pollen assemblages from known vegetation through the use of modern pollen surface-samples (Birks and Gordon, 1985).

The reconstruction of climate from the assemblages relies on ecological knowledge about the modern equivalents. The most direct approach uses multivariate transfer functions to relate modern pollen assemblages to the climate or environment in which they occur. The functions can then be applied to fossil assemblages to reconstruct the past climate or environment (Birks, 1981). This approach is useful in cases where the fossils (pollen) are abundant and can be calculated as percentages (Birks, pp. 107–123 in this volume). In the case of plant macrofossils, the calculation of percentages is not useful (Birks, 2001a) and quantitative transfer functions for macrofossil assemblages cannot be constructed. The advantage of the transfer function approach is that a continuous climate reconstruction can be made corresponding to every level in the pollen diagram. Thus relatively small and short-lived oscillations can, in theory, be detected. Other numerical methods besides multivariate transfer functions have also been developed to relate modern surface-samples to fossil assemblages (e.g. modern analogues, response surfaces; see Birks, pp. 107–123 in this volume), and these are part of the assemblage approach.

23.2 COMPARISON OF POLLEN AND MACROFOSSIL DATA: HOW THEY COMPLEMENT EACH OTHER

Pollen analysis is based on the principal that pollen grains are abundantly produced and freely dispersed to the site of preservation. However, the representation of species by their pollen is a result of many factors, including production, dispersal and preservation (Bennett and Willis, 2001). Wind-pollinated species produce more numerous and widely dispersed pollen than insect-pollinated species. The abundant wind-dispersed types (e.g. *Pinus, Betula, Alnus, Corylus, Artemisia*, Gramineae, Chenopodiaceae, etc.) are well represented in the pollen record. However, these pollen grains can often be blown long distances, and into areas where they do not grow. Although they make up the regional pollen rain, the long-distance or extra-regional component can confuse climate interpretations, especially in areas of low local pollen production such as at tree-lines, in treeless vegetation such as grassland or tundra, and in forest types dominated by insect-pollinated taxa (e.g. mixed temperate forest, tropical forest; Ritchie, 1995; Birks and Birks, 2000). Pollen assemblages are usually presented as percentages of a pollen sum, but concentration and influx estimates can be particularly useful in regions of low or varying local pollen production (Birks and Birks, 1980). There are all degrees of pollen representation within a pollen assemblage (Bennett and Willis, 2001), from overabundant anemophilous tree pollen down to taxa that produce minute amounts of pollen (e.g. apomictic taxa), or pollen that is poorly or not preserved (e.g. Juncaceae, many submerged aquatic plants). Representation factors and modelling approaches have tried to rectify the unevenness of representation in relation to the parent vegetation, especially for tree-pollen types (e.g. Andersen, 1970; Sugita, 1993, 1994; Davis, 2000).

In contrast to pollen, plant macrofossils are much larger and generally produced in smaller numbers. They are usually enumerated as concentrations, rather than percentages or influx (see Birks, 2001a). There is no 'macrofossil rain' representing the regional vegetation comparable to 'pollen rain' (Watts and Winter, 1966). Macrofossils (seeds, fruits, leaves, etc.) are not usually dispersed far from their point of origin, and thus a fossil assemblage reflects the local vegetation around the site of deposition, often aquatic and wetland vegetation with little representation of upland vegetation except in favourable accumulation sites. However, this means that if fossils are found of upland taxa, it is almost certain evidence that the taxa were growing locally. This information is valuable in reconstructing local upland vegetation and in tracing tree-limit changes that are often under climatic control.

Pollen grains are usually identifiable to Family or Genus level. Although some types are species-distinct, it is often difficult or impossible to reach species determinations, especially with the common tree types and many common herb types. Macrofossils on the other hand are often identifiable to species level, and several types of plant part may be preserved from one species (e.g. Birks and Birks, 1980). In many cases, macrofossils have been valuable in determining which tree species were locally present, e.g. *Betula* (van Dinter and Birks, 1996; Freund *et al.*, 2001), *Pinus* (e.g. Watts, 1970), *Picea* (e.g. Watts, 1980; Jackson and Weng, 1999), other conifers of the western USA (e.g. Cwynar, 1987; Dunwiddie, 1987), *Thuja/Juniperus* (Yu, 1997) and *Tilia, Quercus,* etc. (Hannon *et al.*, 2000). Cyperaceae can be distinguished to genus or below (e.g. Birks and Mathewes, 1978) and *Juncus* spp. can be separated (Körber-Grohne, 1964), although Gramineae apart from cereals are more difficult to determine (e.g. Körber-Grohne, 1964). The macrofossil record may also fill in 'blind spots' in the pollen record, as it contains fossils of mosses and Charophyceae, many aquatic plants, and diverse types such as a few lichens, liverworts and Cyanobacteria. Thus, although not a complete record, the combined pollen and plant macrofossil records give a much more detailed picture of palaeoecological and hence environmental and climatic conditions than either record alone.

In addition, plant macrofossils can be used to reconstruct climate parameters not possible with pollen (Birks, 2001a). Stomatal density and index on leaf cuticles of some species respond to the concentration of CO_2 in the air (Woodward, 1987) and the relationship can be used to reconstruct past CO_2 concentrations from fossil cuticles (e.g. Rundgren and Beerling, 1999). $\delta^{13}C$ measurements on macrofossils have also been used as climate responses, for example *Pinus flexilis* (van der Water *et al.*, 1994), mosses and sedges (Figge and White, 1995) and *Salix herbacea* (Rundgren *et al.*, 2000). δD has also been used in tree-rings as a climate proxy (e.g. Dubois and Ferguson, 1985). Most importantly, ^{14}C measured from terrestrial plant macrofossils yields radiocarbon dates that should be free of sedimentary reservoir effects (Birks, 2001a). Any improvement in the reliability of chronologies is essential for the timing of climate changes and thus their comparison between regions.

23.3 COMBINED HOLOCENE POLLEN AND PLANT MACROFOSSIL STUDIES

As mentioned above, plant macrofossils have features that complement pollen analysis. They can indicate the local presence of taxa, they can determine which species were present to a lower taxonomic level, they can fill 'blind spots' in the pollen record, and remains of terrestrial origin can be used for accelerator mass spectrometry (AMS) radiocarbon dating. Combined investigations have been particularly fruitful in the study of the migration and succession of tree taxa during the Holocene, of both alpine and arctic tree-line movements, and in areas with low local pollen production such as grasslands and tundra where long-distance pollen becomes an important confusing factor for climate reconstruction. We illustrate these with a few examples, which by no means are an exhaustive review.

23.3.1 Tree Migration and Vegetation Change

In North America, W.A. Watts pioneered the use of macrofossils as an adjunct to pollen analysis, and for overall palaeoecological purposes. In the midwest USA, pollen analyses indicated that the Mid-Holocene was so warm and dry that forest retreated eastwards and prairie grassland replaced it (e.g. Jacobson and Grimm, 1986). Pollen transfer functions have been used to quantify the changes in temperature and precipitation mainly using changes in the percentages of tree pollen, Gramineae, *Artemisia* and Chenopodiaceae (e.g. Webb and Bryson, 1972; Bartlein *et al.*, 1984). Local information from macrofossil records, particularly at Kirchner Marsh, Minnesota (Watts and Winter, 1966) and Pickerel Lake, South Dakota (Watts and Bright, 1968) showed the extent of lake-drawdown and some complex oscillations in lake level between *c.* 8000 and 4000 ^{14}C yr BP. These indicated changes in the precipitation/evaporation ratio and the amount of temperature and dryness increase necessary to cause the replacement of Boreal forest first by deciduous trees and then by prairie grassland. These and subsequent data from the Great Plains (Barnosky *et al.*, 1987) have been synthesised in climate models to explain the Holocene climate history of North America (e.g. Kutzbach and Webb, 1991). Clearly, more pollen and macrofossil data relating to lake-level changes will be valuable in determining the detailed fine-scale climate changes during the Mid-Holocene of this area and to complement other multi-proxy studies involving diatoms, geochemistry and stable-isotopes (e.g. Engstrom and Nelson, 1991; Haskell *et al.*, 1996; Laird *et al.*, 1998b) and to assess the magnitude and frequency of drought cycles during the Mid-Holocene (Clark *et al.*, 2002).

Plant macrofossils have also been used to determine the timing of local arrival of tree species during the Early Holocene which implies local climate conditions that permit their growth. In Scotland at Abernethy Forest, Birks and Mathewes (1978) showed that the earliest Holocene birch species was *Betula nana*, followed later by *B. pubescens* corresponding to the large rise of *Betula* pollen percentages, suggesting that the mean July temperature had risen above 10 °C. The local arrival of *Pinus sylvestris* shown by macrofossils was *c.* 150 years after the large rise of *Pinus* pollen percentages that indicated the regional arrival of *Pinus*. Subsequently, pollen analysts have been able to identify stomata of conifers in pollen preparations (e.g. Hansen, 1995b; see also review by MacDonald, 2001). Their presence indicates the local occurrence of the taxa, particularly of *Pinus* whose abundant pollen is widely dispersed and can reach high percentages in the pollen sum without being present in the area. Watts (1979) used plant macrofossils in this way to determine the presence of *Picea* and *Pinus* in Appalachia. High percentages of *Picea* pollen at Longswamp, Pennsylvania (Fig. 23.1) were accompanied by large numbers of needles. *Picea* pollen subsequently declined as *Pinus* pollen increased. However, *Picea* needle numbers remained high, showing that the apparent decline of *Picea* pollen percentages had no climatic significance, but was a statistical artifact due to the very high local pollen production by *Pinus*. Needle analysis showed that the *Pinus* species was *P. banksiana*. Only much later in the Holocene did *P. strobus* expand. Watts' analyses from Longswamp also showed that a characteristic Late Glacial *Betula* pollen peak was derived from *B. glandulosa* but that subsequent *Betula* pollen was produced by the tree, *B. populifolia*. During the same period, macrofossils also attested to the local presence *Abies* and *Alnus rugosa*. Interestingly, the Early Holocene replacement of mixed coniferous forest by deciduous forest is shown by the end of pollen and macrofossil deposition of the Boreal types and the expansion of deciduous-tree pollen only, mainly *Quercus*. Macrofossils of these types were not recorded. The seeds of several of the taxa are rare and often large and probably not well preserved as they are designed for animal consumption. Other tree taxa have smaller, lighter seeds that are capable of transport some tens of kilometres outside the forest canopy (Nathan *et al.*, 2002) but the numbers that are widely dispersed are small. Their leaves are fragile and are only preserved in favourable depositional conditions (e.g. Chaney, 1924; Spicer, 1989). Refinement of macrofossil identification has shown that some deciduous-tree budscales can be preserved (Tomlinson, 1985) and the identification of these in Holocene sediments yields information on the local occurrence

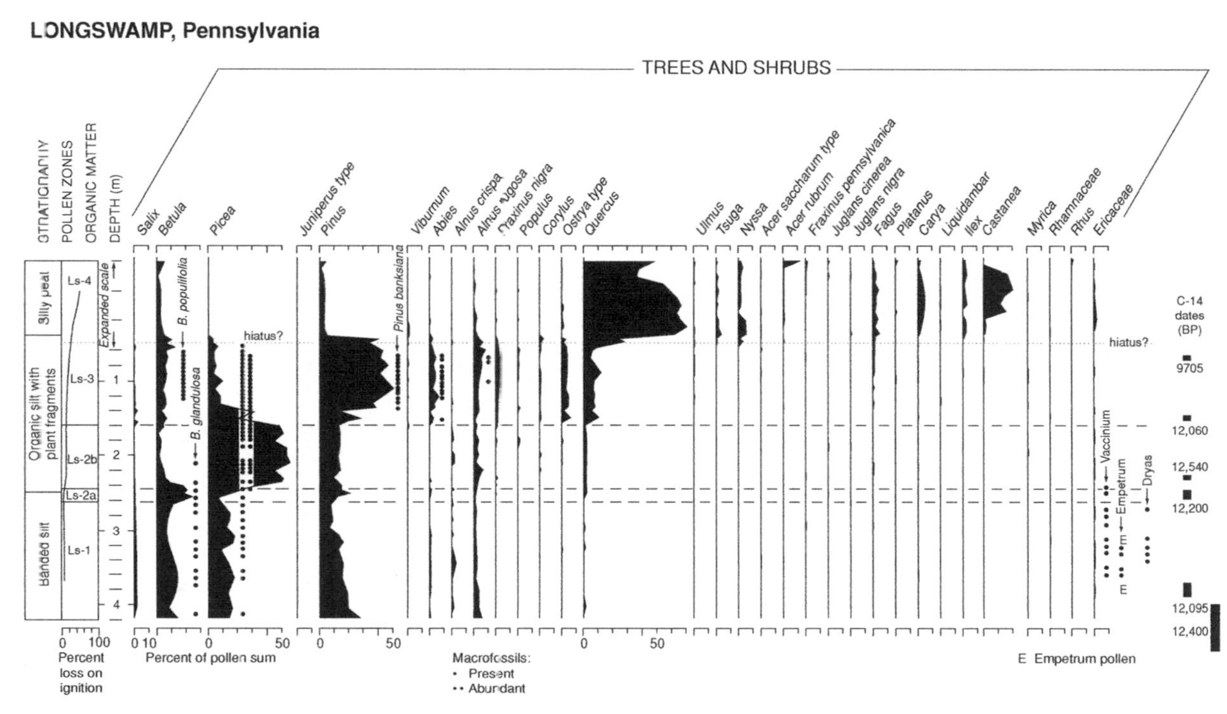

Figure 23.1 Pollen and macrofossil diagram from Longswamp, Pennsylvania, USA by W.A. Watts (1979). Macrofossil records are indicated as black dots adjacent to the appropriate pollen percentage curves.

of these taxa (e.g. Schneider and Tobolski, 1985; Eide *et al.*, 2004). Their identification in fluvial sediments from Iowa (Baker *et al.*, 1996c) has helped towards resolving the conundrum of the presence or absence of thermophilous deciduous trees in the Late Glacial and Early Holoscene of the mid-west USA, and the climatic implications of the alternative conclusions drawn from the low amounts of their pollen (Birks, 2003). The nearest positive records of macrofossils of thermophilous deciduous trees during the Glacial period are from the southeastern USA (Jackson and Givens, 1994; Birks, 2003). Clearly, more sites with the identification of macrofossils can aid in the reconstruction of the local vegetation, the biogeographical history and in the validation or otherwise of climate reconstructions made from pollen data.

23.3.2 Altitudinal Tree-Limits – Alpine Tree-Lines

The position of the forest-limit, tree-line, or tree-limit is hard to detect using pollen percentages or influx. Because of the small geographic separation, pollen is blown or washed into the site from above or below. Maher (1972) and Jalut *et al.* (1996) attempted to locate forest-limits using ratios of pollen types from different vegetation zones. However, the use of macrofossils is more satisfactory. Jackson (1989) and Jackson and Whitehead (1991) tracked the Holocene altitudinal movements of tree taxa in the Adirondack mountains (eastern USA) by analysing pollen and macrofossils from sediments of six lakes at critically positioned altitudes. Macrofossils enabled the separation of tree species within the pollen taxa (e.g. *B. papyrifera* and *B. lutea*) thus allowing more precise interpretation of the pollen data. The macrofossil record traced the local arrival and colonisation of taxa through the Holocene, whereas the pollen record provided a regional picture. Between 8000 and 5000 ^{14}C yr BP most tree taxa occurred at higher altitudes than at present, indicating summer temperatures 1.0–1.6 °C higher. However, their movements were not in parallel, demonstrating that, although there was an overall Holocene climate pattern, factors other than temperature were also important, such as moisture balance, immigration, competition from established vegetation, disturbances (e.g. fire regime), soil conditions and the individual tolerances of the species, as outlined by Iversen (1973). The modern pattern of vegetation and species distribution became established around 2500 years ago after the immigration of *Picea*.

In a detailed study of a small lake, Lago Basso, above tree-line today in the north Italian Alps, Wick and Tinner (1997) compared macrofossils with pollen and conifer stomata influx records (Fig. 23.2). Early Holocene pollen was dominated by *Betula* and *P. cembra,* but macrofossils showed that only *Betula* was locally present, *P. cembra* arriving *c.* 500 years later. *P. cembra* forms the tree-line today, and *Larix* accompanies it at a somewhat lower elevation. The local arrival of *Larix* at Lago Basso was attested by its abundant needles, showing a tree-line at its highest altitude between 8700 and 5000 ^{14}C yr BP. Subsequently, the tree-line descended but there were several well marked phases of tree presence or absence, particularly in the macrofossil records, the later ones associated with the arrival of *P. abies* and the major role of human deforestation and development of treeless meadows on the plateau. The phases of tree-line oscillations could be correlated by Wick and Tinner (1997) with phases of alpine glacial advances and retreats. Macrofossils, mostly of needles, budscales and other vegetative parts, were essential in this elegant study, and provided a clearer pattern of changes than both the stomatal and pollen records.

In Swedish Lapland, Barnekow (1999) used macrofossil and pollen evidence to detect species-limit changes along a 430 m altitudinal transect from boreal forest to alpine tundra. She found

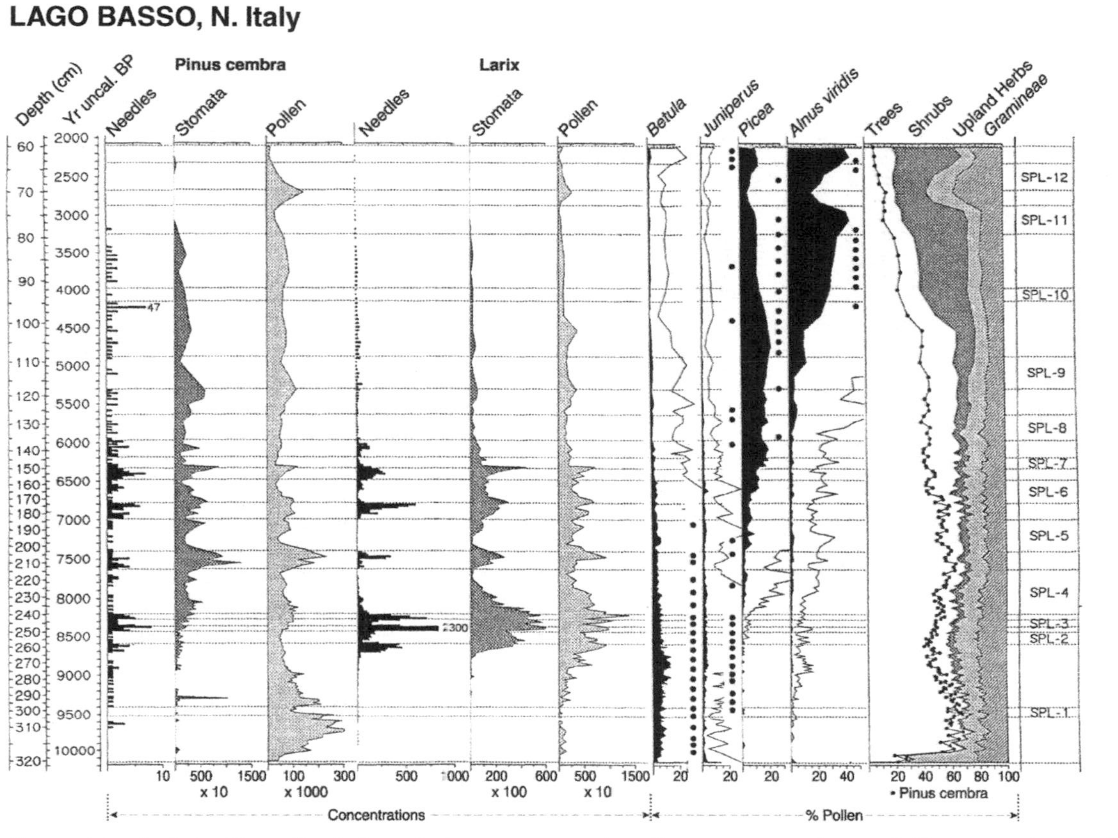

Figure 23.2 Lago Basso, North Italy. Macrofossil (needles), stomata and pollen concentrations of *Pinus cembra* and *Larix*. Macrofossil concentrations are numbers in 45 cm^3 sediment. Stomata and pollen concentrations are number cm^{-3}. Pollen percentages of *Betula*, *Juniperus*, *Picea* and *Alnus viridis*, with their macrofossil records shown as black dots. A summary of the pollen percentage categories is shown on the right. The zones SPL-1 to SPL-12 are interpreted as tree-line depressions. Analysed by L. Wick. After Wick and Tinner (1997).

Early Holocene tree-*Betula* macrofossils 300–400 m higher than today (i.e. 1.5–2.0 °C warmer summer temperatures) in an assemblage suggestive of an oceanic climate. *P. sylvestris* occurred in the valley bottom, but it did not expand until around 5500 ^{14}C yr BP. *Pinus* could not react sooner to the warmer climate because, unlike *Betula,* it cannot withstand deep snow cover. Its expansion suggests the development of a more continental climate with less winter precipitation. It reached about 200 m above its present limit and then descended to its present position after *c.* 4500 ^{14}C yr BP, indicating a mean summer temperature drop of around 1.5 °C over the last 4000 years. At the same time, the *Betula* limit descended around 300 m to its present limit.

Other studies of tree-lines have followed and refined the approach of Blytt and Sernander and used records of tree remains preserved in mountain peats to track altitudinal changes, e.g. Aas and Faarlund (1988, 1996) and Kullman (1995). These so-called megafossils represent spot samples, and each has to be radiocarbon dated. The record depends on the preservation of the tree remains in peat, which probably does not represent the actual altitudinal tree-limit. If climate reconstructions are to be realistic, Birks *et al.* (1996b) point out that it is important to establish the ecological conditions of growth and death of the trees such as moisture balance and peat development, by utilizing detailed peat-stratigraphy, including peat composition, pollen and other macrofossils (e.g. Birks, 1975). In addition, dates on local clusters of stumps or tree-ring measurements can provide overlapping records that can be cross-correlated to reveal population dynamics and patterns of recruitment and mortality that may be related to climate changes (Birks *et al.*, 1996b; Kullman, 1996). *Pinus* remains are better preserved than *Betula* and records of other montane tree and shrub species are sparse or absent (e.g. *Salix*, *Alnus*, *Populus*, *Juniperus*, *Sorbus*). However, large datasets show the former occurrence of trees at higher altitudes in central Scandinavia during the Early to Mid-Holocene suggesting warmer summer temperatures than at present, a steep decline after about 5000 ^{14}C yr BP and then a more gradual decrease to present levels (e.g. Kullman, 1995).

23.3.3 Latitudinal Tree-Limits – Arctic Tree-Line

As for altitudinal studies, the distributions of tree fossils in space and time have been used to track changes in latitudinal tree-limits. The impressive dataset of Kremenetski *et al.* (1998) from across Siberia showed the limits of *Picea obovata, Larix* spp. and *Betula* were already further north than at present by 9500–8000 ^{14}C yr BP, spreading particularly fast at the start of the Holocene, and suggesting summer temperatures *c.* 4–5 °C, or even 7 °C warmer than present. The Early Holocene maximum summer insolation was probably a major influence, resulting in increased continentality, increased penetration of warm North Atlantic waters into the Arctic Seas, and reduced sea-ice cover and lowered albedo, that were combined with eustatically lowered sea-level that moved the coastline some 150 km to the north (MacDonald *et al.*, 2000a). A reversal of these trends resulted in the southward retreat of the tree-line starting about 6000 ^{14}C yr BP and present positions were reached *c.* 4500–3000 ^{14}C yr BP.

The compilation of pine megafossil data in northern Finland (e.g. Eronen and Zetterberg, 1996; Eronen *et al.*, 1999) and the Kola Peninsula (MacDonald *et al.*, 2000b) shows northward and altitudinal extensions of the tree-line 7000–4000 ^{14}C yr BP. The timing of this advance is similar to that found elsewhere in northern Scandinavia, although it is later than in southern Scandinavia and Siberia, indicating regional differences in climate history probably relating to the deglaciation history of the massive Scandinavian ice sheet. The frequency of trees in the

North Finnish 8000-year long ring-matched pine series shows sparse phases which may relate to climatic deteriorations affecting population recruitment and mortality.

In the Mackenzie Delta region of northwestern Canada, Spear (1993) combined pollen and macrofossil data from north of the present tree-line to track tree-line changes during the Holocene. *Picea glauca* and *P. mariana* both grew some 60 km north of their present limits between 9000 and 7000 ^{14}C yr BP. They retreated after 7000 ^{14}C yr BP, partly replaced by *Alnus* scrub and shrub tundra, probably responding to climatic cooling due to decreased summer insolation resulting in moister environmental conditions, peaty soils and increased solifluction.

23.3.4 Treeless Environments

Pollen diagrams from treeless environments often contain considerable amounts of tree pollen derived from long-distance dispersal. Because of the relatively low local pollen production by dwarf shrubs, herbs and grasses, a small influx of tree pollen will register as relatively high percentages. This was demonstrated as long ago as 1940 by Aario in Finland, where *Pinus* pollen reached high percentages in surface samples from treeless vegetation, and more recently by van der Knaap (1987) in surface samples from Spitsbergen, where tree pollen attained values of up to 50 per cent. Thus it is difficult to interpret pollen diagrams from arctic vegetational areas in terms of local vegetation development and thus climate change. For example, Hyvärinen's (1970) Holocene pollen diagrams from Svalbard and Bjørnøya reflect the Early Holocene forest development in northern Fennoscandia by the rise in the AP : NAP (Arboreal Pollen : Non-Arboreal Pollen) ratio and the amount of AP deposited. A small increase in the amount of NAP deposited around 5000 years ago may indicate that the local vegetation and pollen production increased and thus that the climate was more favourable for plant growth. The decline in the AP : NAP ratio together with falling pollen concentrations was interpreted to reflect the overall decline and retreat of forest in northern Fennoscandia as the Late Holocene climate became cooler and wetter. However, the poor taxonomic resolution of the herb-pollen types hindered the interpretation in terms of local floral and vegetational development, and the overall very low pollen concentrations made the whole process of pollen analysis very difficult.

A plant macrofossil study was undertaken on a core from Skardtjørna on the west coast of Spitsbergen by Birks (1991). The record started *c.* 8000 ^{14}C yr BP when macrofossils were at their most abundant, especially remains of *Salix polaris, Dryas octopetala* and *Saxifraga oppositifolia* and seeds of *Silene acaulis, Saxifraga cespitosa* and *S. rivularis.* Remains of *Salix* cf. *glauca, S. herbacea* and *Cassiope hypnoides* were recovered that only grow in inland areas of Spitsbergen today where summer temperatures are *c.* 1.5 °C higher than at the coast, or in areas of greater precipitation such as Hornsund to the south. The vegetation became sparser around 2500 ^{14}C yr BP, as shown in the decline in the concentration of all macrofossils, particularly *S. polaris* and *Dryas*, and the number of taxa per sample. The more warmth-requiring species were no longer recorded, indicating that the vegetation had come to resemble that at the coast today. This interpretation of climate development is more consistent with other Svalbard records from e.g. thermophilous marine molluscs (see Birks, 1991) than with the pattern suggested by Hyvärinen (1970) on the basis of pollen stratigraphy that resembled the climate development in northern Fennoscandia.

The general climatic interpretation of the Holocene pollen and macrofossil records from the largely treeless Great Plains of North America has already been mentioned. The multidisciplinary study at Roberts Creek, Iowa (Baker *et al.*, 1996c) enabled a detailed interpretation to be made of the vegetation around the floodplain sites. The pollen record indicated in broad terms the shift from Late Glacial *Picea*/*Larix* forest to Early Holocene deciduous forest dominated by *Ulmus*, *Quercus*, *Ostrya*/*Carpinus* and *Tilia*. This was replaced around 5500 ^{14}C yr BP by an assemblage dominated by Gramineae (Poaceae) and *Ambrosia* pollen, followed by a return of savanna woodland *c.* 3500 ^{14}C yr BP. Plant macrofossils allowed the reconstruction of the forest vegetation including the understorey in both dry and wet areas, and this was augmented by the bryophyte record. Most striking was the detailed reconstruction made possible of the prairie period vegetation, poorly characterized by the pollen assemblage. A major factor in the conversion to prairie was a large increase in summer fires (see also Grimm, 1983). The Gramineae pollen was produced by prairie grasses, and the shift in δ^{13}C values in the sediments and from a nearby speleothem reflected a change from C_3 to C_4 plants. The detailed macrofossil record allowed the climate change to be interpreted in terms of air-mass changes rather than overall climate warming and drying. The previously prevalent monsoonal pattern bringing moist air to the eastern Midwest broke down at *c.* 5500 ^{14}C yr BP in northeastern Iowa, allowing drier Pacific air to penetrate. The change in seasonality resulted in a shift from the majority of the precipitation falling in the growing season to fewer and more intense rainstorms in spring and drought during the growing season. The latter resulted in more frequent fires which undoubtedly contributed to the demise of regional forests. There is abundant evidence of these fires in the charcoal record and the occurrence of charred and burned seeds and fruits that were produced in summer and autumn. After about 3500 ^{14}C yr BP, oak savanna expanded, accompanied by the return of wetland bryophytes as cooler and moister conditions returned. This multi-proxy study permitted a much more refined reconstruction of climate than from pollen alone, in particular the reconstruction of seasonal changes in precipitation that had large effects on the vegetation rather than gross changes in mean summer temperatures and precipitation. Subsequently, Grimm (2001) has proposed that Mid-Holocene variations in *Ambrosia* pollen and in lake-levels in the northern Great Plains can also be interpreted in terms of changing seasonality of precipitation, with a dryness gradient increasing from east to west.

A Mid-Holocene dry period has been inferred over the whole area (e.g. Barnosky *et al.*, 1987). However, the arid period began later at about 5500 ^{14}C yr BP in northeastern Iowa and southeastern Minnesota whereas some 100 km to the west, e.g. at Kirchner Marsh (Watts and Winter, 1966) prairie vegetation developed about 8000 ^{14}C yr BP. Multi-proxy analyses of floodplain sites in southeastern Minnesota similar to Roberts Creek in northeastern Iowa were used by Baker *et al.* (2002) to position a prairie-forest ecotone to within 100 km here between 8000 and 5500 ^{14}C yr BP.

23.4 VALIDATION OF A POLLEN-BASED CLIMATE RECONSTRUCTION WITH MACROFOSSILS

Climate reconstructions employing the multivariate transfer-function approach (Birks, pp. 107–123 in this volume) using all taxa in a pollen record at a forest-limit situation can be influenced by relatively large amounts of non-locally produced pollen. The transfer functions assume that the total assemblage is locally produced.

Björnfjelltjörn, a small lake in northern Norway at 510 m altitude, lies just above the *Betula* forest-limit today (Birks, pp. 107–123 in this volume). It is surrounded by typical low-alpine plant communities, but a few *B. pubescens* and *Sorbus aucuparia* trees occur in its catchment. The percentage pollen diagram (Fig. 9.3 in Birks, pp. 107–123 in this volume) is dominated by *c.* 40 per cent *Betula* pollen throughout, with higher values of *c.* 50–60 per cent between 10,000 and 8800 cal yr BP. However, at the start of the Holocene, the *Betula* pollen influx is low (Fig. 23.3), suggesting a low local pollen production in which distantly derived *Betula* pollen reached high percentages. Influx increased when organic sedimentation started. The highest influx values occurred 10,800–9500 cal yr BP. *Pinus* pollen percentages were around 10 per cent until 8800 cal yr BP when they rose to a consistent 15–20 per cent, accompanied by *c.* 10 per cent *Alnus* and *Juniperus* pollen. At about 6500 cal yr BP, *Alnus* and *Juniperus* pollen decreased to around 5 per cent and Ericaceae, *Rubus chamaemorus* and *Diapensia lapponica* (not shown on Fig. 9.3 in Birks, pp. 107–123 in this volume) pollen are recorded, accompanied after *c.* 3500 cal yr BP by an increase in *Salix* pollen and the start of a continuous *Picea* pollen record. This diagram is difficult to interpret in terms of local vegetational changes, and one has to infer changes from the minor herb and shrub indicator pollen-types in conjunction with the *Betula* pollen influx.

A mean July temperature reconstruction using transfer functions derived from modern pollen–climate relationships in Scandinavia (discussed by Birks, pp. 107–123 in this volume) shows a consensus pattern (Fig. 23.4) starting above 11 °C and rising to around 11.5–12 °C between 9500 and 6500 cal yr BP. Temperatures fell from the maximum down to 11 °C by 4500 cal yr BP, and then decreased steadily to the present value of about 10.5 °C. The unsmoothed weighted averaging partial least squares (WA-PLS) reconstruction shown on Fig. 23.3 also shows this pattern. The questions arise, was the mean July temperature really greater than 11 °C at the start of the Holocene, was the Early to Mid-Holocene period of maximum warmth applicable locally or was it influenced by the relatively high percentages of long-distance thermophilous taxa such as *Alnus*, and was the Mid-Holocene temperature decrease influenced by lower *Alnus* and *Juniperus* percentages and increased Ericaceae?

These questions can be addressed by an examination of the plant macrofossil record from the same sequence (Fig. 23.3). *Betula pubescens* fruits occurred at the start of the Holocene, implying that *B. pubescens* was growing locally. However, the pollen influx is very low. *Betula* pollen is produced in large quantities for wind dispersal, and the low influx values imply that the pollen was long-distance transported. Winged tree *Betula* fruits are also designed for wind dispersal, and undoubtedly can be blown several kilometres in small quantities, especially over a snow-smoothed winter landscape. Therefore, it is likely that *Betula* was present lower in the valley and the pollen and fruits were blown up above the tree-line. Macrofossil records of *B. nana*, *Dryas octopetala* and *Saxifraga oppositifolia* together with some indicator herb-pollen types (S.M. Peglar, unpublished data) suggest local treeless low-alpine dwarf-shrub heath on relatively base-rich soil. *B. pubescens* fruits and other tree *Betula* macrofossils increased substantially about 10,800 cal yr BP, suggesting that *B. pubescens* actually arrived around the lake then, although it had been in the vicinity over the previous 1500 years. The heavier *B. pubescens* female catkin scales are not recorded until *c.* 9500 cal yr BP. Therefore, the mean July temperature may only have reached 11 °C at about 10,800 cal yr BP, which is the limiting mean July temperature for *B. pubescens* forest growth today in oceanic regions, such as at Björnfjell (Odland, 1996).

The macrofossil record (Fig. 23.3) suggests that *B. pubescens* woodland grew round the lake between 10,800 and 4500 cal yr BP. The understorey was dominated by ferns, suggesting an oceanic western type of birch woodland, containing some *Juniperus*, *B. nana* and *Empetrum*

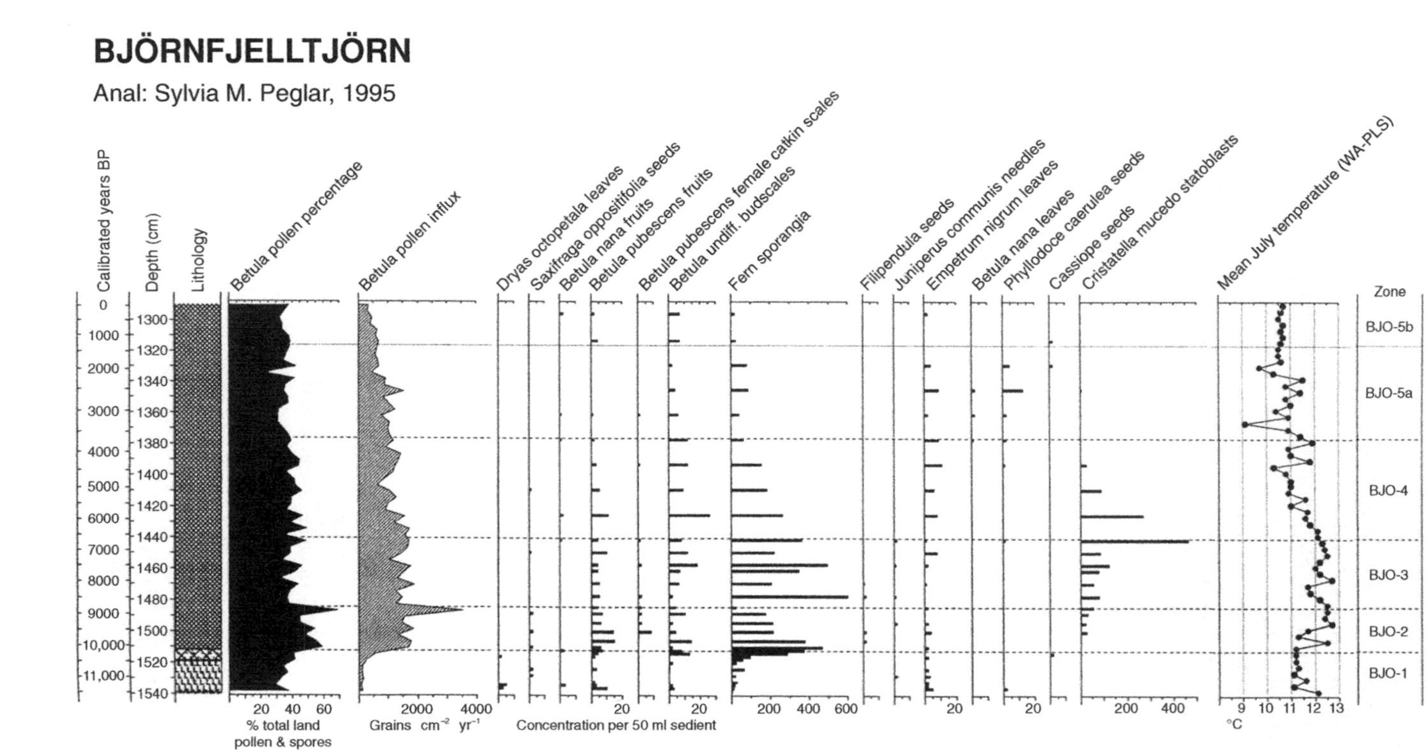

Figure 23.3 Björnfjelltjörn, Nordland, north Norway. Holocene *Betula* pollen percentages and influx, and concentrations of selected macrofossil types plotted against calibrated age. Analysed by S.M. Peglar. Mean July temperature (°C) reconstructed by WA-PLS (Birks, pp. 107–123 in this volume) shown on the right.

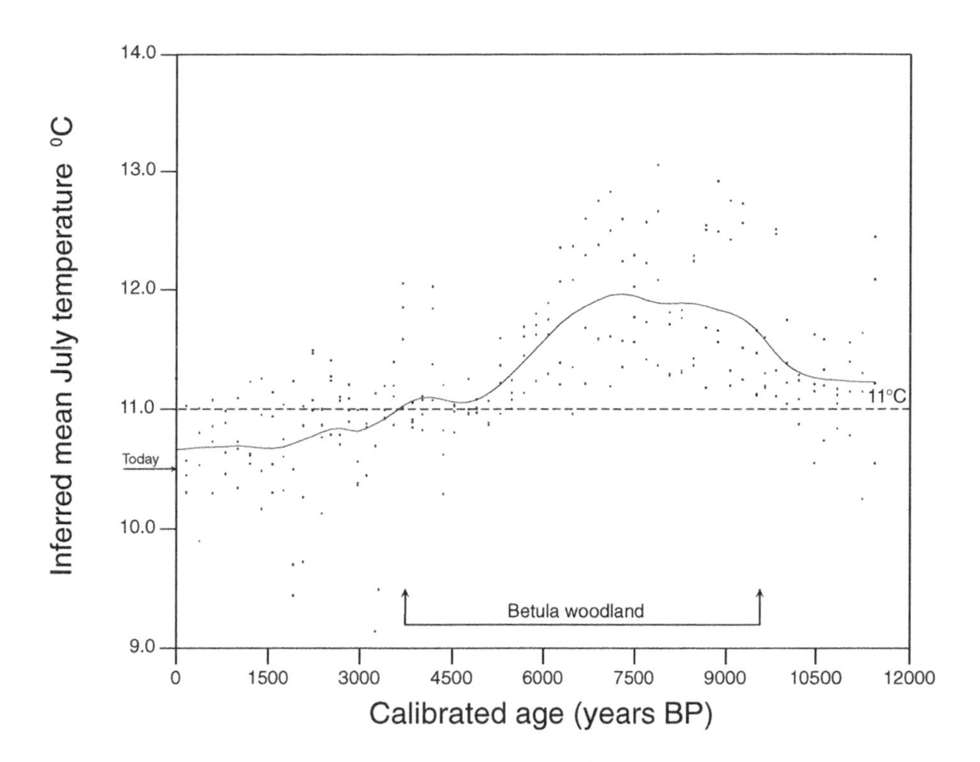

Figure 23.4 Inferred mean July temperature for the Holocene at Björnfjelltjörn. The inferred temperature is based on four different numerical reconstruction methods (weighted averaging, weighted-averaging partial least squares, partial least squares, and modern analogues) (Birks, pp. 107–123 in this volume). The fitted line is a locally weighted least squares regression (LOWESS, span = 0.2, order = 0.2) fitted through all the reconstructed values from the four numerical methods to provide a 'consensus' reconstruction for the Holocene. The reconstructions are plotted against calibrated years BP. The likely extent of *Betula* woodland around Björnfjelltjörn as evidenced by plant macrofossils is indicated, as is 11 °C, the present-day climatic limit of birch woodland in oceanic areas of northern Scandinavia. The present-day mean July temperature of the site is also shown.

(Fig. 23.3). The consensus temperature curve (Fig. 23.4) starts to decrease at about 6500 cal yr BP, reaching 11 °C near 4500 cal yr BP. This is when *Alnus* pollen decreases, suggesting that this change is influenced by the long-distance *Alnus* pollen changes which probably reflected a temperature decrease affecting *Alnus* lower down the valley. At Björnfjell, *Betula* woodland was still present, but probably became more open after 6500 cal yr BP, as shown by a decrease in *Betula* pollen influx and fewer female catkin scales. Ferns decreased and Ericaceae, *B. nana* and *Empetrum* increased, suggesting the development of a heath-rich understorey, perhaps reflecting drier and more continental conditions. An interesting independent line of evidence is provided by the statoblasts of the aquatic bryozoan *Cristatella mucedo* which was common during the period of highest mean July temperatures, but declined after *c.* 6500 cal yr BP and became very rare after 4500 cal yr BP (Fig. 23.3). This is a relatively thermophilous animal which is rare above tree-line in Norwegian lakes today.

Frequent *Betula* remains (including *B. nana*) continued after 6500 cal yr BP, although female catkin scales became rare. Open *Betula* woodland characteristic of the forest-limit probably persisted round the lake until about 3800 cal yr BP. This is the point at which the consensus temperature curve sinks below 11 °C. After this time, *Betula* pollen influx decreased further, although its percentages remained the same. The only marked change in the percentage pollen diagram is the start of a continuous record of *Picea* pollen, although a few alpine plants became better represented. The long-distance *Picea* pollen became prominent in a treeless environment where its pollen was not filtered and local pollen production was low. Tree-birch macrofossils all decreased, suggesting that woodland was no longer present. Macrofossils of *Phyllodoce caerulea* and *Cassiope* are recorded, typical of low-alpine heaths and snowbeds on acid soils. The increased *Salix* pollen percentages may reflect the expansion of low-alpine shrub willows and *S. herbacea*, although there is little macrofossil evidence for this.

In conclusion, the consensus mean July temperature curve based on the pollen stratigraphy (Fig. 23.4) can be validated to a large extent by the macrofossil record (Fig. 23.3) taken in conjunction with pollen influx estimates and the indications of habitat and environmental tolerances of individual pollen and macrofossil taxa. The reconstruction of a mean July temperature over 11 °C for the pioneer phase before *c.* 10,800 cal yr BP is probably an artifact of high *Betula* pollen percentages during a period of very low local pollen production. The following period of temperatures above 11 °C corresponds well with evidence of oceanic fern-rich subalpine birch forest around the lake, that became more open and heathy as the temperature decreased after *c.* 6500 cal yr BP. The reconstructed decrease was probably influenced by the reduced *Alnus* percentages whose origin was relatively local but lower down the valley, but macrofossils show that it also reflected a real summer temperature decrease at Björnfjell. The forest-limit descended below Björnfjelltjörn about 3800 cal yr BP, the few tree *Betula* macrofossils probably originating from scattered local trees at the tree-line and more dense birch forest a few kilometres away at lower altitudes. The temperature curve fell below 11 °C to around 10.5 °C at this time. The agreement from several lines of evidence gives confidence in the quantitative mean July temperature reconstruction from this site. It also shows the advantage of being able to check for false reconstructions, as in the Early Holocene pioneer phase, and to consider the environmental changes interpreted from the other lines of evidence available during times of inferred temperature changes.

23.5 CONCLUSIONS

1. Pollen and plant macrofossils provide complementary evidence about past flora, vegetation and environment.
2. The study of Holocene tree migrations and expansions, of altitudinal and latitudinal tree-line fluctuations and of tundra and prairie history all benefit from combined studies of pollen and plant macrofossils.
3. Plant macrofossils provide an independent means of validating Holocene climate reconstructions based on pollen stratigraphy alone. Macrofossils help to determine what species were actually growing locally. The pollen-based climate reconstruction can be compared with the known ecological tolerances of the species recorded as macrofossils. To be able to use macrofossils, or any other climatic proxy, in this validatory role, it is essential not to combine pollen and plant macrofossils to derive one climate reconstruction

(cf. Huntley, 1994) as the macrofossil data should be independent of the climate reconstruction that requires validation.

Acknowledgements

We thank Sylvia Peglar for the use of her unpublished pollen and macrofossil analyses from Björnfjelltjörn, the Norwegian Research Council for financial support and Heikki Seppä and Arvid Odland for helpful discussions about climate reconstructions and modern biometeorology.

CHAPTER

BIOMARKERS AS PROXIES OF CLIMATE CHANGE

Antoni Rosell-Melé

Abstract: Common applications of biomarkers in palaeoclimate research are briefly reviewed, and the principles behind their use discussed. Biological marker compounds (or biomarkers) are geologically-occurring organic compounds with chemical structures that can be unambiguously linked to natural product precursors. They are thus considered to be molecular fossils. As such, these compounds preserve information regarding the contributions of different sources of biomass to sedimentary organic matter. The use of biomarkers as climate proxies is relatively recent. Their potential in palaeoclimatological reconstruction studies has been increasingly recognized during the last 20 years, but it is only since the 1990s that their application has taken off worldwide, with numerous groups investing in the technology to carry out biomarker measurements. This is probably due to the development of the $U_{37}^{K'}$ index to estimate sea-surface temperature, arguably to date the most successful application of biomarkers. Judging by the publications in journals with highest impact factors, other applications have also generated a great deal of interest, particularly the reconstruction of past variations in atmospheric CO_2, marine primary productivity, relative presence of C_3 versus C_4 plants, and anoxygenic photosynthesis. This information cannot easily be gained by non-geochemical techniques. Thus, biomarkers have earned their own niche in the Earth Sciences and palaeoclimate research.

Keywords: Biomarkers, Climate proxies, Palaeoclimate, Palaeoceanography

24.1 Preliminary Considerations

24.1.1 What Are Biomarkers?

The term biomarker is a contraction of the expression 'biological marker compound' and is used in a variety of scientific disciplines. In palaeoclimate studies it is used to refer to organic molecules, i.e. carbon-based, of relatively low molecular weight (usually less than 1000 mass units), which are found in sediments (see examples in Fig. 24.1). These compounds were initially produced by a variety of organisms either on land or in the aquatic environment. The key factor that determines the value of these molecules (natural products) in palaeoclimate studies is that after their biosynthesis, and the death of the source organisms, they survived deposition to sediments in a recognisable form in terms of their original structure and sterical configuration (i.e. spatial distribution of the atoms). They can thus be considered chemical fossils (Eglinton and Calvin, 1967). The success of these organic components as palaeoproxies largely depends on their resilience to early degradation processes during sedimentation and after incorporation into the sediment.

Molecular organic geochemists have established that the composition of organic matter in sediments is extremely complex, and so are the processes mediating the transformation from the original natural product to the geomolecule. Broadly speaking, there are four major sources of organic matter (and biomarkers), which in order of their importance are: phytoplankton, bacteria, higher plants and zooplankton. Fungi and other organisms, particularly higher animals, do not make significant contributions. All organisms are composed of the same basic chemical classes in different proportions, i.e. carbohydrates, proteins and lipids, with higher plants also containing significant amounts of lignin. It is important to bear in mind that most organic matter is decomposed and recycled. Globally, only 0.3 per cent of the net marine **primary production** (fixed carbon dioxide by plants that it is not respired but available for growth and reproduction) accumulates annually in sediments (cf. Hedges and Oades, 1997). In soils, all organic matter is eventually decomposed and recycled. Preservation of organic matter in terrestrial environments is confined to peat formation in moorland bogs and low-lying swamps and in lakes. Most of the global preservation of carbon occurs in the ocean, in the continental margins, particularly in deltas (Romankevich, 1984; Walsh, 1988). Preservation factors will eventually alter the original proportions of the basic chemical classes. Thus, proteins and carbohydrates are very labile and will be almost completely recycled, whereas lipids and lignins will accumulate preferentially as they are fairly resistant to degradation. However, it is within the category of the lipids that biomarker paleoclimate proxies are found. The term 'lipid' is often defined as any substance produced by organisms that is insoluble in water but extractable by solvents that dissolve fats (see Killops and Killops, 1993). This is a broad definition, which encompasses a wide range of substances used as energy stores by animals and plants (see Fig 24.1) (i.e. **fats**, which are triglycerides formed from the combination of 1,2,3-propanetriol, *I*, so-called glycerol, and three straight chain aliphatic carboxylic acid units, so-called fatty acids; e.g. *II*), cell membrane structure constituents (i.e. **phospholipids**, which are triglycerides containing one phosphoric acid and two fatty acid units combined with glycerol; **glycolipids** in which the phosphate group of a phospholipid has been replaced by a sugar; **ether lipids** in which *n*-alkanols, e.g. *III*, instead of fatty acids are combined with glycerol; e.g. *XIV*), protective coatings (**waxes** comprising a range of substances with high melting points, among these important members are esters of fatty acids with straight chain saturated alcohols, *III*, and hydrocarbons like long-chain *n*-alkanes, *IV*), or photosynthesis (e.g. tetrapyrrole pigments like *XV* and *XVI*). **Terpenoids** are another class of lipids with a variety of functions (e.g. sex pheromones, natural rubber) constructed from C_5 isoprene units. Important terpenoids are the triterpenoids (six isoprene units), which include the steroids, *VII* (in cell membranes act as rigidifiers), and the hopanoids, *X* (control fluidity in bacterial cell membranes); and also the tetraterpenoids of which the most important are the carotenoid pigments, e.g. *XII*.

It should be noted that some lipids appear very resistant to degradation (e.g. aliphatic hydrocarbons), whereas others are more labile (e.g. chlorophylls). The study of the diagenesis of different lipid classes is required to establish which processes affect the accumulation of biomarkers prior to deposition and during the early stages of burial under relatively low temperature and pressure conditions. The overall effect of diagenesis will be a reduction of the absolute amounts within all compound classes with increasing water column and sediment depth. Of course, the rate of degradation will not be the same for all lipids (e.g. Wakeham *et al.*, 1997).

(I) CH_2—CH—CH_2 (OH OH OH) *or* OH OH OH — Glycerol, 1,2,3- Propanetriol

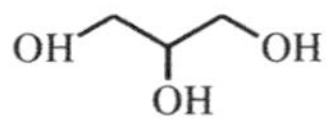

(II) $CH_3(CH_2)_{10}COOH$ — O, OH — Lauric acid, n-Dodecanoic acid

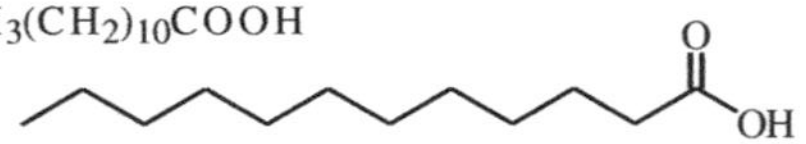

(III) $CH_3(CH_2)_nCH_2OH$ — *n*-alkanol (generally 23< n <29)

(IV) $CH_3(CH_2)_nCH_3$ — *n*-alkane (generally 22< n <34)

(V) *n*-heptacosane

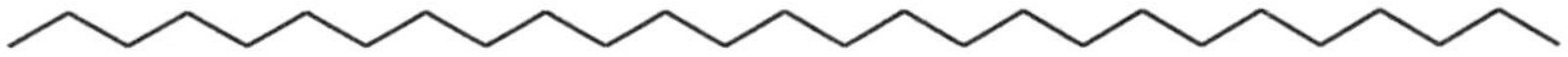

(VI) 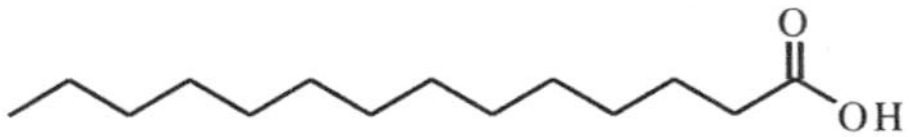 O, OH *n*-tetradecanoic acid

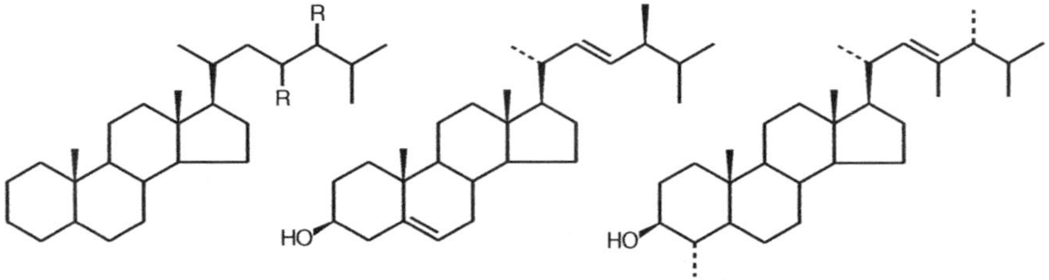

R, R — HO — HO

Steroids basic structure *(VII)*

Diatomsterol *(VIII)*

Dinosterol *(IX)*

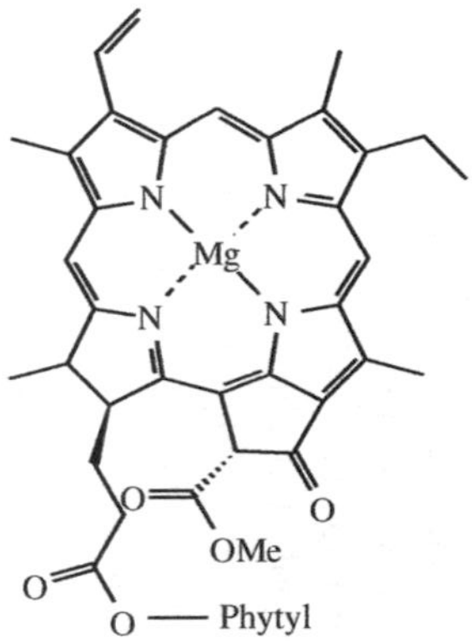

N, N, Mg, N, N — O, OMe, O, O — Phytyl

(XV) Chlorophyll *a*

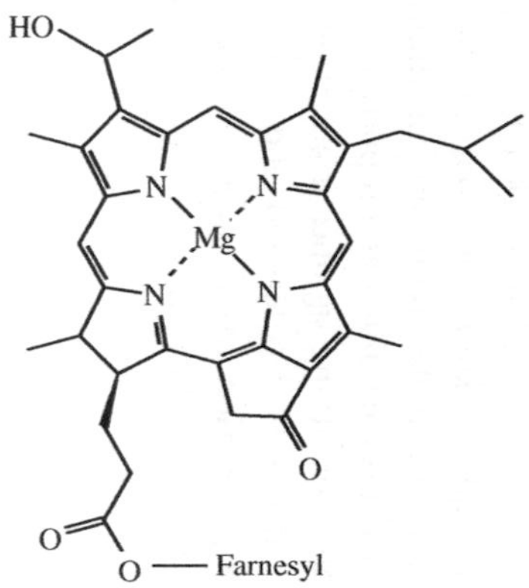

HO — N, N, Mg, N, N — O, O — Farnesyl

(XVI) Bacteriochlorophyll *d*

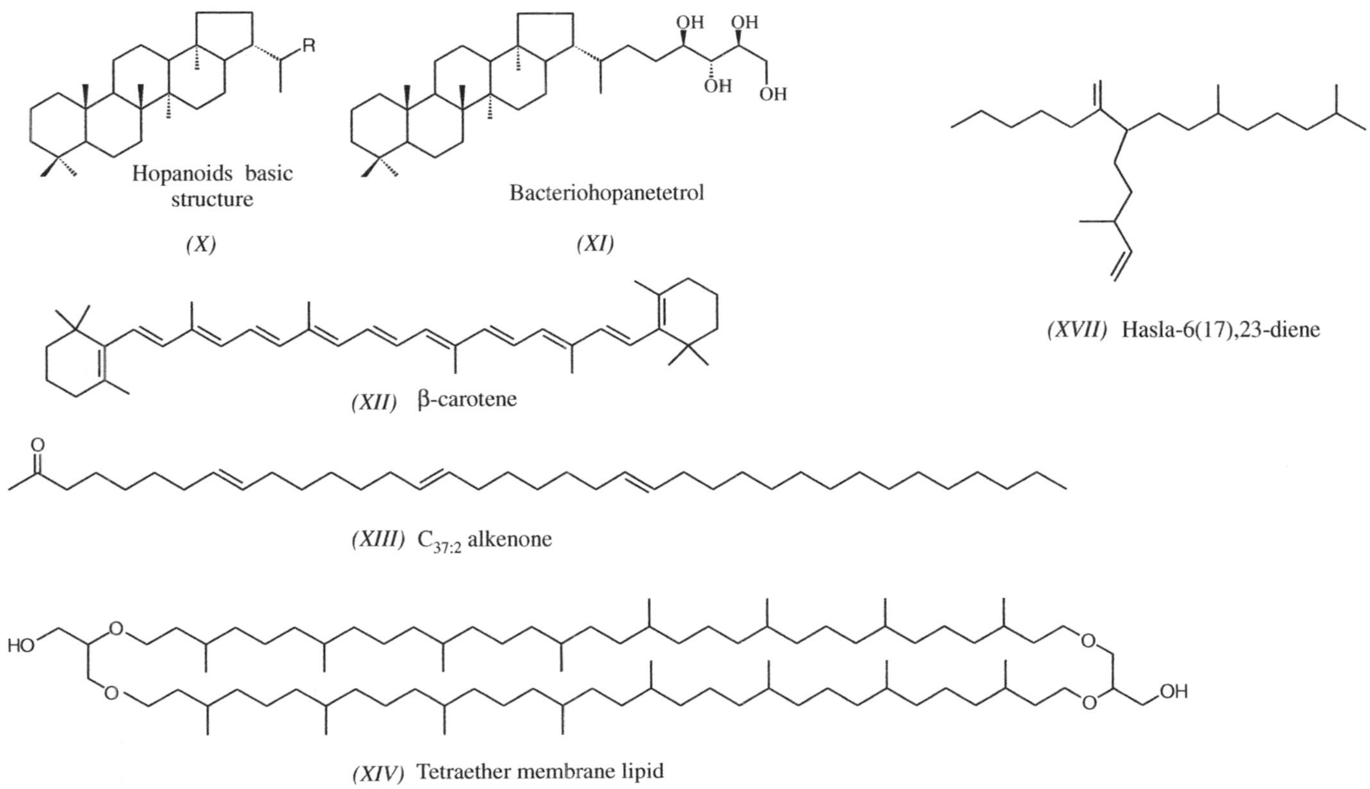

R

Hopanoids basic structure

(X)

OH OH

OH OH

Bacteriohopanetetrol

(XI)

(XVII) Hasla-6(17),23-diene

(XII) β-carotene

O

(XIII) $C_{37:2}$ alkenone

HO

O

O

O

O

OH

(XIV) Tetraether membrane lipid

Figure 24.1 Chemical structures of biomarkers discussed in the text. Structure *(I)* corresponds to glycerol or 1,2,3-propanetriol, and is drawn in two ways to illustrate the procedure followed to represent the structures, where the carbon and hydrogen bonds are omitted for simplicity.

24.1.2 Why Can Biomarkers Be Used as Climate Proxies?

To ensure their viability, organisms have evolved the capability of regulating to different extents the biosynthesis of their constituent biomolecules as a response to changing environmental conditions. Plants and micro-organisms maintain the fluidity of their cell membranes when temperature decreases by lowering the melting point of the constituent lipids. Organisms have also adapted to living under harsh conditions through the production of specific chemicals that protect them from the environment (e.g. plant waxes). They have also developed substances that allow them to extract energy from their surroundings (e.g. photosynthetic pigments) and store it (e.g. triglycerides). Eventually, organisms die and their remains decompose. In the 18th century it was already recognized that soil organic matter is derived from the decomposition of plants and animals (cf. Hedges and Oades, 1997), so that the nature of the organic material deposited in sediments depends on the type of contributing organisms (Didyk *et al.*, 1978). Biomarkers preserve information regarding the contributions of different sources of biomass to sedimentary organic matter. They indicate the type of environment or provide information on an environmental variable at the time that the source organism biosynthesized the component. This information is gained by analysing, in a stratigraphic horizon, for the presence, abundance and isotopic composition of specific biogenic components and their diagenetic transformation products. By studying a biomarker profile through a sedimentary column we can infer how the environment has changed through time. It is the task of the geochemist to figure out how to translate the geochemical data into palaeoenvironmental/climatic information.

24.1.3 How Do Biomarkers Accumulate in Sediments?

This section relates to biomarkers in aquatic environments, and a summary of depositional processes is shown in Fig. 24.2. Biomarkers may have an allochthonous or an autochthonous origin. Autochthonous biomarkers originate close to the site of deposition, whereas allochthonous material is transported from another environment. In both cases the biomarkers are associated to detritus or macromolecular material. This can be part of the original organism (e.g. leaf debris, marine snow), digested remains of the original organism (e.g. faecal pellets), or adsorbed to mineral particles (i.e. **ballast minerals**: silicate and carbonate biominerals and dust), which will eventually settle on the ocean bottom through gravity. In the case of allochthonous materials these are carried to the site of deposition by wind and water (i.e. rivers and currents; e.g. Simoneit *et al.*, 1977; Prahl, 1985; Poynter *et al.*, 1989; Fahl and Stein, 1999). Materials can be further redistributed laterally and vertically by sediment reworking through erosion and redeposition by sediment-gravity flow processes, bottom water currents, tidal movement or contour currents, which are common processes around the continental margins and abyssal depths (see e.g. Weaver *et al.*, 2000). In areas and time periods of intense glaciomarine sedimentological activity, sea-ice and icebergs can also be a vehicle for the transport of continental organic matter to the sea-floor via ice-rafted debris or dropstones, resuspension of marine shelf deposits by iceberg scouring and gravimetric flows caused by advancing ice-sheets (Rosell-Melé *et al.*, 1997; Thomsen *et al.*, 1998; Fahl and Stein, 1999). Clearly, there is a range of modes of transport, which can operate on different time-scales. It cannot be taken for granted that in the same stratigraphic horizon all biomarkers will have the same age, especially those with different origins, as it has been demonstrated through ^{14}C dating of biomarkers (Eglinton *et al.*, 1997). Another complicating factor is that different organic components are associated with particles of different size (Thompson and Eglinton, 1978). During particle dispersion, spatial fractionation of the various chemicals may take place according to the hydrodynamic properties of the carrier particles (Prahl, 1985).

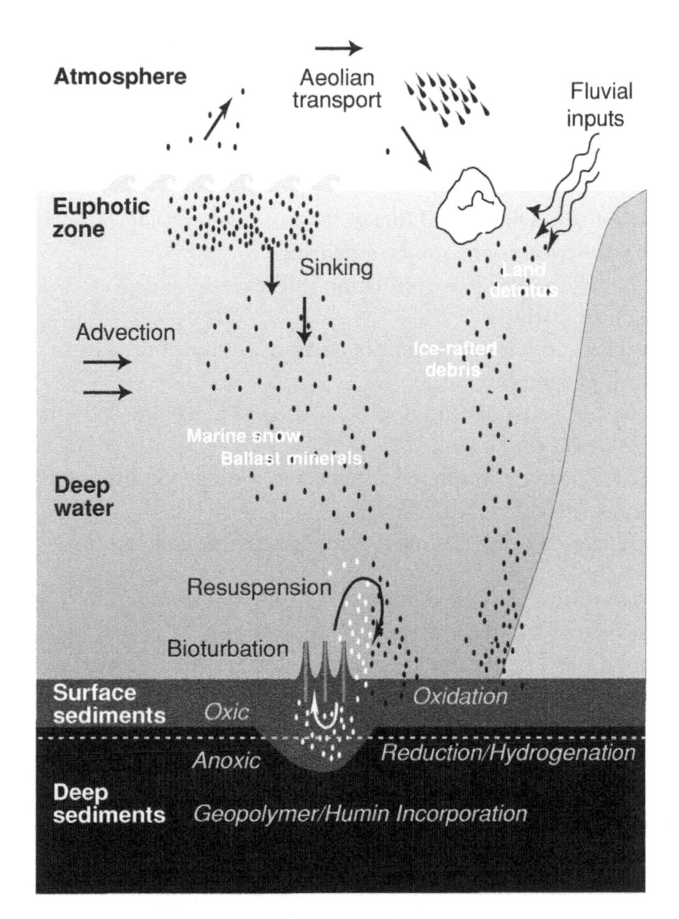

Figure 24.2 Scheme summarizing the depositional processes of biomarkers in aquatic environments. Once deposited in the sediments, biomarkers may undergo further transformations in surface sediments initially in aerobic conditions, and once more deeply buried, in anoxic conditions where free biomarkers may eventually become incorporated into macromolecular materials. In some environments anaerobic conditions prevail in the water column and water/sediment interface, as in the Black Sea or eutrophic lakes.

Consideration must be given to the age of the biomarkers in a sample in relation to the age of other sediment constituents, which are used to investigate other proxy parameters, particularly foraminifera that are typically used to establish the age models of a sediment record. For instance, a study on sediments from the continental shelf and slope off Namibia has found that ages of planktonic biomarkers (alkenones) were 1000–4500 years older than those of foraminifera (Mollenhauer *et al.*, personal communication). Biomarkers and microfossil proxies are associated with particles of different sizes. This separation may lead to a decoupling of different palaeo-records obtained in the same core because bioturbation depends on particle size (Bard, 2001). Lateral advection of (recycled) biomarker-bearing material (e.g. Benthein and Müller, 2000) may also explain age decoupling between biomarkers and other proxy parameters as a function of particle size.

24.1.4 Which Biomarkers Can Be Used as Proxies?

To date only a handful of biomarkers of the thousands present in sediments and soils are used in palaeoclimatic reconstruction. Thus, for a biomarker to qualify as a palaeo-proxy, in principle, it should meet the following requirements:

1. known source organism, which lived in a well-constrained ecological niche
2. known biological function in the source organism
3. resilience to degradation during accumulation in the geosphere
4. established diagenetic pathway
5. established transport mechanisms from decayed organism to burial in the sediment
6. presence over wide geographical areas
7. biological function of the organism and its ecological behaviour have remained unchanged through long geological time spans
8. the biomarker data can be interpreted to relate to a single environmental process (e.g. sea-surface temperature (SST))
9. Amenable to be analysed using routine organic geochemical techniques.

In fact, no biomarker satisfies all these requirements. But this is no serious obstacle for the use of biomarkers as proxies, provided that the interpretation of the data takes into account the caveats of each specific application. The above requirements relate to the study of the absolute content of biomarkers in sediments (or their mass accumulation rates), or the relative abundance/content of a range of biomarker components. If the isotopic composition of a biomarker is the focus of the study, further requirements apply. For instance, carbon isotopic fractionation by phytoplankton is related to changes in ambient CO_2, geometry of the algal cell and growth rates through time, among other factors (see Laws *et al.*, 2002).

Biomarkers are found in sedimentary rocks both 'free' in the bitumen (solvent-extractable) and 'bound' to macromolecular organic matter (solvent-insoluble). Bitumen represents only a small fraction of the total organic matter in sediments (Figs 24.3 and 24.4). The full benefit of

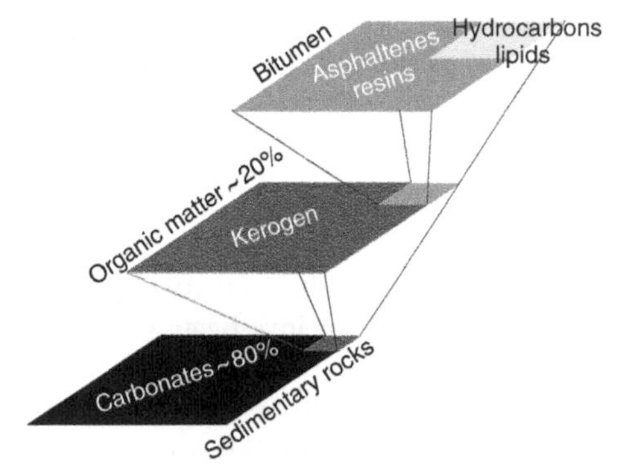

Figure 24.3 Scheme representing the approximate proportion and composition of the various pools of organic matter in sedimentary rocks (after Tissot and Welte, 1984). Biomarkers used in paleoclimate studies belong to the bitumen (fraction soluble in organic solvents), which represents a very small proportion of the total organic matter in sedimentary rocks and is mainly composed by high molecular weight components (i.e. asphaltenes and resins).

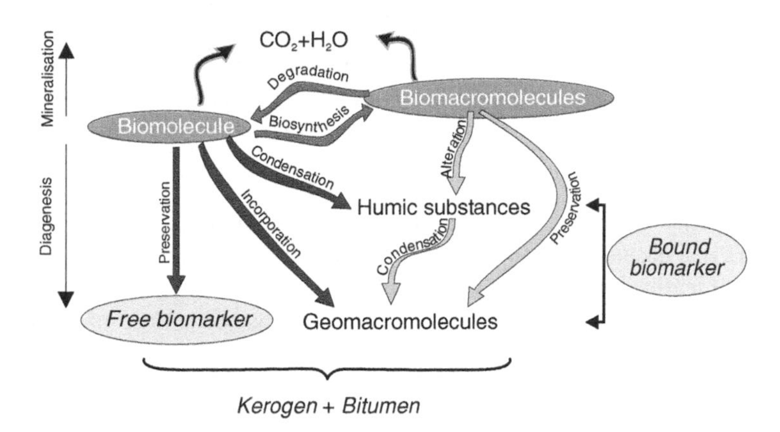

Figure 24.4 Simplified pathways of biomolecules in the environment, reflecting the classical and selective preservation models of kerogen formation. After Tissot and Welte (1984) and Tegelaar *et al*. (1989).

biomarkers as palaeoenvironmental markers is only achieved when the diagenetic reaction pathways from biological precursors to geological products are recognized and understood. However, if the diagenetic transformations of the 'free' sedimentary components are known for some classes of components, those of the potentially dominant 'bound' (into humic materials or kerogen) biomarkers are poorly understood. Hence, the bound biomarker fractions are rarely considered. Nevertheless, the importance of the 'bound' lipids to extend the information provided by the 'free' biomarker distributions for the assessment of depositional environments has been noted, for instance, from sulphur-linked biomarkers released after desulphurisation (Sinninghe-Damsté and de Leeuw, 1990). But these and other attempts have not shown whether the 'bound' distributions are of general significance or only characteristic of very specific ancient sedimentary environments, which are not representative of either past or present global environmental conditions.

24.1.5 Which Biomarkers Are Used as Proxies?

Most biomarker investigations focus on the analysis of marine sediments, whereas there are fewer studies of terrestrial environments, for example studies based on peat-bogs or lake sediments. The most common biomarkers used as climate proxies belong to a few compound types, in particular C_{37}–C_{39} alkenones (e.g. *XIII*), *n*-alkyl lipids (i.e. long-chain *n*-alkanes and to a lesser extent *n*-alkanols and *n*-alkanoic acids; e.g. *II–VI*) and chlorophyll derivatives (i.e. chlorins derived from *XV*). Long-chain alkenones are biosynthesized by a restricted group of phytoplankton belonging to the Haptophyceae (Volkman *et al.*, 1980). This class of algae includes the coccolithophorids and among them the presently dominant species *Emiliania huxleyi*, which is found in all of the world's oceans. In contrast, chlorins can potentially be derived from any photosynthetic organism. In aquatic environments the presence of chlorins in sediments is related to the primary production of phytoplankton (i.e. export net primary production). Chlorins and alkenones are examples of autochthonous biomarkers in marine/aquatic environments, whereas long-chain *n*-alkanes are typical allochthonous components from epicuticular waxes of higher land plants (Eglinton and Hamilton, 1963;

Simoneit, 1977). Other applications of biomarkers have been demonstrated or proposed on compound classes like carotenoid derivatives, hopanoids, fatty acids, sterols, mid-chain diols and alkyl keto-1-ols, *n*-alkenes, phytol and porphyrins among others (e.g. Rosell-Melé and Koç, 1997; Rosell-Melé *et al.*, 1997; Versteegh *et al.*, 1997; Hinrichs *et al.*, 1999; Ishiwatari *et al.*, 1999; Schulte *et al.*, 1999).

The key factors that set apart the most frequently used biomarkers as proxies from other components are that alkenones, *n*-alkyl lipids and chlorins are extremely common in sediments, their sources are identified, and their study provides information on key environmental variables. These three classes of components are biosynthesized by organisms that are spread around the world, in the oceans or on land (e.g. for the alkenones see Müller *et al.*, 1998). Alkenones and *n*-alkanes are also more refractory to degradation than other components (although not perfectly inert), so that sediments become relatively enriched as other compounds are being degraded (e.g. Prahl *et al.*, 1989a; Madureira *et al.*, 1995; Wakeham *et al.*, 1997). The early diagenetic pathways of these three compound classes are reasonably well understood and relatively simple, so that the basic structures of the molecules remain unchanged in immature or shallow sediments. Finally, these biomarkers provide extremely useful information for palaeoclimate research, e.g. on primary productivity (chlorins and alkenones; Summerhayes *et al.*, 1995; Harris *et al.*, 1996; Rosell-Melé and Koç, 1997; Rostek *et al.*, 1997), SST (alkenones; e.g. Brassell *et al.*, 1986; Eglinton *et al.*, 1992; Schneider *et al.*, 1996; Bard *et al.*, 1997; Marlow *et al.*, 2000; Herbert *et al.*, 2001), carbon dioxide concentration (alkenones; Jasper and Hayes, 1990; Pagani *et al.*, 1999), wind intensity (*n*-alkyl lipids; e.g. Poynter *et al.*, 1989; Prahl *et al.*, 1989b; Hinrichs *et al.*, 1999; Ternois *et al.*, 2001) and changes in vegetation patterns on land (*n*-alkyl lipids; e.g. Huang *et al.*, 1999; Yamada and Ishiwatari, 1999).

The opposite can be said for some very abundant families of biomarkers. For example, the hopanoids, *XV–XVI*, are the most abundant complex organic molecules in the geosphere (Ourisson and Albrecht, 1992; Rohmer *et al.*, 1992). Their principal source is a family of prokaryotic bacteria. Despite their widespread occurrence and variety of structures, few viable applications have been proposed so far (e.g. Farrimond *et al.*, 2000). Another example are the haslenes, which are widely distributed C_{25} highly branched isoprenoids (HBI) hydrocarbons (e.g. *XVII*), biosynthesized by a few species of diatoms and potentially very useful to reconstruct environmental conditions of these algae through time (Rowland *et al.*, 2001). Unfortunately HBIs have rarely been found in the stratigraphic column of recent sediments. Sterols are also ubiquitous, and a diverse range of structures have been recognized on a global scale (e.g. *VIII* and *IX*). The complexity of the sterol distributions makes their application as proxies very challenging. Such complexity can be partly attributed to the diversity of their sources including phytoplankton (diatoms, coccolithophorids, dinoflagellates), vascular plants and zooplankton, as well as diagenetic transformations (Volkman, 1986). This diversity of sterol structures has led to the frequent revision of specific sources and to the difficulty in attributing a particular component to a single source. For instance, 24β-methylcholesta-5,22-dien-3β-ol (i.e. diatomsterol, *VIII*) has been wrongly used as marker of unique inputs of algae of the class Bacillariophyceae (i.e. diatoms), because other microalgae also produce them in substantial amounts, including the Haptophyceae (see in Volkman, 1986; Conte *et al.*, 1995). Dinosterol (4α,23,24-trimethyl-5α-cholest-22-en-3β-ol, *IX*) is used as a marker of Dinophyceae (i.e. dinoflagellates; Boon *et al.*, 1979), but not all dinoflagellates synthesizes this sterol (Volkman *et al.*, 1998); and it has also been found in a marine diatom as a minor constituent (2.0–3.6 per cent of total sterols; Volkman *et al.*, 1993).

24.2 INTERPRETATION OF BIOMARKER DATA

Interpretation of biomarker data relies on determining the presence, absolute and relative concentrations and isotopic composition of specific components. In the following sections these analytical parameters are discussed, and examples of applications are given.

24.2.1 Significance of the Presence of Certain Biomarkers

The absence or presence of a specific biomarker in a stratigraphic horizon should be interpreted taking into account:

- the sensitivity of the analytical techniques employed
- the diagenetic processes that may have led to the removal of a biomarker from the environment
- those sedimentary processes leading to its accumulation, ascertaining the autochthonous versus allochthonous origin of the component.

In the interpretation of autochthonous organic matter sources, a particularly useful application of biomarkers is the recognition of anoxygenic photosynthesis in the original water column. At present, the Black Sea is the only large marine water body with an extended euxinic layer reaching into the photic zone (Repeta *et al.*, 1989). Such conditions in the geological past can be recognized in palaeo-water bodies by a molecular approach, using characteristic bacteriochlorophyll (*XVI*) derivatives or certain aromatic isoprenoid hydrocarbons together with their $\delta^{13}C$ signatures (Ocampo *et al.*, 1985; Summons and Powell, 1986; Keely and Maxwell, 1993) that indicate contributions of photosynthetic anaerobic bacteria. It is believed that such conditions occurred widely in the past and may have stimulated the accumulation of organic-matter-rich shales (Passier *et al.*, 1999). For instance, Sinninghe-Damsté *et al.* (1993) investigated the distributions of carotenoid pigments in core samples from the Black Sea and showed that photosynthetic green sulphur bacteria (Clorobiaceae), which are strict anaerobes, had been active in the Black Sea for substantial periods of time during the last 6200 years.

Contributions of lipid matter from vascular land-plants and algae can be distinguished by the chain length of *n*-alkyl lipids, which are generally shorter for algae, i.e. < 20 carbon atoms. For example, land plant contributions are indicated by the presence of C_{27} *(V)*, C_{29} and C_{31} *n*-alkanes, or C_{24}, C_{26} and C_{28} *n*-alkanoic acids (e.g. Eglinton and Hamilton, 1963; Cranwell, 1973; Rieley *et al.*, 1991). In contrast, algal inputs are characterized by C_{17} *n*-alkane or C_{12}, C_{14} *(VI)* and C_{16} *n*-alkanoic acids (e.g. Cranwell *et al.*, 1987), although these components are produced by all plants albeit in lower amounts than in algae. This approach has been used by a number of researchers, for example, to identify wind borne or riverine contributions of land plants in marine sediments (Poynter *et al.*, 1989; Prahl *et al.*, 1989b; Hinrichs *et al.*, 1999; Ternois *et al.*, 2001). In continental environments a similar approach has been applied to investigate changes in English bog vegetation (Nott *et al.*, 2000), or to obtain a measure in an African lakes of the sedimentary input from submerged or floating aquatic macrophytes relative to that from emergent and terrestrial species (Ficken *et al.*, 2000).

24.2.2 Measurement of Absolute Concentrations to Estimate Production Fluxes

The concentration of a biomarker is often related to the flux of the component from its source to the burial site. The objective being to reconstruct changes in the productivity of the source organism (e.g. for chlorins as indicators of primary productivity). Changes in productivity of an algal group can be related to factors such as nutrient abundance and ecological structure, which in turn relate to climatic conditions. The contents of allochthonous biomarkers have been used to establish the occurrence and intensity of a transport process (e.g., *n*-alkyl lipids in deep-sea sediments as indicators of palaeo-wind intensity; e.g. Poynter *et al.*, 1989; ten Haven *et al.*, 1992; Cacho *et al.*, 2000). Such assessments are prone to error unless they take into account other processes which influence the content of a biomarker in the sediment, namely, degradation during transport and burial of the biomarker and dilution by mineral matter. Poynter and Eglinton (1991) defined a basic model to link the concentration of a biomarker in a sediment to its export production below the photic zone:

$$[i]_{sed} = \frac{[Flux_i\,(1 - \alpha_i)]}{BSR} \qquad 1$$

where $[i]_{sed}$ is the concentration of compound i in a sediment; $Flux_i$, the export primary production of compound i from the photic zone; α_i, the fraction of compound i degraded or transformed during transport, deposition and burial; and BSR, the bulk sedimentation rate. It is possible to calculate the latter by measuring the density and age of a sediment unit. Thus, the concentration of a biomarker in a sediment is often expressed in terms of its mass accumulation rate. Unfortunately, to date there is no reliable method to measure or estimate α_i for past environments. Hence, at present the true palaeo-flux of organic matter cannot be determined. This is compounded by the fact that in modern environments degradation rates of biomarkers in the water column and surface sediments can be higher than 99 per cent (e.g. Wakeham and Lee, 1993; Wakeham *et al.*, 1997). Hence, small variations in the processes that control biomarker degradation rates (e.g. oxygen contents, enzymatic reactions, particle surface areas, sedimentation rate) can lead to important variations in accumulation rates, resulting in an unknown degree of uncertainty in the interpretation of the data. Degradation rates are different for different biomarkers. Hence, higher abundance of one species over another may just indicate that some biomarkers have been preferentially preserved, less degraded, rather than a higher production flux. Comparison of data from a range of sites must also account for local factors that may have enhanced preservation. In consequence, extreme care has to be taken when interpreting biomarker mass accumulation rates.

Nevertheless, preservation factors may have not influenced significantly the mass accumulation rates of biomarkers in some settings over extended periods of time. In a number of records from the Atlantic and Indian Ocean, productivity-related proxies (namely alkenones) show a 23-kyr cyclicity. Various authors have argued that this cyclicity reflects changes in export productivity driven by precession during the glacial periods (Rostek *et al.*, 1997; Schulte *et al.*, 1999; Budziak *et al.*, 2000). Given the correlation found between the records of biomarker mass accumulation rates and microfossils, particularly linked to nutrient variability, it seems quite likely that the biomarker records in these studies reflect changes in productivity through time. It is by this process of correlation between proxies that conclusions on changes in export productivity through time can be drawn with a degree of confidence.

24.2.3 Measurement of Relative Concentrations

The calculation of a molecular ratio can avoid some of the caveats discussed in the previous section regarding the use of mass accumulation rates of biomarkers, i.e. not to know the fraction of compound degraded or transformed. Thus, it can be argued that if two components have the same source and are structurally and stereochemically very similar, their diagenetic pathways are bound to be equivalent. Hence, using equation 1 and for two components i and j, where $\alpha_i = \alpha_j$, the ratio of their concentration in the same sediment sample is:

$$\frac{[i]_{sed}}{[j]_{sed}} = \frac{[Flux_i]}{[Flux_j]} \qquad 2$$

where the ratio of two components is equal to their export fluxes (after Poynter and Eglinton, 1991). In practice, it is very difficult to establish that the α values are the same for two different molecules. Even small structural differences may lead to divergent diagenetic pathways or enhanced degradation of one of the components. Eventually, for two components i and j what must be considered is whether the ratio:

$$\frac{(1-\alpha_i)}{(1-\alpha_j)} \approx 1 \qquad 3$$

approaches 1 so that any information derived from equation 2 is climatically significant, rather than related to the relative accumulation efficiency of the molecules in the sediment.

In fact, the most successful application of biomarkers to palaeoceanography is based on this assumption, i.e. the U^{K}_{37} index, based on the relative content in sediments of long chain alkenones (Brassell *et al.*, 1986; Prahl and Wakeham, 1987):

$$U^{K'}_{37} = \frac{37:2}{37:2 + 37:3} \qquad 4$$

(37 : 2 stands for concentration of C_{37} alkenone with 2 double bonds)

U^{K}_{37} has been steadily gaining acceptance for palaeotemperature reconstruction, and an increasing number of research groups worldwide are now successfully measuring it routinely using in-house facilities, as illustrated by the results of an interlaboratory comparison study (Rosell-Melé *et al.*, 2001). The success of the index is related to the resilience of the alkenones to degradation (Prahl *et al.*, 1989a; Teece *et al.*, 1998), and that it provides a representative picture of SSTs in a wide variety of depositional settings (Müller *et al.*, 1998). Since the inception of U^{K}_{37} numerous studies have successfully used this proxy to reconstruct sea-surface paleotemperatures in records spanning tens, thousands, to millions of years (e.g. Eglinton *et al.*, 1992; Schneider *et al.*, 1996; Bard *et al.*, 1997; Emeis *et al.*, 2000; Marlow *et al.*, 2000; Herbert *et al.*, 2001; Marchal *et al.*, 2002).

As argued by Poynter and Eglinton (1991), in some circumstances different α values for two different molecules with the same source and equivalent fluxes can be exploited to infer conditions of deposition leading to enhance preservation of one molecule over the other.

$$\frac{[i]_{sed}}{[j]_{sed}} = \frac{(1-\alpha_i)}{(1-\alpha_j)} \qquad 5$$

In this case the success of the application will depend on the export fluxes of both biomarkers (Flux) being sufficiently close to 1 so that any information derived from equation 5 is representative of the relative accumulation efficiency of the molecules in the sediment. However, Poynter and Eglinton (1991) claimed that no molecules were known that, whilst having different structures, were produced in a constant ratio. Nevertheless they proposed that the ratio of *n*-alkanes to *n*-alkanol from higher plants approached such condition at least in their data set (Gulf of Guinea and Mediterranean sapropels), so that the *n*-alkanol to *n*-alkane ratio was a function of sample degradation. Subsequently, some workers have extrapolated this observation and used such ratio to infer changes in oxygen content in the water column through time (Cacho *et al.*, 2000). However, caution should be exercised with this index as no systematic study has been carried out to establish which are the processes that lead to the preferential degradation of *n*-alkanols. It is not well established either if the relative production rates of *n*-alkanes versus *n*-alkanols in continental settings and their fluxes to the ocean can be considered constant in any particular location and through time. Organic matter reactivity is a function of material matrix as well as inherent lability, so that chemical reactivity does not correspond simply with compound class or structure (e.g. Wakeham and Lee, 1993; Keil and Cowie, 1999; Hedges *et al.*, 2001).

24.2.4 Measurement of Isotopic Composition

Compound-specific isotopic analysis (CSIA) has become one of the most groundbreaking lines of biomarker research in recent years. It can be carried out on any isotopes present in any organic molecule, but it is usually undertaken by measuring the relative abundance of two stable-isotopes of carbon (^{13}C, ^{12}C) in components separated by gas chromatography (Hayes *et al.*, 1990). The measurement is expressed as the ratio of the two isotopes in organic matter relative to a standard (i.e., δ values with units of per mille, ‰):

$$\delta^{13}C = \left(\frac{^{13}C/^{12}C \text{ in sample}}{^{13}C/^{12}C \text{ in standard}} - 1 \right) \times 1000 \qquad 6$$

Carbon compounds of biological origin are enriched in the lighter isotope (^{12}C), while the heavier isotope (^{13}C) is retained in the main forms of inorganic carbon (e.g. carbonate, bicarbonate and carbon dioxide). The amount of carbon isotopic fractionation in the tissue of the plant is dependent on the ratio of the concentration of CO_2 inside the plant to that in the external environment (see Fogel and Cifuentes, 1993). The lightness of biogenic substances results from isotope fractionation processes in the main assimilatory pathways of carbon, and to a lesser extent the ensuing metabolic reactions. A key process is the carbon isotopic fractionation by the enzyme 1,5-biphosphate carboxylase (RuBP carboxylase or RUBISCO), responsible for all carbon dioxide fixation in green plants, algae and most autotrophic bacteria (O'Leary, 1988). Plants that utilize the C_4 pathway for photosynthesis have an alternate enzyme, phosphoenolpyruvate carboxylase (PEP carboxylase) for one of the first steps of carbon dioxide fixation. The isotopic fractionation associated with these enzymes is different, and C_4 plants have a more positive isotope composition in the range of –8 to –18‰ in contrast to the $\delta^{13}C$ of most plants (–20 to –30‰) (O'Leary, 1981; Deleens *et al.*, 1983). A diversity of biogeochemical reactions, like those from chemoautotrophic organisms, also fractionate carbon isotopically in a characteristic manner, so that CSIA provides a method to investigate the sources and transformation of carbon in the environment (e.g. Freeman *et al.*, 1990; Hayes *et al.*, 1990).

One of the most revealing applications of CSIA relies on determining the relative contributions of C_3 and C_4 plants on land to aquatic sediments or peat by determining the $\delta^{13}C$ composition of *n*-alkyl lipids (Ficken *et al.*, 1998; Huang *et al.*, 1999, 2000). C_3 plants include trees, shrubs and cool-climate grasses, whereas C_4 plants include tropical grasses and sedges. Past vegetation changes can be determined from the stratigraphic record of pollen and spores. It is impossible, however, to distinguish between C_3 from C_4 plants using pollen records. Stratigraphic determination of $\delta^{13}C$ in *n*-alkyl lipids can thus be used to determine the emergence of C_4 plants (tropical grassland), for instance, in Africa in relation to changes in partial pressure of CO_2 or rainfall (Ficken *et al.*, 1998; Huang *et al.*, 1999, 2000).

Another useful application of CSIA is to infer oceanic and atmospheric CO_2 partial pressures through the $\delta^{13}C$ values of C_{37} long-chain alkenones (Jasper and Hayes, 1990; Pagani *et al.*, 1999). The reliability of the final pCO_2 estimates depend on the value of the assumptions and proxies used to process the isotopic data. Results of laboratory experiments and field studies imply that the isotopic fractionation of *E. huxleyi*, the main alkenone producer, changes mainly as a function of cellular growth rate, nutrients and ambient CO_2 (see Laws *et al.*, 2002). The coherent downcore variations in estimated pCO_2 of some of the results to date might be taken as an indication of the potential of the approach despite the inherent uncertainties of the method. Some of these uncertainties may be overcome measuring the Sr/Ca ratio in coccoliths from alkenone producers, to infer growth rates and to isolate the CO_2 fractionation effect, and hence to obtain more reliable estimates of past CO_2 values (Stoll *et al.*, 2002).

Measurements of ^{14}C in individual biomarkers by ^{14}C-accelerator mass spectrometry, AMS, is also possible (Eglinton *et al.*, 1996, 1997). Fossil and contemporary sources can be distinguished with ^{14}C AMS, and an apportionment of the different contributions estimated. Components with similar $\delta^{13}C$ values may display different ^{14}C ages (Eglinton *et al.*, 1997). A presumed fossil biomarker should contain no ^{14}C, while an autochthonous compound should contain the $^{14}C/^{12}C$ ratio characteristic of the living biosphere at the time of deposition. Phytoplanktonic biomarkers can be used as a proxy for the ^{14}C concentration of ocean surface waters (Pearson *et al.*, 2000), so an independent sedimentary chronostratigraphy could be constructed based on appropriate marine biomarkers.

24.3 CONCLUDING REMARKS

Changes in key climatic parameters and the carbon cycle can be inferred from studying biomarkers in sediments. The study of biomarkers has afforded novel insights, for instance, into past variability of sea-surface temperature, marine primary productivity and carbon dioxide concentration. It is particularly useful that the information gained is often quantitative, which can be related to climate model reconstructions. These successful applications show that the use of biomarkers in paleoclimatology has ceased to be seen as an application with potential, to be valued by the wider Earth Sciences scientific community as an approach with the same scientific credibility as those granted to classical approaches (e.g. microfossil or palaeomagnetic analyses).

Most advances in the field of biomarkers have been obtained using gas chromatography hyphenated to mass spectrometry, the workhorses of the organic geochemists. As new analytical techniques are developed, and multidisciplinary investigations are conducted, it is likely that new applications will keep being proposed and applied. These may range from the use of the well known C_{37} alkenones to infer changes in surface ocean salinity (Rosell-Melé, 1998), to the lesser known archaeal tetraether lipids to infer marine temperatures and the relative amount of fluvial exported terrestrial carbon in marine sediments (so-called TEX86 and BIT, respectively) (Schouten *et al.*, personal communication). Perhaps CSIA will provide in the future some of the best insights into the climatic evolution of the Earth. For instance, an exciting approach is the use of hydrogen isotopes as proxies of paleotemperature/precipitation (e.g. Anderson *et al.*, 2001). Hopefully, as new analytical techniques are implemented (e.g. hyphenation of liquid chromatography and isotopic ratio mass spectrometry) larger and thermally labile biomolecules will be studied and other isotopes will also be, eventually, routinely measured on biomarkers, like nitrogen isotopes in chlorophyll (Sachs *et al.*, 1999). Finally, perhaps new sources of applications will come up from the relatively less studied continental and, particularly, lake environments. The latter aided by new lake drilling programs like GLAD800 (see http://www.dosecc.org/GLAD800/glad800.html), which may play the same key role that the Ocean Drilling Project played in the study of the oceans.

CHAPTER

25

MULTI-PROXY CLIMATIC RECONSTRUCTIONS

André F. Lotter

Abstract: This chapter presents an overview of the potential of multi-proxy studies for Late Quaternary climatic reconstructions using lacustrine sediment records. Choice of sites, proxies and their climatic sensitivity, leads and lags of different proxies, and chronology are discussed and illustrated by three case studies of Late Glacial, Younger Dryas and Holocene multi-proxy summer temperature reconstructions. Common trends and consistencies in multi-proxy data make climatic reconstructions more credible, whereas differences among multi-proxy records call for a critical evaluation of the strengths and weaknesses of the proxies involved.

Keywords: Climatic reconstruction, Leads and lags, Reaction times, Transfer functions

This chapter deals with multi-proxy climatic reconstructions resulting from the study of biotic and abiotic materials preserved in continental lacustrine deposits. The focus and the selection of examples are the results of a personal bias. Other important types of environmental archives for multi-proxy climatic reconstructions are dealt with elsewhere in this book (e.g. marine deposits, historical records, glaciers and others).

The study of past and present global climatic change and the multiple interactions among geosphere, hydrosphere and biosphere call for a holistic approach to understand these complex and often non-linear relationships. Traditionally, every discipline in palaeoscience has concentrated on developing its own methods to reconstruct past climate. Depending on the specific proxy investigated the focus has usually been on one particular climatic variable. Biotic proxies such as vegetation or aquatic insects have mainly been interpreted in terms of the temperature of the growing season, whereas information from closed lake basins, raised mires, or abiotic archives such as glaciers and river deposits has focussed on precipitation. A wealth of mono-proxy climatic reconstructions go back to the early days of Quaternary science, such as the plant-indicator species approach of Iversen (1944). Even earlier, Gams and Nordhagen (1923) inferred major Late Quaternary climatic changes based on the combination of different types of evidence from European lacustrine deposits, an approach that today would be called a 'multi-proxy' study.

Such multi-proxy studies are becoming increasingly popular, not only in palaeoclimatology but also in palaeoecology and palaeolimnology including biotic assemblages (e.g. pollen, plant macrofossils, diatoms, ostracods, molluscs, chrysophytes, chironomids, cladocerans, coleopterans, see relevant chapters in this book) and/or sedimentary variables (stable-isotopes, grain size, organic and inorganic geochemistry, mineralogy, loss-on-ignition, magnetic sediment properties, etc.). Such studies, however, are very labour-intensive and need careful planning. The major advantage that make multi-proxy studies attractive is the potentially independent lines of evidence they offer for environmental reconstruction. Every proxy has its own strengths and weaknesses, depending on such characteristics as its occurrence and abundance, temporal and spatial resolution, reaction and recovery time, and the climatic variables it approximates (e.g. Guiot, 1991, but see also chapters in this book). In a multi-proxy approach we seek to take advantage of the strengths and identify the weaknesses of each proxy with the aim of building on the consistencies and explaining the disagreement among proxies.

25.1 MULTI-PROXY STUDIES: TYPES AND TROUBLES

In principle, two main approaches may be distinguished: single-site and multi-site studies. Observations and interpretations from single-site studies form the basis of any palaeoclimatic reconstruction. Depending on the sediment-sample size needed for analysis (e.g. <1 cm^3 for most microfossils; several cm^3 for plant macrofossils; up to several kilograms for Coleoptera) the single-site approach may require the use of different cores from the same deposit. Such different stratigraphical sequences then need to be correlated on the basis of either conspicuous stratigraphical marker horizons (e.g. laminae, tephras, etc.) or by using core-correlation techniques such as loss-on-ignition (LOI), magnetic susceptibility, or stable-isotopes. Different proxies may have dissimilar time resolution, so corresponding samples from different cores may represent unequal amounts of time. Depending on the importance of the temporal resolution of the study, this may create problems, e.g. a comparison of rates of change on a high-resolution time scale may be impeded. Preferably, however, all analyses should be carried out on the same core to minimize errors of stratigraphic correlation. Moreover, the sampling intensity and the sample intervals for all proxies should ideally be identical to avoid the above-mentioned difficulties.

Multi-site studies link different proxy data from single-site studies by using a common chronological framework to produce a climatic reconstruction in space and/or time. An example of such a multi-site approach is the reconstruction of European precipitation patterns for Holocene time-intervals by the combined use of lake-level and pollen proxies (Guiot *et al.*, 1993a). In another example, Huijzer and Isarin (1997) assembled botanical, zoological, periglacial, glaciological, aeolian, fluvial, lacustrine and pedological evidence in a multi-proxy database to reconstruct the Weichselian pleniglacial climate in northwestern and central Europe. A reliable chronology is of major importance in such an approach. In many cases, regional or even global overviews and reconstructions are hampered because of poor time control. This is especially the case for the Last Glacial and the Late Glacial periods or for sites in extreme settings such as high-altitude or high-latitude regions, where little suitable dateable organic matter is available for dating, even by accelerator mass spectrometry (AMS).

25.2 ABIOTIC AND BIOTIC PROXIES

The benefit of combining two or more proxies, preferably from the same sediment core, to reconstruct past climate is evident from the wealth of scientific literature. Among abiotic climatic proxies, stable-isotopes play a prominent role (see Leng, pp. 124–139 in this volume). Since the

1970s numerous studies have stressed the potential of combining oxygen isotope analyses with pollen-analytical studies (e.g. Eicher and Siegenthaler, 1976; Lotter and Zbinden, 1989; Ralska-Jasiewiczowa *et al.*, 1992; Andrieu *et al.*, 1993; Hammarlund and Buchardt, 1996; Ahlberg *et al.*, 2001; Tinner and Lotter, 2001). Most of these studies have been carried out in Europe and focus on the Late Glacial period. Nevertheless, there are also such combined studies from North America (e.g. Fritz *et al.*, 1987; Lewis and Anderson, 1992; Yu and Eicher, 1998) as well as from the Holocene of Turkey (e.g. Lemcke, 1996; Reed *et al.*, 1999) and Africa (e.g. Lézine *et al.*, 1990). Close similarities among lacustrine lake marl and ostracod $\delta^{18}O$ records, and oxygen isotope variations in Greenland ice-cores suggest that the lake records are sensitive and respond rapidly to broad climatic shifts (Siegenthaler *et al.*, 1984; von Grafenstein *et al.*, 1999). The major and minor shifts recorded in the Late Glacial $\delta^{18}O$ curves (see Lotter *et al.*, 1992; Björck *et al.*, 1998) reflect changes in the North Atlantic thermohaline circulation that eventually led to climatic oscillations (e.g. Broecker, 2000). It is thus attractive to use the $\delta^{18}O$ signal as proxy for climatic change and then to observe the reaction of biotic and abiotic systems on both sides of the Atlantic (e.g. Björck *et al.*, 1996c; Yu and Wright, 2001). Several studies using stable-isotope records as a proxy for climatic change and pollen as a proxy for vegetation show no substantial lag of vegetation response to major climatic oscillations (e.g. Lotter *et al.*, 1992; Ammann *et al.*, 2000; for further discussion see below).

With biotic climatic proxies, one of the major points of discussion still is whether the organism of interest reacts directly or is triggered by climate-induced changes of habitat or substratum or, mainly in the case of aquatic organisms, is mediated through changes in the lake (e.g. length of ice-cover, mixing regime, anoxia, etc.) as well as in the catchment (e.g. vegetation cover, erosion, nutrient export, etc.). Furthermore, because different organisms have dissimilar ecological and climatic thresholds, it is also important to assess biota-inherent tolerances to particular climatic variables. In recent years, quantification of biotic proxies in terms of climatic variables has helped to elucidate some of these issues. Empirical regression models (transfer functions; see Birks, pp. 107–123 in this volume) infer climatic variables based, for example, on estimates of optima and tolerances of different taxa. Different organisms have been numerically related to one or more environmental variables that are directly (e.g. temperature, precipitation) or indirectly (e.g. lake-water salinity, pH) linked to climate. Applications of such transfer functions to downcore biotic assemblages allow quantitative reconstructions of past climatic variables. Nevertheless, climate as well as other environmental factors can produce similar signals in a proxy record, and it is therefore essential to evaluate the results of such quantitative climatic reconstructions critically. Pienitz *et al.* (2000) give an example of an indirect multi-proxy climatic reconstruction. By applying diatom, fossil-pigment and mineralogical analyses to a sediment core from a lake in the Canadian subarctic, they inferred changes in the salinity of this high-latitude lake that also reflect Holocene changes in effective moisture. Using pollen, plant-macrofossil, diatom and sedimentological analyses in combination with chironomid-inferred temperatures, Levesque *et al.* (1994) and Mayle and Cwynar (1995) estimated summer cooling in Atlantic Canada for the Younger Dryas.

25.3 CASE STUDIES OF LATE QUATERNARY MULTI-PROXY CLIMATIC RECONSTRUCTIONS

In the following section three case studies of multi-proxy climatic reconstructions are discussed. Most multi-proxy climatic reconstructions in the literature deal with the Late Glacial period, as this period gives evidence for the highest amplitude of climatic and environmental change since the end of the last ice-age, whereas Holocene climatic changes are smaller and therefore more difficult to detect. The choice of the three case studies reflects this situation: the first two examples deal with Late Glacial and Early Holocene climatic reconstructions, whereas in the third example multi-proxy climate inferences for the Holocene are discussed.

25.3.1 Lobsigensee: Late Glacial Climatic Change

Lobsigensee is a small kettle-hole lake on the Swiss Plateau where Ammann (1989b) used more than a dozen cores for pollen and other biotic and abiotic proxies. A littoral core was analysed for Coleoptera (beetles) and Trichoptera (caddis flies; Elias and Wilkinson, 1983) as well as stable oxygen isotopes (Eicher and Siegenthaler, 1983). More than three dozen macrofossil samples of terrestrial vegetation radiocarbon-dated by AMS provided an excellent high-resolution Late Glacial and Early Holocene chronology (Ammann and Lotter, 1989).

The sediments older than 12,700 (conventional radiocarbon years) BP contain pollen of heliophilous herbs such as grasses and *Artemisia* as well as pollen and macrofossils of willows and dwarf birch (*Betula nana*). The vegetation during this period consisted of shrub tundra. Beetles and caddis flies of modern boreo-montane distribution such as *Chilostigma siebaldi* were present in these sediments, suggesting mean July temperatures between 10 and 12 °C according to modern analogues (Elias and Wilkinson, 1983). Figure 25.1a shows the similar modern distributions for *C. siebaldi* and *B. nana* (mean July temperature ≥7 °C, see Huijzer and Isarin, 1997 and references therein). At 12,700 BP a first sharp increase in $\delta^{18}O$ values indicates climatic warming (Lotter *et al.*, 1992). Both pollen and macrofossils of dwarf birch decreased as well as remains of cold-indicating caddis flies. Simultaneously, the first records of *Typha latifolia* pollen and of the temperate plant-independent beetle (*Donacia cinerea*) occur (see Fig. 25.1). The Bølling and Allerød vegetation is characterized by open birch woodland and pine-birch woodland, respectively. According to the insect assemblages the mean July temperature was between 14 and 16 °C, whereas *T. latifolia* may grow at temperatures of 12–14 °C today, but is more common when mean July temperatures are ≥15 °C. However, the temperature estimates based on such plant-indicator species have to be considered merely as threshold values, and the occurrence of such indicator plants will not give any information as to the extent to which this threshold had been crossed (see also discussion in Birks, 1981).

Whereas the climatic cooling of the Younger Dryas is evidenced by a decrease in the $\delta^{18}O$ values (Fig. 25.1c) and a renewed increase of heliophilous herb pollen, indicating an opening of the pine-birch woodland around 10,800 BP, the insect assemblages at Lobsigensee give no evidence for cooling (Fig. 25.1b). This contrasts markedly with results from Britain (see Fig. 25.1b) and

Figure 25.1 Multi-proxy climatic reconstruction for the Late Glacial period at Lobsigensee (Central Swiss Plateau). Modified after Ammann (1989a). (a) Comparison of modern biogeographical ranges of the caddis fly *Chilostigma siebaldi*, the beetle *Donacia cinerea*, the dwarf birch (*Betula nana*), and hazel (*Corylus*). (b) Insect-inferred Late Glacial July temperatures for Lobsigensee (grey band, according to Elias and Wilkinson, 1983) and for Britain (solid line, according to Coope, 1977). The timings of occurrence or immigration of constituents of the insect and vegetation assemblages are also shown. (c) Simplified oxygen isotope curve for the Swiss Plateau showing the three major shifts in bulk-carbonate $\delta^{18}O$ (after Lotter *et al.*, 1992). The time-scale refers to conventional (uncalibrated) radiocarbon years BP.

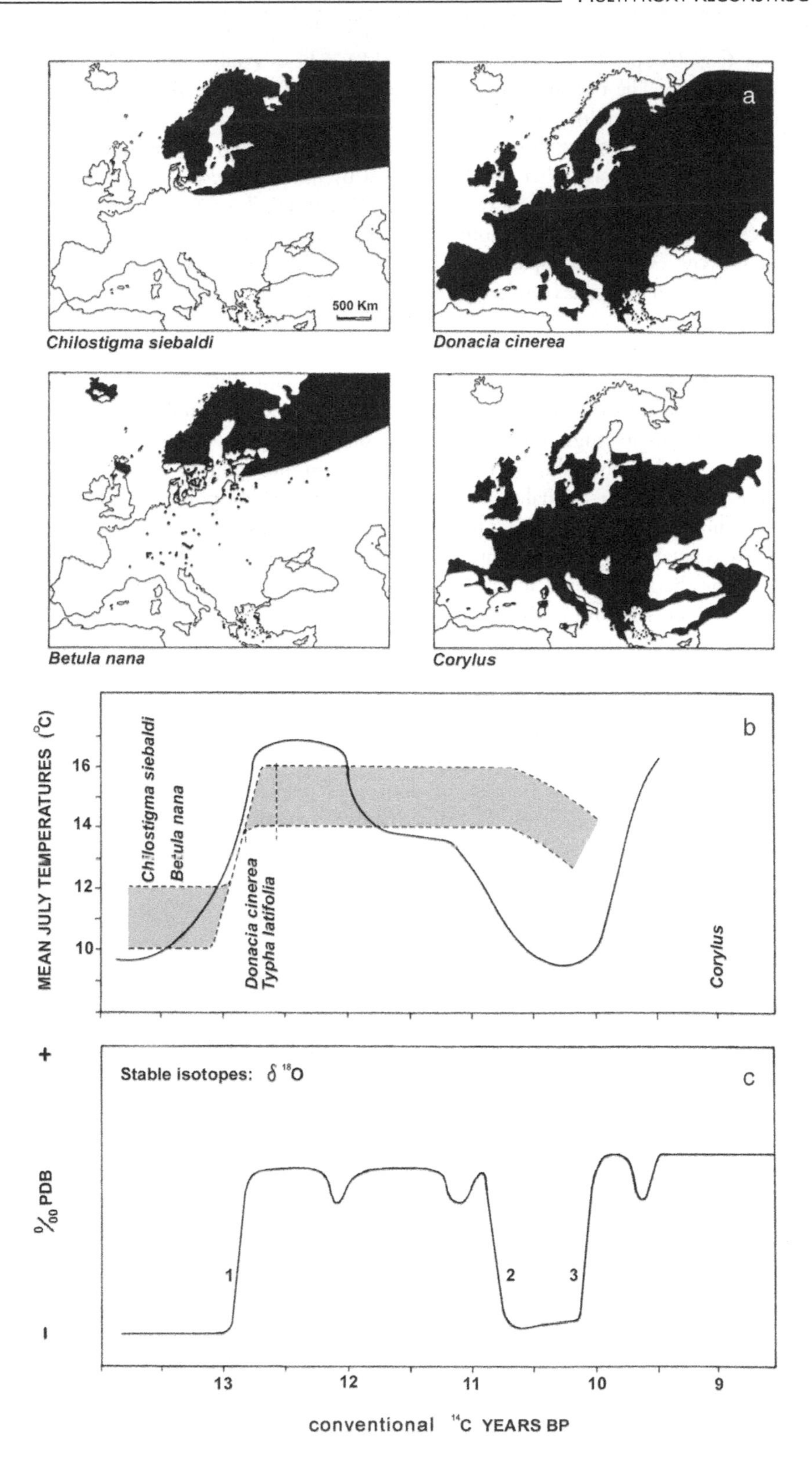
a
500 Km
Chilostigma siebaldi
Donacia cinerea
Betula nana
Corylus
b
MEAN JULY TEMPERATURES (°C)
16
14
12
10
Chilostigma siebaldi
Betula nana
Donacia cinerea
Typha latifolia
Corylus
c
Stable isotopes: δ 18O
+
‰ PDB
-
1
2
3
13
12
11
10
9
conventional 14C YEARS BP

Scandinavia (e.g. Coope, 1977; Coope *et al.*, 1998), where a substantial coleopteran-inferred cooling is indicated. The transition between the Younger Dryas and the Pre-Boreal at 10,000 BP (i.e. the onset of the Holocene) is marked by another major increase in oxygen isotopes (Fig. 25.1c) together with a decrease in herb pollen. During the Pre-Boreal the pine-birch woodland in all of Central Europe was rather rapidly replaced by hazel (*Corylus*) and some centuries later by other temperate deciduous trees. *Corylus* has a modern distribution somewhat similar to that of *D. cinerea* (see Fig. 25.1a). It is noteworthy that the arrival and expansion of this shrub, which today occupies a climate space similar to that of the temperate beetle (*D. cinerea*), took more than 3500 years, whereas the arrival of the reed-plant *T. latifolia* was contemporaneous with that of the beetle.

This example demonstrates the advantage of insects and aquatic organisms as climate proxies, or the possible disadvantages of trees. Once climatic conditions are favourable, the amount of time needed for organisms to migrate depends on factors such as the distance from refugia, their dispersal abilities and the availability of suitable habitats in the region to be colonized. Moreover, changes in productivity (e.g. annual seed or pollen production) in response to climatic change may also play an important role for long-lived organisms such as trees (Ammann *et al.*, 2000). According to Iversen (1954, 1964; but see also Ritchie, 1986) water plants may react fast to climatic change because of their short life-cycle, efficient dispersal and their pedogenic independence. These reasons are also valid for (vegetation-independent) insects (e.g. Coleoptera, Trichoptera, chironomids) and aquatic organisms (e.g. water plants, Cladocera, algae such as diatoms and chrysophytes). These short-lived organisms are therefore most likely to be the best fossil bioindicators for Late Glacial climatic change. If vegetation is used as a climatic proxy, it is important for it to be in equilibrium with climate (see e.g. Birks, 1981; Davis, 1984; Prentice, 1986; Webb, 1986). In the case of the lagged reaction of hazel, Ammann (1989a) favours the hypothesis of a migrational lag because of poor seed dispersal and life-cycle characteristics. Furthermore, increased seasonality and continentality during the Younger Dryas, with very cold winters and late frost, may have inhibited or delayed the expansion of the frost-sensitive hazel, whereas the onset of milder winters during the Holocene may have favoured its rapid expansion throughout Europe during the Early Holocene.

25.3.2 Gerzensee And Kråkenes: High-Resolution Climatic Change of the Younger Dryas

Two independent high-resolution studies from Switzerland and Norway focus on the Younger Dryas as an example of an abrupt, high-amplitude climatic change. Both studies illustrate very elegantly the use of quantitative multi-proxy climatic reconstructions to assess the magnitude and rate of climatic change.

Gerzensee is a small kettle-hole lake located in the sub-montane belt of the Swiss Plateau. An oxygen isotope oscillation of Allerød age was named after this site (Eicher and Siegenthaler, 1976; Lotter *et al.*, 1992). This oscillation has not only been found in the Greenland ice-cores (Björck *et al.*, 1998) but also in LOI stratigraphies and chironomid-inferred temperatures from Atlantic Canada, where it is called the Killarney oscillation (Levesque *et al.*, 1993a, 1993b). A multi-proxy investigation of biotic response to climatic change was undertaken on a core consisting of Late Glacial lake marl (Ammann, 2000). At Kråkenes, a small coastal lake close to a former cirque glacier in west central Norway, a similar study was carried out (Birks *et al.*, 1996a). An extensive series of radiocarbon dates on terrestrial macrofossils was calibrated to produce a calendar-year chronology for Kråkenes (Birks *et al.*, 2000), whereas at Gerzensee a high-resolution chronology was derived by synchronizing the detailed lake-marl oxygen isotope stratigraphy with the GRIP $\delta^{18}O$ record (Schwander *et al.*, 2000). This oxygen isotope stratigraphy is assumed to record the hemispheric temperature changes without lag and is used to indicate the commencement of the climatic change at the beginning and end of the Younger Dryas (see Fig. 25.2).

At both sites quantitative temperature reconstructions based on weighted averaging partial least squares (WA-PLS) regressions and calibrations (see Birks, pp. 107–123 in this volume) were carried out. At Gerzensee pollen and benthic cladoceran WA-PLS reconstructions were used to infer mean summer temperatures (average of June, July and August) (Lotter *et al.*, 2000). At the onset of the Younger Dryas the pollen-inferred mean summer temperatures dropped by 2–3 °C over *c.* 135 years, whereas the cladoceran-inferred temperatures decreased by 2–4 °C over *c.* 225 years. At the end of the Younger Dryas the pollen-inferred temperatures increased by 2–3 °C over *c.* 160 years, while the Cladocera imply an increase of 5–6 °C in *c.* 400 years (see Fig. 25.2). At Gerzensee the inferred rates of temperature change are thus about 0.6 °C per 25 years at the beginning and about 0.3 °C per 25 years at the end of the Younger Dryas. At Kråkenes pollen, cladoceran and chironomid WA-PLS models as well as plant-macrofossil reconstructions based on modern analogues were adopted to infer mean July temperatures (Birks and Ammann, 2000). Although all the investigated organisms exhibited high rates of change (Birks *et al.*, 2000), it is only the plant macrofossils and chironomids that show a marked fall in Younger Dryas mean July temperatures (Fig. 25.2). At the end of the Younger Dryas all organisms imply a steep rise in temperature of 4–6 °C. At Kråkenes the rates of temperature change are about 0.7 °C per 25 years at the onset and between 0.2 (chironomids) and 0.3 °C per 25 years (Cladocera) at the end of the Younger Dryas.

This example demonstrates how independent climatic reconstructions based on terrestrial and aquatic biota may complement each other. The ranges and rates of temperature increase and decrease revealed at these two sites more than a thousand kilometres apart are, within the chronological constraints, surprisingly comparable and reflect rates comparable to those observed in the Greenland ice-cores. However, differences between the two reconstructions do exist. The inferred Allerød summer temperatures at Gerzensee were generally 2–4 °C warmer than at Kråkenes, whereas the Younger Dryas was up to 4 °C colder in western Norway. Here, the comparison with the temperature inferences by aquatic organisms and with plant macrofossils (tree birches established locally around 10,900 cal yr BP) point to the fact that the pioneer Allerød tree-less vegetation was not yet in equilibrium with climate. If the steeply increasing cladoceran-inferred temperatures for the Early Holocene at Gerzensee were real, then the pollen-inferred temperatures that levelled off around 11 °C would point in the same direction.

25.3.3 Sjuodjijaure: Holocene Climatic Change

An increasing number of quantitative reconstructions based on transfer functions give evidence for large temperature changes for the Late Glacial and Early Holocene periods. For the Holocene period, however, only a few quantitative reconstructions and hardly any multi-proxy reconstructions are available. The smaller amplitude of climatic change during the Holocene is usually close to the limits of the prediction errors of the inference models. A multi-proxy approach providing independent reconstructions is therefore of great importance.

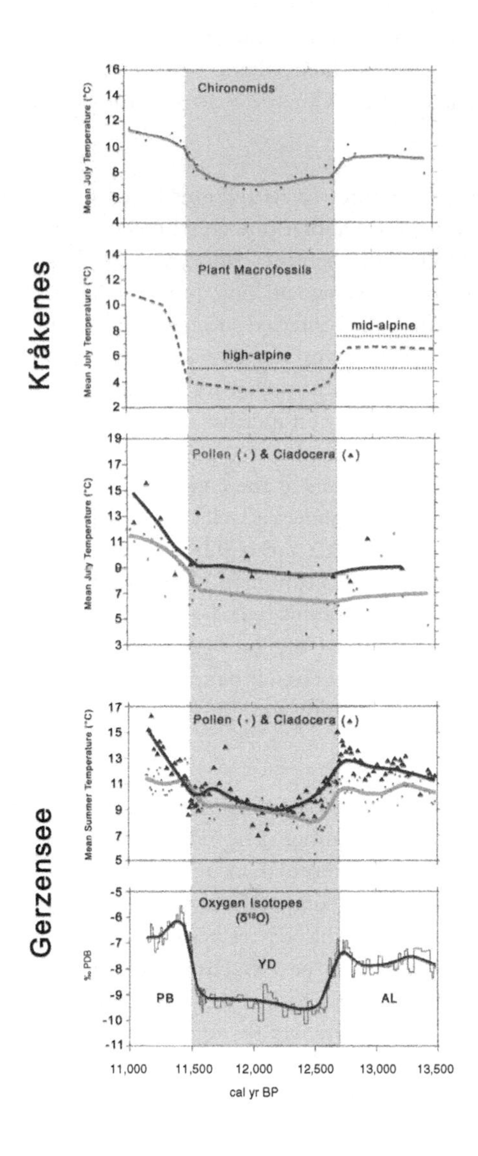

Figure 25.2 Multi-proxy climatic reconstructions at Gerzensee (Switzerland, data from Lotter *et al.*, 2000) and Kråkenes (western Norway, data from Birks and Ammann, 2000). The Gerzensee chronology refers to GRIP ice-core years BP, whereas the Kråkenes chronology is based on calibrated radiocarbon dates. The mean summer (average of June, July, August) and mean July temperatures were inferred from weighted averaging partial least squares regression and calibration models. For plant-macrofossil temperatures modern vegetation analogues were used, and the horizontal dotted lines delimit the mean July temperatures for mid- (5–7.5 °C) and high-alpine vegetation zones (<5 °C). The lines drawn through the individual estimates are LOWESS smoothers (span: 0.25). AL = Allerød; YD = Younger Dryas; PB = Pre-Boreal.

Rosén *et al.* (2001) investigated sediments of Sjuodjijaure, a small lake above tree-line in the Scandes Mountains of northern Sweden. The lake is located in an ecotonal area where small climatic changes may trigger large biotic reactions. The chronology of the core is based on five AMS radiocarbon dates on terrestrial plant remains. Regional WA-PLS transfer functions were used to infer mean July temperatures for the past 9300 years from diatoms, chironomids, pollen and near-infrared spectroscopy (NIRS) of organic sediment. The inferred Holocene temperatures based on these four proxies show a generally similar development (see Fig. 25.3).

Between 9300 and 7300 cal yr BP all reconstructions indicate short-term temperature variations. These are well within the model-inherent errors of prediction (0.9–1.7 °C). However, some of these rapid oscillations are simultaneously manifested by different proxies and can thus be accepted with more confidence: the decreases in LOI values around 8500 and 7600 cal yr BP are concurrent with oscillations in the diatom- and chironomid-inferred temperatures of 0.6–1.7 °C. The LOI decrease around 8200 cal. BP is contemporaneous with the pollen-inferred temperature oscillation (cooling of 0.8 °C and subsequent warming of 1.7 °C) and, to a lesser degree, those evidenced by diatoms and chironomids (0.5–0.6 °C). Depending on the proxy, however, only few or no convincing modern analogues exist in the calibration datasets for the Early Holocene. Rosén *et al.* (2001) are rightly cautious about the interpretation of the reconstructions for this period. The analogues are considerably better after ca. 7300 cal yr BP and may therefore be more reliable.

Around 7300 cal yr BP, a major increase of between 1.5 and 1.7 °C in mean July temperature is indicated by all proxies except the pollen, which may be affected by a migrational lag of the vegetational constituents. Subsequently, the inferred temperatures began to decrease gradually, and lower temperatures then characterize the Late Holocene. Considering the chronological uncertainties, the inference that the climate was warmest during the Mid-Holocene compares well with other investigations from northern Scandinavia.

Correlations among the different proxy-based temperature inferences for the entire Holocene are highest between chironomids and NIRS ($r = 0.92$) and lowest for pollen and NIRS ($r = 0.32$), whereas for the period from 7300 cal yr BP to today they are highest between diatoms and chironomids ($r = 0.84$) and diatoms and pollen ($r = 0.83$) and lowest for NIRS and pollen ($r = 0.45$). The differences in correlation between the different proxy-derived mean July temperatures is not surprising because each of these biotic and abiotic proxies reacts with a different time lag to climatic change. Moreover, July air temperature is not the only controlling factor, for it is correlated with several other climate-related and limnological factors.

To assess Holocene climatic change on a regional or even continental scale, more such multi-proxy studies are needed. By comparing similarities in the different proxy-inferred climatic parameters (see e.g. Rosén, 2001), such studies will eventually not only allow one to assess long-term climatic change but also, given the necessary time-resolution, short-term trends.

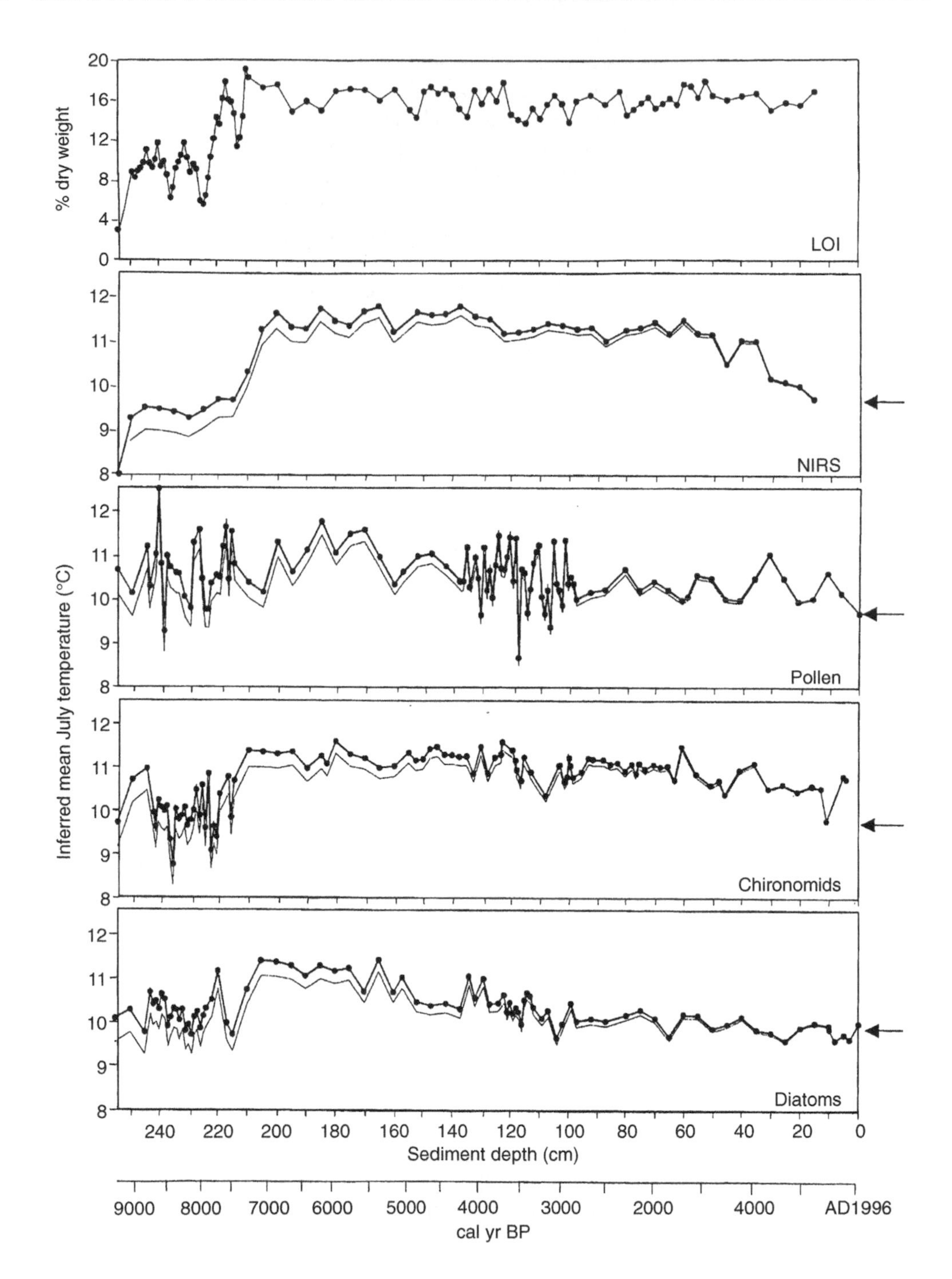

Figure 25.3 Loss-on-ignition (LOI) data and multi-proxy mean July temperature reconstructions from diatoms, chironomids, pollen and near infrared spectroscopy (NIRS) for a sediment core of Sjuodjijaure (northern Sweden). The arrows indicate the modern mean July temperature (9.8 °C) at the lake and the dotted lines represent the inferred July temperatures corrected for land uplift. Data from Rosén *et al.* (2001).

25.4 CONCLUSIONS

Multi-proxy studies are increasingly popular to reconstruct past environmental change at different spatial and temporal scales. As such studies are extremely laborous and time-demanding they need a rigorous design, careful planning and meticulous site selection.

The choice of proxies for such studies is paramount. They should be sensitive climatic indicators (i.e. their variations in the fossil record should be attributable to climate change) and complementary in their ability to reconstruct climate. Fast-reacting proxies such as e.g. stable-isotopes, insects, or aquatic organisms may help to extract the climatic signal directly, whereas long-lived proxies such as upland vegetation and especially trees may show delayed reactions. The combination of these types of proxy may help register leads and lags in the reaction of different biotic and abiotic systems to climatic change and thus enhance our 'understanding of community ecology under a changing climate, including equilibrium or disequilibrium between communities and climate' (Ammann, 1989a). Detecting common trends between independent proxies may make reconstructions more credible, whereas identifying differences calls for critical evaluations.

The chance of detecting past climatic change, especially during the Holocene, is highest at sites that are located at ecotones. Here small changes in climate are likely to lead to biotic response that may be traced in the sedimentary record. However, during times of low-amplitude climatic change such as the Holocene, one of the drawbacks of the multi-proxy approach may be that ecotonal situations are organism-specific, i.e. some biota may react at a site, whereas others are not affected (Lotter and Birks, unpublished data).

Ideally, all proxies should be studied on the same sediment core to minimize the necessity of core correlation and problems of comparison of the observed changes. If fast climatic changes are to be studied, cores with high sediment-accumulation rates that allow fine-resolution studies are preferable. Particular attention must always be paid to chronology. A reliable time-scale (preferably in calendar years) is not only of paramount importance for unambiguous long-distance comparison of climatic reconstructions but also for the estimation of rates of change. Multi-site studies will eventually help in assessing trends and interpretations of climatic reconstructions on a broader spatial scale. More such multi-proxy climatic reconstructions are needed to elucidate geographical patterns of climatic change that may be used as boundary conditions to constrain climate models or as a means of validating Earth System model results.

Acknowledgments

Many of the ideas in this chapter have been drifting around for a long time but discussions in particular with Brigitta Ammann, John Birks, and members of the CHILL-10,000 group helped to crystallized them over the years. Herb Wright Jr and John Birks made valuable comments on the manuscript, Hilary Birks generously provided the data for the Kråkenes reconstructions, and Leonard Bik produced the diagrams. JH patiently endured a sunny 'holiday' in which this chapter was produced. My cordial thanks go to all of you and also to Miguel Torres (10) for company and Cardenal Mendoza for spiritual support.

CHAPTER 26

HOLOCENE CLIMATES OF THE LOWLAND TROPICAL FORESTS

Mark Bush

Abstract: The three major tropical regions have contrasting histories of Holocene climate change. Each region reveals a complex interaction of climatic, environmental and cultural change that prompted modern land use and biogeographic patterns. The onset of the Holocene was marked by some changes of global scale, e.g. rising atmospheric CO_2 concentrations, temperatures and sea-levels, whereas other factors were essentially local, e.g. precipitation, edaphic development and competition. Despite the profound environmental changes associated with the Pleistocene–Holocene transition, in general, lowland tropical forest communities formed modern communities earlier than their temperate forest counterparts. Following the initial warming of the Holocene, temperatures were fairly constant. Consequently, the largest Holocene climatic variables in the lowland tropics were the amount and seasonality of precipitation. Through much of the Holocene the El Niño Southern Oscillation (ENSO) strongly influenced precipitation in the neotropics and Asia. In Africa, orbitally-driven variability in insolation may have outweighed the influence of ENSO. Africa is also different from the other areas as it is the driest of the continents and is the only one to exhibit major biome range changes in the last 11,000 years. In the neotropics, local movement of ecotonal boundaries rather than broad biome change was the norm. In Southeast Asia, Holocene relaxation of island communities isolated by rising sea-level has probably been ecologically more significant than the direct consequences of climate change. The impact of human activities on tropical systems spans the entire Holocene in all the tropical continents. Land clearance with fire, proto-agriculture and hunting, all exerted an influence on systems. The development of larger societies both increased the rate of land disturbance and rendered those societies more susceptible to climate change. Increasing evidence points to a linkage between cultural turnover and periods of extended drought.

Keywords: Climate change, Cultural collapse, Fossil pollen, Holocene, Lake-level, Precipitation, Rain-forest, Temperature, Tropical

The lowland tropics fascinate ecologists because of their mystery, unsurpassed biodiversity and potential to fuel an explosive growth in the comprehension of complex ecosystem dynamics. What makes the lowlands so intriguing is also what makes them such a challenging region for the palaeoecologist. Determining the climatic conditions of the last Ice Age was the target of much of the early work in the lowland tropics (Müller, 1959; van der Hammen, 1963; Absy, 1979; Flenley, 1979). These pioneering researchers faced a bewildering diversity of tropical plants, and a virtually unknown pollen flora. The grail was to test major palaeoclimatic and evolutionary theories such as the presence of Pluvial periods, Pleistocene aridity, Pleistocene cooling and Haffer's (1969) refugial hypothesis (Coetzee, 1964; Livingstone, 1967; van der Hammen, 1974; Servant *et al.*, 1981; Absy, 1982; Colinvaux *et al.*, 1996). The Holocene was almost ignored in these studies as it was assumed to be composed of uniformly 'modern' conditions and therefore of little real interest.

As the probability of impending rapid climate change was realized, interest in Holocene climate dynamics quickened. Lessons could be taken from the past because the forcing functions governing climatic change are the same now as in the Pleistocene, e.g. variation in northern hemispheric insolation, the thermohaline circulation system, and El Niño Southern Oscillation (ENSO)-like phenomena. Suggestions that the southern hemispheric tropics underwent major climatic change before the northern hemisphere has prompted speculation that the tropics may play a significant role in inducing high-latitude climate change (e.g. Harris and Mix, 1999). Consequently, in the past decade, tropical palaeoecology has moved beyond a simple characterization of glacial or interglacial climates to tackle more subtle questions about the timing, nature and extent of climate change, and how those changes are forced.

The importance of the tropics as a driver of climatic change is demonstrated by the observation that the largest single cause of modern inter-annual climatic variability in North America is ENSO (Dettinger *et al.*, 2001). Interhemispheric heat and moisture transport is, and has been, one of the great pumps of climate change. The vast landmass of the tropics (>50 per cent of terrestrial surface area) is thus not only important for supporting most of Earth's biodiversity, but it is also a vital component in generating global climate. Consequently, oceanic and terrestrial records are critical to building and testing models of past and future climate change.

In the last 30 years, many new records of palaeoecological and palaeoclimatic data have emerged from the lowland tropics. Rather than attempt a case-by-case review of a large literature, I will explore the reasons that appear to underlie some commonalities and differences between Holocene lowland tropical records. As my specialization is in neotropical palaeoecology I will draw most examples from that region.

Let us start with some generalities. Deglaciation started *c.* 19,000 yr BP (all ages will be reported in calibrated years before present) and continued steadily, perhaps with periodic reversals, until *c.* 10,000 yr BP (Seltzer *et al.*, 2002, Paduano *et al.*, 2003). Temperatures in the Holocene have been relatively stable, generally varying within the errors of palaeothermometers such as fossil pollen, phytoliths, macrofossils and diatoms. Because the lowland tropics are already the warmest habitats, no biological indicators have been identified that could provide a signal of warmer-than-modern temperatures. The exception to this generalization may be where cold-water upwellings have strengthened or weakened and affected local climate. Nevertheless, such evidence is based on marine faunas, not on terrestrial data (e.g. Sandweiss *et al.*, 1996). Other broad similarities include a trend of increasing fire-frequency from the Pleistocene into the Holocene, and that many coastal locations reflect a rising sea-level that stabilized in the Mid-Holocene.

Despite these broad agreements, the greater the temporal resolution of a record, the more idiosyncratic each record appears to be. Most of that variability lies within the precipitational record. Lake-level, isotopic composition of carbonates and vegetation response, can all be influenced by changes in total precipitation and seasonality, and therein lies the nub of most Holocene palaeoclimatic reconstructions.

26.1 SETTING THE SCENE: THE PRE-HOLOCENE

It is increasingly evident that the Pleistocene and Holocene were not cleanly separated events. At the Last Glacial Maximum (LGM) lowland tropical temperatures dropped at least 5 °C compared with modern (Livingstone, 1967; Flenley, 1979; Liu and Colinvaux, 1985; Bush *et al.*, 2001). However, a general consensus has yet to be reached regarding Ice Age precipitation (e.g. Colinvaux and De Oliveira, 2000; van der Hammen and Hooghiemstra, 2000). The old paradigm of a dry LGM in the tropics was initiated on the basis of studies of African lake-levels (Street-Perrott and Grove, 1976). Because many of the initial records of this aridity were clustered in East Africa and the Saharan region, they were not representative of all of Africa. Sites that contradict a simple pattern of African aridity lie in West Africa where the Pleistocene was clearly mesic (Street-Perrott and Perrott, 1993), and even around Lake Malawi in East Africa there is evidence of moist conditions (De Busk, 1998).

The concept of Pleistocene aridity was extended to encompass South America where it became intertwined with the refugial hypothesis (Haffer, 1969) and attained the status of a paradigm. Not one shred of credible palaeoecological evidence has been offered to substantiate this claim of synchronous Amazon-wide Glacial-age aridity (Räsänen *et al.*, 1978; Salo, 1987; Colinvaux, 1996). There is little doubt that Pleistocene climate change in Amazonia was neither synchronous nor unidirectional (Bush *et al.* 2001). Thus, the start of the Holocene in many neotropical locations was not, as previously supposed, a transition from savanna to forest, but from one forest type to another.

Other ecological changes that carried over from the Pleistocene into at least the Early Holocene included competitive responses to the rising atmospheric CO_2 concentrations. At the LGM, CO_2 concentrations of around 180 ppm would have created physiological stress in many plants. C4 and CAM plants would have held a competitive advantage over similarly sized C3 plants, leading to strong changes in the composition of grassland and shrubland communities (Street-Perrott *et al.*, 1997). Where there was sufficient moisture for forest to have been maintained, the reduced CO_2 concentrations may still have exacted a toll. Unable to support as much leaf area, forest canopies may have been more open and had lower leaf area indices (Cowling and Sykes, 1999; Cowling *et al.*, 2001). Clearly, under such conditions of increased light availability, plus the climatic fluctuations, competitive interactions would have been very different, perhaps allowing cerrado and deciduous elements to invade and exist within 'closed canopy' forests (Pennington *et al.*, 2000).

Further ecological disruption would have followed from the widespread extinction of megafauna that may have been triggered, and was certainly aided, by the range expansion of humans (Coltorti *et al.*, 1998). The loss of megafauna would have impacted plant recruitment through both seed dispersal and seedling grazing. With the loss of the large grazers and foliovores, adult plants would also have been released from grazing pressure. The reduction in predators of small- to medium-sized rodents and artiodactylids (e.g. peccaries), an important group of seed predators and dispersers, may have resulted in further plant community change (Silman *et al.*, in press).

Across the tropics, the climatic oscillations of the Pleistocene–Holocene transition caused unfamiliar conditions that bludgeoned species far faster than evolution could respond. Unfamiliar warmth, new precipitation regimes, increased CO_2 concentrations and fire frequency, altered competition, rising sea-levels and a growing human influence made the onset of the Holocene a harsh shock.

26.2 NEW COMMUNITIES OF THE HOLOCENE

In montane tropical regions, species respond to thermal change by migrating up- or downslope. However, in the lowlands there was no refuge from Ice Age cooling. The cold events of the

Pleistocene simply had to be survived. Some populations may have been unaffected, or thrived when freed from competition, while others may have been reduced to populations surviving in protected microclimates. A resulting reassortment of communities both in relative abundance and the presence/absence of species is predicted; though this remains an untested hypothesis. Although we can observe the arrival of cold-adapted species into lowland Pleistocene floras, demonstrating a change in the composition of the rest of the lowland community is difficult. Thus, we are unsure how much community change was needed at the start of the Holocene to create modern plant assemblages.

Plant migrations on mountain flanks from the peak of the LGM to the warmest time of the Holocene might have resulted in a vertical movement of *c.* 1000–1700 m, according to location. On the flanks of most tropical mountains, this vertical distance can be accomplished in just a few kilometers of horizontal movement. Consequently, the upslope expansion of warm forest taxa at the start of the Holocene was a rapid event (Hansen, 1995a; Hooghiemstra and Cleef, 1995; Colinvaux *et al.*, 1997), occurring much faster than the long overland migration of species in temperate America and Europe (Davis, 1981; Webb, 1987), though see Kullman (1998) for a reappraisal of some migration rates.

With the warming at the start of the Holocene some of these cold-indicator species died out relatively quickly, while others survived as long as 10,000 years. For example, at La Yeguada, Panama (650 m above sea-level), *Quercus* descended *c.* 1000 m elevation to occur at this site in the Pleistocene (Bush *et al.*, 1992). With the onset of the Holocene, *Quercus* lingered in the area until about 8000 yr BP. *Podocarpus* pollen is still a component of lowland records until *c.* 9500–10,000 years ago in Amazonia. Similarly, in Africa the montane element, *Olea*, survived until about 8000 years ago at Barombi Mbo (Maley *et al.*, 1990), and until about 2000 years ago at Lake Malawi (De Busk, 1998).

Cold-adapted species with disjunct distributions in warm settings are well known from the temperate regions, e.g. the Hemlocks of the Appalachian coves. These populations are seen as relicts of ranges that expanded during the last Ice Age. Species of *Podocarpus* and other montane taxa found scattered across Amazonia are probably relicts of the last, or previous, cold-stage populations expansions. Similarly, disjunct floras such as the cerrado elements now restricted to isolated patches of white sand soils in Amazonia, could be the last holdouts of species that were once much more widespread (Prado and Gibbs, 1993).

Whether these cerrado species invaded during the Pleistocene or during phases of Late Miocene or Mid-Pliocene savanna expansion has not been resolved. It is probable that both montane and cerrado elements were able to colonize lowland Amazonia in the turmoil of Pleistocene (or earlier) climate change and have now dwindled into highly restricted ranges. Thus, the Early Holocene in the humid lowland tropics is indicated by a general loss of cold-loving species, rather than by the arrival of warm-loving species (they were already there).

26.2.1 Rising Sea-Level, Isolation and Extinction

The *c.* 120 m lowering of sea level at the LGM exposed much of the continental shelf. Because the continental shelf areas around South America and Africa are narrow, there was little gain in land area suitable for lowland tropical forest development (Fig. 26.1). In contrast, the exposure of the Sunda Shelf in Southeast Asia more than doubled the modern land area for lowland

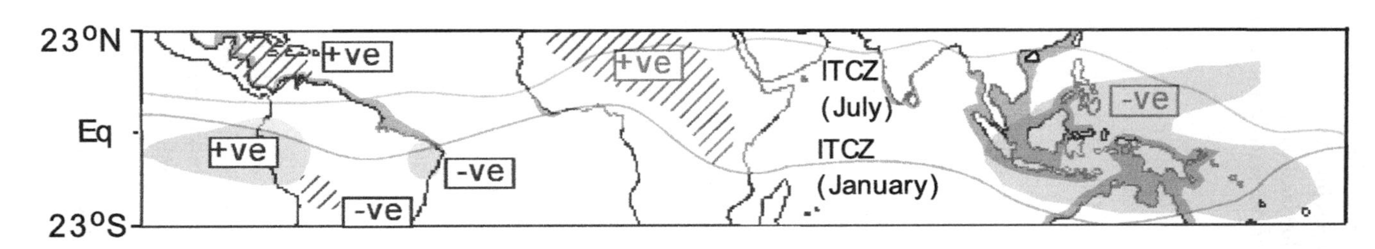

Figure 26.1 Schematic diagram showing characteristic patterns of precipitation response to forcing by ENSO and insolation. Also shown are major areas of land exposed during the LGM low sea-level event, and the ITCZ. Positive and negative signs indicate opposing effects of a particular phase of forcing. Positive sign for ENSO indicates increased precipitation during El Niño events, negative indicates reduced precipitation. This pattern would be reversed during La Niña. Positive and negative signs relating to insolation also represent regions that are characteristically wet when the other is dry as a result of insolation asymmetry and vice versa. Key: [hatched box] Area where insolation is a key climatic driver. [light grey box] Area where ENSO is a key climatic driver. [dark grey box] Areas of continental shelf exposed at the LGM.

forest. Islands presently fragmented by high sea-levels were connected to the mainland with only the deepest channels providing barriers to dispersal, e.g. the channel that maintains Wallace's line. The larger land area would have translated into larger populations and greater mixing of biotas. The rise in sea-level came suddenly. Two phases of rapid, sustained sea-level rise between approximately 16,000 and 12,500 yr BP, and between 11,500 and 8000 yr BP led to the sudden inundation of massive areas of lowland flood forest and mangrove. Harmon *et al.* (1983) estimate that the rate of shoreline advance across the Australian landbridge that connected Australia to Papua New Guinea was about 20 km/millennium, approximately 40 cm/week. For the plants and animals facing this rising tide, the onset of the Holocene saw a massive collapse in populations, a loss of *c.* 90 per cent of lowland habitat. Populations were crammed onto the volcanic slopes of the Sunda Islands, leaving megafauna such as tiger, white rhinoceros, elephant, Banteng cattle, orangutan and tapir, with perilously small habitats. The island populations were sufficiently isolated for speciation to begin in mammals and insects (e.g. Corbet and Pendlebury, 1978). However, a faster process has been relaxation of these communities resulting in the extinction of megafauna from the smaller islands. Human actions have accelerated extinction rates throughout the region, and the prospect for megafaunal survivial on any of the Sunda Islands looks bleak. Biogeographically, this change in Southeast Asian ecology between low and high sea-level was far more profound for that region than any climatic oscillation of the last two million years.

Although the two other continental areas were not as severely affected by sea-level change, from a palaeoecological standpoint the rising sea-level was significant as it caused the hydrology of low lying areas to change. During the Pleistocene low stand, many lakes dried out, not because of reduced precipitation but because of lowered water-tables. The rising sea-level caused bottomlands to flood and is clearly reflected in the formation of flood forests, and lakes in coastal regions between 8000 and 6000 years ago, e.g. the *varzeas* of the lower Amazon.

26.3 INSOLATION VARIABILITY AS A PREDICTOR OF PRECIPITATION

That orbital forcing underlies the timing of glaciations is now well established (Broecker and Denton, 1989), though the mechanisms amplifying that weak signal into massive climate change are perhaps less well understood. The familiar flip-flop from Pleistocene to Holocene conditions is probably causally related to a strengthening thermohaline circulation, reduced volumes of dust entering the atmospheric circulation, and reduced planetary albedo as ice caps shrank. All of these events are tied to changes in the Earth's orbit around the Sun, and all influenced tropical climates.

Another factor of climatic significance that is controlled by insolation is the inter-tropical convergence zone (ITCZ) (Fig. 26.1). Insolation influences the ITCZ in a number of ways, but two of them are important in inducing an asymmetric response north and south of the equator. First, insolation which is positively correlated with the strength of the ITCZ is asymmetric, so that on an annual basis when insolation is strong in the southern hemisphere it is weak in the northern hemisphere. Second, long-term precessional variation in insolation is similarly anti-phased between the hemispheres (Fig. 26.2).

Thus, on first principles, we would expect to see opposing patterns of precipitation change north and south of the equator (Fig. 26.1) (Fritz *et al.*, 2001; Baker, 2002). Some evidence exists for this pattern as North Africa and the Caribbean exhibit a dry late Pleistocene and a wet Early Holocene (Salgado-Labouriau, 1980; Hodell *et al.*, 1991), whereas South America appears to have been wet in the terminal Pleistocene and dry in the Early Holocene (Baker, 2002). In addition to empirical data, an atmosphere-only model (Prell and Kutzback, 1987) predicted that Saharan Africa's summer precipitation was highly sensitive to insolation.

The model suggests that a 1 per cent increase in summer radiation increased precipitation by 5 per cent. De Menocal *et al.* (2000a) suggest that this relationship is not linear and that it only responds to insolation departures of >4.2 per cent. Consequently, the increased precipitation would turn on and off abruptly. As the insolation departure in the northern hemisphere exceeded 4.2 per cent between 14,800 and 5500 yr BP, these times bound the period when insolation would drive African precipitation. Given a maximum Holocene insolation departure of +8 per cent, the model predicts a precipitation increase in Africa of 40 per cent at *c.* 10,000 yr BP (Fig. 26.2). The predictions of this model are supported by observations of increasing lake depth in Saharan and eastern Africa in the Early Holocene. The predicted Mid- to Late Holocene reduction in precipitation is also supported by falling lake-levels from about 7000 years ago.

Despite broad trends that can be predicted by insolation, local factors are superimposed adding complexity to the history. During the peak of the Early Holocene humid episode, part of the Sahara was vegetated. Yet even in the Sahara, the Mid-Holocene desiccation did not proceed evenly, as the western Sahara remained wet longer than eastern Sahara. Such details of climate pattern cannot be explained easily by insolation variation and reflect this basic pattern being overridden by local factors (Gasse, 2000).

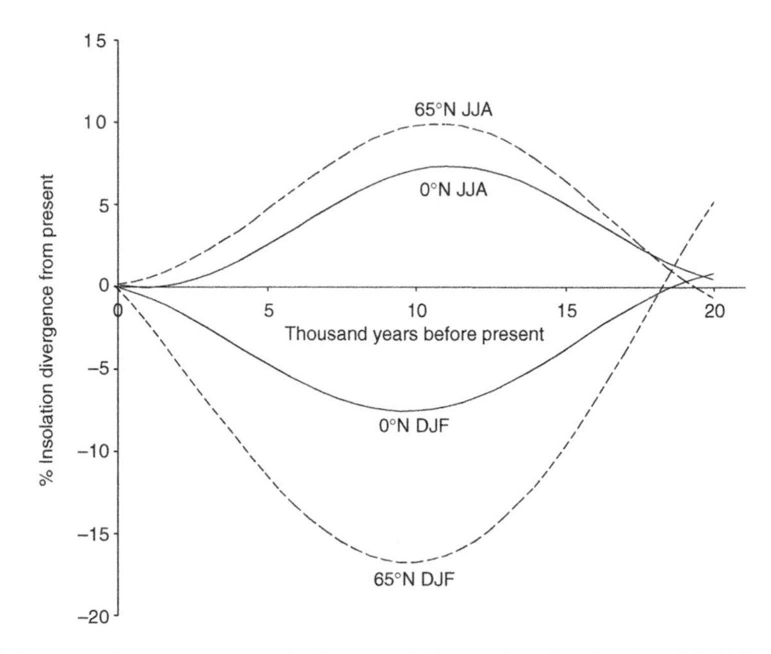

Figure 26.2 Insolation departures for June and December for the last 20,000 years at the equator and 65°N. Data from *Analyseries* (Laskar, 1990; Paillard *et al.*, 1996).

Other indications of local climatic variability are evident in lake-levels from east and west equatorial Africa. Throughout the Early Holocene, the Rift Valley lakes of eastern African were deeper than the modern lakes (Street-Perrott and Perrott, 1993). In the Mid-Holocene, starting about 7000 years ago lake-level at many sites started a decline that lasted until the present (De Busk, 1998; Johnson *et al.*, 2002). At the West African site of Lake Sinnda, which lies close to the modern forest/grassland ecotone, a Mid-Holocene change in vegetation is evident. Vincens *et al.* (1998) report falling lake-levels as early as 4800 yr BP, with substantial replacement of forest by grassland and a dry lake bed between *c.* 3500 and 2800 yr BP. The lake refilled at 1300 yr BP, although human disturbance prevented a full regrowth of forest. The basic pattern of this study is present at other inland West African sites (Elenga *et al.*, 1994). However, in wetter coastal locations, maritime air systems maintained moist tropical forests throughout the Holocene, e.g. Bosumtwi and Barombi-Mbo (Talbot *et al.*, 1984; Maley *et al.*, 1990). These sites suggest that a wetter-than-present phase lasted longer in West Africa than in East Africa and caused Barombi Mbo to overflow between *c.* 8000 and 4000 yr BP (Talbot *et al.*, 1984).

Southern Amazonia has a broadly opposite pattern of precipitation change to that of Africa. Servant *et al.* (1981) suggested that Mid-Holocene drying of Amazonia was so intense that lowland tropical forests were fragmented. The initial suggestion of South American Mid-Holocene aridity was supported by two sites, Carajas (Absy *et al.*, 1991) and Titicaca (Wirrmann and De Oliveira Almeida, 1987; Baker *et al.*, 2001). In the lowlands, the best evidence of Mid-Holocene drying is found in ecotonal areas (such as Carajas), particularly from the cerrado/savanna transition (Behling and Hooghiemstra, 1999). Most mesic sites suggest a very wet start to the Holocene followed by reduced Mid-Holocene precipitation, but there is little accord regarding the timing of this event. Indeed, two sites in the ecotonal area (seasonal forest to moist forest) of eastern Amazonia show increasing lake-levels throughout the Holocene (Bush *et al.*, 2000).

26.3.1 A Caution on the Interpretation of Lake-Level and Vegetation as Proxies for Precipitation

An important distinction must be made between lake-level and forest cover. A lake can dry out without a coincidental loss of regional forest cover (Colinvaux *et al.*, 1996; Colinvaux and De Olivera, 2000), especially if the reduction in precipitation is during the wet season rather than the dry season. In many areas it is the duration and intensity of the dry season that sets the vegetation type rather than the amount of rain received in the wet season (Sternberg, 2001). Consequently, an area can become significantly drier, and its lake-levels fall, without necessarily impacting vegetation. Thus, inferences drawn from lake-level alone would exaggerate the biological significance of a reduction in wet season precipitation, while inferences drawn from pollen alone would underestimate change in precipitation.

Another potentially misleading relationship exists between lake size and apparent vegetation change. As a lake dries and open-water area is reduced, an increased proportion of pollen arrives from local sources, e.g. the marginal swamp (Jacobson and Bradshaw, 1981; Prentice, 1985; Sugita, 1993). Therefore, if a region is dry enough that a large, modern lake contracted to a marsh or a small lake, there would be a shift in the palynological signal from a primarily arboreal input to a primarily swamp (non-arboreal) input that is independent of regional vegetation changes (Bush and Colinvaux, 1994; Bush, 2002). Consequently, basing interpretations of regional forest cover on arboreal/non-arboreal pollen representation must be done with great caution.

Pennington *et al.* (2000) suggest that reconstructions should not be based on a simple view of 'forested', which is assumed to be similar to modern forests of the area, or 'savanna', which is assumed to be a C4 grass-dominated community. Rather than this dichotomy, we should consider a wide range of intermediate forest types that would differ in species composition, structure, habit and leaf area index (Cowling and Sykes, 1999; Pennington *et al.*, 2000).

During periods of reduced Holocene precipitation, forest composition may well have oscillated across a wide amplitude of forest types. In the areas that currently receive most precipitation, the range of variation might be from a semi-deciduous forest to a true rain forest. Whereas in the driest areas there may have been replacement of cerrado with savanna. No Holocene site in the neotropics indicates an oscillation from wet forest to savanna. A core from a seasonally inundated swamp in Noel Kempf Mercado Park, Bolivia, however, shows a swing from savanna to mesic forest at *c.* 3000 yr BP (Mayle *et al.*, 2000). The modern vegetation of Noel Kempf is a mosaic of forest and savanna, in which small changes in hydrology or climate could induce an ecotonal migration. Thus this record reflects a local movement of vegetation, rather than a regional replacement of one biome with another.

As more detailed records with better ^{14}C dating control are established, it is probable that the concept of 'wet' or 'dry' phases in the Holocene will be replaced by the recognition of many rapid and frequent climate changes. What we are observing at the coarser scale is a set of tendencies that are of significance to those of us trying to understand the 'big picture', but are of much less relevance culturally or ecologically than the short-term variations in climate.

26.4 ENSO-LIKE PHENOMENA AND THE CLIMATE RECORD

In the late 20th century ENSO phase (positive = El Niño, negative = La Niña) and strength was the single largest cause of interannual climatic variability (Diaz and Markgraf, 1992). In general, positive ENSO phases cause the export of moisture from the tropics to temperate regions. Strong positive ENSO events led to droughts in northeastern Brazil and in Southeast Asia. However, a closer examination reveals that a large element of variability exists between ENSO events and how they are manifested regionally (Diaz and Markgraf, 1992). Although positive ENSO phases are usually correlated with drought, in western Amazonia a more complex pattern is associated with these events. During the strong El Niño event of 1982/83, the dry season was extremely dry but this was followed by unusually heavy rains in the wet season. The average result was a slightly wetter than average year in Amazonian Ecuador and Peru. But averages never killed a plant or snuffed out a population. The important event was the increase in seasonality. Both the drought phase and the flood phase of this El Niño event could have been important factors that shaped local habitats.

The periodicity of ENSO events, 3–7 years, is too fine a resolution to be revealed by lowland tropical palynological studies. The best record to date of ENSO periodicity comes from the Galapagos which experience strong droughts during El Niño events. At least 435 moderate to very strong El Niño events have occurred in the last 7100 years (Riedinger *et al.*, 2002). El Niño was frequent between *c.* 7100 and 5900 yr BP, rather quiet between 5900 and 4600 yr BP and then active once more between 4600 and 3100 yr BP and between 1900 and 900 yr BP. Sandweiss *et al.* (1999) attributed warmer and more settled climates between *c.* 8500 and 5800 yr BP to a weakened ENSO. On the basis of fish and molluscan remains in Peruvian coastal middens,

Sandweiss *et al.* (1999) suggest the onset of stronger ENSO conditions at 5800 yr BP. Nevertheless, there is broad agreement that by about 5000 yr BP El Niño was a significant factor in neotropical climate systems.

Mörner (1992) has proposed that there are also mega-ENSO events that occur on a centennial- to millennial-scale. It is evident that the Holocene has seen some major changes in ENSO on a millennial-scale, but no evidence of a regular long-term oscillation has yet been found. Although ENSO is suggested as the proximate cause of much Holocene climatic change, no ultimate cause has been identified and it is not understood why ENSO should vary on a millennial time-frame.

As we gain a better grasp of the history of ENSO, at the centennial- to millennial-scale, it provides us with a framework against which regional climate change can be tested. Modern observations of ENSO climatic patterns can lead to predictions regarding positive and negative ENSO phases on a regional basis. For example, in negative phase (La Niña) most of Northern Australia, Indonesia and Papua New Guinea would be wet, the Galapagos would be dry and eastern Amazonia would be wet (Fig. 1). East Africa north of the equator would be dry, whereas further south there would be increased precipitation. India, southern China and peninsula Malaysia would be cool. Whether this pattern is evident indicating the presence of mega-ENSO events has yet to be resolved, but it certainly provides a framework for future analyses.

26.5 HUMANS AND FIRE

Humans and fire have a linked record of transforming landscapes. The expansion of Australian fire-tolerant *Eucalyptus*-dominated communities appears to coincide with human occupation of the land. Singh and Geissler (1985) have suggested that this expansion was a direct consequence of human manipulation of fire to improve grazing and drive game. Humans are known to have used fire as a tool for modifying lowland tropical forests throughout the Holocene. However, it is not known to what extent fire is also a natural phenomenon within all lowland tropical communities. Obviously, grasslands and scrub woodlands are fire-prone and their species are fire-adapted. The return period of fires in these settings is generally controlled by fuel availability, so that decadal-scale fire cycles are typical. In more mesic habitats, such as semi-deciduous forest and everwet forest, the role of fire is less well known. Although fire has been recorded in wet forest settings in Southeast Asian systems as long ago as 50,000 yr BP (Haberle, 1993), these burns were associated with human occupation.

The general assumption that rainforests do not burn naturally was challenged by Saldarriaga and West (1986) who found charcoal in soil pits near São Gabriel in northern Amazonia. The fire events dated to before 6000 yr BP, at that time considered too early for human-induced fires. In the years since Saldarriaga and West's (1986) discovery of Amazonian charcoal, archaeological evidence points to a longer use of Amazonia than was previously expected. Rock paintings dating to *c.* 12,500 yr BP were found near Monte Allegre, Brazil (Roosevelt *et al.*, 1996) and a thriving centre of occupation at Santarem, Brazil, dates to 8000 years ago (Roosevelt *et al.*, 1991). Although Piperno and Stothert (2003) have now demonstrated squash agriculture in coastal Ecuador at about 10,000 yr BP, the earliest evidence of Amazonian cultivation was from Ecuadorian Amazonia where charcoal and maize pollen co-occur for the last 6000 years (Bush *et al.* 1989). These data admit the possibility that the São Gabriel charcoal was the product of human disturbance. Another spark that prompted debate over the susceptibility of lowland forests to fire, was provided by the massive fires that swept across Kalimantan, Indonesia, in the wake of 1982/83 and 1997/98 El Niño droughts. All kinds of forest were burned, and the fact that the fires were initially set by humans did not detract from the observation that wet forest can burn. However, this debate is far from settled.

Because of the long record of human occupation in Africa and Asia, Australia and the neotropics are the best places to test the hypothesis that humans modified the landscape through fire. The suggestion that early Australian settlers transformed mesic woodland to fire-resistant *Eucalyptus* forests, has received some support from paleoecological data. The few sites with long records suggest that there was a turnover of forest type associated with fire within the approximate time-frame of human occupation of Australia (Kershaw *et al.*, 1997). However, whether such a vast transformation was truly attributable to human activity is unclear. Singh and Geissler (1985) suggests that the last interglacial may hold the key to this puzzle. If fire was a regular component of moist tropical landscapes at a time when humans had not arrived in Australia, it would be a sure sign that fires are a natural part of the ecosystem. This hypothesis remains to be tested satisfactorily.

What is evident in most neotropical records is that fire frequency is much greater in the Holocene than in the Pleistocene (Bush *et al.*, 1992; Kershaw *et al.*, 1997). In some locations increased seasonality at the start of the Holocene, with more convective activity and lightning strikes, may have caused these fires. Holocene increases in fire frequency in dry regions, such as Australia, may reflect increased precipitation, which in turn increased fuel loads and allowed fires to be carried across the landscape (Walker and Chen, 1987). However, in other places the most plausible explanation of the charcoal abundance is increased human activity and the onset of agriculture (Piperno *et al.*, 1990).

By the Mid-Holocene, fire is a regular event in many of the watersheds analysed for fossil pollen. That the increase in fire is associated with a general Mid-Holocene drying may not be a coincidence, but in many areas the fires are accompanied by the presence of cultivars such as maize in the neotropics, or sweet potato in Asia. Indeed, among Mid-Holocene records from closed-basin systems that lie close to rivers in the neotropics, fire is so prevalent that it is more usual to obtain a record with substantial amounts of charcoal than without it. The response of mesic tropical communities to the increased frequency of fire is another area awaiting research.

26.6 PALAEOCLIMATE AND THE RISE AND FALL OF CULTURES

Poor land management, overexploitation of the land and growing human populations combine to form a socio-economic system that teeters on the brink of collapse (Sheets, 2001). Although environmental determinism has had its champions and critics (e.g. Meggers, 1971; Robertshaw, 1988; Hassan, 2000). An increasing body of evidence points to a repeated pattern of cultural growth terminating rapidly during times of severe drought. The great civilizations of Mesopotamia, the Indus valley and Egypt, all show a strong cultural collapse or turnover between 4800 and 4600 yr BP which is coincident with strong regional drought (Dalfes *et al.*, 1997).

The rise and fall of Mesoamerican cultures has long elicited speculation as to their causes (e.g. Meggers, 1954; Adams, 1973). Erosion of soil, lack of wood and environmental determinism have all been blamed, but without any real evidence. Studies of lake sediment in Mexico, Honduras and Guatemala, have shown that Mayan farming practices did indeed accelerate soil erosion. However, the collapse of the Mayan culture was not the gradual decay projected by some, but a sudden abandonment of a way of life. Using pollen, lake chemistry and $\delta^{18}O$ as proxies for lake-level, Curtis *et al.* (1998) demonstrated the abandonment of Mayan farms and cities in Guatemala. In Mexico, the Mayan collapse coincided with the most severe and prolonged drought experienced in more than 1000 years (Brenner *et al.*, 2001). The Maya were not alone in their demise, good evidence has been found linking a number of cultural collapses to sudden climate change. In a recent review, Weiss and Bradley (2001) outlined a series of such events that included the demise of the Natufian cultures (*c.* 8400 yr BP), the Tiwanaku civilization of the Andes (*c.* 1000 yr BP), the Moche of coastal Peru (*c.* 1400 yr BP), the Anasazi of the southwestern USA (*c.* 750 yr BP) and the Jamestown Colony, USA (AD 1606–1612) (Stahle *et al.*, 1998). The collapse of indigenous peoples at the time of European conquest of the Americas is often related to epidemics of typhus, measles, smallpox, influenza and the common cold (e.g. Denevan, 1976). The possibility that these bouts of disease were further exacerbated by drought, flood or famine is another exciting area of future research.

26.6.1 A Footnote on Conservation Areas and Global Climate Change

One of the most important practical legacies that Holocene palaeoecologists may have to offer is in helping to formulate conservation policy in a world of rapidly changing climate. Most conservation plans are essentially static, in that they do not accommodate future climate change. Traditionally, a community or species is 'protected' by setting aside a nature reserve. The boundaries of the reserve are fixed, but the community is not. As climate change induces species to migrate, we should expect community turnover and local extinctions (Davis and Zabinski, 1992). Active management of small reserves may delay or even halt this process, but in larger areas, a new conservation approach is needed. Identification of biodiversity hotspots and areas of significant threat (e.g. Myers *et al.*, 2000) are a first but incomplete step toward protection. Understanding the resilience of those communities to change, and the probable response to change become major determinants of a successful conservation strategy (Hannah *et al.*, 2002). As the vast majority of global biodiversity is essentially tropical, there is an important role for tropical palaeoecologists to play in this debate.

Holocene climate change has shaped natural communities and the development of human cultures. Understanding these changes leads us to a better understanding of our own history, and makes us aware that the contingency of history (Gould, 1989) applies to cultural as well as to organismal evolution. After 50 years of research, the lowland tropics remain an undersampled set of ecosystems, and are still full of challenge. In some ways, the challenges are growing as the most attainable lakes have already been cored and their pollen histories described. Although our knowledge of tropical plant types has improved somewhat, only a tiny fraction of pollen types are identifiable to species and, with one notable exception (Roubik and Moreno, 1991), comprehensive pollen atlases are still a distant dream. Tracing the interaction of humans, climate, and natural communities of the lowland tropics throughout the Holocene is just beginning; it is a truly exciting time to be studying these systems.

Acknowledgements

This work was funded through NSF DEB-9732951.

CHAPTER

27

THE HOLOCENE OF MIDDLE LATITUDE ARID AREAS

Louis Scott

Abstract: The methods used in Holocene research are numerous and they vary in different desert regions. Continuous high-resolution records of Holocene palaeoenvironments in the deserts of the world are difficult to construct because biomass production, which can serve as source of proxy records, is low under dry open-air conditions. Where available organic material is formed in exposed sites like lakes it is unlikely to be preserved in the long term in arid zones. Desert lake sediments are often interrupted and increased productivity during wet phases is often responsible for poor uniformity and continuity of biological data records. In deserts like the Sahara, lake sediments have nevertheless received much attention. Alternatively, many palaeoenvironmental studies in arid areas focus on cave shelter inclusions, e.g. in the North American Southwest where urine-impregnated packrat middens have been studied extensively. By responding to the orbital precession cycle, deserts in different hemispheres reacted out of phase with each other during the Early and Mid-Holocene. Smaller scale variability superimposed on this broad pattern is observed, but global correlation of these cycles has up to now not been possible.

Keywords: Atacama, Deserts, Middens, Namib, Palaeolakes, Sahara

Significant development of more humid climates in deserts took place during the Holocene. Current human population growth, agriculture and industry increasingly impoverish natural dry land ecosystems affecting their beauty and sustainability. In order to put the current scale of desertification of dry lands into perspective and to understand global change of deserts and areas bordering on them, long-term rainfall records of desert growth and decline are needed. Desert boundary shifts should be useful to consider in construction of global change models.

Moist areas generally contain more proxy records of Holocene climates than dry areas. Fossils are not easily preserved in open-air sites in desert regions due to low primary production, generally oxygen-rich conditions and the lack of anoxic lake and swamp deposits. The shortage of lakes and groundwater also limits the development of useful non-organic sediments like stalagmites. The evidence for environmental change in deserts or semi-deserts is therefore limited. Desert environments, however, can be good for preservation under certain conditions; for example, mummies in dry areas. Fossil or proxy record preservation in deserts can be aided by desiccation that prevents microbial action and decomposition. High time-resolution is, however, rarely achievable in the available records. Most are not continuous and often only represent specific conditions, e.g. moist events or dry events. They are usually not capable of reflecting sub-millennial-scale variations.

Examples of palaeoenvironmental research presented here are restricted to typical sub-tropical deserts like the Sahara, the Namib and central southern Africa, the Australian region, the North American Southwest and the Atacama. They are centred on the Tropics of Capricorn and Cancer between *c.* 15 and 32°N and S (Fig. 27.1). Continental and altitude configuration of these deserts varies considerably, but due to their positions on the same latitudes, they are comparable with each other. With up to 400-mm rainfall and strong evaporation, the chosen regions are influenced by sub-tropical high-pressure belts or cold ocean currents and typically are situated on continental west-sides or interiors. The methods of palaeoenvironmental study in these deserts can be applied more widely in other dry areas, e.g. those surrounding deserts with rainfall of up to 600 mm, or tropical or temperate arid regions like Eritrea or central Asian deserts, respectively.

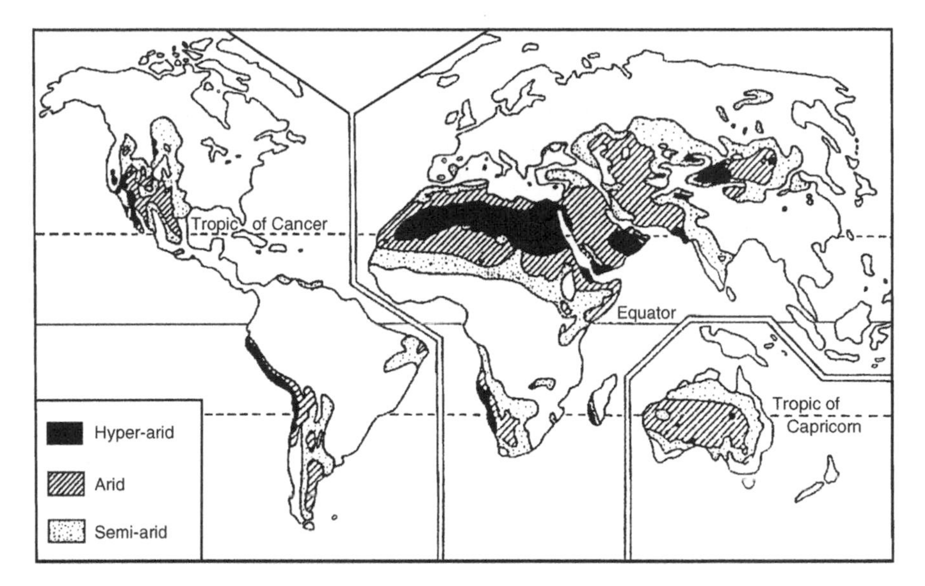

Figure 27.1 Map showing the arid regions of the world. Redrawn from Thomas (1989) with kind permission from John Wiley and Sons.

27.1 EXAMPLES OF PALAEOENVIRONMENTAL RECORDS IN DRY AREAS

The available material in a particular region usually dictates what is studied and resources vary widely from desert to desert due to local physiographic differences or varying biodiversity. Low primary production reduces the potential for radiocarbon dating. However, more recently, TL and OSL methods have revolutionized dating of sands or aeolian deposits in dry areas, despite a potential for a certain degree of error (Bradley, 1999).

27.1.1 Caves and Shelters

Different types of biological and mineral inclusions in caves and shelters are good sources of proxy data in dry areas and some shelters can be very productive (e.g. Sandelowsky, 1977). Although clastic cave floor sediments can reflect climatic conditions, (e.g. aeolian Kalahari sands, dateable by OSL dating) (Robbins *et al.*, 2000), their biotic contents have more often been used as environmental indicators. Charcoal sequences derived from firewood in floor sediments can give indications of palaeoenvironments, e.g. in central Australia and the Namib Desert (Smith, *et al.*, 1995; Albrecht *et al.*, 2001). Pollen can be preserved in low numbers in the cave stalagmites when they are sealed in carbonate precipitate after evaporation of drip water (Fig. 27.2) (Burney *et al.*, 1994; see also Lauritzen, pp. 242–263 in this volume). Pollen in cave-floor sediments in the sub-humid regions is usually only found in the modern surface and not in deeper layers (e.g. Border Cave, Rose Cottage Cave, etc., in South Africa) (Scott unpublished data). It is destroyed with time due to damp oxidizing conditions, which allows microbial action (Fig. 27.2). In contrast, cave deposits in deserts can be expected to be productive as result of desiccation (Davis, 1987, 1990). Whereas unconsolidated cave floor deposits are often poor in fossils (pollen, macro-plants, insects, hair, etc.), coprolites and urine-impregnated middens in caves and shelters are richer (Scott, 1987; Betancourt *et al.*, 1990). Urine impregnation and cementing of midden inclusions by 'amberrat' or 'hyracium' (Betancourt, *et al.*, 1990; Scott, 1990) protect organic inclusions from oxidation, biological decomposition or disintegration. These materials are available in suitable dry rocky shelters and can be dated by radiocarbon dating. Kuch *et al.* (2002) have demonstrated preservation in middens from the Atacama Desert is good enough to protect DNA. The most intensive and notable application of urine-impregnated middens in Holocene studies was carried out by several researchers from the American Southwest in the study of packrat (*Neotoma*) nest accumulations (Betancourt *et al.*, 1990; Cole, 1990, Spaulding, 1990; Thompson, 1990; Thompson, *et al.*, 1993; Thompson and Anderson, 2000). Packrat middens studies are mainly based on macro-botanical rather than pollen remains, while pollen from lake deposits complements the results. The disadvantages of pollen in middens include possible anomalous trapping by rodent activity (Betancourt *et al.*, 1990; Thompson, 1990). The unpredictable architecture of middens make stratigraphical studies difficult, but macro remains in middens have the advantage of species-level identifications. Pollen in lakes or middens, is partly wind-transported and represents a wider region than macro remains.

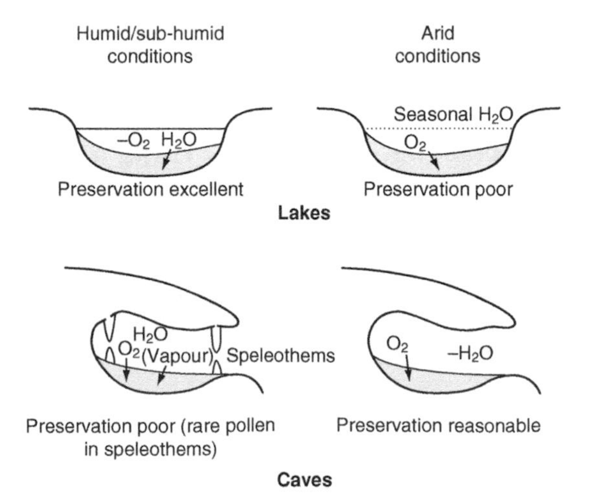

Figure 27.2 Preservation quality of microscopic organic material like pollen in caves and lake basins under humid/sub-humid and arid conditions.

Nests of extinct stick-rats (*Leporillus*) in Australia are similar in appearance to packrat middens (McCarthy *et al.*, 1996). These accumulations and those made by other fauna-like rock wallabies can be found over large parts of the Australian desert region (Nelson, *et al.*, 1990; McCarthy *et al.*, 1996; Pearson and Betancourt, 2002). Relatively few studies on Australian midden inclusions are available (e.g. Pearson and Dodson, 1993) but numerous deposits were dated (Pearson *et al.*, 1999). Nest middens in the Atacama Desert in Chile are built by a variety of species (*Lagidium*, *Phyllotis*, *Abrocoma* and *Octodontomys*) (Betancourt *et al.*, 2000; Holmgren, *et al.*, 2001). Their macrofossil inclusions make an important contribution in reconstruction of Holocene environments in a high-lying part of this desert. In the Southern African desert region, dassie rats (*Petromus*) build middens similar to those of packrats, but no detailed studies on their plant contents have yet been made. Some samples gave unpublished dates ranging as far back as *c.* 7000 yr BP (Scott, 1990; Scott and Cooremans, 1992; Scott, unpublished data).

Procavia capensis (hyrax) urine and faecal accumulations occur in the dry parts of Africa and the Middle East including samples as old as 20,000 years (Pons and Quézel; 1958; Scott, 1990, 1994, 1996; Fall *et al.* 1990; Thinon *et al.*, 1996) (Fig. 27.3). The studies on hyrax midden materials in Southern Africa and the Sahara focussed on pollen analysis but the investigation from Jordan dealt with macrobotanical rests and pollen (Fall *et al.*, 1990). The African middens therefore appear to contain fewer macrofossils than those from Jordan. Macrofossil extraction from Late Holocene hyrax middens from the Namib Desert yielded mainly finely chewed plant material, small seeds (*Acacia* spp. and unidentified types) and insect remains (Scott, 1996). The advantage of pollen in hyrax deposits in comparison with that in lowland basins, which are usually over-represented by

Figure 27.3 (a) Hyrax (*Procavia capensis*). Reprinted with kind permission from National Museum, Bloemfontein. (b) Solidified hyrax dung heaps from the Southwestern Cape province. Reprinted with kind permission from Y. Fernandez-Jalvo.

local swamp elements, is that it gives a more direct reflection of vegetation on mountain slopes. Porcupine urine accumulations are not as easily available. An example from Wonderwerk Cave, South Africa, preserved pollen and macro-inclusions like chewed bone (Scott *et al.*, 1995). Mid- to Late Holocene bat dung from Arnhem Cave on the edge of the Namib Desert also contains pollen (E. Marais and L. Scott, unpublished data).

Pollen assemblages are found in *Parahyena brunnei* (brown hyena) and *Crocuta crocuta* (spotted hyena) coprolites in Equus Cave (southern Kalahari Desert) (Scott, 1987) and Las Ventanas Cave in the dry part of Spain (Carrión *et al.*, 2001), respectively. The sediments, in which they were found, were barren or poor in pollen. The rich hyena accumulated fauna in Equus Cave has been used in palaeoenvironmental reconstruction by Klein *et al.* (1991). $\delta^{15}N$ and $\delta^{18}O$ from ostrich eggshells from that cave also provided paleoenvironmental data (E. Lee-Thorp and L. Beaumont, 1995; Johnson *et al.*, 1997).

Micro-faunal accumulations by owls in caves provided reconstructions of Holocene environments, e.g. in the Namib (Brain, 1974; Brain and Brain, 1977; Avery, 1991). Owl diets in fossil pellets in Mirabeb Shelter include an increasing gecko-to-gerbil ratio as moisture conditions and grass abundance deteriorated (Brain, 1974; Brain and Brain, 1977) (Fig. 27.4).

A shortage in ground water in desert areas may limit the availability of stalagmites but they are available in desert fringe areas with karstic geology, e.g. the Kalahari (Burney *et al.*, 1994; Brook *et al.*, 1999a). Remarkably detailed, high-resolution isotopic records have been produced in sub-humid areas, e.g. from the Cango Caves and the Makapansgat Valley (Talma and Vogel, 1992; Holmgren *et al.*, 1999), and this can potentially be achieved in deserts. Brook *et al.* (1999a) dated drowned stalagmites in karstic solution cavities in Namibia that must have formed under conditions of lower water tables and therefore during times of low rainfall. In order to identify moist phases they have documented increased phases of stalagmite growth in dry areas (Brook *et al.*, 1999a).

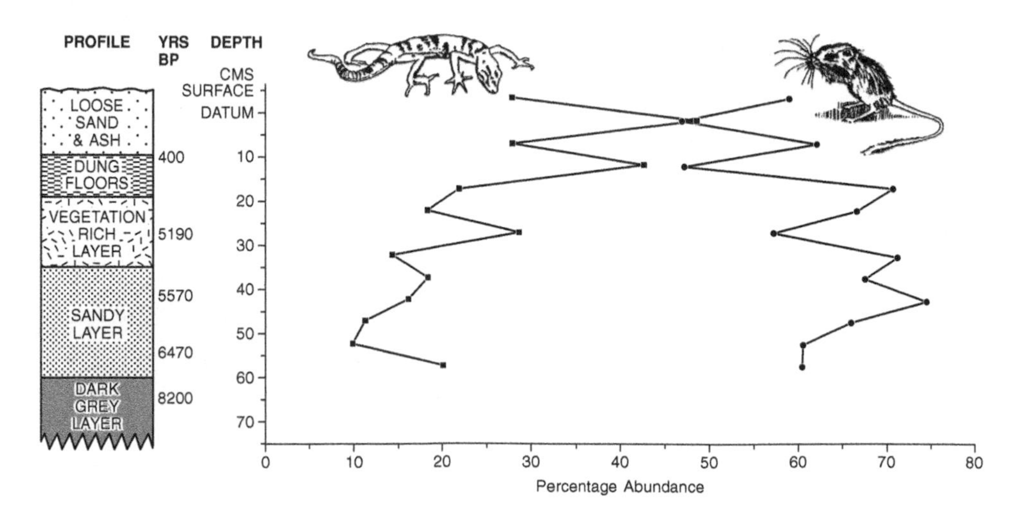

Figure 27.4 Percentage abundance of geckos and gerbils from Mirabib Shelter deposits, Namib Desert. Reprinted from Brain and Brain (1977) with kind permission from the Nature Conservation and Tourism Branch of the South West Africa Administration.

27.1.2 Palaeolakes and Springs

A Late Holocene core of the deep karstic Lake Otjikoto on the edge of the Namib Desert has been taken by divers under 50 m of water, and provided a Late Holocene pollen sequence (Scott *et al.*, 1991). Records in shallower lake basins in arid regions are, however, more unlikely to be preserved (Fig. 27.2), only leaving evidence of their fresh and brackish water phases (Teller, 1998). Dramatic evidence for the existence of large paleolakes occurs in the Sahara (Fig. 27.5) including macrofaunal, pollen, diatom, algal, cyanobacterial and other remains (e.g. Petit-Maire, 1986; Pachur and Hoelzmann, 2000). Typical problems with lake palynology in deserts include that they may contain pollen that has been wind-transported over great distances due to the paucity of local vegetation, and that pollen may have been destroyed after deposition. When drier conditions and seasonal desiccation set in and lakes begin to experience repeated seasonal desiccation and aeration, pollen or fine organic matter will be destroyed down to a depth corresponding with the water table (Fig. 27.2). If the deposits are covered and sealed off from the elements, pollen might presumably be protected (Ritchie and Haynes, 1987). Possible input of pollen via rivers from distant regions, e.g. from the West African forest into Lake Chad (Maley, 1983), may also complicate environmental interpretations. Desert lake radiocarbon dating may suffer from the 'reservoir effect' of deposits (Betancourt *et al.*, 2000; Grosjean *et al.*, 2001). Also, the input of glacial melt-water in high-altitude Atacama lakes during phases of warming can leave evidence of 'too full' lakes (Grosjean and Núñez, 1994). Exposed deposits of diatomites in Southern Peru and palaeo-wetland deposits in Chile indicate higher lake-levels and timing of spring formation indicative of hydrological changes (Placzek *et al.*, 2001; Rech *et al.*, 2002). The pollen contents of desert springs, e.g. at Eksteenfontein and Windhoek in the Namibian region, gave indications of vegetation during times of spring activity (Scott *et al.*, 1991, 1995). While tufa growth may be impeded by increased moisture availability in sub-humid areas, in extremely dry areas it can be indicative of increased flow events of springs or streams. 'Waterfall' tufas are common in the Namib Desert. A massive example from the Naukluft area at Blässkranz, is under investigation (G.A. Brook and E. Marais, personal communication).

27.1.3 Alluvial Records

Desert streams may drop their silt loads before river mouths are reached. The position of silt accumulations along a river course may reflect stream energy (Vogel and Rust, 1987; Vogel, 1989). Cycles of arid aggradation and humid incision can be indicative of cyclic changes in moisture

Figure 27.5 Cover of *Episodes* 9(1), 1986, showing eroded lake deposits from Mali in the Sahara (Petit-Maire, 1986). Reprinted with kind permission from the International Union of Geological Sciences.

regime. Palaeoflood or slack water deposits with dateable organic rests that are to be found in major drainage systems like the Orange are indicative of the distant inland catchment areas and not necessarily of desert conditions (Zavada, 2000).

27.1.4 Ground Water

Sonntag *et al.* (1980) investigated the frequency and origin of Pleistocene and Holocene groundwater dates and isotopes in the Sahara and identified times of recharge. East–west depletion of D and ^{18}O of groundwater of Atlantic origin resulting of progressive rainout towards Egypt can be used as standard for past variations in westerly flow (Sonntag *et al.*, 1980). Water chemistry, including noble gas and isotopic composition, has provided clues about soil conditions and the origin of weather systems of rain in the Namib and Kalahari (Stute and Talma, 1998).

27.1.5 Aeolian Deposits

The luminescence dating of sand grains provides new information about Holocene dune mobilization in the Kalahari region and spreading of desert conditions in north-western Namibia (Thomas, *et al.*, 1997; Eitel, *et al.*, 2001). Swezey (2001) compiled radiacarbon dates of non-eolian materials associated with eolian evidence in the Sahara and suggested that widespread dune mobilization occurred after 5000 cal yr BP.

27.1.6 Soils and Wood

Periods of pedogenesis in desert soils indicate changing climatic conditions (Vogel and Rust, 1987). Dead *Acacia erioloba* trees still standing in soil now devoid of groundwater, grew 500 yr BP ago in the Namib (Vogel, 1989) and testifies to wetter conditions than at present.

27.1.7 Marine Deposits Adjacent to Deserts

Offshore deposits have provided pollen and other inclusions like dust and clay minerals that indicate Holocene conditions along desert coastlines and their interiors (Sirocko, 1996; Gingele, 1996). Pollen in deposits from core GeoB1023-5 along the Namib Desert coast yielded a high-resolution Holocene sequence (Shi *et al.*, 1998). Part of the pollen contents must have been derived from the interior upstream catchment of the Kunene River including the wooded Angolan highlands.

27.2 HOLOCENE ENVIRONMENTAL RECONSTRUCTION IN TYPICAL DESERT AREAS

Despite logistical difficulties, scientists have studied the environmental history of the Saharan/Sahelian and Near Eastern regions intensely (Petit-Maire, 1986). Street-Perrott and Perrott (1993) and Gasse (2000) have recently discussed the general climate history of the Sahara in detail. Marked hydrological shifts have taken place during the Holocene in this region and yielded a great variety of multi-disciplinary records (Petit-Maire and Riser, 1981; Petit-Maire *et al.*, 1981; Petit-Maire, 1986). The Holocene climatic pattern seemed to have affected the development of ancient civilizations profoundly (Hassan, 1997). The greatest part of available research on the Sahara deals with widespread sedimentary and cultural evidence of moist conditions during the Early to Mid-Holocene (*c.* 10,000–6000 cal yr BP). Biological data including pollen and isotopic records indicate a major transformation of the region. The northern Sahelian vegetation boundary was displaced by at least a 500 km northward from the east to the west across the continent, up to *c.* 21°N (Servant and Servant-Vidary, 1980; Petit-Maire and Riser, 1981; Maley, 1983; Ritchie and Haynes, 1987; Hoelzmann *et al.*, 1998; Abell and Hoelzmann, 2000; Pachur and Hoelzmann, 2000). Dramatic evidence for equitable conditions in the now hyper-arid Sahara includes rock art depicting cattle farming, other cultural remains, old lake deposits, crocodiles, fish hippopotami, etc. (Petit-Maire, 1986; Gasse, 2000) (Fig. 27.5). The productivity in the desert was much higher than at present and it supported dense populations of people and fauna (Petit-Maire, 1986; Pachur and Wünneman, 1996; Pachur and Peters, 2001). This moisture amelioration seems to have been strongly regulated by orbital precessional forcing that increases monsoon strength (Street-Perrott and Perrott, 1993), lighter isotopic contents in groundwater (Gasse, 2000) and mollusc shells (Abell and Hoelzmann, 2000). However, some additional global factors must have been active to account for anomalies not corresponding to this forcing, e.g. an arid interruption at *c.* 8200 cal yr BP, which is recorded in several sites in the region (Servant and Servant-Vidary, 1980; Gasse, 2000). After another dry event *c.* 4000 cal yr BP moisture continued to fluctuate but conditions never returned to the optimal levels of the Early to Mid-Holocene (Gasse, 2000).

Land pollen in offshore marine core GeoB1023-5, west of northern Namibia, suggests the Early Holocene in the northern Namib and southern Angola was relatively dry between 11,000 and 9000 cal yr BP (Shi *et al.*, 1998, 2000). Sedimentological analysis of silty deposits in valley basins from northern Namibia supports this (Eitel *et al.*, 2001). Currently submerged speleothems from Aikab, representing lower water tables than at present, may suggest aridity between *c.* 10,000 and 7500 cal yr BP (Brook *et al.*, 1999a). Relatively dry conditions during the Early Holocene in the semi-arid Kalahari region of Botswana are indicated by pollen in a stalagmite *c.* 12,000–8000 cal yr BP (Burney *et al.*, 1994). Pollen in spring deposits from Eksteenfontein, south of the Orange River, suggest that the dry succulent Karoo biome was established around 12,600 cal yr BP and well developed by 9400 yr BP (Scott *et al.*, 1995). Offshore pollen in sediments in marine core GeoB1023-5 shows a gradual replacement from after *c.* 9000 cal yr BP until 7000 cal yr BP of desert elements by woodland and forest types. The latter were probably derived from the interior highlands (Shi *et al.*, 1998, 2000). The gerbil-to-gecko ratios in owl pellet remains from Mirabib suggest that this extremely arid site received slightly more moisture *c.* 7000–6000 cal yr BP (Brain and Brain, 1977) (Fig. 27.4). Spring deposits from Windhoek in central Namibia indicate a relatively warm, moist phase with grass and woodland pollen between 7900 and 6400 cal yr BP (Scott *et al.*, 1991). Excess of air in an aquifer from Stampriet in the western Kalahari may suggest rising water tables at *c.* 7000 cal yr BP (Stute and Talma, 1998). Vogel (1989), however, presents evidence of the deposition of silt-loads at Natab along the Kuiseb River showing that flood energy was not strong enough to reach the Atlantic Ocean. The above mentioned pollen sequences indicate drier conditions after 6400 cal yr BP. After 5000 cal yr, conditions varied but became gradually drier. This trend finds support in the presence of pollen of the pioneer plant, *Tribulus*, and other dryness indicators in karstic Lake Otjikoto before *c.* 3000 cal yr BP (Scott *et al.*, 1991). High numbers of grass pollen in the hyrax dung in the Kuiseb River (Scott, 1996) suggest relatively moist but fluctuating conditions *c.* 2000 cal yr BP, while $\delta^{13}C$ analysis points to higher C_4 dietary contributions (Scott and Vogel, 2000). Conditions deteriorated but they remained favourable until *c.* 1400 cal yr BP when sheep were kept at Mirabib Shelter (Sandelowsky, 1977). Between *c.* 1200 and 800 cal yr BP dry but fluctuating climates are suggested in Kuiseb River until the development of relatively dry climate of the present day (Scott, 1996).

Most of central Australia falls into a desert zone but much more is known about Holocene palaeoclimates of the mesic fringes of the continent (Markgraf *et al.*, 1992; Harrison and Dodson, 1993). Geomorphological evidence from the eastern desert margin (western New South Wales) indicate dune movements that imply climate change in the latter half of the Holocene (Williams *et al.*, 1991). Except for some Late Holocene pollen-records indicating fluctuations in wooded vegetation over the last thousand years, few analyses of the biological contents of *Leporillus* middens have been done (Pearson and Dodson, 1993). A large number of them have been dated showing a decrease in frequency of older dates in the Holocene (Pearson *et al.*, 1999). This is probably due to natural decay of middens with time. Strong fluctuations in the frequency curve possibly partly reflect productivity of stick rats and therefore moisture conditions. However, Pearson *et al.* (1999) show caution with palaeoclimatic inferences in view of other possible factors that might have caused the fluctuations.

Holocene palaeoclimatic reconstructions in the American Southwest are based on a large body of evidence based on packrat middens, pollen and lake beds. The results are reasonably consistent with simulation models (Thompson *et al.*, 1993). The biomization method where fossil biological data are related to modern biomes has been successfully applied in this region (Thompson and Anderson, 2000). At *c.* 10,000 cal yr BP, palaeoclimate data suggest wetter conditions in the Southwest and Rocky Mountains. Down-slope extension of forest trees in the mountains and the presence of mesophytic plants are due to increased summer monsoonal precipitation (Thompson *et al.*, 1993). By 6800 cal yr BP, the North American ice sheet had disappeared completely and conditions in the Southwest were still strongly under influence of summer insolation but became more like present conditions. Wetter conditions occurred in the southern Rocky Mountains and temperatures were greater than at present but in contrast, it was drier in the northern Great Basin (Thompson and Anderson, 2000). An enhanced summer monsoon in the south probably accounts for this anomaly. After *c.* 5700 cal yr BP a gradual increase in the dominance of desert plants are recorded in south central New Mexico and northern Mexico (Betancourt *et al.*, 2001).

Increased precipitation of 400 mm, in contrast to 200 mm today, in the Altiplano of Chile was responsible for high lake-levels in the Early Holocene until 9500 cal yr BP (Grosjean and Núñez, 1994). Soil formation prior to 8100 cal yr BP is indicated but the contribution of melting of permafrost is still under debate. However, Lake Aguas Calientes is reported to have receded by 9400 cal yr BP and Laguna Miscanti by 8700 cal yr BP suggesting arid conditions towards the Mid-Holocene until *c.* 3800 cal yr BP (Grosjean and Núñez, 1994; Grosjean *et al.*, 2001). On the basis of water table fluctuations of wetlands plant and the contents of fossil rodent middens, a dry phase between 9000 and 8000 cal yr BP is indicated, and slightly more mesic conditions prevailed from after *c.* 8000 cal yr BP until the Mid-Holocene (Betancourt *et al.*, 2000; Rech *et al.*, 2002). This appears to be in conflict with lake-levels dates but discrepancies could be coming from the 'reservoir effect' (Betancourt *et al.*, 2000; Grosjean *et al.*, 2001). Modern dry conditions were reached after *c.* 3000 cal yr BP (Betancourt *et al.*, 2000).

27.3 THE GLOBAL PATTERN OF CHANGE IN DESERTS

On a millennial-scale, interhemispheric asymmetry in moisture fluctuations is generally found between desert regions in the northern and southern hemispheres. More high-resolution data are, however, needed to confirm the pattern and to investigate smaller scale anomalies like the 8200 cal yr BP dry event in the Sahara. Palaeodata suggest that the Early Holocene of the Namib Desert appeared to be relatively dry between *c.* 11,000 and 8000 cal yr BP in response to minimum summer insolation at 20°S becoming wetter in the Mid-Holocene (Shi *et al.*, 2000). In comparison the Atacama was relatively wet between 11,800 and 10,500 cal yr BP and showed the lowest water tables and dryness between 9000 and 8000 cal yr BP (Betancourt *et al.*, 2000). Although all the data from the Atacama Desert are not consistent, a low intensity wet phase between 8000 and 3000 cal yr BP (Betancourt *et al.*, 2000; Placzek *et al.*, 2001) corresponds fairly well with the climate development in the Namib Desert. An interesting question is how the El Niño Southern Oscillation (ENSO) phenomenon might have distorted intrahemispheric parallelism during especially the Late Holocene. A broadly parallel Holocene development seems to be indicated between the American Southwest and the Sahara. It seems reasonable to expect that the changing latitudinal effect of precessional forcing in the Holocene would create broadly parallel climate sequences on different continents on the same latitude within limitations of the asymmetric continental and mountain range configurations. The great difference in altitude between the proxy sources in the high-lying Andes or Rockies and the lower-lying Namib Desert or Sahara and the difference in continental and oceanic shapes and processes, including anomalies of the ENSO type, may eventually prove to have caused dissimilarities.

On a millennial-scale, global desert growth and decline during the Holocene as reflected by environmental data, seems to correspond well with moisture fluctuations in immediate longitudinal surroundings but may be the opposite along latitudinal gradients away from the equator. On sub-millennial-scales the nature of this relationship is far from clear and should be addressed in current research. New high-resolution speleothem data suggest that sub-tropical areas experience very marked variability on a centennial-scale (Holmgren, *et al.*, 1999). It is a challenge to produce similar Holocene records from arid lands in order to allow correlation with other regions. Where conventional techniques have failed, creative new approaches and methods are required in order to probe climate and vegetation change and to understand desertification processes more fully.

Acknowledgements

I thank the following persons for information: Julio Betancourt, John Dodson, Martin Iriondo, Eugene Marais and Stuart Pearson.

CHAPTER

28

UNRAVELLING CLIMATIC INFLUENCES ON LATE HOLOCENE SEA-LEVEL VARIABILITY

Ian D. Goodwin

Abstract: Predictions of future coastal change in response to 'Greenhouse Warming' have been based on a historical, secular trend in sea-level rise, with an estimated total rise of up to 0.88 m, and an average estimate of 0.4 ± 0.1 m over the next 100 years (Houghton *et al.*, 2001). However, our knowledge of past decadal to centennial sea-level change and its influence on coastal evolution over the last few hundred to 1000 years is scant. This chapter outlines the climatic and non-climatic factors influencing relative sea-level (RSL) change, and discusses the advancements in Late Holocene studies that are enabling a multi-decadal to centennial examination of coastal evolution and the influence of sea-level changes over the 2000 years. High-resolution proxy sea-level data are being interpreted from fixed biological indicators, such as intertidal coral microatolls and encrusting tubeworms and barnacles together with saltmarsh and chenier sedimentary sequences. Three case studies on the resolution of last millennial sea-level and climate changes are presented for the Southwest Pacific Ocean, South Atlantic coast of South Africa, and the northeast Atlantic Ocean coastline of USA. A fourth case study is presented on the application of multicentennial coral $\delta^{18}O$ derived sea-surface temperature (SST) time-series to temporally extending RSL trends from tide-gauge measurements. Broadly, the recent studies described above depict some agreement in regional sea-level history between oceanic regions; however, each region displays a characteristic signal. There is considerable evidence from both the northern and southern hemispheres that sea-level was at least 0.2–0.5 m lower than late 20th century sea-level during the period AD 1400–1850 (including the Little Ice Age). Estimates of higher sea-level during the prior period AD 700–1300 are less regionally consistent in both hemispheres, and in the southern hemisphere, may principally reflect the influence of the post-glacial global isostatic adjustment processes, rather than climatic factors alone. This general hypothesis requires rigorous examination using developing high-resolution palaeo-methods in both coastal evolution and palaeoceanographic studies.

Keywords: Climate, Late Holocene, Microatoll, Saltmarsh, Sea-level, Sea-surface temperature

In the last decade, there has been a growing interest in research on high-resolution sea-level records over the last 2000 years. An expanding archive of palaeoclimate data spanning the last few thousand years has enabled researchers to conclude that significant regional, and perhaps hemispherical, climatic fluctuations have occurred on multi-decadal and centennial time-scales during this period. The extent to which these climatic fluctuations have affected decadal to centennial sea-level over the past 2000 years is critical to future climate and sea-level predictions. Palaeo sea-level studies, summarized in this chapter, indicate that climatic-driven absolute sea-level anomalies are likely to be on the order of *c.* ±0.5 m on multi-decadal to centennial time-scales. These magnitudes are critical to the habitation of many small island states and low-lying continental regions. Reciprocally, spatial and temporal variation in high-resolution palaeo sea-level records is relevant to the study of ocean–atmosphere interactions (Goodwin *et al.*, 2000).

The millennial-scale, regional relative sea-level RSL envelopes (sea-level change determined relative to a fixed datum on the land) since 6000 yr BP have been modelled for numerous global locations, based on the geophysical relationships that describe the post-glacial glacio- and hydro-isostatic adjustments and the resultant redistribution of global ocean water mass around the globe (Peltier *et al.*, 1978; Clark and Lingle, 1979; Peltier, 1998). There is general agreement between the field-interpreted and modelled RSL curves on millennial time-scales at most sites (Peltier, 2001). A recent study by Nunn and Peltier (2001) demonstrates the agreement for data from the Fiji Islands, in the southwest Pacific Ocean. However, on centennial time-scales, the field-interpreted RSL curves (Pirazzoli, 1991) often describe non-linear, sea-level behaviour which oscillates around the millennial linear trend. Factors contributing to any non-linear RSL behaviour include: local vertical movements of the land through tectonic processes; small eustatic contributions from fluctuations in global ice volume; and the impact of small climate fluctuations on ocean temperatures and salinity, which affect the ocean density and steric sea-level. In addition, errors in the interpretation and spatial reliability of proxy sea-level datums, together with difficulties in obtaining chronological control, may also have contributed to the differences between the interpreted and modelled data. However, the recognition that the field-derived RSL records describe variability not explained by the post-glacial isostatic and local tectonic adjustments alone, indicates that climate forcing of multi-decadal to centennial sea-level fluctuations may indeed be significant, and preserved in the coastal palaeo-record. This is entirely consistent with cyclical sea-level and oceanographic circulation behaviour in response to atmosphere–ocean interactions on seasonal to multi-decadal time-scales, as observed in data from instrumental measurements over the last 100 years (Douglas, 2001).

The notion of oscillating sea-level at centennial-scale has been debated in the scientific literature for at least the last 40 years, commencing with papers by Shephard (1963) on smooth sea-level curves and Fairbridge (1961, 1987, 1995) on oscillating sea-level curves, that were recently reviewed by Kearney (2001). The initial papers lead to a vigorous debate on whether synchronous global sea-level changes existed or whether Holocene sea-level was more accurately characterized by regional sea-level changes on the centennial to multi-centennial time-scales. However, over the last 25 years, palaeo sea-level studies and analysis of the instrumental tide-gauge data have unequivocally supported the case for distinct regional RSL changes, on decadal, centennial and millennial time-scales, as do the geophysical models of the regional response to the glacial-hydro-isostatic adjustment on millennial time-scales (Peltier, 2001). It is of interest that discussions on present or future sea-level rise associated with 'Greenhouse Warming' are often based on a linear sea-level trend with global expression until AD 2100 and beyond. This is largely due to conflicting signals in the instrumental sea-level records on inter-decadal time-scales for the same global regions (Douglas, 1991).

This chapter provides an up-to-date summary of the recent progress in geological studies, towards the resolution of both the variability and non-linearity of regional sea-level changes from high-resolution (multi-decadal and centennial or better) sea-level records spanning the past 2000 years. This work is playing a major role in extending the temporal and spatial coverage of the more recent instrumental records, and allowing the investigation of multi-decadal variability over longer time-series. The chapter does not address advancements in the interpretation of decadal variability in the instrumental tide-gauge records. This has recently been summarized by Sturges and Hong (2001) and demonstrates the intrinsic role of climatic factors, particularly the open-ocean wind field on decadal sea-level variability. It is important for researchers on the Late

Holocene sea-level history to become more acquainted with the lessons learnt from the analysis of the instrumental data.

28.1 CLIMATIC AND NON-CLIMATIC FACTORS INFLUENCING RSL CHANGE OVER THE LATE HOLOCENE

28.1.1 Non-Climatic-Driven Factors

The influence of geodynamics on RSL history around the globe is better resolved than the influence of climate. The principal geodynamic influence on Late Holocene sea-level change is the global glacio-isostatic adjustment (GIA) process. This accounts for the secular (millennial) supply of post-glacial melt-water to the oceans, and the subsequent redistribution of the ocean-water masses around the globe, together with the response of the Earth's lithosphere and mantle to the globally, redistributed water and ice loads, at both mid-ocean and continental shelf locations (Peltier, 1998). Clark and Lingle (1979) established six distinct GIA responses across the globe, which reflected the distance from the former Late Quaternary ice sheets. A compilation of the RSL history for the equator to mid-latitudes, defined as the far-field zones from the former ice sheets, was reported in Grossman *et al.* (1998).

Excellent state-of-the-art summaries on the GIA process and its application to the resolution of regional sea-level trends in the instrumental record, are provided in Peltier (1998, 2001). Other geodynamic influences on RSL include: the possible secular changes in the Earth's equipotential surface, known as the geoid, and the migration of the topographic features on the geoidal surface (±100 m) (Mörner, 1976, Grossman *et al.*, 1998); the possible migration of the location of ocean tidal amphidromic nodes and changes in tidal range associated with coastal and inner shelf bathymetric changes associated with geodynamic and sedimentological processes (Hinton, 1998). Any change in the geological configuration and volume of the ocean basins can be considered negligible over the Late Holocene. The above geodynamic processes affect sea-level variability on millennial time-scales and longer, and are not discussed further in this chapter. However, both instantaneous and persistent geodynamic processes such as tsunamis, tectonism and subsidence do affect sea-level variability on annual to centennial time-scales, and need to be thoroughly investigated before any consideration of climatic influences on sea-level at a site can be made. Local and regional tectonic uplift and subsidence, are best resolved for each site, according to gradual or rapid impact on the RSL (Pirazzoli, 1996). In addition, the catastrophic impacts of tsunamis on coastal morphology and the preservation of the geological record of coastal sea-level changes (Bryant *et al.*, 1996) need to be considered for much of the world's coastline.

28.1.2 Climatic-Driven Factors

Palaeoclimate data from ice-cores, corals, deep-sea sediments and tree-rings have enabled a fundamental understanding of the Quaternary Ice Age cycle of ice sheet growth and decay, and sea-level highstands and lowstands on a millennial time-scale. Atmospheric and oceanic variability influences both the ocean mass and sea-level in two distinct ways. Changes in ocean mass are principally driven by the fluctuating distribution of ice and water on the continents, which is principally controlled by temperature and solar insolation. Precipitation patterns over the continents can result in long-term changes to terrestrial water storage and, hence, sea-level. Indirectly, anthropogenic practices such as surface-water impoundment and groundwater extraction can influence sea-level in this way (Sahagian *et al.*, 1994). Ice volume changes due to snow accumulation variability in Antarctica, Greenland and the non-polar glaciers instantaneously affect the eustatic sea-level through the effect on ocean mass. In contrast, ice sheet dynamical responses to climate change are significantly lagged: up to 35,000 years for the interior of East Antarctica to decades for small glaciers (Whillans, 1981; Drewry and Morris, 1992; Goodwin, 1998). Ice sheet surface and remote sensing studies have been directed at resolving the modern global ice volumes in Antarctica, Greenland and the non-polar glaciers (Meier, 1984; Wingham *et al.*, 1998). Comparative studies on secular palaeo ice-volume have been based on the combined study of ice-core data and glacial geological data on former ice sheet configurations (Denton and Hughes, 1981). A trend in increasing East Antarctic ice volume over the last 4000 years has been interpreted by Goodwin (1998) on the basis of increasing surface elevation as determined from ice-core gas concentrations, and from sedimentological and glaciological studies showing progradation of the East Antarctic ice margin. It was determined that the component increase in ice volume in East Antarctica may have contributed as much as 1 m of equivalent eustatic sea-level lowering during the Late Holocene. This is attributed to Holocene climatic warming and an associated increase in precipitation over the continent. In contrast, Conway *et al.* (1999) suggested that the marine-based West Antarctic Ice Sheet may have continued to retreat, supplying melt-water to the ocean during the Late Holocene due to a lagged ice dynamic response to Late Pleistocene climate change. This was a view also supported by Stirling *et al.* (1995) and Lambeck (2002) on the basis of geophysical modelling on millennial time-scales. However, Peltier (2001) argues that GIA theory requires a cessation of extra melt-water added to the global ocean during the Late Holocene (since 4000 yr BP). If melt-water was continued to be added, then a higher than present sea-level stand during the Mid- to Late Holocene, would not have occurred over much of the low to mid-latitudes regions in the southern hemisphere. Field evidence supports the Peltier (2001) view, with an extensive dataset showing that RSL on far-field islands and continental coasts was ~1–2 m higher than present during the Mid- to Late Holocene (Grossman *et al.*, 1998). It is highly probable that the cumulative Antarctic contribution to millennial-scale sea-level changes during the Late Holocene is small, with the lowering contribution from increased ice volume in East Antarctica cancelled out by the rise contribution from melting in West Antarctica. Recent numerical modelling work by Mitrovica *et al.* (2001) has compared the global spatial patterns of sea-level change for continuing ice mass variations in Greenland, Antarctica and the non-polar glaciers with the global distribution of sea-level change determined from tide-gauge sites corrected for the GIA adjustment. The Mitrovica *et al.* (2001) study demonstrates that ~0.6 mm/year sea-level rise trend over the last century, can be explained by increased ice ablation and melt-water contributions from the Greenland ice sheet under a warming climate, rather than by significant contributions from Antarctica or the non-polar glaciers.

Sea-level variability on seasonal to interannual time-scales, also occurs with or without ocean mass variation, as a result of variations in the spatial distribution of water density throughout the surface mixed-layer of the ocean. An increase in the surface ocean temperature and or a decrease in salinity, decreases the ocean density and hence forces a rise in steric sea-level (Church *et al.*, 1991). This is known as thermal expansion. The converse is known as saline contraction. On monthly to interannual time-scales the wind-driven ocean circulation results in dynamic variability in regional sea-level due principally to temporal variability in the direction and strength of ocean currents. The temporal and spatial patterns of dynamic sea-level are a marked characteristic of coupled ocean–atmosphere phenomena such as the El Niño Southern Oscillation (ENSO) (Wyrtki, 1975), the North Atlantic Oscillation (NAO) and the Arctic and Antarctic Annular Oscillations (AO and AAO, respectively). Sea-level variability produced by both water density and dynamic changes are known as steric changes (Patullo *et al.*, 1955). Recently, Hong *et al.* (2000a) and Sturges and Hong (2001) have established that the decadal variability in steric sea-level is produced by wind-forcing over the open ocean and propagates by long low-frequency Rossby waves. Pacific Ocean sea-level oscillates throughout the ENSO cycle. On monthly to bi-annual time-scales, equatorial sea-level ranges up to ±40 cm at the eastern and western basin boundaries (shown in Fig. 28.1), due to fluctuating surface ocean wind stress and the resulting ocean circulation (after Wyrtki, 1985). Temporal changes in regional steric sea-level over the last 10 years can be readily observed in real-time and archived satellite altimetry data available on the internet (http://www-ccar.colorado.edu/~leben/research.html). Researching sea-level variability on decadal time-scales is difficult since only a few, fragmentary oceanographic or hydrographic data records exist for periods more than a few decades (Wunsch, 1992). However, indices of low-frequency, coupled atmospheric and oceanic phenomena in the high to mid-latitudes, such as the recently discovered, Pacific Decadal Oscillation (PDO) in SST (Mantua *et al.*, 1997) may prove to be applicable to investigating decadal sea-level variability. As Sturges and Hong (2001) have shown, long-term data on open-ocean, wind-stress curl (torque) may be the most significant for the resolution of centennial sea-level trends. Predicted scenarios of global-scale sea-level rise, forced by 'Accelerated Greenhouse Warming', are based on models of the ocean's steric response to warming and changes in the thermohaline circulation, in addition to eustatic effects contributed by potential ice volume changes in Greenland, and the small non-polar glaciers (Houghton *et al.*, 2001). The predicted global sea-level rise trend for AD 2000–2100 is in the range of 0.2–0.7 m (Houghton *et al.*, 2001).

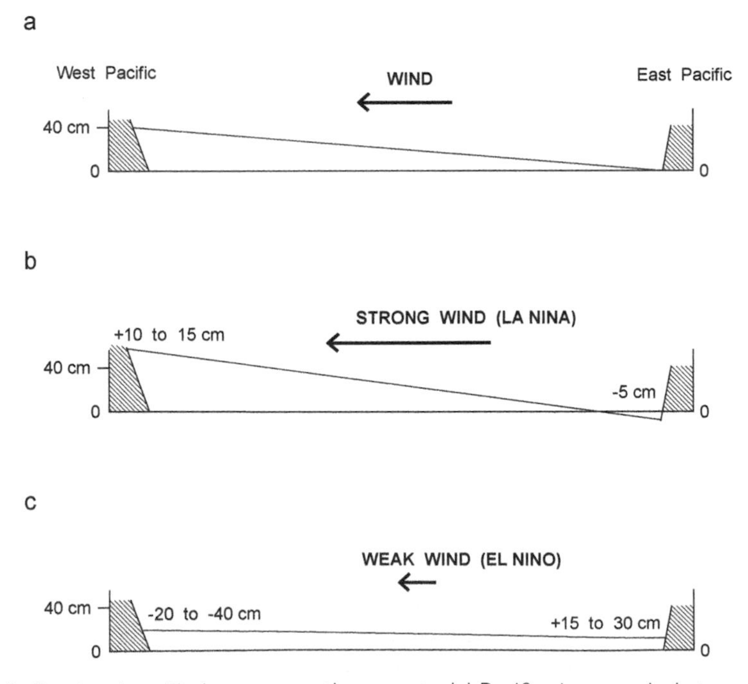

Figure 28.1 Sea-level oscillations across the equatorial Pacific due to wind stress and oceanic circulation variability associated with ENSO. (a) Normal conditions; (b) La Niña conditions; (c) El Niño conditions. After Wyrtki (1985) and McGregor and Nieuwolt (1998).

In reality, the typical regional and decadal variability in sea-level is of the same order of magnitude. Consequently, some oceanic regions where the decadal variability is high, will experience enhanced impacts due to the overall rise in sea-level, and to the contrary, some regions will experience little impact.

High-resolution, palaeo sea-level studies can provide longer-term constraints on both the centennial and millennial trends in sea-level behaviour, together with an understanding of the regional oceanic expression of multi-decadal variability. Scientific techniques designed to unravel multi-decadal sea-level variability over the Late Holocene are of paramount importance to future global change predictions, by providing longer-term records containing both sea-level trends and a quantification of decadal variability. In particular, palaeo-studies can investigate the timing of the onset and trend in recent sea-level rise with respect to proxy air temperature and SST datasets derived from ice-cores, corals and tree-rings, such as the global and hemispherical time-series derived by Mann *et al.* (1998, 2000). Of particular importance to palaeo sea-level studies in the Pacific basin are long-term shifts in the frequency of ENSO as summarized by Quinn *et al.* (1987), which may be potentially resolved in high-resolution palaeo sea-level studies (see Section 28.3).

28.2 PROGRESS IN RESOLVING HIGH-RESOLUTION SEA-LEVEL PROXY DATA FOR COMPARISON WITH THE INSTRUMENTAL SEA-LEVEL AND CLIMATE RECORDS

28.2.1 Fixed Biological Sea-Level Indicators: Coral Microatolls, Encrusting Tubeworms and Barnacles

The most common fixed biological indicators of former and present sea-level are corals, tubeworms, barnacles and bivalves, all of which occupy a distinct vertical niche within the tidal range (Pirrazoli, 1996). Recent papers on Late Holocene sea-level studies have established the use of coral microatolls (Smithers and Woodroffe, 2000) and tubeworms (Baker and Haworth, 2000a) as high-resolution datums. It is not possible to include an in-depth review of all the biological sea-level indicators in this section. Hence, this section will provide an example on progress using coral microatolls as sea-level indicators.

Live coral microatolls are typically single coral colonies (*Porites* sp.), commonly annular in plan with a dead, predominantly flat upper surface and a living outside margin. They are often massive structures, a metre or more in diameter. Microatolls grow upwards until the upper surface is constrained by water level of between mean low water neap (MLWN) and mean low water springs (MLWS) depending upon reef environment. Smithers and Woodroffe (2000) surveyed 282 living microatolls (*Porites* sp.) on intertidal environments (open reef flats, inter-island passages and lagoons) around Cocos (Keeling) Islands (12°05′S, 96°50′E), a mid-ocean atoll in the Indian Ocean. They found that the upper surface elevation of microatolls spanned a vertical 40 cm, or 30 per cent of the spring tidal range. The variability was largely due to surface ponding during lower tides in areas influenced by inter-island hydrodynamics. Microatolls in open reef flat environments are the most suitable as sea-level datums. These were found to have upper surfaces in a small datum range of 8 cm, lying between MLWN and MLWS (Smithers and Woodroffe, 2000). Once the upper surfaces of microatolls reach the constraining water level they continue to grow laterally, with a distinct annual band of skeletal calcium carbonate, that can be identified by X-radiography. It has been demonstrated that during periods of higher or lower mean sea-level the radial microtopography of the upper surface of the microatolls reliably traces sea-level fluctuations over time (Woodroffe and McLean, 1990). Coral microatolls were used by Spencer *et al.* (1997) to reconstruct sea-level change on Tongareva (Penryn) Atoll (8°59'S, 158°02'W) in the Northern Cook Islands (see Fig. 28.2). Spencer *et al.* (1997) established that it was also necessary to understand the variability in annual growth area in addition to variations in the upper surface microtopography of microatolls, to determine the true sea-level record accurately. Correction of the surface microtopography using annual microatoll growth area is particularly relevant following the erosion of the microatoll surface during prolonged interannual periods of low sea-level (in this case, due to El Niño events). Other factors such as high SST and coral bleaching may influence the ability of individual microatolls reliably to record sea-level history continuously over multidecadal periods. As both modern and fossil microatolls of the same *Porites* sp. are found on modern reef flats and or adjacent raised reef flats, there is the potential to obtain a sea-level record using a number of different age microatolls. A composite sea-level record for the locality can be assembled using TIMS Th/U dating, and annual growth ring counting as the dating control (Woodroffe and McLean, 1990).

There is also the potential that coral microatolls can provide a coupled sea-level and SST. Woodroffe and Gagan (2000) measured sub-annual oxygen isotope ($\delta^{18}O$) composition on both a modern and Mid-Holocene age coral microatolls collected from Christmas Island (2°00'N, 157°30'W) in the central Pacific Ocean. They found that the microatolls reliably record and preserve SST variations associated with ENSO, in addition to sea-level variations recorded by the microatoll surface microtopography. The fossil microatoll showed no signs of diagenetic alteration

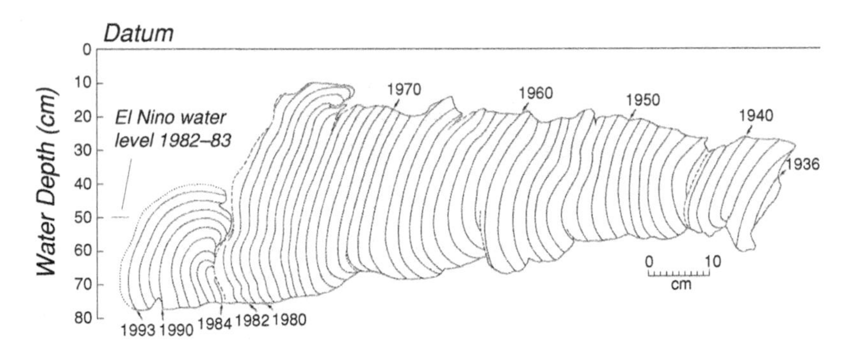

Figure 28.2 Annual coral growth bands mapped from X-ray positive prints of a coral slice from a study on microatoll at Tongareva Atoll, Northern Cook Islands by Spencer *et al.* (1997). A sea-level datum relative to the Tongareva tide-gauge that was used for the microatoll surface microtopography measurements, is shown above the microatoll surface. The datum is 1.25 m on the nearby tide-gauge at Omoka, Tongareva. After Spencer *et al.* (1997). Note that significant sub-aerial erosion of the microatoll surface occurred during the low (–40 cm) (Wyrtki, 1985) sea-level conditions during the 1982–1983 El Niño, and to a lesser extent during the 1940–1941 and 1942–1943 El Niño. Spencer *et al.* (1997) attribute the 1.51 mm/year trend in the annual microatoll surface elevation from 1949 to 1979, to regional sea-level rise.

and the integrity of the coral $\delta^{18}O$ at subannual resolution appeared to be unaltered by subaerial processes including exposure to rainfall.

28.2.2 Saltmarsh, Chenier and Mangrove Sequences as High-Resolution Sea-Level Indicators

RSL records have been obtained from coring sequences of saltmarsh and chenier deposits in meso and macrotidal regions. The methods for RSL studies in these depositional environments are outlined in van de Plassche (1986). This technique has recently been applied in South Australia, and used to interpret tectonic effects and separate out the secular trend in regional sea-level in the modern tide-gauge records (Harvey *et al.*, 1999). The relationship between tidal level and the lateral coastal-ecological succession was established in the modern environment extending from the upper surface of subtidal *Posidonia* sp. seagrass at 0.25 ± 0.25 m above Tidal Datum (approximately Mean Low Water Spring), to the upper surface of the intertidal to supratidal samphire marsh at 3.0 ± 0.4 m above Tidal Datum. Holocene RSL changes were interpreted from vertical facies stratigraphy in a series of vibrocores drilled across the modern supratidal to intertidal zone. Using these techniques Harvey *et al.* (1999) were able to separate the land-level rise and secular sea-level rise components of RSL change in a nearby instrumental RSL record. The overall RSL trend in the historic tide-gauge record showed minimal long-term change (–0.02 mm/year). Harvey *et al.* (1999) subtracted the land-level rise trend (0.33 mm/year due to GIA processes established from the paleo sea-level data) from the overall tide-gauge RSL trend to establish that secular sea-level rise since 1930 was 0.31 mm/yr.

Along the coastlines where post-glacial subsidence has occurred during the Late Holocene (e.g. northeastern USA and southern Britain and North Sea coasts) sea-level history has been often interpreted from vertical saltmarsh accretion, that is assumed to be in equilibrium with sea-level rise (van de Plassche, 1986; Reed, 1990; Allen, 1991). However, marsh-accretion curves do not adequately reflect sea-level changes at some sites due to variations in sediment supply, land subsidence/uplift rates, anthropogenic influences, accommodation space and autocompaction (Haslett *et al.*, 2001). Recent work by Gehrels (1999), van de Plassche (2000), Edwards and Horton (2000) and Haslett *et al.* (2001) have described a more comprehensive methodology for determining proxy sea-level from saltmarsh peat cores, by using *Foraminifera*-based biostratigraphy as tidal indicators. Biostratigraphy using *Foraminifera* enables the interpretation of distinct tidal relationships (Scott and Medioli, 1978) for former marsh surfaces in downcore saltmarsh sequences (Haslett *et al.*, 2001). Van de Plassche (2000) summarises the methodology into three stages as follows:

- a marsh accretion record is determined from marsh chronostratigraphy, where dates are usually determined from radiocarbon or ^{210}Pb methods
- a marsh-surface elevation record is established relative to a tidal datum using known *Foraminifera*-tidal relationships
- the RSL record is determined from the subtraction of the latter from the former.

In addition to these methods, Haslett *et al.* (2001) and Edwards and Horton (2000) establish a series of sea-level index points (SLIPs) based on accurately surveyed litho- and bio-stratigraphic markers in the modern environment that can be used to describe water-level variations from vertical saltmarsh facies changes.

28.3 DISCUSSION ON SEA-LEVEL HISTORY OF THE LAST MILLENNIUM FROM PROXY DATA

The above sections have outlined the advancements in the interpretation of depositional environments and fixed biological sea-level indicators that are associated with the development of high-resolution dating techniques. In the following sections, three case studies are summarized to illustrate the progress that has recently been made in unravelling palaeo sea-level history at centennial or multi-decadal resolution. In palaeoclimatic terms, the last millennium has often been defined in the literature by two periods, principally based on proxy temperature history in the northern hemisphere. These two periods are

- the Medieval Warm Period (MWP) (or Little Climatic Optimum, LCO) with surface temperature warmer than present between ~1050 and 690 yr BP
- a cooler than present period, the Little Ice Age (LIA) between ~575 and 150 yr BP (Lamb, 1972a).

There has been considerable debate on the global proliferation of these climate events, and on the regional variability in SST and air temperature, ocean and atmospheric circulation during this period (Mann *et al.*, 1998; Crowley and Lowery, 2000; Jones *et al.*, 2001b). Bradley and Jones (1993) found that geographical and temporal variations existed around the globe during the LIA and Hughes and Diaz (1994) and Crowley and Lowery (2000) debated the existence of a MWP or Medieval Optimum. Mann *et al.* (2000) argue that the last few years of the last century experienced the warmest air temperatures in the northern hemisphere in the last millennium, including the so-called MWP. They also concluded that climatic conditions appear to have been warmer during the 11th–14th centuries, than the subsequent period from the 15th–19th centuries. Crowley and Lowery (2000) using 15 palaeoclimate records from the northern hemisphere concluded that the MWP comprised three relatively narrow intervals: AD 1010–1040, AD 1070–1105 and AD 1155–1190 where temperatures were within ±0.05 °C of the mid-20th century warm period, and the MWP was on average only 0.2 °C warmer than the LIA. Few high-resolution proxy temperature records covering the last millennium exist for sites in the southern hemisphere. However, from the scant evidence it is possible to conclude that the peak warming in the southern hemisphere during the last millennium may have lagged the northern hemisphere (Hughes and Diaz, 1994). Hence the interpretation of last millennial steric and eustatic sea-level changes, and the inherent influence of multi-decadal to centennial climate fluctuations is complicated by both the regionality of climate change and the complexity of regional coastal evolution. The latter is fashioned by the regional sediment budget, embayment accommodation space and antecedent geology, wave and wind climate fluctuations, the global isostatic adjustment, tectonics, coral reef evolution (tropics), in addition to sea-level variability. However, bearing this complexity in mind, researchers have reported evidence for sea-level variability on the order of at least ±0.5 m over the last millennium, although much of the former sea-level information is more qualitative than quantitative.

28.3.1 The South-West Pacific Region

Nunn (1998) compiled a summary of published evidence of former sea-levels in the equatorial and southwest Pacific regions, over the last millennium. The proxy sea-level indicators used in the compilation are not described in detail, but include both erosional notches, elevated storm deposits, coral on emerged reef flats and archaeological information. Unfortunately, the Nunn

(1998) compilation was derived from sporadic spatial evidence without reference to temporal sequences of former sea-level at one site. On the basis of this dataset, Nunn (1998) argued for a distinct pattern of last millennial sea-level variability across a very wide region of the Pacific. This is summarized as follows: a rise in sea-level by ~0.9 m from near the present level during the MWP; a subsequent fall in sea-level of ~0.9 m coincident with the LIA; and a subsequent rise in sea-level in the last 200 years. The latter interpretation is consistent with the instrumental records of the last century, although the timing of the onset of sea-level rise has not been established because both the instrumental records are too short and the palaeo sea-level data are both spatially and temporally sparse in the South Pacific.

However, Gehrels (2001) points out in his criticism of the Nunn (1998) paper that the sea-level envelope is constructed using uncalibrated radiocarbon ages, and age mid-points rather than calendar age ranges were shown. This distorts any possible temporal interpretations. Perhaps more importantly, the vertical uncertainties in the proxy sea-level data with respect to mean sea-level or relative tidal level were not included. Gehrels (2001) replotted the Nunn (1998) data with error boxes and his interpretation on the locus of Southwest Pacific sea-level over the last millennium is shown in Fig. 28.3.

Despite the shortcomings of Nunn (1998), the sea-level overview provides a reasonable hypothesis for testing, as the relationship between changes in ocean thermal structure, salinity and/or

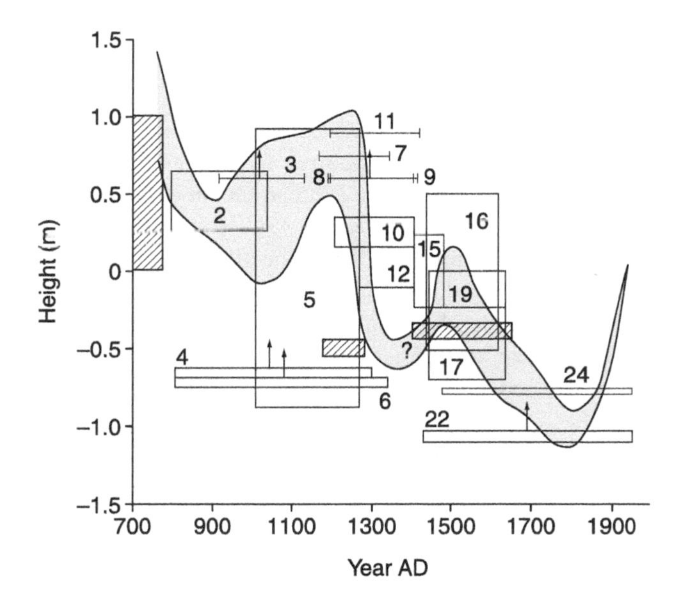

Figure 28.3 The relative sea-level envelope (shaded) for the last 1300 years from Nunn (1998) data. Gehrels (2001) has reworked the Nunn (1998) data and has shown the range of calibrated radiocarbon ages, together with the vertical uncertainty in the proxy sea-level indicators, as boxes. The numbers refer to the samples in Nunn (1998). Relative sea-level is expressed as an anomaly in metres with respect to present mean sea-level (0 m). The envelope shows that relative sea-level was lower than present by up to 1 m for some period in the last few hundred years. Also plotted in Fig. 28.3 to allow regional comparison, are the relative sea-level data for Langebaan Lagoon, South Africa from Compton (2001), which are shown in the boxes with diagonal hatching.

circulation and the sea-level response is supported by clear physical relationships in modern coupled ocean–atmosphere climate studies. However, our knowledge of decadal to centennial climate fluctuations over the last millennium in the southern hemisphere is scant. With this in mind, sea-level research in the Pacific basin, needs to systematically focus on first, resolving the influence of multi-decadal to centennial climate variability associated with ENSO (Quinn *et al.*, 1987; Dunbar and Cole, 1999) on coastal evolution, before high-resolution records of sea-level variability can be derived. There have been many interpretations of coastal sedimentary sequences that require a more energetic wave climate, and/or major storm events during the LIA, to explain their deposition. Allen (1998) interprets a sequence of storm deposits within a prograded beach ridge plain, along the Aitutaki coast (18°50′S, 160°45′W), Cook Islands, Southwest Pacific, as indicative of an increase in storm events during the last 1000 years, when compared with the period prior to 900–1040 cal yr AD. Similarly, Thom (1978) interpreted a renewed phase of dune transgression along the east coast of Australia commencing at around 300–500 yr BP which he attributed to a more energetic storm wind and wave climate, in the Tasman Sea.

28.3.2 The South African Region

A recent study on the evolution of Langebaan Lagoon (33°15′S, 18°15′E) on the southwest coast of South Africa by Compton (2001) produced a sub-millennial RSL history along the South Atlantic coast for the last 7000 years. A RSL curve was developed from radiocarbon-dated sediments and shell obtained from vibrocores taken in salt pan and saltmarsh sediments surrounding the Langebaan Lagoon. Compton (2001) concluded that RSL was up to 0.7 m higher than present at 1300 cal yr BP (AD 700) and that sea-level fell to –0.5 m by 700 cal yr BP (AD 1300). RSL had commenced rising by 500 cal yr BP (AD 1500), with an elevation of –0.2 m with respect to present sea-level, and continued to rise to the present. The RSL interpretations for Langebaan Lagoon are supported by the interpretations from many other sites in South Africa, from both the Atlantic and Indian Ocean coasts, which are summarized in Compton (2001). The RSL data spanning the last 1300 years for the Langebaan Lagoon are shown plotted on Fig. 28.3 for comparison with the RSL data from the Southwest Pacific region. Despite the uncertainties in the age and relationship of the former sea-level indicators, RSL data from both regions support the existence of a lower than present sea-level between AD 1500 and 1700. However, there are considerable differences in the interpretation of sea-level variability between AD 1150 and 1300.

28.3.3 The North Atlantic Region

In contrast to the above examples, field investigations in microtidal saltmarshes in Connecticut and mesotidal saltmarshes in Maine, along the northeast coast of the USA have produced higher-resolution palaeo sea-level records with decimeter accuracy. The records from Long Island Sound, Connecticut (41°16′N, 72°31′W) describe sea-level oscillations of up to a few decimeters on a centennial time-scale since AD 800. The sea-level records were obtained by subtracting a palaeo mean high-water record, derived from foraminiferal analysis, from a saltmarsh accumulation record, obtained by acceleration mass spectrometry (AMS) ^{14}C dating of paleo marsh-surface indicators (van de Plassche *et al.*, 1998; Varekamp *et al.*, 1992). The relative marsh-elevation data has a vertical error in interpretation of ±0.1 m, based on the identification of foraminiferal biostratigraphic zones. The relative to Mean High Water (MHW) sea-level curve for the period since AD 800 was presented in van de Plassche *et al.* (1998) and the AD 750–950 portion upgraded in van de Plassche (2000) together with a correlation of the sea-level curve and proxy climate time-series from the northern hemisphere. Gehrels (1999) reported a high-resolution relative sea-level curve for the Machiasport area (44°41′N, 67°24′W) of eastern Maine based on foraminiferal saltmarsh stratigraphy corrected for sediment autocompaction. Whilst the relative sea-level curve extends back to 6000 yr BP, the period from AD 900 (1100 cal yr BP) to the present is of most relevance to this discussion. After both chronological and sea-level indicator errors were considered, Gehrels (1999) interpreted low-amplitude fluctuations in a millennial-scale period of relative sea-level (mean tide level, MTL) stability between ~850 AD and 1550 AD. The rate of relative sea-level rise during this period was at a minimum in the last 2000 years. Subsequently, sea-level rose by 0.5 m during the last 300 years.

Van de Plassche (2000) argued that the MHW oscillations lagged northern hemisphere temperature fluctuations, as described by time-series from central Greenland, northern Eurasia and Fennoscandia, and the northern hemisphere summer temperature anomaly time-series of Bradley and Jones (1993), by 0–100 years. Gehrels (1999) plotted both the Maine and Connecticut relative sea-level curves against the paleo SST record from the Sargasso Sea. There was reasonably good agreement between the Maine and Connecticut relative sea-level fluctuations at a centennial-scale during the last 1100 years, and both demonstrate a lower than present sea-level between AD 1000 and the present. Gehrels (1999) noted some correspondence between the sea-level fluctuations and the Sargasso Sea SST record, which prompted him to conclude that sea-level fluctuations may reflect the steric ocean response to climate change. The Connecticut MHW sea-level curve is shown in Fig. 28.4, and is plotted against the most recent northern hemisphere summer temperature anomaly time-series of Mann *et al.* (1998). Whilst the temporal resolution of the MHW sea-level curve is much coarser than the filtered (40-year Gaussian moving average) annual temperature anomaly series, there is good agreement between the two series on a centennial level. At a diagnostic rather than mechanistic level, this implies that sea-level has varied by a few decimeters in concert with air temperature anomalies of 0.2–0.6 °C. However, Kelley *et al.* (2001), in reply to the van de Plassche (2000) paper, argue that the applicability of the interpreted MHW curve as indicative of hemispherical changes may

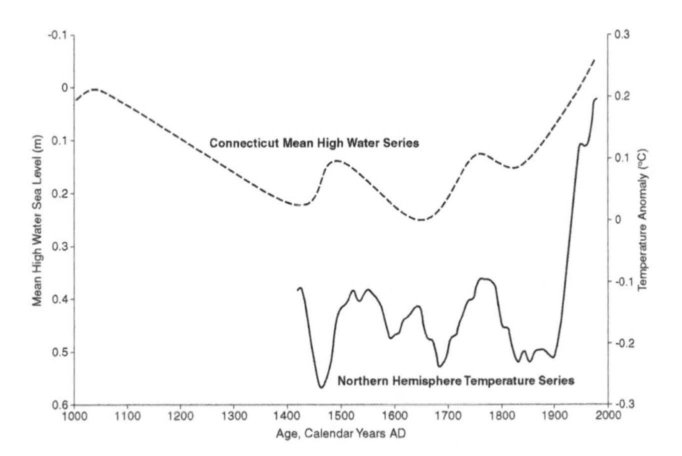

Figure 28.4 The sea-level curve relative to Mean High Water for Connecticut coast, northeast USA from van de Plassche et *al.* (1998), plotted against the annual summer temperature anomaly for the northern hemisphere (40-year smoothing). After Mann et *al.* (1998). Data are available online from ftp://ftp.ngdc.noaa.gov/paleo/paleocean/by_contributor/mann1998/mannhem.dat.

be marred by local effects which could result in more than one sea-level interpretation of the marsh accretion sequence. These local effects include variability in marsh autocompaction and peat accumulation in near equilibrium with sea-level such as increased marsh accumulation rates in equilibrium with new accommodation space afforded by sea-level rise. Van de Plassche (2001) responded to the Kelley *et al.* (2001) criticism by demonstrating that the robustness of the combined approach using both marsh accumulation and a high water record determined from foraminiferal analysis warrants a more regional and perhaps hemispherical consideration of the MWH sea-level record. Kelley *et al.* (2001) reinforce the potential of sea-level studies based on foraminiferal biostratigraphy, as foraminifera assemblages respond to water-level changes more rapidly than marsh plants, and may record more rapid sea-level fluctuations. In summary, both the Gehrels (1999) and van de Plassche (2000) studies indicate that sea-level was lower by at least 0.2–0.3 m along the northeast coast of the USA during the LIA.

28.3.4 Extending the Modern Instrumental Tide-Gauge Records of Mean Sea-Level Change Using High-Resolution Proxy SST Records from Corals

There are now 11 or more, long proxy records of SST from the tropical Indian, Pacific and Atlantic Oceans (Dunbar and Cole, 1999). In the tropical Indian Ocean, proxy SST interpreted from oxygen isotope ($\delta^{18}O$) measurements on coral cores, indicate that SST decreased by 1 °C between AD 1700 and 1800 and then increased to the present by 1.5–2 °C since AD 1800 (R. Dunbar, personal communication; Dunbar, 2001). These rates of change in proxy SST are comparable to the instrumentally derived rates of change for this century. The question of interest is, to what degree has sea-level fluctuated in accordance with these SST changes over the last few hundred years and when did the modern sea-level rise trend that is observed in the instrumental record commence?

One method to resolve this issue may be to develop a coupled approach to derive both proxy SST and sea-level time-series in regional coral palaeoclimate studies. In section 28.2.1 above, reference has been made to the advancements in measuring both SST and RSL variability using microatolls (Woodroffe and Gagan, 2000). In addition there is the potential to use SST time-series from studies of reef slope *Porites* corals, for sea-level studies. Kühnert *et al.* (1999) reported a 200-year stable-isotope record from *Porites lutea* corals drilled on the reef slope of the Houtman Abrolhos Islands (29°S, 114°E), on the West Australian continental shelf. The Houtman Abrolhos Island reefs are the southernmost major reef complex in the Indian Ocean. The reef complex is sustained at the high latitude site between 28 and 29° S by the Leeuwin current which transports warm, tropical water poleward along the shelf edge, sometimes meandering with incursions of warm water onto the shelf in the vicinity of the islands (Hatcher, 1991). The Leeuwin Current is coupled to the Indonesian Throughflow, and hence the SSTs and current strength along the West Australian coast experience considerable temporal variability associated with ENSO fluctuations. Pariwono *et al.* (1986) established that long-period sea-level signals propagate north to south along the Western Australian coast in association with the Leeuwin Current, mainly during the southern hemisphere's summer (Godfrey and Ridgeway, 1985). Kühnert *et al.* (1999) examined the sub-annually resolved, proxy SST record from the Houtman Abrolhos Islands, and instrumental SSTs from the Comprehensive Ocean Atmosphere Data Set (COADS) with the nearby Geraldton (28°50′S, 114°36′E) tide-gauge record of RSL variability, to assess the influence of the Leeuwin Current strength on the SSTs. The overall coherence was surprisingly weak between the coral $\delta^{18}O$ and the RSL when compared to the coherence between the coral $\delta^{18}O$ and the COADS SSTs. The most coherence between the coral $\delta^{18}O$ and the RSL series was at frequencies of 0.09–0.21 per year (11–4.8 years). However, despite these shortcomings, Kühnert *et al.* (1999) suggested that the analysis of coral $\delta^{18}O$ SST's and relative sea-level trends on multi-decadal time-scales deserved more investigation.

Whilst the Geraldton RSL record spans only 1965 to present, the relative sea-level record at Fremantle (also a stable continental site, 370 km further south along the West Australian coast, 32°02′S, 115°45′E), spans from 1897 to present (Mitchell *et al.*, 2001). These two annual RSL records are highly correlated ($r = 0.930$, $n = 33$, significant at the 99.9-per cent interval). This suggests that both records are largely free of effects due to local subsidence or uplift and anthropogenic influences. However, both RSL records contain a component of GIA which is estimated from the global models of Peltier (1998) to be a RSL lowering component in the range of 0.2–0.3 mm/year. This is not adjusted for in the following analysis as a more detailed study, beyond the scope of this chapter, is required, following the methods of Harvey *et al.* (2002). Linear regression of the Fremantle RSL time-series revealed a relative sea-level rise trend of 1.38 mm/year for the 96-year period, and a total rise of 0.131 m which is equivalent to that determined for the Fremantle site by Mitchell *et al.* (2000). The annual coral $\delta^{18}O$ SST time-series and the annual Fremantle RSL series are determined in this study to be moderately correlated ($r = 0.430$, $n = 87$, significant at the 99.9-per cent interval). Consequently, linear regression of the coral $\delta^{18}O$ SST time-series and the Fremantle RSL series was applied for the common time period 1897–1993. Linear regression was used to represent the centennial-scale trend in the data and does not account for fluctuating sea-level behaviour and SST on annual to decadal time-scales in the data. The comparative trend in the coral $\delta^{18}O$ SST time-series shows a decrease in $\delta^{18}O$ of –0.131‰, which is equivalent to an increase in SST of +0.79 °C (using the Kühnert *et al.*, 1999) $\delta^{18}O$ to SST transfer function of –0.1‰ for +0.6 °C). The coral $\delta^{18}O$ SST and Fremantle RSL time-series are shown in Fig. 28.5.

Kühnert *et al.* (1999) determined that the SST had increased by +1.4 °C between 1795 and 1993. Accordingly, by extrapolating the coupled sea-level rise and increased SST relationship for the last century over the full SST record from 1795 to 1993, it is calculated that RSL rose by a total 0.232 m. This assumes that sea-level rise is coupled to increased SST over the full record. This is based on the temperature dependence of both instrumental sea-level trends over the last century and sea-level predictions for this century. These results are encouraging and outline a possible methodology for investigating interdecadal and perhaps annual sea-level change at many sites around the globe covering the last two to three centuries depending upon the age of living *Porites* corals and the equivalent length of $\delta^{18}O$ measurements. However, it is stressed that the component of RSL change due to the GIA, geological and anthropogenic influences must be accounted for at each site, using both site studies and modelling, before the climatic influence on secular sea-level change can be determined.

28.4 CONCLUSIONS AND FUTURE RESEARCH DIRECTIONS

The existing body of coastal and sea-level research indicates that multi-decadal sea-level changes on the order of up to 1 m, but typically ±0.5 m or less, can be detected in the coastal sedimentary record covering the last few hundred years. Broadly, the recent northern and southern hemisphere studies described above, depict a common pattern of lower (0.2–0.5 m) than late 20th century sea-level during the period AD 1400–1850 which encompasses the LIA.

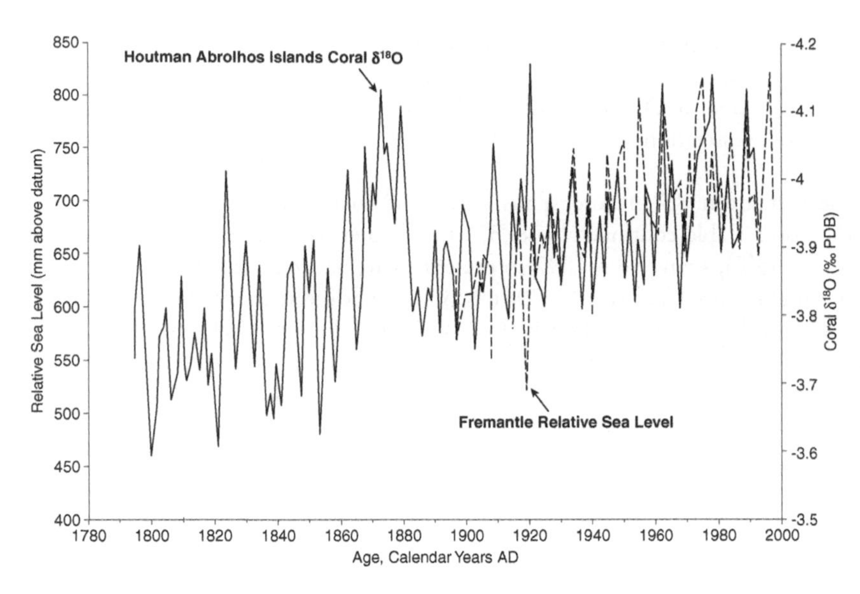

Figure 28.5 Annual mean relative sea-level for Fremantle (dashed line), South Western Australia plotted against the annual mean coral $\delta^{18}O$ (proxy SST) (stippled line) for a coral core drilled at the Houtman Abrolhos Islands, offshore from the south Western Australian coast, for the 1795–1993 period. The coral $\delta^{18}O$ (proxy SST) data are from the paper by Kuhnert *et al.* (1999) and are available from http://www.ngdc.noaa.gov/paleo/corals/html. The Fremantle tide-gauge relative sea-level data were obtained from the National Tidal Facility, The Flinders University of South Australia, Australia. The two datasets are moderately correlated ($r = 0.430$, $n = 87$) for the common time period. The diagram shows the concomitant rise in relative sea-level with proxy SST (positive SST is negative $\delta^{18}O$) on a multi-decadal trend analysis.

The evidence for higher (0.5–1.0 m) than late 20th century sea-level during the period AD 1000–1300, and its relationship to climate variability is more circumspect. It is highly probable that at southern hemisphere sites where prominent emergence has occurred since the AD 1000–1300 period, the relative sea-level change may be primarily explained by the GIA process. However, a small component of the interpreted higher sea-level during this period may be consistent with warm climate or circulation changes during the MWP. The onset of the recent phase of sea-level rise is not known, but the palaeo sea-level research constrains it to the last 200 years. These sea-level changes may be related to changes in ocean thermal structure, salinity and/or wind-driven circulation. However, there is considerable debate amongst coastal scientists on the interpretation of coastal evolution and high-resolution sea-level signals, the application of different dating methods and whether the signals are regionally representative. Nevertheless, this debate is enabling scientists to close the gaps and develop more robust methodologies to extract data on coastal evolution at decadal or multidecadal resolution over the last millennium. These methodologies have the potential to enable the separation of climate-forced sea-level changes from the cumulative or integrated sea-level curve at many global locations, which includes a signal from the GIA, geological and anthropogenic influences.

Because the fluctuations in sea-level over the last few millennia are the same magnitude as those predicted for the future under 'Greenhouse conditions', future research should contain coupled objectives to achieve an integrated data base on coastal response to climate change. Much attention has been focussed on the coastal response to sea-level rise according to basic shore profile translations such as the Bruun Rule (Bruun, 1962). However, a major focus for coupled coastal and palaeoclimate research is to establish the response of coastlines to a suite of climate forced variables. These include wave climate (significant and maximum height, period and direction), major storm events, regional shifts in sea-level due to changes in ocean basin wind stress and circulation, in addition to steric and eustatic sea-level changes.

Most of the predictions of future regional and global sea-level change have been based on tide-gauge data corrected for the global isostatic adjustment. An improved understanding of the tide-gauge data can be made with signal to noise data obtained from palaeo-records. This has the potential to enable better interpretation of cyclical ocean circulation patterns and the discrimination of local tectonic influences such as subsidence or uplift on sea-level records. However, the establishment of robust dating techniques and multiple chronologies is fundamental to any advancement in the interpretation and resolution of coastal evolution and the discrimination of the role of climate, anthropogenic and geological variables in sea-level change.

Acknowledgements

The Fremantle tide-gauge data were provided by the National Tidal Facility, Flinders University of South Australia, Adelaide, South Australia, Australia. The author acknowledges the IGBP Past Global Changes (PAGES) programme and the Land–Ocean Interactions in the Coastal Zone (LOICZ) programme, for assisting with funds to conduct the workshop on 'Late Holocene Sea-Level and Climate Change' in 1999, and in their support for an international research activity on 'Coastal Records of Sea-level and Climate Variability over the Last 2000 Years'.

CHAPTER

SIMULATION OF HOLOCENE CLIMATE CHANGE USING CLIMATE-SYSTEM MODELS

Martin Claussen

Abstract: Holocene climate variability provides an excellent test ground for climate-system models. It appears that interpretation of palaeoclimatic records in terms of atmospheric or oceanic phenomena only fails to successfully explain the full range of Holocene climate dynamics. Instead, theories which take into account interactions between the atmosphere, the oceans and the biosphere are needed. This review addresses transient simulations of Holocene climate changes with focus on the dynamics at high northern latitudes and Northern Africa.

Keywords: Atmosphere–biosphere interaction, Biome paradox, CGCMs, Green Sahara, Multiple equilibria, Vegetation dynamics

Following the traditional concept of Hann (1908), climate has been considered as the sum of all meteorological phenomena which characterize the mean state of the atmosphere at any point of the Earth's surface. This classical definition has proven to be useful for climatology, the descriptive view of climate. However, for understanding climate dynamics, i.e. the processes which govern the mean state of the atmosphere, the classical definition appears to be too restrictive since the mean state of the atmosphere is affected by more than just atmospheric phenomena. In modern text books on climate dynamics, therefore, climate is described in a wider sense in terms of state and ensemble statistics of the climate-system (e.g. Peixoto and Oort, 1992). The climate-system, according to the modern concept, consists of the **abiotic world**, or physical climate-system, and the living world, called the **biosphere**. The physical climate-system is further subdivided into open systems, namely the atmosphere, the hydrosphere (mainly the oceans but also rivers), the cryosphere (inland ice, sea-ice, permafrost and snow cover), the pedosphere (the soils) and, if long time-scales of many millennia are concerned, the lithosphere (the Earth's crust and the more flexible upper Earth's mantle).

The physical concept of climate is reflected in discussions of climate models only to some extent. For example, coupled general circulation models of the atmosphere and the ocean (CGCMs) are considered as 'the most complete type of climate models currently available' (McGuffie and Henderson-Sellers, 1997). Recently, Grassl (2000) stated that CGCMs will become a basis for Earth System models that describe the feedbacks of societies to climate anomaly predictions. Grassl (2000) further suggested that parameterization of vegetation and other processes and boundary conditions in CGCMs has to be improved. This view potentially underestimates the role of vegetation dynamics and biogeochemical cycles in affecting the climate-system. Only recently, climate-system models of various degree of complexity have been built in which not only atmospheric and oceanic, but also vegetation dynamics are explicitly simulated (e.g. Petoukhov *et al.*, 2000; Cox *et al.*, 2000; for a detailed discussion on climate models, specification of boundary conditions in climate models etc., the reader is referred to the chapter on climate model simulations by Paul Valdes, pp. 20–35 in this volume).

Holocene climate variability provides an excellent test ground for climate-system models. It appears that interpretation of palaeoclimatic records in terms of atmospheric or oceanic phenomena only fails to successfully explain the full range of Holocene climate dynamics. Instead, theories that take into account interactions between the atmosphere, the oceans and the biosphere are needed. As examples, I present two cases: climate changes at high northern latitudes which are associated with the so-called biome paradox and climate changes in Northern Africa associated with the 'green Sahara'.

29.1 CHANGES AT HIGH NORTHERN LATITUDES: THE BIOME PARADOX

Palaeobotanical evidence indicates that during the Early to Mid-Holocene, some 9000 to 6000 years ago, Boreal forests extended north of the modern tree line at the expense of tundra (see Prentice *et al.*, 2000 for an overview of a special issue on Mid-Holocene and Glacial Maximum vegetation geography of the northern continents and Africa). It is assumed that this migration was triggered by changes in the Earth's orbit which led to stronger insolation in northern hemisphere summer (Harrison *et al.*, 1998). Insolation during northern hemisphere winter, however, was weaker than today. Hence when assuming that changes in insolation are the dominant cause for climate change one would expect warmer summers, but colder winters than today during the Early and Mid-Holocene. However, not only summers but also winters were presumably warmer than today in many regions of the northern hemisphere as reported, for example, for Europe by Cheddadi *et al.* (1997). Most climate models in which only atmospheric dynamics were considered show a wintertime cooling in Europe. Some atmospheric models reveal a wintertime warming which is too weak though to induce some vegetation changes as seen in the pollen record (Texier *et al.*, 1997; Harrison *et al.*, 1998). The enhanced winter warming was attributed to shifts in tree-line (e.g. Otterman *et al.*, 1984), the so called taiga–tundra feedback. Therefore, the Mid-Holocene wintertime warming at some high northern regions is sometimes called the biome paradox (Berger, 2001).

The taiga–tundra feedback is a feedback between migration of vegetation and near-surface atmospheric energy fluxes. The taiga–tundra feedback was first discussed by Otterman *et al.* (1984) and Harvey (1988, 1989a, 1989b). It is based on the observation that the albedo of snow-covered vegetation is much lower for forests, such as taiga, than for low vegetation, such as tundra. For snow-covered grass, albedo values of some 0.75 were measured (Betts and Ball, 1997), and for snow-covered forests, 0.2–0.4. Hence the darker, snow-covered taiga receives more solar energy than snow-covered tundra mainly in spring and early summer, thereby warming the near-surface atmosphere which, in turn, favours the growth of taiga. This feedback is sketched in Fig. 29.1a. The taiga–tundra feedback is modified by changes in evaporation. An increase in evaporation, perhaps owing to an increase in plant growth, tends to reduce the sensible heat flux. Thereby it cools the near-surface atmosphere. On the other hand, stronger evaporation increases atmospheric water vapour which could lead to a stronger atmospheric

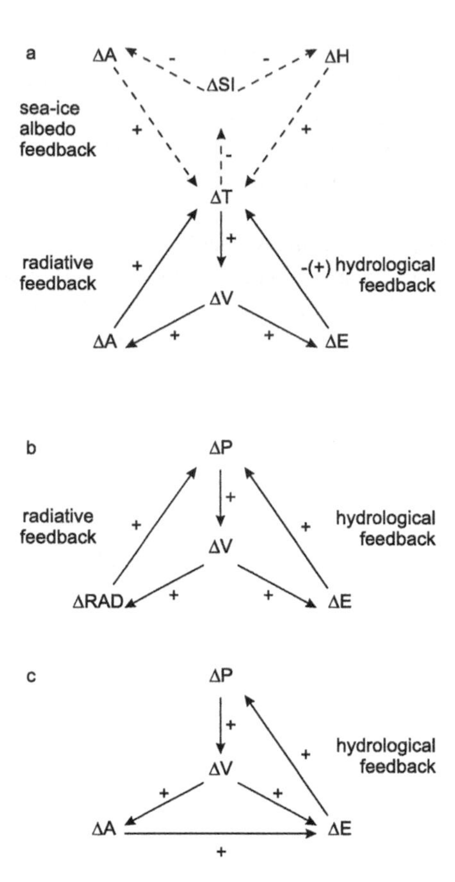

Figure 29.1 Sketch of atmosphere–vegetation interactions at (a) northern high latitudes and (b) sub-tropics. In the sketch, not all possible feedbacks are drawn, but only those which I consider the most important ones on a large scale. The arrows, for example from ΔT to ΔV, or ΔT to ΔSI in (a), have to be interpreted as relations, e.g. $\Delta V \sim \Delta T$, or $\Delta SI \sim \Delta T$. The dashed arrows in (a) indicate the sea-ice albedo feedback (left hand side) and the sea-ice heat flux feedback (right-hand side), both of which enhance the biogeophysical feedback. T refers to temperature or temperature sum; *V*, to Boreal forest fraction or vegetation density or area covered by vegetation; *P*, to annually or seasonally averaged precipitation; *A*, to absorption of solar energy at the surface; *E*, to evaporation and transpiration; *SI*, to sea-ice area; *H*, to heat fluxes from the ocean into the atmosphere; *RAD*, to the divergence of the net radiative flux in the atmosphere. Reprinted from Claussen (2003) with kind permission from Springer-Verlag.

radiation, i.e. stronger near-surface warming. If cloudiness becomes stronger owing to increased evaporation, then near-surface warming is diminished by the reduction of insolation (e.g. Betts, 1999). Hence, the hydrological branch of the taiga–tundra feedback can work in both directions, amplification or attenuation of the initial signal. Model experiments indicate that the hydrological feedback seems to dominate in summer, but with respect to annual mean changes, it appears to be smaller than the surface albedo feedback.

Foley *et al.* (1994) analysed the sensitivity of northern hemisphere climate in detail. By imposing an increase of forest area by some 20 per cent as a surface condition, they find that changes in land surface conditions give rise to an additional warming of some 4 °C at regions north of 60°N in spring and about 1 °C in the other seasons. Orbital forcing would produce only some 2 °C. The additional warming is mainly caused by a reduction of snow and sea-ice volume by nearly 40 per cent which, in turn, reduced the surface albedo in the Arctic region. Further simulations using similar experiment set up, but different models (TEMPO, 1996) corroborate the earlier results.

To analyse the biogeophysical feedback, i.e. the **inter**action between changes in vegetation structure and the atmosphere, coupled atmosphere–vegetation models were designed. For example, Gutman *et al.* (1984) and Gutman (1984, 1985) explored the idea of relating the surface parameters of an atmospheric model (e.g. albedo and water availability) to climatic variables. They used the Budyko (1974) radiative index of dryness D to characterize the geobotanic type of a climate zone and proposed a simple relation between albedo, water availability and D. Henderson-Sellers (1993) and Claussen (1994) coupled comprehensive atmospheric circulation models with (diagnostic) biome models, i.e. with models of macro ecosystems assuming an equilibrium with climate. These asynchronously coupled atmosphere-biome models can be used to assess equilibrium solutions of the system, but not system dynamics. Vegetation dynamics was included later in the models used by Ganopolski *et al.* (1998a) and Levis *et al.* (1999a, 1999b), for example.

Let us return to the problem of the vegetation–snow–albedo feedback at high northern latitudes, i.e. the taiga–tundra feedback. One would expect that this feedback is a positive one: a reduction in surface albedo increases near-surface temperatures which, in turn, favours growth of taller vegetation, reducing surface albedo further (see Otterman *et al.*, 1984). The feedback is limited by topographical constraints, e.g. coast lines, or by the insolation. The studies of Claussen and Gayler (1997) and Texier *et al.* (1997), when using different atmospheric models, but the same biome model of Prentice *et al.* (1992), confirm the earlier assertion of a positive vegetation–snow–albedo feedback. However, both models show a rather small northward expansion of Boreal forests. This is not surprising as the annual cycle of sea-surface temperatures (SST) and Arctic sea-ice volume is kept constant. Hence in the atmosphere–vegetation models, no feedback between changing sea-ice and near-surface energy fluxes was considered. The sea-ice–albedo feedback operates in a similar way as the taiga–tundra feedback: an increase in near-surface warming reduces the extent of sea-ice thereby lowering the surface albedo, enhancing the absorption of solar radiation which eventually leads to further melting of sea-ice, and *vice versa* (see Fig. 29.1a).

To explore the specific contribution of the taiga–tundra feedback and the sea-ice–albedo feedback to the biome paradox, Ganopolski *et al.* (1998a) used a coupled atmosphere–ocean–vegetation model. They find a summer warming on average over northern hemisphere continents of some 1.7 °C (in comparison with present-day climate) owing to orbital forcing of the atmosphere alone. Inclusion of ocean–atmosphere feedbacks (but keeping vegetation structure constant in time) reduces this signal to some 1.2 °C, because the thermal inertia of the ocean leads to a slight damping of the direct atmospheric response to insolation changes. The taiga–tundra feedback (but now without any oceanic feedback) enhances the summer warming to 2.2 °C. In the full system (including all feedbacks) this additional warming is not reduced, as one would expect from linear reasoning, but it is increased to 2.5 °C as a result of a synergism between the taiga–tundra feedback and the Arctic sea-ice–albedo feedback. For Boreal winter, orbital forcing induces a cooling of some –0.8 °C in the model. The pure biogeophyscial feedback reduces wintertime cooling to

–0.7 °C, and the atmosphere–ocean interaction, to –0.5 °C. The synergism between the two feedbacks, however, causes a winter warming of some 0.4 °C (see Plate 11). Obviously the results of Ganopolski *et al.* (1998a) indicate that the so called biome paradox is presumably caused not by a pure biospheric feedback, but by the synergism between the taiga–tundra feedback and the sea-ice–albedo feedback.

29.2 CHANGES IN THE SUBTROPICS: THE GREEN SAHARA

Palaeoclimatic reconstructions indicate that during the Early and Mid-Holocene, Northern African summer monsoon was stronger than today according to lake-level reconstructions (Yu and Harrison, 1996), estimates of aeolian dust fluxes (deMenocal *et al.*, 2000a) and distribution sand dunes (Sarnthein, 1978). Moreover, palaeobotanic data (Jolly *et al.*, 1998) reveal that the Sahel reached at least as far north as 23°N, presumably more so in the western than in the eastern part. (The present boundary extends up to 18°N.) Hence, there is an overall consensus that during the Holocene optimum, the Sahara was greener than today (Prentice *et al.*, 2000) – albeit estimates of Mid-Holocene vegetation distribution in Northern Africa vary (Frenzel *et al.*, 1992; Hoelzmann *et al.*, 1998; Anhuf *et al.* 1999) with respect to vegetation coverage and geographical detail.

The greening was tentatively attributed to changes in orbital forcing (e.g. Kutzbach and Guetter, 1986) that was suggested to amplify the Northern African summer monsoon. However, this amplification in summer rain did not seem to be large enough to explain a large-scale greening. Models of atmospheric circulation in which present-day land cover was prescribed did not yield an increase in monsoon rainfall strong enough to sustain widespread vegetation growth (Joussaume *et al.*, 1999). By modifying surface conditions in Northern Africa (increased vegetation cover, increased area of wetlands and lakes), Kutzbach *et al.* (1996) obtained some change in their model which leads to an increase in precipitation in the southeastern part of the Sahara, but almost none in the western part (See Fig. 29.2a). Broström *et al.* (1998) obtained a stronger, but still unrealistically small, northward migration of savanna.

When coupling an atmospheric model with an equilibrium vegetation model of Prentice *et al.* (1992), Texier *et al.* (1997) yielded a positive feedback between vegetation and precipitation in this region which is, however, too weak to get any substantial greening (see Fig. 29.2b). They suggested an additional (synergistic) feedback between SSTs and land surface changes. Claussen and Gayler (1997) found a strong feedback between vegetation and precipitation and an almost complete greening in the western Sahara and some in the eastern part (see Fig. 29.2c). Claussen and Gayler (1997) and Claussen *et al.* (1998) explained the positive feedback by an interaction between high albedo of Saharan sand deserts and atmospheric circulation as hypothesized by Charney (1975).

Charney (1975) argued that the high albedo of sand deserts causes a negative deviation of the atmospheric radiation budget. Above sand deserts, more solar radiation is reflected to space and more thermal radiation leaves the surface into space because of reduced cloudiness over the dry deserts than over darker, vegetated surfaces. This is indicated in Fig. 29.1b, left-side branch, i.e. ΔRAD - ΔV. Hence above sand deserts, the atmosphere cools more strongly than over vegetation, and this radiative cooling is compensated by adiabatic sinking and heating of

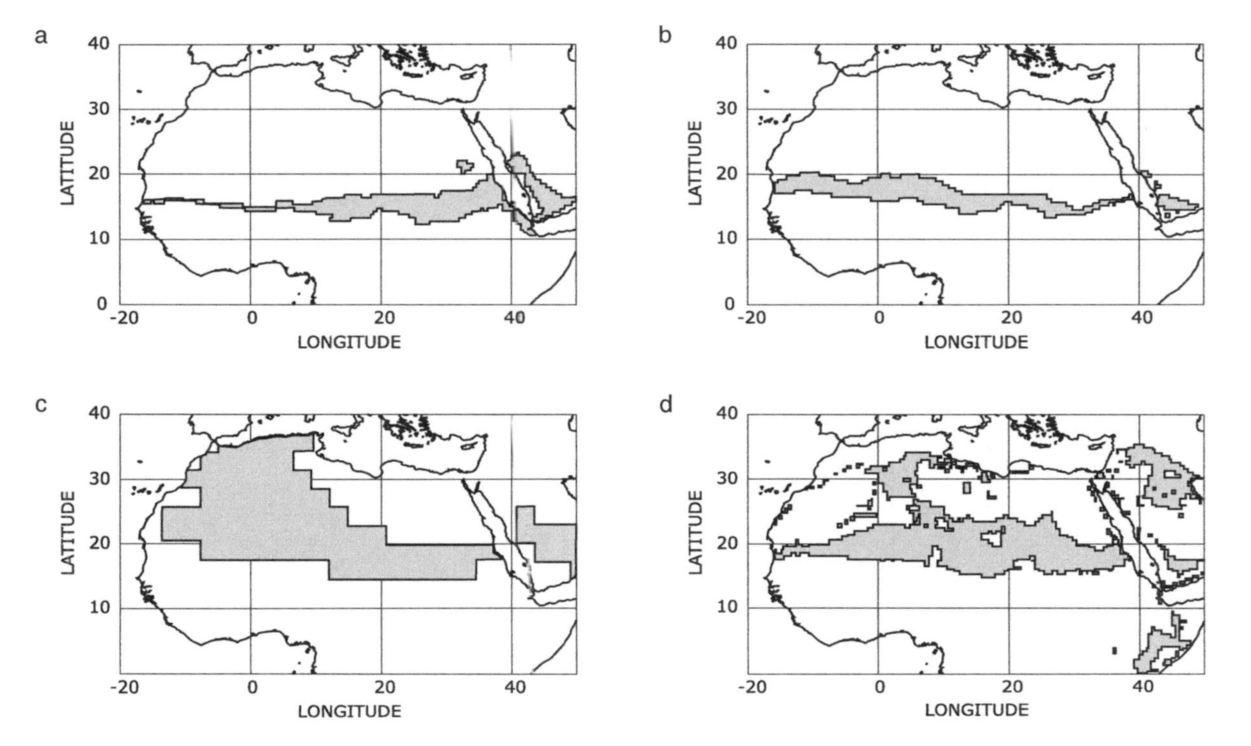

Figure 29.2 Reduction of Sahara (grey shade) from present-day climate to Mid-Holocene climate simulated (a) by the model of Kutzbach *et al.* (1996) in which changes in land-surface conditions were prescribed, (b) by the atmosphere–vegetation model of Texier *et al.* (1997), (c) Claussen and Gayler (1997) and (d) Doherty *et al.* (2000).

air – just as in Föhn situations in the leeward side of mountains. Sinking motion further reduces cloudiness and the possibility of convective rain (ΔP in Fig. 29.1b). When comparing Mid-Holocene climate and present-day climate, then the reverse argument applies, i.e. a reduction in albedo by an increase in vegetation decreases the negative deviation of the atmospheric radiation budget which leads to a decrease in the large-scale sinking motion. Convective rain is not inhibited.

As demonstrated by Dickinson (1992), Charney's theory of albedo-induced aridification is not generally valid. It is incomplete as it ignores any interaction with changes in soil moisture, which is sketched in Fig. 29.1b, right-side branch. (In a later paper, Charney *et al.*, 1977, prescribed large soil moisture for vegetation and little soil moisture, in the case of desert in order to simulate hydrological feedbacks to a first approximation.) Moreover, Eltahir and Gong (1996) argued that the variability in SST off the coast of West Africa and changes in land cover in the coastal region of West Africa may play a more important role in the regional climate than do changes in land cover near the desert border as suggested by Charney. However for the interior of Northern Africa, Charney's theory seems to apply when extended to include the interaction with soil moisture and evaporation (Claussen, 1997). It appears that the combined subtropical albedo-vegetation-precipitation feedback loop operates in an amplifying, i.e. positive way (see right-side branch in Fig. 29.1b with ΔE as change in evaporation.)

Recently, Doherty *et al.* (2000) used a coupled atmosphere–dynamic vegetation model to assess the problem of the 'green' Sahara. They report a reduction of Saharan desert by some 50 per cent (see Fig. 29.2d) which seems to agree best with Anhuf *et al.*'s (1999) reconstruction, but is too low in comparison with the estimate by Hoelzmann *et al.* (1998). The result by Doherty *et al.* (2000) is not directly taken from the coupled atmosphere–dynamic vegetation model. Doherty *et al.* (2000) used the so-called anomaly approach (see below), i.e. they interpreted the difference between simulations of Mid-Holocene and present-day climate by using an equilibrium vegetation model BIOME-3 (Haxeltine and Prentice, 1996). The original coupled model, as shown by Doherty *et al.* (2000), has a rather strong bias towards a green Sahara in present-day climate.

To assess critically the differences between model results, deNoblet-Ducoudre *et al.* (2000) compared the 'extreme' – concerning the magnitude of Saharan greening – models of Claussen and Gayler (1997) and Texier *et al.* (1997). Both groups used the same biome model, but different atmospheric models. Moreover, the atmospheric model and biome model were asynchronously coupled in different manners: Claussen and Gayler (1997) used the output of the climate model directly to drive the biome model, while Texier *et al.* (1997) took the difference between model results and a reference climate as input to the biome model. The latter, so-called anomaly approach prevents the coupled model to drift to an unrealistic climate which could be induced by some positive feedbacks between biases in either model. Hence this method is similar to the 'flux correction' in coupled atmosphere–ocean models. It turns out that the difference between coupling procedures affects the results of the coupled atmosphere–biome model only marginally. Hence, deNoblet-Ducoudre *et al.* (2000) concluded that the differences in Northern Africa greening cannot be attributed to the coupling procedure; it could be traced back to different representations of the atmospheric circulation in the tropics. The atmospheric model of Claussen and Gayler (1997) somewhat overestimates the duration of the Northern African monsoon, while the other model of Texier *et al.* (1997) yielded an unrealistic near-surface pressure distribution and, therefore, a too zonal circulation. The authors demonstrated why the one model yielded an unrealistically arid climate and they considered the other model as more realistic as far as the simulation of the Mid-Holocene greening was concerned. But they could not exclude that the model of Claussen and Gayler (1997) might have produced the strong biogeophysical feedback for the wrong reason. This issue certainly needs further consideration.

In all model experiments, the change in surface albedo was considered an important external (in the case of sensitivity studies by Kutzbach *et al.*, 1996, and Broström *et al.*, 1998) or internal (in the case of atmosphere–vegetation experiments) forcing. By use of new satellite measurements, Pinty *et al.* (2000a, 2000b) estimated that the surface albedo of sand desert in Northern Africa can be above 0.5 in between 18°N and 20°N. On average over the Sahara (taken from 17°N to 31°N) the albedo reaches 0.39 – a value much larger than conventionally used in atmospheric models. The value of 0.35 used in the experiments by Claussen (1997) and Ganopolski *et al.* (1998a) comes closest to the new estimate. Knorr *et al.* (2001) found that implementation of the new albedo values into their atmospheric model considerably improved simulated rainfall over Northern Africa. Moreover, Knorr *et al.* (2001) argued that for an albedo change of some 0.15 assumed for the Southern Sahara in several previous experiments on Holocene climate change (deNoblet-Ducoudre *et al.*, 2000) was too small. It should be larger by a factor of 2 according to the new estimates. This study is corroborated by Bonfils *et al.* (2001) who demonstrate that the choice of desert albedo not only influences the present-day simulated rainfall, but also the simulated climate change during the Mid-Holocene.

A further argument concerns the interaction with the ocean. Kutzbach and Liu (1997) provided simulations using an asynchronously and partially coupled atmosphere–ocean model (no fresh water fluxes, no dynamic sea-ice model). They found an increase in Northern African monsoon precipitation as a result of increased SST bringing the model in closer agreement with palaeo data. Similarly, Hewitt and Mitchell (1998), using a fully coupled atmosphere–ocean model, observed an increase in precipitation over Northern Africa, but still not as intense as data suggest. They assumed that missing biospheric feedbacks caused their model 'failure'. All models yield only small changes in annual mean SST and small changes in deep-ocean circulation of some 1–2 × 10^6 m^3/s (e.g. Weber, 2001). However, as mentioned by Kutzbach and Liu (1997) and Hewitt and Mitchell (1998), it is not the change in annual mean SST but the change in their annual cycle. In their model, Hewitt and Mitchell (1998) find enhanced precipitation in the early part of the wet season over Northern Africa, but towards the end of the summer over Southeast Asia.

Ganopolski *et al.* (1998a) have readdressed this issue using a coupled atmosphere–vegetation–ocean model in different combinations (as an atmosphere-only model, atmosphere–vegetation model, atmosphere–ocean model and fully coupled model). They concluded that in the subtropics, the biospheric feedback dominates, while the synergism between this feedback and an increase in monsoon precipitation owing to increased SST adds only little. Braconnot *et al.* (1999) found a stronger impact of SST than Ganopolski *et al.* (1998a) did, albeit the biospheric feedback still seemed to dominate. Unfortunately, Braconnot *et al.* (1999) did not perform a systematic feedback and synergism analysis or factor-separation technique. Hence, the relative magnitude of feedbacks could not be quantified in their simulation.

29.3 ABRUPT CLIMATE CHANGES

The interaction between components of the climate-system is non-linear. Therefore one might expect multiple equilibrium solutions. Gutman *et al.* (1984) and Gutman (1984, 1985) found only unique, steady-state solutions in their zonally averaged model. The possibility of multiple equilibria in the three-dimensional atmosphere–vegetation system was discovered later by Claussen (1994) and subsequently analysed in detail by Claussen (1997, 1998) for present-day climate, i.e. present-day insolation and SST. Two solutions of the atmosphere–vegetation system appear: the arid, present-day climate and a humid solution resembling more the Mid-Holocene climate, i.e. with a Sahara greener than today, albeit less green than in the Mid-Holocene. The two solutions differ mainly in the sub-tropical areas of Northern Africa and, but only slightly, in Central East Asia. The possibility of multiple equilibria in the atmosphere–vegetation system of North-West Africa has recently been corroborated by Wang and Eltahir (2000a, 2000b) and Zeng and Neelin (2000) by using completely different models of the tropical atmosphere and dynamic vegetation.

Interestingly, the stability of the atmosphere–vegetation system seems to change with time: experiments with Mid-Holocene vegetation yield only one solution, the green Sahara (Claussen and Gayler, 1997), while for the Last Glacial Maximum (LGM), two solutions exist (Kubatzki and Claussen, 1998) (see Plate 12).

So far, no other regions on Earth have been identified in which multiple equilibria could evolve on a large scale. Levis *et al.* (1999b) have sought multiple solutions of the atmosphere–vegetation–sea-ice system at high northern latitudes. Their model converges to one solution in this region corroborating the earlier assertion (Claussen, 1998) that multiple solutions manifest themselves in the subtropics, mainly in Northern Africa.

Why do we find multiple solutions in the subtropics, but none at high latitudes – and why for present-day and LGM climate but not for Mid-Holocene climate? Claussen *et al.* (1998) analysed large-scale atmospheric patterns in present-day, Mid-Holocene and LGM climate. They found that velocity potential patterns, which indicate divergence and convergence of large-scale atmospheric flow, differ between arid and humid solutions mainly in the tropical and sub-tropical regions. It appears that the Hadley–Walker circulation slightly shifts to the west. This is consistent with Charney's (1975) theory of albedo-induced aridification in the sub-tropics. Moreover, changes in surface conditions directly influence vertical motion, and thereby large-scale horizontal flow, in the tropics (Eltahir, 1996), but hardly at middle and high latitudes (e.g. Lofgren 1995a, 1995b). For the Mid-Holocene climate, the large-scale atmospheric flow was already close to the humid mode – in the simulations – even if one prescribes present-day land surface conditions in the model owing to changes in insolation. During the LGM, insolation differed little from present-day conditions.

A more ecological interpretation of multiple equilibria was given by Brovkin *et al.* (1998). They developed a conceptual model of vegetation–precipitation interaction in the western Sahara which is applied to interpret the results of comprehensive models. Conceptual models are used to explore the consequences of simple assumptions in a physically consistent way. Hence these models are constructed for building hypotheses. Their predictive capacity is very limited. The conceptual model (see Fig. 29.3) finds three solutions for present-day and LGM climate; one of these, however, is unstable to infinitesimally small perturbation. The humid solution is shown to be less probable than the arid solution, and this could explain the existence of the Sahara desert as it is today. For Mid-Holocene climate, only one solution is obtained. Application of the conceptual model to biospheric feedbacks at high latitudes (Levis *et al.*, 1999b) yields only one solution for present-day conditions.

The discussion of multiple equilibria seems to be somewhat academic. However, the existence of these can explain abrupt transitions in vegetation structure (Brovkin *et al.*, 1998; Claussen *et al.*, 1998). If global stability changes in the sense that one equilibrium solution becomes less stable to finite amplitude perturbation than the others, then an abrupt change of the system from the less stable to a more stable equilibrium is to be expected. Brovkin *et al.* (1998) found in their box model that the green solution becomes less stable around 3600 years ago. Keeping in mind that the variability of precipitation is larger in humid regions than in arid regions of Northern Africa (e.g. Eischeid *et al.*, 1991), one would expect a transition roughly in between 6000 and 4000 years ago.

In fact, there is evidence that the Mid-Holocene wet phase in Northern Africa ended around 5000–4500 years ago even in the high continental position of the East Sahara (Pachur and Wünnemann, 1996; Pachur and Altmann, 1997). Petit-Maire and Guo (1996) presented data which suggest that the transition to the present-day arid climate did not occur gradually, but in two steps with two arid periods, at 6700–5500 years ago and 4000–3600 years ago. Other reconstructions indicate that freshwater lakes in the Eastern Sahara began to disappear from 5700 to 4000 years ago, when recharge of aquifers ceased at the end of the wet phase (Pachur and Hoelzmann, 1991). Pachur (1999) suggested that climate change at the end of the Mid-Holocene was faster in the western than in the Eastern Sahara. Recently, deMenocal *et al.* (2000a) found an abrupt change in terrigenous material dated at 5500 years ago. The record was obtained from marine cores drilled near the Atlantic coast of Northern Africa.

Claussen *et al.* (1999) have analysed the transient structures in global vegetation pattern and atmospheric characteristics by using the model of Petoukhov *et al.* (2000), which was also used by Ganopolski *et al.* (1998a). They demonstrated (see Fig. 29.3) that subtle changes in orbital forcing could have triggered changes in Northern African climate which were than strongly amplified by biogeophysical feedbacks in this region. The timing of the transition, which started at around 5500 years ago in the model, was affected by changes in tropical SST and by the large-scale meridional temperature gradient over Eurasia. The latter was modified in part by processes at high northern latitudes, in particular by the synergism between the taiga–tundra feedback and the sea-ice–albedo feedback, discussed in the previous section. Hence the abrupt aridification – abrupt in comparison with the subtle change in orbital forcing – was presumably a regional effect, the timing of it, however, was likely to be governed by global processes.

Interestingly, further analysis of the model of Petoukhov *et al.* (2000) revealed that this model did not yield two distinct equilibrium solutions in Northern Africa prior to the Mid-Holocene aridification. Thus, the rapid change simulated by Claussen *et al.* (1999) was attributed to a rapid, steady change of one equilibrium solution, but not to a jump between two solutions. The authors attribute the apparently smooth behaviour in phase space on the coarse spatial resolution of their in which Northern Africa is represented by just three grid boxes, Sahara, Sudan and tropical Northern Africa. Close inspection of Plate 12 reveals that the strongest changes in Saharan vegetation are to be expected in the Western Sahara. Therefore, the authors concluded

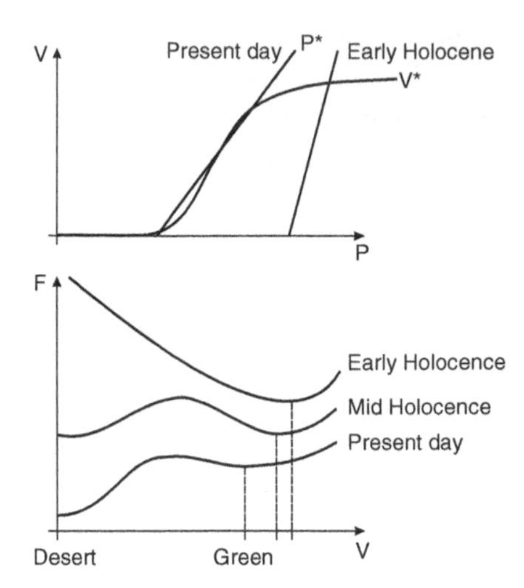

Figure 29.3 Stability diagram of a conceptual model of atmosphere–vegetation interaction in the sub-tropics. Upper figure: Given some arbitrary, but fixed, precipitation value *P*, sub-tropical vegetation achieves an equilibrium V^* with *P*. The curve is obtained by using a comprehensive atmosphere–(equilibrium) vegetation model. *Vice versa*, given an arbitrary, but fixed value of vegetation fraction *V*, the precipitation attains an equilibrium value P^*. $P^*(V)$ depends, however, on external forcing. Hence two curves $P^*(V)$ for present-day and Early Holocene climate are depicted. The intersects of $P^*(V)$ and $V^*(P)$ are the equilibrium solutions of the dynamic model $V(P)$. For present day conditions, three equilibrium solutions are obtained: two stable equilibria and one – the intermediate solution – which is unstable to infinitesimal perturbation. Lower figure: The Lyapounov potential $F(V)$ visualizes the stability of equilibrium solutions: a maximum indicates an unstable solution and a minimum, a stable solution. A local minimum appears to be only locally stable, the absolute minimum the most stable solution.

that only a model which resolves the geographical differences between Western and Eastern Sahara could, in principle, produce multiple equilibria.

29.4 CONCLUSIONS AND PERSPECTIVES

By and large, palaeoclimatic reconstructions reveal a warming trend during the Early and Mid-Holocene, mainly in the northern hemisphere, and a cooling trend thereafter. Boreal forests were presumably more widespread in the Early and Mid-Holocene than in the Late Holocene. In Northern Africa, the climate during the Early Holocene was wetter than today and the Sahara was much greener. Changes in oceanic circulation were presumably small, at least in the Mid- and Late Holocene. Hence, Holocene climate variability was rather modest in comparison with changes during the Last Glacial or changes between glacial and interglacial climate. Nonetheless, several aspects, in particular the so-called biome paradox, i.e. the wintertime warming despite decreased insolation during the Mid-Holocene, and the disproportional response of palaeo-monsoon to orbital forcing, are interesting challenges for climate-system modellers. Other challenges which were not discussed in this overview include the response of the climate-system to the small changes in solar luminosity, volcanic activity, anthropogenic land-cover change and greenhouse gas emissions.

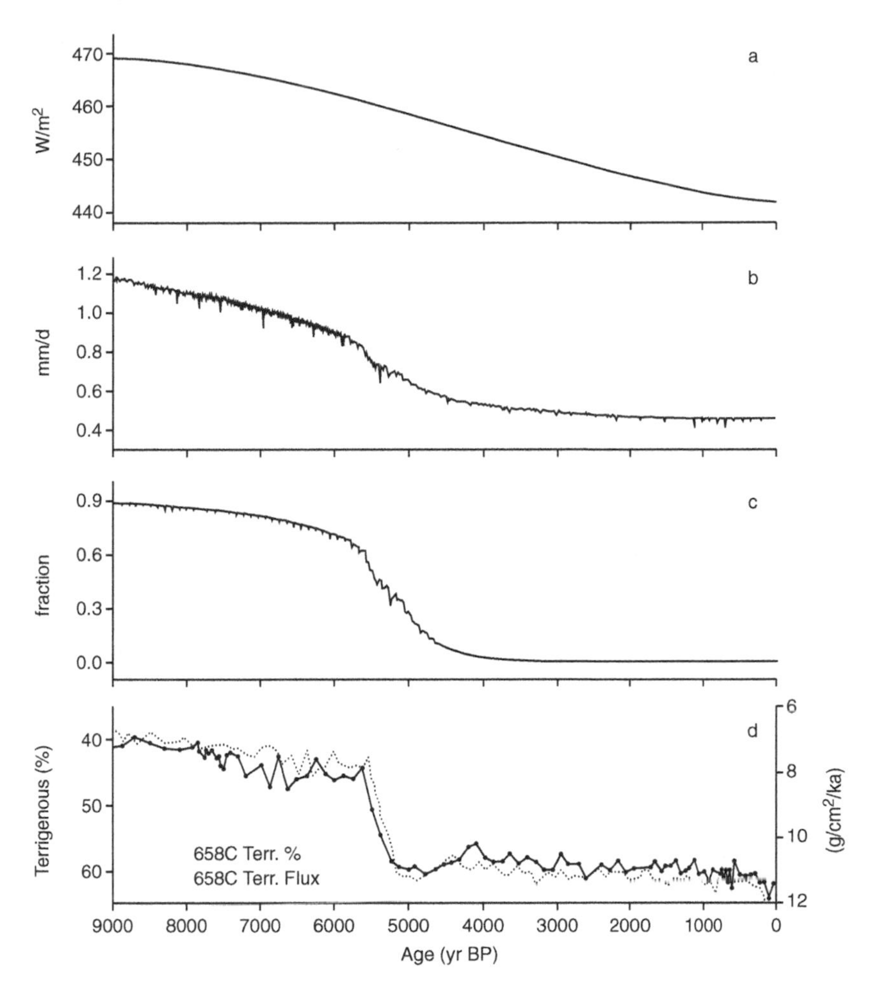

Figure 29.4 Simulation of transient development of precipitation (b) and vegetation fraction (c) in the Sahara as response to changes in insolation changes (a) insolation changes on average over the northern hemisphere during Boreal summer are depicted). These results obtained by Claussen *et al.* (1999) are compared with data of terrigenous material (left ordinate, d) and estimates of fluxes of terrigeneous material (right ordinate, d) from North Atlantic cores off the North African coast. This figure is taken with modifications from deMenocal *et al.* (2000a) and reprinted from Claussen (2003) with kind permission from Springer-Verlag.

Most climate simulations have focussed on Mid-Holocene climate around 6000 years ago, because changes in inland ice from glacial to interglacial climate have most likely come to an end. So far, we have learnt that simulations with atmosphere–ocean models fail to successfully describe all aspects of Mid-Holocene climate, and it is concluded that biospheric dynamics have to be taken into account in climate-system simulations. Transient simulations of last several thousand years reveal that Mid-Holocene climate around 6000 years ago is close to a strong change in phase space – in particular, if Northern African climate is concerned. Therefore, it seems advisable for climate modellers to not focus on 6000 years, but on 7000 or 8000 years before present – when changes in inland ice were presumably restricted to small areas of North America.

In this chapter, only atmosphere–ocean–vegetation interaction has been discussed without taking into account changes in biogeochemical cycles. Ice-core data from the Vostok (Barnola *et al.*, 1993) and the Taylor Dome (Indermühle *et al.*, 1999) show that the atmospheric CO_2 concentration in the Early Holocene, some 9000–8000 years ago, was approximately 20 ppmv lower than the pre-industrial value of 280 ppmv. Therefore a complete simulation of Holocene climate changes requires inclusion of biogeochemical models. Brovkin *et al.* (2002) present a first attempt in this direction by implementing a model of terrestrial and oceanic carbon cycle into a climate-system model. However, to properly represent Holocene changes in atmospheric CO_2, they have to make some *ad hoc* assumptions about changes in oceanic biogeochemistry. This result demonstrates the necessity of detailed palaeobotanic reconstruction of Holocene carbon pools as well as oceanic biogeochemistry.

Ocean and terrestrial and oceanic biosphere were presumably not in equilibrium during the Early and Mid-Holocene, and perhaps it will require investigation of climate-system dynamics beyond the Holocene to fully understand these changes. The climate-system has certainly been pushed away from equilibrium – which presumably has existed throughout the last several thousand years before the industrial revolution – by human-induced land-cover change and changes in atmospheric chemistry. To explore the consequences of natural and anthropogenic perturbation of the climate-system and, eventually, on our society it seems necessary to study the resilience of the natural Earth System in its current geological epoch – the Holocene. We have just begun to tackle this task.

Acknowledgements

The author wishes to acknowledge the constructive comments by an anonymous reviewer. Last, but not least this paper could not have been written without the fruitful discussion within the CLIMBER-2 group, in particular with Victor Brovkin, Andrey Ganopolski, Claudia Kubatzki, Stefan Rahmstorf, Vladimir Petoukhov, Eva Bauer and Anja Hünerbein.

REFERENCES

Aaby, B. 1976. Cyclic climatic variations in climate over the past 5500 years reflected in raised bogs. *Nature*, 263, 281–284.

Aario, L. 1940. Waldgrenzen und subrezente Pollenspektren in Petsamo, Lappland. *Annales Academiae Scientiarum Fennicae A*, 54, 1–120.

Aas, T. and Faarlund, T. 1988. Post-glacial forest limits in central south Norwegian mountains. Radiocarbon dating of subfossil pine and birch stumps. *Norsk Geografisk Tidsskrift*, 42, 25–61.

Aas, B. and Faarlund, T. 1996. The present and the Holocene subalpine birch belt in Norway. *Paläoklimaforschung*, 20, 19–42.

Abbott, M.B., Seltzer, G.O., Kelts, K.R. and Southon, J. 1997. Holocene paleohydrology of the tropical Andes from lake records. *Quaternary Research*, 47, 70–80.

Abbott, M.B., Finney, B.P., Edwards, M.E. and Kelts, K.R. 2000a. Lake-level reconstruction and paleohydrology of Birch Lake, central Alaska, based on seismic reflection profiles and core transects. *Quaternary Research*, 53, 154–166.

Abbott, M.B., Wolfe, B.B., Aravena, R., Wolfe, A.P. and Seltzer, G.O. 2000b. Holocene hydrological reconstructions from stable-isotopes and palaeolimnology, Cordillera Real, Bolivia. *Quaternary Science Reviews*, 19, 17–18.

Abell, P.I. 1985. Oxygen isotope ratios in modern African gastropod shells: a database for palaeoclimatology. *Chemical Geology Isotope Geoscience Section*, 58, 183–193.

Abell, P.I. and Hoelzmann, P. 2000. Holocene palaeoclimates in northwestern Sudan: stable-isotope studies on molluscs. *Global and Planetary Change*, 26, 1–12.

Abell, P.I. and Williams, M.A.J. 1989. Oxygen and carbon isotope ratios in gastropods shells as indicators of paleoenvironments in the Afar region of Ethiopia. *Palaeogeography, Palaeolimnology, Palaeoecology*, 74, 265–278.

Absolon, A. 1973. Ostracoden aus einigen Profilen spät-und postglazialer Karbonatablagerungen in Mitteleuropa. *Mitteilungen der Bayerischen Staatsammlung für Palaeontologie und Historische Geologie*, 13, 47–94.

Absy, M.L. 1979. *A palynological study of Holocene sediments in the Amazon basin*. PhD Thesis. University of Amsterdam, Amsterdam.

Absy, M.L. 1982. Quaternary palynological studies in the Amazon basin. In G.T. Prance (ed.), *Biological diversification in the tropics*. Columbia University Press, New York, pp. 66–73.

Absy, M.L., Clief, A., Fournier, M., *et al.* 1991. Mise en évidence de quatre phases d'ouverture de la forêt dense dans le sud-est de L'Amazonie au cours des 60,000 dernières années. Première comparaison avec d'autres régions tropicales. *Comptes Rendus Academie des Sciences Paris*, Series II 312, 673–678.

Adams, R.E.W. 1973. The collapse of maya civilization: a review of previous theories. In T.P. Culbert (ed.), *The Classic Maya Collapse*. University of New Mexico Press, Albuquerque, pp. 21–34.

Adkins, J.F., Cheng, H., Boyle, E.A., Druffel, E.R.M. and Edwards, R.L. 1998. Deep-sea coral evidence for rapid change in ventilation of the deep North Atlantic 15,400 years ago. *Science*, 280, 725–728.

Agustí-Panareda, A. Thompson, R. and Livingstone D.M. 2000. Reconstructing temperature variations at high elevation lake sites in Europe during the instrumental period. *Verhandlungen Internationalen Vereinigung Limnologie*, 27, 479–483.

Ahlberg, K., Almgren, E., Wright, H.E. and Ito, E. 2001. Holocene stable-isotope stratigraphy at Lough Gur, County Limerick, western Ireland. *The Holocene*, 11, 367–372.

Ahlstrom, R.V.N., Van West, C.R. and Dean, J.S. 1995. Environmental and chronological factors in the Mesa Verde–Northern Rio Grande migration. *Journal of Anthropological Archaeology*, 14, 125–142.

Aitken, M.J. 1998. *An introduction to optical dating*. Oxford University Press, Oxford.

Al-Aasm, I.S., Taylor, B.E. and South, B. 1990. Stable-isotope analysis of multiple carbonate samples using selective acid extraction. *Chemical Geology*, 80, 119–125.

Albrecht, M., Hubert, B., Eichhorn, B.F., *et al.* 2001. Oruwanje 95/1: a late Holocene stratigraphy in north-western Namibia. *Cimbebasia*, 17, 1–22.

Alibert, C. and McCulloch, M.T. 1997. Strontium/calcium ratios from modern *Porites* corals from the Great Barrier Reef as a proxy for sea surface temperature: calibration of the thermometer and monitoring of ENSO. *Paleoceanography*, 12, 345–364.

Allan, R.J. and Haylock, M.R. 1993. Circulation features associated with the winter rainfall decrease in south-western Australia. *Journal of Climate*, 7, 1356–1367.

Allen, J.R.L. 1991. Saltmarsh accretion and sea-level movement in the inner Severn Estuary, southwest Britain: the archaeological and historical contribution. *Journal of Geological Society, London*, 148, 485–494.

Allen, J.R.M., Brandt, U., Brauer, A., *et al.* 1999. Rapid environmental changes in southern Europe during the last glacial period. *Nature*, 400, 740–743.

Allen, J.R.M., Watts, W.A. and Huntley, B. 2000. Weichselian palynostratigraphy, palaeovegetation and palaeoenvironment: the record from Lago Grande di Monticchio, southern Italy. *Quaternary International*, 73/74, 91–110.

Allen, J.R.M., Watts, W.A., McGee, E. and Huntley, B. 2002. Holocene environmental variability: the record from Lago Grande di Monticchio, Italy. *Quaternary International*, 88, 69–80.

Allen, M.S. 1998. Holocene sea-level change on Aitutaki, Cook Islands: landscape change and human response. *Journal of Coastal Research*, 14, 10–22.

Alley, R.B., Meese, D.A., Shuman, C.A., *et al.* 1993. Abrupt increase in snow accumulation at the end of the Younger Dryas event. *Nature*, 362, 527–529.

Alley, R.B. and Anandakrishnan, S. 1995. Variations in melt-layer frequency in the GISP2 ice core: implications for Holocene summer temperatures in central Greenland. *Annals of Glaciology*, 21, 64–70.

Alley, R.B., Mayewski, P.A., Sowers, T., Stuiver, M., Taylor, K.C. and Clark, P.U. 1997. Holocene climatic instability: a prominent, widespread event 8200 yr. ago. *Geology*, 25, 483–486.

Alley, R.B., Anandakrishnan, S. and Jung, P. 2001. Stochastic resonance in the North Atlantic. *Paleoceanography*, 16, 190–198.

Almquist, H., Dieffenbacher-Krall, A.C., Flanagan-Brown, R. and Sanger, D. 2001. The Holocene record of lake-levels at Mansell Pond, central Maine, USA. *The Holocene*, 11, 189–201.

Almquist-Jacobson, H. 1995. Lake-level fluctuations at Ljustjarnen, central Sweden and their implications for the Holocene climate of Scandinavia. *Palaeogeography, Palaeoclimatology, Palaeoecology* 118, 269 – 290.

Alverson K., Oldfield, F. and Bradley, R. 2000. Past global changes and their significance for the future. *Quaternary Science Reviews*, 19, 479.

Alverson, K., Bradley, R., Briffa, K., *et al.* 2001. A global paleoclimate observing system. *Science*, 293, 47–48.

Alverson K., Bradley, R.S. and Pedersen, T.F. (eds) 2003. *Paleoclimate, global change and the future. A PAGES Project Synthesis*. Springer-Verlag, Berlin.

Ammann, B. 1989a. Response times in bio- and isotope-stratigraphies to Late-Glacial climatic shifts: an example from lake deposits. *Eclogae geologicae Helvetiae*, 82, 183–190.

Ammann, B. 1989b. Late-Quaternary palynology at Lobsigensee. Regional vegetation history and local lake development. *Dissertationes Botanicae*, 137, 1–157.

Ammann, B. 2000. Biotic responses to rapid climatic changes: introduction to a multidisciplinary study of the Younger Dryas and minor oscillations on an altitudianl transect in the Swiss Alps. *Palaeogeography, Palaeoclimatology, Palaeoecology*, 59, 191–201.

Ammann, B., and Lotter, A.F. 1989. Late-Glacial radiocarbon- and palynostratigraphy on the Swiss Plateau. *BOREAS*, 18, 109–126.

Ammann, B., Birks, H.J.B., Brooks, S.J., *et al.* 2000. Quantification of biotic responses to rapid climatic changes around the Younger Dryas: a synthesis. *Palaeogeography, Palaeoclimatology, Palaeoecology*, 159, 313–347.

Ammann, C.M., Kiehl, J.T., Otto-Bliesner, B.L., Zender, C.S. and Bradley, R.S. 2003. Coupled simulations of the 20th century including external forcing. *Journal of Climate*, in press.

Anadón, P., De Deckker, P. and Julià, R. 1986. The Pleistocene lake deposits of the NE Baza Basin, Spain): salinity variations and ostracod succession. *Hydrobiologia*, 14, 199–208.

Anadón, P., Gliozzi, E. and Mazzini, I. 2002. Paleoenvironmental reconstruction of marginal marine environments from combined palaeoecological and geochemical analyses on ostracods. In J.A. Holmes and A.R. Chivas (eds), *The Ostracoda: Applications in Quaternary Research*, Vol. 131, AGU Geophysical Monograph Series, Washington, DC.

Andersen, N., Paul, H.A., Bernasconi, S.M., *et al.* 2001. Large and rapid climate variability during the Messinian salinity crisis: evidence from deuterium concentrations of individual biomarkers. *Geology*, 29, 799–802.

Andersen, S.T. 1966. Interglacial vegetational succession and development in Denmark. *The Palaeobotanist*, 15, 117–127.

Andersen, S.T. 1969. Interglacial vegetation and soil development. *Meddelelser fra Dansk Geologisk Forening*, 19, 90–102.

Andersen, S.T. 1970. The relative pollen productivity and pollen representation of north European trees, and correction factors for tree pollen spectra. *Danmarks Geologiske Undersøgelse II*, 96, 96.

Anderson, D.E. 1998. A reconstruction of Holocene climatic changes from peat bogs in north-west Scotland. *BOREAS*, 27, 208–224.

Anderson, L., Abbott, M.B. and Finney, B.P. 2001. Holocene climate inferred from oxygen isotope ratios in lake sediments, central Brooks Range, Alaska. *Quaternary Research*, 55, 313–321.

Anderson, N.J. 2000. Diatoms, temperature and climatic change. *European Journal of Phycology*, 35, 307–314.

Anderson, R.Y. 1993. The varve chronometer in Elk Lake: record of climatic variability and evidence for solar-geomagnetic ^{14}C climate connections. In J.P. Bradbury and W.E. Dean (eds), *Elk Lake, Minnesota: evidence for rapid climate change in the north-central United States.* Geological Society of America, Denver, Colorado, pp. 45–67.

Anderson, R.Y. and Kirkland, D.W. 1960. Origin, varves and cycles of Jurassic Todilto Formation, New Mexico. *Bulletin of the American Association of Petrology and Geology*, 44, 37–52.

Anderson, R.Y., Dean, W.E., Bradbury, J.P. and Love, D. 1985a. Meromictic lakes and varved sediments in North America. *United States Geological Survey Bulletin*, 1607, 1–19.

Anderson, R.Y., Nuhfer, E.B. and Dean, W.E. 1985b. Sedimentation in a blast-zone lake at Mount St. Helens, Washington: implications for varve formation. *Geology*, 13, 348–352.

Andersson, G. 1902. Hasseln i Sverige fordom och nu. Hazel in Sweden, past and present. *Svenska Geologiske Undersøgelse Ca*, 3, 168.

Andrews, J.E., Riding, R. and Dennis, P.F. 1993. Stable-isotope compositions of recent freshwater cyanobacterial carbonates from the British Isles: local and regional environmental controls. *Sedimentology*, 40, 303–314.

Andrews, J.E., Riding, R. and Dennis, P.F. 1997. The stable-isotope record of environmental and climatic signals in modern terrestrial microbial carbonates from Europe. *Palaeogeography, Palaeoclimatology, Palaeoecology*, 129, 171–189.

Andrews, J.T. 1975. Glacial systems: an approach to glaciers and their environments. Duxbury Press, North Scituate, MA.

Andrews, J.T., Kristjansdottir, G.B., Geirsdottir, A., *et al.* 2001a. Late Holocene ~ 5 cal ka trends and century-scale variability of N. Iceland marine records. In D. Seidov, B. Haupt and M. Maslin (eds), *The oceans and rapid climate change: past, present and future*, Vol. 126. AGU Geophysical Monograph Series, Washington, DC, pp. 69–81.

Andrews, J.T., Helgadottir, G., Geirsdottir, A. and Jennings, A. 2001b. Multicentury-scale records of carbonate (hydrographic?) variability on N. Iceland margin over the last 5000 years. *Quaternary Research*, 56, 199–206.

Andrieu, V., Eicher, U. and Reille, M. 1993. La fin du dernier Pléniglaciaire dans les Pyrénées France: donnée pollinique, isotopique et radiométrique. *Comptes Rendus de L Academie des Sciences Paris*, 316, 245–250.

Andrus, C.F., Crowe, D.E., Sandweiss, D.H., Reitz, E.J. and Romanek, C.S. 2002. Otolith $\delta^{18}O$ record of mid-Holocene sea surface temperatures in Peru. *Science*, 295, 1508–1511.

Anhuf, D., Frankenber, P. and Lauer, W. 1999. Die postglaziale Warmphase vor 8000 Jahren. *Geologische Rundschau*, 51, 454–461.

Aranuvachapun, S. and Brimblecombe, P. 1978. Extreme ebbs in the Thames. *Weather*, 33, 126–131.

Aranuvachapun, S. and Brimblecombe, P. 1979. *Tawatodsamad Klongdun*: an old Thai weather poem. *Weather*, 34, 459–464.

Arbogast, R.M., Magny, M. and Pétrequin, P. 1996. Climat, cultures céréalieres et densité de population au néolithique: le cas des lacs du Jura français de 3500 à 2500 av. J.-C. *Archäologisches Korrespondenzblatt*, 26, 121–144.

Ariztegui, D., Farrimond. P. and Mckenzie, J.A. 1996. Compositional variations in sedimentary lacustrine organic matter and their implications for high alpine holocene environmental changes: Lake St Moritz, Switzerland. *Organic Geochemistry*, 24, 453–461.

Armitage, P.D., Cranston, P.S. and Pinder, L.C.V. (eds), 1995. *The Chironomidae. The biology and ecology of non-biting midges.* Chapman and Hall, London.

Ashley, G.M. 1975. Rhythmic sedimentation in glacial Lake Hitchcock, Massachusetts, Connecticut. *Special Publication Society of Economic Paleontologists and Mineralogists*, 23, 304–320.

Atkinson, T.C. 1983. Growth mechanisms of speleothems in Castleguard Cave, Columbia Icefields, Alberta, Canada. *Arctic and Alpine Research*, 15, 523–536.

Atkinson, T.C., Briffa, K.R. and Coope, G.R. 1987. Seasonal temperatures in Britain during the past 22,000 years, reconstructed from beetle remains. *Nature*, 325, 587–592.

Austin, J. 1983. Krakatoa sunsets. *Weather*, 38, 226–231.

Avery, M.D. 1991. Micromammals and environmental change at Zebrarivier Cave, Central Namibia. *SWA Wissenschaftliche Gesellschaft*, XXXVIII, 79–86.

Avsejs, L.A., Nott, C.J., Xie, S., Maddy, D., Chambers, F.M. and Evershed, R.P. 2002. 5-*n*-Alkylresorcinols as biomarkers of sedges in an ombrotrophic peat section. *Organic Geochemistry*, 33, 861–867.

Ayalon, A., Bar-Matthews, M. and Kaufman, A. 1999. Petrography, strontium, barium, and uranium concentration, and strontium and uranium isotope ratios as palaeoclimatic proxies: Soreq Cave, Israel. *The Holocene*, 9, 715–722.

Bagnato, S., Linsley, B.K., Wellington, G.H. and Howe, S.S. 2001. Developing the use of the massive coral genus *Diploastrea* for paleoclimate reconstruction. GSA Abstracts. Online at: http://gsa.confex.com/gsa/2001AM/finalprogram/abstract_27188.htm

Baillie, M.G.L. 1994 Dendrochronology raises questions about the nature of the AD 536 dust-veil event. *The Holocene*, 4, 212–217.

Baillie, M.G.L. 1995. A slice through time: dendrochronology and precision dating. Routledge, London.

Baillie, M.G.L. 1996. Extreme environmental events and the linking of the tree-ring and ice-core records. In J.S. Dean, D.M. Meko and T.W. Swetnam (eds), *Tree-rings, environment and humanity: proceedings of the International Conference, Tucson, Arizona, 17–21 May 1994. Radiocarbon*, 703–711.

Baillie, M.G.L. 1999a. Putting abrupt environmental change back into human history. In P. Slack (ed.), *Environments and historical change: the Linacre Lectures 1998.* Oxford University Press, Oxford.

Baillie, M.G.L. 1999b. Exodus to Arthur: catastrophic encounters with comets. Batsford, London.

Baillie, M.G.L. and Munro, M.A.R. 1998. Irish tree-rings, Santorini and volcanic dust veils. *Nature,* 332, 344–346.

Baker, A. and Smart, P.L. 1995. Recent flowstone growth rates: field measurements in comparison to theoretical predictions. *Chemical Geology*, 122, 121–128.

Baker, A., Smart, P.L., Edwards, R.L. and Richards, D.A. 1993a. Annual growth banding in a cave stalagmite. *Nature,* 364, 518–520.

Baker, A., Smart, P.L. and Ford, D.C. 1993b. Northwest European palaeoclimate as indicated by growth frequency variations of secondary calcite deposits. *Palaeogeography, Palaeoclimatology, Palaeoecology*, 100, 291–301.

Baker, A., Smart, P.L. and Edwards, R.L. 1995. Paleoclimate implications of mass spectrometric dating of a British flowstone. *Geology Boulder*, 23, 309–312.

Baker, A., Barnes, W.L. and Smart, P.L. 1996a. Luminescence and discharge variations in stalagmite drip waters, Bristol, England. *Karst Waters Institute Special Publication*, 2, 4–6.

Baker, A., Smart, P.L. and Barnes, W.L. 1996b. Speleothem luminescence intensity and spectral characteristics: signal calibration and a record of paleovegetation change. *Chemical Geology*, 130, 65–76.

Baker, A., Barnes, W.L. and Smart, P.L. 1997a. Variations in the discharge and organic matter content of stalagmite drip waters in Lower Cave, Bristol. *Hydrological Processes*, 11, 1541–1555.

Baker, A., Caseldine, C.J., Hatton, J., Hawkesworth, C.J. and Latham, A.G. 1997b. A cromerian complex stalagmite from the Mendip Hills, England. *Journal of Quaternary Science*, 12, 533–537.

Baker, A., Genty, D., Dreybrodt, W., Barnes, W.L., Mockler, N.J. and Grapes, J. 1998. Testing theoretically predicted stalagmite growth rate with recent annually laminated samples: implications for past stalagmite deposition. *Geochimica Cosmochimica Acta*, 62, 393–404.

Baker, A., Proctor, C.J. and Barnes, W.L. 1999. Variations in stalagmite luminescence laminae structure at Poole's Cavern, England, AD 1910–1996: calibration of a paleoprecipitation proxy. *The Holocene*, 9, 683–688.

Baker, A., Proctor, C.J., Lauritzen, S.E. and Lundberg, J. 2000. SPEP: high-resolution stalagmite records of NE Atlantic climate in the last millennium. *PAGES Newsletter*, 8, 14.

Baker, P.A 2002. Paleoclimate: trans-Atlantic climate connections. *Science*, 296, 67–68.

Baker, P.A., Seltzer, G.O., Fritz, S.C., *et al.* 2001. The history of South American tropical precipitation for the past 25,000 years. *Science*, 291, 640–643.

Baker, R.G.V. and Haworth, R.J. 2000a. Smooth or oscillating late Holocene sea-level curve? Evidence from paleo-zoology of fixed biological indicators in east Australia and beyond. *Marine Geology*, 163, 367–386.

Baker, R.G.V. and Haworth, R.J. 2000b. Smooth or oscillating late Holocene sea-level curve? Evidence from cross-regional statistical regressions of fixed biological indicators. *Marine Geology*, 163, 353–365.

Baker, R.G., Bettis, E.A., Schwert, D.P., *et al.* 1996c. Holocene paleoenvironments of northeast Iowa. *Ecological Monographs*, 66, 203–234.

Baker, R.G., Bettis, E.A., Denniston, R.F., Gonzalez, L.A., Strickland, L.E. and Kreig, J.R. 2002. Holocene paleoenvironments in southeastern Minnesota: chasing the prairie-forest ecotone. *Palaeogeography, Palaeoclimatology, Palaeoecology*, 177, 103–122.

Baldini, J.U.L., McDermott, F. and Fairchild, I.J. 2002. Structure of the 8200-year cold event revealed by a speleothem trace element record. *Science*, 296, 2203–2206.

Baldwin, M.P. and Dunkerton, T.J. 2001. Stratospheric harbingers of anomalous weather regimes. *Science*, 294, 581–584.

Ballantyne, C.K. 1989. The Loch Lomond readvance on the Island of Skye, Scotland: glacier reconstruction and palaeoclimate implications. *Journal of Quaternary Science*, 4, 95–108.

Ballantyne, C.K. 2002. Paraglacial geomorphology. *Quaternary Science Reviews*, 21, 1935–2017.

Bangs, M., Battarbee, R.W., Flower, R.J., *et al.* 2000. Climate change in Lake Baikal: diatom evidence in an area of continuous sedimentation. *International Journal of Earth Sciences*, 89, 251–259.

Barber, D.C., Dyke, A., Hillaire-Marcel, C., *et al.* 1999a. Forcing of the cold event of 8200 years ago by catastrophic drainage of Laurentide lakes. *Nature,* 400, 344–348.

Barber, K.E. 1981. *Peat stratigraphy and climatic change.* A.A. Balkema, Rotterdam.

Barber, K.E. 1994. Deriving Holocene palaeoclimates from peat stratigraphy: some misconceptions regarding the sensitivity and continuity of the record. *Quaternary Newsletter*, 72, 1–9.

Barber K.E. and Langdon, P.G. 2001. Testing the palaeoclimatic signal from peat bogs: temperature or precipitation forcing? Abstracts, PAGES-PEPIII/ESF-HOLIVAR International Conference: Past climate variability through Europe and Africa. *ECRC/CEREGE*, pp. 58–59.

Barber, K.E., Chambers, F.M., Maddy, D., Stoneman, R. and Brew, J.S. 1994. A sensitive high resolution record of late Holocene climatic change from a raised bog in northern England. *The Holocene*, 4, 198–205.

Barber, K.E., Dumayne-Peaty, L., Hughes, P.D.M., Mauquoy, D. and Scaife, R.G. 1998. Replicability and variability of the recent macrofossil and proxy-climate record from raised bogs: field stratigraphy and macrofossil data from Bolton Fell Moss and Walton Moss, Cumbria, England. *Journal of Quaternary Science*, 13, 515–528.

Barber, K.E., Maddy, D., Rose, N., Stevenson, A.C., Stoneman, R. and Thompson, R. 2000. Replicated proxy-climate signals over the last 2000 yr from two distant UK peat bogs: new evidence for regional palaeoclimate teleconnections. *Quaternary Science Reviews*, 19, 481–487.

Bard, E. 2001. Paleoceanographic implications of the difference in deep-sea sediment mixing between large and fine particles. *Paleoceanography*, 16, 235–239.

Bard, E., Hamelin, B., Fairbanks, R.G. and Zindler, A. 1990. Calibration of the ^{14}C time-scale over the past 30,000 years using mass spectrometric U-Th ages from Barbados corals. *Nature*, 345, 405–410.

Bard, E., Rostek, F. and Sonzogni, C. 1997. Interhemispheric synchrony of the last deglaciation inferred from alkenone paleothermometry. *Nature*, 385, 707–710.

Bard, E., Raisbeck, G., Yiou, Y. and Jouzel, J. 2000. Solar irradiance during the last 120 years based on cosmogenic nuclides. *Tellus*, 52B, 985–992.

Barker, P.A., Street-Perrott, F.A., Leng, M.J., *et al.* 2001. A 14 ka oxygen isotope record from diatom silica in two alpine tarns on Mt Kenya. *Science*, 292, 2307–2310.

Barlow, L.K., Sadler, J.P., Ogilvie, A.E.J., *et al.* 1997. Interdisciplinary investigations of the end of the Norse western settlement in Greenland. *The Holocene*, 7, 489–499.

Bar-Matthews, M. and Ayalon, A. 1997. Late Quaternary paleoclimate in the eastern mediterranean region from stable-isotope analysis of speleothems at Soreq Cave, Israel. *Quaternary Research*, 47, 155–168.

Bar-Matthews, M., Ayalon, A. and Kaufman, A. 1998. Middle to Late Holocene 6500 yr period. paleoclimate in the Eastern Mediterranean region from stable isotopic composition of speleothems from Soreq Cave, Israel. In A.S. Issar and N. Brown (eds), *Water, environment and society in times of climatic change*. Kluwer, Dordrecht, pp. 204–214.

Bar-Matthews, M., Ayalon, A., Kaufman, A. and Wasserburg, G.J. 1999. The eastern Mediterranean paleoclimate as a reflection of regional events, Soreq Cave, Israel. *Earth and Planetary Science Letters*, 166, 85–95.

Bar-Matthews, M., Ayalon, A. and Kauffman, A. 2000. Timing andhydrological conditions of sapropel events in the eastern mediterranean, as evident from speleothems, Soreq cave, Israel. *Chemical Geology*, 169, 145–156.

Barnekow, L. 1999. Holocene tree dynamics in the Abisko area, northern Sweden, based on pollen and macrofossil records, and the inferred climatic changes. *The Holocene*, 9, 253–265.

Barnes, D.J. and Lough, J.M. 1993. On the nature and causes of density banding in massive coral skeletons. *Journal of Experimental Marine Biology and Ecology*, 167, 91–108.

Barnola, J.M., Raynaud, D., Lorius, C., Korotkevich, Y.S. 1993. Vostok ice core atmospheric CO_2 concentrations records. In Trends '93 Database. *CDIAC, ORNL*, Oak Ridge.

Barnosky, C.W., Grimm, E.C. and Wright, H.E. 1987. Towards a post-glacial history of the northern Great Plains: a review of the paleoecologic problems. *Annals of Carnegie Museum*, 56, 259–273.

Bartlein, P.J. and Whitlock, C. 1993. Paleoclimatic interpretation of the Elk Lake pollen record. *Geological Society of America Special Paper*, 276, 275–293.

Bartlein, P.J., Webb, T. and Fleri, E.C. 1984. Holocene climatic change in the northern Midwest: pollen-derived estimates. *Quaternary Research*, 22, 361–374.

Bartlein, P.J., Prentice, I.C. and Webb, T. 1986. Climatic response surfaces from pollen data for some eastern North American taxa. *Journal of Biogeography*, 13, 35–57.

Baruch,U., and Bottema, S. 1999. A new pollen diagram from Lake Hula. Vegetational, climatic, and anthropogenic implications. In H. Kawanabe, G.W. Coulter, and A.C. Roosevelt (eds), *Ancient lakes: their cultural and biological diversity*. Kenobi Productions, Belgium, pp. 75–86.

Bar-Yosef, O. 2000. The impact of radiocarbon dating on Old World archaeology: past achievements and future expectations. *Radiocarbon*, 42, 23–39.

Bar-Yosef, O., and Belfer-Cohen, A. 1992. From foraging to farming in the Mediterranean Levant. In A.B. Gebauer and T.D. Price (eds), Transitions to agriculture in prehistory. Prehistory Press, Madison, Wisconsin, pp. 21–48.

Basnett, T.A. and Parker, D.E. 1997. Development of the global mean sea level pressure data set. *Hadley Centre Climate Research Technical Note. CRTN 79*. Met Office, London.

Bastin, B. 1978. L'analyse pollinique des stalagmites: Une nouvelle possibilité d'approche des fluctuations climatiques du quaternaire. *Annales de la Societé Geologique de Belgue*, 101, 13–19.

Battarbee, R.W. 2000. Palaeolimnological approaches to climate change, with special regard to the biological record. *Quaternary Science Reviews*, 19, 197–124.

Battarbee, R.W., Juggins, S., Gasse, F., Anderson, N.J., Bennion, H. and Cameron, N.G. 2000. European Diatom Database EDDI. An information system for palaeoenvironmental reconstruction. *European Climate Science Conference, Vienna City Hall, Vienna, Austria, 19–23 October 1998*, pp. 1–10.

Battarbee, R.W., Cameron, N.G., Golding, P., *et al.* 2001a. Evidence for Holocene climate variability from the sediments of a Scottish remote mountain lake. *Journal of Quaternary Science*, 16, 339–346.

Battarbee, R.W., Jones, V.J., Flower, R.J., *et al.* 2001b. Diatoms. In J.P. Smol and H.J.B. Birks (eds), *Tracking environmental change using lake sediments. Vol. 3. Terrestrial, algal and siliceous indicators*. Kluwer Academic Publishers, Dordrecht, pp. 155–202.

Battarbee, R.W., Grytnes, J.A., Thompson, R., *et al.* 2002. Comparing palaeolimnological and instrumental evidence of climate change for remote mountain lakes over the last 200 years. *Journal of Paleolimnology*, 28, 161–179.

Bauer, E., Claussen, M., Huenerbein, A. and Brovkin, V. 2002. Assessment of climate forcing contributing to the Late Maunder Minimum. Abstracts, AGU Spring Meeting. AGU, Washington, DC.

Bayliss, A., Ramsey, B. and McCormac, F.C. 1997. Dating Stonehenge In B. Cunliff and C. Renfrew (eds), *Science and Stonehenge. Proceedings of the British Academy*, 92, 39–59.

Beck, J.W., Edwards, R.L., Ito, E., *et al.* 1992. Sea-surface temperature from coral skeletal strontium/calcium ratios. *Science*, 257, 644–647.

Beck, J.W., Richards, D.A., Edwards, R.L., *et al.* 2001. Extremely large variations of atmospheric ^{14}C concentration during the last glacial period. *Science*, 292, 2453–2458.

Becker, B. and Schirmer, W. 1977. Palaeoecological study on the Holocene valley development of the River Main, Southern Germany, *BOREAS*, 6, 303–321.

Beer, J., Mende, W., Stellmacher, R. and White, O.R. 1996. Intercomparisons of proxies for past solar variability. In P.D. Jones, R.S. Bradley and J. Jouzel (eds), *Climate variations and forcing factors of the last 2000 years*. Springer-Verlag, Berlin, pp. 501–517.

Behl, R.J., and Kennett, J.P. 1996. Brief interdstadial events in the Santa Barbara basin, NE Pacific, during the past 60 kyr. *Nature*, 379, 243–246.

Behling, H. and Hooghiemstra, H. 1999. Environmental history of the Colombian savannas of the Llanos Orientales since the last glacial maximum from lake records El Pinal and Carimagua. *Journal of Paleolimnology*, 21, 461–476.

Behre, K.-E. 1967. The late glacial and early post-glacial history of vegetation and climate in northwestern Germany. *Review of Palaeobotany and Palynology*, 4, 147–161.

Bell, W.T. and Ogilvie, A.E.J. 1978. Weather compilations as a source of data for the reconstruction of European climate during the medieval period. *Climate Change*, 1, 331–348.

Belyea, L.R. and Warner, B.G. 1994. Dating of the near-surface layer of a peatland in northwestern Ontario, Canada. *BOREAS*, 23, 259–269.

Benn, D.I. and Evans, D.J.A. 1998. *Glaciers and glaciation*. Arnold, London.

Bennett, J.R., Cumming, B.F., Leavitt, P.R., Chiu, M., Smol, J.P. and Szeicz, J., 2001. Diatom, pollen, and chemical evidence of post-glacial climatic change at Big Lake, South-Central British Columbia, Canada. *Quaternary Research*, 55, 332–343.

Bennett, K.D. 2002. Comment: the Greenland 8200 cal. yr BP event detected in loss-on-ignition profiles in Norwegian lacustrine sediment sequences. *Journal of Quaternary Science*, 17, 97–99.

Bennett, K.D. and Willis, K.J. 2001. Pollen. In J.P. Smol, H.J.B. Birks and W.M. Last (eds), *Tracking environmental changes using lake sediments. Vol. 3. Terrestrial and siliceous indicators*, Kluwer Academic Publishers, Dordrecht, pp. 5–32.

Benthien, A. and Müller, P.J. 2000. Anomalously low alkenone temperatures caused by lateral particle and sediment transport in the Malvinas Current region, western Argentine Basin. *Deep-Sea Research*, I47, 2369–2393.

Berger, A. 1978. Long-term variations of caloric insolation resulting from Earth's orbital elements. *Quaternary Research*, 9, 139–167.

Berger, A. 2001. The role of CO_2, sea-level and vegetation during the Milankovitch-forced glacial–interglacial cycles. In L.O. Bengtsson and C.U. Hammer (eds), *Geosphere–biosphere interactions and climate*. Cambridge University Press, New York, pp. 119–146.

Berger, A. and Loutre, M.F. 1991. Insolation values for the climate of the last 10 million years. *Quaternary Science Reviews*, 10, 297–317.

Berglund, B.E. 1966. Late-Quaternary vegetation in eastern Blekinge, south-east Sweden. II. Post-glacial time. *Opera Botanica*, 12, 190.

Beschel, R.E. 1950. Flechten als Altersmasstab rezenter Moränen. *Zeitschrift für Gletscherkunde und Glazialgeologie*, 1, 152–161.

Beschel, R.E. 1957. Lichenometrie im Gletschervorveld. *Jahrbuch, Verein zum Schutz der Alpenplanzen und-tierre München*, 22, 164–185.

Beschel, R.E. 1961. Dating rock surfaces by lichen growth and its application to glaciology and physiography lichenometry. In G.O. Raasch (ed.), *Geology in the Arctic 2*. University of Toronto Press, Toronto, pp. 1044–1062.

Betancourt, J.L., van Devender, T. and Martin, P.S. 1990. *Packrat middens. The last 40,000 years of biotic change*. University of Arizona Press, Tucson.

Betancourt, J.L., Latorre, C., Rech, J.A., Quade, J. and Rylander, K.A. 2000. A 22-year record of monsoonal precipitation from Northern Chile's Atacama Desert. *Science*, 289, 1542–1546.

Betancourt, J.L., Rylander, K.A., Peñalba, C. and McVickar, J.L. 2001. Late Quaternary vegetation history of Rough Canyon, south-central New Mexico, USA. *Palaeogeography, Palaeoclimatology, Palaeoecology*, 165, 71–95.

Betts, A.K. and Ball, J.H. 1997. Albedo over the boreal forest. *Journal of Geophysical Research*, 102D, 28901–28909.

Betts, R.A. 1999. The impact of land use on the climate of the present day. In H. Ritchie (ed.), *Research activities in atmospheric and oceanic modelling. CAS/JSC WGNE Report 28*. WMO, Geneva, pp. 7.11–7.12.

Bianchi, G.G. and McCave, I.N. 1999. Holocene periodicity in North Atlantic climate and deep-ocean flow south of Iceland, *Nature*, 397, 515–513.

Bickerton, R.W. and Matthews, J.A. 1993. 'Little Ice Age' variations of outlet glaciers from the Jostedalsbreen ice-cap, southern Norway: a regional lichenometric-dating study of ice-marginal moraine sequences and their climatic significance. *Journal of Quaternary Science*, 8, 45–66.

Bigler, C. and Hall, R.I. 2002. Diatoms as indicators of climatic and limnological change in Swedish Lapland: a 100-lake calibration set and its validation for palaeoecological reconstruction. *Journal of Paleolimnology*, 27, 97–115.

Bigler, C. and Hall, R.I. 2003. Diatoms as quantitative indicators of July temperature: a century-scale validation with meteorological data from northern Sweden. *Palaeogeography, Palaeoclimatology, Palaeoecology*, 189,147–160.

Bigler, C., Larocque, I., Peglar, S.M., Birks, H.J.B. and Hall, R.I. 2002. Quantitative multiproxy assessment of long-term patterns of Holocene environmental change from a small lake near Abisko, northern Sweden. *The Holocene*, 12, 481–496.

Billamboz, A. 1992. Tree-ring analysis from an archaeodendrochronological perspective. The structural timber from the South West German Lake Dwellings. *Lundqua*, 34, 34–40.

Billamboz, A. 1995. Proxyséries dendrochronologiques et occupation néolithique des bords du lac de Constance. *Palynosciences*, 3, 69–81.

Billamboz, A. 1996. Tree-rings and pile-dwellings in southwestern Germany: following in the footsteps of Bruno Huber. In J.S. Dean, D.M. Meko and T.W. Swetnam (eds), *Tree-rings, environment and humanity.* Proceedings of the International Conference, Tucson, Arizona, 17–21 May 1994. *Radiocarbon*, 471–483.

Billman, B.R., Lambert, P.M. and Leonard, B.L. 2000. Cannibalism, warfare and drought in the Mesa Verde region during the 12th century AD. *American Antiquity*, 65, 145–178.

Bin, L., Daoxian, Y., Yushi, L. and Lauritzen, S.E. 1997. Luminescence and palaeoenvironmental record in a stalagmite in Panlong Cave, Guilin. *Acta Geoscientia Sinica*, 18, 400–406.

Binford, L R. 1968. Post Pleistocene adaptations. In S.B. Binford and L.R. Binford (eds), *New perspectives in archeology.* Aldine, Chicago, IL, pp. 313–342.

Binford, M., Kolata, A.L., Brenner, M., *et al.* 1997. Climate variation and the rise and fall of an Andean civilization. *Quaternary Research*, 47, 235–248.

Bird, M.I., Ayliffe, L.K., Fifield, L.K., *et al.* 1999. Radiocarbon dating of 'old' charcoal using a wet oxidation-stepped combustion procedure. *Radiocarbon*, 41, 127–140.

Birks, C.J.A. and Koç, N. 2002. A high-resolution diatom record of late-Quaternary sea-surface temperatures and oceanographic conditions from the eastern Norwegian Sea. *Boreas*, 31, 323–344.

Birks, H.H. 1975. Studies in the vegetational history of Scotland IV. Pine stumps in Scottish blanket peats. *Philosophical Transactions of the Royal Society of London B*, 270, 181–226.

Birks, H.H. 1991. Holocene vegetational history and climatic change in west Spitsbergen: plant macrofossils from Skardtjørna, an arctic lake. *The Holocene*, 1, 209–218.

Birks, H.H. 2001a. Plant macrofossils. In J.P. Smol, H.J.B. Birks and W.M. Last (eds), *Tracking environmental changes using lake sediments. Vol. 3. Terrestrial and siliceous indicators.* Kluwer Academic Publishers, Dordrecht, pp. 49–74.

Birks, H.H. 2003. The importance of plant macrofossils in the reconstruction of late glacial vegetation and climate: examples from Scotland, western Norway, and Minnesota USA. *Quaternary Science Reviews*, 22, 453–473.

Birks, H.H. and Ammann, B. 2000. Two terrestrial records of rapid climatic change during the glacial–Holocene transition 14,000–9000 calendar years BP from Europe. *Proceedings of the National Academy of Sciences USA*, 97, 1390–1394.

Birks, H.H. and Birks, H.J.B. 2000. Future uses of pollen analysis must include plant macrofossils. *Journal of Biogeography*, 27, 31–35.

Birks, H.H. and Mathewes, R.W. 1978. Studies in the vegetational history of Scotland V. Late Devensian and Early Flandrian pollen and macrofossil stratigraphy at Abernethy Forest, Inverness-shire. *New Phytologist*, 80, 455–484.

Birks, H.H., Battarbee, R.W., Beerling, D.J., *et al.* 1996a. The Kråkenes late-glacial palaeoenvironmental project. *Journal of Paleolimnology*, 15, 281–286.

Birks, H.H., Vorren, K.-D. and Birks, H.J.B. 1996b. Holocene treelines, dendrochronology and palaeoclimate. *Paläoklimaforschung*, 20, 1–18.

Birks, H.H., Battarbee, R.W. and Birks, H.J.B. 2000. The development of the aquatic ecosystem at Kråkenes Lake, western Norway, during the late glacial and early Holocene: a synthesis. *Journal of Paleolimnology*, 23, 91–114.

Birks, H.J.B. 1981. The use of pollen analysis in the reconstruction of past climates: a review. In T.M.L. Wigley, M.J. Ingram and G. Farmer (eds), *Climate and history.* Cambridge University Press, Cambridge, pp. 111–138.

Birks, H.J.B. 1986. Late-Quaternary biotic changes in terrestrial and lacustrine environments, with particular reference to north-west Europe. In B.E. Berglund (ed.), *Handbook of Holocene palaeoecology and palaeohydrology.* John Wiley, Chichester, pp. 3–65.

Birks, H.J.B. 1995. Quantitative palaeoenvironmental reconstructions. In D. Maddy and J.S. Brew (eds), *Statistical modelling of Quaternary science data*. Quaternary Research Association, Cambridge, pp. 161–254.

Birks, H.J.B. 1998. Numerical tools in palaeolimnology - progress, potentialities, and problems. *Journal of Paleolimnology*, 20, 307–332.

Birks, H.J.B. 2001b. Maximum likelihood environmental calibration and the computer program WACALIB: a correction. *Journal of Paleolimnology*, 25, 111–115.

Birks, H.J.B. and Birks, H.H. 1980. *Quaternary palaeoecology*. Edward Arnold, London.

Birks, H.J.B. and Gordon, A.D. 1985. *Numerical methods in Quaternary pollen analysis*. Academic Press, London.

Birks, H.J.B., Line, J.M., Juggins, S., Stevenson, A.C. and ter Braak, C.J.F. 1990. Diatoms and pH reconstruction. *Philosophical Transactions of the Royal Society, London B*, 327, 263–278.

Björck, S., Kromer, B., Johnsen, S., *et al.* 1996c. Synchronized terrestrial–atmospheric deglacial records around the North Atlantic. *Science*, 274, 1155–1160.

Björck, S., Rundgren, M., Ingólfsson, Ó. and Funder, S. 1997. The Pre-Boreal oscillation around the Nordic Seas: terrestrial and lacustrine responses. *Journal of Quaternary Science*, 12, 455–465.

Björck, S., Walker, M.J.C., Cwynar, L.C., *et al.* 1998. An event stratigraphy for the Last Termination in the North Atlantic region based on the Greenland ice-core record: a proposal by the INTIMATE group. *Journal of Quaternary Science*, 13, 283–292.

Björck, S., Muscheler, R., Kromer, B., *et al.* 2001. High-resolution analyses of an early Holocene climate event may imply decreased solar forcing as an important climate trigger. Geology 29, 1107–1110.

Black, D.E., Peterson, L.C., Overpeck, J.T., Kaplan, A., Evans, M.N. and Kashgarian, M. 1999. Eight centuries of North Atlantic Ocean atmosphere variability. *Science*, 286 1709–1713.

Blackford, J.J. and Chambers, F.M. 1991. Proxy records of climate from blanket mires: evidence for a Dark Age 1400BP. climatic deterioration in the British Isles. *The Holocene*, 1, 63–67.

Blackford, J.J. and Chambers, F.M. 1993. Determining the degree of peat decomposition in peat-based palaeoclimatic studies. *International Peat Journal*, 5, 7–24.

Blackford, J.J. and Chambers, F.M. 1995. Proxy climate record for the last 1000 years from Irish blanket peat and a possible link to solar variability. *Earth and Planetary Science Letters*, 133, 145–150.

Blais-Stevens, A., Bornhold, B.D., Kemp, A.E.S., Dean, J.M., and Vaan, A.A. 2001, Overview of Late Quaternary stratigraphy in Saanich Inlet, British Columbia: results of ocean drilling prorgam leg 169S. *Marine Geology*, 174, 3–26.

Blake, Jr, W. 1993. Holocene emergence along the Ellesmere Island coasts of northernmost Baffin Bay. *Norsk Geologisk Tidsskrift*, 73, 147–160.

Bloomfield, P. 1976. *Fourier analysis of time series. An introduction*. Wiley, New York.

Blunier, T., Chappellaz, J.A., Schwander, J., Stauffer, B. and Raynaud, D. 1995. Variations in atmospheric methane concentration during the Holocene epoch. *Nature*, 374, 46–49.

Blunier, T., Chappellaz, J., Schwander, J., *et al.* 1998. Asynchrony of Antarctic and Greenland climate change during the last glacial period. *Nature*, 384, 739–743.

Bluszcz, A., Goslar, T., Hercman, H., Pazdur, M.F. and Walanus, A. 1988. Comparison of TL, ESR and ^{14}C dates of speleothems. *Quaternary Science Reviews*, 7, 417–421.

Blytt, A. 1876. Essay on the immigration of the Norwegian flora during alternating rainy and dry periods. Kristiana, Cammermeyer.

Bonadonna, F. and Leone, G. 1995. Palaeoclimatological reconstruction using stable-isotope data on molluscs from Valle di Castiglione, Roma, Italy. *The Holocene*, 5, 461–469.

Bond, G., Showers, W., Cheseby, M. *et al.* 1997. A pervasive millennial-scale cycle in North Atlantic Holocene and glacial climates. *Science*, 278, 1257–1266.

Bond, G., Kromer, B., Beer, J., *et al.* 2001. Persistent solar influence on North Atlantic climate during the Holocene. *Science*, 294, 2130–2136.

Bonfils, C., deNoblet-Ducoudre, N., Braconnot, P., Joussaume, S. 2001. Hot desert albedo and climate change: mid-Holocene monsoon in North Africa. *Journal of Climate*, 14, 3724–3737.

Boninsegna, J.A. and Holmes, R.L. 1985. Fitzroya Cupressoides yields 1534-year long South American chronology. *Tree-Ring Bulletin*, 45, 37–42.

Bonsal, B.R., Zhang, X., Vincent, L.A. and Hogg, W.D. 2001. Characteristics of daily and extreme temperatures over Canada. *Journal of Climate*, 14, 1959–1976.

Boomer, I. and Eisenhauer, G. 2002. Ostracod faunas as palaeoenvironmental indicators in marginal marine environments. In J.A. Holmes and A.R. Chivas (eds), *The Ostracoda: Applications in Quaternary Research*, Vol. 131, AGU Geophysical Monograph Series, Washington, DC, pp. 135–149.

Boon, J.J., Rijpstra, W.I.C., de Lange, F. and de Leeuw, J.W. 1979. Black Sea sterol. A molecular fossil for dinoflagellate blooms. *Nature*, 277, 125–126.

Boone, J.L. 2002. Subsistence strategies and early human population history: an evolutionary ecological perspective. *World Archaeology*, 34, 6–25.

Bornhold, B.D. and Kemp, A.E.S. 2001. Late Quaternary sedimentation in Saanich Inlet, British Columbia, Canada: ocean drilling program leg 169S, Preface. *Marine Geology*, 174, 1.

Boto, K. and Isdale, P. 1985. Fluorescent bands in massive corals result from terrestrial fulvic acid inputs to nearshore zone. *Nature*, 315, 396–397.

Bottema, S. 1995. The Younger Dryas in the eastern Mediterranean. *Quaternary Science Reviews*, 14, 883–891.

Boubée, J.A.T. 1983. Past and present fauna of Lake Maratoto with special reference to the Chironomidae, PhD thesis. University of Waikato, New Zealand.

Bourgeois, J.B., Koerner, R.M., Gayewski, K. and Fisher, D.A. 2000. A Holocene pollen record from Ellesemere Island, Nunavut, Canada. *Quaternary Research*, 54, 275–283.

Bouwmann, A.F. 1990. Global distribution of the major sols and land cover types. In A.F. Bouwmann (ed.), *Soils and the greenhouse effect*. John Wiley, Chichester, pp. 33–59.

Bowen, R. 1988. *Isotopes in the earth sciences*. Elsevier Applied Science, Amsterdam.

Bowman, S. 1990. *Radiocarbon dating*. British Museum Press, London.

Braconnot, P., Joussaume, S., Marti, O. and de Noblet, N. 1999. Synergistic feedbacks from ocean and vegetation on the african monsoon response to mid-Holocene insolation. *Geophysical Research Letters*, 26, 2481–2484.

Braconnot, P., Marti, O., Joussaume, S. and Leclainche, Y. 2000. Ocean feedback in response to 6 kyr BP insolation. *Journal of Climate*, 13, 1537–1553.

Bradbury, J.P. 1988. A climatic-limnological model of diatom succession for paleolimnological interpretation of varved sediments at Elk Lake, Minnesota. *Journal of Paleolimnology*, 1, 115–131

Bradbury, J.P. and Dean, W.E. 1993. Elk Lake, Minnesota: evidence for rapid climate change in the north-central United States. *Geological Society of America Special Paper*, 276, 336.

Bradbury, J.P. and Dieterich-Rurup, K.V. 1993. Holocene diatom paleolimnology of Elk Lake, Minnesota. *Geological Society of America Special Paper*, 276, 215–237.

Bradbury, J.P., Leyden, B., Salgado-Labouriau, M., *et al.* 1981. Late Quaternary history of Lake Valencia, Venezuela. *Science*, 214, 1299–1305.

Bradbury, J.P., Forester, R.M., Bryant, W.A. and Covich, A.P. 1990. Paleolimnology of Laguna de Cocos, Albio Island, Rio Hondo, Belize. In M.D. Pohl (ed.), *Ancient Maya wetland agriculture. Excavations on Albion Island, Northern Belize*. Westview Press, Boulder, CO, pp. 119–154.

Bradbury, J.P., Bezrukova, Y.V., Chernyaeva, G.P., *et al.* 1994. A synthesis of post-glacial diatom records from Lake Baikal. *Journal of Paleolimnology*, 10, 213–252.

Bradbury, J.P., Cumming, B. and Laird, K. 2002. A 1500-year record of climatic and environmental change in Elk Lake, Minnesota III: measures of past primary production. *Journal of Paleolimnology*, 27, 321–340.

Bradley, R.S. 1988. The explosive volcanic eruption signal in northern hemisphere continental temperature records. *Climatic Change*, 12, 221–243.

Bradley, R.S. 1996. The Taconite Inlet lakes project. *Journal of Paleolimnology*, 16, 97–255.

Bradley, R.S. 1999. Paleoclimatology: reconstructing climates of the Quaternary. International Geophysics Series, 64. Harcourt Academic Press, San Diego.

Bradley, R.S. and Jones, P.D. 1985. Data bases for isolating the effects of increasing carbon dioxide concentrations. In M.C. MacCracken and F.M. Luther (eds), *Detecting the climatic effects of increasing carbon dioxide, DOE/ER-0235*. US Department of Energy, Carbon Dioxide Research Division, Washington, DC, pp. 29–53.

Bradley, R.S. and Jones, P.D. 1993. 'Little Ice Age' summer temperature variations: their nature and relevance to recent global warming trends. *The Holocene*, 3, 367–376.

Braidwood, R.J. 1952. From cave to village. *Scientific American*, 187, 62–66.

Braidwood, R.J. 1960. The agricultural revolution. *Scientific American*, 203, 130–141.

Braidwood, R.J. and Howe, B. (eds). 1960. *Prehistoric investigations in Iraqi Kurdistan*. Oriental Institute Studies in Ancient Oriental Civilization, 31. University of Chicago Press, Chicago.

Brain, C.K. 1974. The use of microfaunal remains as habitat indicators in the Namib. *South African Archaeological Society, Goodwin Series*, 2, 55–60.

Brain, C.K. and Brain, V. 1977. Microfaunal remains from Mirabib: some evidence of palaeo-ecological changes in the Namib. *Madoqua*, 10, 285–293.

Brandriss, M.E., O'Neil, J.R., Edlund, M.B. and Stoermer, E.F. 1998. Oxygen isotope fractionation between diatomaceous silica and water. *Geochimica et Cosmochimica Acta*, 62, 1119–1125.

Brassell, S.C., Eglinton, G., Marlowe, I.T., Pflaumann, U. and Sarnthein, M. 1986. Molecular stratigraphy: a new tool for climatic assessment. *Nature*, 320, 129–133.

Brauer, A. 1994. Weichselzeitliche Sedimente des Holzmaares - Warvenchronologie des Hochglazials und Nachweis von Klimaschwankungen. *Documenta naturae*, 85, 1–210.

Brauer, A., Litt, T., Negendank, J.F.W. and Zolitschka, B. 2001. Lateglacial varve chronology and biostratigraphy of lakes Holzmaar and Meerfelder Maar, Germany. *BOREAS*, 30, 83–88.

Brenner, M., Hodell, D.A., Curtis, J.H., Rosenmeier, M.F., Binford, M.W. and Abbott, M.B. 2001. Abrupt climate change and Pre-Columbian cultural collapse. In V. Markgraf (ed.), *Interhemispheric climate linkages*. Academic Press, San Diego, pp. 87–104.

Bridgewater, N.D. Heaton, T.H.E. and O'Hara, S.L. 1999. A late Holocene palaeolimnological record from central Mexico, based on faunal and stable-isotope analysis of ostracod shells. *Journal of Paleolimnology*, 22, 383–397.

Briffa, K.R. 1999. Analysis of dendrochronological variability and associated natural climates in Eurasia - the last 10,000 years (Advance 10 K). PAGES (Post Global Changes) 7,1, 6–8.

Briffa, K.R., Jones, P.D., Schweingruber, F.H. and Osborn, T.J. 1998. Influence of volcanic eruptions on Northern Hemisphere summer temperature over the last 600 years. *Nature*, 393, 450–455

Briffa, K.R., Osborn, T.J., Schweingruber, F.H., *et al.* 2001. Low-frequency temperature variations from a northern tree-ring density network. *Journal of Geophysical Research*, 106, 2929–2941.

Brimblecombe, P. 1982. Long term trends in London fog. *Science of the Total Environment*, 22, 19–29.

Brimblecombe, P. 1987. *The big smoke*. Methuen, London.

Brimblecombe, P. 1990. Writing on smoke. In H. Bradby (ed.), *Dirty words*. Earthscan Publications, London, pp. 93–114.

Brimblecombe, P. 1995. A meteorological service in fifteenth century Sandwich. *Environment and History*, 1, 241–249.

Brodersen, K.P. and Anderson, N.J. 2000. Subfossil insect remains Chironomidae and lake-water temperature inference in the Sisimiut - Søndre Strømfjord region, southern West Greenland. In P.R. Dawes and A.K. Higgins (eds), *Review of Greenland activities 1999. Geology of Greenland Survey Bulletin*, 186, 78–82.

Brodersen, K.P. and Lindegaard, C. 1999. Mass occurrence and sporadic distribution of *Corynocera ambigua* Zetterstedt Diptera, Chironomidae in Danish lakes. Neo- and palaeolimnological records. *Journal of Paleolimnology*, 22, 41–52.

Brodin, Y.W. 1986. The post-glacial history of Lake Flarken, southern Sweden, interpreted from subfossil insect remains. *Internationale Revue der Gesamten Hydrobiologie*, 71, 371–432.

Brodin, Y-W. and Gransberg, M. 1993. Responses of insects, especially Chironomidae Diptera, and mites to 130 years of acidification in a Scottish lake. *Hydrobiologia*, 250, 201–212.

Broecker, W.S. 1995. *The glacial world according to Wally*. Eldigio Press, New York.

Broecker, W.S. 1998. Paleocean circulation during the last deglaciation: a bipolar seesaw? *Paleoceanography*, 13, 119–121.

Broecker, W.S. 2000. Abrupt climate change: causal constraints provided by the paleoclimate record. *Earth-Science Reviews*, 51, 137–154.

Broecker, W.S. and Denton, G.H. 1989. The role of ocean–atmosphere reorganizations in glacial cycles. *Geochimica et Cosmochimica Acta*, 53, 2465–2501.

Broecker, W.S. and Olson, E.A. 1960. Radiocarbon measurements and annual rings in cave formations. *Nature,* 185, 93–94.

Broecker, W.S., Peng, T.-H., Östlund, G. and Stuiver, M. 1985. The distribution of bomb radiocarbon in the ocean. *Journal of Geophysical Research*, 90, 6953–6970.

Brook, G.A., Burney, D.A. and Cowart, J.B. 1990. Desert paleoenvironmental data from cave speleothems with examples from the Chahuahuan, Somali-Chalib, and Kalahari deserts. *Palaeogeography, Palaeoclimatology, Palaeoecology*, 76, 311–329.

Brook, G.A., Marais E. and Cowart, J.A. 1999a. Evidence of wetter and drier conditions in Namibia from tufas and submerged speleothems. *Cimbebasia*, 15, 29–39.

Brook, G.A., Rafter, M., Railsback, L.B., Sheen, S.W. and Lundberg, J. 1999b. A high-resolution proxy record of rainfall and ENSO since AD 1550 from layering in stalagmites from AnjohibeCave, Madagascar. *The Holocene*, 9, 695–706.

Brooks, S.J. 1996. Three thousand years of environmental history in a Cairngorms lochan revealed by analysis of non-biting midges Insecta: Diptera: Chironomidae. *Botanical Journal of Scotland*, 48, 89–98.

Brooks, S.J. 2000. Lateglacial fossil midge Insecta: Diptera: Chironomidae. stratigraphies from the Swiss Alps. *Palaeogeography, Palaeoclimatology, Palaeoecology*, 159, 261–279.

Brooks, S.J. and Birks, H.J.B. 2000. Chironomid-inferred late-glacial air temperatures at Whitrig Bog, south-east Scotland. *Journal of Quaternary Science*, 15, 759–764.

Brooks, S.J. and Birks, H.J.B. 2001. Chironomid-inferred air temperatures from late-glacial and Holocene sites in north-west Europe: progress and problems. *Quaternary Science Reviews*, 20, 1723–1741.

Brooks, S.J. and Birks, H.J.B. In press. The dynamics of Chironomidae populations in response to environmental change during the past 300 years in Spitsbergen. *Journal of Paleolimnology*,

Broström, A., Coe, M., Harrison, S.P., *et al.* 1998. Land surface feedbacks and palaeomonsoons in northern Africa. *Geophysical Research Letters*, 25, 3615–3618.

Brovkin, V., Claussen, M., Petoukhov, V., Ganopolski, A. 1998. On the stability of the atmosphere-vegetation system in the Sahara/Sahel region. *Journal of Geophysical Research*, 103, 31613–31624.

Brovkin, V., Bendtsen, J., Claussen, M., *et al.* 2002. Carbon cycle, vegetation and climate dynamics in the Holocene: experiments with the CLIMBER-2 model. *Global Biogeochemical Cycles*, 16, 1139.

Brundin, L. 1956. Die bodenfaunistischen Seetypen und ihre Anwendbarkeit auf die Südhalbkugel. Zugleich eine Theorie der produktionsbiologischen Bedeutung der glazialen Erosion, Vol. 37. Institute of Freshwater Research, Drottningholm, pp. 186–235.

Brundin L. 1966. Transantarctic relationships and their significance, as evidenced by chironomid midges, with a monograph of the subfamilies Podonominae and Aphroteniinae and the austral Heptagyiae. *Kungl Svenska Vetenskapsakademiens Handlingar*, 11, 1–472.

Bruun, P. 1962. Sea-level rise as a cause of shore erosion. Proceedings of the American Society of Civil Engineers, Waterways and Harbour Division, 88, 117–130.

Bryant, E.A., Young, R.W. and Price, D.M. 1996. Tsunami as a major control on coastal evolution, Southeastern Australia. *Journal of Coastal Research*, 12, 831–840.

Buckland, P.C., Dugmore A.J. and Edwards, K.J. 1997. Bronze Age myths? Volcanic activity and human response in the Mediterranean and North Atlantic regions. *Antiquity*, 71, 581–593.

Budyko, M.I. 1974. Climate and Life. International Geophysical Series, 18. Academic Press, New York.

Budziak, D., Schneider, R.R., Rostek, F., Muller, P.J., Bard, E. and Wefer, G. 2000. Late Quaternary insolation forcing on total organic carbon and C_{37} alkenone variations in the Arabian Sea. *Paleoceanography*,15, 307–321.

Buhmann, D. and Dreybrodt, W. 1985. The kinetics of calcite dissolution and precipitation in geologically relevant situations of karst areas. 1. Open system. *Chemical Geology*, 48, 189–211.

Burnaby, T.P. 1953. A suggested alternative to the correlation coefficient for testing the significance of agreement between pairs of time series and its application to geological data. *Nature,* 172, 210–211.

Burney, D.A., Brook, G.A. and Cowart, J.B. 1994. A Holocene pollen record for the Kalahari Desert of Botswana from a U-series dated speleothem. *The Holocene*, 4, 225–232.

Burns, S. and Maslin, M.A. 1999. Composition and circulation of bottom water in the western Atlantic Ocean during the last glacial, based on pore-water analyses from the Amazon Fan. *Geology*, 27, 1011–1014.

Busenberg, E. and Plummer, L.N. 1986. A comparative study of the dissolution and crystal growth kinetics of calcite and aragonite. In F.A. Mumton (ed.), *Studies in diagenesis*. US Geological Survey, VA, pp. 139–168.

Bush, M.B. 2002. On the interpretation of fossil Poaceae pollen in the humid lowland neotropics. *Palaeogeography, Palaeoclimatology, Palaeoecology*, 177, 5–17.

Bush, M.B. and Colinvaux, P.A. 1994. A paleoecological perspective of tropical forest disturbance: records from Darien, Panama. *Ecology*, 75, 1761–1768.

Bush, M.B., Piperno, D.R. and Colinvaux, P.A. 1989. A 6000-year history of Amazonian maize cultivation. *Nature*, 340, 303–305.

Bush, M.B., Piperno, D.R., Colinvaux, P.A., *et al.* 1992. A 14,300-year paleoecological profile of a lowland tropical lake in Panama. *Ecological Monographs*, 62, 251–276.

Bush, M.B., Miller, M.C., De Oliveira, P.E. and Colinvaux, P.A. 2000. Two histories of environmental change and human disturbance in eastern lowland Amazonia. *The Holocene*, 10, 543–554.

Bush, M.B., Stute, M., Ledru, M.-P., *et al.* 2001. Paleotemperature estimates for the lowland Americas between 30°S and 30°N at the last glacial maximum. In V. Markgraf (ed.), *Interhemispheric climate linkages: present and past interhemispheric climate linkages in the americas and their societal effects*. Academic Press, New York, pp. 293–306.

Cacho, I., Grimalt, J.O., Sierro, F.J., Shackleton, N. and Canals, M. 2000. Evidence for enhanced Mediterranean thermohaline circulation during rapid climatic coolings. *Earth and Planetary Science Letters*, 183, 417–429.

Campbell, C. 1998. Late Holocene lake sedimentology and climate change in Southern Alberta, Canada. *Quaternary Research*, 49, 96–101.

Campbell, I.D., Campbell, C., Apps, M.J., Rutter, N.W. and Bush, A. 1998. Late Holocene ~1500 yr periodicities and their implications. *Geology* 26, 471–473.

Campbell, I.D., Campbell, C., Yu, Z.C., Vitt, D.H. and Apps, M.J. 2000. Millennial-scale rhythms in peatlands in the western interior of Canada and in the global carbon cycle. *Quaternary Research*, 54, 155–158.

Camuffo, D. 1993. Analysis of the sea surges at Venice from AD782 to 1990. *Theoretical and Applied Climatology*, 47, 1–14.

Camuffo, D. and Sturaro, G. 2004. Use of proxy-documentary and instrumental data to assess the risk factors leading to sea flooding in Venice. *Global and Planetary Change*, 40, 93–103.

Camuffo, D., Cocheo, C., and Enzi, S. 2000a. Seasonality of instability phenomena hailstorms and thunderstorms. in Padova, northern Italy, from archive and instrumental sources since AD 1300. *The Holocene*, 10, 635–642.

Camuffo. D., Demarée, G., Davies, T.D., *et al.* 2000b. Improved understanding of past climatic variability from early daily European instrumental sources IMPROVE. *European Climate Science Conference, Vienna, 19–23 October 1998.*

Cane, M.A., Clement, A.C., Kaplan, A., Kushnir, Y., Murtugudde, R. and Zebiak, S. 1997. Twentieth-century sea surface temperature trends. *Science*, 275, 957–960.

Cappers, R.T.J. 2001. Problems in correlating pollen diagrams of the Near East: a preliminary report. In A.B. Damania (ed.), *Origins of agriculture and crop domestication*. Genetic Resources Conservation Program, University of California, Davis, CA,160–169.

Card, V.M. 1997. Varve-counting by the annual pattern of diatoms accumulated in the sediment of Big Watab Lake, Minnesota, AD 1837–1990. *BOREAS*, 26, 103–112.

Cardinal, D., Hamelin, B., Bard, E. and Patzold, J. 2001. Sr/Ca, U/Ca and $\delta^{18}O$ records in recent massive corals from Bermuda: relationships with sea surface temperature. *Chemical Geology*, 176, 213–233.

Carney, T. F. 1972. Content analysis: a technique for systematic inference from communications. B.T. Batsford, London.

Carrión, J.S., Riquelme, J.A., Navarro, C. and Munuera, M. 2001. Pollen in coprolites reflects the late glacial landscape in southern Spain. *Palaeogeography, Palaeoclimatology, Palaeoecology*, 176, 193–205.

Caseldine, C.J., Baker, A., Charman, D.J. and Hendon, D. 2000. A comparative study of optical properties of NaOH peat extracts: implications for humification studies. *The Holocene*, 10, 649–658.

Catchpole, A. J. W. and Moodie, D. W. 1975. *Environmental data from historical accounts by contents analysis.* Department of Geography, University of Manitoba, Winnipeg.

Chambers, F.M. and Blackford, J.J. 2001. Mid- and Late-Holocene climatic changes: a test of periodicity and solar forcing in proxy-climate data from blanket peat bogs. *Journal of Quaternary Science*, 16, 329–338.

Chambers, F.M., Barber, K.E., Maddy, D. and Brew, J. 1997. A 5500-year proxy-cliamte and vegetation record from blanket mire at Talla Moss, Borders, Scotland. *The Holocene*, 7, 391–399.

Chambers, F.M., Mauquoy, D. and Todd, P.A. 1999. Recent rise to dominance of Molinia caerulea in environmentally sensitive areas: new perspectives from palaeoecological data. *Journal of Applied Ecology*, 36, 719–733.

Chaney, R.W. 1924. Quantitative studies of the Bridge Creek Flora. *American Journal of Science*, 8, 127–144.

Changnon, S.A. 1999. Assessment of historical thunderstorm data for urban effects: the Chicago case. *Climate Change*, 49, 161–169.

Chapman, M.R. and Shackleton, N.J. 2000. Evidence of 550 year and 1000 years cyclicities in North Atlantic pattern during the Holocene. *The Holocene*, 10, 287–291.

Chappellaz, J.A., Fung, I.Y. and Thompson, A.M. 1993. The atmospheric CH_4 increase since the last glacial maximum. Part 1. Source estimates. *Tellus, Series B*, 45b, 228–241.

Chappellaz, J.A., Brook, E., Blunier, T. and Malaizé, B. 1997. CH_4 and $\delta^{18}O$ of O_2 records from Antarctic and Greenland ice: a clue for stratigraphic disturbance in the bottom part of the GRIP and GISP2 ice cores. In *Greenland summit ice cores*, Vol. 102, No. C12. JGR Special Publication, pp. 26547–26558.

Charles, C.D., Hunter, D.E. and Fairbanks, R.G. 1997. ENSO and the Monsoon in a coral record of Indian Ocean surface temperature. *Science*, 277, 925–928.

Charleson, R.J., Lovelock, J.E., Meinrat, O., Warren, A. and Warren, S.G. 1987. Oceanic phytoplankton, atmospheric sulphur, cloud albedo and climate. *Nature*, 326, 655–661.

Charman, D.J. 1992. Relationship between testate amoebae Protozoa: rhizopoda. and microenvironmental parameters on a forested peatland in northeastern Ontario. *Canadian Journal of Zoology*, 70, 2474–2482.

Charman, D.J. 1997. Modelling hydrological relationships of testate amoebae Protozoa : Rhizopoda. on New Zealand peatlands. *Journal of the Royal Society of New Zealand*, 27, 465–483.

Charman, D.J. 2002. *Peatlands and environmental change.* John Wiley, Chichester.

Charman, D.J. and Hendon, D. 2000. Long-term changes in soil water tables over the past 4500 years: relationships with climate and North Atlantic atmospheric circulation and sea surface temperature. *Climatic Change*, 47, 45–59.

Charman, D.J. and Warner, B.G. 1997. The ecology of testate amoebae Protozoa: Rhizopoda in oceanic peatlands in Newfoundland, Canada: modelling hydrological relationships for palaeoenvironmental reconstruction. *Ecoscience*, 4, 555–562.

Charman, D.J., Hendon, D. and Packman, S. 1999. Multiproxy surface wetness records from replicate cores on an ombrotrophic mire: implications for Holocene palaeoclimate records. *Journal of Quaternary Science*, 14, 451–463.

Charman, D.J., Hendon, D. and Woodland, W.A. 2000. *The identification of peatland testate amoebae. Technical Guide.* Quaternary Research Association, London.

Charman, D.J., Caseldine, C., Baker, A., Gearey, B.R. and Hatton, J. 2001. Palaeohydrological records from peat profiles and speleothems in Sutherland, northwest Scotland. *Quaternary Research*, 55, 223–234.

Charman, D.J., Brown, A.D., Hendon, D., Kimmel, A. and Karofeld, E. In press. Testing the relationship between Halocene peatland and palaeoclimate reconstructions and instrumental data. *Quaternary Science Reviews.*

Charney, J.G. 1975. Dynamics of deserts and droughts in the Sahel. *Quarterly Journal of the Royal Meteorological Society*, 101, 193–202.

Charney, J., Stone, P.H., Quirk, W.J. 1977. Drought in the Sahara: a biogeophysical feedback mechanism. *Science*, 187, 434–435.

Chaudhuri, P. and Marron, J.S. 1999. Sizer for exploration of structures in curves. *Journal of the American Statistical Association*, 94, 807–823.

Cheddadi, R., Yu, G., Guiot, J., Harrison, S.P., Prentice, I.C. 1997. The climate of Europe 6000 years ago. *Climate Dynamics*, 13, 1–9.

Cheddadi, R., Lamb, H.F., Guiot, J. and van der Kaars, S. 1998. Holocene climatic change in Morocco: a quantitative reconstruction from pollen data. *Climate Dynamics*, 14, 883–890.

Childe, V.G. 1952. *New light on the most ancient East.* Routledge and Kegan Paul, London.

Chinn, T.J. 1999. New Zealand glacier response to climate change of the past 2 decades. *Global and Planetary Change*, 22, 155–168.

Chivas, A.R., De Deckker, P. and Shelley, J.M.G. 1986a. Magnesium and strontium in non-marine ostracod shells as indicators of palaeosalinity and palaeotemperature. *Hydrobiologia*, 143, 135–142.

Chivas, A.R., De Deckker, P. and Shelley, J.M.G. 1986b. Magnesium content of non-marine ostracod shells: a new palaeosalinometer and palaeothermometer. *Palaeogeography, Palaeoclimatology, Palaeoecology*, 54, 43–61.

Christensen, C.J., Gorsline, D.S., Hammond, D.E. and Lund, S.P. 1994. Non-annual laminations and expansion of anoxic basin-floor conditions in Santa Monica Basin, California Borderland, over the past four centuries. *Marine Geology*, 116, 399–418.

Church, J.A., Godfrey, J.S., Jackett, D.R. and McDougall, T.J. 1991. A model of sea level rise caused by ocean thermal expansion. *Journal of Climate*, 4, 438–456.

Clague, J.J. and Mathewes, R.W. 1996. Neoglaciation, glacier-dammed lakes, and vegetation change in northwestern British Colombia, Canada. *Arctic, Antarctic and Alpine Research*, 28, 10–24.

Clark, I. and Fritz, I. 1997. *Environmental isotopes in hydrogeology*. Lewis, Boca Raton, LA.

Clark, J.A. and Lingle, C.S. 1979. Predicted RSL changes 18,000 years BP to present. caused by late-glacial retreat of the Antarctic ice sheet. *Quaternary Research*, 11, 279–298.

Clark, J.S., Grimm, E.C., Donovan, J.J., Fritz, S.C., Engstrom, D.R. and Almendinger, J.E. 2002. Drought cycles and landsape responses to past aridity on prairies of the Northern Great Plains, USA. *Ecology*, 83, 595–601.

Clausen, H.B., Hammer, C.U., Christensen, J., *et al.* 1995. 1250 years of global volcanism as revealed by central Greenland ice cores. In R.J. Delmas (ed.), *Ice core studies of global biogeochemical cycles, NATO ASI Series I, Vol. 30*. Springer-Verlag, New York, pp. 517–532.

Clausen, H.B., Hammer, C.U., Hvidberg, C.S., Dahl-Jensen, D. and Steffensen, J.P. 1997 A comparison of the volcanic records over the past 4000 years from the Greenland Ice Core Project and Dye 3 Greenland ice cores. *Journal of Geophysical Research*, 102, 26707–26723.

Claussen, M. 1994. On coupling global biome models with climate models. *Climate Research*, 4, 203–221.

Claussen, M. 1997. Modelling biogeophysical feedback in the African and Indian Monsoon region. *Climate Dynamics*, 13, 247–257.

Claussen, M. 1998. On multiple solutions of the atmosphere-vegetation system in present-day climate. *Global Change in Biology*, 4, 549–559.

Claussen, M. 2003. Does landsurface matter in weather and climate? In P. Kabat, M. Claussen, P.A. Dirmeyer, *et al.* (eds), *Vegetation, water, humans and the climate: a new perspective on an interactive system*. Springer-Verlag, Berlin.

Claussen, M. and Gayler, V. 1997. The greening of Sahara during the mid-Holocene: results of an interactive atmosphere-biome model. *Global Ecology and Biogeography Letters*, 6, 369–377.

Claussen, M., Brovkin, V., Ganopolski, A., Kubatzki, C. and Petoukhov, V. 1998. Modelling global terrestrial vegetation–climate interaction. *Philosophical Transactions of the Royal Society of London B*, 353, 53–63.

Claussen, M., Kubatzki, C., Brovkin, V., Ganopolski, A., Hoelzmann, P. and Pachur, H.J. 1999. Simulation of an abrupt change in Saharan vegetation at the end of the mid-Holocene. *Geophysical Research Letters*, 24, 2037–2040.

Claussen, M., Mysak, L.A., Weaver, A.J., *et al.* 2002. Earth system models of intermediate complexity: closing the gap in the spectrum of climate system models. *Climate Dynamics*, 18, 579–586.

Clement, A.C., Seager, R. and Cane, M.A. 1999. Orbital controls on ENSO and the tropical climate. *Paleoceanography*, 14, 441–456.

Clement, A.C., Seager, R. and Cane, M.A. 2000. Suppression of El Niño during the mid-Holocene by changes in the earth's orbit. *Paleooceanography*, 15, 731–737.

Cleveland, W.S. 1993. *Visualizing data.* Hobart Press, Summit.

CLIMAP. 1981. Geological Society of America Map and Chart Series, MC-36. Geological Society of America.

CLIMAP Project Members. 1976. The surface of the ice-age earth. *Science*, 191, 1131–1137.

Clow, G.D. and Waddington, E.D. 2003. Reconstructing climatic changes at Summit, Greenland and Taylor Dome, Antarctica using borehole paleothermometry. Proceedings of International Symposium on Ice Cores and Climate, Kangerlussuaq, Greenland, August 2001. *Annals of Glaciology*, 35, special issue.

Clymo, R.S., Oldfield, F., Appleby, P.G., Pearson, G.W., Ratnesar, P. and Richardson, N. 1990. The record of atmospheric deposition on a rainwater-dependent peatland. *Philosophical Transactions of the Royal Society of London Series B- Biological Sciences*, 327, 331–338.

Cobb, K.M., Charles, C.D. and Hunter, D.E. 2001. A central tropical Pacific coral demonstrates Pacific, Indian, and Atlantic decadal climate connections. *Geophysical Research Letters*, 28, 2209–2212.

Coe, M.T. 1997. Simulating continental surface waters: an application to Holocene northern Africa. *Journal of Climate*, 10, 1680–1689.

Coe, M.T. and Bonin, G. 1997. Feedbacks between climate and surface water in northern Africa during the middle Holocene. *Journal of Geophysical Research*, 102D, 11087–11101.

Coetzee, J.A. 1964. Evidence for a considerable depression of vegetation belts during the upper Pleistocene on the east African mountains. *Nature*, 204, 564–566.

Cohen, A.L. and Hart, S.R. 1997. The effect of colony topography on climate signals in coral skeleton. *Geochimica et Cosmochimica Acta*, 61, 3905–3912.

Cohen, A.L., Layne, G.D., Hart, S.R. and Lobel, P.S. 2001. Kinetic control of skeletal Sr/Ca in a symbiotic coral: implications for the paleotemperature proxy. *Paleoceanography*, 16, 20–26.

Cohen, A.L., Owens, K.E., Layne, G.D. and Shimizu, N. 2002. The effect of algal symbionts on the accuracy of Sr/Ca paleotemperatures from coral. *Science*, 296, 331–333.

COHMAP Members, 1988. Climatic changes of the last 18,000 years: observations and model simulations. *Science*, 241, 1043–1052.

Cole, J.E. 2001. A slow dance for El Niño. *Science*, 291, 1496–1497.

Cole, J.E. and Cook, E.R. 1998. The changing relationship between ENSO variability and moisture balance in the continental United States. *Geophysical Research Letters*, 25, 4529–4532.

Cole, J.E. and Fairbanks, R.G. 1990. The Southern Oscillation recorded in the oxygen isotopes of corals from Tarawa Atoll. *Paleoceanography*, 5, 669–683.

Cole, J.E., Fairbanks, R.G. and Shen, G.T. 1993. The spectrum of recent variability in the Southern Oscillation: results from a Tarawa Atoll coral. *Science*, 262, 1790–1793.

Cole, J.E., Dunbar, R.B., McClanahan, T.R. and Muthiga, N.A. 2000. Tropical Pacific forcing of decadal SST variability in the western Indian Ocean over the past two centuries. *Science*, 287, 617–619.

Cole, J.E., Overpeck, J.T. and Cook, E.R. 2002. Multiyear La Niñas and prolonged US drought. *Geophysical Research Letters*, 29, 25-1–25-4.

Cole, K. 1990. Late Quaternary vegetation gradients through the Grand Canyon. In J.L. Betancourt, T. van Devender and P.S. Martin (eds), *Packrat middens. The last 40,000 years of biotic change.* University of Arizona Press, Tucson, pp. 240–258.

Cole-Dai, J., Mosley-Thompson, E., Wight, S.P. and Thompson, L.G. 2000. A 4100-year record of explosive volcanism from an East Antarctica ice core. *Journal of Geophysical Research*, 105, 24431–24441.

Coleman, D.C. and Fry, B. 1991. *Carbon isotope techniques.* Academic Press, New York.

Colinvaux, P.A. 1996. Quaternary environmental history and forest diversity in the neotropics. In J.B.C. Jackson, A.F. Budd and A.G. Coates (ed.), *Evolution and environment in tropical America.* University of Chicago Press, Chicago, pp. 425.

Colinvaux, P.A. and De Oliveira, P.E. 2000. Palaeoecology and climate of the Amazon basin during the last glacial cycle. *Journal of Quaternary Science,* 15, 347–356.

Colinvaux, P.A., Liu, K.B., De Oliveira, P.E., Bush, M.B., Miller, M.C. and Steinitz-Kannan, M. 1996. Temperature depression in the lowland tropics in glacial times. *Climate Change,* 32, 19–33.

Colinvaux, P.A., Bush, M.B., Steinitz-Kannan, M. and Miller, M.C. 1997. Glacial and post-glacial pollen records from the Ecuadorian Andes and Amazon. *Quaternary Research,* 69–78.

Colleta, P., Pentecost, A. and Spiro, B. 2001. Stable-isotopes in charophyte incrustations: relationships with climate and water chemistry. *Palaeogeography, Palaeoclimatology, Palaeoecology,* 173, 9–19.

Collins, M., Osborn, T. J., Tett, S. F. B., Briffa, K. R. and Schweingruber, F. H. 2002. A comparison of the variability of a climate model with paleotemperature estimates from a network of tree-ring densities. *Journal of Climate,* 15, 1497–1515.

Colman, S.M. 1981. Rock-weathering rates as a function of time. *Quaternary Research,* 15, 250–264.

Coltorti, M., Ficcarelli, G. and Torre, D. 1998. The last occurrence of Pleistocene megafauna in the Ecuadorian Andes. *Journal of South American Earth Sciences,* 11, 581.

Compton, J.S. 2001. Holocene sea-level fluctuations inferred from the evolution of depositional environments of the southern Langebaan Lagoon salt marsh, South Africa. *The Holocene,* 11, 395–406.

Conolly, A.P. and Dahl, E. 1970. Maximum summer temperature in relation to the modern and Quaternary distributions of certain arctic-montane species in the British Isles. In D. Walker and R.G. West (eds), *Studies in the Vegetational History of British Isles.* Cambridge University Press, Cambridge, pp. 159–223.

Conrad, V. and Pollak, L.D. 1962. *Methods in climatology.* Harvard University Press, Cambridge, MA.

Conte, M.H., Thompson, A., Eglinton, G. and Green, J.C. 1995. Lipid biomarker diversity in the coccolithophorid *Emiliania huxleyi* Prymnesiophyceae and the related species *Gephyrocapsa oceanica. Journal of Phycology,* 31, 272–282.

Conway, H., Hall, B.L., Denton, G.H., Gades, A.M. and Waddington, E.D. 1999. Past and future grounding-line retreat of the West Antarctic ice sheet. *Science,* 286, 280–283.

Cook, E.R. and Peters, K. 1997 Calculating unbiased tree-ring indices for the study of climatic and environmental change. *The Holocene,* 7, 361–370.

Cook, E.R., Meko, D.M., Stahle, D.W. and Cleaveland, M.K. 1999. Drought reconstructions for the continental United States. *Journal of Climate,* 12, 1145–1162.

Coope, G.R. 1977. Fossil coleopteran assemblages as sensitive indicators of climatic change during the Devensian last cold stage. *Philosophical Transactions of the Royal Society London B,* 280, 313–340.

Coope, G.R., Lemdahl, G., Lowe, J.J. and Walkling, A. 1998. Temperature gradients in northern Europe during the last glacial–Holocene transition 14–9 ^{14}C kyr BP interpreted from coleopteran assemblages. *Journal of Quaternary Science,* 13, 419–433.

Coplan, T.B., Kendall, C. and Hopple, J. 1982. Intercomparison of stable-isotope reference samples. *Nature,* 302, 236–238.

Corbet, A.S. and Pendlebury, H.M. 1978. *The butterflies of the Malay Peninsula.* Malayan Nature Society, Kuala Lumpur.

Correge, T., Delcroix, T., Recy, J., Beck, W., Cabioch, G. and Le Cornec, F. 2000. Evidence for stronger El Niño/Southern Oscillation ENSO events in a mid-Holocene massive coral. *Paleoceanography,* 15, 465–470.

Cowling, S.A. and Sykes, M.T. 1999. Physiological significance of low atmospheric CO_2 for plant-climate interactions. *Quaternary Research,* 52, 237–242.

Cowling, S.A., Maslin, M.A. and Sykes, M.T. 2001. Paleovegetation simulations of lowland Amazonia and implications for neotropical allopatry and speciation. *Quaternary Research,* 55, 140–149.

Cox, P.M., Betts, R.A., Jones, C.D., Spall, S.A. and Totterdell, I.J. 2000. Will carbon cycle feedbacks amplify global warming in the 21st century? *Nature,* 404, 184–187.

Craddock, J. M. 1976. Annual rainfall in England since 1725. *Quarterly Journal of the Royal Meteorological Society,* 102, 823–840.

Craig, H. 1957. Isotopic standards for carbon and oxygen and correction factors for mass spectromatic analysis of carbon dioxide. *Geochimica et Cosmochimica Acta*, 12, 133–149.

Craig, H. 1961. Isotopic variations in meteoric waters. *Science*, 133, 1702–1703.

Craig, H. 1965. The measurement of oxygen isotope palaeotemperatures. In E. Tongiorgi (ed.), *Stable-isotopes in oceanographic studies and palaeotemperatures.* Consiglio Nazionale delle Ricerche Laboratorio di Geologia Nucleare, Pisa, pp. 161–182.

Cranston, P.S. 1997. Identification guide to the Chironomidae of New South Wales. AWT Identification Guide Number 1. Australian Water Technologies Pty Ltd, West Ryde, pp. 1–376.

Cranston, P.S. and Martin, J. 1989. Family Chironomidae. In N.L. Evenhuis (ed.), *Catalog of the Diptera of the Australasian and Oceanic regions.* E.J. Brill and Bishop Museum Press, Leiden, pp. 252–274.

Cranwell, P.A. 1973. Chain-length distribution of n-alkanes from lake sediments in relation to post-glacial environmental change. *Freshwater Biology*, 3, 259–265.

Cranwell, P.A., Eglinton, G. and Robinson, N. 1987. Lipids of aquatic organisms as potential contributors to lacustrine sediments-II. *Organic Geochemistry*, 11, 513–527.

Crary, A.P. 1968. Presidential Address. International Symposium on Antarctic Glaciological Exploration (ISAGE), Hanover, New Hampshire, September 1968, x–xvi. W. Heffer and Sons, Cambridge, UK.

Cressman, G.P. 1959. An operational objective analysis system. *Monthly Weather Review*, 87, 347–364.

Criss, R.E. 1999. *Principles of stable-isotope distribution.* Oxford University Press, Oxford.

Cromack, M. 1991. Interpretation of laminated sediments from glacier-fed lakes, northwest Spitsbergen. *Norsk Geologiske Tidskrift*, 71, 129–132.

Crowley, T.J. 2000. Causes of climate change over the past 1000 years. *Science*, 289, 270–277.

Crowley, T.J. and Kim, K.-Y. 1999. Modeling the temperature response to forced climate change over the last six centuries. *Geophysical Research Letters*, 26, 1901–1904.

Crowley, T.J. and Lowery, T.S. 2000. How warm was the Medieval Warm Period? *Ambio*, 29, 51–54.

Crowley, T.J. and North, G.R. 1992. *Paleoclimatology.* Oxford University Press, New York.

CRU. 1987. Flavit et dissipati sunt – it blew and they were scattered: the Spanish Armada storms. *Chinook*, 9, 28–43.

Cubasch, U. 2002. Modelling the late Maunder minumum. *Abstracts, AGU Spring Meeting.* AGU, Washington, DC.

Cuffey, K.M. and Clow, G.D. 1997. Temperature, accumulation and ice sheet elevation in central Greenland through the last deglacial transition. *Journal of Geophysical Research*, 102, 26383–26396.

Cullen, H.M., deMenocal, P.D, Hemming, S., *et al.* 2000. Climate change and the collapse of the Akkadian empire: evidence from the deep-sea. *Geology*, 28, 379–82.

Cumming, B.F. and Smol, J.P. 1993. Development of diatom-based salinity odels for paleoclimatic research from lakes in British Columbia Canada. *Hydrobioloiga*, 269/270, 179–196.

Cunningham, W.L., Leventer, A., Andrews, J.T., Jennings, A.E. and Licht, K.J. 1999. Late Pleistocene–Holocene marine conditions in the Ross Sea, Antarctica: evidence from the diatom record. *The Holocene*, 9, 129–139.

Curry, B.B. 1999. An environmental tolerance index for ostracodes as indicators of physical and chemical factors in aquatic habitats. *Palaeogeography, Palaeoclimatology, Palaeoecology*, 148, 51–63.

Curtis, J.H., Brenner, M., Hodell, D.A., Balser, R.A., Islebe, G.A. and Hooghiemstra, H. 1998. A multi-proxy study of Holocene environmental change in the Maya lowlands of Peten, Guatemala. *Journal of Paleolimnology*, 19, 139–159.

Curtis, J.H., Brenner, M. and Hodell, D.A. 2001. Climate change in the Circum–Caribbean Late Pleistocene to present and implications for regional biogeography. In C.A. Woods and F.E. Sergile (eds), *Biogeography of the West Indies: patterns and perspectives*, 2nd Edn. CRC Press, Boca Raton, LA, pp. 35–54.

Cwynar, L.C. 1987. Fire and the forest history of the North Cascade range. *Ecology*, 68, 791–802.

Dahl, E. 1998. The phytogeography of northern Europe British Isles, Fennoscandia and adjacent areas. Cambridge University Press, Cambridge.

Dahl, S.O. and Nesje, A. 1992. Paleoclimatic implications based on equilibrium-line altitude depressions of reconstructed Younger Dryas and Holocene cirque glaciers in inner Nordfjord, western Norway. *Palaeogeography, Palaeoclimatology, Palaeoecology*, 94, 87–97.

Dahl, S.O. and Nesje, A. 1994. Holocene glacier fluctuations at Hardangerjøkulen, central southern Norway: a high-resolution composite chronology from lacustrine and terrestrial deposits. *The Holocene*, 4, 269–277.

Dahl, S.O. and Nesje, A. 1996. A new approach to calculating Holocene winter precipitation by combining glacier equilibrium-line altitudes and pine-tree limits: a case study from Hardangerjøkulen, central southern Norway. *The Holocene*, 6, 381–398.

Dahl, S.O., Nesje, A. and Øvstedal, J. 1997. Cirque glaciers as morphological evidence for a thin Younger Dryas ice sheet in east-central southern Norway. *Boreas*, 26, 161–180.

Dahl-Jensen, D. 1997. Reconstruction of the past climate from the GRIP temperature profile by Monte Carlo inversion. *Eos. Transactions AGU Fall Meeting, San Francisco, USA, 78, F6, oral presentation U12B–5.* AGU, Washington, DC.

Dahl-Jensen D., Johnsen, S.J., Hammer, C.U., Clausen, H.B. and Jouzel, J. 1993. Past accumulation rates derived from observed annual layers in the GRIP ice core from Summit, Central Greenland. In W.R. Peltier (ed.), *Ice in the climate system. Nato ASI series 1: Global Environmental Change 12.* Springer-Verlag, Berlin, pp. 517–532.

Dahl-Jensen, D., Gundestrup, N.S., Keller, K.R., *et al.* 1997. A search in north Greenland for a new ice-core drilling site. *Journal of Glaciology*, 43, 300–3–6.

Dahl-Jensen, D., Mosegaard, K., Gundestrup, N., *et al.* 1998. Past temperatures directly from the Greenland ice sheet. *Science*, 282, 268–271.

Dalfes, H.N., Kukla, G. and Wiess, H. (eds). 1997. *Third millennium BC climate change and old world collapse. NATO ASI series 1, 49.* Springer-Verlag, Berlin, pp. 723–733.

Damnati, B. and Taieb, M. 1995. Solar and ENSO signatures in laminated deposits from Lake Magadi Kenya during the Pleistocene/Holocene transition. *Journal of African Earth Sciences*, 21, 373–382.

Damon, P.E. and Peristykh, A.N. 2000. Radiocarbon calibration and application to geophysics, solar physics and astrophysics. *Radiocarbon*, 42, 137–150.

Dansgaard, W. 1964. Stable-isotopes in precipitation. *Tellus*, 16, 436–468.

Dansgaard, W., Johnsen, S.J., Clausen, H.B. and Langway Jr, C.C. 1971. Climate revealed by the Camp century ice core. In K.K. Turekian (ed.), *The Late Cenozoic glacial ages*. Yale University Press, New Haven.

Dansgaard, W., Johnsen, S.J., Clausen, H.B. and Gundestrup, N. 1973. Stable Isotope Glaciology. Meddelelser om Grønland, 197(2).

Dansgaard, W., Johnsen, S.J., Reeh, N., Gundestrup, N., Clausen, H.B. and Hammer, C.U. 1975. Climatic change, Norsemen and modern man. *Nature*, 255, 24–28.

Dansgaard, W., Clausen, H.B., Gundestrup, N., *et al.* 1982. A new Greenland deep ice core. *Science*, 218, 1273–1277.

Dansgaard, W., White, J.W.C. and Johnsen, S.J. 1989. The abrupt termination of the Younger Dryas climate event. *Nature*, 339, 532–534.

Dansgaard, W., Johnsen, S.J., Clausen, H.B., *et al.* 1993. Evidence of general instability of past climate from a 250-kyr ice-core record. *Nature*, 364, 218–220.

D'Arrigo, R., Frank, D., Jacoby, G. and Pederson, N. 2001. Spatial response to major volcanic events in or about AD 536, 934 and 1258: frost rings and other dendrochronological evidence from Mongolia and Northern Siberia. *Climatic Change*, 49, 239–246

D'Arrigo, R. D., Davi, N., Jacoby, G. and Wiles, G. 2002. A tree-ring temperature reconstruction from the Wrangell Mountains, Alaska (1593–1992): evidence for pronounced regional cooling during the Maunder Minimum. *Abstracts, AGU Spring Meeting*. AGU, Washington, DC.

Dauta, A., Devaux, J., Piquemal, F. and Boumnich, L. 1990. Growth rate of four freshwater algae in relation to light and temperature. *Hydrobiologia*, 207, 221–226.

David B. and Lourandos, H. 1998. Rock art and socio-demography in northeastern Australian prehistory. *World Archaeology*, 30, 193–219.

Davies, S.J., Metcalfe, S.E., Caballero, M.E. and Juggins, S. 2002. Developing diatom-based transfer functions for Central Mexican lakes. *Hydrobioloiga*, 467, 199–213.

Davis, M.B. 1981. Quaternary history and the stability of forest communities. In D.C. West, H.H. Shugart and D.B. Botkin (eds), *Forest succession: concepts and application*. Springer-Verlag, New York, pp. 132–154.

Davis, M.B. 1984. Climatic instability, time lags, and community disequilibrium. In J. Diamond and T.J. Case (eds), *Community ecology*. Harper and Row, New York, pp. 269–284.

Davis, M.B. 2000. Palynology after Y2K - understanding the source area of pollen in sediments. *Annual Review of Earth and Planetary Sciences*, 28, 1–18.

Davis, M.B. and Zabinski, C. 1992. Changes in geographic range resulting from greenhouse warming: effects on biodiversity in forests. In R. Peters and T. Lovejoy (eds), *Global warming and biodiversity*. Yale University Press, New Haven, pp. 297–308.

Davis, O.K. 1987. Recent developments in the study of arid lands. *Episodes*, 10, 41–42.

Davis, O.K. 1990. Caves as sources of biotic remains in arid western North America. *Palaeogeography, Palaeoclimatology, Palaeoecology*, 76, 331–348.

Dean, J.A., Gumerman, G.J., Epstein, J.M., *et al.* 1999a. Understanding Anasazi culture change through agent-based modeling. In T.A. Kohler and G.J. Gumerman (eds), *Dynamics of human and primate societies: agent-based modeling of social and spatial processes*. Oxford University Press, New York, pp. 179–205.

Dean, J.M., Kemp, A.E.S., Bull, D., Pike, J., Petterson, G. and Zolitschka, B. 1999b. Taking varves to bits: scanning electron microscopy in the study of laminated sediments and varves. *Journal of Paleolimnology*, 22, 121–136.

Dean, J.M., Kemp, A.E.S. and Pearce, R.B. 2001. Paleo-flux records from electron microscope studies of Holocene laminated sediments, Saanich Inlet, British Columbia. *Marine Geology*, 174, 139–158.

Dean, W.E. 1997. Rates, timing, and cyclicity of Holocene eolian activity in north-central United States: evidence from varved lake sediments. *Geology*, 25, 331–334.

Dean, W., Anderson, R., Bradbury, J.P. and Anderson, D. 2002. A 1500-year record of climatic and environmental change in Elk Lake, Minnesota I: varve thickness and gray-scale density. *Journal of Paleolimnology*, 27, 287–299.

De Busk, G.H.J. 1998. A 37,500-year pollen record from Lake Malawi and implications for the biogeography of Afromontane forests. *Journal of Biogeography*, 25, 479–500.

De Deckker, P. 1982a. Late Quaternary ostracods from Lake George, New South Wales. *Alcheringa*, 6, 305–318.

De Deckker, P. 1982b. Holocene ostracods, other invertebrates and fish remains from cores of four maar lakes in southeastern Australia. *Proceedings of the Royal Society of Victoria*, 94, 183–220.

De Deckker, P. 2002. Ostracod palaeoecology. In J.A. Holmes and A.R. Chivas (eds), *The Ostracoda: Applications in Quaternary Research*, Vol. 131, AGU Geophysical Monograph Series, Washington, DC, pp. 121–134.

De Deckker, P., Chivas, A.R. and Shelley, J.M.G. 1999. Uptake of Mg and Sr in the euryhaline ostracod *Cyprideis* determined from *in vitro* experiments. *Palaeogeography, Palaeoclimatology, Palaeoecology*, 148, 105–116.

DeGeer, G. 1912. A geochronology of the last 12,000 years. *Compte Rendu XI Session du Congres Geologique International Stockholm*, 1910, 241–257.

Deines, P. 1970. Mass spectrometer correction factors for the detection of small isotopic variation of carbon and oxygen. *International Journal of Mass Spectrometry and Ion Physics*, 4, 283–295.

Deleens, E., Ferhi, A. and Queiroz, O. 1983. Carbon isotope fractionation by plants using the C_4 pathway. *Physiologie Vegetale*, 21, 897–905.

Delorme, L. D. 1978. Distribution of freshwater ostracodes in Lake Erie. *Journal of Great Lakes Research*, 4, 216–220.

deMenocal, P. 2001. Cultural response to climate change during the Holocene. *Science*, 292, 667–673.

deMenocal, P., Oriz, J., Guilderson, T., *et al.* 2000a. Abrupt onset and termination of the African Humid Period: rapid climate responses to gradual insolation forcing. *Quaternary Science Reviews*, 19, 347–361.

deMenocal, P., J. Ortiz, T. Guilderson and M. Sarnthein 2000b. Coherent high- and low-latitude climate variability during the Holocene warm period. *Science*, 288, 2198–2202.

Denevan, W.M. 1976. The aboriginal population of Amazonia. In W.M. Denevan (ed.), *The native population of the Americas in 1492*. University of Wisconsin Press, Madison, WI, pp. 205–234.

Dennis, P.F., Rowe, P.J. and Atkinson, T.C. 1998. Stable-isotope composition of palaeoprecipitation and palaeogroundwaters from speleothem fluid inclusions. In: P. Murphy (ed) *Isotope techniques in the study of environmental change*. IAEA, Vienna, pp. 663–671.

Dennis, P.F., Rowe, P.J. and Atkinson, T.C. 2001. The recovery and isotopic measurement of water from fluid inclusions in speleothem. *Geochimica Cosmochimica Acta*, 65, 871–884.

Denniston, R.F., Gonzàles, L.A., Baker, R.G., *et al.* 1999. Speleothem evidence for Holocene fluctuations of the praire-forest ecotone, north-central USA. *The Holocene*, 9, 671–676.

Denniston, R.F., Gonzàlez, L.A., Asmerom, Y., Reagan, M.K. and Recelli-Snyder, H. 2000. Speleothem carbon isotopic records of Holocene environments in the Ozark Highlands, USA. *Quaternary International*, 67, 21–27.

deNoblet-Ducoudre, N., Claussen, M. and Prentice, I.C. 2000. Mid-Holocene greening of the Sahara: first results of the GAIM 6000-year BP experiment with two asynchronously coupled atmosphere/biome models. *Climate Dynamics*, 16, 643–659.

Denton, G.H. and Hughes, T.J. 1981. *The last great ice sheets*. Wiley, New York.

Dettinger, M.D., Battisti, D.S., Garreaud, R.D., McCabe, G.J.J. and Bitz, C.M. 2001. Interhemsipheric effects of interannual and decadal ENSO-like climate variations on the Americas. In V. Markgraf (ed.), *Interhemispheric climate linkages*. Academic Press, San Diego, pp. 1–16.

Deutsch, K.W., Platt, J. and Senghaas, D. 1971. Conditions favoring major advances in social science. *Science*, 171, 450–459.

Diaz, H.F. and Markgraf, V. 1992. *El Niño: historical and paleoclimatic aspects of the southern oscillation*. Cambridge University Press, Cambridge.

Dickinson, R.E. 1992. Changes in land use. In K.E. Trenberth (ed.), *Climate system modelling*. Cambridge University Press, Cambridge, pp. 698–700.

Dickson, J.A.D. 1978. Length-slow and length-fast calcite: a tale of two elongations. *Geology*, 6, 560–561.

Dickson, J.A.D. 1993. Inorganic calcite morphology: roles of fluid chemistry and fluid flow. Response to reply of discussion. *Journal of Sedimentary Petrology*, 63, 1160–1162.

Didyk, B.M., Simoneit, B.R.T., Brassell, S.C. and Eglinton, G. 1978. Organic geochemical indicators of palaeoenvironmental conditions of sedimentation. *Nature*, 272, 216–221.

Digerfeldt, G. 1988. Reconstruction and regional correlation of Holocene lake-level fluctuations in lake Bysjön, South Sweden. *BOREAS*, 17, 165–182.

Dimitriadis, S. and Cranston, P.S. 2001. An Australian Holocene climate reconstruction using Chironomidae from a tropical volcanic maar lake. *Palaeogeography, Palaeoclimatology, Palaeoecology*, 176, 109–131.

Doherty, R., Kutzbach, J., Foley, J. and Pollard, D. 2000. Fully coupled climate/dynamical vegetation model simulations over Northern Africa during the mid-Holocene. *Climate Dynamics*, 16, 561–573.

Domack, E., Leventer, A., Dunbar, R., *et al.* 2001. Chronology of the Palmer Deep site, Antarctic Peninsula: a Holocene palaeoenvironmental reference for the circum-Antarctic. *The Holocene*, 11, 1–9.

Dominguez, S. 2002. Optimal gardening strategies: maximizing the input and retention of water in prehistoric gridded fields in North Central New Mexico. *World Archaeology*, 34, 131–163.

Donnelly, J.P., Bryant, S.S., Butler, J., *et al.* 2001. 700-yr sedimentary record of intense hurricane landfalls in southern New England. *Geological Society of America Bulletin*, 113, 714–727.

Dorale, J.A., Gonzalez, L.A. and Reagan, M.K. 1992. A high-resolution record of Holocene climate change in speleothem calcite from Cold Water Cave, Northeast Iowa. *Science*, 258, 1626–1630.

Doran, P.T., Priscu, J.C., Lyons, W.B., *et al.* 2002. Antarctic climate cooling and terrestrial ecosystem response. *Nature*, 415, 517–520.

Douglas, B.C. 1991. Global sea level rise. *Journal of Geophysical Research*, 96, 6981–6992.

Douglas, B.C. 2001. Sea level change in the era of the recording tide gauge. In B.C. Douglas, M.S. Kearney and S.P. Leatherman (eds), *Sea level rise: history and consequences*. Academic Press, San Diego, pp. 37–64.

Douglas, M.S.V., Smol, J.P. and Blake Jr, W. 1994. Marked post-18th century environmental change in high-arctic ecosystems. *Science*, 266, 416–419.

Douglass, A.E. 1919. *Climatic cycles and tree growth I*. Carnegie Institute, Washington, DC.

Dowdeswell, J.A, Hagen, J.O., Bjornsson, H., *et al.* 1997. The mass balance of circum-Arctic glaciers and recent climate change. *Quaternary Research*, 48, 1–14.

Drewry, D.J. and Morris, E.M. 1992. The response of large ice sheets to climatic change. *Philosophical Transactions of the Royal Society of London B*, 338, 235–242.

Dreybrodt, W. 1988. Processes in karst systems. Physics, chemistry and geology. Springer-Verlag, Berlin.

Dreybrodt, W. 1999. Chemical kinetics, speleothem growth and climate. *Boreas*, 28, 347–356.

Druffel, E.M. 1981. Radiocarbon in annual coral rings from the eastern tropical Pacific Ocean. *Geophysical Research Letters*, 8, 59–62.

Druffel, E.R.M. 1985. Detection of El Niño and decade time-scale variations of sea surface temperature from banded coral records: implications for the carbon dioxide cycle. In W.S. Broecker and E.T. Sundquist (eds), *The carbon cycle and atmospheric CO_2: natural variations Archean to present. Geophysical monograph.* American Geophysical Union, Washington DC, pp. 111–122.

Druffel, E.R.M. 1987. Bomb radiocarbon in the Pacific: annual and seasonal timescale variations. *Journal of Marine Research*, 45, 667–698.

Druffel, E.R.M. 1997. Pulses of rapid ventilation in the North Atlantic surface ocean during the past century. *Science*, 275, 1454–1457.

Druffel, E.R.M. and Griffin, S. 1993. Large variations of surface ocean radiocarbon: evidence of circulation changes in the southwestern Pacific. *Journal of Geophysical Research*, 98, 20249–20259.

Druffel, E.R.M. and Griffin, S. 1999. Variability of surface ocean radiocarbon and stable-isotopes in the southwestern Pacific. *Journal of Geophysical Research*, 104, 23607–23613.

Druffel, E.R.M. and Linick, T.W. 1978. Radiocarbon in annual coral rings of Florida. *Geophysical Research Letters*, 5, 913–916.

Drummond, C.N., Patterson, W.P. and Walker, J.C.G. 1995. Climatic forcing of carbon-oxygen isotopic covariance in temperate-region marl lakes. *Geology*, 23, 1031–1034.

Dubois, A.D. and Ferguson, D.K. 1985. The climatic history of pine in the Cairngorms based on radiocarbon dates and stable-isotope analyses, with an account of events leading up to its colonisation. *Review of Palaeobotany and Palynology*, 46, 55–80.

Dugmore, A.J. 1989. Tephrachronological studies of Holocene glacier fluctuations in south Iceland. In J. Oerlemans (ed.), *Glacier fluctuations and climatic change*. Kluwer Academic Publishers, Dordrecht, pp. 37–55.

Dugmore, A.J., Larsen, G. and Newton, A.J. 1995. Seven Tephra isochrones in Scotland. *The Holocene*, 5, 257–266.

Dulinski, M. and Rozanski, K. 1990. Formation of $^{13}C/^{12}C$ isotope ratios in speleothems: a semi-dynamic model. *Radiocarbon*, 32, 7–16.

Dunbar, R. 2001. Kenya corals record western Indian Ocean variability since 1697 AD. Abstracts of the First ARTS Open Sciences Meeting, IRD Research Center, Noumea, 4–7 November, 2001, 24–27.

Dunbar, R. and Cole, J. 1999. Annual Records of Tropical Systems (ARTS). Recommendations for research. *PAGES Workshop Report*, Series 99–1.

Dunbar, R.B., Wellington, G.M., Colgan, M.W. and Glynn, P.W. 1994. Eastern Pacific sea surface temperature since 1600 A.D.: the $\delta^{18}O$ record of climate variability in Galapagos corals. *Paleoceanography*, 9, 291–316.

Dunwiddie, P.W. 1987. Macrofossil and pollen representation of coniferous trees in modern sediments from Washington. *Ecology*, 68, 1–11.

Dupont, L.M. 1986. Temperature and rainfall variation in the Holocene based on comparative palaeoecology and isotope geology of a hummock and a hollow Boutangerveen, The Netherlands. *Review of Palaeobotany and Palynology*, 48, 71–159.

Dupont, L.M. and Brenninkmeijer, C.A.M. 1984. Palaeobotanic and isotopic analysis of late subboreal and early subatlantic peat from Engbertsdijksveen-VII, the Netherlands. *Review of Palaeobotany and Palynology*, 41, 241.

Dyke, A.S. and Morris, T.F. 1997. Post-glacial history of the Bowhead whale and of driftwood penetration: implications for paleoclimate, Central Canadian Arctic. Paper 89–24. Geological Survey of Canada, Ottawa.

Dyke, A.S., Hooper, J. and Savelle, J.M. 1996. A history of sea-ice in the Canadian Arctic Archipelago based on the post-glacial remains of the bowhead whale *Balaena mysticetus*. *Arctic*, 49, 235–255.

Easterling, D.R., Horton, B., Jones, P.D., *et al.* 1997. A new look at maximum and minimum temperature trends for the globe. *Science*, 277, 364–367.

Edlund, M.B., Stoermer, E.F. and Pilskaln, C.H. 1995. Siliceous microfossil succession in the recent history of two basins in Lake Baikal, Siberia. *Journal of Paleolimnology*, 14, 165–184.

Edwards, R.J. and Horton, B.P. 2000. Reconstructing RSL change using UK salt-marsh foraminifera. *Marine Geology*, 169, 41–56.

Edwards, R.L., Chen, J.H., Ku, T.-L. and Wasserburg, G.J. 1987. Precise timing of the last interglacial period from mass spectrometric determination of thorium-230 in corals. *Science*, 236, 1547–1553.

Edwards, T.W.D., Wolfe, B.B. and MacDonald, G.M. 1996. Influence of changing atmospheric circulation on precipitation $\delta^{18}O$ relations in Canada during the Holocene. *Quaternary Research*, 46, 211–218.

Efron, B. and Tibshirani, R.J. 1993. *An introduction to the bootstrap*. Chapman and Hall, London.

Eglinton, G. and Calvin, M. 1967. Chemical fossils. *Scientific American*, 261, 32–43.

Eglinton, G. and Hamilton, R.J. 1963. The distribution of alkanes. In T. Swain (ed.), *Chemical plant taxonomy*, Academic Press, London, pp.187–208.

Eglinton, G., Bradshaw, S.A., Rosell, A., Sarnthein, M., Pflaumann, U. and Tiedemann, R. 1992. Molecular record of secular sea surface temperature changes on 100-year time-scales for glacial terminations I, II and IV. *Nature*, 356, 423–426.

Eglinton, T.I., Aluwihare, L.I., Bauer, J.E., Druffel, E.R.M. and McNichol, A.P. 1996. Gas chromatographic isolation of individual compounds from complex matrices for radiocarbon dating. *Analytical Chemistry*, 68, 904–912.

Eglinton, T.I., Benitez-Nelson, B.C., Pearson, A., Nichol, A.P., Bauer, J.E. and Druffel, E.R.M. 1997. Variability in radiocarbon ages of individual organic compounds from marine sediments. *Science*, 277, 796–799.

Eicher, U. and Siegenthaler, U. 1976. Palynological and oxygen isotope investigation on Late-Glacial sediment cores from Swiss lakes. *BOREAS*, 5, 109–117.

Eicher, U. and Siegenthaler, U. 1983. Stable-isotopes in lake marl and mollusc shells from Lobsigensee Swiss Plateau. Studies in the late Quaternary of Lobsigensee 6. *Revue de Paléobiologie*, 2, 217–220.

Eide, W., Bigelow, N.H., Peglar, S.M., Birks, H.H. and Birks, H.J.B. 2004. Holocene tree migrations in the Setesdalen valley, southern Norway, reconstructed from macrofossil and pollen evidence. *Vegetation History and Archaeobotany*, in press.

Eischeid, J.D., Diaz, H.F., Bradley, R.S. and Jones, P.D. 1991. *A comprehensive precipitation data set for global land areas. DOE/ER-6901T-H1, TR051.* US Department of Energy, Carbon Dioxide Research Program, Washington, DC.

Eitel, B., Blümel, W.D., Hüser, K. and Mauz, B. 2001. Dust and loessic alluvial deposits in Northwestern Namibia Damaraland, Kaokoveld: sedimentology and palaeoclimatic evidence based on luminescence data. *Quaternary International*, 76/77, 57–65.

Elenga, H., Schwartz, D. and Vincens, A. 1994. Pollen evidence for Late Quaternary vegetation and inferred climate changes in Congo. *Palaeogeography, Palaeoclimatology, Palaeoecology*, 109, 345–346.

Elias, S.A. 1997. The mutual climatic range method of palaeoclimatic reconstruction based on insect fossils: new applications and interhemispheric comparisons. *Quaternary Science Reviews*, 16, 1217–1225.

Elias, S.A. and Wilkinson, B. 1983. Lateglacial insect fossil assemblages from Lobsigensee Swiss Plateau. Studies in the late Quaternary of Lobsigensee 3. *Revue de Paléobiologie*, 2, 189–204.

Eltahir, E.A.B. 1996. Role of vegetation in sustaining large-scale atmospheric circulation in the tropics. *Journal of Geophysical Research*, 101(D2), 4255–4268.

Eltahir, E.A.B. and Gong, C. 1996. Dynamics of wet and dry years in West Africa. *Journal of Climate*, 9, 5, 1030–1042.

Emeis, K.C., Struck, U., Schulz, H.M., *et al.* 2000. Temperature and salinity variations of Mediterranean Sea surface waters over the last 16,000 years from records of planktonic stable oxygen isotopes and alkenone unsaturation ratios. *Palaeogeography, Palaeoclimatology, Palaeoecology*, 158, 259–280.

Engeset, R.V., Elvehoy, H., Andreassen, L.M., *et al.* 2000. Modelling of historic variations and future scenarios of the mass balance of Svartisen ice cap, northern Norway. *Annals of Glaciology*, 31, 97–103.

Engstrom, D.R. and Nelson, S. 1991. Paleosalinity from trace metals in fossil ostracodes compared with observational records at Devils Lake, North Dakota. *Palaeogeography, Palaeoclimatology, Palaeoecology*, 83, 295–312.

Engstrom, D.R., Fritz, S.C., Almendinger, J.E. and Juggings, S. 2000. Chemical and biological trends during lake evolution in recently deglaciated terrain. *Nature*, 408, 161–166.

Enmar, R., Stein, M., Bar-Matthews, M., Sass, E., Katz, A. and Lazar, B. 2000. Diagenesis in live corals from the Gulf of Aqaba. I. The effect on paleo-oceanography tracers. *Geochimica Et Cosmochimica Acta*, 64, 3123–3132.

Epstein, S., Buchsbaum, R., Lowenstam, H.A. and Urey, H.C. 1953. Revised carbonate water isotopic temperature scale. *Geological Society American Bulletin*, 64, 1315–1326.

Erez, J. and Luz, B. 1983. Experimantal paleotemperature equation for planktonic foraminifera. *Geochimica et Cosmochimica Acta*, 47, 1025–1031.

Eronen, M. and Zetterberg, P. 1996. Expanding megafossil-data on Holocene changes at the polar/alpine pine limit in northern Fennoscandia. *Paläoklimaforschung*, 20, 127–134.

Eronen, M., Hyvärinen, H. and Zetterberg, P. 1999. Holocene humidity changes in northern Finnish Lapland inferred from lake sediments and submerged Scots pines dated by tree-rings. *The Holocene*, 9, 569–580.

Eugster, H.P. and Jones, B.F. 1979. Behavior of major solutes during closed-basin brine evolution. *American Journal of Science*, 279, 609–631.

Evans, J.G., Limbrey, S., Máté, J., *et al.* 1993. An environmental history of the Upper Kennet Valley, Wiltshire, for the last 10,000 years. *Proceedings of the Prehistoric Society*, 59, 139–195.

Evans, M.N., Fairbanks, R.G. and Rubenstone, J.L. 1999. The thermal oceanographic signal of ENSO reconstructed from a Kiritimati Island coral. *Journal of Geophysical Research*, 104, 13409–13421.

Evans, M., Kaplan, A. and Cane, M.A. 2000. Intercomparison of coral oxygen isotope data and historical sea surface temperature SST: potential for coral-based SST field reconstructions. *Paleoceanography*, 15, 551–563.

Evans, M., Cane, M.A., Schrag, D.P., *et al.* 2001. Support for tropically-driven Pacific decadal variability based on paleoproxy evidence. *Geophysical Research Letters*, 28, 3689–3692.

Evans, M., Kaplan, A. and Cane, M. 2002. Pacific sea surface temperature field reconstruction from coral $\delta^{18}O$ data using reduced space objective analysis. *Paleoceanography*, 17, 7-1–7-13.

Fahl, K. and Stein, R. 1999. Biomarkers as organic-carbon and environmental indicators in the Late Quaternary Arctic Ocean: problems and perspectives. *Marine Chemistry*, 63, 293–309.

Fairbanks, R.G. 1989. A 17,000-year glacio-eustatic sea-level record: influence of glacial melting rates on the Younger Dryas event and deep-ocean circulation. *Nature*, 342, 637–642.

Fairbanks, R.G. and Dodge, R.E. 1979. Annual periodicity of the $^{18}O/^{16}O$ and $^{13}C/^{12}C$ ratios in the coral *Montastrea annularis*. *Geochimica et Cosmochimica Acta*, 43, 1009–1020.

Fairbanks, R.G., Evans, M.N., Rubenstone, J.L., *et al.* 1997. Evaluating climate indices and their geochemical proxies measured in corals. *Coral Reefs*, 16, S93–S100.

Fairbridge, R.W. 1961. Eustatic changes in sea level. In L.H. Ahrens, F. Press, K. Raukawa and S.K. Runcorn (eds), *Physics and Chemistry of the Earth*, Vol. 4. Pergamon, New York, pp. 99–185.

Fairbridge, R.W. 1987. The spectre of sea level in a Holocene time frame. In M.R. Rampino, J.E. Sanders, W.S. Newman and L.K. Kongsson (eds), *Climate: history, periodicity and predictability*, Van Nostrand, New York, pp.127–142.

Fairbridge, R.W. 1995. Some personal reminiscences on of the idea of cycles, especially in the Holocene. In C.W. Finkl Jr (ed.), *Holocene cycles. Journal of Coastal Research Special Issue*, 17, 11–20.

Fairchild, I.J., Borsato, A., Tooth, A.F., *et al.* 2000. Controls on trace element Sr-Mg. compositions of carbonate cave waters: implications for speleothem climatic records. *Chemical Geology*, 166, 255–269.

Fall, P.L, Lindquist, C.A. and Falconer, S.E.1990. Fossil hyrax middens from the Middle East: a record of palaeovegetation and human disturbance. In J.L. Betancourt, T. van Devender and P.S. Martin (eds), *Packrat middens. The last 40,000 years of biotic change*. University of Arizona Press, Tucson, pp. 409–427.

Fallon, S.J., McCulloch, M.T., van Woesik, R. and Sinclair, D.J. 1999. Corals at their latitudinal limits: laser ablation trace element systematics in *Porites* from Shirigai Bay, Japan. *Earth and Planetary Science Letters*, 172, 221–238.

Fallon, S.J., White, J.C. and McCulloch, M.T. 2002. *Porites* corals as recorders of mining and environmental impacts: Misima Island, PapuaNew Guinea. *Geochimica Cosmochimica Acta*, 66, 45–62.

Farrimond, P., Head, I.M. and Innes, H.E. 2000. Environmental influence on the biohopanoid composition of recent sediments. *Geochimica et Cosmochimica Acta*, 64, 2985–2992.

Felis, T., Patzold, J., Loya, Y., Fine, M., Nawar, A.H. and Wefer, G. 2000. A coral oxygen isotope record from the northern Red Sea documenting NAO, ENSO, and North Pacific teleconnections on Middle East climate variability since the year 1750. *Paleoceanography*, 15, 679–694.

Ferguson, C.W. and Graybill, D.A. 1983, Dendrochronology of Bristlecone Pine: a progress report. *Radiocarbon*, 25, 287–288.

Ficken, K.J., Street-Perrot, F.A., Perrott, R.A., Swain, D.L., Olago, D.O. and Eglinton, G. 1998. Glacial/interglacial variations in carbon cycling revealed by molecular and isotope stratigraphy of Lake Nkunga, Mt Kenya, East Africa. *Organic Geochemistry*, 29, 1701–1719.

Ficken, K.J., Li, B., Swain, D.L. and Eglinton, G. 2000. An *n*-alkane proxy for the sedimentary input of submerged/floating freshwater aquatic macrophytes. *Organic Geochemistry*, 31, 745–749.

Figge, R. and White, J.W.C. 1995. High-resolution Holocene and late glacial atmospheric CO_2 record: variability tied to changes in thermohaline circulation. *Global Biogeochemical Cycles*, 9, 391–403.

Fischer, D.H. 1996. *The great wave: price revolutions and the rhythm of history*. Oxford University Press, Oxford.

Fischer, G. and Wefer, G. (eds), 1999. *Uses of proxies in paleoceanography: examples from the South Atlantic*. Springer-Verlag, Berlin, pp. 735.

Fischer, H. 2001. Imprint of large-scale atmospheric transport patterns on sea-salt records in Northern Greenland ice cores. *Journal of Geophysical Research*, 106, 23977–23984.

Fisher, D.A. 1990. A zonally averaged isotope model coupled to a regional variable elevation stable-isotope model. *Annals of Glaciology*, 14, 65–71.

Fisher, D.A. 1992. Possible ice-core evidence for a fresh melt-water cap over the Atlantic Ocean in the early Holocene. In E. Bard and W.S. Broecker (eds), *The Last Deglaciation: Absolute and Radiocarbon Chronologies. Nato ASI series 1: Global Environmental Change 12.* Springer-Verlag, Berlin, pp. 267–293.

Fisher, D.A. 2002. High resolution multiproxy climatic records from ice cores, tree-rings, corals and documentary sources using eigenvector techniques and maps: assessment of recovered signal and errors. *The Holocene*, 12, 401–419.

Fisher, D.A., Reeh, N. and Clausen, H.B. 1985. Stratigraphic noise in time series derived from ice cores. *Annals of Glaciology*, 7, 76–86.

Fisher, D.A., Koerner, R.M, Kuivinen, K., *et al.* 1996. Inter-comparison of ice core $\delta^{18}O$. and precipitation records from sites in Canada and Greenland over the last 3500 years and over the last few centuries in detail using EOF techniques. In P.D. Jones, R.S. Bradley and J. Jouzel (eds), *Climate Variations and Forcing Mechanisms of the Last 2000 Years. NATO ASI Series I, Vol. 41.* Springer-Verlag, Berlin, pp. 297–328.

Fisher, D.A., Koerner, R.M., Bourgeois, J.C., *et al.* 1998. Penny Ice Cap cores, Baffin Island, Canada, and the Wisconsinan Foxe Dome connection: two states of Hudson Bay ice cover. *Science*, 279, 692–695.

Fisher, D.A., Koerner, R.M. and Reeh, N. 1995. Holocene climate records from Agassiz Ice Cap, Ellesmere Island, NWT, Canada. *The Holocene*. 5, 19–24.)

Fitzhugh, B. 2001. Risk and invention in human technological evolution. *Journal of Anthropological Archaeology*, 20, 125–167.

Flannery, K.V. 1965. The ecology of early food production in Mesopotamia. *Science*, 147, 1247–1256.

Fleitmann, D., Burns, S.J., Mudelsee, M., *et al.* 2002. Holocene variability in the Indian Ocean Monsoon: a stalagmite-based, high-resolution oxygen isotope record from southern Oman. *PAGES Newsletter*, 10, 7–8.

Flenley, J.R. 1979. *A geological history of tropical rainforest*. Butterworths, London.

Flint, R.F. 1959. Pleistocene climates in eastern and southern Africa. *Geological Society of America Bulletin*, 70, 343–374.

Flower, R.J. 1993. Diatom preservation: experiments and observations on dissolution and breakage in modern and fossil material. *Hydrobiologia*, 269/270, 473–484.

Foged, N. 1964. Freshwater diatoms from Sptizbergen. Tromsö Museums Skrifter, No. 11. Universitetsforlaget, Tromsö/Oslo, pp. 205.

Fogel, M.L. and Cifuentes, L.A. 1993. Isotope fractionation during primary production. In M.H. Engel and S.A. Macko (eds), *Organic geochemistry principles and applications*. Plenum Press, New York, pp. 73–98.

Foley, J.A. 1994. The sensitivity of the terrestrial biosphere to climate change: a simulation of the middle Holocene. *Global Biogeochemical Cycles*, 8, 505–525.

Foley, J.A., Kutzbach, J.E., Coe, M. and Levi, S. 1994. Feedbacks between climate and boreal forests during the Holocene epoch. *Nature*, 371, 52–53.

Folk, R. and Assereto, R. 1976. Comparative fabrics of length-slow and length-fast calcite and calcitized aragonite in a Holocene speleothem, Carlsbad Caverns, New Mexico. *Journal of Sedimentary Petrology*, 56, 486–496.

Ford, D.C. and Williams, P.W. 1989. *Karst geomorphology and hydrology*. Unwin Hyman, London.

Forester, R.M. and Brouwers, E.M. 1985. Hydrochemical parameters governing the occurrence of estuarine and marginal esturine ostracodes: an example from south-central Alaska. *Journal of Paleontology*, 59, 344–369.

Forester, R.M. and Smith, A.J. 1994. Late Glacial climate estimates for southern Nevada: the ostracode fossil record. In *The 5th Annual International High-Level Radioactive Waste Management Conference and Exposition, Las Vegas*, pp. 2553–2561.

Forester, R.M., Delorme, L.D. and Bradbury, J.P. 1987. Mid-Holocene climate in northern Minnesota. *Quaternary Research*, 28, 263–273.

Forman, R.T.T. 1964. Growth under controlled conditions to explain the hierarchical distributions of a moss, *Tetraphis pellucida*. *Ecological Monographs*, 34, 1–25.

Forsythe, W., Breen, C., Callaghan, C. and McConkey, R. 2000. Historic storms and shipwrecks in Ireland: a preliminary survey of severe synoptic conditions as a causal factor in underwater archaeology. *International Journal of Nautical Archaeology*, 29, 247–259.

Forti, P. 2001. Speleology in the next milennium. Presidents Address, *13th International Speleological Congress (UIS), Brasilia, July 2001*.

Fotheringham, A.S., Brundson, C. and Charlton, M. 2000. *Quantitative geography: perspectives on spatial data analysis*. Sage Publications, London.

Francus, P. 1998. An image-analysis technique to measure grain-size variation in thin sections of soft clastic sediments. *Sedimentary Geology*, 121, 289–298.

Francus, P. and Asikainen, C.A. 2001. Sub-sampling unconsolidated sediments: a solution for the preparation of undisturbed thin-sections from clay-rich sediments. *Journal of Paleolimnology*, 26, 323–326.

Francus, P., Bradley, R.S., Abbott, M.B., Patridge, W. and Keimig, F. 2002. Palaeoclimate studies of minerogenic sediments using annually resolved textural parameters. *Geophysical Research Letters*, 29, 20, 1998

Francus, P., Keimig, F. and Besonen, M. 2002. A algorithm to aid varve counting and measurement from thin sections. *Journal of Paleolimnology*, 28, 283–286

Frank, A.H.E. 1969. Pollen stratigraphy of the lake of Vico central Italy. *Palaeogeography, Palaeoclimatology, Palaeoecology*, 6, 67–85.

Free, M. and Robock, A. 1999. Global warming in the context of the Little Ice Age. *Journal of Geophysical Research*, 104, 19057–19070.

Freeman, K.H., Hayes, J.M., Trendel, J.-M. and Albrecht, P. 1990. Evidence from carbon isotope measurements for diverse origins of sedimentary hydrocarbons. *Nature*, 343, 254–256.

Frenzel, B., Pesci, M., Velichko, A.A. (eds), 1992. Atlas of paleoclimates and paleoenvironments of the Northern Hemisphere: Late Pleistocene–Holocene. Geographical Research Institute, Budapest.

Freund, H., Birks, H.H. and Birks, H.J.B. 2001. The identification of wingless *Betula* fruits in Weichselian sediments in the Gross Todshorn borehole Lower Saxony, Germany – the occurrence of *Betula humilis* Schrank. *Vegetation History and Archaeobotany*, 10, 107–115.

Frich, P., Alexander, L.V., Della-Martin, P., *et al.* 2002. Observed coherent changes in climatic extremes during the second half of the twentieth century. *Climate Research*, 19, 193–212.

Friedman, M. and O'Neil J.R. 1977. Compilation of stable-isotope fractionation factors of geochemical interest. In M. Fleischer (ed.), *Data of Geochemistry*, 6th Edn. Geological Survey Professional Paper 440-KK.

Friedrich, M., Kromer, B., Spurk, M., Hofmann, J. and Kaiser, K.F. 1999 Palaeo-environment and radiocarbon calibration as derived from Late glacial/Early Holocene tree-ring chronologies. *Quaternary International*, 61, 27–39

Friis-Christensen, E., Frohlich, C., Haigh, J.D., Schussler, M. and von Steiger, R. 2000. Solar variability and climate. Kluwer, Dordrecht (reprinted from *Space Science Reviews*, 94).

Frisia, S. 1996. TEM and SEM investigation of speleothem carbonates: another key to the interpretation of environmental parameters. *Karst Waters Institute Special Publication*, 2, 33–34.

Frisia, S., Borsato, A., Fairchild, I.J. and McDermott, F. 2000. Calcite fabrics, growth mecanisms, and environments of formation in speleothems from the Italian Alps and soutwestern Ireland. *Journal of Sedimentary Research*, 70, 1183–1196.

Fritts, H.C. 1991. *Reconstructing large-scale climatic patterns from tree-ring data.* University of Arizona Press, Tucson.

Fritz, P. and Poplawski, S. 1974. ^{18}O and ^{13}C in the shells of freshwater molluscs and their environments. *Earth and Planetary Science Letters*, 24, 91–98.

Fritz, P., Morgan, A.V., Eicher, U. and McAndrews, J.H. 1987. Stable-isotope, fossil Coleoptera and pollen stratigraphy in late Quaternary sediments from Ontario and New York State. *Palaeogeography, Palaeoclimatology, Palaeoecology*, 58, 183–202.

Fritz, S.C. 1996. Paleolimnological records of climate change in North America. *Limnology and Oceanography*, 41, 882–889.

Fritz, S.C., Juggins, S., Battarbee, R.W. and Engstrom, D.R. 1991. Reconstruction of past changes in salinity and climate using a diatom-based transfer function. *Nature,* 352, 706–708.

Fritz, S.C., Juggins, S. and Battarbee, R.W. 1993. Diatom assemblages and ionic characterisation of lakes of the Northern Great Plains, North America: a tool for reconstructing past salinity and climate fluctuations. *Canadian Journal of Fisheries and Aquatic Sciences*, 50, 1844–1856.

Fritz, S.C., Engstrom, D.R. and Haskell, B.J. 1994. 'Little Ice Age' aridity in the North American Great Plains: a high-resolution reconstruction of salinity fluctuations from Devils Lake, North Dakota, USA. *The Holocene*, 4, 69–73.

Fritz, S.C., Cumming, B.R., Gasse, F. and Laird, K.L. 1999. Diatoms as indicators of hydrologic and climatic change in saline lakes. In E.F. Stoermer and J.P. Smol (eds), *The diatoms: applications for the environmental and earth sciences.* Cambridge University Press, Cambridge, pp. 41–72.

Fritz, S.C., Ito, E., Yu, Z., Laird, K.R. and Engstrom, D.R. 2000. Hydrologic variation in the northern Great Plains during the last two millennia. *Quaternary Research*, 53, 175–184.

Fritz, S.L., Metcalfe, S.E. and Dean, W. 2001. Holocene climate patterns in the Americas inferred from paleolimnological records. In V. Markgraf (ed.), *Interhemispheric climate linkages.* Academic Press, San Diego, pp. 241–263.

Frogley, M.R., Tzedakis, P.C. and Heaton, T.H.E. 1999. Climate variability in northwest Greece during the last interglacial. *Science*, 285, 1886–1889.

Fröhlich, C. 2000. Observations of irradiance variations. *Space Science Reviews*, 94, 15–24.

Fröhlich, C. and Lean, J. 1998. The sun's total irradiance: cycles and trends in the past two decades and associated climate change uncertainties. *Geophysical Research Letters*, 25, 4377–4380.

Frumkin, A., Carmi, I., Gopher, A., Ford, D.C., Schwarcz, H.P. and Tsuk, T. 1999. A Holocene millennial-scale climatic cycle from a speleothem in Nahal Qanah Cave, Israel. *The Holocene*, 9, 677–682.

Fujii, Y., Kamiyama, K., Kawamura, T., *et al.* 1990. 6000-year climate records in an ice core from the Høghetta ice dome in northern Spitsbergen. *Annals of Glaciology*, 14, 85–89.

Furbish, D.J. and Andrews, J.T. 1984. The use of hypsometry to indicate long-term stability and response of valley glaciers to change in mass transfer. *Journal of Glaciology*, 30, 199–211.

Gagan, M.K., Chivas, A.R. and Isdale, P. 1994. High-resolution isotopic records from corals using ocean temperature and mass spawning chronometers. *Earth and Planetary Sciences*, 121, 549–558.

Gagan, M.K., Chivas, A.R. and Isdale, P.J. 1996. Timing coral-based climatic histories using ^{13}C enrichments driven by synchronized spawning. *Geology*, 24, 1009–1012.

Gagan, M.K., Ayliffe, L.K., Hopley, D., *et al.* 1998. Temperature and surface-ocean water balance of the mid-Holocene tropical western Pacific. *Science*, 279, 1014–1018.

Gagan, M.K., Ayliffe, L.K., Beck, J.W., *et al.* 2000. New views of tropical paleoclimates from corals. *Quaternary Science Reviews*, 19, 45–64.

Gaillard, M.-J. 1985. Post-glacial palaeoclimatic changes in Scandinavia and Central Europe. A tentative correlation based on studies of lake-level fluctuations. *Ecologia Mediterranea*, 11, 159–175.

Gajewski, K., Vance, R., Sawada, M., *et al.* 2000. The climate of North America and adjacent ocean waters ca. 6 ka. *Canadian Journal of Earth Sciences*, 37, 661–681.

Gallée, H., van Ypersele, J. P., Fichefet, T., Tricot, C. and Berger, A. 1991. Simulation of the last glacial cycle by a coupled, sectorially averaged climate-ice sheet model. I. The climate model. *Journal of Geophysical Research-Atmospheres*, 96, 13139–13161.

Gams, H. and Nordhagen, R. 1923. Postglaziale Klimaänderungen und Erdkrustenbewegungen in Mitteleuropa. *Landeskundliche Forschungen*, 25.

Ganopolski, A. and Rahmstorf, S. 2001. Rapid changes of glacial climate simulated in a coupled climate model. *Nature*, 409, 153–158.

Ganopolski, A., Kubatzki, C., Claussen, M., Brovkin, V. and Petoukhov, V. 1998a. The influence of vegetation–atmosphere–ocean interaction on climate during the mid-Holocene. *Science*, 280, 1916–1919.

Ganopolski, A., Rahmstorf, S., Petoukhov, V. and Claussen, M. 1998b. Simulation of modern and glacial climates with a coupled global model of intermediate complexity. *Nature*, 391, 351–356.

Garreaud, R.D. and Battisti, D.S. 1999. Interannual ENSO and interdecadal ENSO-like variability in the Southern Hemisphere tropospheric circulation. *Journal of Climate*, 12, 2113–2123.

Gascoyne, M. 1983. Trace-element partition coefficients in the calcite-water system and their palaeoclimatic significance in cave studies. *Journal of Hydrology*, 61, 213–222.

Gascoyne, M. 1992a. Geochemistry of the actinides and their daughters. In Ivanovich, M. and Harmon, R.S. (eds), *Uranium-series disequilibrium: applications to earth, marine, and environmental sciences*. Clarendon Press, Oxford, pp. 34–61.

Gascoyne, M. 1992b. Palaeoclimate determination from cave calcite deposits. *Quaternary Science Reviews*, 11, 609–632.

Gascoyne, M. and Nelson, D.E. 1983. Growth mechanisms of recent speleothems from Castleguard Cave, Columbia Icefields, Alberta, Canada, inferred from a comparison of uranium-series and carbon-14 age data. *Arctic and Alpine Research*, 15, 537–542.

Gasse, F. 1987. Diatoms for reconstructing palaeoenvironments and palaeohydrology in tropical semi-arid zones. Example of some lakes from Niger since 12,000 BP. *Hydrobiologia*, 154, 127–163

Gasse, F. 2000. Hydrological changes in the African tropics since the last glacial maximum. *Quaternary Science Reviews*, 19, 189–211.

Gasse, F. and Van Campo, E. 1994. Abrupt post-glacial climate events in west Asia and north Africa monsoon domains. *Earth and Planet Science Letters*, 126, 435–465.

Gasse, F., Fontes, J.C., Plaziat, J.C., *et al.* 1987. Biological remains, geochemistry and stable-isotopes for the reconstruction of environmental and hydrological changes in the Holocene lakes from North Sahara. *Palaeogeography, Palaeoclimatology, Palaeoecology*, 60, 1–46.

Gasse, F., Barker, P. Gell, P. Fritz S.C. and Chalie, F. 1997. Diatom-inferred salinity in palaeolakes: an indirect tracer of climate change. *Quaternary Science Reviews*, 16, 547–563.

Gat, J.R. 1971. Comments on stable-isotope method in regional groundwater investigations. *Water Resource Research*, 7, 980–993.

Gat, J.R. 1980. The isotopes of hydrogen and oxygen in precipitation. In P. Fritz and J.C. Fontes (eds), *Handbook of evironmental isotope geochemistry*. Elsevier Scientific, Amsterdam, pp. 21–48.

Gehrels, W.R. 1999. Middle and Late Holocene sea-level changes in eastern Maine reconstructed from foraminiferal saltmarsh stratigraphy and AMS ^{14}C dates on basal peats. *Quaternary Research*, 52, 350–359.

Gehrels, W.R. 2001. Discussion on Nunn, P.D. 1998 Sea-level changes over the past 1000 years in the Pacific (*Journal of Coastal Research*, 14, 23–30). *Journal of Coastal Research*, 17, 244–245.

Gell, P.A. 1997. The development of a diatom database for inferring lake salinity, western Australia: towards a quantitative approach for reconstructing past climates. *Australian Journal of Botany*, 45, 389–423

Genty, D. and Quinif, Y. 1996. Annually laminated sequences in the internal structure of some Belgian stalagmites: importance for palaeoclimatology. *Journal of Sedimentary Research*, 66, 275–288.

Genty, D. and Massault, M. 1997. Bomb ^{14}C recorded in laminated speleothems: calculation of dead carbon proportion. *Radiocarbon*, 39, 33–48.

Genty, D., Baker, A., Barnes, W.L. and Massault, M. 1996. Growth rate, grey level and luminescence of stalagmite laminae. *Karst Waters Institute Special Publication*, 2, 36–39.

Genty, D., Baker, A. and Barnes, W.L. 1997. Comparison of annual luminescent and visible laminae in a recently developed stalagmite: environmental significance. *Comptes Rendus de l'Academie des Sciences*, 325, 193–200.

Genty, D., Massault, M., Gilmour, M., Baker, A., Verheyden, S. and Keppens, E. 1999. Caculation of past dead carbon proportion and variability by thecomparison of AMS ^{14}C and TIMS U/Th ages on two Holocene stalagmites. *Radiocarbon*, 41, 251–270.

Genty, D., Baker, A., Massault, M., *et al.* 2001. Dead carbon in stalagmites: Carbonate bedrock paleodissolution vs. ageing of soil organic matter. Implications for ^{13}C variations in speleothem. *Geochimica Cosmochimica Acta*, 65, 3443–3457.

Gersonde, R. and Zielinski, U. 2000. The reconstruction of late Quaternary Antarctic sea-ice distribution: the use of diatoms as a proxy for sea ice. *Palaeogeography, Palaeoclimatology, Palaeoecology*, 162, 263–286.

Geyh, M.A. and Franke, H.W. 1970. Zur Wachstumsgeschwindigkeit von Stalagmiten. *Atompraxis*, 16, 46–48.

Gingele, F.X. 1996. Holocene climatic optimum in southwest Africa: evidence from the marine clay mineral record. *Palaeogeography, Palaeoclimatology, Palaeoecology*, 122, 77–78.

Glaser, P.H., Siegel, D.I., Romanowicz, E.A. and Shen, Y.P. 1997. Regional linkages between raised bogs and the climate, groundwater, and landscape of north-western Minnesota. *Journal of Ecology*, 85, 3–16.

Godfrey, J.S. and Ridgway, K.R. 1985. The large-scale environment of the poleward-flowing Leeuwin Current, Western Australia: longshore steric height gradients, wind stresses and geostrophic flow. *Journal of Physical Oceanography*, 15, 481–495.

Godwin, H. 1981. *The archives of the peat bogs*. Cambridge University Press, Cambridge.

Goede, A. and Vogel, J.C. 1991. Trace element variations and dating of a late Pleistocene Tasmanian speleothem. *Palaeogeography, Palaeoclimatology, Palaeoecology*, 88, 121–131.

Goede, A., Veeh, H.H. and Ayliffe, L.K. 1990. Late Quaternary palaeotemperature records for two Tasmanian speleothems. *Australian Journal of Earth Sciences*, 37, 267–278.

Goede, A., McCulloch, M., McDermott, F. and Hawkesworth, C.J. 1998. Aeolian contribution to strontium isotope variations in a Tasmanian speleothem. *Chemical Geology*, 149, 37–50.

Goedicke, H. 1975. *The report of Wenamun*. Johns Hopkins University Press, Baltimore.

Gonfiantini, R. 1986. Environmental isotopes in lake studies. In P. Fritz and J.-Ch. Fontes (eds), *Handbook of environmental isotope geochemistry, Vol. 2, The terrestrial environment.*, B. Elsevier, Amsterdam, pp. 113–168.

Goñi, M.A., Nelson, B., Blanchette, R.A. and Hedges, J.I. 1993. Fungal degradation of wood lignins: Geochemical perspectives from CuO-derived phenolic dimers and monomers. *Geochimica Cosmochimica Acta*, 57, 3985–4002.

Gonzàles, L.A., Carpenter, S.J. and Lohmann, K.C. 1992. Inorganic calcite morphology: roles of fluid chemistry and fluid flow. *Journal of Sedimentary Petrology*, 62, 382–399.

Goodwin, I. D. 1998. Did changes in Antarctic ice volume influence Late Holocene sea-level lowering? *Quaternary Science Reviews*, 17, 319–332.

Goodwin, I.D., Harvey, N., van de Plassche, O., Oglesby, R. and Oldfield, F. 2000. Research targets 2000 years of climate-induced sea-level fluctuations. *EOS*, 81, 311–312.

Gordon, D., Smart, P.L., Ford, D.C., *et al.* 1989. Dating of late Pleistocene interglacial and interstadial periods in the United Kingdom from speleothem growth frequency. *Quaternary Research*, 31, 14–26.

Gore A.J.P. (ed.) 1983. Ecosystems of the world 4B. mires: swamp, bog, fen and moor. Regional studies. Elsevier Science, Amsterdam.

Gorsline, G.S., Nava-Sanchez, E. and Murillo de Nava, J. 1996. A survey of occurrences of Holocene laminated sediments in California Borderland Basins: products of a variety of depositional processes. In A.E.S. Kemp (ed.), *Palaeoclimatology and palaeoceanography from laminated sediments, 116.* The Geological Society, London, pp. 93–110.

Goslar, T. 1992. Possibilities for reconstructing radiocarbon level changes during the Late Glacial by using a laminated sequence of Gosciaz Lake. *Radiocarbon*, 34, 826–832.

Gosselin, D.C., Nebelek, P.E., Peterman, Z.E. and Sibray, S. 1997. A reconnaissance study of oxygen, hydrogen and strontium isotopes in geochemically diverse lakes, Western Nebraska, USA. *Journal of Paleolimnology*, 17, 51–65.

Gotschalk, M.K.E. 1971. Stormvloeden en rivieroverstormingen in Nederland Deel I: de period voor 1400. van Gorcum, Assen.

Gotschalk, M.K.E. 1975. Stormvloeden en rivieroverstormingen in Nederland Deel I: de period 1400–1600. van Gorcum, Assen.

Gotschalk, M.K.E. 1977. Stormvloeden en rivieroverstormingen in Nederland Deel I: de period 1600–1700. van Gorcum, Assen.

Gould, S.J. 1989. *Wonderful life*. Hutchinson Radius, London.

Gove, H.E. 2000. Some comments on accelerator mass spectrometry. *Radiocarbon*, 42, 127–135.

Granados, I. and Toro, M. 2000. Recent warming in a high mountain lake Laguna Cimera, Central Spain. inferred by means of fossil chironomids. *Journal of Limnology*, 59 (Suppl. 1), 109–119.

Grassl, H. 2000. Status and improvements of coupled general circulation models. *Science*, 288, 1991–1997.

Greuell, W. 1992. Hintereisferner, Austria: mass balance reconstruction and numerical modelling of historical length variation. *Journal of Glaciology*, 38, 233–244.

Grichuk, V.P. 1969. Opyt rekonstruktusü nekotorykh elementov klimata severnogo polusharuä v Atlanticheskü period golotsena. In M.I. Neishtadt (ed.), *Golotsen VIII Kongressu INQUA, Paris*. Izd-vo Nauka, Moscow, pp. 41–57.

Griffiths, H.I. 1998/9. Freshwater Ostracoda from the Mesolithic lake site at Lough Boora, Co. Offaly, Ireland. *Irish Journal of Earth Sciences*, 17, 39–49.

Griffiths, H.I. and Horne, D.J. 1998. Fossil distribution of reproductive modes in non-marine ostracods. In K. Martens (ed.), *Sex and parthenogenesis. Evolutionary ecology of reproductive modes in non-marine Ostracods.* Backhuys Publishers, Leiden, pp. 101–118.

Griffiths, H.I., Pillidge, K., Hill, C.J., Learner, M.A. and Evans, J.G. 1996. Ostracod gradients in a calcareous coastal stream: implications for the interpretation of tufas and travertines. *Limnologica*, 26, 49–61.

Griffiths, H.I., Reed, J.M. Leng, M.J., Ryan, S. and Petkovski, S. 2002. The recent palaeoecology and conservation status of Balkan Lake Dojran. *Biological Conservation*, 104, 35–49.

Grimm, E.C. 1983. Chronology and dynamics of vegetation change in the prairie-woodland region of southern Minnesota, USA. *New Phytologist*, 93, 311–350.

Grimm, E.C. 2001. Trends and palaeoecological problems in the vegetation and climate history of the northern Great Plains, USA. *Proceedings of the Royal Irish Academy*, 101B, 47–64.

Groisman, P. Y. 1992. Possible regional climate consequences of the Pinatubo eruption: an empirical approach. *Geophysical Research Letters*, 19, 1603–1606.

Grosjean, M. and Núñez, L.A. 1994. Lateglacial, early and Middle Holocene environments, human occupation, and resource use in the Atacama Northern Chile. *Geoarchaeology*, 9, 271–286.

Grosjean, M., Valero-Garces, B.L., Geyh, M.A., *et al.* 1997. Mid- and late-Holocene limnogeology of Laguna del Negro Francisco, northern Chile, and its palaeoclimatic implications. *The Holocene*, 7, 151–159.

Grosjean, M., van Leeuwen, J.F.N., van der Knaap, W.O *et al.* 2001. A 22,000 ^{14}C year BP sediment and pollen record of climate change from Laguna Miscanti 23°S, northern Chile. *Global and Planetary Change*, 28, 35–31.

Gross, E., Jacomet, S.T. and Schibler, J. 1990. Stand und Ziele der wirtschaftsarchäologischen Forschung an neolithischen Ufer- und Inselsiedlungen im unteren Zürichseeraum Kt Zürich, Schweiz. In J. Schibler, J. Sedlmaier and H. Spycher (eds), *Beiträge zur Archäozoologie, Archäologie, Anthropologie, Geologie und Paläontologie*. Festschrift Hans R. Stampfli, Basle, pp. 77–100.

Gross, M.G., Gucluer, S.M., Creager, J.S. and Dawson, W.A. 1963. Varved marine sediments in a stagnant fjord. *Science*, 141, 918–919.

Gross-Klee, E. and Maise, C. 1997. Sonne, Vulkane und Seeufersiedlungen. *Jahrbuch der Schweizerischen Gesellschaft für Ur- und Frühgeschichte*, 80, 85–94.

Grossman, E.E., Fletcher, C.P. III, Richmond, B.M. 1998. The Holocene sea-level highstand in the equatorial Pacific: analysis of the insular paleosea-level database. *Coral Reefs*, 17, 309–327.

Grottoli, A.G. 1999. Variability of stable-isotopes and maximum linear extension in reef-coral skeletons at Kaneohe Bay, Hawaii. *Marine Biology*, 135, 437–449.

Grottoli, A.G. 2002. Effect of light and brine shrimp on skeletal delta ^{13}C in the Hawaiian coral *Porites compressa*: A tank experiment. *Geochimica et Cosmochimica Acta*, 66, 1955–1967.

Grottoli, A.G. and Wellington, G.M. 1999. Effect of light and zooplankton on skeletal delta ^{13}C values in the eastern Pacific corals *Pavona clavus* and *Pavona gigantea*. *Coral Reefs*, 18, 29–41.

Grove, J.M. 1988. *The Little Ice Age*. Methuen, New York.

Grove, J.M. and Battagel, A. 1983. Tax records from western Norway, as an index of Little Ice Age environmental and economic deterioration. *Climatic Change*, 5, 265–282.

Grumet, N.S., Wake, C.P., Mayewski, P.A., *et al.* 2001. Variability of sea-ice in Baffin Bay over the last millennium. *Climate Change*, 49, 129–145.

Guilderson, T.P. and Schrag, D.P. 1998. Abrupt shifts in subsurface temperatures in the tropical Pacific associated with changes in El Niño. *Science*, 281, 240–243.

Guilderson, T.P. and Schrag, D.P. 1999. Reliability of coral records from the western Pacific warm pool: a comparison using age-optimized records. *Paleoceanography*, 4, 457–464.

Guilderson, T.P., Schrag, D.P., Kashgarian, M. and Southon, J. 1998. Radiocarbon variability in the western equatorial Pacific inferred from a high-resolution coral record from Nauru Island. *Journal of Geophysical Research-Oceans*, 103, 24641–24650.

Guilderson, T.P., Caldeira, K. and Duffy, P.B. 2000. Radiocarbon as a diagnostic tracer in ocean and carbon cycle modeling. *Global Biogeochemical Cycles*, 14, 887–902.

Guiot, J. 1990. Methodology of the last climatic cycle reconstruction in France from pollen data. *Palaeogeography, Palaeoclimatology, Palaeoecology*, 80, 49–69.

Guiot, J. 1991. Structural characteristics of proxy data and methods for quantitative climate reconstruction. In B. Frenzel, A. Pons and B. Gläser (eds), *Evaluation of climate proxy data in relation to the European Holocene*. Akademie der Wissenschaften und der Literatur Mainz and G. Fischer Verlag, Stuttgart, pp. 271–284.

Guiot, J., Reille, M., de Bealieu, J.L. and Pons, A. 1992. Calibration of the climatic signal in a new pollen sequence from La Grand Pile. *Climate Dynamics*, 6, 259–264.

Guiot, J., de Beaulieu, J.L., Cheddadi, R., David, F., Ponel, P. and Reille, M. 1993a. The climate in western Europe during the last glacial/interglacial cycle derived from pollen and insect remains. *Palaeogeography, Palaeoclimatology, Palaeoecology*, 103, 73–93.

Guiot, J., Harrison, S.P. and Prentice, I.C. 1993b. Reconstruction of Holocene precipitation patterns in Europe using pollen and lake-level data. *Quaternary International*, 40, 139–149.

Gulliksen, S. and Scott, M. 1995. Report of the TIRI workshop, Saturday 13 August 1994. *Radiocarbon*, 37, 820–821.

Gutman, G. 1984. Numerical experiments on land surface alterations with a zonal model allowing for interaction between the geobotanic state and climate. *Journal of Atmospheric Science*, 41, 2679–2685.

Gutman, G. 1985. On modelling dynamics of geobotanic state-climate interaction. *Journal of Atmospheric Science*, 43, 305–306.

Gutman, G., Ohring, G. and Joseph, J.H. 1984. Interaction between the geobotanic state and climate: a suggested approach and a test with a zonal model. *Journal of Atmospheric Science*, 41, 2663–2678.

Guzman, H.M. and Tudhope, A.W. 1998. Seasonal variation in skeletal extension rate and stable isotopic $^{13}C/^{12}C$ and $^{18}O/^{16}O$ composition in response to several environmental variables in the Caribbean reef coral *Siderastrea siderea. Marine Ecology-Progress Series*, 166, 109–118.

Haakensen, N. 1989. Akkumulasjon på breene i Sør-Norge vinteren 1988–89. *Været*, 13, 91–94.

Haberle, S.G. 1993. *Late Quaternary environmental history of the Tari Basin, Papua New Guinea*, PhD Thesis. Australian National University, Canberra.

Haffer, J. 1969. Speciation in Amazonian forest birds. *Science*, 165, 131–137.

Haigh, J.D. 1996. The impact of solar variability on climate. *Science*, 272, 981–984.

Hajdas, I., Zolitschka, B., Ivy-Ochs, S.D., *et al.* 1995. AMS radiocarbon dating of annually laminated sediments from Lake Holzmaar, Germany. *Quaternary Science Reviews*, 14, 137–143.

Hall, N.M.J. and Valdes, P.J. 1997. A GCM simulation of the climate 6000 years ago. *Journal of Climate*, 10, 3–17.

Halsey, L.A., Vitt, D.H. and Bauer, I.E. 1998. Peatland initiation during the Holocene in continental western Canada. *Climatic Change*, 40, 315–342.

Hammarlund, D. and Buchardt, B. 1996. Composite stable-isotope records from a Late Weichselian lacustrine sequence at Graenge, Lolland, Denmark: evidence of Allerød and Younger Dryas environments. *BOREAS*, 25, 8–22.

Hammarlund, D., Barnekow, L., Birks, H.J.B., Buchardt, B. and Edwards, T.W.D. 2002. Holocene changes in atmospheric circulation recorded in the oxygen-isotope stratigraphy of lacustrine carbonates from northern Sweden. *The Holocene*, 12, 339–351.

Hammer, C.U. 1980. Acidity of polar ice cores in relation to absolute dating, past volcanism, and radio echos. *Journal of Glaciology*, 25, 359–372.

Hammer, C.U., Clausen, H.B., Neftel, A., Kristinsdottir, P. and Johnson, E. 1985. Continuous impurity analysis along the Dye 3 deep core. In Greenland Ice Core: *Geophysics, Geochemistry and the Environment, Geophysical Monographs Series*, vol 33, edited by C.C. Langway Jr., H. Oeschger and W. Dansgaard, AGU, Washington DC, 90–94.

Hann, B.J., Warner, B.G. and Warwick, W.F. 1992. Aquatic invertebrates and climate change: a comment on Walker *et al.* 1991. *Canadian Journal of Fisheries and Aquatic Sciences*, 49, 1274–1276.

Hann, J. 1908. Handbuch der Klimatologie, Band I, 3. Auflage.

Hannah, L., Midgley, G.F., Lovejoy, T., *et al.* 2002. Conservation of biodiversity in a changing climate. *Conservation Biology*, 16, 264–268.

Hannon, G.E. and Gaillard, M.-J. 1997. The plant-macrofossil record of past lake-level changes. *Journal of Palaeolimnology*, 18, 15–28.

Hannon, G.E., Bradshaw, R.G. and Emborg, J. 2000. 6000 years of forest dynamics in Suserup Skov, a seminatural Danish woodland. *Global Ecology and Biogeography*, 9, 101–114.

Hansen, B.C.S. 1995a. A review of lateglacial pollen records from Ecuador and Peru with reference to the Younger Dryas Event. *Quaternary Science Reviews*, 14, 853–865.

Hansen, B.C.S. 1995b. Conifer stomate analysis as a paleoecological tool: an example from the Hudson Bay Lowland. *Canadian Journal of Botany*, 73, 244–252.

Hansen, J. and Lebedeff, S. 1987. Global trends of measured surface air temperature. *Journal of Geophysical Research*, 92, 13345–13372.

Hansson, M.E. 1994. The Renland ice core. A northern hemisphere record of aerosol comparison over 120,000 years. *Tellus*, 46B, 390–418.

Hardy, D.R., Bradley, R.S. and Zolitschka, B. 1996. The climatic signal in varved sediments from Lake C2, northern Ellesmere Island, Canada. *Journal of Paleolimnology*, 16, 227–238.

Harmon, R.S., Mitterer, R.M., Kriausakul, N., *et al.* 1983. U-series and amino-acid racemization geochronology of Bermuda: implications for eustatic sea-level fluctuation over the past 250,000 years. *Palaeogeography, Palaeoclimatology, Palaeoecology*, 44, 41–70.

Harris, P.G., Zhao, M., Rosell-Melé, A., Tiedemann, R., Sarnthein, M. and Maxwell, J.R. 1996. Chlorin accumulation rate as a proxy for Quaternary marine primary productivity proxy. *Nature*, 383, 63–65.

Harris, S.E. and Mix, A.C. 1999. Pleistocene precipitation balance in the Amazon basin recorded in deep-sea sediments. *Quaternary Research*, 51, 14–26.

Harrison, S.P. and Dodson, J. 1993. Climates of Australia and New Guinea since 18,000 yr BP. In H.E. Wright, J.E. Kutzbach, T. Webb III, W.F. Ruddiman, F.A. Street-Perrott and P.J. Bartlein (eds), *Global climates since the last glacial maximum.* University of Minnesota Press, Minneapolis, pp. 265–293.

Harrison, S.P., Prentice, I.C. and Guiot, J. 1993. Climatic controls on Holocene lake-level changes in Europe. *Climate Dynamics*, 8, 189–200.

Harrison, S.P., Jolly, D., Laarif, F., *et al.* 1998. Intercomparison of simulated global vegetation distributions in response to 6 kyr BP orbital forcing. *Journal of Climate*, 11, 2721–2742.

Harvey, L.D.D. 1988. A semianalytic energy balance climate model with explicit sea ice and snow physics. *Journal of Climate*, 1, 1065–1085.

Harvey, L.D.D. 1989a. An energy balance climate model study of radiative forcing and temperature response at 18 ka. *Journal of Geophysical Research*, 94, 12873–12884.

Harvey, L.D.D. 1989b. Milankovitch forcing, vegetation feedback, and North Atlantic deep-water formation. *Journal of Climate*, 2, 800–815.

Harvey, N., Barnett, E.J., Bourman, R.P. and Belperio, A.P. 1999. Holocene sea-level change at Port Pirie, South Australia: A contribution to global sea-level rise estimates from tide gauges. *Journal of Coastal Research*, 15, 607–615.

Harvey, N., Belperio, A.P., Bourman, R.P. and Mitchell, B. 2002. Isostatic and anthropogenic sgnals affecting sea-level records at tide guage sites in Southern Australia. *Global and Planetary Change*, 32, 1–11.

Haskell, B.J., Engstrom, D.R. and Fritz, S.C. 1996. Late Quaternary paleohydrology in the North Amrican Great Plains inferred from the geochemistry of endogenic carbonate and fossil ostracodes from Devils Lake, North Dakota, USA. *Palaeogeography, Palaoeclimatology, Palaeoecology*, 124, 179–193.

Haslam, C.J. 1988. Late Holocene peat stratigraphy and climatic change: a macrofossil investigation from the raised mires of north western Europe, PhD Thesis. University of Southampton, Southampton.

Haslett, S.K., Strawbridge, F., Martin, N.A. and Davies, C.F.C. 2001. Vertical saltmarsh accretion and its relationship to sea-level in the Severn Estuary, UK: an investigation using foraminifera as tidal indicators. *Estuarine, Coastal and Shelf Science*, 52, 143–153.

Hassan, F.A.1997. Nile floods and political disorder in early Egypt. In H.N. Dalfes, G. Kukla and H. Weiss (eds), *NATO ASI Series 149*, Springer-Verlag, Berlin, pp. 1–23.

Hassan, F.A. 2000. Environmental perception and human responses in history and prehistory. In R.J. McIntosh, J.A. Tainter and S.K. McIntosh (eds), *The way the wind blows: climate, history and human action.* Colombia University Press, Washington, DC.

Hatcher, B.G. 1991. Coral reefs in the Leeuwin Current: an ecological perspective. *Journal of the Royal Society of Western Australia*, 74, 115–127.

Haug, G.H., Hughen, K.A., Sigman, D.M., Peterson, L.C. and Röhl, U. 2001. Southward migration of the intertropical convergence zone through the Holocene. *Science*, 293, 1304–1308.

Hausmann, S., Lotter, A.F., van Leeuwen, J.F.N., *et al.* 2002. Interactions of climate and land use documented in the varved sediments of Seebergsee in the Swiss Alps. *The Holocene*, 12, 279–289.

Haxeltine, A. and Prentice, I.C. 1996. BIOME3: an equilibrium biosphere model based on ecophysiological constraints, resource availability and competition among plant funtional types. *Global Biogeochemical Cycles*, 10, 693–709.

Hayes, J.M., Freeman, K.H., Popp, B.N. and Hoham, C.H. 1990. Compound-specific isotopic analyses, a novel tool for reconstruction of ancient biogeochemical processes. *Organic Geochemistry*, 16, 1115–1128.

Hays, P.D. and Grossman, E.L. 1991. Oxygen isotopes in meteoric calcite cements as indicators of continental paleoclimate. *Geology*, 19, 441–444.

Heaton, T.H.E., Holmes, J.A. and Bridgewater, N.D. 1995. Carbon and oxygen isotope variations among lacustrine ostracods: implications for palaeoclimatic studies. *The Holocene*, 5, 428–434.

Hedges, J.I. and Oades, J. 1997. Comparative organic geochemistries of soils and marine sediments. *Organic Geochemistry*, 27, 319–361.

Hedges, J.I., Baldock, J.A., Gelinas, Y., Lee, C., Peterson, M. and Wakeham, S.G. 2001. Evidence for non-selective preservation of organic matter in sinking marine particles. *Nature*, 409, 801–804.

Heer, O. 1865. *Die Urwelt der Schweiz.* Zürich,

Hegmon M., Nelson, M.C. and Ennes, M.J. 2000. Corrugated pottery, technological style, and population movement in the Mimbres region of the American southwest. *Journal of Anthropological Research*, 56, 217–240.

Heim, C., Nowaczyk, N.R. and Negendank, J.F.W. 1997. Near East desertification: evidence from the Dead Sea. *Naturwissenschaften*, 84, 398–401.

Heinrichs, M.L., Wilson, S.E., Walker, I.R., Smol, J.P., Mathewes, R.W. and Hall, K.J. 1997. Midge- and diatom-based paleosalinity reconstructions for Mahoney Lake, Okanagan Valley, British Columbia, Canada. *International Journal of Salt Lake Research*, 6, 249–267.

Heiri, O. and Lotter, A.F. In press. 9000 years of chironomid population dynamics in an Alpine lake: long-term trends, sensitivity to disturbance and resilience of the fauna. *Journal of Paleolimnology*.

Heiri, O., Lotter, A.F., Hausmann, S, and Kienast, F. 2003. A chironomid-based Holocene summer air temperature reconstruction from the Swiss Alps. *The Holocene*, 13, 477–484.

Henderson-Sellers, A. 1993. Continental vegetation as a dynamic component of global climate model. a preliminary assessment. *Climatic Change*, 23, 337–378.

Hendon, D., Charman, D.J. and Kent, M. 2001. Palaeohydrological records derived from testate amoebae analysis from peatlands in northern England: within-site variability, between-site comparability and palaeoclimatic implications. *The Holocene*, 11, 127–148.

Hendy, C.H. 1971. The isotopic geochemistry of speleothems. Part 1. The calculation of the effects of different modes of formation on the isotopic composition of speleothems and their applicability as paleoclimatic indicators. *Geochimica Cosmochimica Acta*, 35, 801–824.

Hendy, E.J., Gagan, M.K., Alibert, C.A., McCulloch, M.T., Lough, J.M. and Isdale, P.J. 2002. Abrupt decrease in tropical Pacific Sea surface salinity at end of Little Ice Age. *Science*, 295, 1511–1514.

Henrikson, L. and Oscarson, H.G. 1985. History of the acidified Lake Gårdsjön: the development of chironomids. *Ecological Bulletin*, 37, 58–63.

Herbert, T.D., Schuffert, J.D., Andreasen, D., 2001. Collapse of the California Current during glacial maxima linked to climate change on land. *Science*, 293, 71–76.

Hercman, H. and Lauritzen, S.E. 1996. Comparison of speleothem dating by the TL, ESR, ^{14}C and ^{230}Th/^{234}U methods. *Karst Waters Institute Special Publication*, 2, 47–50.

Hewitt, C.D. and Mitchell, J.F.B. 1998. A fully coupled GCM simulation of the climate of the mid-Holocene. *Geophysical Research Letters*, 25, 361–364.

Hibler III, W.D. and Johnsen, S J. 1979. The 20-yr cycle in Greenland ice core records. *Nature*, 280, 481–483

Hicks, D.M., McSaveney, M.J. and Chinn, T.J. H. 1990. Sedimentation in pro-glacial Ivory Lake, Southern Alps, New Zealand. *Artic and Alpine Research*, 22, 26–42.

Hicks, S., Miller, U. and Saarnisto, M. (eds) 1994. Laminated sediments. Rixensart, Belgium, Council of Europe. Journal of the European Study Group on Physical, Chemical, Biological and Mathematical Techniques Applied to Archaeology, 41.

Hill, C. and Forti, P. 1997. *Cave Minerals of the World.* National Speleological Society, Huntsville, AL.

Hillman, G.C. 1996. Late Pleistocene changes in wild plant food available to hunter-gatherers of the northern Fertile Crescent: possible preludes to cereal cultivation. In D. Harris (ed.), *The origins and spread of agriculture and pastoralism in Eurasia.* UCL Press, London, pp. 159–203.

Hinrichs, K.U., Schneider, R.R., Müller, P.J. and Rullkötter, J. 1999. A biomarker perspective on paleoproductivity variations in two Late Quaternary sediment sections from the Southeastern Atlantic Ocean. *Organic Geochemistry*, 30, 341–366.

Hintikka, V. 1963. Über das Grossklima einigor Pflanzenmareale in zwei Klimakoordinatensystemen Dargestelt. *Annales Botanici Societatis Zoologicae Botanicae Fennicae "Vanamo"*, 34, 1–64.

Hinton, A. 1998. Tidal changes. *Progress in Physical Geography*, 22, 282–294.

Hodell, D.A. and Schelske, C.L. 1998. Production, sedimentation, and isotopic composition of organic matter in Lake Ontario. *Limnology and Oceanography*, 43, 200–214.

Hodell, D.A., Curtis, J.H., Jones, G.A., *et al.*1991. Reconstruction of Caribbean climate change over the past 10,500 years. *Nature*, 352, 790–793.

Hodell, D.A., M. Brenner, J.H. Curtis and T. Guilderson, 2001. Solar forcing of drought frequency in the Maya Lowlands. *Science*, 292, 1367–1370.

Hodgkinson, R.L. 1991. Microfossil processing: a damage report. *Micropaleontology*, 37, 320–326.

Hoefs, J. 1997. *Stable-isotope geochemistry*. Springer, Berlin.

Hoelzmann, P., Jolly, D., Harrison, S.P., Laarif, F., Bonnefille, R., Pachur, H.-J. 1998. Mid-Holocene land-surface conditions in northern Africa and the arabian peninsula: A data set for the analysis of biogeophysical feedbacks in the climate system. *Global Biogeochemical Cycles*, 12, 35–51.

Hoffmann, G. 2002. Modelling the water isotope signal in the Quaternary. *PAGES Newsletter*, 10, 12–13.

Hofmann, W. 1971. Die postglaziale Entwicklung der Chironomiden- und Chaoboriden-Fauna Dipt. des Schöhsees. *Archiv für Hydrobiologie Supplement*, 40, 1–74.

Hofmann, W. 1987. Stratigraphy of Cladocera Crustacea. and Chironomidae Insecta: Diptera in three sediment cores from the Central Baltic Sea as related to paleo-salinity. *Internationale Revue der Gesamten Hydrobiologie*, 72, 97–106.

Hofmann, W. 1991. Stratigraphy of Chironomidae Insecta: Diptera and Cladocera Crustacea in Holocene and Würm sediments from Lac du Bouchet Haute Loire, France. *Documents du CERLAT*, 2, 363–386.

Hollstein, E. 1980. *MittelEuropaische Eichenchronologie*. Phillip Von Zabern, Mainz am Rhein.

Holmes, J.A. 1996. Trace element and stable-isotope geochemistry of non-marine ostracod shells in Quaternary palaeoenvironmental reconstruction. *Journal of Paleolimnology*, 15, 223–235.

Holmes, J.A. 1997. Recent non-marine Ostracoda from Jamaica, West Indies. *Journal of Micropalaeontology*, 16, 137–143.

Holmes, J.A. 1998. A late Quaternary ostracod record from Wallywash Great Pond, a Jamaican marl lake. *Journal of Paleolimnology*, 19, 115–128.

Holmes, J.A. 2001. Ostracoda. In J.P. Smol, H.J.B. Birks and W.M. Last (eds), *Tracking environmental change using lake sediments, Vol. 4. Zoological indicators*. Kluwer, Dordrecht, pp. 125–151.

Holmes, J.A., and Chivas, A.R. 2002. Ostracod shell chemistry - overview. In J.A. Holmes and A.R. Chivas (eds), *The Ostracoda: Applications in Quaternary Research*, Vol. 131, AGU Geophysical Monograph Series, Washington, DC., pp. 185–204.

Holmes, J.A. and Griffiths, H.I. 1998. Ostracoda from Star Carr. In P. Mellars and S.P. Dark (eds), *Star Carr in context: new archaeological and palaeoecological investigations at the early Mesolithic site of Star Carr, North Yorkshire*. The Macdonald Institute for Archaeological Research, Cambridge, pp. 175–178.

Holmes, J.A., Street-Perrott, F.A., Allen, M.J., *et al.* 1997. Holocene palaeolimnology of Kajemarum Oasis, Northern Nigeria: an isotopic study of ostracodes, bulk carbonate and organic carbon. *Journal of Geological Society*, 154, 311–319.

Holmes, J.A., Fothergill, P.A., Street-Perrott, F.A. and Perrott, R.A. 1998. A high-resolution Holocene ostracod record from the Sahel zone of Northeastern Nigeria. *Journal of Paleolimnology*, 20, 369–380.

Holmes, R.L., Adams, R.K. and Fritts, H.C. 1986. Tree-ring chronologies of Western North America: California, Eastern Oregon and Northern Great Basin. Chronology Series VI. University of Arizona, Tucson.

Holmgren, C.A., Betancourt, J.L., Rylander K.A., *et al.* 2001. Holocene history from fossil rodent middens near Arequipa, Peru. *Quaternary Research*, 56, 242–251.

Holmgren, K., Lauritzen, S.E. and Possnert, G. 1994. $^{230}Th/^{234}U$ and ^{14}C dating of a late Pleistocene stalagmite in Lobatse II Cave, Botswana. *Quaternary Geochronology*, 13, 111–119.

Holmgren, K., Karlén, W. and Shaw, P.A. 1995. Paleoclimatic significance of the stable-isotope composition and petrology of a late Pleistocene stalagmite from Botswana. *Quaternary Research*, 43, 320–328.

Holmgren, K., Karlén, W., Lauritzen, S.E., *et al.* 1999. A 3000-year high-resolution stalagmite-based record of palaeoclimate for northeastern South Africa. *The Holocene*, 9, 295–309.

Holmlund, P., Karlén, W. and Grudd, H. 1996. Fifty years of mass balance and glacier front observations at the Tarfala Research Station. *Geografiska Annaler*, 78A, 105–114.

Holzhauser, H. 1984. Zur Geschichte der Aletschgletscher und des Fieschergletschers. *Physische Geographie*, 13, 448.

Hong, B.G., Sturges, W. and Clarke, A.J. 2000a. Sea level on the US East Coast: decadal variability caused by open ocean wind-curl forcing. *Journal of Physical Oceanography*, 30, 2088–2098.

Hong, Y.T., Jiang, H.B., Liu, T.S., *et al.* 2000b. Response of climate to solar forcing recorded in a 6000-year $\delta^{18}O$ time-series of Chinese peat cellulose. *The Holocene*, 10, 1–7.

Hong, Y.T., Wang, Z.G., Jiang, H.B., *et al.* 2001. A 6000-year record of changes in drought and precipitation in northeastern China based on a delta C-13 time series from peat cellulose. *Earth and Planetary Science Letters*, 185, 111–119.

Hooghiemstra, H. and Cleef, A. 1995. Pleistocene climatic change and environmental and generic dynamics in the north Andean montane forest and paramo. In S.P. Churchill, H. Balsev and E. Forero (eds), *Biodiversity and conservation of neotropical montane forests.* The New York Botanical Garden, New York, pp. 35–49.

Horne, D.J., Baltanás, A. and Paris, G. 1998. Geographical distribution of reproductive modes in living non-marine ostracods. In K. Martens (ed.), *Sex and parthenogenesis. Evolutionary ecology of reproductive modes in non-marine Ostracods.* Backhuys Publishers, Leiden, pp. 77–99.

Horne, D.J., Cohen, A. C. and Martens, K. 2002. Taxonomy, morphology and biology of Quaternary and living Ostracoda. In J.A. Holmes and A.R. Chivas (eds), *The Ostracoda: Applications in Quaternary Research,* Vol. 131, AGU Geophysical Monograph Series, Washington, DC, pp. 5–36.

Horner, R.A. 1985. Ecology of sea ice microalgae. In R.A. Horner (ed.), *Sea ice biota.* CRC Press, Boca Raton, FL. pp. 83–100.

Hostetler, S.W., Bartlein, P.J., Clark, P.U., Small, E.E. and Solomon, A.M. 2000. Simulated influences of Lake Agassiz on the climate of central North America 11,000 years ago. *Nature*, 405, 334–337.

Houghton, J.T., Ding, Y., Griggs, D.J., Noguer, M., van der Linden, P.J. and Xiaosu D. (eds) 2001. *Climate change: the scientific basis. Contribution of Working Group I to the Third Assessment Report of the Intergovernmental Panel on Climate Change (IPCC).* Cambridge University Press, Cambridge.

Hoyt, D.V. and K.H. Schatten, 1997. *The role of the sun in climate change.* Oxford University Press, Oxford.

Hu, F.S., Ito, E., Brubaker, L.B. and Anderson, P.M. 1998. Ostracode geochemical record of Holocene climatic change and implications for vegetational response in the northwestern Alaska Range. *Quaternary Research*, 49, 86–95.

Hu, F.S., Slawinski, D., Wright Jr, H.E., *et al.* 1999. Abrupt changes in North American climate during early Holocene times. *Nature*, 399, 437–440.

Huang, Y. and Fairchild, I.J. 2001. Partitioning of Sr^{2+} and Mg^{2+} into calcite under karst-analogue experimental conditions. *Geochimica Cosmochimica Acta*, 65, 47–62.

Huang, Y.S., Street-Perrott, F.A., Perrot, R.A., Metzger, P. and Eglington, G. 1999. Glacial–interglacial environmental changes inferred from molecular and compound-specific $\delta^{13}C$ analyses of sediments from Sacred Lake, Mt Kenya. *Geochimica et Cosmochimica Acta*, 63, 1383–1404.

Huang, Y., Dupont, L., Sarnthein, M., Hayes, J.M. and Eglinton, G. 2000. Mapping of C_4 plant input from North West Africa into North East Atlantic sediments. *Geochimica et Cosmochimica Acta*, 64, 3505–3513.

Hudson, J.H., Shinn, E.A., Halley, R.B. and Lidz, B. 1976. Sclerochronology: a tool for interpreting past environments. *Geology*, 4, 361–364.

Hughen, K.A., Overpeck, J.T., Peterson, L.C. and Anderson, R.F. 1996. The nature of varved sedimentation in the Cariaco Basin, Venezuela, and its palaeoclimatic significance. In A.E.S. Kemp (ed.), *Palaeoclimatology and palaeoceanography from laminated sediments.* Geological Society Special Publication, London, pp. 171–183.

Hughen, K.A., Overpeck, J.T. and Anderson, R.F. 2000. Recent warming in a 500-year palaeotemperature record from varved sediments, Upper Soper Lake, Baffin Island, Canada. *The Holocene*, 10, 9–20.

Hughes, M.K. and Brown, P.M. 1992 Drought frequency in Central California since 101 BC recorded in Giant Sequoia tree-rings. *Climate Dynamics*, 6, 161–167.

Hughes, M.K. and Diaz, H.F. 1994. Was there a 'Medieval Warm Period' and if so where and when? *Climate Change*, 26, 109–142.

Hughes, P.D.M., Mauquoy, D., Barber, K.E. and Langdon, P.G. 2000. Mire development pathways and palaeoclimatic records from a full Holocene peat archive at Walton Moss, Cumbria, England. *The Holocene*, 10, 465–479.

Huijzer, A.S. and Isarin, R.F.B. 1997. The reconstruction of past climates using multi-proxy evidence: an example of the Weichselian pleniglacial in northwest and central Europe. *Quaternary Science Reviews*, 12, 513–533.

Hulme, M. 1996. Recent climatic change in the world's drylands. *Geophysical Research Letters*, 23, 61–64.

Huntley, B. 1993. The use of climate response surfaces to reconstruct palaeoclimate from Quaternary pollen and plant macrofossil data. *Philosophical Transactions of the Royal Society of London B*, 341, 215–223.

Huntley, B. 1994. Late Devensian and Holocene palaeoecology and palaeoenvironments of the Morrone Birkwoods, Aberdeenshire. *Journal of Quaternary Science*, 9, 311–336.

Huntley, B. 1996. Quaternary Palaeoecology and ecology. *Quaternary Science Reviews*, 15, 591.

Huntley, B. 2001. Reconstructing past environments from the Quaternary palaeovegetation record. *Proceedings of the Royal Irish Academy*, 101B, 3–18.

Hurrell, J.W. 1995. Decadal trends in the North Atlantic Oscillation: Regional temperatures and precipitation. *Science*, 269, 676–679.

Hutson, W.H. 1977. Transfer functions under no-analog conditions: experiments with Indian Ocean planktonic foraminifera. *Quaternary Research*, 8, 355–367.

Hyvärinen, H. 1970. Flandrian pollen diagrams from Svalbard. *Geografiska Annaler*, 52A, 213–222.

Imbrie, J. and Kipp, N.G. 1971. A new micropaleontological method for quantitative paleoclimatology: application to a late Pleistocene Caribbean core. In K.K. Turekian (ed.), *The Late Cenozoic Glacial Ages*. Yale University Press, New Haven, pp. 71–181.

Imbrie, J. and Webb, T. 1981. Transfer functions: calibrating micropaleontological data in climatic terms. In A. Berger (ed.), *Climate variations and variability: factors and theories*, D. Reidel, Dordrecht, pp. 125–134.

Immirzi, C.P., Maltby, E. and Clymo, R.S. 1992. *The global status of peatlands and their role in carbon cycling*. A report for Friends of the Earth by the Wetland Ecosystems Research Group, Department of Geography, University of Exeter. Friends of the Earth, London, pp. 1–145.

Indermühle, A., Stocker, T.F., Joos, F., *et al.* 1999. Holocene carbon cycle dynamics based on CO_2 trapped in ice at Taylor Dome, Antarctica. *Nature*, 398, 121–126.

Ingram, H.A.P. 1978. Soil layers in mires: function and terminology. *Journal of Soil Science*, 29, 224–227.

Ingram, M., Underhill, D.J. and Wigley, T.M.L. 1978. Historical climatology. *Nature*, 276, 329–334.

Innes, J.L. 1984. Relative dating of Neoglacial moraine ridges in North Norway. *Zeitschrift für Gletscherkunde und Glazialgeologie*, 20, 53–63.

Innes, J.L. 1985a. Lichenometry. *Progress in Physical Geography*, 9, 187–254.

Innes, J.L. 1985b. A standard Rhizocarpon nomenclature. *Boreas*, 14, 83–85.

Innes, J.L. 1986a. The use of percentage cover measurements in lichenometric dating. *Arctic and Alpine Research*, 18, 209–216.

Innes, J.L. 1986b. Influence of sampling design on lichen size-frequency distribution and its effect on derived lichenometric indices. *Arctic and Alpine Research*, 18, 201–208.

International Study Group. 1982. An inter-laboratory comparison of radiocarbon measurements in tree-rings. *Nature*, 198, 619–623.

IPCC. 2001. Climate change 2001: the scientific basis. In J.T. Houghton, Y. Ding, D.J. Griggs, M. Noguer, P.J. van der Linden and D. Xiaosu (eds), Working Group 1 contribution to the IPCC Third Assessment Report (Summary for Policy Makers). Cambridge University Press, Cambridge.IPCC 2001. see website: http://www.ipcc.ch

Isdale, P. 1984. Fluorescent bands in massive corals record centuries of coastal rainfall. *Nature*, 310, 578–579.

Isdale, P.J., Stewart, B.J. and Lough, J.M. 1998. Palaeohydrological variation in a tropical river catchment: a reconstruction using fluorescent bands in corals of the Great Barrier Reef, Australia. *The Holocene*, 8, 1–8.

Ishiwatari, R., Yamada, K., Matsumoto, K., Houtatsu, M. and Naraoka, H. 1999. Organic molecular and carbon isotopic records of the Japan Sea over the past 30kyr. *Paleoceanography*, 14, 260–270.

Itkonen, A., Marttila, V., Meriläinen, J.J. and Salonen, V.P. 1999. 8000-year history of palaeoproductivity in a large boreal lake. *Journal of Paleolimnology*, 21, 271–294.

Ito, E. 2002. Mg/Ca, Sr/Ca, $\delta^{18}O$ and $\delta^{13}C$ chemistry of Quaternary lacustrine ostracode shells from North American continental interior. In J.A. Holmes and A.R. Chivas (eds), *The Ostracoda: Applications in Quaternary Research*, Vol. 131, AGU Geophysical Monograph Series, Washington, DC, pp. 267–278.

Ivanovich, M. and Harmon, R.S. 1992. Uranium-series disequilibrium: applications to earth, marine, and environmental sciences. Clarendon Press, Oxford.

Iversen, J. 1944. *Viscum, Hedera* and *Ilex* as climate indicators. A contribution to the study of the post-glacial temperature climate. *Geologiske Föreningen i Stockholm Förhandlingar*, 66, 463–483.

Iversen, J. 1954. The Late-Glacial flora of Denmark and its relation to climate and soil. *Danmarks Geologiske Undersogelse*, II, 87–119.

Iversen, J. 1958. The bearing of glacial and interglacial epochs on the formation and extinction of plant taxa. In O. Hedberg (ed.), *Systematics of today*. Uppsala Universitets Årsskrift 6, pp. 210–215.

Iversen, J. 1964. Plant indicators of climate, soil, and other factors during the Quaternary. *Report of the Sixth International Congress on the Quaternary, Warsaw 1961, Palaeobotanical Section*, 2, 421–428.

Iversen, J. 1973. The development of Denmark's nature since the last glacial. *Danmarks Geologiske Undersøgelse V Series*, No. 7-C, 7–126.

Iwakuma, T. 1986. *Ecology and production of* Tokunagayusurika akamusi *Tokunaga. and* Chironomus plumosus *L. Diptera: Chironomidae. in a shallow eutrophic lake*, PhD Thesis, Kyushu University, Japan.

Jackson, S.T. 1989. Post-glacial vegetation changes along an elevational gradient in the Adirondack Mountains New York: a study of plant macrofossils. *New York State Museum and Science Service Bulletin 465*. NY State Museum and Science Service, Albany, NY.

Jackson, S.T. and Givens C.R. 1994. Late Wisconsinan vegetation and environment of the Tunica Hills region, Louisiana/Mississippi. *Quaternary Research*, 41, 316–325.

Jackson, S.T. and Weng, C. 1999. Late Quaternary extinction of a tree species in eastern North America. *Proceedings of the National Academy of Sciences*, 96, 13847–13852.

Jackson, S.T. and Whitehead, D.R. 1991. Holocene vegetation patterns in the Adirondack Mountains. *Ecology*, 72, 631–653.

Jacobson, G.L. and Bradshaw, R.H.W. 1981. The selection of sites for paleovegetational studies. *Quaternary Research*, 16, 80–96.

Jacobson, G.L and Grimm, E.C. 1986. A numerical analysis of Holocene forest and prairie vegetation in central Minnesota. *Ecology*, 67, 958–966.

Jalut, G., Galop, D., Aubert, S. and Belet, J.M. 1996. Late-glacial and post-glacial fluctuations of the tree limits in the Mediterranean Pyrenees: the use of pollen ratios. *Paläoklimaforschung*, 20, 189–201.

Jasper, J.P. and Hayes, J.M. 1990. A carbon isotope record of CO_2 levels during the late Quaternary. *Nature*, 347, 462–464.

Jennings, A.E. and N.J. Weiner 1996. Environmental change on eastern Greenland during the last 1300 years: evidence from formainifera and lithofacies in Nansen Fjord. *The Holocene*, 6, 179–191.

Jennings, A.E., Knudsen, K.L., Hald, M., Hansen, C.V. and Andrews, J.T. 2002. A Mid-Holocene shift in Arctic sea ice varibility on the East Greenland shelf. *The Holocene*, 12, 49–58.

Jewson, D.H. and Granin, N.G. 2000. How can present day studies of diatoms help in understanding past climatic change in Baikal? *Terra Nostra*, 9, 29–33.

Jiang, H., Björck, S. and Svensson, N.-O. 1998. Reconstruction of Holocene sea-surface salinity in the Skagerrak-Kattegat: a climatic and environmental record of Scandinavia. *Journal of Quaternary Science*, 13, 107–114.

Jiang, H., Seidenkrantz, M.-S., Knudsen, K.L. and Eiríksson, J. 2001. Diatom surface sediment assemblages around Iceland and their relationships to oceanic environmental variables. *Marine Micropaleontology*, 41, 73–96.

Jiang, H., Seidenkrantz, M.-S., Knudsen, K.L. and Eiríksson, J. 2002. Late-Holocene summer sea-surface temperatures based on a diatom record from the north Icelandic shelf. *The Holocene*, 12, 137–147.

Johannesson, T., Raymond, C. and Waddington, E. 1989. Time-scale for adjustment of glaciers to changes in mass balance. *Journal of Glaciology*, 35, 355–369.

Johnsen, S.J., Dahl-Jensen, D., Dansgaard, W. and Gundestrup, N. 1995. Greenland palaeotemperatures derived from GRIP bore hole temperature and ice core isotope profiles. *Tellus, Series B*, 47, 624–629.

Johnsen, S.J., Dahl-Jensen, D., Gundestrup, N.S., *et al.* 2001. Oxygen isotope and palaeotemperature records from six Greenland ice-core stations: Camp Century, Dye-3, GRIP, GISP2, Renland and NorthGRIP. *Journal of Quaternary Science*, 16, 299–307.

Johnson, A. and Earle, T.K. 1987. *The evolution of human societies: from foraging group to agrarian state.* Stanford University Press, Stanford.

Johnson, B.J., Miller, G.H., Fogel, M.L. and Beaumont, P.B. 1997. The determination of Late Quaternary palaeoenvironments at Equus Cave, South Africa, using stable-isotopes and amino acid racemization in ostrich eggshell. *Palaeogeography, Palaeoclimatology, Palaeoecology*, 136, 121–137.

Johnson, T.C., Barry, S.L., Chan, Y. and Wilkinson, P. 2001. Decadal record of climate variability spanning the past 700 yr in the southern tropics of East Africa. *Geology*, 29, 83–86.

Johnson, T.C., Brown, E.T., McManus, J., Barry, S., Barker, P. and Gasse, F. 2002. A high resolution paleoclimate record spanning the past 25,000 years in southern East Africa. *Science*, 296, 113–116.

Jolly, D., Harrison, S.P., Damnati, B. and Bonnefille, R. 1998. Simulated climate and biomes of Africa during the late quarternary: comparison with pollen and lake status data. *Quaternary Science Reviews*, 17, 629–657.

Jones, M., Leng, M.J., Eastwood. W., Keen, D. and Turney, C. 2002. Interpreting stable-isotope records from freshwater snail-shell carbonate: a Holocene case study from Lake Gölhisar, Turkey. *The Holocene*, 12, 629–634.

Jones, M., Leng, M.J., Eastwood, W., Keen, D. and Turney, C. In press. Stable-isotope records from freshwater snail shell carbonate. *The Holocene.*

Jones, P.D. 1995. Land surface temperatures: is the network good enough? *Climatic Change*, 31, 545–558.

Jones, P.D. and Briffa, K.R. 1995. Growing season temperatures over the former Soviet Union. *International Journal of Climatology*, 15, 943–959.

Jones, P.D. and Lister, D.H. 1998. Riverflow reconstructions and their analysis on 15 catchments over England and Wales. *International Journal of Climatology*, 18, 999–1013.

Jones, P.D.,Wigley, T.M.L. and Briffa, K.R. 1987. *Monthly mean pressure reconstructions for Europe 1780–1980 and North America 1858–1980.* NDP-025 Carbon Dioxide Information Analysis Centre, Oak Ridge National Laboratory, Oak Ridge, TN.

Jones, P.D., Bradley, R.S. and Jouzel, J. (eds). 1996. *Climatic variations and forcing mechanisms over the last 2000 years.* Springer-Verlag, Berlin.

Jones, P.D., Osborn, T.J. and Briffa, K.R. 1997a. Estimating sampling errors in large-scale temperature averages. *Journal of Climate*, 10, 2548–2568.

Jones, P.D., Jonsson, T. and Wheeler, D. 1997b. Extension to the North Atlantic Oscillation using early instrumental pressure observations from Gibraltar and South-west Iceland. *International Journal of Climatology*, 17, 1433–1450.

Jones, P.D., Briffa, K.R., Barnett, T.P. and Tett, S.F.B. 1998. High-resolution palaeoclimatic records for the last millennium: interpretation, integration and comparison with General Circulation Model control run temperatures. *The Holocene*, 8, 455–471.

Jones, P.D., Davies, T.D., Lister, D.H., *et al.* 1999a. Monthly mean pressure reconstructions for Europe for the 1780–1995 period. *International Journal of Climatology*, 19, 347–364.

Jones, P.D., New, M., Parker, D.E., Martin, S. and Rigor, I.G. 1999b. Surface air temperature and its variations over the last 150 years. *Reviews of Geophysics*, 37, 173–199.

Jones, P.D., Osborn, T.J., Briffa, K.R., *et al.* 2001a. Adjusting for sampling density in grid-box land and ocean surface temperature time series. *Journal of Geophysical Research*, 106, 3371–3380.

Jones, P.D., Osborn, T.J. and Briffa, K.R. 2001b. The evolution of climate over the last millennium. *Science*, 292, 662–667.

Jones, P.D., Briffa, K.R., Osborn, T.J., Moberg, A. and Bergström, H. 2002. Relationships between circulation strength and the variability of growing season and cold season climate in northern and central Europe. *The Holocene*, 12, 643–656.

Jones, T.L., G.M. Brown, L.M. Raab, *et al.* 1999c. Environmental imperatives reconsidered: demographic crises in western North America during the Medieval climatic anomaly. *Current Anthropology*, 40, 137–170.

Jones, V.J., Flower, R.J., Appleby, P.G., *et al.* 1993. Palaeolimnological evidence for the acidification and atmospheric contamination of lochs in the Cairngorm and Lochnagar areas of Scotland. *Journal of Ecology*, 81, 3–24.

Joussaume, S. and Taylor, K. E. 1995. Status of the Paleoclimate Modeling Intercomparison Project (PMIP). *Proceedings of the first international AMIP scientific conference WCRP-92*, Monterey, CA, pp. 425–430.

Joussaume, S., Taylor, K.E., Braconnot, P., *et al.* 1999. Monsoon changes for 6000 years ago: Results of 18 simulations from the Paleoclimate Modeling Intercomparison Project (PMIP). *Geophysical Research Letters*, 26, 859–862.

Jouzel, J., Alley, R.B., Cuffey, K.M., *et al.* 1997. Validity of the temperature reconstruction from water isotopes in ice cores. *Journal Geophysical Research Special Publication*, 102, 26471–26448.

Jouzel, J., Hoffmann, G., Koster, R. D. and Masson, V. 2000. Water isotopes in precipitation: data/model comparison for present-day and past climates. *Quaternary Science Reviews*, 19, 363–379.

Joynt, E.H. and Wolfe, A.P. 2001. Paleoenvironmental inference models from sediment diatom assemblages in Baffin Island lakes Nunavut, Canada. and reconstruction of summer water temperature. *Canadian Journal of Fisheries and Aquatic Sciences*, 58, 1222–1243.

Juggins, S. 1997. *MAT, a C++ computer program for the modern analogue technique.* See website: http://www.campus.ncl.ac.uk/staff/Stephen.Juggins/.

Juggins, S. and ter Braak, C.J.F. 1997a. CALIBRATE, a C++ program for analysing and visualising species-environment relationships and for predicting environmental values from species assemblages. University of Newcastle, Newcastle, UK.

Juggins, S. and ter Braak, C.J.F. 1997b. *WAPLS, a C++ computer program for weighted-averaging partial least squares regression and calibration.* See website: http://www.campus.ncl.ac.uk/staff/Stephen.Juggins/.

Juggins, S., Battarbee, R.W., Frtiz, S.C. and Gasse, F. 1994. The CASPIA project: diatoms, salt lakes and environmental change. *Journal of Paleolimnology*, 12, 191–196.

Juillet-Leclerc, A. and Labeyrie, L. 1987. Temperature dependence of the oxygen isotopic fractionation between diatom silica and water. *Earth and Planetary Science Letters*, 84, 69–74.

Kaiser, K.F. 1993 Growth rings as indicators of glacial advances, surges and floods. *Dendrochronologia*, 11, 101–122.

Kalnay, E., Kanamitsu, M., Kistler, R. *et al.*, 1996. The NMC/NCAR 40-year reanalysis project. *Bulletin of the American Meteorological Society*, 77, 437–471.

Kapsner, W.R., Alley, R.B., Shuman, C.A., Anandakrishnan, S. and Grootes, P.M. 1995. Dominant influence of atmospheric circulation on snow accumulation in Greenland over the past 18,000 years. *Nature*, 373, 52–54.

Karlén, W. 1976. Lacustrine sediments and tree-limit variations as indicators of Holocene climatic fluctuations in Lappland: Northern Sweden. *Geografiska Annaler*, 58A, 1–34.

Karlén, W. 1981. Lacustrine sediment studies. *Geografiska Annaler*, 63A, 273–281.

Karlén, W. 1988. Scandinavian glacial and climatic fluctuations during the Holocene. *Quaternary Science Reviews*, 7, 199–209.

Karlén, W. and Matthews, J.A. 1992. Reconstructing Holocene glacier variations from glacial lake sediments: Studies from Nordvestlandet and Jostedalsbreen-Jotunheimen, southern Norway. *Geografiska Annaler*, 74A, 327–348.

Kaufman, A., Wasserburg, G.J., Porcelli, D., Bar-Matthews, M., Ayalon, A. and Halicz, L. 1998. U-Th isotope systematics from the Soreq cave, Israel and climate correlations. *Earth and Planetary Science Letters*, 156, 141–155.

Kearney, M.S. 2001. Late Holocene sea level variation. In B.C. Douglas, M.S. Kearney and S.P. Leatherman (eds), *Sea level rise, history and consequences*. Academic Press, San Diego, pp. 13–36.

Keatings, K.W., Heaton, T.H.E. and Holmes, J.A. 2002. Carbon and oxygen isotope fractionation in non-marine ostracods: results from a 'natural culture' environment. *Geochimica et Cosmochimica Acta*, 66, 1701–1711.

Keely, B.J. and Maxwell, J.R. 1993. The Mulhouse basin evidence from porphyrin distributions for water column anoxia during deposition of marls. *Organic Geochemistry* 20, 1217–1225.

Keigwin, L.D. 1996. The Little Ice Age and Medieval Warm Period in the Sargasso Sea. *Science*, 274, 1504–1507.

Keil, R.G. and Cowie, G.L. 1999. Organic matter preservation through the oxygen-deficient zone of the NE Arabian Seas as discerned by organic carbon:mineral surface area ratios. *Marine Geology*, 161, 13–22.

Kelley, J.T., Belnap, D.F. and Daly, J.F. 2001. Comment on North Atlantic climate-ocean variations and sea level in Long Island Sound, Connecticut, since 500 cal yr AD. *Quaternary Research*, 55, 105–107.

Kelly, D.E. 1997. Convection in ice-covered lakes: effects on algal suspension. *Journal of Plankton Research*, 19, 1859–1880.

Kelts, K. and K.J. Hsü, 1978. Freshwater carbonate sedimentation. In A. Lerman (ed.), *Lakes: chemistry, geology, physics*. Springer, Berlin, pp. 295–323.

Kemp, A.E.S. (ed.) 1996. *Palaeoclimatology and Palaeoceanography from laminated sediments*. Geological Society of London Special Publication 116. The Geological Society, London.

Kemp, A.E.S., Pearce, R.B., Koizumi, I., Pike, J. and Rance, S.J. 1999. The role of mat-forming diatoms in formation of the Mediterranean sparopels. *Nature*, 398, 57–61.

Kemp, A.E.S., Pike, J., Pearce, R.B. and Lange, C.B. 2000. The 'Fall dump': a new perspective on the role of a 'shade flora' in the annual cycle of diatom production and export flux. *Deep-Sea Research Part II*, 47, 2129–2154.

Kershaw, A.P., Bush, M.B., Hope, G.S., Weiss-K.-F., Goldammer, J.G. and Sanford, R. 1997. The contribution of humans to past biomass burning in the tropics. In J.S. Clark, H. Cachier, J.G. Goldammer and B. Stocks (eds), *Sediment records of biomass burning and global change*. Springer-Verlag, Berlin, pp. 413–442.

Key, R.M. 1996. WOCE Pacific Ocean radiocarbon program. *Radiocarbon*, 38 3. 415–425.

Keys, D. 1999. *Catastrophe: an investigation into the origins of the modern world*. Century, London.

Keyser, D. 1977. Ecology and zoogeography of recent brackish-water Ostracoda Crustacea from South-west Florida. In H. Löffler, and D. Danielopol (eds), *Aspects of the ecology and zoogeography of recent and fossil Ostracoda*. Dr. W. Junk Publishers, The Hague, pp. 207–222.

Kilian, M.R., van der Plicht, J. and van Geel, B. 1995. Dating raised bogs: new aspects of AMS ^{14}C wiggle-matching, a reservoir effect and climatic change. *Quaternary Science Reviews*, 14, 959–966.

Kilian, M.R., van Geel, B. and van der Plicht, J. 2000. ^{14}C AMS wiggle-matching of raised bog deposits and models of peat accumulation. *Quaternary Science Reviews*, 19, 1011–1033.

Killops, S.D. and Killops, V.J. 1993. *An introduction to organic geochemistry*. Longman Scientific and Technical, Harlow.

Kislev, M.E., Nader, D. and Carmi, I. 1992. Epipaleolithic 19,000 BP. cereal and fruit diet at Ohalo II, Sea of Galilee, Israel. *Review of Palaeobotany and Palynology*, 73, 161–186.

Kistler R., Kalnay, E., Collins, W. *et al.* 2001. The NCEP/NCAR 50-Year Reanalysis: Monthly Means, CD-ROM and Documentation. *Bulletin of the American Meteorological Society*, 82, 247–267.

Kitagawa, H. and van der Plicht, J. 1998. Atmospheric radiocarbon calibration to 45,000 yr BP: Late Glacial fluctuations and cosmogenic isotope production. *Science*, 279, 1187–1190.

Kivinen, E. and Pakarinen, P. 1981. Geographical distribution of peat resources and major peatland complex types in the world. *Annales Academiae Scientorum Fennicae AIII*,132, 1–28.

Kjøllmoen, B. 1998. Glasiologiske undersøkelser i Norge 1996 og 1997. Rapport 20, Norges vassdrags-og energiverk, Oslo, 139.

Klein, R.G., Cruz-Uribe, K. and Beaumont, P.B. 1991. Environmental, ecological and palaeoanthropological implications of the Late Pleistocene mammalian fauna from Equus Cave, Northern Cape Province, South Africa. *Quaternary Research*, 36, 94–119.

Kleypas, J.A., Buddemeier, R.W., Archer, D., *et al.* 1999. Geochemical consequences of increased atmospheric CO_2 on corals and coral refs. *Science*, 284, 118–120.

Knorr, W., Schnitzler, K.-G., Govaerts, Y. 2001. The role of bright desert regions in shaping North African climate. *Geophysical Research Letters*, 28, 3489–3492.

Knox, J.C. 2000. Sensitivity of modern and Holocene floods to climate change. *Quaternary Science Reviews*, 19, 439–457.

Knutson, D.K., Buddemeier, R.W. and Smith, S.V. 1972. Coral chronometers: seasonal growth bands in reef corals. *Science*, 177, 270–272.

Koerner, R.M.1979. Accumulation, ablation, and oxygen isotope variations on the Queen Elizabeth Islands ice caps, Canada. *Journal of Glaciology*, 22, 25–41.

Koerner, R.M. 1989. Ice core evidence for extensive melting of the Greenland ice sheet during the last interglacial. *Science*, 244, 964–968.

Koerner, R.M. 1997. Some comments on climatic reconstructions from ice cores drilled in areas of high melt. *Journal of Glaciology*, 43, 90–97.

Koerner, R.M. 2001. Sudden climatic change. Proceedings of the second International Symposium on Environmental Research in the Arctic and Fifth Ny-Ålesund Scientific Seminar, 21–25 February, NIPR, Tokyo. *Memoirs of National Institute of Polar Research, Special Issue*, 54, 203–207.

Koerner, R.M. and Fisher, D.A. 1990. A record of Holocene summer climate from a Canadian high-Arctic ice core. *Nature*, 343, 630–631.

Koerner, R.M. and Fisher, D.A. 2002. Ice core evidence for massive glacier retreat in the last interglacial and early Holocene. *Annals of Glaciology*, 35, 19–24.

Koerner, R.M. and Russell, R.D. 1979. $\delta(^{18}O)$ variations in snow on the Devon Island Ice Cap, Northwest Territories, Canada. *Canadian Journal of Earth Science*, 16, 1419–1427.

Kohfeld, K.E. and Harrison S.P. 2000. How well can we simulate past climates? Evaluating the models using global palaeoenvironmental datasets. *Quaternary Science Reviews*, 19, 321–346.

Koinig, K.A, Sommaruga-Wögrath, S., Schmidt, R., Tessadri, R. and Psenner, R. 1998a. Acidification processes in high alpine lakes. In M.J. Haigh, J. Krecek, G.S. Raijwar, M.P. Kilmartin (eds), *Headwaters: water resources and soil conservation*. A.A. Balkema Publishers, Rotterdam, pp. 45–54.

Koinig, K.A, Schmidt, R., Sommaruga-Wögrath, S., Tessadri, R. and Psenner, R. 1998b. Climate change as the primary cause for pH shifts in a high alpine lake. *Water, Air and Soil Pollution*, 104, 167–180.

Kon, M. 1984. Swarming and mating of *Chironomus yoshimatsui* Diptera: Chironomidae: seasonal change in the timing of swarming and mating. *Journal of Ethology*, 2, 37–45.

Körber-Grohne, U. 1964. Bestimmungsschlussel für subfossile Juncus-samen und Gramineen-Fruchte. Probleme der Kustenforschung im Sudlichen Nordseegebiet 7. August Lax, Hildesheim.

Korhola, A. 1995. Holocene climatic variations in southern Finland reconstructed from peat-initiation data. *The Holocene*, 5, 43–58.

Korhola, A. 1996. Initiation of a sloping mire complex in southwestern Finland: autogenic versus allogenic controls. *Ecoscience*, 3, 216–222.

Korhola, A., Weckström, Holmström, L. and Erästö, P. 2000. A quantitative Holocene climatic record from diatoms in northern Fennoscandia. *Quaternary Research*, 54, 284–294.

Korhola, A., Birks, H.J.B., Olander, H. and Blom, T. 2001. Chironomids, temperature and numerical methods: a reply to Seppälä. *The Holocene*, 11, 615–622.

Korhola, A., Vasko, K., Toivonen, H.T.T. and Olander, H. 2002. Holocene temperature changes in northern Fennoscandia reconstructed from chironomids using Bayesian modelling. *Quaternary Science Reviews*, 21, 1841–1860.

Kotlyakov, V.M., Nikolayev, V.M., Korotkov, I.M. and Klementyev, O.L. 1991. Climate-stratigraphy of Severnaya Zemlya ice dome in the Holcene. In G.I. Khudyakov (ed.), *Stratigraphy and correlation of Quaternary deposits of East Asia and the Pacific region*. Nauka, Moscow, pp.100–112. (In Russian.)

Kremenetski, C.V., Sulerzhitsky, L.D. and Hantemirov, R. 1998. Holocene history of the northern range limits of some trees and shrubs in Russia. *Arctic and Alpine Research*, 30, 317–333.

Krippendorff, K. 1980. *Content analysis : an introduction to its methodology*. Sage Publications, Beverly Hills.

Kruss, P. 1983. Climate change in East Africa: numerical simulation from the 100 years of terminus record at Lewis Glacier, Mount Kenya. *Zeitschrift für Gletscherkunde und Glazialgeologie*, 19, 43–60.

Kubatzki, C. and Claussen, M. 1998. Simulation of the global biogeophysical interactions during the last glacial maximum. *Climate Dynamics*, 14, 461–471.

Kuch, M., Rohland, N., Betancourt, J.L., Latorre, C., Steppan, S. and Poinar, H.M. 2002. Molecular analysis of a 11 700-year-old rodent midden from the Atacame desert, Chile. *Molecular Ecology*, 11, 913–924.

Kühnert, H., Patzold, J., Hatcher, B., *et al.* 1999. A 200-year coral stable oxygen isotope record from a high-latitude reef off Western Australia. *Coral Reefs*, 18, 1–12.

Kühnert, H., Pätzold, J., Wyrwoll, K.-H. and Wefer, G. 2000. Monitoring climate variability over the past 116 years in coral oxygen isotopes from Ningaloo Reef, Western Australia. *International Journal of Earth Sciences*, 88, 725–732.

Kullman, L. 1995. Holocene tree-limit and climate history from the Scandes mountains, Sweden. *Ecology*, 76, 2490–2502.

Kullman, L. 1996. Structural population dynamics of pine *Pinus sylvestris* L. during the past 500 years close to the tree-limit in northern Sweden. *Paläoklimaforschung*, 20, 75–82.

Kullman, L. 1998. Palaeoecological, biogeographical and palaeoclimatological implications of early Holocene immigration of *Larix sibirica* Ledeb. into the Scandes Mountains, Sweden. *Global Ecology and Biogeography Letters*, 7, 181–188.

Kuniholm, P.I. 1996 The Prehistoric Aegean: dendrochronological progress as of 1995. In K. Randsborg (ed.), Absolute Chronology: Archaeological Europe 2500–500 BC. *Acta Archaeologica Supplementum*, 1, 327–335.

Kureck, A. 1979. Two circadian eclosion times in *Chironomus thummi*; ecological adjustment to different temperature levels and the role of temperature cycles in Chironomidae. In D.A. Murray (ed.), *Chironomidae: ecology, systematics, cytology and physiology*. Pergamon Press, New York.

Kutzbach, J.E. 1996. Vegetation and soil feedbacks on the response of the African monsoon to orbital forcing in the Early to Mid-Holocene. *Nature*, 384, 623–626.

Kutzbach, J.E. and Guetter, P.J. 1986. The influence of changing orbital parameters and surface boundary conditions on climate simulations for the past 18,000 years. *Journal of Atmospheric Science*, 43, 1726–1759.

Kutzbach, J.E. and Liu, Z. 1997. Response of the African monsoon to orbital forcing and ocean feedbacks in the middle Holocene. *Science*, 278, 440–444.

Kutzbach, J.E. and Webb, T. 1991. Late Quaternary climatic and vegetational change in eastern North America: concepts, models, and data. In L.C.K. Shane and E.J. Cushing (eds), *Quaternary Landscapes*. University of Minnesota Press, Minneapolis, MN, pp. 175–217.

Kutzbach, J.E., Guetter, P.J., Behling, P.J. and Selin, R. 1993. Simulated climatic changes: Results of the COHMAP climate model experiments. In H.E. Wright Jr., J.E. Kutzbach, T. Webb III, W.F. Ruddiman, F. A. Street-Perrott, and P.J. Bartlein (eds), *Global climates since the last glacial maximum*. University of Minnesota Press, Minneapolis, MN, pp. 24–93.

Kutzbach, J.E., Bonan, G., Foley, J., Harrison, S.P. 1996. Vegetation and soil feedbacks on the response of the African monsoon to orbital forcing in the early to middle Holocene. *Nature*, 384, 623–626.

Kutzbach, J.E., Gallimore, R., Harrison, S., Behling, P., Selin, R. and Laarif, F. 1999. Climate and Biome simulations for the past 21,000 years. *Quaternary Science Reviews*, 17, 473–506.

Labeyrie, L. 1974. New approach to surface seawater palaeotemperatures using $^{18}O/^{16}O$ ratios in silica of diatom frustules. *Nature*, 248, 40–42.

Labeyrie, L.D. and Juillet, A. 1982. Oxygen isotopic exchangeability of diatom valve silica; interpretation and consequences for paleoclimatic studies. *Geochimica et Cosmochimica Acta*, 46, 967–975.

Labeyrie, L., Cole, J.E., Alverson, K. and Stocker, T.F. 2003. The history of climate dynamics. In K. Alverson, R.S. Bradley and T. Pedersen (eds), *Paleoclimate, global change and the future*. Springer, Berlin, 33–61.

Laird, K.R., Fritz, S.C., Grimm, E.C. and Mueller, P.G. 1996. Century-scale paleoclimatic reconstruction from Moon Lake, a closed basin in the northern Great Plains. *Limnology and Oceanography*, 41, 890–902.

Laird, K.R., Fritz, S.C. and Cumming, B.F. 1998a. A diatom-based reconstruction of drought intensity, duration, and frequency from Moon Lake, North Dakota: a sub-decadal record of the last 2300 years. *Journal of Paleolimnology*, 19, 161–179.

Laird, K.R., Fritz, S.F., Cumming, B.F. and Grimm, E.C. 1998b. Early-Holocene limnological and climate variability in the Northern Great Plains. *The Holocene*, 8, 275–285.

LaMarche Jr, V.C. and Fritts, H.C. 1972. Tree-rings and sunspot numbers. Tree-*Ring Bulletin*, 32, 19–32.

Lamb, A. L., Leng, M.J., Lamb, H.F. and Mohammed, M.U. 2000. A 9000-year oxygen and carbon isotope record of hydrological change in a small Ethiopian crater lake. *The Holocene*, 10, 167–177.

Lamb, H.F., Roberts, N., Leng, M.J., Barker, P., Benkaddour, A. and van der Kaars, S. 1999. Lake evolution in a semi-arid montane environment: responses to catchment change and hydroclimatic variation. *Journal of Paleolimnology*, 21, 325–343.

Lamb, H.H. 1963. What can we learn about the trend of our climate? *Weather*, 18,194–216.

Lamb, H.H. 1972a. *Climate: past, present and future*. Methuen, London.

Lamb, H.H. 1972b. British Isles weather types and a register of the daily sequence of circulation patterns 1861–1971. *Geophysical Memoir No.116*. HMSO for the Meteorological Office, London.

Lamb, H.H. 1977. *Climate: past, present and future*, Vol. 2. Methuen, London.

Lamb, H., Kebede, S., Leng. M.J., Ricketts, D., Telford, R. and Umer, M. 2002. Origin and stable-isotope composition of aragonite laminae in an Ethiopian crater lake. In E. Odada and D. Olago (eds), *The East African Great Lakes Region: limnology, palaeoclimatology and biodiversity, Advances in Global Research Series*. Kluwer, Dordrecht.

Lambeck, K. 2002. Sea level change from Mid Holocene to recent time: An Australian example with global implications. *Ice Sheets, Sea Level and the Dynamic Earth*. American Geophysical Union, Washington, D.C. Geodynamic Series, 29, 35–50.

Lamoureux, S. 1994. Embedding unfrozen lake sediments for thin section preparation. *Journal of Paleolimnology*, 10, 141–146.

Lamoureux, S. 1999. Spatial and interannual variations in sedimentation patterns recorded in nonglacial varved sediments from the Canadian High Arctic. *Journal of Paleolimnology*, 21, 73–84.

Lamoureux, S. 2001. Varve chronology techniques. In W.M. Last and J.P. Smol (ed.), *Tracking environmental change using lake sediments: Physical and geochemical techniques.* Developments in Paleoenvironmental Research, Vol. 1. Kluwer Academic Publishers, Dordrecht, pp. 247–260.

Lamoureux, S.F. and R. S. Bradley 1996. A late Holocene varved sediment record of environmental change from northern Ellesmere Island, Canada. *Journal of Paleolimnology*, 16, 239–255.

Lang, B. accepted. The use of ultrasound in the preparation of carbonate sediments. *Journal of Paleolimnology*.

Langdon, C., Takahashi, T., Sweeney, C., *et al.* 2000. Effect of calcium carbonate saturation state on the calcification rate of an experimental coral reef. *Global Biogeochemical Cycles*, 14, 639–954.

Langdon, P.G. 1999. Reconstructing Holocene climate change in Scotland utilising peat stratigraphy and tephrochronology, PhD thesis. University of Southampton, Southampton.

Langdon, P.G. and Barber, K.E. 2001. New Holocene tephras and a proxy climate record from a blanket mire in northern Skye, Scotland. *Journal of Quaternary Science*, 16, 753–759.

Langdon, P.G., Barber, K.E. and Hughes, P.D.M. 2003. A 7500-year peat-based palaeoclimatic reconstruction and evidence for an 1100-year cyclicity in bog surface wetness from Temple Hill Moss, Pentland Hills, Southeast Scotland. *Quaternary Science Reviews*, 22, 259–274.

Langdon, P.G. and Barber, K.E. In press. Snapshots in time: precise correlations of peat based proxy climate records in Scotland using tephras. *The Holocene*, 14.

Lappalainen, E. (ed.) 1996. *Global peat resources*. International Peat Society, Finland.

Lara, A. and Villalba, R. 1993 A 3620-year temperature record from *Fitzroya cupressoides* tree-rings in Southern South America. *Science*, 260, 1104–1106.

Larkin, A., Haigh, J.D. and Djavidnia, S. 2000. The effect of solar UV irradiance variations on the earth's atmosphere. *Space Science Reviews*, 94, 199–214.

Larocque, I., Hall, R.I. and Grahn, E. 2001. Chironomids as indicators of climate change: a 100-lake training set from a subarctic region of northern Sweden Lapland. *Journal of Paleolimnology*, 26, 307–322.

Larson, D.O. and Michaelson, J. 1990 Impacts of Climatic Variability and Population Growth on Virgin branch Anasazi Cultural. *American Antiquity* 55, 227–249.

Larson, D.O., Neff, H., Graybill, D.A., Michaelsen, J. and Ambos, E. 1996. Risk, climatic variability and the study of southwestern prehistory: an evolutionary perspective. *American Antiquity*, 61, 217–241.

Laskar, J. 1990. The chaotic motion of the solar system: A numerical estimate of the chaotic zones. *Icarus*, 88, 266–291.

Last, W.M. and Smol, J.P. (ed.) 2001a. *Tracking environmental change using lake sediments, Vol. 1: Basin analysis, coring and chronological techniques*. Developments in Paleoenvironmental Research, 1. Kluwer Academic Publishers, Dordrecht.

Last, W.M. and Smol, J.P. (ed.) 2001b. *Tracking environmental change using lake sediments, Vol. 2: Physical and geochemical methods*. Developments in Paleoenvironmental Research, 2. Kluwer Academic Publishers, Dordrecht.

Last, W.M., Vance, R.E., Wilson, S. and Smol, J.P. 1998. A multi-proxy limnologic record of rapid early-Holocene hydrologic change on the northern Great Plains, southwestern Saskatchewan, Canada. *The Holocene*, 8, 503–520.

Laumann, T. and Reeh, N. 1993. Sensitivity to climate change of the mass balance of glaciers in southern Norway. *Journal of Glaciology*, 39, 656–665.

Lauritzen, S.E. 1993. Natural environmental change in Karst: the Quaternary record. *Catena Supplement*, 25, 21–40.

Lauritzen, S.E. 1995. High-resolution paleotemperature proxy record during the last interglaciation in Norway from speleothems. *Quaternary Research*, 43, 133–146.

Lauritzen, S.E. 1996. Calibration of speleothem stable-isotopes against historical records: a Holocene temperature curve for north Norway? *Karst Waters Institute Special Publication* 2, 78–80.

Lauritzen, S.E. and Lundberg, J. 1998. Rapid temperature variations and volcanic events during the Holocene from a Norwegian speleothem record. *PAGES Open Science Meeting, University of London, April 20–23, 1998*, pp. 88.

Lauritzen, S.E. and Lundberg, J. 1999a. Calibration of the speleothem delta function: an absolute temperature record for the Holocene in northern Norway. *The Holocene*, 9, 659–670.

Lauritzen, S.E. and Lundberg, J. 1999b. Speleothems and climate: a special issue of *The Holocene*. *The Holocene*, 9, 643–647.

Lauritzen, S.E. and Onac, P.B. 1996. Uranium-series dating of some speleothems from Romania. *Theoretical and Applied Karstology*, 8, 25–36.

Lauritzen, S.E., Ford, D.C. and Schwarcz, H.P. 1986. Humic substances in speleothem matrix-paleoclimatic significance. *Proceedings of the 9th International Speleological Congress, Barcelona*, 1, 77–79.

Lauritzen, S.E., Løvlie, R., Moe, D. and Østbye, E. 1990. Paleoclimate deduced from a multidiciplinary study of a half-million-year-old stalagmite from Rana, Northern Norway. *Quaternary Research*, 34, 306–316.

Lauritzen, S.E., Haugen, J.E., Gilje-Nilsen, H. and Løvlie, R. 1994. Geochronological potential of isoleucine epimerization in calcite speleothem. *Quaternary Research*, 41, 52–58.

Lauritzen, S.E., Hercman, H. and Glazek, J. 1996. Preliminary comparison between Norwegian and Polish speleothem growth frequencies. *Karst Waters Institute Special Publication*, 2, 81–83.

Lawrence, J.K. and A.A. Ruzmaikin, 1998. Transient solar influence on terrestrial temperature fluctuations. *Geophysical Research Letters*, 25, 159–162.

Laws, E.A., Popp, B.N., Cassar, N. and Tanimoto, J. 2002. ^{13}C discrimination patterns in oceanic phytoplankton: likely influence of CO_2 concentrating mechanisms, and implications for paleoreconstructions. *Functional Plant Biology*, 29, 323–333.

Lea, D.W., Boyle, E.A. and Shen, G.T. 1989. Coralline barium records temporal variability in equatorial Pacific upwelling. *Nature*, 340, 373–376.

Lean, J.L. 1996. Reconstructions of past solar variability. In P.D. Jones, R.S. Bradley and J. Jouzel (eds), *Climate variations and forcing factors of the last 2000 years*. Springer, Berlin, pp. 519–532.

Lean, J.L. 2000. Evolution of the Sun's spectral irradiance since the Maunder Minuimum. *Geophysical Research Letters*, 27, 2425–2428.

Lean, J.L., Skumanich, A. and White, O.R. 1992. Estimating the Sun's radiative output during the Maunder Minimum. *Geophysical Research Letters*, 19, 1591–1594.

Lean, J., Beer, J. and Bradley, R.S. 1995. Reconstruction of solar irradiance since AD 1600 and implications for climate change. *Geophysical Research Letters*, 22, 3195–3198.

Le Bec, N., Juillet-Leclerc, A., Correge, T., Blamart, D. and Delcroix, T. 2000. A coral delta ^{18}O record of ENSO driven sea surface salinity variability in Fiji south-western tropical Pacific. *Geophysical Research Letters*, 27, 3897–3900.

Le Cornec, F. and Correge, T. 1996. Determination of uranium to calcium and strontium to calcium ratios in corals by inductively coupled plasma mass spectrometry. *Journal of Analytical Atomic Spectrometry*, 12, 969–973.

Leder, J.J., Swart, P.K., Szmant, A. and Dodge, R.E. 1996. The origin of variations in the isotopic record of scleractinian corals: I. Oxygen. *Geochimica Cosmochimica Acta*, 60, 2857–2870.

Leemann, A. and Niessen, F. 1994a. Varve formation and the climatic record in an Alpine proglacial lake: Calibrating annually laminated sediments against hydrological and meteorological data. *The Holocene*, 4, 1–8.

Leemann, A. and Niessen, F. 1994b. Holocene glacial activity and climatic variations in the Swiss Alps: reconstructing a continuous record from proglacial lake sediments. *The Holocene*, 4, 259–268.

Lee-Thorp, J.A. and Beaumont, P.B. 1995. Vegetation and seasonality shifts during the late Quaternary deduced from $^{13}C/^{12}C$ ratios of grazers at Equus Cave, South Africa. *Quaternary Research*, 43, 426–432.

Lee-Thorp, J., Holmgren, K. and Lauritzen, S.E. 2002. Rapid climate shifts in the southern African interior throughout the mid to late Holocene. *Geophysical Research Letters*, 27, 1–15.

Legrand, M. 1995. Sulphur-derived species in polar ice: a review. In R. Delmas (ed.), *Ice Core Studies of Global Biogeochemical Cycles, NATO ASI series, Series I, Vol. 30.* Springer, Berlin, pp. 91–119.

Legrand, M., Feniet-Saigne, C., Saltzman, E.S., Germain, C., Barkov, N.I. and Petrov, V.N.1991. Ice core record of oceanic emissions of dimethyl sulphide during the last climate cycle. *Nature*, 350, 144–146.

Legrand, M., Hammer, C., De Angelis, M., *et al.* 1997. Sulfur-containing species methanesulfonate and SO_4. over the last climatic cycle in the Greenland ice core project central Greenland. In Greenland summit ice cores. *Journal of Geophysical Research, special publication*, 102, 26663–26679.

Lemcke, G. 1996. Paläoklimarekonstruktion am Van See Ostanatolien, Türkei. ETH, Zürich, pp. 11786.

Lemcke, G. and Sturm, M. 1997. $\delta^{18}O$ and trace element measurements as proxy for the reconstruction of climate changes at Lake Van Turkey: preliminary results. In H.N. Dalfez, G. Kukla and H. Weiss (eds), *Third millennium BC climate change and the Old World collapse, Brussels: NATO ASI Series I 49.* Springer-Verlag, Berlin, pp. 653–678.

Leng, M.J., Lamb, A.L., Lamb, H.F. and Telford, R.J. 1999a. Palaeoclimatic implications of isotopic data from modern and early Holocene shells of the freshwater snail *Melanoides tuberculata*, from lakes in the Ethiopian Rift Valley. *Journal of Paleolimnology*, 21, 97–106.

Leng, M.J., Roberts, N., Reed, J. M. and Sloane, H.J. 1999b. Late Quaternary climatic and limnological variations based on carbon and oxygen isotope data from authigenic and Ostracod carbonate in the Konya Basin, Turkey. *Journal of Paleolimnology*, 22, 187–204.

Leng, M.J., Barker, P., Greenwood, P, Roberts, N. and Reed, J.M. 2001. Oxygen isotope analysis of diatom silica and authigenic calcite from Lake Pinarbasi, Turkey. *Journal of Paleolimnology*, 25, 343–349.

Lent, R.M. and Lyons, W.B. 2001. Biogeochemistry of silica in Devil's Lake: implications for diatom preservation. *Journal of Paleolimnology* 26, 53–66.

Leonard, E.M. 1986. Use of lacustrine sedimentary sequences as indicators of Holocene glacial history, Banff National Park, Alberta, Canada. *Quaternary Research*, 26, 218–231.

Leventer, A., Dunbar, R.B. and DeMaster, D.J. 1993. Diatom evidence for late Holocene climatic events in Granite Harbor, Antarctica: *Paleoceanography*, 8, 373–386.

Leventer, A., Domack, E.W., Ishman, S.E., Brachfield, S., McClennen, C.E. and Manley, P. 1996. Productivity cycles of 200–300 years in the Antarctic Peninsula region: understanding linkages among the sun, atmosphere, oceans, sea ice, and biota. *Geological Society of America Bulletin*, 108, 1626–1644.

Levesque, A.J., Mayle, F.E., Walker, I.R. and Cwynar, L.C. 1993a. A previously unrecognised late-glacial cold event in eastern North America. *Nature,* 361, 623–626.

Levesque, A.J., Mayle, F.E., Walker, I.R., and Cwynar, L.C. 1993b. The Amphi-Atlantic Oscillation: a proposed late-glacial climatic event. *Quaternary Science Reviews*, 12, 629–643.

Levesque, A.J., Cwynar, L.C. and Walker, I.R. 1994. A multiproxy investigation of Late-Glacial climate and vegetation change at Pine Ridge Pond, Southwest New Brunswick, Canada. *Quaternary Research*, 42, 316–327.

Levesque, A.J., Cwynar, L.C. and Walker, I.R. 1997. Exceptionally steep north–south gradients in lake temperatures during the last deglaciation. *Nature,* 385, 423–426.

Levis, S., Foley, J.A., Pollard, D. 1999a. Potential high-latitude vegetation feedbacks on CO_2-induced climate change. *Geophysical Research Letters*, 26, 747–750.

Levis, S., Foley, J.A., Brovkin, V., Pollard, D. 1999b. On the stability of the high-latitude climate-vegetation system in a coupled atmosphere–biosphere model. *Global Ecology and Biogeography*, 8, 489–500.

Lewes, V.B. 1910. Smoke and its prevention. *Nature,* 85, 290–295.

Lewis, C.F.M. and Anderson, T.W. 1992. Stable-isotope O and C. and pollen trends in eastern Lake Erie, evidence for a locally-induced climatic reversal of Younger Dryas age in the Great Lakes basin. *Climate Dynamics*, 6, 241–250.

Lézine, A.M., Casanova, J., and Hillaire-Marcel, C. 1990. Across an early Holocene humid phase in western Sahara: pollen and isotope stratigraphy. *Geology*, 18, 264–267.

Li, H.-C. and Ku, T.-L. 1997. $\delta^{13}C$–$\delta^{18}O$ covariance as a paleohydrological indicator for closed-basin lakes. *Palaeogeography, Palaeoclimatology, Palaeoecology*, 133, 69–80.

Li, H.-C., Bischoff, J. L., Ku, T. -L., Lund, S.P. and Stott, L.D. 2000. Climate variability in east-central California during the past 1000 years reflected by high-resolution geochemical and isotopic records from Owens Lake sediments. *Quaternary Research*, 54, 189–197.

Libby, W. 1965. *Radiocarbon dating.* University of Chicago Press, Chicago, pp. 175.

Lindegaard, C. 1995. Classification of water-bodies and pollution. In P.D. Armitage, P.S. Cranston and L.C.V. Pinder (eds), *The Chironomidae. The biology and ecology of non-biting midges*. Chapman and Hall, London.

Line, J.M., ter Braak, C.J.F. and Birks, H.J.B. 1994. WACALIB version 3.3 - a computer program to reconstruct environmental variables from fossil assemblages by weighted averaging and to derive sample-specific errors of predition. *Journal of Paleolimnology*, 10, 147–152.

Linge, H.C., Lauritzen, S.E., Lundberg, J. and Berstad, I.M. 2001a. Isotope stratigraphy of Holocene speleothems: examples from a cave system in Rana, northern Norway. *Palaeogeography, Palaeoclimatology, Palaeoecology*, 167, 209–224.

Linge, H.C., Lundberg, J. and Lauritzen, S.E. 2001b. Isotope stratigraphy of Holocene speleothems: an example from Fauske, northern Norway. *The Holocene*, submitted.

Linsley, B.K., Dunbar, R.B., Wellington, G.M. and Mucciarone, D.A. 1994. A coral-based reconstruction of Intertropical Convergence Zone variability over Central America since 1707. *Journal of Geophysical Research*, 99, 9977–9994.

Linsley, B.K., Messier, R.G. and Dunbar, R.B. 1999. Assessing between-colony oxygen isotope variability in the coral *Porites lobata* at Clipperton Atoll. *Coral Reefs*, 18, 13–27.

Linsley, B.K., Ren, L., Dunbar, R.B. and Howe, S.S. 2000a. El Niño Southern Oscillation ENSO and decadal-scale climate variability at 10N in the eastern Pacific from 1893 to 1994: a coral-based reconstruction from Clipperton Atoll. *Paleoceanography*, 15, 322–335.

Linsley, B.K., Wellington, G.M. and Schrag, D.P. 2000b. Decadal sea surface temperature variability in the subtropical South Pacific from 1726 to 1997 AD. *Science*, 290, 1145–1148.

Liu, K.-B. and Colinvaux, P.A. 1985. Forest changes in the Amazon basin during the last glacial maximum. *Nature*, 318, 556–557.

Liu, Y. and He, J. 1990. Translated title: The study on the palaeoclimatic long-term variation of the speleothem stalagmite. collected from Luohuzi Cave in Guilin, China. Earth Science. *Journal of China University of Geoscience,* 15, 689–696.

Liu, Z., Kutzbach, J.E. and Wu, L. 2000. Modeling climate shift of El Niño variability in the Holocene. *Geophysical Research Letters*, 27, 2265–2268.

Livingstone, D.A. 1967. Post-glacial vegetation of the Ruwenzori Mountains in equatorial Africa. *Ecological Monographs*, 37, 25–52.

Livingstone, D.M. 1999. Ice break-up on southern Lake Baikal and its relationship to local and regional air temperatures in Siberia and to the North Atlantic Oscillation. *Limnology and Oceanography*, 44, 1486–1497.

Livingstone, D.M. and Hajdas, I. 2001. Climatically relevant periodicities in the thicknesses of biogenic carbonate varves in Soppensee, Switzerland 9740–6870 calender yr BP. *Journal of Paleolimnology*, 25, 17–24.

Livingstone, D.M. and Lotter, A.F. 1998. The relationship between air and water temperature in lakes of the Swiss Plateau: a case study with palaeolimnological implications. *Journal of Paleolimnology*, 19, 181–198.

Livingstone, D.M., Lotter, A.F. and Walker, I.R. 1999. The decrease in summer surface water temperature with altitude in Swiss Alpine lakes: a comparison with air temperature lapse rates. *Arctic, Antarctic and Alpine Research*, 31, 341–352.

Loewe, F. 1971. Considerations of the origin of the Quaternary ice sheet in North America. *Arctic and Alpine Research*, 3, 331–344.

Lofgren, B.M. 1995a. Sensitivity of land–ocean circulations, precipitation, and soil moisture to perturbed land surface albedo. *Journal of Climate*, 8, 2521–2542.

Lofgren, B.M. 1995b. Surface albedo-climate feedback simulated using two-way coupling. *Journal of Climate*, 8, 10, 2543–2562.

Loope, D.B., Swinehart, J.B. and Mason, J.P. 1995. Dune-dammed paleovalleys of the Nebraska Sand Hills: intrinsic versus climatic controls on the accumulation of lake and marsh sediments. *Geological Society of America Bulletin*, 107, 396–406.

Lorius, C., Merlivat, L., Jouzel, J. and Pourchet, M. 1979. A 30,000-yr isotope climatic record from Antarctic ice. *Nature*, 280, 644–648.

Lotter, A.F. 1989. Evidence of annual layering in Holocene sediments of Soppensee, Switzerland. *Aquatic Sciences*, 51, 19–30.

Lotter, A.F. 1998. The recent eutrophication of Baldeggersee Switzerland, as assessed by fossil diatom assemblages. *The Holocene*, 8, 395–405.

Lotter, A.F. and Bigler, C. 2000. Do diatoms in the Swiss Alps reflect the length of ice cover. *Aquatic Sciences*, 62, 125–141.

Lotter, A.F. and G. Lemke, 1999. Methods for preparing and counting biochemical varves. *BOREAS*, 28, 243–255.

Lotter, A.F. and Zbinden, H. 1989. Late-Glacial pollen analysis, oxygen-isotope record, and radiocarbon stratigraphy from Rotsee Lucerne, Central Swiss Plateau. *Eclogae geologicae Helvetiae*, 82, 191–202.

Lotter, A.F., Eicher, U., Birks, H.J.B. and Siegenthaler, U. 1992. Late-glacial climatic oscillations as recorded in Swiss lake sediments. *Journal of Quaternary Science*, 7, 187–204.

Lotter, A.F., Birks, H.J.B., Hofmann W. and Marchetto, A. 1997a. Modern diatom, Cladocera, chironomid and chrysophyte cyst assemblages as quantitative indicators for the reconstruction of past environmental conditions in the Alps. I. Climate. *Journal of Paleolimnology*, 18, 395–420.

Lotter, A.F., Sturm, M., Teranes, J.L. and Wehrli, B. 1997b. Varve formation since 1885 and high-resolution varve analyses in hypertrophic Baldegger See Switzerland. *Aquatic Sciences*, 59, 304–325.

Lotter, A.F., Walker, I.R., Brooks, S.J. and Hofmann, W. 1999. An intercontinental comparison of chironomid palaeotemperature inference models: Europe versus North America. *Quaternary Science Reviews*, 18, 717–735.

Lotter, A.F., Birks, H.J.B., Eicher, U., Hofmann, W., Schwander, J., and Wick, L. 2000. Younger Dryas and Alleröd summer temperatures at Gerzensee Switzerland. inferred from fossil pollen and cladoceran assemblages. *Palaeogeography, Palaeoclimatology, Palaeoecology*, 159, 349–361.

Lough, J.M. and Barnes, D.J. 1997. Several centuries of variation in skeletal extension, density, and calcification in massive *Porites* colonies from the Great Barrier Reef: A proxy for seawater temperature and a background of variability against which to identify unnatural change. *Journal of Experimental Marine Biology and Ecology*, 211, 29–67.

Lough, J.M. and Barnes, D.J. 2000. Environmental controls on growth of the massive coral *Porites*. *Journal of Experimental Marine Biology and Ecology*, 245, 225–243.

Lowe, J.J. 1993. Isolating the climatic factors in Early- and Mid-Holocene palaeobotanical records from Scotland. In F.M. Chambers (ed.), *Climatic change and human impact on the landscape*. Chapman and Hall, London, pp. 67–82.

Lowe, J.J. and Walker, M.J.C. 1997. *Reconstructing Quaternary environments*. Longman, Harlow.
Lowe, J.J. and Walker, M.J.C. 2000. Radiocarbon dating of the last glacial–interglacial transition ca. 14-9 ^{14}C ka BP in terrestrial and marine records: the need for new quality assurance protocols. *Radiocarbon*, 42, 53–68.
Luckman, B. 1986. Reconstruction of Little Ice Age events in the Canadian Rocky Mountains. *Géographie Physique et Quaternaire*, 40, 17–28.
Luckman, B.H. 1995 Calendar-dated, early Little Ice Age glacier advance at Robson Glacier, British Columbia, Canada. *The Holocene*, 5, 149–159.
Lüder, B. and Zolitschka, B. 2001. Jahreszeitlich geschichtete Sedimente des Sacrower Sees Brandenburg als Archiv zur Rekonstruktion von Paläoumweltbedingungen. *Die Erde*, 132, 381–397.
Lukas, R. and Lindstrom, E.J. 1988. The mixed layer of the Western Equatorial Pacific, Aha Huliko, a Hawaiian winter workshop on the mixed layer. Hawaii Institute of Geophysics, University of Hawaii, Honolulu, HI.
Lumley S.H. and Switsur, R. 1993. Late Quaternary chronology of the Taitao Peninsula, southern Chile. *Journal of Quaternary Science*, 8, 161–165.
Luterbacher, J., Rickli, R., Xoplaki, E., Tinguely, C., Beck, C., Pfister, C. and Wanner, H. 2001. The Late Maunder Minimum 1675–1715. A key period for studying decadal-scale climatic change in Europe. *Climatic Change*, 49, 441–462.
Maasch, K.A. and Saltzman, B. 1990. A low-order dynamical model of global climatic variability over the full Pleistocene. *Journal of Geophysical Research-Atmospheres*, 95, 1955–1963.
MacAyeal, D.R. 1993. Binge/Purge oscillations of the Laurentide Ice Sheet as a cause of the North Atlantic's Heinrich Events. *Paleoceanography*, 8, 775–784.
McCarroll, D. 1989. Potential and limitations of the Schmidt hammer for relative-age dating: field tests on Neoglacial moraines, Jotunheimen, southern Norway. *Arctic and Alpine Research*, 21, 268–275
McCarroll, D. 1994. A new approach to lichenometry: dating single-age and diachronous surfaces. *The Holocene*, 4, 383–396.
McCarroll, D. and Nesje, A. 1993. Vertical extent of ice sheets in Nordfjord, western Norway: measuring degree of rock surface weathering using Schmidt hammer and rock surface roughness. *Boreas*, 22, 255–265.
McCarthy, L. Head, L. and Quade, J. 1996. Holocene palaeoecology of the northern Flinders Range, South Australia, based on stick-nest rat *Leporillus* spp. middens: a preliminary overview. *Palaeogeography, Palaeoclimatology, Palaeoecology*, 123, 205–218.
McClung, D.M. and Armstrong, R.L. 1993. Temperate glacier time response from field data. *Journal of Glaciology*, 39, 323–326.
McConnaughey, T.A. 1986. Oxygen and carbon isotope disequilibria in Galápagos corals: isotopic thermometry and calcification physiology. University of Washington, Seattle, WA.
McConnaughey, T.A. 1989. ^{13}C and ^{18}O isotopic disequilibria in biological carbonates: I. Patterns. *Geochimica et Cosmochimica Acta*, 53, 151–162.
McCormac, F.G., Thompson, M. and Brown, D. 2001. Characterisation, optimisation, and standard measurements for two small-sample high-precision radiocarbon counters. *Centre for Archaeology Report 8*. English Heritage, London.
McCormac, F.G., Hogg, A.G., Reimer, P.J., *et al.* 2002. High precision calibration of the radiocarbon time-scale for the southern hemisphere AD 1850–950. *Radiocarbon*, 44, 641–651.
McCorriston, J. and Hole, F. 1991. The ecology of seasonal stress and the origin of agriculture in the Near East. *American Anthropologist*, 91, 46–69.
McCrea, J.M. 1950. On the isotopic chemistry of carbonates and palaeo-temperature scale. *Journal of Chemical Physics*, 18, 849–857.
McDermott, F., Frisia, S., Huang, Y., *et al.* 1999. Holocene climate variability in Europe: evidence from $\delta^{18}O$, textural and extension-rate variations in three speleothems. *Quaternary Science Reviews*, 18, 1021–1038.
McDermott, F., Mattey, D.P. and Hawkesworth, C.J. 2001. Centennial-scale Holocene climate variability revealed by a high-resolution spelothem $\delta^{18}O$ record from SW Ireland. *Science*, 294, 1328–1331.

MacDonald, G.M. 2001. Conifer stomata. In J.P. Smol, H.J.B. Birks and W.M. Last (eds), *Tracking environmental changes using lake sediments, Vol. 3: Terrestrial and siliceous indicators*. Kluwer Academic Publishers, Dordrecht, pp. 33–48.

MacDonald, G. M., Edwards, T.W.D., Moser, K.A., Pienitz, R. and Smol, J.P. 1993. Rapid response of treeline vegetation and lakes to past climate warming. *Nature,* 361, 243–246.

MacDonald, G.M., Velichko, A.A., Kremenetski, C.V. *et al.* 2000a. Holocene treeline history and climate change across northern Eurasia. *Quaternary Research*, 53, 302–311.

MacDonald, G.M., Gervais, B.R., Snyder, J.A., Tarasov, G.A. and Borisova, O.K. 2000b. Radiocarbon dated *Pinus sylvestris* L. wood from beyond tree-line on the Kola Peninsula, Russia. *The Holocene*, 10, 134–147.

McGlone, M.S. and Wilmshurst, J.M. 1999. A Holocene record of climate, vegetation change and peat bog development, east Otago, South Island, New Zealand. *Journal of Quaternary Science*, 14, 239–254.

McGlone, M.S., Kershaw, A.P. and Markgraf, V. 1992. El Niño/Southern Oscillation and climatic variability in Australasian and South American paleoenvironmental records. In H.F. Diaz and V. Markgraf (eds), *El Niño: Historical and paleoclimatic aspects of the Southern Oscillation*. Cambridge University Press, Cambridge, pp. 435–462.

McGregor, G.R. and Nieuwolt, S. 1998. *Tropical Climatology*. John Wiley and Sons, Chichester.

McGuffie, K. and Henderson-Sellers, A. 1997. *A climate modelling primer*, 2nd edition. J Wiley and Sons, Chichester.

McIntosh, R.J., Tainter, J.A. and McIntosh, S.K. 2000. Climate, history and human action. In R.J. McIntosh, J.A. Tainter and S.K. McIntosh (eds), *The way the wind blows: climate, history and human action*. Columbia University Press, New York, pp. 1–42.

Mackay, A.W., Flower, R.J., Kuzmina, A.E. *et al.* 1998. Diatom succession trends in recent sediments from Lake Baikal and their relationship to atmospheric pollution and to climate change. *Philisophical Transactions of the Royal Society of London B*, 353, 1011–1055.

Mackay, A.W., Battarbee, R.W., Flower, R.J., *et al.* 2000. The deposition and accumulation of endemic planktonic diatoms in the sediments of Lake Baikal and an evaluation of their potential role in climate reconstruction during the Holocene. *Terra Nostra*, 9, 34–48.

Mackay, A.W., Battarbee, R.W., Flower, R.J., *et al.* 2003. Assessing the potential for developing internal diatom-based inference models in Lake Baikal. *Limnology and Oceanography*, 48, 1183–1192.

McKenzie, G.M and Busby, J.R. 1992. A quantitative estimate of Holocene climate using a bioclimatic profile of *Nothofagus cunninghamii* Hook. *Oerst. Journal of Biogeography*, 19, 531–540.

McKenzie, J.A. 1985. Carbon isotopes and productivity in the lacustrine and marine environment. In W. Stumm (ed.), *Chemical processes in lakes*. Wiley, New York.

McKenzie, J.A. and Hollander, D.J. 1993. Oxygen-isotope record in recent carbonate sediments from Lake Greifen, Switzerland 1750–1986: application of continental isotopic indicator for evaluation of changes in climate and atmospheric circulation patterns. Climate change in Continental Isotopic Records. *Geophysical Monograph*, 78, 101–111.

McMullen, J.A., Barber, K.E. and Johnson, B. 2000. Contributions of palaeoecology to the conservation management of raised bogs in the United Kingdom. *O'Dell Memorial Monograph 27*, University of Aberdeen, Aberdeen, pp. 1–109.

McNichol, A.P., Ertel, J.R. and Eglinton, T.I. 2000. The radiocarbon content of individual lignin-derived phenols: technique and initial results. *Radiocarbon*, 42, 219–227.

McNish, A.G. and E.A. Johnson, 1938. Magnetization of unmetamorphosed varves and marine sediments. *Terrestrial Magnetism and Athmospheric Electricity*, 43, 401–407.

Madureira, L.A.S., Conte, M.H. and Eglinton, G. 1995. Early diagenesis of lipid biomarker compounds in North Atlantic sediments. *Paleoceanography*, 10, 627–642.

Magnuson, J.J., Robertson, D.M., Benson, B.J., *et al.* 2000. Historical trends in lake and river ice cover in the Northern Hemisphere. *Science*, 289, 1743–1746.

Magny, M. 1993a. Holocene fluctuations of lake-levels in the French Jura and sub-Alpine ranges, and their implications for past general patterns. *The Holocene*, 3, 306–313.

Magny, M. 1993b. Solar influences on Holocene climatic changes illustrated by correlations between past lake-level fluctuations and the atmospheric ^{14}C record. *Quaternary Research*, 40, 1–9.

Magny, M. 1995. Paleohydrological changes in Jura, France, and climatic oscillations around the North Atlantic from Allerød to Preboreal. *Géographie Physique et Quaternaire*, 49, 401–408.

Mahaney, W.C. 1987. Lichen trimlines and weathering features as indicators of mass balance changes and successive retreat stages of the Mer de Glace in the western Alps. *Zeitschrift für Geomorphologie N.F.*, 31, 411–418.

Maher, L.J. 1972. Absolute pollen diagram of Redrock Lake, Boulder County, Colorado. *Quaternary Research*, 2, 531–553.

Maise, C. 1998. Archäoklimatologie - vom Einfluss nacheiszeitlicher Klimavariabilität in der Ur- und Frühgeschichte. *Jahrbuch der Schweizerischen Gesellschaft für Ur- und Frühgeschichte*, 81, 197–235.

Maley, J. 1983. Histore de la végétation et du climat de l'Afrique nord-tropicale au Quaternaire récent. *Bothalia*, 14, 377–389.

Maley, J., Livingstone, D.A., Giresse, P., *et al.* 1990. Lithostratigraphy, volcanism, palaeomagnetism and palynology of Quaternary lacustrine deposits from barombi Mbo west Cameroon: preliminary results. *Journal of Volcanic and Geothermal Research*, 42, 319–335.

Mallory, J.P., McCormac, F.G., Reimer, P.J. and Marsadolov, L.S. 2002. The date of Pazyryk. In K. Boyle, C. Renfrew and M. Levine (eds), *Ancient interactions: east and west in Eurasia*. McDonald Institute Monographs, Cambridge, pp. 199–211.

Malmgren, B.A. 1978. Comparison of visual and statistical correlation in time series curves. *Mathematical Geology*, 10, 103–106.

Malmgren, B.A. and Nordlund, U. 1997. Application of artificial neural networks to paleoceanographic data. *Palaeogeography, Palaeoclimatology, Palaeoecology*, 136, 359–373.

Malmgren, B.A., Kucera, M., Nyberg, J. and Waelbroeck, C. 2001. Comparison of statistical and artificial neural network techniques for estimating past sea surface temperatures from planktonic foraminifer census data. *Paleoceanography*, 16, 520–530.

Malmgren, B.A., Winter, A. and Chen, D. 1998. El-Niño-Southern Oscillation and North Atlantic Oscillation control of climate in Puerto Rico. *Journal of Climate*, 11, 2713–2717.

Manabe, S. and Stouffer, R. J. 1997. Coupled ocean–atmosphere model response to freshwater input: comparison to Younger Dryas event. *Paleoceanography*, 12, 321–336.

Manabe, S. and Stouffer, R. J. 1999. Are two modes of thermohaline circulation stable? *Tellus Series A: Dynamic Meteorology and Oceanography*, 51, 400–411.

Mangerud, J., Andersen, S.T., Berglund, B.E. and Donner, J.J. 1974. Quaternary stratigraphy of Norden, a proposal for terminology and classification. *BOREAS*, 3, 109–128.

Manley, G. 1974. Central England temperatures: monthly means 1659 to 1973. *Quarterly Journal of the Royal Meteorological Society*, 100, 389–405.

Mann, M.E., Bradley, R.S. and Hughes, M.K. 1998. Global-scale temperature patterns and climate forcing over the past six centuries. *Nature*, 392, 779–787.

Mann, M.E., Bradley, R.S. and Hughes, M.K. 1999. Northern Hemisphere temperatures during the past millenium: inferences, uncertainties and limitations. *Geophys. Res. Letts.*, 26, 759–762.

Mann, M.E., Gille, E., Bradley, R.S., *et al.* 2000. Global temperature patterns in past centuries: an interactive presentation. *Earth Interactions*, 4, 1–29, see website: http://EarthInteractions.org

Mantua, N.J., Hare, S.R., Zhang, Y., Wallace, J.M. and Francis, R.C. 1997. A Pacific interdecadal climate oscillation with impacts on salmon production. *Bulletin of the American Meteorological Society*, 78, 1069–1079.

Marchal, O., Cacho, I., Stocker, T.F., *et al.* 2002. Apparent long-term cooling of the sea surface in the northeast Atlantic and Mediterranean during the Holocene. *Quaternary Science Reviews*, 21, 455–483.

Markgraf, V. and Diaz, H.F. 2001. The past ENSO record: a synthesis. In H.F. Diaz and V. Markgraf (eds), *El Niño and the Southern Oscillation: Multiscale variability and global impacts*. Cambridge University Press, Cambridge, pp. 465–488.

Markgraf, V., Bradbury, J.P. and Busby, J.R. 1986. Paleoclimates in southwestern Tasmania during the last 13,000 years. *Palaios*, 1, 368–380.

Markgraf, V., Dodson, J.R., Kershaw, A.P. McGlone, M.S. and Nicholls, N. 1992. Evolution of late Pleistocene and Holocene climates in the circum-South pacific land areas. *Climate Dynamics*, 6, 193–211.

Marlow, J.R., Lange, C., Wefer, G. and Rosell-Melé, A. 2000. Upwelling intensification as part of the Pliocene–Pleistocene climate transition. *Science*, 290, 2288–2291.

Marshall, S.J. and Clarke, G.K.C. 1999. Modeling North American freshwater runoff through the last glacial cycle. *Quaternary Research*, 52, 300–315.

Martens, H. and Næs, T. 1989. *Multivariate calibration*. J. Wiley and Sons, Chichester.

Martens, K., De Deckker, P. and Marples, T.G. 1985. Life History of *Mytilocypris henricae* Chapman. Crustacea: Ostracoda. in Lake Bathurst, New South Wales. *Australian Journal of Marine and Freshwater Research*, 36, 807–819.

Martens, K., Horne, D. J. and Griffiths, H. I. 1998. Age and diversity of non-marine ostracods. In K. Martens (ed.), *Sex and parthenogenesis. evolutionary ecology of reproductive modes in non-marine Ostracods*. Backhuys Publishers, Leiden, pp. 37–55.

Martinez-Cortizas, A., Pontevedra-Pombal, X., Garcia-Rodeja, E., Novoa-Munoz, J.C. and Shotyk, W. 1999. Mercury in a Spanish peat bog: archive of climate change and atmospheric metal deposition. *Science*, 284, 939–942.

Masefield, J. 1926. *Odtaa, one damn thing after another: a novel.* William Heinemann, London.

Maslin, M.A. and Burns, S.J. 2000. Reconstruction of the Amazon Basin effective moisture availability over the last 14,000 years. *Science*, 290, 2285–2287.

Maslin M.A. and Tzedakis, C. 1996. *Sultry last interglacial gets sudden chill. EOS*, 77, 353–354.

Maslin M.A., Mikkelsen, N., Vilela, C. and Haq, B. 1998. Sea-level- and gas-hydrate-controlled catastrophic sediment failures of the Amazon Fan. *Geology*, 26, 1107–1110.

Maslin, M.A., Seidov D. and Lowe, J. 2001. Synthesis of the nature and causes of sudden climate transitions during the Quaternary. In D. Seidov, B. Haupt and M. Maslin (eds), *The oceans and rapid climate change: past, present and future*, Vol. 126. AGU Geophysical Monograph Series, Washington, DC, pp. 9–52.

Mason, I.M., Guzkowska, M.A.J. and Rapley, C.G. 1994. The response of lake-levels and areas to climatic change. *Climatic Change*, 27, 161–197.

Massaferro, J. and Brooks, S.J. 2002. The response of chironomids to Late Quaternary climate change in the Taitao Peninsula, southern Chile. *Journal of Quaternary Science*, 17, 101–111.

Masson, V., Cheddadi, R., Braconnot, P., Joussaume, S. and Texier, D. 1999. Mid-Holocene climate in Europe: what can we infer from PMIP model-data comparisons? *Climate Dynamics*, 15, 163–182.

Matheney, R.K. and Knauth, L.P. 1989. Oxygen-isotope fractionation between marine biogenic silica and seawater. *Geochimica et Cosmochimica Acta*, 53, 3207–3214.

Matthews, J.A. 1985. Radiocarbon dating of surface and buried soils: principles, problems and prospects. In K.S. Richards, R.R. Arnett and S. Ellis (eds), *Geomorphology and Soils*. George Allen and Unwin, London, pp. 269–288.

Matthews, J.A. 1992. *The ecology of recently deglaciated terrain*. Cambridge University Press, Cambridge.

Matthews, J.A. and Caseldine, C.J. 1987. Arctic-alpine Brown Soils as a source of palaeoenvironmental information: further ^{14}C dating and palynological evidence from Vestre Memurubreen, Jotunheimen, Norway. *Journal of Quaternary Science*, 2, 59–71.

Matthews, J.A. and Dresser, P.Q. 1983. Intensive ^{14}C dating of a buried palaeosol horizon. *Geologiska Föreningen i Stockholm Förhandlingar*, 105, 59–63.

Matthews, J.A. and Karlén, W. 1992. Asynchronous neoglaciation and Holocene climatic change reconstructed from Norwegian glaciolacustrine sedimentary sequences. *Geology*, 20, 991–994.

Matthews, J.A. and Shakesby, R.A. 1984. The status of the 'Little Ice Age' in southern Norway: relative-age dating of neoglacial moraines with Schmidt hammer and lichenometry. *BOREAS*, 13, 333–346.

Matthews, J.A., Dahl, S.O., Nesje, A., Berrisford, M.S. and Anderson, C. 2000. Holocene glacier variations in central Jotunheimen, southern Norway, based on distal glaciolacustrine sediment cores. *Quaternary Science Reviews*, 19, 1625–1647.

Matzke-Karasz, R., Horne, D. C., Janz, H., Griffiths, H. I., Hutshinson, W. F. and Preece, R. C. 2001. 5000 year-old spermatozoa in Quaternary Ostracoda Crustacea. *Naturwissenschaften*, 88, 268–272.

Mauquoy, D. and Barber, K. 1999a. A replicated 3000 yr proxy-climate record from Coom Rigg Moss and Felecia Moss, the Border Mires, northern England. *Journal of Quaternary Science*, 14, 263–275.

Mauquoy, D. and Barber, K. 1999b. Evidence for climatic deteriorations associated with the decline of *Sphagnum imbricatum* Hornsch ex Russ. in six ombrotrophic mires from northern England and the Scottish Borders. *The Holocene*, 9, 423–437.

Mauquoy, D., van Geel, B., Blaauw, M. and van der Plicht, J. 2002. Evidence from northwest European bogs shows 'Little Ice Age' climatic changes driven by variations in solar activity. *The Holocene*, 12, 1–6.

Mayewski, P. A., Meeker, L. D., Twickler, M. S., *et al.* 1997. Major features and forcing of high-latitude northern hemisphere atmospheric circulation using a 110,000-year-long glaciochemical series. *Journal of Geophysical Research*, 102, 26345–26366.

Mayle, F.E. and Cwynar, L.C. 1995. A review of multi-proxy data for the Younger Dryas in Atlantic Canada. *Quaternary Science Reviews*, 14, 813–821.

Mayle, F.E., Burbridge, R. and Killeen, T.J. 2000. Millennial-scale dynamics of southern Amazonian rain forests. *Science*, 290, 2291–2294.

Mees, F., Verschuren, D., Nijs, R. and Dumont, H. 1991. Holocene evolution of the crater lake at Malha, Northwest Sudan. *Journal of Paleolimnology*, 5, 227–253.

Meggers, B.J. 1954. Environmental limitation on the development of culture. *American Anthropologist*, 56, 801–824.

Meggers, B.J. 1971. *Amazonia, man and culture in a counterfeit paradise.* Aldine, Chicago.

Meier, M.F. 1984. Contribution of small glaciers to global sea level. *Science*, 226, 1418–1421.

Merkt, J. 1971. Zuverlässige Auszählungen von Jahresschichten in Seesedimenten mit Hilfe von Groß-Dünnschliffen. *Archiv für Hydrobiologie*, 69, 145–154.

Merkt, J. and H. Müller, 1999. Varve chronology and palynology of the Lateglacial in Northwest Germany from lacustrine sediments of Hämelsee in Lower Saxony. *Quaternary International*, 61, 41–59.

Merkt, J., Müller, H., Knabe, W., Müller, P. and Weiser, T. 1993. The early Holocene Saksunarvatn tephra found in lake sediments in NW Germany. *BOREAS*, 22, 93–100.

Metcalfe, S.E. 1995. Holocene environmental change in the Zacapu Basin, Mexico: a diatom-based record. *The Holocene*, 5, 196–208.

Metcalfe, S.E., O'Hara, S.L., Caballero, M. and Davies, S.J. 2000. Records of Late Pleistocene–Holocene climatic change in Mexico: a review. *Quaternary Science Reviews*, 19, 699–721.

Meyers, P.A. and Lallier-Verges, E. 1999. Lacustrine sedimentary organic matter records of Late Quaternary paleoclimates. *Journal of Paleolimnology*, 21, 345–372.

Mezquita, F., Tapia, G. and Roca, J. R. 1999. Ostracoda from springs on the eastern Iberian Peninsula: ecology, biogeography and palaeolimnological implications. *Palaeogeography, Palaeoclimatology, Palaeoecology*, 148, 65–85.

Miano, T.M., Sposito, G. and Martin, J.P. 1988. Fluorescence spectroscopy of humic substances. *Soil Science America Journal*, 52, 1016–1019.

Miller, N.G. and Futyma, R.P. 1987. Palaeohydrological implications of Holocene peatland development in northern Michigan. *Quaternary Research*, 27, 297–311.

Min, G.R., Edwards, R.L., Taylor, F.W., Recy, J., Gallup, C.D. and Beck, J.W. 1995. Annual Cycles of U/Ca in Coral Skeletons and U/Ca Thermometry. *Geochimica Et Cosmochimica Acta*, 59, 2025–2042.

Minobe, S. 1999. Resonance in bidecadal and pentadecadal climate oscillations over the North Pacific: role in climatic regime shifts. *Geophysical Research Letters*, 26, 855–858.

Mitchell, E.A.D., Van der Knaap, W.O., van Leeuwen, J.F.N., Buttler, A., Warner, B.G. and Gobat, J.M. 2001. The palaeoecological history of the Praz-Rodet bog Swiss Jura. based on pollen, plant macrofossils and testate amoebae Protozoa. *The Holocene*, 11, 65–80.

Mitchell, J.F.B. 1990. Greenhouse warming: is the Mid-Holocene a good analogue? *Journal of Climate*, 3, 1177–1192.

Mitchell, J.F.B. and Karoly, D.J. 2001. Detection of climate change and attribution of causes. In J.T. Houghton, Y. Ding, D.J. Griggs, M. Noguer, P.J. van der Linden and D. Xiaosu (eds), *Climate Change 2001: The Scientific Basis*. Cambridge University Press, Cambridge, pp. 695–738.

Mitchell, J.M. 1976. An overview of climatic variability and its causal mechanisms. *Quaternary Research*, 6, 481–493.

Mitchell, W., Chittleborough, J., Ronai, B. and Lennon, G.W. 2000. Sea level rise in Australia and the Pacific. In M. Grzechnik and J. Chittleborough (eds), *Proceedings of the Pacific Islands Conference on Climate Change, Climate Variability and Sea Level Rise, Rarotonga, Cook Islands, 3–7 April, 2000*. National Tidal Facility, Adelaide, South Australia, pp. 47–57.

Mitrovica, J.X., Tamisiea, M.E., Davis, J.L. and Milne, G.A. 2001. Recent mass balance of polar ice sheets inferred from patterns of global sea-level change. *Nature*, 409, 1026–1029.

Mitsuguchi, T., Matsumoto, E., Abe, O., Uchida, T. and Isdale, P. 1996. Mg/Ca thermometry in coral skeletons. *Science*, 274, 961–963.

Mitsuguchi, T., Uchida, T., Matsumoto, E., Isdale, P.J. and Kawana, T. 2001. Variations in Mg/Ca, Na/Ca, and Sr/Ca ratios of coral skeletons with chemical treatments: implications for carbonate geochemistry. *Geochimica Et Cosmochimica Acta*, 65, 2865–2874.

Moine, O., Rousseau, D.-D., Jolly, D. and Vianey-Lund, M. 2002. Paleoclimatic reconstruction using mutual climatic range on terrestrial mollusks. *Quaternary Research*, 57, 162–172.

Moodie, D.W. and Catchpole, A.J.W. 1976. Valid climatological data from historical sources by content analysis. *Science*, 193, 51–53.

Moore, A.M.T. 1985. The development of Neolithic societies in the Near East. *Advances in World Archaeology*, 4, 1–69.

Moore, A.M.T. and Hillman, G.C. 1992. The Pleistocene to Holocene transition and human economy in southwest Asia: the impact of the Younger Dryas. *American Antiquity*, 57, 482–494.

Moore, G.W. 1956. Aragonite speleothems as indicators of paleotemperature. *American Journal of Science*, 254, 746–753.

Moore, J.J., Hughen, K.A., Miller, G.H. and Overpeck, J.T. 2001. Little Ice Age recorded in summer temperature reconstruction from varved sediments of Donard Lake, Baffin Island, Canada. *Journal of Paleolimnology*, 25, 503–517.

Moore, M.D., Schrag, D.P. and Kashgarian, M. 1997. Coral radiocarbon constraints on the source of the Indonesian throughflow. *Journal of Geophysical Research-Oceans*, 102, 12359–12365.

Moore, P.D. 1975. Origin of blanket mires. *Nature*, 256, 267–269.

Moore, P.D. 1993. The origin of blanket mire, revisited. In F.M.Chambers (ed.), *Climate Change and Human Impact on the Landscape*. Chapman and Hall, London, pp. 217–224.

Morimoto, M., Abe, O., Kayanne, H., Kurita, N., Matsumoto, E. and Yoshida, N. 2002. Salinity records for the 1997–1998 El Niño from western Pacific corals. *Geophysical Research Letters*, 29, 35-1–4.

Mörner, N.-A. 1976. Eustasy and geoid changes. *Journal of Geology*, 84, 123–151.

Mörner, N.-A. 1992. Present El Nino-ENSO events and past super-ENSO events effect of changes in the Earth's rate of rotation. *Paleo ENSO Records International Symposium 1992*, Lima, Peru, pp. 201–206.

Mossman, R.C. 1897. The non-instrumental meteorology of London 1713–1896. *Quarterly Journal of the Royal Meteorological Society*, 23, 287–298.

Mourguiart, P. and Carbonel, P. 1994. A quantitative method of palaeolake-level reconstruction using ostracod assemblages - an example from the Bolivian altiplano. *Hydrobiologia*, 288, 183–193.

Mourguiart, P. and Montenegro, M.E. 2002. Climate changes in the Lake Titicaca area: evidence from ostracod ecology. In J.A. Holmes and A.R. Chivas (eds), *The Ostracoda: Applications in Quaternary Research,* Vol. 131, AGU Geophysical Monograph Series, Washington, DC, pp. 151–166.

Mourguiart, P., Wirrmann, D., Fournier, M. and Servant, M. 1992. Reconstruction quantitative des niveaux du petit lac Titicaca au cours de l'Holocene. *Comptes Rendus de L Academie des Sciences Paris*, 315, 875–880.

Mourguiart, P., Corrège, T., Wirrmann, D., *et al.* 1998. Holocene palaeohydrology of Lake Titicaca estimated from an ostracod-based transfer function. *Palaeogeography, Palaeoclimatology, Palaeoecology*, 143, 51–72.

Müller, A., Gagan, M.K. and McCulloch, M.T. 2001. Early marine diagenesis in corals and geochemical consequences for paleoceanographic reconstructions. *Geophysical Research Letters*, 28, 4471–4474.

Muller, J. 1959. Palynology of recent Orinoco River and shelf sediments: reports of the Orinoco shelf expedition. *Micropaleontology*, 5, 1–32.

Müller, P.J., Kirst, G., Ruhland, G., von Storch, I. and Rosell-Melé, A. 1998. Calibration of the alkenone paleotemperature index Uk37 based on core-tops from the eastern South Atlantic and the global ocean 60°N–60°S. *Geochimica et Cosmochimica Acta*, 62, 1757–1772.

Myers, N., Mittermeier, R.A., Mittermeier, C.G., da Fonseca, G.A.B. and Kent, J. 2000. Biodiversity hotspots for conservation priorities. *Nature*, 403, 853–858.

NAS. 2002. *Abrupt climate change: inevitable surprises.* National Academy of Sciences Press, Washington, DC.

Nathan, R., Katul, G.G. Horn, H.S., *et al.* 2002. Mechanisms of long-distance dispersal of seeds by wind. *Nature,* 418, 409–413.

National Research Council. 1994. *Solar influences on global change.* National Academy Press, Washington, DC.

Neale, J. W. 1988. Ostracods and palaeosalinity reconstruction. In P. De Deckker, J.-P. Colin and J.-P. Peypouquet (eds), *Ostracoda in the earth sciences.* Elsevier, Amsterdam, pp. 125–155.

Nederbragt, A.J. and Thurow, J.W. 2001. A 6000-yr varve record of Holocene climate in Saanich Inlet, British Columbia, from digital sediment colour analysis of ODP Leg 169S cores. *Marine Geology*, 174, 95–110.

Neff, U., Burns, S.J., Mangini, A., Mudelsee, M., Fleitmann, D. and Matter, A. 2001. Strong coherence between solar variability and the monsoon in Oman between 9 and 6kyr Fago. *Nature,* 411, 290–293.

Nelson, D.J., Webb, R.H. and Long, A. 1990. Analysis of stick-nest rat *Leporillus*: Muridae. middens from central Australia. In J.L. Betancourt, T. van Devender and P.S. Martin (eds), *Packrat middens. The last 40, 000 years of biotic change.* University of Arizona Press, Tucson, pp. 428–434.

Nesje, A. 1992. Topographical effects on the equilibrium-line altitude on glaciers. *GeoJournal,* 27.4, 383–391.

Nesje, A. and Dahl, S.O. 2000. *Glaciers and environmental change.* Arnold, London.

Nesje, A., Kvamme, M., Rye, N. and Løvlie, R. 1991. Holocene glacial and climate history of the Jostedalsbreen region, western Norway: evidence from lake sediments and terrestrial deposits. *Quaternary Science Reviews*, 10, 87–114.

Nesje, A., Johannessen, T. and Birks, H.J.B. 1995. Briksdalsbreen, western Norway: climatic effects on the terminal response of a temperate glacier between AD 1901 and 1994. *The Holocene*, 5, 343–347.

Nesje, A., Lie, Ø. and Dahl, S.O. 2000. Is the North Atlantic Oscillation reflected in Scandinavian glacier mass balance records? *Journal of Quaternary Science*, 15, 587–601.

Nesje, A., Matthews, J.A., Dahl, S.O., Berrisford, M.S. and Andersson, C. 2001. Holocene glacier fluctuations of Flatebreen and winter precipitation changes in the Jostedalsbreen region, western Norway, based on glaciolacustrine records. *The Holocene*, 11, 267–280.

New, M., Hulme, M. and Jones, P.D. 1999. Representing twentieth-century space–time climate variability. Part I: Development of a 1961–1990 mean monthly terrestrial climatology. *Journal of Climate*, 12, 829–856.

New, M., Hulme, M. and Jones, P.D. 2000. Representing twentieth century space–time climate variability. II: Development of 1901–1996 monthly grids of terrestrial surface climate. *Journal of Climate*, 13, 2217–2238.

Niemi, T.M., Ben-Avraham, Z. and Gat, J.R. 1997. The Dead Sea, the lake and its setting. *Monographs on Geology and Geophysics, No. 36.* Oxford University Press, New York.

Niklewski, I. and van Zeist, W. 1970. A late Quaternary pollen diagram from northwestern Syria. *Acta Botanica Neerlandica*, 19, 737–754.

Noon, P.E., Leng, M.J. and Jones, V.J. 2003. Oxygen isotope $\delta^{18}O$: evidence of hydrological change through the mid- to late- Holocene c. 6000 ^{14}C BP to present day at Signey Island, maritime Antarctic. *The Holocene*, 13, 153–160.

Nott, C.J., Xie, S.C., Avsejs, L.A., Maddy, D., Chambers, F.M. and Evershed, R.P. 2000. *n*-Alkane distributions in ombrotrophic mires as indicators of vegetation change related to climatic variation. *Organic Geochemistry*, 31, 231–235.

Nunn, P.D. 1998. Sea-level changes over the past 1,000 years in the Pacific. *Journal of Coastal Research*, 14, 23–30.

Nunn, P.D. and Peltier, W.R. 2001. Far-field test of the ICE-4G model of global isostatic response to deglaciation using empirical and theoretical Holocene sea-level reconstructions for the Fiji Islands, southwestern Pacific. *Quaternary Research*, 55, 203–214.

Nydal, R. and Gislefoss, J.S. 1996. Further application of bomb ^{14}C as a tracer in the atmosphere and ocean. *Radiocarbon*, 38, 389–407.

Nye, J.F. 1960. The response of glaciers and ice sheets to seasonal and climatic changes. *Proceedings of the Royal Society of London, Series A*, 256, 559–584.

O'Brien, P.E., Cooper, A.K., Richter, C., *et al.* 2001. *Proceedings of ODP, Initial Reports, 188 CD ROM*. Available from: Ocean Drilling Program, Texas A&M University, College Station, TX.

O'Brien, S.R., Mayewski, P.A., Meeker, L.D., Meese, D.A., Twickler, M.S., and Whitlow, S.I. 1996. Complexity of Holocene climate as reconstructed from a Greenland ice core. *Science*, 270, 1962–1964.

Ocampo, R., Callot, H.J. and Albrecht, P. 1985. Occurrence of bacterioporphyrins in oil shale. *Journal of the Chemical Society: Chemical Communications*, 200–201.

Odgaard, B.V. 1994. The Holocene vegetation history of northern West Jutland, Denmark. *Opera Botanica*, 123, 1–171.

Odland, A. 1996. Differences in the vertical distribution pattern of *Betula pubescens* in Norway and its ecological significance. *Paläoklimaforschung*, 20, 43–59.

Oerlemans, J. 1988. Simulation of historic glacier variations with a simple climate–glacier model. *Journal of Glaciology*, 34, 333–341.

Oerlemans, J. and Fortuin, J.P.F. 1992. Sensitivity of glaciers and small ice caps to greenhouse warming. *Science*, 258, 115–118.

Oeschger, H. 2000. Perspectives on Global Change Science: isotopes in the Earth System, past and present. In K. Alverson, F. Oldfield and R.S. Bradley. 2000. Past global changes and their significance for the future. *Quaternary Science Reviews*, 19, 37–44.

O'Hanlon, L. 2002. Making waves. *Nature*, 415, 360–362.

Ohlendorf, C., Niessen, F. and Weissert, H. 1997. Glacial varve thickness and 127 years of instrumental climate data: a comparison. *Climatic Change*, 36, 391–411.

Ohlendorf, C., Bigler, C., Goudsmit, G.H., *et al.* 2000. Causes and effects of long periods of ice cover on a remote high Alpine lake. *Journal of Limnology*, 59, 65–80.

Ohmura, A., Kasser, P. and Funk, M. 1992. Climate at the equilibrium line of glaciers. *Journal of Glaciology*, 38, 397–411.

Ojala, A. 2001. Varved lake sediments in southern and central Finland: long varve chronologies as a basis for Holocene palaeoenvironmental reconstructions, PhD Thesis, University of Turku. Geological Survey of Finland, Espoo.

Ojala, A. and Francus, P. 2002. X-ray densitometry versus BSE-image analysis of thin-sections: a comparative study of varved sediments of Lake Nautajärvi, Finland. *BOREAS*, 31, 57–64.

Ojala, A. and Saarnisto, M. 1999. Comparative varve counting and magnetic properties of a 8400 years sequence of an annually laminated sediment in Lake Valkiajärvi, Central Finland. *Journal of Palaeolimnology*, 22, 335–348.

Ojala, A., Saarinen, T. and Salonen, V.-P. 2000. Preconditions for the formation of annually laminated lake sediments in southern and central Finland. *Boreal Environment Research*, 5, 243–255.

Olander, H., Korhola, A., Birks, H.J.B. and Blom, T. 1999. An expanded calibration model for inferring lake-water temperatures from chironomid assemblages in northern Fennoscandia. *The Holocene*, 9, 279–294.

Olausson, E. and Olsson, I.U. 1969. Varve stratigraphy in a core from the Gulf of Aden. *Palaegeography, Palaeoclimatology, Palaeoecology*, 6, 87–103.

Oldfield, F. and Alverson, K. 2002. The societal relevance for past global change research. In K. Alverson, R.S. Bradley and T.F. Pedersen (eds), *Paleoclimate, global change and the future.* Springer, Berlin, pp. 1–11.

Oldfield, F., Richardson, N. and Appleby, P.G. 1995. Radiometric dating ^{210}Pb, ^{137}Cs, ^{241}Am. of recent ombrotrophic peat accumulation and evidence for changes in mass-balance. *The Holocene*, 5, 141–148.

O'Leary, M.H. 1981. Carbon isotope fractionation in plants. *Phytochemistry*, 20, 553–567.

O'Leary, M.H. 1988. Carbon isotopes in photosynthesis. *Bioscience*, 38, 328–336.

Oliver, D.R. 1971. Life history of Chironomidae. *Annual Review of Entomology*, 16, 211–230.

Onac, B.P. 1997. Crystallography of speleothems. In C. Hill and P. Forti (eds), *Cave minerals of the world.* National Speleological Society, Huntsville, AL, pp. 230–236.

O'Neil, J.R., Clayton, R.N. and Mayeda, T.K. 1969. Oxygen isotope fractionation in divalent metal carbonates. *Journal of Chemical Physics*, 51, 5547–5558.

Open University. 1989. *Ocean chemistry and deep-sea sediments.* Open University and Pergamon, Milton Keynes, pp. 134.

O'Sullivan, P.E. 1983. Annually laminated lake sediments and the study of Quaternary environmental changes. *Quaternary Science Reviews*, 1, 245–313.

Osvald, H. 1923. Die vegetation des hochmoores Komosse. *Svenska Vaxtsociologiska Sallskapets Handlingar*, 1, 436.

Otterman, J., Chou, M.-D., Arking, A. 1984. Effects of non-tropical forest cover on climate. *Journal of Climate and Applied Meterology*, 23, 762–767.

Otto-Bliesner, B. 1999. El Niño/La Niña and Sahel precipitation during the middle Holocene. *Geophysical Research Letters*, 26, 87–90.

Ourisson, G. and Albrecht, P. 1992. Hopanoids. 1. Geohopanoids: the most abundant natural products on Earth? *Accounts of Chemical Research*, 25, 398–402.

Overpeck, J.T., Webb, T. and Prentice, I.C. 1985. Quantitative interpretation of fossil pollen spectra: dissimilarity coefficients and the method of modern analogs. *Quaternary Research*, 23, 87–108.

Overpeck, J., Hughen, K., Hardy, D., *et al.* 1997. Arctic environmental change of the last four centuries. *Science*, 278, 1251–1256.

Pachur, H.-J. 1999. Paläo-Environment und Drainagesysteme der Ostsahara im Spätpleistozän und Holozän. In Nordost-AfriKa: Strukturen und Resourcen; Ergebnisse aus dem Sonderforschungsbereich. In E. Klitzsch and U. Thorweihe (eds), *Geowissenschaftliche Probleme in Ariden und Semiariden Gebieten.* Wiley-VCH, Weinheim, pp. 366–455.

Pachur, H.-J. and Altmann, N. 1997. The Quaternary Holocene, ca. 8000a BP. In H. Schandelmeier and P.-O. Reynolds (eds), *Palaeogeographic-Palaeotectonic atlas of North-Eastern Africa, Arabia, and adjacent areas – Late Neoproterozoic to Holocene.* Balkema, Rotterdam, pp. 111–125.

Pachur, H.-J. and Hoelzmann, P. 1991. Paleoclimatic implications of Late Quaternary lacustrine sediments in Western Nubia, Sudan. *Quaternary Research*, 36, 257–276.

Pachur, H.-J. and Hoelzmann, P. 2000. Late Quaternary palaeoecology and palaeoclimates in the eastern Sahara. *Journal of African Earth Sciences*, 30, 929–939.

Pachur, H.-J. and Peters, J. 2001. The position of the Mursuq sand sea in the palaeodrainage system of the eastern Sahara. *Palaeoecology of Africa*, 27, 259–290.

Pachur, H.-J. and Wünnemann, B. 1996. Reconstruction of the palaeoclimate along 30°E in the eastern Sahara during the Pleistocene/Holocene transition. In K. Heine (ed.), *Palaeoecology of Africa and the surrounding islands.* A.A.Balkema, Rotterdam/Brookfied, pp.1–32.

Paduano, G.M., Bush, M.B., Baker, P.A., Fritz, S.L. and Seltzer, G.O. 2003. The late Quaternary vegetation history of Lake Titicaca, Peru/Bolivia. Paleogeography, Paleoclimatology, Paleoecology, 194, 259–279.

Pagani, M., Arthur, M.A. and Freeman, K.H. 1999. Miocene evolution of atmospheric carbon dioxide. *Paleoceanography*, 14, 273–292.

Page, M.J., Trustrum, N.A. and DeRose, R.C. 1994. A high resolution record of storm-induced erosion from lake sediments, New Zealand. *Journal of Palaeolimnology*, 11, 333–348

Paillard, D., Labeyrie, L. and Yiou, P. 1996. Macintosh program performs time-series analysis. *EOS Transactions AGU*, 77, 379.

Palacios-Fest, M. R. 1997. Continental ostracode paleoecology from the Hohokam Pueblo Blanco area, central Arizona. *Journal of Archaeological Science*, 24, 965–983.

Pariwono, J.I., Bye, J.A.T. and Lennon, G.W. 1986. Long-period variations in sea-level in Australasia. *Geophysical Journal of the Royal Astronomical Society*, 87, 43–54.

Parker, D.E. 1994. Effects of changing exposure of thermometers at land stations. *International Journal of Climatology*, 14, 1–31.

Passier, H.F., Bosch, H.J., Nijenhuis, I.A., *et al.* 1999. Sulphidic Mediterranean surface waters during Pliocene sapropel formation. *Nature*, 397, 146–149.

Paterson, C.G. and Walker, K.F. 1974. Recent history of *Tanytarsus barbitarsus* Freeman Diptera: Chironomidae. in the sediments of a shallow, saline lake. *Australian Journal of Marine and Freshwater Research*, 25, 315–325.

Paterson, W.S.B. 1994. *The physics of glaciers*, 3rd Edn. Pergamon and Elsevier Science, Tarrytown, NY.

Paterson, W.S.B., Koerner, R.M., Fisher, D., *et al.* 1977. An oxygen-isotope climatic record from the Devon Island ice cap Arctic Canada. *Nature*, 266, 508–511.

Pattullo, J., Munk, W., Revelle, R. and Strong, E. 1955. The seasonal oscillation in sea level. *Journal of Marine Research*, 14, 88–156.

Pätzold, J. 1986. Temperature and CO_2 changes in tropical surface waters of the Phillippines during the last 120 years: record in the stable-isotopes of hermatypic corals, 12. Geologische-Palaeontologische Institut, Christian-Albrechts-Universität, Kiel, Germany.

Pawley, A., Fritz, S.C., Baker, P.A., Seltzer, G.O. and Dunbar, R. 2001. The biological, chemical, and physical limnology of Lake Titicaca, Bolivia/Peru. In M. Munawar and R.E. Hecky (eds), *The Great Lakes of the World (GLOW): food-web, health and integrity*. Backhuys, Leiden, The Netherlands, pp. 195–215.

Payette, S. 1984. Peat inception and climatic change in northern Quebec. In N.A. Morner and W. Karlen (eds), *Climatic change on a yearly to millenial basis*. Reidel, Netherlands, pp. 173–179.

Pearce, R.B., Kemp, A.E.S., Koizumi, I., Pike, J., Cramp, A. and Rowland, S.J. 1998. A lamina-scale, SEM-based study of a late Quaternary diatom-ooze sapropel from the Mediterranean ridge, Site 971. In A.H.F. Robertson, K.-C. Emeis, C. Richter and A. Camerlenghi (eds), *Proceedings of the Ocean Drilling Program, Scientific Results*, 160, 349–363.

Pearson, A., Eglinton, T.I. and McNichol, A.P. 2000. An organic tracer for surface ocean radiocarbon. *Paleoceanography*, 15, 541–550.

Pearson, G.W. 1986. Precise calendrical dating of known growth-period samples using a 'curve fitting' technique. *Radiocarbon*, 28, 292–299.

Pearson, G.W. and Stuiver, M. 1986. High-precision calibration of the radiocarbon time-scale, 500–2500 BC. *Radiocarbon*, 28, 839–862.

Pearson, S. and Dodson, J. 1993. Stick-nest rat middens as source of palaeoecological data in Australian deserts. *Quaternary Research*, 39, 347–354.

Pearson, S. and Betancourt, J.L. 2002. Understanding arid environments using fossil rodent middens. *Journal of Arid Environments*, 50, 499–511.

Pearson, S., Lawson, E., Head, L., McCarthy, L. and Dodson, J. 1999. The spatial and temporal patterns of stick-nest rat middens in Australia. *Radiocarbon*, 41, 295–308.

Peglar, S.M. 1993. The mid-Holocene *Ulmus* decline at Diss Mere, Norfolk, UK: a year-by-year pollen stratigraphy from annual laminations. Holocene 3, 1–13.

Peglar, S.M., S.C. Fritz, T. Alapieti, M. Saarnisto and H.J.H. Birks, 1984. Composition and formation of laminated sediments in Diss Mere, Norfolk, England. *BOREAS*, 13, 13–28.

Peiser, B.J. 1998. Comparative analysis of late Holocene environmental and social upheaval: evidence for a disaster around 4000 BP. In B.J. Peiser, T. Palmer and M. Bailey (eds), *Natural Catastrophes during Bronze Age Civilisations*. BAR International Series, 728, 117–139.

Peixoto, J.P., Oort, A.H. 1992. *Physics of climate*. American Institute of Physics, New York, pp. 111–125.

Pellatt, M.G., Smith, M.J, Mathewes, R.W., Walker, I.R. and Palmer, S.L. 2000. Holocene treeline and climate change in the Subalpine Zone near Stoyoma Mountain, Cascade Mountains, south-western British Columbia, Canada. *Arctic, Antarctic and Alpine Research*, 32, 73–83.

Peltenburg, E. 2000. From nucleation to dispersal. Late third millennium BC settlement pattern transformations in the Near East and Aegean. *Subartu VII*, 183–206.

Peltier, W.R. 1998. Post-glacial variations in the level of the sea: implications for climate dynamics and solid-earth geophysics. *Reviews of Geophysics*, 36, 603–689.

Peltier, W.R. 2001. Global glacial isostatic adjustment and modern instrumental records of relative sea level history. In B.C. Douglas, M.S. Kearney and S.P. Leatherman (eds), *Sea level rise, history and consequences.* Academic Press, San Diego, pp. 65–95.

Peltier, W.R., Farrell, W.E. and Clark, J.A. 1978. Glacio isostasy and relative sea level: a finite element model. *Tectonophysics*, 50, 81–110.

Pennington, R.T., Prado, D.E. and Pendry, C.A. 2000. Neotropical seasonally dry forests and Quaternary vegetation changes. *Journal of Biogeography*, 27, 261–273.

Perfiliev, B.W. 1929. Zur Mikrobiologie der Bodenablagerungen. *Verhandlungen der Internationalen Vereinigung für Limnologie*, 4, 107–143.

Perrette, Y., Delannoy, J.J., Bolvin, H., *et al.* 2000. Comparative study of a stalagmite sample by stratigraphy, laser induced fluorescence spectroscopy, EPR spectrometry and reflectance imaging. *Chemical Geology*, 162, 221–243.

Peterson, T.C., Easterling, D.R., Karl, T.R., *et al.* 1998a. Homogeneity adjustments of in situ atmospheric climate data: a review. *International Journal of Climatology*, 18, 1493–1517.

Peterson, T.C., Karl, T.C., Jamason, P.F., Knight, R. and Easterling, D.R. 1998b. The first difference method: Maximising station density for the calculation of long-term global temperature change. *Journal of Geophysical Research*, 103, 25967–25974.

Petit, J.R., Jouzel, J., Raynaud, D., *et al.* 1999. Climate and atmospheric history of the past 420,000 years from the Vostok ice core, Antarctica. *Nature,* 399, 429–436.

Petit-Maire, N. 1986. Palaeoclimates of the Sahara in Mali. *Episodes*, 9, 7–16.

Petit-Maire, N. 1994. Natural variability of the Asian, Indian and African monsoons over the last 130 ka. In M. Desbois (ed.), *Global precipitation and climate change. NATO ASI Series, Vol. 126,* Springer-Verlag, Berlin, pp. 1–26.

Petit-Maire, N. and Guo, Z. 1996. Mise en évidence de variations climatiques holocènes rapides, en phase dans les déserts actuels de Chine et du Nord de l'Afrique. *Sciences de la Terre et des Planètes*, 322, 847–851.

Petit-Maire, N. and Riser, J. 1981. Holocene lake deposits and palaeoenvironments in central Sahara, northeastern Mali. *Palaeogeography, Paeoclimatology, Palaeoecology*, 35, 45–61.

Petit-Maire, N., Celles, J.C., Commelin, D., Delibrias, and Raimbault, M. 1981. The Sahara in northern Mali: man and his environment between 10,000 and 3500 years bp. Preliminary results. *African Archaeological Review*, 1, 105–125.

Petoukhov, V., Ganopolski, A., Brovkin, V., *et al.* 2000. CLIMBER-2: a climate system model of intermediate complexity. Part I: Model description and performance for present climate. *Climate Dynamics*, 16, 1–17.

Pétrequin, P. 1997. Management of architectural woods and variations in population density in the fourth and third millennia BC Lakes Chalain and Clairvaux, Jura, France. *Journal of Anthropological Archaeology*, 15, 1–19.

Pétrequin, P., Arbogast, R.M., Bourquin-Mignot, C., Lavier, C. and Viellet, A. 1998. Demographic growth, environmental changes and technical adaptations: responses of an agricultural community from the 32nd to the 30th centuries BC. *World Archaeology*, 30, 181–192.

Petterson, G., Odgaard, B.V. and Renberg, I. 1999. Image analysis as a method to quantify sediment components. *Journal of Palaeolimnology*, 22, 443–455.

Pfadenhauer, J., Schneekloth, H., Schneider, R. and Schneider, S. 1993. Mire distribution. In A.L. Heathwaite and Kh. Göttlich (eds), *Mires: process, exploitation and conservation.* John Wiley, Chichester, pp. 77–121.

Pfister, C. and Brazdil, R. 1999. Climatic variability in sixteenth-century Europe and its social dimensions: a synthesis. *Climatic Change*, 43, 5–53.

Pfister, C., SchwarzZanetti, G. and Wegmann, M. 1996. Winter severity in Europe: the fourteenth century. *Climate Change*, 34, 91–108.

Pfister, C., Luterbacher, J., Schwarz-Zanetti, G., and Wegmann, M. 1998. Winter air temperature variations in western Europe during the Early and High Middle Ages AD 750–1300. *The Holocene*, 8, 535–552.

Pfister, C., Brázdil, R., Glaser, R., *et al.* 1999. Documentary evidence on climate in sixteenth-century Europe. *Climate Change*, 43, 55–110.

Picaut, J., Ioualalen, I., Menkes, C., Delcroix, T. and McPhaden, M.J. 1996. Mechanism of the zonal displacements of the Pacific warm pool: implications for ENSO. *Science*, 274, 1486–1489.

Pichon, J.-J., Labeyrie, L.D., Bareille, G., Labracherie, M., Duprat, J. and Jouzel, J. 1992. Surface water temperature changes in the high latitudes of the southern hemisphere over the last glacial–interglacial cycle. *Paleoceanography*, 7, 289–318.

Pienitz, R., Walker, I.R., Zeeb, B.A. Smol, J.P. and Leavitt, P.R. 1992. Biomonitoring past salinity changes in an athalassic subarctic lake. *International Journal of Salt Lake Research*, 1, 91–123.

Pienitz, R., Smol, J.P. and Birks, H.J.B. 1995. Assessment of freshwater diatoms as quantitative indicators of past climatic change in the Yukon and Northwest Territories, Canada. *Journal of Paleolimnology*, 13, 21–49.

Pienitz, R., Smol, J.P. and MacDonald, G. 1999. Paleolimnological reconstruction of the Holocene climatic trends from two boreal treeline lakes, Northwest Territories, Canada. *Arctic, Antarctic and Alpine Research*, 31, 82–93.

Pienitz, R., Smol, J.P., Last, W.M., Leavitt, P.R., and Cumming, B. 2000. Multi-proxy Holocene palaeoclimatic record from a saline lake in the Canadian Subarctic. *The Holocene*, 10, 673–686.

Pigott, C.D. 1970. The response of plants to climate and *Climate Change*. In F.H. Perring (ed.), *The Flora of a changing Britain*. Classey, Faringdon, pp. 32–44.

Pigott, C.D. 1981. Nature of seed sterility and natural regeneration of *Tilia cordata* near its northern limit in Finland. *Annals Botanicae Fennici*, 18, 255–263.

Pigott, C.D. 1992. Are the distributions of species determined by failure to set seed? In C. Marshall and J. Grace (eds), *Fruit and Seed Prodution, Society for Experimental Biology Seminar Series 47*. Cambridge University Press, Cambridge, pp. 203–216.

Pigott, C.D. and Huntley, J.P. 1981. Factors controlling the distribution of *Tilia cordata* at the northern limits of its geographical range III. Nature and causes of seed sterility. *New Phytologist*, 87, 817–839.

Pike, J. and Kemp, A.E.S. 1996. Preparation and analysis techniques for studies of laminated sediments. In A.E.S. Kemp (ed.), *Palaeoclimatology and palaeoceanography from laminated sediments*. Geological Society of London Special Publication, 116. The Geological Society, London, pp. 37–48.

Pike, J. and Kemp, A.E.S. 1997. Early Holocene decadal-scale ocean variability recorded in Gulf of California laminated sediments. *Paleoceanography*, 12, 227–238.

Pilcher, J.R. 1991a. Radiocarbon dating. In P.L. Smart and P.D. Frances (eds), *Quaternary dating methods: a user's guide*. Quaternary Research Association Technical Guide 4. Quaternary Research Association, Cambridge, UK, pp. 16–36.

Pilcher, J.R. 1991b. Radiocarbon dating for the Quaternary scientist. In J.J. Lowe (ed.), *Radiocarbon dating: recent applications and future potential, Quaternary Proceedings No. 1*. Quaternary Research Association, Cambridge, UK, pp. 27–33.

Pilcher, J.R. 1993. Radiocarbon dating and the palynologist: a realistic approach to precision and accuracy. In F.M. Chambers (ed.), *Climate change and human impact on the landscape*. Chapman and Hall, London, pp. 23–32.

Pilcher, J.R. and Baillie, M.G.L. 1980. Six modern oak chronologies from Ireland. *Tree-Ring Bulletin*, 40, 23–34.

Pilcher, J.R., Baillie, M.G.L., Schmidt, B. and Becker, B. 1984. A 7272-Year tree-ring chronology for Western Europe. *Nature*, 312, 150–152.

Pilcher, J.R., Hall, V.A. and McCormac, F.G. 1995. Dates of Holocene Icelandic volcanic eruptions from tephra layers in Irish peats. *The Holocene*, 5, 103–110.

Pilcher, J.R., Hall, V.A. and McCormac, F.G. 1996. An outline tephrochronology for the north of Ireland. *Journal of Quaternary Science*, 11, 485–494.

Pinty, B., Roveda, F., Verstaete, M.M., *et al.* 2000a. Surface albedo retrieval from Meteosat. 1: Theory. *Journal of Geophysical Research*, 105, 18099–18112.

Pinty, B., Roveda, F., Verstaete, M.M., *et al.* 2000b. Surface albedo retrieval from Meteosat. 2: Applications. *Journal of Geophysical Research*, 105, 18113–18134.
Piperno, D.R. and Stothert, K.E. 2003. Phytolith evidence for early Holocene Cucurbita domestication in southwest Ecuador. *Science*, 299, 1054–1057.
Piperno, D.R., Bush, M.B. and Colinvaux, P.A. 1990. Paleoenvironments and human settlement in late-glacial Panama. *Quaternary Research*, 33, 108–116.
Pirazzoli, P. A. 1991. *World atlas of Holocene sea-level changes.* Elsevier Oceanography Series 58. Elsevier, Amsterdam.
Pirazzoli, P.A. 1996. *Sea-level changes. The last 20,000 years.* Wiley, Chichester.
Pisias, N., Dauphin, J.P. and Sancetta, C. 1973. Spectral analysis of late Pleistocene–Holocene sediments. *Quaternary Research*, 3, 3–9.
Placzek, C., Quade, J. and Betancourt, J.L. 2001. Holocene lake-level fluctuations of lake Aricota, Southern Peru. *Quaternary Research*, 56, 181–190.
Plog, S. 1990. Agriculture, sedentism and environment in the evolution of political systems. In S. Upham (ed.), *The evolution of political systems.* Cambridge University Press, Cambridge, pp. 177–202.
Polyak, V.J. and Asmerom, Y. 2001. Late Holocene climate and cultural changes in the southwestern united states. *Science*, 294, 148–151.
Polyak, V.J., Cokendolpher, J.C., Norton, R.A. and Asmeron, Y. 2001. Wetter and cooler late Holocene climate in the southwestern United States from mites preserved in stalagmites. *Geology Boulder*, 29, 643–646.
Pons, A. and Quézel P. 1958. Premières remarques sur l'étude palynologique d'un guano fossile du Hoggar. *Comptes Rendus Séances Academie Sciences*, 244, 2290–2292.
Pope, V.D., Gallani, M.L., Rowntree, P.R. and Stratton, R.A. 2000. The impact of new physical parametrizations in the Hadley Centre climate model - HadAM3. *Climate Dynamics*, 16, 123–146.
Porinchu, D.F. and Cwynar, L.C. 2000. The distribution of freshwater Chironomidae Insecta: Diptera. across treeline near the lower Lena River, northeast Siberia, Russia. *Arctic, Antarctic and Alpine Research*, 32, 429–437.
Porinchu, D.F. and Cwynar, L.C. 2002. Late-Quaternary history of midge communities and climate from a tundra site near the lower Lena River, Northeast Siberia. *Journal of Paleolimnology*, 27, 59–69.
Porter, S.C. 1975. Equilibrium-line altitudes of late Quaternary glaciers in the Southern Alps, New Zealand. *Quaternary Research*, 5, 27–47.
Porter, S.C. 1981. Glaciological evidence of Holocene climatic change. In T.M.L. Wigley, M.J. Ingram and C. Fermer (eds), *Climate and history.* Cambridge University Press, Cambridge, pp. 82–110.
Poynter, J.G. and Eglinton, G. 1991. The biomarker concept-strength and weaknesses. *Fresenius Journal of Analytical Chemistry*, 339, 725–731.
Poynter, J.-G., Farrimond, P., Robinson, N. and Eglinton, G. 1989. Aeolian derived higher plant lipids in the marine sedimentary record: links with paleoclimate. In M. Leinen and M. Sarnthein (eds), *Paleoclimatology and paleometerology: modern and past patterns of global atmospheric transport.* Kluwer Academic Press, Dordrecht, pp. 435–462.
Prado, D.E. and Gibbs, P.E. 1993. Patterns of species distributions in the seasonal dry forests of South America. *Annals of the Missouri Botanical Garden*, 80, 902–927.
Prahl, F.G. 1985. Chemical evidence of differential particle dispersal in the southern Washington coastal environment. *Geochimica et Cosmochimica Acta*, 49, 2533–2539.
Prahl, F.G. and Wakeham, S.G. 1987. Calibration of unsaturation patterns in long-chain ketone compositions for palaeotemperature assessment. *Nature*, 320, 367–369.
Prahl, F.G., de Lange, G.J., Lyle, M. and Sparrow, M.A. 1989a. Post-depositional stability of long-chain alkenones under contrasting redox conditions. *Nature*, 341, 434–437.
Prahl, F.G., Muehlhausen, L.A. and Lyle, M. 1989b. An organic geochemical assessment of oceanographic conditions at MANOP site over the past 26,000 years. *Paleoceanography*, 4, 495–510.
Prell, W.L. and Kutzbach, J.E. 1987. Monsoon variability over the past 150,000 years. *Journal of Geophysical Research*, 92, 8411–8425.
Prentice, I.C. 1985. Pollen representation, source area, and basin size: Toward a unified theory of pollen analysis. *Quaternary Research*, 23, 76–86.

Prentice, I.C. 1986. Vegetation responses to past climatic variation. *Vegetatio*, 67, 131–141.

Prentice, I.C., Bartlein, P.J. and Webb, T. 1991. Vegetation and climate change in eastern North America since the last glacial maximum. *Ecology*, 72, 2038–2056.

Prentice, I.C., Cramer, W., Harrison, S.P., Leemans, R., Monserud, R.A. and Solomon, A.M. 1992. A global biome model based on plant physiology and dominance, soil properties and climate. *Journal of Biogeography*, 19, 117–134.

Prentice, I.C., Sykes, M.T., Lautenschlager, M., Harrison, S.P., Dennissenko, O. and Bartlein, P.J. 1993. Modelling the global vegetation patterns and terrestrial carbon storage at the last glacial maximum. *Global Ecology and Biogeography Letters*, 3, 67–76.

Prentice, I.C., Jolly, D. and BIOME 6000 members. 2000. Mid-Holocene and glacial-maximum vegetation geography of the northern continents and Africa. *Journal of Biogeography*, 27, 507–519.

Price, G.D., McKenzie, J.E., Pilcher, J.R. and Hoper, S.T. 1997. Carbon-isotope variation in *Sphagnum* from hummock-hollow complexes: implications for Holocene climate reconstruction. *The Holocene*, 7, 229–233.

Pross, J., Klotz, S. and Mosbrugger, V. 2000. Reconstructing palaeotemperatures for the Early and Middle Pleistocene using the mutual climatic range method based on plant fossils. *Quaternary Science Reviews*, 19, 1785–1799.

Przbylowicz, W., Schwarcz, H.P. and Latham, A.G. 1991. Dirty calcites; 2, Uranium-series dating of artificial calcite–detritus mixtures. *Chemical Geology*, 86, 161–178.

Psenner, R. and Schmidt, R. 1992. Climate driven pH control of remote alpine lakes and effects of acid deposition. *Nature*, 356, 781–783.

Qin, X., Tan, M., Liu, T., Wang, X., Li, T. and Lu, J. 1999. Spectral analysis of a 1000-year stalagmite lamina-thickness record from Shihua Cavern, Bejing, China, and its climatic significance. *The Holocene*, 9, 689–694.

Quinn, T.M., Crowley, T.J. and Taylor, F.W. 1996. New stable-isotope records from a 173-year coral record from Espiritu Santo, Vanuatu. *Geophysical Reserch Letters*, 23, 3413–3416.

Quinn, T.M., Crowley, T.J., Taylor, F.W., Henin, C., Joannot, P. and Join, Y. 1998. A multicentury stable-isotope record from a New Caledonia coral: Interannual and decadal sea surface temperature variability in the southwest Pacific since 1657 AD. *Paleoceanography*, 13, 412–426.

Quinn, W. H., Neal, V. T. and Antunez de Mayolo, S. E. 1987, El Niño occurrences over the past four and a half centuries. *Journal of Geophysical Research*, 92, 14449–14461.

Racca, J.M.J., Philipert, A., Racca, R. and Prairie, Y.T. 2001. A comparison between diatom-based pH inference models using Artificial Neural Networks (ANN), Weighted Averaging (WA) and Weighted Averaging Partial Least Squares (WA-PLS). regressions. *Journal of Paleolimnology*, 26, 411–422.

Racca, J.M.J., Wild, M.D., Birks, H.J.B. and Prairie, Y.T. 2003. Separating wheat from chaff: diatom taxon selection using an artificial neural network pruning algorithm. *Journal of Paleolimnology*, 29, 123–133.

Rackham O. 1980. *Ancient woodland: its history, vegetation and uses in England.* Arnold, London.

Rahmstorf, S. and R. Alley, 2002. Stochastic resonance in glacial climate. *EOS Transactions AGU*, 83, 129 and 135.

Rahmstorf, S., Marotzke, J., and Willebrand, J. 1996. Stability of the Thermohaline Circulation. In W. Kraus (ed.), *The Warmwatersphere of the North Atlantic Ocean.* Gebrüder Bornträger, Berlin, pp. 129–157.

Ralska-Jasiewiczowa, M., van Geel, B., Goslar, T., and Kuc, T. 1992. The record of the Late Glacial/Holocene transition in the varved sediments of lake Gosciaz, central Poland. *Sveriges Geologiska Undersökning Ca*, 81, 257–268.

Ralska-Jasiewiczowa, M., T. Goslar, T. Madeyska and L. Starkel (eds) 1998. *Lake Gosciaz, central Poland – A monographic study, Part 1.* W. Szafer Institute of Botany, Krakow.

Ramrath, A., Sadori, L. and Negendank, J.F.W. 2000. Sediments from Lago di Mezzano, central Italy: a record of Lateglacial/Holocene climatic variations and anthropogenic impact. *The Holocene*, 10, 87–96.

Ramseyer, K., Miano, T., D'Orazio, V., Wildberger, A., Wagner, T. and Geister, J. 1997. Nature and origin of organic matter in carbonates from speleothems, marine cements and coral skeletons. *Organic Geochemistry*, 26, 361–378.

Raper, S.C.B., Briffa, K.R. and Wigley, T.M.L. 1996. Glacier change in northern Sweden from AD 500: a simple geometric model of Storglaciären. *Journal of Glaciology*, 42, 341–351.

Räsänen, M., Salo, J. and Kalliola, R. 1987. Fluvial perturbance in the western Amazon basin regulation by long term sub-Andean tectonics. *Science*, 238, 1398–1401.

Rathburn, A.E., Pichon, J.-J., Ayress, M.A. and de Deckker, P. 1997. Microfossil and stable-isotope evidence for changes in Late Holocene palaeoproductivity and palaeoceanographic conditions in the Prydz Bay region of Antarctica. *Palaeogeography, Palaeoclimatology, Palaeoecology*, 131, 585–510.

Raynaud, D., Chappellaz, J., Ritz, C. and Martinerie, P. 1997. Air content along the Greenland Ice core. a record of surface climatic parameters and elevation in central Greenland. Special Issue *Journal of Geophysical Research*, 102, 26607–26614.

Reale, O. and Dirmeyer, P. 2000. Modeling the effects of vegetation on Mediterranean climate during the Roman Classical Period Part I. Climate history and model sensitivity. *Global and Planetary Change*, 25, 163–184.

Reale, O. and Shukla, J. 2000. Modeling the effects of vegetation on Mediterranean climate during the Roman Classical Period Part II. Model simulation. *Global and Planetary Change*, 25, 185–214.

Rech, J.A., Quade, J. and Betancourt, J.L. 2002. Late Quaternary paleohydrology of the central Atacama desert lat 22° – 24° S. Chile. *Geological Society of America Bulletin*, 114, 334–348.

Reed, D.J. 1990. The impact of sea-level rise on coastal salt marshes. *Progress in Physical Geography*, 14, 465–481.

Reed, J.M. 1998. A diatom-conductivity transfer function for Spanish salt lakes. *Journal of Paleolimnology*, 19, 399–416.

Reed, J.M., Roberts, N. and Leng, M.J. 1999. An evaluation of the diatom response to Late Quaternary environmental change in two lakes in the Konya Basin, Turkey, by comparison with stable-isotope data. *Quaternary Science Reviews*, 18, 631–646.

Reichert, B.K., Bengtsson, L. and Oerlemans, J. 2001. Midlatitude forcing mechanisms for glacier mass balance investigated using general circulation models. *Journal of Climate*, 14, 3767–3784.

Renberg, I. 1981a. Formation, structure and visual appearance on iron-rich, varved lake sediments. *Verh. Internat. Verein. Limnology*, 21, 94–101.

Renberg, I. 1981b. Improved methods for sampling, photographing and varve-counting of varved lake sediments. *BOREAS*, 10, 255–258.

Renssen, H., Goosse, H., Fichefet, T. and Campin, J. M. 2001. The 8.2 kyr BP event simulated by a global atmosphere–sea–ice–ocean model. *Geophysical Research Letters*, 28, 1567–1570.

Repeta, D.J., Simpson, D.J., Jorgensen, B.B. and Jannasch, H.W. 1989. Evidence for anoxygenic photosynthesis from the distribution of bacteriochlorophylls in the Black Sea. *Nature*, 342, 69–72.

Retelle, M.J. and J. Child, 1996. Suspended sediment transport and deposition in a High Arctic meromictic lake. *Journal of Paleolimnology*, 16, 151–167.

Richards, M., H. Corte-Real, P. Forster, *et al.* 1996. Palaeolithic and Neolithic lineages in the European mitochondrial gene pool. *American Journal of Human Genetics*, 59, 185–203.

Richardson, T.L., Gibson, C.E. and Heaney, S.I. 2000. Temperature growth and seasonal succession of phytoplankton in Lake Baikal, Siberia. *Freshwater Biology*, 44, 431–440.

Richter-Bernburg, G. 1964. Solar cycle and other climatic periods in varvitic evaporites. In A.E.M. Nairn (ed.), *Problems in palaeoclimatology*. Springer, New York, 510–519.

Ricketts, R.D. and Anderson, R.F. 1998. A direct comparison between the historical record of lake-level and the d18O signal in carbonate sediments from Lake Turkana, Kenya. *Limnology and Oceanography*, 43, 811–822.

Ricketts, R.D and Johnson, T.C. 1996. Early Holocene changes in lake-level and productivity in Lake Malawi as interpreted from oxygen and carbon isotopic measurements from authigenic carbonates. In T.C. Johnson and E.O. Odada (eds), *The limnology, climatology and palaeoclimatology of the East Africa lakes*. Gordon and Breach, Amsterdam.

Riedinger, M.A., Steinitz-Kannan, M., Last, W.M. and Brenner, M. 2002. A ~6100 ^{14}C yr record of El Nino activity from the Galapagos Islands. *Journal of Paleolimnology*, 27, 1–7.

Rieley, G., Collier, R.J., Jones, D.M. and Eglinton, G. 1991. The biogeochemistry of Ellesmere Lake, UK-I: source correlation of leaf wax inputs to the sedimentary lipid record. *Organic Geochemistry*, 17, 901–912.

Rieradevall, M. and Brooks, S.J. 2001. An identification guide to subfossil Tanypodinae larvae Insecta: Diptera: Chironomidae. based on cephalic setation. *Journal of Paleolimnology*, 25, 81–99.

Rietti-Shati, M., Shemesh, A. and Karlen, W. 1998. A 300-year climate record from biogenic silica oxygen isotopes in an equatorial high-altitude lake. *Science*, 281, 980–982.

Rind, D. 1998. Latitudinal temperature gradients and climate change. *Journal of Geophysical Research*, 103D, 5943–5971.

Rind, D. 2000. Relating paleoclimate data and past temperature gradients: some suggestive rules. *Quaternary Science Reviews*, 19, 381–390.

Rind, D. 2002. The Sun's role in climate variations. *Science*, 296, 673–677.

Rind, D., Lean, J. and Healy, R. 1999. Simulated time-dependent climate response to solar radiative forcing since 1600. *Journal of Geophysical Research*, 104D, 1973–1990.

Ritchie, J.C. 1986. Climatic change and vegetation response. *Vegetatio*, 67, 65–67.

Ritchie, J.C. 1995. Current trends in studies of long-term plant community dynamics. *New Phytologist*, 130, 469–494.

Ritchie, J.C. and Haynes, C.V. Jr. 1987. Holocene vegetation zonation in the eastern Sahara. *Nature*, 330, 645–647.

Robbins, L. H., Murphy, M.L., Brook, G. A., *et al.* 2000. Archaeology, Palaeoenvironment, and Chronology of the Tsodilo Hills White Paintings Rock Shelter, Northwest Kalahari Desert, Botswana. *Journal of Archeological Science*, 27, 1085–1113.

Roberts, D. and McMinn, A. 1998. A weighted averaging regression and calibration model for inferring lakewater salinity from fossil diatom assemblages in saline lakes of the Vestfold Hills: a new tool for interpreting Holocene lake histories in Antarctica. *Journal of Palaeolimnology*, 19, 99–113.

Roberts, J.H., Holmes, J.A. and Swan, A.R.H. 2002. Ecophenotypy in *Limnocythere inopinata* Ostracoda. from the late Holocene of Kajemarum Oasis North-eastern Nigeria. *Palaeogeography, Palaeoclimatology, Palaeoecology*, 185, 41–52.

Roberts, M.S., Smart, P.L. and Baker, A. 1998. Annual trace element variations in a Holocene speleothem. *Earth and Planetary Science Letters*, 154, 237–246.

Roberts, N., Reed, J., Leng, M.J., *et al.* 2001. The tempo of Holocene climatic change in the eastern Mediterranean region: new high-resolution crater-lake sediment data from central Turkey. *The Holocene*, 11, 721–736.

Robertshaw, P.T. 1988. Environment and culture in the Late Quaternary of Eastern Africa: a critique of some correlations. In J. Bower and D. Lubell (eds), *Prehistoric cultures and environments in the Late Quaternary of Africa*. Cambridge University Press, Cambridge.

Robertson, A., Overpeck, J., Rind, D. *et al.* 2001. Hypothesized climate forcing time series for the last 500 years. *Journal of Geophysical Research*, 106D, 14783–14804.

Robertson, I., Lucy D., Baxter, L., *et al.* 1999. A kernel-based Bayesian approach to climatic reconstruction. *The Holocene*, 9, 495–500.

Robinson, W.J. 1976. Tree-ring dating and archaeology in the American Southwest. *Tree-Ring Bulletin*, 36, 9–20.

Robinson, W.R. and Cameron, C.M. 1991. *A directory of tree-ring dated prehistoric sites in the American Southwest*. The University of Arizona, Tucson, Arizona.

Robock, A. 2000. Volcanic eruptions and climate. *Reviews of Geophysics*, 38, 191–219.

Robock, A. and J. Mao, 1992. Winter warming from large volcanic eruptions. *Geophysical Research Letters*, 19, 2405–2408.

Robock, A. and J. Mao, 1995. The volcanic signal in surface temperature observations. *Journal of Climate*, 8, 1086–1103.

Rodbell, D.T., Seltzer, G.O., Anderson, D.M., Abbott, M.B., Enfield, D.B. and Newman, J.H. 1999. An ~15,000-year record of El Niño-driven alluviation in southwestern Ecuador. *Science*, 283, 516–520.

Rohmer, M., Bisseret, P. and Neunlist, S. 1992. The hopanoids, prokaryotic triterpenoids and precursors of ubiquitous molecular fossils. In J.M. Moldowan, P. Albrecht and R.P. Philp (eds), *Biological markers in sediments and petroleum*. Prentice Hall, New Jersey, pp. 1–17.

Rom, W., Golser, R., Kutschera, W., Priller, A., Steier, P. and Wild, E.M. 1999. AMS ^{14}C dating of equipment from the iceman and of spruce logs from the prehistoric salt mines of Hallstatt. *Radiocarbon*, 41, 183–197.

Romankevich, E. 1984. *Geochemistry of organic matter in the ocean*. Springer-Verlag, Berlin.

Roosevelt, A.C., Housley, R.A., Imazio da Silveira, M., Maranca, S. and Johnson, R. 1991. Eighth millennium pottery from a prehistoric shell midden in the Brazilian Amazon. *Science*, 254, 1621–1624.

Roosevelt, A.C., Lima da Costa, M., Lopes Machado, C., *et al.* 1996. Paleoindian cave dwellers in the Amazon: the peopling of the Americas. *Science*, 272, 373–384.

Rösch, M. 1993. Prehistoric land use as recorded in a lake-shore core at Lake Constance. *Vegetation History and Archaeobotany*, 2, 213–232.

Rose, M.R., Dean, J.S. and Robinson, W.J. 1981. *The past climate of Arroyo Hondo, New Mexico, reconstructed from tree-rings*. School of American Research Press, New Mexico.

Rosell-Melé, A. 1998. Interhemispheric appraisal of the value of alkenone indices as temperature and salinity proxies in high-latitude locations. *Paleoceanography*, 13, 694–703.

Rosell-Melé, A. and Koç, N. 1997. Paleoclimatic significance of the stratigraphic occurrence of photosynthetic biomarker pigments in the Nordic seas. *Geology*, 25, 49–52.

Rosell-Melé, A., Maslin, M., Maxwell, J.R. and Schaeffer, P. 1997. Biomarker evidence for 'Heinrich' events. *Geochimica et Cosmochimica Acta*, 61, 1671–1678.

Rosell-Melé, A., Bard, E., Emeis, K.C., *et al.* 2001. Precision of the current methods to measure the alkenone proxy UK37 and absolute alkenone abundance in sediments: Results of an interlaboratory comparison study. *Geochemistry Geophysics Geosystems 2*, Paper no. 2000GC000141.

Rosen, A.M. 1995. The social response to environmental change in Early Bronze Age Canaan. *Journal of Anthropological Archaeology*, 14, 26–44.

Rosen, A.M. 1998. Early to mid-Holocene environmental changes and their impact on human communities in southeastern Anatolia. In A.S. Issar and N. Brown (eds), *Water, Environment and Society in Times of Climatic Change*.Kluwer, Dordrecht, pp. 215–240.

Rosén, P. 2001. Holocene climate history in northern Sweden reconstructed from diatom, chironomid and pollen records and near-infrared spectroscopy of lake sediments, PhD Thesis. University of Umeå, Umeå.

Rosén P., Hall, R, Korsman, T. and Renberg, I. 2000. Diatom-transfer functions for quantifying past air temperature, pH and total organic carbon concentration from lakes in northern Sweden. *Journal of Paleolimnology*, 24, 109–123

Rosén, P., Segerström, U., Erikson, L., Renberg, I. and Birks, H.J.B. 2001. Holocene climatic change reconstructed from diatoms, chironomids, pollen and near-infrared spectroscopy at an alpine lake Sjuodjijaure. in northern Sweden. *The Holocene*, 11, 551–562.

Rosenmeier, M.F., Hodell, D.A., Brenner, M. *et al.* 2002. Influence of vegetation change on watershed hydrology: implications for paleoclimatic interpretation of lacustrine ^{18}O records. *Journal of Paleolimnology*, 27, 117–131.

Rosqvist, G.C., Rietti-Shati, M. and Shemesh, A. 1999. Late glacial to middle Holocene climatic record of lacustrine biogenic silica oxygen isotopes from a Southern Ocean island. *Geology*, 27, 967–970.

Rossby, C.G. 1939. Relation between variations in the intensity of the zonal circulation of the atmosphere and the displacements of the semi-permanent centers of action. *Journal Marine Research*, 2, 38–55.

Rossignol-Strick, M. 1985. Mediterranean Quaternary sapropels, an immediate response of the African monsoon to variation of insolation. *Palaeogeography, Palaeoclimatology, Palaeoecology*, 49, 237–263.

Rossignol-Strick, M. 1995. Sea–land correlation of pollen records in the eastern Mediterranean for the glacial–interglacial transition: biostratigraphy versus radiometric time-scale. *Quaternary Science Reviews*, 14, 893–915.

Rostek, F., Bard, E., Beaufort, L., Sonzogni, C. and Ganssen, G. 1997. Sea surface temperature and productivity records for the past 240 kyr in the Arabian Sea. *Deep-Sea Research*, 44, 1461–1480.

Roubik, D.W. and Moreno, E. 1991. Pollen and Spores of Barro Colorado Island. Monographs in Systematic Botany, v. 36, Missouri Botanical Garden, St Louis, MO.

Rowland, S.J., Belt, S.T., Wraige, E.J., Masse, G., Roussakis, C. and Robert, J.M. 2001. Effects of temperature on polyunsaturation in cytostatic lipids of *Haslea ostrearia. Phytochemistry*, 56, 597–602.

Rozanski, K., Stichler. W., Gonfiantini, R., *et al.* 1992. The IAEA ^{14}C intercomparison exercise 1990. *Radiocarbon*, 34, 506–519.

Rundgren, M. and Beerling, D. 1999. A Holocene CO_2 record from the stomatal index of subfossil *Salix herbacea* L. leaves from northern Sweden. *The Holocene*, 9, 509–513.

Rundgren, M., Loader, N.J. and Beerling, D.J. 2000. Variations in the carbon isotope composition of late-Holocene plant macrofossils: a comparison of whole-leaf and cellulose trends. *The Holocene*, 10, 149–154.

Ruzmaikin, A.A. 1999. Can El Niño amplify the solar forcing of climate? *Geophysical Research Letters*, 26, 2255–2258.

Rymer, L. 1978. The use of uniformitarianism and analogy in palaeoecology, particularly pollen analysis. In D. Walker and J.C. Guppy (eds), *Biology and Quaternary environments.* Australian Academy of Sciences, Canberra, pp. 245–257.

Ryves, D.B., Juggins, S., Fritz, S.C. and Battarbee, R.W. 2001. Experimental diatom dissolution and the quantification of microfossil preservation in sediments. *Palaeogeography, Palaeoclimatology, Palaeoecology*, 172, 99–113.

Ryves, D.B., McGowan, S. and Anderson, N.J. 2002. Development and evaluation of a diatom-conductivty model from lakes in West Greenland. *Freshwater Biology*, 47, 995–1014.

Ryves, D.B., Jewson, D.H., Sturm, M. 2003. Quantitative and qualitative relationships between planktonic diatom communities and diatom assemblages in sedimenting material and surface sediments in Lake Baikal, Siberia. *Limnology and Oceanography*, 48, 1183–1192.

Saarinen, T. 1999. Palaeomagnetic dating of Late Holocene sediments in Fennoscandia. *Quarternary Science Reviews*, 18, 889–897.

Saarinen, T. and Petterson, G. 2001. Image analysis techniques. In W. Last and J.P. Smol (ed.), Tracking environmental change using lake sediments: Physical and Geochemical Methods. Developments in Paleoenvironmental Research, 2. Kluwer Academic Publishers, Dordrecht, pp. 23–39.

Saarnisto, M. 1986. Annually laminated lake sediments. In B.E. Berglund (ed.), *Handbook of Holocene palaeoecology and palaeohydrology.* Wiley and Sons, Chichester, pp. 343–370.

Saarnisto, M. and Kahra, A. (ed.) 1992. *Laminated sediments.* Geological Survey of Finland, Special Paper 14.

Saarnisto, M., Huttunen, P. and Tolonen, K. 1977. Annual lamination of sediments in Lake Lovojärvi, southern Finland, during the past 600 years. *Annales Botanicae Fennici*, 14, 35–45.

Sachs, J.P., Repeta, D.J. and Goericke, R. 1999. Nitrogen and carbon isotopic ratios of chlorophyll from marine phytoplankton. *Geochimica et Cosmochimica Acta*, 63, 1431–1441.

Sadler, J.P. and Jones, J.C. 1997. Chironomids as indicators of Holocene environmental change in the British Isles. In A.C. Ashworth, P.C. Buckland and J.P. Sadler (eds), *Studies in Quaternary entomology. Quaternary Proceedings*, 5, 219–232.

Sahagian, D.L., Scwartz, F.W. and Jacobs, D.K. 1994. Direct anthropogenic contributions to sea level rise in the twentieth century. *Nature*, 367, 54–57.

Sakai, K. and Peltier, W.R. 1996. A Multibasin reduced model of the global thermohaline circulation: paleoceanographic analyses of the origins of ice-age climate variability. *Journal of Geophysical Research-Oceans*, 101, 22535–22562.

Saldarriaga, J.G. and West, D.C. 1986. Holocene fires in the northern Amazon basin. *Quaternary Research*, 26, 358–366.

Salgado-Labouriau, M.L. 1980. A pollen diagram of the Pleistocene–Holocene boundary of Lake Valencia, Venezuela. *Review of Palaeobotany and Palynology*, 30, 297–312.

Salo, J. 1987. Pleistocene forest refuges in the Amazon: Evaluation of the biostratigraphical, lithostratigraphical and geomorphological data. *Annales Zoologici Fennici*, 24, 203–211.

Saltzman, E.S. 1995. Ocean/atmosphere cycling of dimethylsulfide . In R. Delmas (ed.), *Ice Core Studies of Global Biogeochemical Cycles, NATO ASI Series, Series I, Vol. 30.* Springer-Verlag, New York, pp. 65–89.

Saltzman, E.S., Whung, P.-Y. and Mayewski, P.A. 1997. Methanesulfonate in the Greenland Ice Sheet Project 2 ice core. *Special Issue Journal of Geophysical Research*, 102, 26649–26658.

Sandelowsky, B.H. 1977. Mirabib: an archaeological study in the Namib. *Madoqua*, 10, 221–283.
Sandweiss, D.H., Richardson, J.B., Reitz, E.J., Rollins, H.B. and Maasch, K.A. 1996. Geoarcheological evidence from Peru for a 5000 years BP onset of El Niño. *Science*, 273, 1531–1533.
Sandweiss, D.H., Maasch, K.A. and Anderson, D.G. 1999. Climate and culture: transitions in the Mid-Holocene. *Science*, 283, 499–500.
Sarkanen, K.V. and Ludwig, C.H. 1971. *Lignins*. Wiley-Interscience, New York
Sarnthein, M. 1978. Sand deserts during glacial maximum and climatic optimum. *Nature*, 272, 43–46.
Sarnthein, M., *et al.* 2001. Fundamental modes and abrupt changes in North Atlantic circulation and climate over the last 60 kyr, In P. Schaefer *et al.* (eds), *The Northern North Atlantic: A changing environment*. Springer-Verlag, New York, 365–410.
Saros, J.E. and Fritz, S.C. 2000. Changes in the growth rates of saline-lake diatoms in response to variation in salinity, brine type and nitrogen form. *Journal of Plankton Research*, 22, 1071–1083.
Saucer, P.E., Miller, G.H. and Overpeck, J.T. 2001. Oxygen isotope ratios of organic matter in arctic lakes as a paleoclimate proxy: field and laboratory investigations. *Journal of Paleolimnolgy*, 25, 43–64.
Schakau, B. 1986. Preliminary study of the development of the subfossil chironomid fauna Diptera. of Lake Taylor, South Island, New Zealand, during the younger Holocene. *Hydrobiologia*, 143, 287–291.
Schakau, B. 1990. Stratigraphy of the fossil Chironomidae Diptera. from Lake Grasmere, South Island, New Zealand, during the last 6000 years. *Hydrobiologia*, 214, 213–221.
Scharf, BW., Pirrung, M., Boehrer, B. *et al.* 2001. Limnogeological studies of maar lake Ranu Klindungan, East Java, Indonesia. *Amazoniana*, 16, 487–516.
Schlesinger, W.H. 1997. *Biogeochemistry: an analysis of global change*. Academic Press, San Diego.
Schlüchter, C. (ed.) 1979. *Moraines and varves*. A.A. Balkema, Rotterdam.
Schmidt, M., Botz, R, Stoffers, P., Anders, T. and Bohrmann, G. 1997. Oxygen isotopes in marine diatoms: a comparative study of analytical techniques and new results on the isotope composition of recent marine diatoms. *Geochimica et Cosmochimica Acta*, 61, 2275–2280.
Schmitz Jr, W.J. 1995. On the interbasin-scale thermohaline circulation. *Reviews of Geophysics*, 33, 151–173.
Schneider, R. and Tobolski, K. 1985. Lago di Ganna: late-glacial and Holocene environments of a lake in the southern Alps. *Dissertationes Botanicae*, 1985, 229–271.
Schneider, R.R., Müller, P.J., Ruhland, G., Meinecke, G., Schmidt, H. and Wefer, G. 1996. Late Quaternary surface temperatures and productivity in the East-equatorial south Atlantic: response to changes in Trade/Monsoon wind forcing and surface water advection. In G. Wefer, W.H. Berger, G. Siedler and D.J. Webb (eds), *The South Atlantic: present and past circulation*. Springer-Verlag, Berlin, pp. 527–551.
Schnell, Ø.A. and Willassen, E. 1996. The chironomid Diptera. communities in two sediment cores from Store Hovvatn, S. Norway, an acidified lake. *Annales of Limnologie*, 32, 45–61.
Schostakowitsch, W.B. 1936. Geschichtete Bodenablagerungen der Seen als Klima-Annalen. *Meteorologische Zeitschrift*, 5, 176–182.
Schrag, D.P. 1999. Rapid analysis of high-precision Sr/Ca ratios in scleractinian corals and other marine carbonates. *Paleoceanography*, 14, 97–102.
Schrag, D.P. and Linsley, B.K. 2002. Paleoclimate: corals, chemistry, and climate. *Science*, 296 5566), 277–278.
Schrag, D.P., Hampt, G. and Murray, D.W. 1996. The temperature and oxygen isotopic composition of the glacial ocean. *Science*, 272, 1930–1932.
Schuenemeyer, J.H. 1978. Reply to comparison of visual and statistical correlation in time series curves. *Mathematical Geology*, 10, 106–108.
Schulte, S., Rostek, F., Bard, E., Rullkötter, J. and Marchal, O. 1999. Variations of oxygen-minimum and primary productivity recorded in sediments of the Arabian Sea. *Earth and Planetary Science Letters*, 173, 205–221.
Schulz, H., von Rad, U. and Stackelberg, U. 1996. Laminated sediments from the oxygen-minimum zone of the northeastern Arabian Sea. In A.E.S. Kemp (ed.), *Palaeoclimatology and palaeoceanography from laminated sediments*, The Geological Society, London, pp. 185–207.
Schulz, H., von Rad, U. and Erlenkeusser, H. 1998, Correlation between Arabian Sea and Greenland climate oscillations of the past 110,000 years. *Nature*, 393, 54–57.

Schulz, M. and Mudelsee, M. 2002. REDFIT: estimating red-noise spectra directly from unevenly spaced paleoclimatic time series. *Computers and Geosciences*, 28, 421–426.

Schulz, M. and Stattegger, K. 1997. SPECTRUM: spectral analysis of unevenly spaced paleoclimatic time series. *Computers and Geosciences*, 23, 929–945.

Schwalb, A., Burns, S.J. and Kelts, K. 1999. Holocene environments from stable-isotope stratigraphy of ostracods and authigenic carbonate in Chilean Altiplano lakes. *Palaeogeography, Palaeoclimatology, Palaeoecology*, 148, 153–168.

Schwander, J., Eicher, U. and Ammann, B. 2000. Oxygen isotopes of lake marl at Gerzensee and Leysin Switzerland, covering the Younger Dryas and two minor oscillations, and their correlation to the GRIP ice core. *Palaeogeography, Palaeoclimatology, Palaeoecology*, 159, 203–214.

Schwarcz, H.P. 1980 Absolute age determination of archaeological sites by uranium-series dating of travertines. *Archaeometry*, 22, 3–24.

Schwarcz, H.P. 1986. Geochronology and isotope geochemistry in speleothems. In P. Fritz and J. Fontes (eds), *Handbook of environmental isotope geochemistry*. Elsevier, Amsterdam, pp. 271–303.

Schwarcz, H.P. and Blackwell, B.A. 1992. Archaeological applications. In M. Ivanovich and R.S. Harmon (eds), *Uranium-series disequilibrium: applications to earth, marine, and environmental sciences*. Clarendon Press, Oxford, pp. 513–552.

Schwarcz, H.P. and Latham, A.G. 1989. Dirty calcites I: Uranium series dating of contaminated calcite using leachates alone. *Isotope Geoscience*, 80, 35–43.

Schwarcz, H.P. and Yonge, C. 1983. Isotopic composition of Paleowaters as inferred from speleothem and its fluid inclusions. In *Paleoclimates and paleowaters; a collection of environmental isotope studies*. International Atomic Energy Agency, Vienna, Austria.

Schweingruber, F.H. 1988. Tree-rings: basics and applications of dendrochronology. Kluwer, Dordrecht.

Scott, D.B. and Medioli, F.S. 1978. Vertical zonations of marsh foraminifera as accurate indicators of former sea levels. *Nature*, 272, 528–531.

Scott, E.M., Aitchison, T.C., Harkness, D.D., Cook, G.T. and Baxter, M.S. 1990a. An overview of all three stages of the international radiocarbon intercomparison. *Radiocarbon*, 32, 309–319.

Scott, E.M., Long, A. and Kra, R.S. (eds) 1990b. Proceedings of the International Workshop on Intercomparison of Radiocarbon Laboratories. *Radiocarbon*, 32, 253 397.

Scott, L. 1987. Pollen analysis of hyena coprolites and sediments from Equus Cave, Taung, southern Kalahari South Africa. *Quaternary Research*, 28,144–156.

Scott, L. 1990. Hyrax Procaviidae. and dassie rat Petromuridae. middens in paleoenvironmental studies in Africa. In J.L. Betancourt, T.R. van Devender and P.S. Martin (eds), *Packrat middens: the last 40,000 years of biotic change*. University of Arizona Press, Tucson, pp. 398–407.

Scott, L. 1994. Palynology of late Pleistocene Hyrax middens, southwestern Cape Province, South Africa: a preliminary report. *Historical Biology*, 9, 71–81.

Scott, L. 1996. Palynology of hyrax middens: 2000 years of palaeo-environmental history in Namibia. *Quaternary International*, 33, 73–79.

Scott, L. and Cooremans, B. 1992. Pollen in recent *Procavia* hyrax, *Petromus* dassie rat. and bird dung in South Africa. *Journal of Biogeography*, 19, 205–215.

Scott, L. and Vogel. JC. 2000. Evidence for environmental conditions during the last 20,000 years in Southern Africa from ^{13}C in fossil hyrax dung. *Global and Planetary Change*, 26, 207–215.

Scott, L., Cooremans, B., de Wet, J.S. and Vogel., J.C. 1991. Holocene environmental changes in Namibia inferred from pollen analysis of swamp and lake deposits. *The Holocene*, 1, 8–13.

Scott, L., Steenkamp, M. and Beaumont, P.B. 1995. Palaeoenvironmental conditions in South Africa at the Pleistocene–Holocene transition. *Quaternary Science Reviews*, 14, 937–947.

Scott, M. 2000. Bayesian methods: what can we gain and at what cost? *Radiocarbon*, 42, 181.

Scuderi, L.A. 1990, Tree-ring evidence for climatically effective volcanic eruptions. *Quaternary Research*, 34, 67–85.

Seibold, E. 1958. Jahreslagen in Sedimenten der mittleren Adria. *Geologische Rundschau*, 47, 100–117.

Seidov, D. and Maslin, M. 1999. North Atlantic Deep Water circulation collapse during the Heinrich events. *Geology*, 27, 23–26.

Seidov, D. and Maslin, M.A. 2001. Atlantic Ocean heat piracy and the bipolar climate seesaw during Heinrich and Dansgaard-Oeschger events. *Journal of Quaternary Science*, 16, 321–328.

Seidov, D., Barron, E., Haupt B.J. and Maslin, M.A. 2001. Ocean bi-polar seesaw and climate: southern versus northern meltwater impacts. In D. Seidov, B. Haupt and M. Maslin (eds), *The oceans and rapid climate change: past, present and future*, Vol. 126. AGU Geophysical Monograph Series, Washington, DC, pp. 147–167.

Seltzer, G.O., Baker, P.A., Cross, S., Dunbar, R. and Fritz, S.C. 1998. High resolution seismic reflection profiles from Lake Titicaca, Peru-Bolivia: evidence for Holocene aridity in the tropical Andes. *Geology*, 26, 167–170.

Seltzer, G.O., Rodbell, D.T., Baker, P.A., Fritz, S.C., Topia, P.M., Rowe, H.D. and Dunbar, R.B. 2002. Early warming of tropical South America at the last glacial-interglacial transition. *Science*, 296, 1685–1686.

Seltzer, G., Rodbell, D. and Burns, S. 2000. Isotopic evidence for late Quaternary climatic change in tropical South America. *Geology*, 28, 35–38.

Seppä, H. and Birks, H.J.B. 2001. July mean temperature and annual precipitation trends during the Holocene in the Fennoscandian tree-line area: pollen-based climate reconstructions. *The Holocene*, 11, 527–539.

Sernander, R. 1908. On the evidence of post-glacial changes of climate furnished by the peat-mosses of Northern Europe. *Geologiska Föreningens i Stockholm Förhandlingar*, 30, 467–478.

Servant, M. and Servant-Vidary, S.1980. L'environnement quaternaire du bassin du Tchad. In M.A.J Williams and H. Faure (eds), *The Sahara and the Nile. Quaternary environments and prehistoric occupation in northern Africa*. Balkema, Rotterdam, pp. 133–162.

Servant, M., Fontes, J.-C., Rieu, M. and Saliège, X. 1981. Phases climatiques arides holocènes dans le sud-ouest de l'Amazonie Bolivie. *Comptes Rendus Academie Scientifique Paris, Series II*, 292, 1295–1297.

Severinghaus, J.P. and Brook, E.J. 1999. Abrupt climate change at the end of the last glacial period inferred from trapped air in polar ice. *Science*, 286, 930–934.

Sheets, P. 2001. The effects of explosive volcanism on simple to complex societies in ancient middle America. In V. Markgraf (ed.), *Interhemispheric climate linkages*. Academic Press, San Diego, pp. 73–86.

Shemesh, A., Charles, C.D. and Fairbanks, R.G. 1992. Oxygen isotopes in biogenic silica – global changes in ocean temperature and isotopic composition. *Science*, 256, 1434–1436.

Shemesh, A., Burckle, L.H. and Hays, J.D. 1995. Late Pleistocene oxygen isotope records of biogenic silica from the Atlantic sector of the Southern Ocean. *Paleoceanography*, 10, 179–196.

Shemesh, A., Rosqvist, G., Riett-Shati, M., Rubensdotter, L., Bigler, C., Yam, R. and Karlen, W. 2001. Holocene climatic changes in Swedish Lapland inferred from an oxygen-isotope record of lacustrine biogenic silica. *The Holocene*, 11, 447–454.

Shen, C.-C., Lee, T., Chen, C.-Y., Wang, C.-H., Dai, C.-F. and Li, L.-A. 1996. The calibration of Δ[Sr/Ca] versus sea surface temperature relationship for *Porites* corals. *Geochimica et Cosmochimica Acta*, 60, 3849–3858.

Shen, G.T. and Dunbar, R.B. 1995. Environmental controls on uranium in reef corals. *Geochimica et Cosmochimica Acta*, 59, 2009–2024.

Shen, G.T., Boyle, E.A. and Lea, D.W. 1987. Cadmium in corals as a tracer of historical upwelling and industrial fallout. *Nature*, 328, 794–796.

Shen, G.T., Campbell, T.M., Dunbar, R.B., Wellington, G.M., Colgan, M.W. and Glynn, P.W. 1991. Paleochemistry of manganese in corals from the Galapagos Islands. *Coral Reefs*, 10, 91–101.

Shen, G.T., Cole, J.E., Lea, D.W., Linn, L.J., McConnaughey, T.A. and Fairbanks, R.G. 1992a. Surface ocean variability at Galápagos from 1936–1982: calibration of geochemical tracers in corals. *Paleoceanography*, 563–588.

Shen, G.T., Linn, L.J., Campbell, T.M., Cole, J.E. and Fairbanks, R.G. 1992b. A chemical indicator of trade wind reversal in corals from the western tropical Pacific. *Journal of Geophysical Research*, 97, 12,689–12,698.

Shennan, S.J. 2000. Population, culture history and the dynamics of culture change. *Current Anthropology*, 41, 811–835.

Shennan, S.J. 2001. Demography and cultural innovation: a model and some implications for the emergence of modern human culture. *Cambridge Archaeological Journal*, 11.1, 5–16.

Shephard, F.P. 1963. Thirty-five thousand years of sea level. In T. Clements (ed.), *Essays in Marine Geology in honour of K.O. Emery*. University of California Press, Los Angeles, pp. 1–10.

Shi, N., Dupont, L.M., Beug, H.-J. and Schneider, R. 1998. Vegetation and climate changes during the last 21,000 years in SW Africa based on a marine pollen record. *Vegetation History and Archaeobotany*, 7, 127–140.

Shi, N., Dupont, L., Beug, H.J. and Schneider, R.R. 2000. Correlation between vegetation in southwestern Africa and oceanic upwelling in the past 21,000 years, *Quaternary Research*, 54, 72–80.

Shindell, D.T., Rind, D., Balachandran, N., Lean, J. and Lonergan, P. 1999. Solar cycle variability, ozone and climate. *Nature*, 284, 305–308.

Shindell, D.T., Schmidt, G.A., Mann, M.E., Rind, D. and Waple, A. 2001. Solar forcing of regional climate change during the Maunder Minimum. *Science*, 294, 2149–2152.

Shindell, D.T., Schmidt, G.A., Mann, M.E., Rind, D., Waple, A. and Miller, R. 2002. Solar and volcanic forcing of climate change during the Maunder Minimum. *Abstracts, Spring AGU Meeting*. AGU, Washington, DC.

Shopov, Y. 1997. Luminescence of cave minerals. In C. Hill and P. Forti (eds), *Cave minerals of the world*. National Speleological Society, Huntsville, AL, pp. 244–248.

Shopov, Y., Dermendjiev, V. and Buykliev, G. 1989. Investigation on the old variations of the climate and solar activity by a new method (LLMZA) of cave flowstone from Bulgaria. *Proceedings of the 10th International Congress of Speleology*, 1, 95–97.

Shopov, Y., Ford, D.C. and Schwarcz, H.P. 1994. Luminescent microbanding in speleothems: High-resolution chronology and paleoclimate. *Geology*, 22, 407–410.

Shopov, Y., Tsankov, L.T., Yonge, C.J., Krouse, H.P.R. and Jull, A.J.T. 1997. Influence of the bedrock CO_2 on stable-isotope records in cave calcites. *Proceedings of the 12th International Congress of Speleology*, 1, 65–68.

Shore, J.S., Bartley, D.D. and Harkness, D.D. 1995. Problems encountered with the ^{14}C dating of peat. *Quaternary Science Reviews*, 14, 373–383.

Shotyk, W., Weiss, D., Appleby, P.G., *et al.* 1998. History of atmospheric lead deposition since 12,370 ^{14}C yr BP from a peat bog, Jura Mountains, Switzerland. *Science*, 281, 1635–1640.

Shulmeister, J. and Lees, B.G. 1995. Pollen evidence from tropical Australia for the onset of an ENSO-dominated climate at c. 4000 BP. *The Holocene*, 5, 10–18.

Shuman, B., Bravo, J., Kaye, J., Lynch, J.A., Newby, P. and Webb, T. 2001. Late Quaternary water-level variations and vegetation history at Crooked Pond, Southeastern Massachusetts. *Quaternary Research*, 56, 401–410.

Siegenthaler, U. and Eicher, U. 1986. Stable oxygen and carbon isotope analyses. In B.E. Berglund (ed.), *Handbook of Holocene paleoecology and palaeohydrology*. Wiley, London, pp. 407–422.

Siegenthaler, U., Eicher, U., Oeschger, H., and Dansgaard, W. 1984. Lake sediments as continental $\delta^{18}O$ records from the glacial/post-glacial transition. *Annals of Glaciology*, 5, 149–152.

Silliman, J.E., Meyers, P.A. and Bourbonniere R.A. 1996. Record of post-glacial organic matter delivery and burial in sediments of Lake Ontario. *Organic Geochemistry*, 24, 463–472.

Silman, M.R., Terborgh, J.T. and Kiltie, R.A. 2003. Population regulation of a rainforest dominant tree by a major seed predator. *Ecology*, 84, 431–438.

Simoneit, B.R.T. 1977. Organic matter in eolian dusts over the Atlantic Ocean. *Marine Chemistry*, 5, 443–464.

Simoneit, B.R.T., Chester, R. and Eglinton, G. 1977. Biogenic lipids in particulates from the lower atmosphere over the eastern Atlantic. *Nature*, 267, 682–685.

Sinclair, D.J., Kinsley, L.P.J. and McCulloch, M.T. 1998. High resolution analysis of trace elements in corals by laser ablation ICP-MS. *Geochimica Et Cosmochimica Acta*, 62, 1889–1901.

Singh, G. and Geissler, E.A. 1985. Late Cainozoic history of fire, lake-levels, and climate at Lake George, New South Wales, Australia. *Philosophical Transactions of the Royal Society of London*, 311, 379–447.

Sinka, K.J. and Atkinson, T.C. 1999. A mutual climatic range method for reconstructing palaeoclimate from plant remains. *Journal of Geological Society, London*, 156, 381–396.

Sinninghe-Damsté, J.S. and de Leeuw, J.W. 1990. Analysis, structure and geochemical significance of organically-bound sulphur in the geosphere: state of the art and future research. *Organic Geochemistry*, 16, 1077–1101.

Sinninghe-Damsté, J.S., Kohnen, M.E.L. and de Leeuw, J.W. 1993. Thiophenic biomarkers for palaeoenvironmental assessment and molecular stratigraphy. *Nature*, 345, 609–611.

Sirocko, F. 1996. The evolution of the monsoon climate over the Arabian Sea during the last 24,000 years. *Palaeoecology of Africa*, 24, 53–69.

Sirocko, F., Sarnthein, M., Erlenkeuser, H., Lange, H., Arnold, M. and Duplessy, J.-C. 1993. Century-scale events in monsoonal climate over the past 24,000 years. *Nature*, 364, 322–324.

Sissons, J.B. 1979. Palaeoclimatic inferences from former glaciers in Scotland and the Lake District. *Nature*, 278, 518–521.

Skabitchevsky, A.P. 1929. On the biology of *Melosira baicalensis* K Meyer. *Wisl. Rus. Hydrobiol. Zh.*, 8, 93–114 (in Russian).

Smith, A.G. and Cloutman, E.W. 1988. Reconstruction of Holocene vegetation history in three dimensions at Waun-Fignen-Felen, an upland site in South Wales. *Philosophical Transactions of the Royal Society of London B*, 322, 159–219.

Smith, A.J. 1993. Lacustrine ostracodes as hydrochemical indicators in lakes of the north-central United States. *Journal of Paleolimnology*, 8, 121–134.

Smith, A.J. and Forester, R.M. 1994. Estimating past precipitation and temperature from fossil ostracodes. *The 5th Annual International High-level Radioactive Waste Management Conference and Exposition, Las Vegas*, pp. 2545–2552.

Smith, A.J. and Horne, D. J. 2002. Ecology of marine, marginal marine and nonmarine ostracods. In J.A. Holmes and A.R. Chivas (eds), *The Ostracoda: Applications in Quaternary Research,* Vol. 131, AGU Geophysical Monograph Series, Washington, DC, pp. 37–64.

Smith, A.J., Delorme, L.D. and Forester, R.M. 1992. A lake's solute history from ostracodes: comparison of methods. In Y.K. Kharaka and A.S. Maest (eds), *Water-Rock Interaction: Proceedings of the 7th International Symposium on Water-Rock Interaction – WRI-7, Park City, Utah, USA*. AA Balkema, Rotterdam, pp. 677–680.

Smith, A.J., Donovan, J.J., Ito, E. and Engstrom, D.R. 1997. Ground-water processes controlling a prairie lake's response to middle Holocene drought. *Geology*, 25, 391–394.

Smith, A.J., Donovan, J.J., Ito, E., Engstrom, D.R. and Panek, V.A. 2002. Climate-driven hydrologic transients in lake sediment records: multiproxy record of mid-Holocene drought. *Quaternary Science Reviews*, 21, 625–646.

Smith, L.C., MacDonald, G.A., Frey, K.E., *et al.* 2000. US–Russia venture probes Siberian peatlands' sensitivity to climate. *EOS*, 81, 497–504.

Smith, M.A., Vellen, L. and Pask, J. 1995. Vegetation history from archaeological charcoals in central Australia: The late Quaternary record from Purijarra rock shelter. *Vegetation History and Archaeobotany*, 4, 171–177.

Smith, N.D. 1978. Sedimentation processes and patterns in a glacier-fed lake with low sediment input. *Canadian Journal of Earth Sciences*, 15, 741–756.

Smith, N.E. 1995. Decadal climate variability in the western tropical Atlantic and the potential for proxy records from the coral Siderastrea siderea, Masters Thesis. University of Colorado, Boulder, CO.

Smith, S.V., Buddemeier, R.W., Redalje, R.C. and Houck, J.E. 1979. Strontium-calcium thermometry in coral skeletons. *Science*, 204, 404.

Smith, T.M., Reynolds, R.W., Livezey, R.E. and Stokes, D.C. 1996. Reconstruction of historical sea surface temperatures using empirical orthogonal functions. *Journal of Climate*, 9, 1403–1420.

Smithers, S.G. and Woodroffe, C.D. 2000. Microatolls as sea-level indicators on a mid-ocean atoll. *Marine Geology*, 168, 61–78.

Smithers, S.G. and Woodroffe, C.D. 2001. Coral microatolls and 20th century sea level in the eastern Indian Ocean. *Earth and Planetary Science Letters*, 191, 173–184.

Smol, J.P. 1988. Paleoclimate proxy data from freshwater arctic diatoms. *Verh. Internat. Verein. Limnology*, 23, 837–844.

Snyder, J.A., Wasylik, K., Fritz, S.C. and Wright, H.E., Jr. 2001. Diatom-based conductivity reconstruction and palaeoclimatic interpretation of a 40-ka record from Lake Zeribar, Iran. *The Holocene*, 11, 737–745.

Sommaruga-Wögrath, S., Koinig, K.A., Schmidt, R., Sommaruga, R., Tessadri, R. and Psenner, R. 1997. Temperature effects on the acidity of remote alpine lakes. *Nature*, 387, 64–67.

Sonntag, C., Thorweihe, U.L.F., Rudolph, J., *et al.* 1980. Paleoclimatic evidence in apparent ^{14}C ages of Saharan groundwaters. *Radiocarbon*, 22, 871–178.

Sorvari, S. Korhola, A. and Thompson, R. 2002. Lake diatom response to recent Arctic warming in Finnish Lapland. *Global Change Biology*, 8, 153–163.

Sowers, T., Brooks, E., Etheridge, D., *et al.* 1997. An interlaboratory comparison of techniques for extracting and analyzing trapped gases in ice cores. Greenland Summit Ice Cores. *Journal of Geophysical Research special publication*, 102, 26527–26538.

Sparks, T.H. and Carey, P.D. 1995. The responses of species to climate over two centuries: an analysis of the Marsham phenological record, 1736–1947. *Journal of Ecology*, 83, 321–329.

Spaulding, W.G. 1990. Packrat middens: their composition and methods of analysis. In J.L. Betancourt, T.R. van Devender and P.S. Martin (eds), *Packrat middens: the last 40,000 years of biotic change*. University of Arizona Press, Tucson, pp. 59–84.

Spear, R.W. 1993. The palynological record of Late-Quaternary arctic tree-line in northwest Canada. *Review of Palaeobotany and Palynology*, 79, 99–111.

Spencer, T., Tudhope, A.W., French, J.R., Scoffin, T.P. and Utanga, A. 1997. Reconstructing sea level change from coral microatolls, Tongareva Penryn. Atoll, Northern Cook Islands. *Proceedings of the 8th International Coral Reef Symposium*, 1, 489–494.

Spicer, R.A. 1989. The formation and interpretation of plant fossil assemblages. *Advances in Botanical Research*, 16, 95–191.

Sprowl, D.R. 1993. On the precision of the Elk Lake varve chronology. In J.P. Bradbury and W.E. Dean (eds), Elk Lake, Minnesota: evidence for rapid climate change in the north-central United States. Geological Society of America Special Paper, 276, 69–74.

Stager, J.C. and Mayewski, P.A. 1997. Abrupt early to mid-Holocene climatic transition registered at the equator and the poles. *Science*, 276, 1834–1836.

Stager, J.C., Cumming, B. and Meeker, L. 1997. A high-resolution 11,400-yr diatom record from Lake Victoria, East Africa. *Quaternary Research*, 47, 81–89.

Stahle, D.W., Van Arsdale, R.B. and Cleaveland, M.K. 1992. Tectonic signal in baldcypress trees at Reelfoot Lake, Tennessee. *Seismological Research Letters*, 63, 439–447.

Stahle, D.W., Cleaveland, M.K., Blanton, D.B., Therrell, M.D. and Gay, D.A. 1998. The lost colony and the Jamestown droughts. *Science*, 280, 564–567.

Standley, L.J. and Kaplan, L.A. 1998. Isolation and analysis of lignin-derived phenols in aquatic humic substances: improvements on the procedures. *Organic Geochemistry*, 28, 689–697.

Steier, P and Rom, W. 2000. The use of Bayesian statistics for ^{14}C dates of chronologically ordered samples: a critical analysis. *Radiocarbon*, 42 2. 183–198.

Steig, E.J. 2000. Wisconsinan and Holocene climate history from an ice core at Taylor Dome, Western Ross Embayment, Antarctica. *Geog. Annaler*, 82A, 213–235.

Sternberg, L.D.S.L. 2001. Savanna–forest hysteresis in the tropics. *Global Ecology and Biogeography*, 10, 369–378.

Stevens, L.R., Wright, H.E. and Ito, E. 2001. Proposed changes in seasonality of climate during the Lateglacial and Holocene at Lake Zeribar, Iran. *The Holocene*, 11, 747–755.

Stevenson, F.J. 1982. *Humus chemistry*. Wiley Interscience, Chichester.

Stine, S. 1994. Extreme and persistent drought in California and Patagonia during mediaeval time. *Nature*, 369, 546–549.

Stirling, C.H., Esat, T.M., McCulloch, M.T. and Lambeck, K. 1995. High-precision U-series dating of corals from western Australia and implications for the timing and duration of the Last Interglacial. *Earth and Planetary Science Letters*, 135, 115–130.

Stocker, T.F. 1998. The seesaw effect. *Science*, 282, 61–62.

Stockhausen, H. 1998. Geomagnetic secular variation 0–13,000 yr BP as recorded in sediments from three maar lakes from the West Eifel Germany. *Geophysical Journal International*, 135, 898–910.

Stoll, H.M. and Schrag, D.P. 1998. Effects of Quaternary sea level cycles on strontium in seawater. *Geochimica Et Cosmochimica Acta*, 62, 1107–1118.

Stoll, H.M., Rosenthal, Y. and Falkowski, P. 2002. Climate proxies from Sr/Ca of coccolith calcite: Calibrations from continuous culture of *Emiliania huxleyi*. *Geochimica et Cosmochimica Acta*, 66, 927–936.

Stothers, R.B. 1999 Volcanic dry fogs, climate cooling, and plague pandemics in Europe and the Middle East. *Climatic Change*, 42, 713–723.

Stothers, R.B. 2000 Climatic and Demographic Consequences of the massive Volcanic Eruption of 1258. *Climatic Change*, 45, 361–374.

Stott, P.A., Tett, S.F.B., Jones, G.S., Allen, M.R., Ingram, W.J. and Mitchell, J.F.B. 2001. Attribution of Twentieth century climate change to natural and anthropogenic causes. *Climate Dynamics*, 17, 1–21.

Street-Perrott, F.A. and Grove, A.T. 1976. Environmental and climatic implications of late Quaternary lake-level fluctuations in Africa. *Nature*, 261, 385–390.

Street-Perrott A.F. and Perrott R.A. 1993. Holocene vegetation, lake-levels, and climate of Africa, In H.E. Wright, J.E. Kutzbach, T. Webb, T., W.E. Ruddiman, F.A. Street-Perrott and P.J. Bartlein (eds), *Global climates since the Last Glacial Maximum*. University of Minnesota Press, Minneapolis, pp. 318–356.

Street-Perrott, F.A., Huang, Y., Perrott, R.A., *et al.* 1997. Impact of lower atmospheric carbon dioxide on tropical mountain ecosystems. *Science*, 278, 1422–1426.

Street-Perrott, F.A., Holmes, J.A., Waller, M.P. *et al.* 2000. Drought and dust deposition in the West African Sahel: a 5500-year record from Kajemurum Oasis, northeastern Nigeria. *The Holocene*, 10, 293–302.

Strestik, J., and Vero, J. 2000. Reconstruction of the spring temperatures in the 18th century based on the measured lengths of grapevine sprouts. *Idojaras*, 104, 123–136.

Stuiver, M. and Braziunas, T.F. 1993a. Modeling atmospheric ^{14}C influences and ^{14}C ages of marine samples to 10,000 BC. *Radiocarbon*, 35 1. 191–200.

Stuiver, M. and Braziunas, T.F. 1993b. Sun, ocean, climate and atmospheric $^{14}CO_2$: an evaluation of causal and spectral relationships. *The Holocene*, 3, 289–305.

Stuiver, M. and Reimer, P.J. 1993. Extended ^{14}C data base and revised CALIB 3.0 ^{14}C age calibration program. *Radiocarbon*, 35, 215–230.

Stuiver, M., Heusser, C.J. and Yang, I.C. 1978. North American glacial history extended to 75,000 years ago. *Science*, 200, 16–21.

Stuiver, M., Braziunas, T.F., Becker, B. and Kromer, B. 1991. Climatic, solar, oceanic and geomagnetic influences on late glacial and $^{14}C/^{12}C$ change. *Quaternary Research*, 35, 1–24.

Stuiver, M., Reimer, P.J., Bard, E., *et al.* 1998. INTCAL98 radiocarbon age calibration, 24,000–0 BP. *Radiocarbon*, 40, 1041–1077.

Sturges W. and Hong, B.G. 2001. Decadal variability of sea level. In B.C. Douglas, M.S. Kearney and S.P. Leatherman (eds), *Sea level rise: history and consequences*. Academic Press, San Diego, pp. 165–180.

Sturm, M. 1979. Origin and composition of clastic varves. In C. Schlüchter (ed.), *Moraines and varves*. A.A. Balkema, Rotterdam, pp. 281–285.

Stute, M. and Talma, A.S. 1998. Glacial temperatures and moisture transport regimes reconstructed from noble gases and $\delta^{18}O$, Stampriet aquifer, Namibia. Isotope techniques in the Study of Environmental Change. *Proceedings of the Symposium of the International Atomic Energy Agency, Vienna*, pp. 307–328.

Suess, H.E. 1965. Secular variations of the cosmic-ray-produced C14 in the atmosphere and their interpretations. *Journal of Geophysical Research*, 70, 5937–5950.

Sugita, S. 1993. A model of pollen source area for an entire lake surface. *Quaternary Research*, 39, 239–244.

Sugita, S. 1994. Pollen representation of vegetation in Quaternary sediments: theory and method in patchy vegetation. *Journal of Ecology*, 82, 881–897.

Summerhayes, C.P., Kroon, D., Rosell-Melé, A., *et al.* 1995. Variability in the Benguela Current upwelling system over the past 70,000 years. *Progress in Oceanography*, 35, 207–251.

Summons, R.E. and Powell, T.G. 1986. Chlorobiaceae in Palaeozoic seas revealed by biological markers, isotopes and geology. *Nature,* 319, 763–765.

Sun, Z.C., Feng, X. J., Li, D. M., Yang, F., Qu, Y. H. and Wang, H. J. 1999. Cenozoic ostracoda and palaeoenvironments of the northeastern Tarim Basin, western China. *Palaeogeography, Palaeoclimatology, Palaeoecology*, 148, 37–50.

Sutherland, D.G. 1984. Modern glacier characteristics as a basis for inferring former climates with particular reference to the Loch Lomond Stadial. *Quaternary Sciences Reviews*, 3, 291–309

Svendsen, J.I. and J. Mangerud, 1997. Holocene glacial and climatic variations on Spitsbergen, Svalbard. *The Holocene*, 7, 45–57.

Swart, P.K. 1983. Carbon and oxygen isotope fractionation in scleractinian corals: a review. *Earth-Science Reviews*, 19, 51–80.

Swart, P.K., Leder, J.J., Szmant, A.M. and Dodge, R.E. 1996. The origin of variations in the isotopic record of scleractinian corals .2. Carbon. *Geochimica Et Cosmochimica Acta*, 60, 2871–2885.

Swetnam, T.W. 1993, History and climate change in Giant Sequoia groves. *Science*, 262, 885–889.

Swezey, C. 2001. Eolian sediment responses to late Quaternary climate changes: temporal and spatial patterns in the Sahara. *Palaeogeography, Plaeoclimatology, Palaeoecology*, 167, 119–155.

Talbot, M.R. 1990. A review of the palaeohydrological interpretation of carbon and oxygen isotopic ratios in primary lacustrine carbonates. *Chemical Geology Isotopes Geoscience Section*, 80, 261–279.

Talbot, M.R. and Kelts, K. 1986. Primary and diagenetic carbonates in the anoxic sediments of Lake Bosumtwi, Ghana. *Geology*, 14, 912–916.

Talbot, M.R., Livingstone, D.A., Palmer, P.G., *et al.* 1984. Preliminary results from sediment cores from Lake Bosumtwi, Ghana. *Palaeoecology of Africa*, 16, 173–192.

Talma, A.S. and Vogel, J.C. 1992. Late Quaternary palaeotemperatures derived from a speleothem from Cango Caves, Cape Province, South Africa. *Quaternary Research*, 37, 203–213.

Talma, A.S., Vogel, J.C. and Partridge, T.C. 1974. Isotopic contents of some Transvaal speleothems and their palaeoclimatic significance. *South African Journal of Science*, 70, 135–140.

Taylor, D.M., Griffiths, H.I., Pedley, H.M. and Prince, I. 1994. Radiocarbon-dated Holocene pollen and ostracod sequences from barrage tufa-dammed fluvial systems in the White Peak, Derbyshire, UK. *The Holocene*, 4, 356–364.

Taylor, F. and McMinn, A. 2001. Evidence from diatoms for Holocene climate fluctuation along the East Antarctic margin. *The Holocene*, 11, 455–466.

Taylor, F. and McMinn, A. 2002. Late Quaternary diatom assemblages from Prydz Bay, eastern Antarctica. *Quaternary Research* 57, 151–161.

Taylor, F., Whitehead, J. and Domack, E. 2001. Holocene paleo climate change in the Antarctic Peninsula: evidence from the diatom, sedimentary and geochemical record. *Marine Micropaleontology*, 41, 25–43.

Teece, M.A., Getliff, J.M., Leftley, J.W., Parkes, R.J. and Maxwell, J.R. 1998. Microbial degradation of the marine prymnesiophyte *Emiliania huxleyi* under oxic and anoxic conditions as a model for early diagenesis: long chain alkadienes, alkenones and alkyl alkenoates. *Organic Geochemistry*, 29, 863–880.

Tegelaar, E.W., de Leeuw, J., Derenne, S. and Largeau, C. 1989. A reappraisal of kerogene formation. *Geochimica et Cosmochimica Acta* 53, 3103–3106.

Teller, J.T.1987. Proglacial lakes and the southern margin of the Laurentide Ice Sheet. In W.F. Ruddiman and H.E. Wright (eds), *North America and Adjacent Oceans during the Last Deglaciation. The Geology of North America, Vol. K-3.* Geological Society of America, Boulder CO, pp. 39–69.

Teller, J.T. 1998. Fresh water lakes in arid regions. *Palaeoecology of Africa*, 25, 241–253.

TEMPO. 1996. Potential role of vegetation feedback in the climate sensitivity of high-latitude regions: a case study at 6000 years BP. *Global Biogeochemical Cycles*, 10, 727–736.

ten Haven, H.L., Eglinton, G., Farrimond, P., *et al.* 1992. Variations in the content and composition of organic matter in sediments underlying active upwelling regimes: a study from ODP Legs 108, 112, and 117. In C.P.Summerhayes, W.L. Prell and K.C. Emeis (eds), *Upwelling systems: evolution since the early Miocene. Geological Society Special Publication*, 64, 229–246.

Teranes, J.L. and McKenzie, J.A. 2001. Lacustrine oxygen isotope record of 20th-century climate change in central Europe: evaluation of climatic controls on oxygen isotopes in precipitation. *Journal of Paleolimnology*, 26, 131–146.

ter Braak, C.J.F. 1995. Non-linear models for multivariate statistical calibration and their use in palaeoecology: a comparison of inverse *k*-nearest neighbours, partial least squares and weighted averaging partial least squares. and classical approaches. *Chemometrics and Intelligent Laboratory Systems*, 28, 165–180.

ter Braak, C.J.F. 1996. *Unimodal models to relate species to environment.* DLO-Agricultural Mathematical Group, Wageningen.

ter Braak, C.J.F. and Barendregt, L.G. 1986. Weighted averaging of species indicator values: its efficiency in environmental calibration. *Mathematical Biosciences*, 75, 57–72.

ter Braak, C.J.F. and Juggins, S. 1993. Weighted averaging partial least squares regression (WA-PLS): an improved method for reconstructing environmental variables from species assemblages. *Hydrobiologia*, 269/270, 485–502.

ter Braak, C.J.F. and Looman, C.W.N. 1986. Weighted averaging, logistic regression and the Gaussian response model. *Vegetatio*, 65, 3–11.

ter Braak, C.J.F. and Šmilauer, P. 2002. CANOCO reference manual and user's guide to Canoco for Windows: Software for Canonical Community Ordination version 4.5. Microcomputer Power, Ithaca, NY.

ter Braak, C.J.F., Juggins, S., Birks, H.J.B. and van der Voet, H. 1993. Weighted averaging partial least squares regression WA-PLS): definition and comparison with other methods for species–environment calibration. In G.P. Patil and C.R. Rao (eds), *Multivariate Environmental Statistics.* Elsevier Science Publishers, Amsterdam, pp. 525–560.

Ternois, Y., Kawamura, K., Keigwin, L., Ohkouchi, N. and Nakatsuka, T. 2001. A biomarker approach for assessing marine and terrigenous inputs to the sediments of Sea of Okhotsk for the last 27,000 years. *Geochimica et Cosmochimica Acta*, 65, 791–802.

Texier, D., de Noblet, N., Harrison, S.P., *et al.* 1997. Quantifying the role of biosphere–atmosphere feedbacks in climate change: coupled model simulations for 6000 years BP and comparison with palaeodata for northern Eurasia and northern Africa. *Climate Dynamics*, 13, 865–882.

Thinon, M., Ballouche, A. and Reille, M. 1996. Holocene vegetation of the central Saharan mountains: the end of a myth. *The Holocene*, 6, 457–462.

Thom, B.G. 1978. Coastal sand deposition in southeast Australia during the Holocene. In J.L. Davies and M.A.G. Williams (eds.), *Landform evolution in Australasia.* ANU Press, Australia, 197–214.

Thomas, D.S.G. 1989. The nature of arid environments. In D.S.G. Thomas (ed.), *Arid zone geomorphology.* Halsted Press, Toronto.

Thomas, D.S.G., Stokes, S. and Shaw, P.A. 1997. Holocene aeolian activity in the southwestern Kalahari Desert, southern Africa: significance and relationships to late Pleistocene dune building events. *The Holocene*, 7, 273–281.

Thompson, L.G. 1996. Climate changes for the last 2000 years inferred from ice core evidence in tropical ice cores. In D. Jones, R.S. Bradley and J. Jouzel (eds), *Climate Variations and Forcing Mechanisms of the Last 2000 Years. NATO ASI Series I, Vol. 41.* AGU, Washington, DC, pp. 281–297.

Thompson, L.G. and Mosley-Thompson, E. 1992. Tropical ice core paleoclimatic records, Quelccaya Ice Cap, Peru, AD 470–1984. *Byrd Polar Research Center Miscellaneous Publication # 321.* University Printing Services, The Ohio State University, Columbus, OH.

Thompson, L.G., Mosely-Thompson, E., Davis, M., *et al.* 1989. Holocene–Late Wisconsin Pleistocene climatic ice core records from Qinghai–Tibetan Plateau. *Science*, 242, 474–477.

Thompson, L.G., Davis, M.E., Mosley-Thompson, E., *et al.* 1998. A 25,000-year tropical climate history from Bolivian ice cores. *Science*, 282, 1858–1864.

Thompson, R. 1995. Complex demodulation and the estimation of the changing continentality of Europe's climate. *International Journal of Climatology*, 15, 175–185.

Thompson, R. 1999. A time-series analysis of the changing seasonality of precipitation in the British Isles and neighbouring areas. *Journal of Hydrology*, 224, 169–183.

Thompson, R.S.1990. Late Quaternary vegetation and climate in the Great Basin. In J.L. Betancourt, T. van Devender and P.S. Martin (eds), *Packrat Middens. The last 40,000 years of biotic change*. Tucson, University of Arizona Press, 200–239.

Thompson, R.S. and Anderson, K.H. 2000. Biomes of western North America at 18,000, 6000 and 0 ^{14}C yr BP reconstructed from pollen and packrat midden data. *Journal of Biogeography*, 27, 555–584.

Thompson, R.S, Whitlock, C, Bartlein, P.J., Harrison, S.P. and Spaulding, W.G. 1993. Climatic changes in the western United States since 18,000 yr BP. In H.E. Wright, J.E. Kutzbach, T. Webb III, W.F. Ruddiman, F.A. Street-Perrott and P.J. Bartlein (eds), *Global climates since the last Glacial Maximum*. University of Minnesota Press, Minneapolis, pp. 468–513.

Thompson, S. and Eglinton, G. 1978. The fractionation of a recent sediment for organic geochemical analysis. *Geochimica et Cosmochimica Acta*, 42, 199–207.

Thomsen, C., Schulz-Bull, D.E., Petrick, G. and Duinker, J.C. 1998. Seasonal variability of the long-chain alkenone flux and the effect on the UK37 index in the Norwegian Sea. *Organic Geochemistry*, 28, 311–323.

Tiljander, M., Ojala, A,, Saarinen, T. and Snowball, I. 2002. Documentation of the physical properties of annually laminated varved. sediments at a sub-annual to decadal resolution for environmental interpretation. *Quaternary International*, 88, 5–12.

Tinner, W. and Lotter, A.F. 2001. Central European vegetation response to abrupt climate change at 8.2 ka. *Geology*, 29, 551–554.

Tissot, B. and Welte, D.H. 1984. Petroleum formation and occurrence. Berlin: Springer-Verlag.

Todd, M.C. and Mackay, A.W. 2003. Large-scale climate controls on Lake Baikal ice cover. *Journal of Climate*, 16, 3186–3199.

Toggweiler, J.R., Dixon, K. and Broecker, W.S. 1991. The Peru upwelling and the ventilation of the south Pacific thermocline. *Journal of Geophysical Research*, 96, 20467–20497.

Toivonen, H.T.T., Mannila, H., Korhola, A. and Olander, H. 2001. Applying Bayesian statistics to organism-based environmental reconstruction. *Ecological Applications*, 11, 618–630.

Tomlinson, P. 1985. An aid to the identification of fossil buds, bud-scales and catkin-bracts of British trees and shrubs. *Circaea*, 3, 45–130.

Torrence, C. and Campo, G.P. 1998. A practical guide to wavelet analysis. *Bulletin of the American Meteriological Society*, 79, 61–78.

Torsnes, I., Rye, N. and Nesje, A. 1993. Modern and Little Ice Age equilibrium-line altitudes on outlet valley glaciers from Jostedalsbreen, western Norway: an evaluation of different approaches to their calculation. *Arctic and Alpine Research*, 25, 106–116.

Trenberth, K.E. 1992. *Climate system modelling*. Cambridge University Press, Cambridge, UK.

Trenberth, K.E. 1995. Atmospheric circulation climate changes. *Climatic Change*, 31 427–453.

Trichet, J., Défarge, C., Tribble, J., Tribble, G. and Sansone, F. 2001. Christmas Island lagoonal lakes, models for the deposition of carbonate–evaporite–organic laminated sediments. *Sedimentary Geology*, 140, 177–189.

Tudhope, A.W., Shimmield, G.B., Chilcott, C.P., Jebb, M., Fallick, A.E. and Dalgleish, A.N. 1995. Recent changes in climate in the far western equatorial Pacific and their relationship to the Southern Oscillation: oxygen isotope records from massive corals, Papua New Guinea. *Earth and Planetary Science Letters*, 136, 575–590.

Tudhope, A.W., Chilcott, C.P., McCulloch, M.T., *et al.* 2001. Variability in the El Niño-Southern Oscillation through a glacial–interglacial cycle. *Science*, 291, 1511–1517.

Turney, C.S.M., Coope, G.R., Harkness, D.D., Lowe, J.J. and Walker, M.J.C. 2000. Implications for the dating of Wisconsinan Weichselian. late-glacial events of systematic radiocarbon age differences between terrestrial plant macrofossils from a site in SW Ireland. *Quaternary Research*, 53, 114–121.

Tvede, A. and Laumann, T. 1997. Glacier variations on a meso-scale: examples from glaciers in the Aurland Mountains, southern Norway. *Annals of Glaciology*, 24, 130–134.

Tyson, R.V. 1995. *Sedimentary organic matter*. Chapman and Hall, London.

UNESCO. 1970. Combined heat, ice and water balance at selected glacier basins. *Technical Papers in Hydrology*, 5, 1–20.

Urban, F.E., Cole, J.E. and Overpeck, J.T. 2000. Modification of tropical Pacific variability by its mean state inferred from a 155 year coral record. *Nature*, 407, 989–991.

Urey, H.C., Lowenstam, H.A., Epstein, S. and McKinney, C.R. 1951. Measurement of palaeotemperatures and temperatures of the Upper Cretaceous of England, Denmark and Southeastern United States. *Geological Society of America Bulletin*, 62, 399–416.

Usdowski, E. and Hoefs, J. 1990. Kinetic ^{13}C ^{12}C and ^{18}O ^{16}O effects upon dissolution and outgassing of CO_2 in the system CO_2–H_2O. *Chemical Geology*, 80, 109–118.

Van Campo, E. and Gasse, F. 1993. Pollen and diatom-inferred climatic and hydrological changes in Shumixi Co basin western Tibet. since 13,000 yr BP. *Quaternary Research*, 39, 300–313.

van de Plassche, O. (ed.) 1986. *Sea-level research*. Geo Books, Norwich.

van de Plassche, O. 2000. North Atlantic climate–ocean variations and sea level in Long Island Sound, Connecticut, since 500 cal yr AD. *Quaternary Research*, 53, 89–97.

van de Plassche, O. 2001. Reply to Kelley *et al. Quaternary Research*, 55, 108–111.

van de Plassche, O., van der Borg, K. and de Jong A.F.M. 1998. Sea level–climate correlation during the past 1400 yr. *Geology*, 26, 319–322.

van de Wal, R.S.W. and Oerlemans, J. 1995. Response of valley glaciers to climate change and kinematic waves: a study with a numerical ice-flow model. *Journal of Glaciology*, 41, 142–152.

van den Dool, H.M., Krijnen, H.J. and Schuurmans, C.J.E. 1978. Average winter temperatures at De Bilt Netherlands: 1634–1977. *Climate Change*, 1, 319–330.

van der Hammen, T. 1963. A palynological study on the Quaternary of British Guiana. *Leidse Geologische Mededelingen*, 29, 125–180.

van der Hammen, T. 1974. The Pleistocene changes of vegetation and climate in tropical South America. *Journal of Biogeography*, 1, 3–26.

van der Hammen, T. and Hooghiemstra, H. 2000. Neogene and Quaternary history of vegetation, climate and plant diversity in Amazonia. *Quaternary Science Reviews*, 19, 725–742.

van der Knaap, W.O. 1987. Long-distance transported pollen and spores on Spitsbergen and Jan Mayen. *Pollen et Spores*, 24, 449–453.

van der Plicht, J., Jansma E. and Kars, H. 1995. The 'Amsterdam Castle': a case study of wiggle matching and the proper calibration curve. *Radiocarbon*, 37, 965–968.

van der Water, P.K., Leavitt, S.W. and Betancourt, J.L. 1994. Trends in stomatal density and $^{13}C/^{12}C$ ratios of *Pinus flexilis* needles during last glacial–interglacial cycle. *Science*, 264, 239–243.

van Dinter, M. and Birks, H.H. 1996. Distinguishing fossil *Betula nana* and *B. pubescens* using their wingless fruits: implications for the late-glacial vegetational history of western Norway. *Vegetation History and Archaeobotany*, 5, 229–240.

van Geel, B. 1978. A palaeoecological study of Holocene peat bog sections in Germany and the Netherlands. *Review of Palaeobotany and Palynology*, 25, 1–120.

van Geel, B. and Middeldorp, A. A. 1988. Vegetational history of Carbury Bog Co. Kildare, Ireland. during the last 850 years and a test of the temperature indicator value of 2H/1H measurements of peat samples in relation to historical sources and meteorological data. *New Phytologist*, 109, 377–392.

van Geel, B. and Mook, W.G. 1989. High resolution ^{14}C dating of organic deposits using natural ^{14}C variations. *Radiocarbon*, 31, 151–155.

van Geel, B., Buurman, J. and Waterbolk, H.T. 1996. Archaeological and palaeoecological indications of an abrupt climate change in The Netherlands, and evidence for climatological teleconnections around 2650 BP. *Journal of Quaternary Science*, 11, 451–460.

van Geel, B., van der Plicht, J., Kilian, M.R., *et al.* 1998. The sharp rise of Delta ^{14}C ca. 800 cal BC: Possible causes, related climatic teleconnections and the impact on human environments. *Radiocarbon*, 40, 535–550.

van Geel, B., C.J. Heusser and J.E. Schuurmans, 2000. Climatic change in Chile at around 2700 B.P. and global evidence for solar forcing: a hypothesis. *The Holocene*, 10, 659–664.

van Harten, D. 2000. Variable noding in *Cyprideis torosa* Ostracoda Crustacea: an overview, experimental results and a model from catastrophe theory. *Hydrobiologia*, 419, 131–139.

van Heteren, S., Huntley, D.J., van de Plassche, O. and Lubberts, R.K. 2000. Optical dating of dune sand for the study of sea-level change. *Geology*, 28, 411–414.

van Kreveld, Sarnthein M., Erlenkeuser H., Grootes P. (see website: http://www.geo.vu.nl/users/isotopen/;ages/2_people/Former_mem.html) In M.J. Nadeau, U. Pflaumann and A. Voelker 2000. Potential links between surging ice sheets, circulation changes and the Dansgaard-Oeschger cycles in the Irminger Sea 60–18 kyr. *Paleoceanography*, 15, 425–442.

van Loon, H. and Rogers, J.C. 1978. The Seesaw in Winter temperatures between Greenland and Northern Europe. Part I: General description. *Monthly Weather Reviews*, 106, 296–310.

van Zeist, W. 1967. Late Quaternary vegetation history of western Iran. *Review of Palaeobotany and Palynology*, 2, 301–311.

van Zeist, W. and Bottema, S. 1977. Palynoplogical investigations in western Iran. *Palaeohistoria*, 19, 19–85.

van Zeist, W. and Wright Jr, H.E. 1961. Preliminary pollen studies at Lake Zeribar, Zagros Mountains, southwestern Iran. *Science*, 140, 61–67.

Varekamp, J.C., Thomas, E. and van de Plassche, O. 1992. Relative sea level rise and climate change over the past 1500 years Clinton, CT, USA. *Terra Nova*, 4, 293–304.

Vasko, K., Toivonen, H.T.T. and Korhola, A. 2000. A Bayesian multinomial Gaussian response model for organism-based environmental reconstruction. *Journal of Paleolimnology*, 24, 243–250.

Vassiljev, J., Harrison, S.P. and Guiot, J. 1998. Simulating the Holocene lake-level record of Lake Bysjon, Southern Sweden. *Quaternary Research*, 49, 62–71.

Velle, G. 1998. A paleoecological study of chironomids Insecta: Diptera. with special reference to climate, MSc Thesis. University of Bergen, Bergen, Norway.

Verbolov, V.I., Sokol'nikov, V.N. and Shimaraev, M.N. 1965. *Hydrometeorological conditions and heat balance of Lake Baikal.* Nauka, Moscow.

Verheyden, S., Keppens, E., Fairchild, I.J., McDermott, F. and Weis, D. 2000. Mg, Sr, and Sr isotope geochemistry of a Belgian Holocene speleothem: implications for paleoclimate reconstructions. *Cave Geology*, 169, 131–144.

Verschuren, D. 1994. Sensitivity of tropical-African aquatic invertebrates to short-term tends in lake-level and salinity: a paleolimnological test at Lake Oloidien, Kenya. *Journal of Paleolimnology*, 10, 253–263.

Verschuren, D. 1997. Taxonomy and ecology of subfossil Chironomidae Insecta, Diptera from Rift Valley lakes in central Kenya. *Archiv für Hydrobiologie Supplement*, 107, 467–512.

Verschuren, D., Laird, K.R. and Cumming, B.F. 2000. Rainfall and drought in equatorial east Africa during the past 1100 years. *Nature*, 403, 410–414.

Versteegh, G.J.M., Bosch, H.J. and De Leeuw, J.W. 1997. Potential palaeoenvironmental information of C_{24} to C_{36} mid- chain diols, keto-ols and mid-chain hydroxy fatty acids; a critical review. *Organic Geochemistry*, 27, 1–13.

de Villiers, S., Shen, G.T. and Nelson, B.K. 1994. Sr/Ca thermometry: coral skeletal uptake and surface ocean variability in the eastern equatorial Pacific upwelling area. *Geochimica Cosmochimica Acta*, 58, 197–208.

de Villiers, S., Nelson, B.K. and Chivas, A.R. 1995. Biological-controls on coral Sr/Ca and $\delta^{18}O$ reconstructions of sea-surface temperatures. *Science*, 269, 1247–1249.

Vincens, A., Schwartz, D., Bertaux, J., Elenga, H. and de Namur, C. 1998. Late Holocene climatic changes in West Equatorial Africa inferred from pollen from Lake Sinnda, Southern Congo. *Quaternary Research*, 50, 34–45.

Vogel, J.C. 1989. Evidence of past climatic change in the Namib Desert. *Palaeogeography, Palaeoclimatology, Palaeoecology*, 70, 355–366.

Vogel, J.C. and Rust, U. 1987. Environmental changes in the Kaokoland Namib Desert during the present millenium. *Madoqua*, 15, 1–16.

Volkman, J.K. 1986. A review of sterol markers for marine and terrigenous organic matter. *Organic Geochemistry*, 9, 83–99.

Volkman, J.K., Eglinton, G., Corner, E.D.S. and Forsberg, T.E.V. 1980. Long-chain alkenes and alkenones in the marine coccolithophorid *Emiliania huxleyi*. *Phytochemistry*, 19, 2619–2622.

Volkman, J.K., Barrett, S.M., Dunstan, G.A. and Jeffrey, S.W. 1993. Geochemical significance of the occurrence of dinosterol and other 4-methyl sterols in a marine diatom. *Organic Geochemistry*, 20, 7–15.
Volkman, J.K., Barrett, S.M., Blackburn, S.I., Mansour, M.P., Sikes, E.L. and Gelin, F. 1998. Microalgal biomarkers: A review of recent research developments. *Organic Geochemistry*, 29, 1163–1179.
von Grafenstein, U. 2002. Oxygen isotope studies of ostracods from deep lakes. In J.A. Holmes and A.R. Chivas (eds), *The Ostracoda: Applications in Quaternary Research*, Vol. 131, AGU Geophysical Monograph Series, Washington, DC, pp. 249–266.
von Grafenstein, U., Erlenkeuser, H., Kleinmann, H., Muller, J. and Trimborn, P. 1994. High-frequency climate oscillations during the last deglaciation as revealed in oxygen isotope records of benthic organisms (Ammersee, southern Germany). *Journal of Paleolimnology*, 11, 344–357.
von Grafenstein, U., Erlenkeuser, H., Muller, J., Trimborn, P., and Alefs, J. 1996. A 200-year mid-European air temperature record preserved in lake sediments: an extension of the $\delta^{18}O_p$–air temperature relation into the past. *Geochimica et Cosmochimica Acta*, 60, 4025–4036.
von Grafenstein, U., Erlenkeuser, H, Muller, J., Jouzel, J. and Johnsen, S. 1998. The cold event 8200 years ago documented in oxygen isotope records of Holocene atmospheric precipitation in Europe and Greenland. *Climate Dynamics*, 14, 73–81.
von Grafenstein, U., Erlenkeuser, H., Brauer, A., Jouzel, J. and Johnsen, S.J. 1999. A mid-European decadal isotope-climate record from 15,500 to 5000 years BP. *Science*, 284, 1654–1657.
Vyverman, W. and Sabbe, K. 1995. Diatom-temperature transfer functions based on the altitudinal zonation diatom assemblages in Papua New Guinea: a possible tool in the reconstruction of regional palaeoclimatic changes. *Journal of Paleolimnology*, 13, 65–77.
Wakeham, S.G. and Lee, C. 1993. Production, transport, and alteration of particulate organic matter in the marine water column. In M.H. Engel and S.A. Macko (eds), *Organic geochemistry, principles and applications*. Plenum Press, New York, pp. 145–169.
Wakeham, S.G., Hedges, J.I., Lee, C., Peterson, M.L. and Hernes, P.J. 1997. Compositions and transport of lipid biomarkers through the water column and surficial sediments of the equatorial Pacific Ocean. *Deep-Sea Research II*, 44, 2131–2162.
Walker, D. and Chen, Y. 1987. Palynological light on tropical rainforest dynamics. *Quaternary Science Reviews*, 6, 77–92.
Walker, D. and Walker, P.M. 1961. Stratigraphic evidence of regeneration in some Irish bogs. *Journal of Ecology*, 49, 169–185.
Walker, D., Head, M., Hancock, G. and Murray, A. 2000. Establishing a chronology for the last 1000 years of laminated sediment accumulation at Lake Barrine, a tropical upland maar lake, northeastern Australia. *The Holocene*, 10, 415–427.
Walker, G.T. and Bliss, E.W. 1932. World Weather V. *Memoirs of the Royal Meteorological Society*, 4, 53–84.
Walker, I.R. 2001. Midges: Chironomidae and related Diptera. In J.P. Smol, H.J.B. Birks and W.M. Last (eds), *Tracking environmental change using lake sediments, Vol. 4: Zoological indicators*. Kluwer Academic Publishers, Dordrecht.
Walker, I.R. and MacDonald, G.M. 1995. Distributions of Chironomidae Insecta: Diptera and other freshwater midges with respect to treeline, Northwest Territories, Canada. *Arctic and Alpine Research*, 27, 258–263.
Walker, I.R. and Mathewes, R.W. 1987. Chironomidae Diptera and post-glacial climate change at Marion Lake, British Colombia, Canada. *Quaternary Research*, 27, 89–102.
Walker, I.R. and Mathewes, R.W. 1988. Late Quaternary fossil Chironomidae (Diptera) from Hippa Lake, Queen Charlotte Islands, British Colombia, with special reference to Corynocera Zett. *Canadian Entomologist*, 120, 739-751.
Walker, I.R. and Paterson, C.G. 1983. Post-glacial chironomid succession in two small, humic lakes in the New Brunswick–Nova Scotia, Canada border area. *Freshwater Invertebrate Biology*, 2, 61–73.
Walker, I.R., Smol, J.P., Engstrom, D.R. and Birks, H.J.B. 1991. An assessment of Chironomidae as quantitative indicators of past climatic change. *Canadian Journal of Fisheries and Aquatic Sciences*, 48, 975–987.

Walker, I.R., Smol, J.P., Engstrom, D.R. and Birks, H.J.B. 1992. Aquatic invertebrates, climate, scale, and statistical hypothesis testing: a response to Hann, Warner, and Warwick. *Canadian Journal of Fisheries and Aquatic Sciences*, 49, 1276–1280.

Walker, I.R., Wilson, S.E. and Smol, J.P. 1995. Chironomidae (Diptera): quantitative palaeosalinity indicators for lakes of western Canada. *Canadian Journal of Fisheries and Aquatic Sciences*, 52, 950–960.

Walker, I.R., Levesque, A.J., Cwynar, L.C. and Lotter, A.F. 1997. An expanded surface–water palaeotemperature inference model for use with fossil midges from eastern Canada. *Journal of Paleolimnology*, 18, 165–178.

Walsh, J. 1988. *On the nature of continental shelves*. Academic Press, London.

Walters, S.D. 1970. Water for Larsa; an Old Babylonian archive dealing with irrigation. Yale University Press, New Haven.

Walther, G-R., Post, E., Menzel, A., *et al.* 2002. Ecological responses to recent climate change. *Nature*, 416, 389–395.

Wang, C.H. and Yeh, H.W. 1985. Oxygen isotopic composition of DSDP Site 480 diatoms: implications and applications. *Geochimica et Cosmochimica Acta*, 49, 1469–1478.

Wang, G. and Eltahir, E.A.B. 2000a. Biosphere–atmosphere interactions over West Africa. Part I. Development and validation of a coupled dynamic model. *Quarterly Journal of the Royal Meteorological Society*, 126, 1239–1260.

Wang, G. and Eltahir, E.A.B. 2000b. Biosphere–atmosphere interactions over West Africa. 2. Multiple Equilibria. *Quarterly Journal of the Royal Meteorological Society*, 126, 1261–1280.

Wang, K. 1989. Paleo-temperature study for some cave deposits in Guilin (translated title). *Carsologica Sinica*, 8, 222–225.

Wang, L., Sarnthein, M., Erlenkeuser, H., *et al.* 1999. Holocene variations in Asian monsoon moisture: a bidecadal sediment record from the South China Sea. *Geophysical Research Letters*, 26, 28890–2892.

Wansard, G. 1996. Quantification of paleotemperature changes during isotopic stage 2 in the La Draga continental sequence NE Spain, based on the Mg/Ca ratio of freshwater ostracods. *Quaternary Science Reviews*, 15, 237–245.

Waple, A.M., Mann, M.E. and Bradley, R.S. 2001. Long-term patterns of solar irradiance forcing in model experiments and proxy-based surface temperature reconstructions. *Climate Dynamics*, 18, 563–578.

Warner, B.G. and Charman, D.J. 1994. Holocene changes on a peatland in northwestern Ontario interpreted from testate amebas Protozoa, analysis. *BOREAS*, 23, 270–279.

Warner, B.G.and Hann, B.J. 1987. Aquatic invertebrates as paleoclimatic indicators? *Quaternary Research*, 28, 427–430.

Warwick, W.F. 1980. Palaeolimnology of the Bay of Quinte, Lake Ontario: 2800 years of cultural influence. *Canadian Journal of Fisheries and Aquatic Sciences*, 206, 1–117.

Warwick, W.F. 1989. Chironomids, lake development and climate: a commentary. *Journal of Paleolimnology*, 2, 15–17.

Wasson, R.J. and Claussen, M. 2002. Earth system models: a test using the mid-Holocene in the southern hemisphere. *Quaternary Science Reviews*, 21, 819–824.

Watanabe, O., Fujii, Y., Kamiyama, K. *et al.* 1999. Basic analyses of Dome Fuji ice core. Part 1: stable oxygen and hydrogen isotope ratios, major chemical compositions and dust concentration. *Polar Meteorology and Glaciology, National Institute of Polar Research*, 13, 83–89.

Watanabe, T., Gagan, M.K., Correge, T., Scott-Gagan, H., Coewley, J. and Hantoro, W. 2003. Oxygen isotope systematics in *Diploastrea heliopora*: New coral archive of tropical paleoclimate. *Geochimica Et Cosmochimica Acta*, 67, 1349–58.

Watts, W.A. 1970. The full-glacial vegetation of northwestern Georgia. *Ecology*, 51, 17–33.

Watts, W.A. 1979. Late Quaternary vegetation of central Appalachia and the New Jersey coastal plain. *Ecological Monographs*, 49, 427–469.

Watts, W.A. 1980. Late Quaternary vegetation history of White Pond on the inner Coastal Plain of South Carolina. *Quaternary Research*, 13, 187–199.

Watts, W.A. and Bright, R.C. 1968. Pollen, seed, and mollusk analysis of a sediment core from Pickerel Lake, northeastern South Dakota. *Geological Society of America Bulletin*, 79, 855–876.

Watts, W.A. and Winter, T.C. 1966. Plant macrofossils from Kirchner Marsh, Minnesota: a paleoecological study. *Geological Society of America Bulletin*, 77, 1339–1360.

Watts, W.A., Allen, J.M.R., Huntley, B, and Fritz, S.C. 1996. Vegetation history and climate of the last 15,000 years at Laghi di Monticchio, southern Italy. *Quaternary Science Reviews*, 15, 113–132.

Weaver, A.J., Eby, M., Wiebe, E.C., *et al.* 2001. The UVic Earth System Climate Model: model description, climatology and application to past, present and future climates. *Atmosphere–Ocean*, 39, 361–428.

Weaver, P.P.E., Wynn, R.B., Kenyon, N.H. and Evan, J. 2000. Continental margin sedimentation, with special reference to the north-east Atlantic margin. *Sedimentology*, 47, 239–256.

Webb, J.D.C., Elsom, D.M. and Reynolds, D.J. 2001. Climatology of severe hailstorms in Great Britain. *Atmospheric Research*, 56, 291–308.

Webb, T. 1986. Is vegetation in equilibrium with climate? How to interpret late-Quaternary pollen data. *Vegetatio*, 67, 75–91.

Webb, T. 1987. The appearance and disappearance of major vegetational assemblages: long-term vegetational dynamics in eastern North America. *Vegetation* 69, 177–188.

Webb, T. and Bryson, R.A. 1972. Late and post-glacial climatic change in the Northern Midwest, USA: quantitative estimates derived from fossil pollen spectra by multivariate statistical analysis. *Quaternary Research*, 2, 70–115.

Weber, J.N. and Woodhead, P.M.J. 1972. Temperature dependence of oxygen-18 concentration in reef coral carbonates. *Journal of Geophysical Research*, 77, 463–473.

Weber, S.L. 2001. The impact of orbital forcing on the climate of an intermediate-complexity coupled model. *Global and Planetary Change*, 30, 7–12.

Webster, K.E., Soranno, P.A., Baines, S.B. *et al.* 2000. Structuring features of lake districts: landscape controls on lake chemical responses to drought. *Freshwater Biology*, 43, 499–515.

Wefer, G., Berger, W.H., Bijma, J. and Fischer, G. 1999. Clues to ocean history: a brief overview of proxies. In G. Fischer and G. Wefer (eds), *Uses of proxies in Paleoceanography: examples from the South Atlantic.* Springer, Berlin, pp. 735.

Wehrli, B. 1997. High resolution varve studies in Baldeggersee. *Aquatic Sciences*, 59, 283–375.

Weiss, H. and Bradley, R.S. 2001. What Drives Societal Collapse? *Science*, 291, 988.

Weiss, H., Courty, M.A., Wetterstrom, W. *et al.* 1993. The genesis and collapse of third millenium north Mesopotamian civilization. *Science*, 261, 995–1004.

Wellington, G.M. and Glynn, P.W. 1983. Environmental influences on skeletal banding in eastern Pacific Panama corals. *Coral Reefs*, 1, 215–222.

Wellington, G.M., Dunbar, R.B. and Merlen, G. 1996. Calibration of stable oxygen isotope signatures in Galápagos corals. *Paleoceanography*, 11, 467–480.

Welten, M. 1944. Pollenanalytische, stratigraphische und geochronologische Untersuchungen aus dem Faulenseemoos bei Spiez. *Veröffentlichungen des Geobotanischen Institutes Rübel*, 21, 1–201.

Werner, M., Heimann, M. and Hoffmann, G. 2001. Isotopic composition and origin of polar precipitation in present and glacial climate simulations. *Tellus Series B-Chemical and Physical Meteorology*, 53, 53–71.

Westman, P. and Sohlenius, G. 1999. Diatom stratigraphy in five offshore sediment cores from the northwestern Baltic proper implying large scale circulation changes during the last 8500 years. *Journal of Paleolimnology*, 22, 53–69.

Whatley, R. 1983. The application of Ostracoda to palaeoenvironmental analysis. In R.F. Maddocks (ed.), *Applications of Ostracoda.* University of Houston, Houston, pp. 51–77.

Whatley, R.C. 1988. Population structure of ostracods: some general principles for the recogntion of palaeoenvironments. In P. De Deckker, J.P. Colin and J.P. Peypouquet (ed.), *Ostracoda in the Earth Sciences.* Elsevier, Amsterdam, pp. 245–256.

Whillans, I.M. 1981. Reaction of accumulation zone portions of glaciers to climatic change. *Journal of Geophysical Research*, 86, C5, 4274–4282.

White, W.B. 1997. Color of speleothems. In C. Hill and P. Forti (eds), *Cave minerals of the world.* National Speleological Society, Huntsville, AL, pp. 239–243.

Whittaker, E.J. 1922. Bottom deposits of McKay Lake, Ottawa Ontario. *Proceedings and Transactions of the Royal Society of Canada*, 16, 141–157.

Wick, L. and Tinner, W. 1997. Vegetation changes and timberline fluctuations in the Central Alps as indicators of Holocene climatic oscillations. *Arctic and Alpine Research*, 29, 445–458.

Wick, L, Lemcke, G. and Sturm, M. 2003. Evidence for Late Glacial and Holocene climatic change and human impact in eastern Anatilia: high-resolution pollen, charcoal, isotopic, and geochemical records from the laminated sediments of Lake Van, eastern Turkey. *The Holocene*, 13, 97–107.

Wiederholm, T. 1980. Effects of dilution on the benthos of an alkaline lake. *Hydrobiologia* 68, 199–207.

Wiederholm, T. (ed.) 1983. Chironomidae of the Holarctic region. Keys and diagnoses. Part 1. Larvae. *Entomologica Scandinavica Supplement*, 19.

Wiederholm, T. 1984. Responses of aquatic insects to environmental pollution. In V.H. Resh and D.M. Rosenberg (eds), *The ecology of aquatic insects*. Praeger, New York, pp. 508–557.

Wigley, T.M.L. and Brown, M.C. 1976. The physics of caves. In T.D. Ford and C.H.D. Cullingford (eds), *The science of speleology*. Academic Press, London, pp. 329–358.

Wigley, T.M.L., Jones, P.D. and Raper, S.C.B. 1997. The observed global warming record: what does it tell us? *Proceedings of the National Academy of Sciences USA*, 94, 8314–8320.

Wilkinson, C.R. 1999. Global and local threats to coral reef functioning and existence: review and predictions. *Marine and Freshwater Research*, 50, 867–878.

Willemse, N.W. and Tornqvist, T.E. 1999. Holocene century-scale temperature variability from West Greenland lake records. *Geology*, 27, 580–584.

Williams, M.A.J., De Deckker, P., Adamson, D.A. and Talbot, M.R. 1991. Episodic fluviatile, lacustrine and aeolian sedimentation in the late Quaternary desert margin system, central western new South Wales. In M.A.W. Williams, P. De Deckker and P. Kershaw (eds), *The Cainozoic in Australia: a reappraisal of the evidence*. Special Publication 18, Geological Society of Australia, pp. 258–287.

Williams, P.W., Marshall, A., Ford, D.C. and Jenkinson, A.V. 1999. Paleoclimatic interpretation of stable-isotope data from Holocene speleothems of the Waitomo district, North Island, New Zealand. *The Holocene*, 9, 649–658.

Willis, J.H. 1944. *Weatherwise*. George Allen and Unwin, London.

Wilmshurst, J.M., McGlone, M.S. and Charman, D.J. 2002. Holocene vegetation and climate change in southern New Zealand: linkages between forest composition and quantitative surface moisture reconstructions from an ombrogenous bog. *Journal of Quaternary Science*, 17, 653–666.

Wilson, D.L., Smith, W.O. and Nelson, D.M. 1986. Phytoplankton bloom dynamics of the western Ross Sea ice edge: 1. Primary productivity and species specific production. *Deep-Sea Research*, 33, 1375–1387.

Wilson, S.E., Cumming, B.F. and Smol, J.P. 1994. Diatom-salinity relationships in 111 lakes from the Interior Plateau of British Columbia, Canada: the development of diatom-based models for paleosalinity reconstructions. *Journal of Paleolimnology*, 12, 197–221.

Wilson, S.E., Smol, J.P. and Sauchyn, D.J. 1997. A Holocene paleosalinity diatom record from southwestern Saskatchewan, Canada: Harris Lake revisited. *Journal of Paleolimnology*, 17, 23–31

Wingham, D., Ridout, A.J., Scharroo, R., Arthern, R.J. and Shum, C.K. 1998. Antarctic elevation change from 1992 to 1996. *Science*, 282, 456–458.

Wirrmann, D. and De Oliveira Almeida, L.F. 1987. Low Holocene level 7700–3650 years ago. of Lake Titicaca Bolivia. *Palaeogeography, Palaeoclimatology, Palaeoecology*, 59, 315–323.

Wirrmann, D. and Mourguiart, P. 1995. Late Quaternary spatiotemporal limnological variations in the altiplano of Bolivia and Peru. *Quaternary Research*, 43, 344–354.

Wohlfarth, B., Björck, S. and Possnert, G. 1995. The Swedish time-scale: a potential calibration tool for the radiocarbon time-scale during the late Weichselian. *Radiocarbon*, 37, 347–359.

Wohlfarth, B., Holmquist, B., Cata, I. and Linderson, H. 1998. The climatic significance of clastic varves in the Angermanalvan Estuary, northern Sweden, AD 1860 to 1950. *The Holocene*, 8, 521–534.

Wolfe, A.P. 2002. Climate modulates the acidity of Arctic lakes on millennial time-scales. *Geology*, 30, 215–218.

Wolfe, A.P. 2003. Diatom community responses to lake Holocene climatic variability, Baffin Island, Canada: a comparison of numerical approaches. *The Holocene*, 13, 29–37.

Wolfe, B.B., Edwards, T.W.D., Aravena, R. and MacDonald, G.M. 1996. Rapid Holocene hydrologic change along boreal tree-line revealed by delta ^{13}C and delta ^{18}O in organic lake sediments, Northwest Territories, Canada. *Journal of Paleolimnology*, 15, 171–181.

Wolfe, B.B., Edwards, T.W.D., Aravena, R. *et al.* 2000. Holocene paleohydrology and paleoclimate at treeline, north-central Russia, inferred from oxygen isotope records in lake sediment cellulose. *Quaternary Research*, 53, 319–329.

Wolfe, B.B., Edwards, T.W.D., Jiang, H.B., MacDonald, G.M., Gervais, B.R. and Snyder, J.A. 2003. Effect of varying oceanicity on early- to mid-Holocene palaeohydrology, Kola Peninsula, Russia: isotopic evidence from treeline lakes. *The Holocene*, 13, 153–160.

Wolter, K. and Timlin, M.S. 1998. Measuring the strength of ENSO – how does 1997/98 rank? *Weather*, 53, 315–324.

Woodland, W.A., Charman, D.J. and Sims, P.C. 1998. Quantitative estimates of water tables and soil moisture in Holocene peatlands from testate amoebae. *The Holocene*, 8, 261–273.

Woodroffe, C.D. and Gagan, M.K. 2000. Coral microatolls from the central Pacific record late Holocene El Niño. *Geophysical Research Letters*, 27, 1511–1514.

Woodroffe, C. and McLean, R. 1990. Microatolls and recent sea-level change on coral atolls. *Nature*, 344, 531–534.

Woodward, F.I. 1987. Stomatal numbers are sensitive to increases in CO_2 from pre-industrial levels. *Nature*, 327, 617–618.

Wright Jr, H.E. 1960. Climate and prehistoric man in the eastern Mediterranean. In R.J. Braidwood and B. Howe (eds), *Prehistoric Investigations in Iraqi Kurdistan*. University of Chicago Press, Chicago, pp. 71–97.

Wright Jr, H.E. 1961. Pleistocene glaciation in Kurdistan. *Eiszeitalter und Gegenwart*, 12, 134–164.

Wright Jr, H.E. 1968. Natural environment of early food production north of Mesopotamia. *Science*, 161, 334–339.

Wright Jr, H.E. 1976. The environmental setting for plant domestication in the Near East. *Science*, 194, 385–389.

Wright Jr, H.E. 1993. Environmental determinism in Near Eastern prehistory. *Current Anthropology*, 34, 458–489.

Wunsch, C. 1992. Decade to century changes in the ocean circulation. *Oceanography*, 5, 99–106.

Wyrtki, K. 1975. El Niño, the dynamic response of the equatorial Pacific Ocean to atmospheric forcing. *Journal of Physical Oceanography*, 5, 572–584.

Wyrtki, K. 1985. Sea level fluctuations in the Pacific during the 1982–1983 El Niño. *Geophysical Research Letters*, 12, 125–128.

Xia, J., Engstrom, D.R. and Ito, E. 1997a. Geochemistry of ostracode calcite: Part 2. The effects of water chemistry and seasonal temperature variation on Candona rawsoni. *Geochimica et Cosmochimica Acta*, 61, 383–391.

Xia, J., Haskell, B.J., Engstrom, D.R. and Ito, E. 1997b. Holocene climate reconstructions from tandem trace-element and stable-isotope composition of ostracodes from Coldwater Lake, North Dakota, USA. *Journal of Paleolimnology*, 17, 85–100.

Xia, J., Ito, E. and Engstrom, D.R. 1997c. Geochemistry of ostracode calcite: Part 1: An experimental determination of oxygen isotope fractionation. *Geochimica et Cosmochimica Acta*, 61, 377–382.

Xie, S., Nott, C.J., Avsejs, L.A., *et al.* 2000. Palaeoclimate records in compound-specific δD values of a lipid biomarker in ombrotrophic peat. *Organic Geochemistry*, 31, 1053–1057.

Xiong, L. and Palmer, J.G. 2000. *Libocedrus bidwillii* tree-ring chronologies in New Zealand. *Tree-Ring Bulletin*, 56, 1–16.

Yamada, K. and Ishiwatari, R. 1999. Carbon isotopic compositions of long-chain *n*-alkanes in the Japan sea sediments: implications for palaeoenvironmental changes over the past 85 kyr. *Organic Geochemistry*, 30, 367–377.

Yamaguchi, D.K., Atwater, B.F., Bunker, D.E., Benson, B.E. and Reid, M.S. 1997. Tree-ring dating the 1700 Cascadia earthquake. *Nature*, 389, 922–923.

Yan, Z., Ye, L., Zhao, S., Liu, M., Liu, R. and Zhao, D. 1984. Oxygen isotope composition, paleotemperature and ^{230}Th/^{234}U dating of speleothem from the fourth cave of Peking Man site. In Z. Su (ed.), *Developments in geoscience; contribution to the 27th International Geological Congress.* Academica Sinica, Beijing, pp. 177–183.

Yang, H., Rose, N.L. and Battarbee, R.W. 2001. Dating of recent catchment peats using spheroidal carbonaceous particle SCP. concentration profiles with particular reference to Lochnagar, Scotland. *The Holocene,* 11, 593–597.

Yarnal, B. 1993. *Synoptic climatology in environmental analyses: a primer.* Belhaven Press, London.

Yasuda,Y., Kitigawa, H. and Nakagawa, T. 2000. The earliest record of major anthropogenic deforestation in the Ghab Valley, northwest Syria: a palynological study. *Quaternary International,* 73/74, 127–136.

Yu, G. and Harrison, S.P. 1996. An evaluation of the simulated water balance of Eurasia and northern Africa at 6000 y BP using lake status data. *Climate Dynamics,* 12, 723–735.

Yu, Z. 1997. Late Quaternary paleoecology of *Thuja* and *Juniperus* Cupressaceae. at Crawford Lake, Ontario, Canada: pollen, stomata and macrofossils. *Review of Palaeobotany and Palynology,* 96, 241–254.

Yu, Z. and Eicher, U. 1998. Abrupt climate oscillations during the Last Deglaciation in Central North America. *Science,* 282, 2235–2238.

Yu, Z. and Ito, E. 1999. Possible solar forcing of century scale drought frequency in the northern Great Plains. *Geology,* 27, 263–266.

Yu, Z. and Wright, H.E. 2001. Response of interior North America to abrupt climate oscillations in the North Atlantic region during the last deglaciation. *Earth-Science Reviews,* 52, 333–369.

Yu, Z., Ito, E., Engstrom, D.R. and Fritz, S.C. 2002. A 2100-year decadal-resolution trace-element and stable-isotope record from Rice Lake in the northern Great Plains, USA. *The Holocene,* 12, 605–617.

Yurtsever, Y. and Gat, J.R. 1981. Atmospheric waters. In J.R. Gatand R.Gonfiantini (ed.), *Stable-isotope Hydrology. IAEA Technical report Series 210,* pp. 103–142.

Zanchetta, G., Bonadonna, F. and Leone, G. 1999. A 37-meter record of paleoclimatological events from stable-isotope data on continental molluscs in Valle di Castiglione, Near Rome, Italy. *Quaternary Research,* 52, 293–299.

Zavada, P.K. 2000. Slackwater sediments and paleofloods. Their significance for Holocene palaeoclimatic reconstruction and flood prediction. In T.C. Partridge and R.R. Maud (eds), *The Cenozoic of Southern Africa.* Oxford University Press, New York, pp. 198–206.

Zdanowicz, C.M., Zielinski, G.A., Wake, C., Fisher, D.A. and Koerner. R.M. 2000. A Holocene record of eolian dust deposition on the Penny Ice Cap, Baffin Island, Canadian Arctic and its paleoenvironmental significance. *Quaternary Research,* 53, 62–69.

Zeebe, R.E. 1999. An explanation of the effect of seawater carbonate concentration on foraminiferal oxygen isotopes. *Geochimica et Cosmochimica Acta,* 63, 2001–2001.

Zeng, N. and Neelin, J.D. 2000. The role of vegetation–climate interaction and interannual variability in shaping the African savanna. *Journal of Climate,* 13, 2665–2670.

Zhang, Y., Wallace, J.M. and Battisti, D.S. 1997. ENSO-like interdecadal variability: 1900–93. *Journal of Climate,* 10, 1004–1020.

Zielinski, G.A., Mayewski, P.A., Meeker, L.D., *et al.* 1994. Record of explosive volcanism since 7000 B.C. from the GISP2 Greenland ice core and implications for the volcano–climate system. *Science,* 264, 948–952.

Zielinski, G.A., Mayewski, P.A., Meeker, L.D., *et al.* 1997. Volcanic aerosol records and tephrochronology of the Summit, Greenland ice cores. *Special Issue Journal of Geophysical Research,* 102, 26625–26640.

Zolitschka, B. 1996a. High resolution lacustrine sediments and their potential for paleoclimatic reconstruction. In P.D. Jones, R.S. Bradley and J. Jouzel (eds), *Climatic variations and forcing mechanisms of the last 2000 years.* Springer-Verlag, Berlin, pp. 453–478.

Zolitschka, B. 1996b. Recent sedimentation in a high arctic lake, northern Ellesmere Island, Canada. *Journal of Paleolimnology,* 16, 169–186.

Zolitschka, B. 1998. A 14,000-year sediment yield record from Western Germany based on annually laminated sediments. *Geomorphology,* 22, 1–17.

Zolitschta, B. and Negendank, J.F.W. 1996. Sedimentology, dating and palaeoclimatic interpretation of a 76.3 ka record from Lago Grande di Monticchio, Southern Italy. *Quaternary Science Reviews*, 15, 101–112.

Zolitschka, B., Negendank, J.F.W. and Lottermoser, BG. 1995. Sedimentological proof and dating of the early Holocene volcanic eruption of Ulmener Maar Vulkaneifel, Germany. *Geologische Rundschau*, 84, 213–219.

Zolitschka, B., Brauer, A., Negendank, J.F.W., Stockhausen, H. and Lang, A. 2000. Annually dated late Weichselian continental paleoclimate record from the Eifel, Germany. *Geology*, 28, 783–786.

Zolitschka, B., Mingram, J., van der Gaast, S., Jansen, J.H.F. and Naumann, R. 2001. Sediment logging techniques. In W.M. Last and J.P. Smol (eds), *Tracking environmental change using lake sediments: physical and geochemical methods. Developments in Paleoenvironmental Research, 1.* Kluwer Academic Publishers, Dordrecht, pp. 137–154.

Zoltai, S.C. and Vitt, D.H. 1990. Holocene climatic change and the distribution of peatlands in western interior Canada. *Quaternary Research*, 33, 231–240.

INDEX

Lightning Source UK Ltd.
Milton Keynes UK
UKOW05f1020080716

277867UK00018B/377/P